Lecture Notes in Computer Science   10173

Commenced Publication in 1973
Founding and Former Series Editors:
Gerhard Goos, Juris Hartmanis, and Jan van Leeuwen

More information about this series at http://www.springer.com/series/7407

Heike Trautmann · Günter Rudolph
Kathrin Klamroth · Oliver Schütze
Margaret Wiecek · Yaochu Jin
Christian Grimme (Eds.)

Evolutionary Multi-Criterion Optimization

9th International Conference, EMO 2017
Münster, Germany, March 19–22, 2017
Proceedings

 Springer

Editors

Heike Trautmann
University of Münster
Münster
Germany

Günter Rudolph
TU Dortmund University
Dortmund
Germany

Kathrin Klamroth
University of Wuppertal
Wuppertal
Germany

Oliver Schütze
CINVESTAV-IPN
Mexico City
Mexico

Margaret Wiecek
Clemson University
Clemson, SC
USA

Yaochu Jin
University of Surrey
Guildford
UK

Christian Grimme
University of Münster
Münster
Germany

ISSN 0302-9743   ISSN 1611-3349 (electronic)
Lecture Notes in Computer Science
ISBN 978-3-319-54156-3   ISBN 978-3-319-54157-0 (eBook)
DOI 10.1007/978-3-319-54157-0

Library of Congress Control Number: 2017932125

LNCS Sublibrary: SL1 – Theoretical Computer Science and General Issues

Printed on acid-free paper

This Springer imprint is published by Springer Nature
The registered company is Springer International Publishing AG
The registered company address is: Gewerbestrasse 11, 6330 Cham, Switzerland

Preface

The acronym EMO stands for a well-established, biennial international conference series devoted to the theory and practice of evolutionary multi-criterion optimization. The first EMO conference took place in 2001 in Zürich (Switzerland), with subsequent conferences taking place in Faro (Portugal), Guanajuato (Mexico), Matsushima (Japan), Nantes (France), Ouro Preto (Brazil), Sheffield (UK), and Guimarães (Portugal).

The 9th International Conference on Evolutionary Multi-Criterion Optimization (EMO 2017) took place in Münster, Germany, during March 19–22, 2017. Today, Münster is the regional metropolis of the Westfalia region hosting numerous administrative, public service, cultural, and scientific authorities. In the past, the city of Münster witnessed momentous events in European history: Münster was the location of the Anabaptist rebellion during the Protestant Reformation and the site of the signing of the Treaty of Westphalia ending the Thirty Years' War in 1648. The castle of Münster was built 1767–1787 and severely damaged during World War II. After restoration it became the official domicile of the University of Münster in 1954, providing the opportunity to host the EMO 2017 conference in this historic building.

EMO 2017 received 72 submissions, which were evaluated in a single-blind peer-review process by 102 reviewers, with a minimum of three referees per paper.

As a result, a total of 46 papers were accepted for presentation and publication in this volume, from which 33 were chosen for oral and 13 for poster presentation.

The conference profited from the presentations of three keynote speakers on research subjects fundamental to the EMO field: Michael Emmerich (LIACS Leiden University, The Netherlands) reviewed "Indicator-Based Multiobjective and Set-Oriented Optimization," Thomas Stützle (IRIDIA, Université libre de Bruxelles, Belgium) shed light on the "Automated Design of Algorithms from Flexible Algorithm Frameworks," and Julian Togelius (New York University, USA) discussed the "Challenges for Multi-criterion Optimization in Games."

Needless to say, the success of such a conference depends on authors, reviewers, and organizers. We are grateful to all authors for submitting their best and latest work, to all the reviewers for the generous way they spent their time and provided their valuable expertise in preparing these reviews, to the program chairs for their hard work in compiling an ambitious scientific program, to the keynote speakers for delivering impressive talks, to the competition chairs for creating the framework for the Black Box Optimization Challenge, to the proceedings and publicity chairs for taking care about EMO proceedings and website, and to the local organizers who helped to make EMO 2017 happen.

Last but not least, we would like to thank the Deutsche Forschungsgemeinschaft (DFG), the European Research Center for Information Systems (ERCIS), and the University of Münster for their generous financial support.

March 2017

Günter Rudolph
Heike Trautmann

Organization

EMO 2017 was organized by the Department of Information Systems and Statistics, University of Münster, Germany, and the Computational Intelligence Group of TU Dortmund University, Germany.

General Chairs

Heike Trautmann	University of Münster, Germany
Günter Rudolph	TU Dortmund University, Germany

Program Chairs

Kathrin Klamroth	University of Wuppertal, Germany
Oliver Schütze	CINVESTAV, Mexico
Margaret Wiecek	Clemson University, USA
Yaochu Jin	University of Surrey, UK

Proceeding and Publicity Chairs

Christian Grimme	University of Münster, Germany
Jakob Bossek	University of Münster, Germany

Competition Chairs

Tobias Glasmachers	University of Bochum, Germany
Michael Emmerich	LIACS Leiden University, The Netherlands

Local Organizers

Pascal Kerschke	University of Münster, Germany
Mike Preuss	University of Münster, Germany

Program Committee

Ajith Abraham	Machine Intelligence Research Labs (MIR Labs)
Richard Allmendinger	University College London, UK
Adiel Teixeira De Almeida	Federal University of Pernambuco, Brazil
Maria João Alves	INESC Coimbra, Portugal
Helio Barbosa	Laboratório Nacional de Computação Científica, Brazil
Mickael Binois	Ecole des Mines de Saint-Etienne, France
Bernd Bischl	LMU Munich, Germany

Juergen Branke	University of Warwick, UK
Dimo Brockhoff	Inria Lille, Nord Europe, France
Matthias Carnein	University of Münster, Germany
Marco Chiarandini	University of Southern Denmark, Denmark
Sung-Bae Cho	Yonsei University, South Korea
João Clímaco	University of Coimbra, Portugal
Leandro Coelho	Pontifícia Universidade Católica do Parana, Brazil
Carlos Coello Coello	CINVESTAV-IPN, Mexico
Fernanda Costa	Universidade do Minho, Portugal
Yves De Smet	Université Libre de Bruxelles, Belgium
Kalyanmoy Deb	Michigan State University, USA
Alexandre Delbem	University of Sao Paulo, Brazil
Clarisse Dhaenens	Université Lille 1 (Polytech Lille, CRIStAL, Inria), France
Michael Doumpos	Technical University of Crete, Greece
Rolf Drechsler	University of Bremen, Germany
Matthias Ehrgott	Lancaster University, UK
Michael T.M. Emmerich	Leiden University, The Netherlands
Andries Engelbrecht	University of Pretoria, South Africa
A. Ismael F. Vaz	University of Minho, Portugal
Jonathan Fieldsend	University of Exeter, UK
Peter Fleming	University of Sheffield, UK
Carlos M. Fonseca	University of Coimbra, Portugal
Katrin Franke	Norwegian University of Science and Technology, Norway
Xavier Gandibleux	Université de Nantes, France
Antonio Gaspar-Cunha	I3N, University of Minho, Portugal
Martin Josef Geiger	Helmut Schmidt University, Germany
Bernard Grabot	ENIT, France
Christian Grimme	University of Münster, Germany
Jerzy Grzymala-Busse	University of Kansas, USA
Jussi Hakanen	University of Jyväskylä, Finland
Jin-Kao Hao	University of Angers, France
Carlos Henggeler Antunes	University of Coimbra, Portugal
Masahiro Inuiguchi	Osaka University, Japan
Hisao Ishibuchi	Osaka Prefecture University, Japan
Johannes Jahn	University of Erlangen-Nuremberg, Germany
Andrzej Jaszkiewicz	Poznan University of Technology, Poland
Laetitia Jourdan	Inria/LIFL/CNRS, France
Miłosz Kadziński	Poznan University of Technology, Poland
Timoleon Kipouros	University of Cambridge, UK
Joshua Knowles	University of Manchester, UK
Mario Koeppen	Kyushu Institute of Technology, Japan
Jana Krejčí	University of Trento, Italy
Dario Landa-Silva	University of Nottingham, UK
Bill Langdon	University College London, UK

Contents

Contents XIII

On the Effect of Scalarising Norm Choice
in a ParEGO implementation

Naveed Reza Aghamohammadi[1], Shaul Salomon[1,2],
Yiming Yan[1], and Robin C. Purshouse[1(✉)]

[1] Department of Automatic Control and Systems Engineering,
University of Sheffield, Sheffield, UK
naveed.aghamohammadi@ieee.org, {yiming.yan,r.purshouse}@sheffield.ac.uk
[2] Department of Mechanical Engineering,
Ort Braude College of Engineering, Karmiel, Israel
shaulsal@braude.ac.il
http://www.sheffield.ac.uk/acse

Abstract. Computationally expensive simulations play an increasing
role in engineering design, but their use in multi-objective optimization
is heavily resource constrained. Specialist optimizers, such as ParEGO,
exist for this setting, but little knowledge is available to guide their configuration. This paper uses a new implementation of ParEGO to examine three hypotheses relating to a key configuration parameter: choice
of scalarising norm. Two hypotheses consider the theoretical trade-off
between convergence speed and ability to capture an arbitrary Pareto
front geometry. Experiments confirm these hypotheses in the bi-objective
setting but the trade-off is largely unseen in many-objective settings.
A third hypothesis considers the ability of dynamic norm scheduling
schemes to overcome the trade-off. Experiments using a simple scheme
offer partial support to the hypothesis in the bi-objective setting but no
support in many-objective contexts. Norm scheduling is tentatively recommended for bi-objective problems for which the Pareto front geometry
is concave or unknown.

Keywords: Expensive optimization · Surrogate-based optimization ·
Performance evaluation

1 Introduction

1.1 Motivation

The use of modelling and simulation plays an increasingly important role in the
evaluation of designs for complex engineered products. However these simulations can often require run-times of hours or days, even using high-performance
computing resources. In this setting, perhaps only a few hundred candidate
designs can be explored using high-fidelity models. Traditional multi-objective
optimizers are unsuitable in this environment due to the typically large number

© Springer International Publishing AG 2017
H. Trautmann et al. (Eds.): EMO 2017, LNCS 10173, pp. 1–15, 2017.
DOI: 10.1007/978-3-319-54157-0_1

(thousands) of solution evaluations that they require to achieve approximate convergence. This has led to the development of specialist algorithms, such as the Pareto Efficient Global Optimization (ParEGO) algorithm [11] that are able to achieve successful results on a smaller budget of function evaluations. Despite the popularity of ParEGO and related optimizers, there is little understanding of how these, usually quite complex, algorithms can be configured to ensure an effective optimization process. The present study focuses on one aspect of a ParEGO configuration – the choice of scalarising norm – and considers the effect of this component on both the speed and quality of optimizer convergence.

1.2   Related Works

In recent years, the use of surrogate modelling to replace expensive function evaluations has become more widespread and has enabled efficient application of multi-objective optimization to real-world problems [7]. This review focuses on optimizers that are related to ParEGO, in that they use Kriging (or one of its variants) to build the surrogate model. For a review of wider methods refer to [16]. ParEGO uses Latin Hypercube Sampling (LHS) [15] to initialise a set of designs; it scalarises the vector objective function and builds a single surrogate (known as a *DACE* model); the model is searched for a design that maximises the single objective of Expected Improvement (EI); this solution is evaluated and the process then iterates; the algorithm stops when the budget of high fidelity evaluations is exhausted. MOEA/D-EGO [21] uses LHS for initialisation, but fits a DACE model to each objective; then, instead of generating a single solution, it creates a batch of solutions by multi-objective maximisation of EI according to a set of scalarised functions using different weighting vectors. Multi-EGO [9] uses LHS for its initialisation, builds DACE models for each objective function and then, instead of using scalarisation, uses the vector EI of the objective functions. Voutchkov and Keane's algorithm [18], uses Sobol sampling for initialisation and works directly with the estimates of objective values (rather than EI). MSPOT [20] is similar but uses LHS for initialisation and selects only one solution at a time using hypervolume contribution as a metric. SMS-EGO [14] creates a model for each objective then instead of using EI, the algorithm uses lower confidence bounds to identify a decision vector which optimises the expected hypervolume indicator. ϵ-EGO [19] is very similar to SMS-EGO except that instead of using the hypervolume indicator, it uses the additive ϵ-indicator. ParEGO has also been extended to use a double Kriging strategy and a modified EI criterion which jointly accounts for the objective function approximation error and the probability of finding Pareto optimal solutions [2].

Despite the set of variants available, until very recently [7] there had been few attempts to compare the performance of the algorithms. In particular, there has been little attempt to analyse and understand how the different mechanisms within these complex algorithms affect performance in the context of a given problem landscape. A notable exception is the work of Cristescu and Knowles [1], which assesses strategies for selecting solutions to use in updating the DACE model in ParEGO. Our paper aims to make a further contribution in this direction, with a focus on scalarising norms.

The paper is organised as follows. In Sect. 2, the original ParEGO algorithm is described in more detail, together with our new implementation. In Sect. 3, we set out the hypotheses relating to scalarisation that will be examined in the study and define the experiments that will be performed to test these hypotheses. Results for static and dynamic choices of scalarising norms are presented in Sects. 4 and 5 respectively. Section 6 concludes the paper.

2 ParEGO implementation

2.1 Knowles' ParEGO

In this section, we elaborate a little further on ParEGO – for full details refer to [11]. The algorithm is an extension to Jones et al.'s EGO algorithm [10] which deals with single-objective problems of a similarly expensive nature. ParEGO is a decomposition-based multi-objective solver, meaning the multi-objective problem is decomposed into a set of single-objective problems using weight vectors and scalarising functions; the augmented Tchebycheff function being used in the original paper. ParEGO begins by creating an initial population of $11k - 1$ candidate solutions using LHS, then evaluates and normalises the k number of objectives. At each iteration of the algorithm, a direction vector is randomly generated from a set of weight vectors generated according to Simplex Lattice Design [15]. The number of directions $|D|$ is defined according to $\binom{s+k-1}{k-1}$, where s is the configurable parameter. ParEGO employs the weighted scalarisation function to re-value all previously visited solutions and uses all or part of these solutions to estimate a DACE model according to maximum likelihood. To find a new solution, ParEGO uses a simple genetic algorithm to maximise EI. The new solution is evaluated and the procedure continues until the evaluation budget is expended.

2.2 A New ParEGO Implementation

In the current work, ParEGO is treated as a framework rather than a specific algorithm. The implementation of ParEGO differs slightly from the original in ways outlined below. The implementation has been undertaken in *Tigon*, which is an open-source C++ library that has been developed to support the *Liger* open-source optimization workflow software [6]. In Tigon, an optimization algorithm is assembled from a set of components according to the Decorator design pattern. Figure 1 depicts a workflow diagram for the new implementation of ParEGO.

This implementation of ParEGO begins by defining an optimization problem. Next, an initial population is created using LHS and is evaluated on the high-fidelity evaluation functions. Next, the Direction Iterator creates a set of weight vectors according to generalised decomposition [5] and randomly chooses from amongst them. Subsequently, the Scalarization operator applies the weights

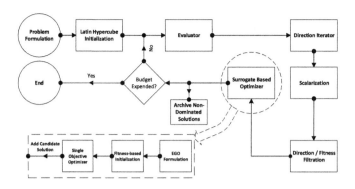

Fig. 1. Implemented framework for ParEGO

to the L_p-norm scalarization function $(\sum_{i=1}^{k} \omega_i |f_i(\mathbf{x}) - z_i^*|^p)^{\frac{1}{p}}$, where w_i is the weight of the ith objective, $f_i(\mathbf{x})$ is the evaluation of a solution $\mathbf{x}$ on the ith objective, z_i^* relates to the ideal point, and p is the order of the scalarisation norm. Next the Direction/Fitness Filtration operator sorts the solutions according to their scalarised fitness and direction. In the Surrogate Based Optimizer the DACE model is constructed and searched over using the efficient single-objective optimizer ACROMUSE [12], which is the default evolutionary optimizer implemented in Tigon. A new solution is identified, evaluated, and the optimizer iterates until the evaluation budget is expended. All solutions are archived off-line as the search progresses, with the archive filtered for non-dominated solutions.

3   Experiments with scalarising norms

We employ a formal hypothesis testing framework to explore the effect of different scalarising norm choices on the performance of ParEGO.

3.1   Hypotheses

1. To obtain good coverage of the Pareto front, the scalarising norm must be of a higher-order than the shape of the front (see [13] for the geometric analysis relating scalarisation functions to Pareto dominance relations).
2. In the case where two different scalarising norms are of sufficient order to capture the front, the lower-order norm will converge faster than the higher-order norm (see [5] for underpinning argumentation and experiments).
3. By increasing the order of the norm dynamically during an optimization run, enhanced convergence can be achieved for problems where higher-order norms are needed to capture the shape of the front.

3.2   Test Problems and Performance Indicators

In order to test the hypotheses we use two test functions: a scalable concave test function – DTLZ2 [3] – and a scalable convex version that replaces the shape of DTLZ2 with the convex shape from the WFG test suite [8]. We refer to this latter problem as DTLZ2CX. To assess the quality of the solution obtained, two performance metrics have been used: hypervolume (HV) [4] and inverse generational distance (IGD) [17]. The hypervolume reference point is defined as the anti-ideal. Raw hypervolume values are normalised with respect to the hypervolume of the global Pareto front. Metrics are applied to the off-line archive.

3.3   Experimental Set-Up

To implement the hypothesis tests, we consider a set of experiments that are repeated for optimizers using a range of static and dynamic scalarising norms.

Static norm experiments: These tests try to validate or reject hypotheses 1 and 2. The experiment has 6 categories of tests (outlined in Table 1) and in each of those tests three norms of $p \in \{1, 2, \infty\}$ are tested with the same initial population on convex and concave versions of DTLZ2. The budget for evaluation for these tests, including the initial population defined using LHS, is the following: 1400 evaluations for two-objective tests (CAT1-2), 1800 evaluations for four-objective tests (CAT3-4) and 2000 evaluations for seven-objective tests (CAT5-6). These budgets are larger than allowable for some industrial problems but enable long-run convergence to be observed. The expected outcome for the convex problem is that all norms can capture the Pareto front given a sufficiently high budget, with the lower-order norms achieving this result more quickly with norm $p = 1$ expected to be fastest. As for the concave tests, norm $p = \infty$ will capture the Pareto front, norm $p = 2$ will partly capture the Pareto front and norm $p = 1$ will only capture the extremes of the front.

Dynamic norm experiments: These tests implement ParEGO with a pre-defined schedule of changing norms in order to confirm or reject hypothesis 3. Again six test categories were created with the parameters following the same setup as in Table 1; however budget size and implementation of norms in these experiments is different. For a given budget of 500 evaluations, which excludes the initial population size, the budget is divided into four quarters: the first quarter begins with the $p = 1$ norm, at the beginning of the second quarter the norm is switched to $p = 2$, at the beginning of the third quarter norm $p = 50$ comes into effect, and at the beginning of the fourth quarter the Tchebycheff norm ($p = \infty$) is operating. The expected outcome for concave geometries is for the norm scheduled version to outperform all static cases. For convex geometries, performance is expected to be better than static $p = \infty$ but worse than $p = 1$.

Both types of experiment are repeated for $k \in \{2, 4, 7\}$. In all tests, the number of decision variables that comprise the distance function 'g' in DTLZ2 and DTLZ2CX is fixed at 5, so that the total number of decision variables is $d = k + 4$. The number of weight vectors is determined by $s = 3$. The initial population size for LHS follows the original guidelines set by Knowles of $11d - 1$ and the maximum number of solutions chosen to estimate the DACE model is 80 (again, following Knowles). All experiments were replicated 11 times.

Table 1. Experimental configuration for 6 categories of test

Test category	Decision vector size	Objective vector size	Number of directions	Initial population size	DACE surrogate size	Pareto front geometry
CAT1	6	2	10	65	80	Concave
CAT2	6	2	10	65	80	Convex
CAT3	8	4	20	87	80	Concave
CAT4	8	4	20	87	80	Convex
CAT5	11	7	84	120	80	Concave
CAT6	11	7	84	120	80	Convex

4   Results—Static Norms

In this section, results are presented for the static norm experiments. The median HV is identified at evaluations $i = 500$ and $i = 1000$, with convergence trajectories shown for $i = [1, 500]$ and $i = [501, 1000]$ to indicate trends on alternative budgets. Box plots for IGD metrics summarising performance across the 11 runs are shown for $i = 500$ and $i = 1000$. Scatterplots and parallel coordinates plots are also included as needed to support understanding of the results.

2-objective instances: Results for the concave CAT1 test are shown in Fig. 2. $p = \infty$ outperforms $p = 1$ and $p = 2$ in terms of both median HV and the IGD distribution. Scatterplots of median HV runs (Fig. 3) show that $p = 1$ has found good quality solutions only at the edges of the global front (the latter is indicated by the dashed line in the plots). For the convex CAT2 tests, $p = 1$ and $p = 2$ show more rapid convergence than $p = \infty$ for the median run, although the IGD results are inconclusive (see Fig. 4). Scatterplots for median runs (not shown) all indicate good convergence to the global front.

4-objective instances: As shown in Fig. 5, for the 4-objective concave CAT3 test, $p = \infty$ does not exhibit the clear outperformance seen previously for $k = 2$. This appears to be as a result of slower convergence, but this result is curiously not replicated on the convex CAT4 (see Fig. 6), where $p = \infty$ exhibits similar convergence to $p = 1$ and $p = 2$.

7-objective instances: The situation for the 7-objective concave CAT5 test is similar to the finding for $k = 4$ and is not shown. For the convex CAT6 case (shown in Fig. 7), the median run for $p = 1$ suffers slow convergence but the IGD results indicate substantial inter-run variability. Parallel coordinates plots corresponding to the median HV results (see Fig. 8) suggest that all norms are struggling to converge to the global front.

The 2-objective results offer support for hypotheses 1 and 2 – with $p = \infty$ offering better coverage of concave fronts, but with a convergence speed penalty compared to $p = 1$ and $p = 2$ for convex fronts. The 4-objective and 7-objective

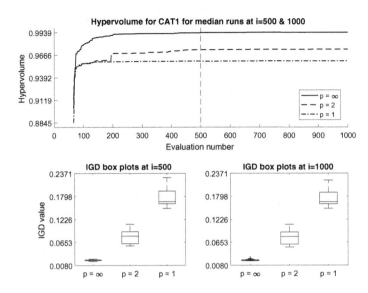

Fig. 2. HV convergence profiles and IGD box plots for static CAT1 tests

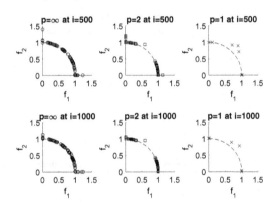

Fig. 3. Scatterplots relating to median HV runs for static CAT1 tests

results, by contrast, offer less support to these hypotheses, with all norms suffering convergence issues within the available limited budget.

5 Results – Dynamic Norms

Many different possibilities exist for applying dynamic norms within ParEGO. Here, for simplicity, after initialisation the remaining budget is divided into quarters and the norm is increased over the course of the optimization process. $p = 1$, $p = 2$, $p = 50$, and $p = \infty$ are used in each successive quarter. Experiments are

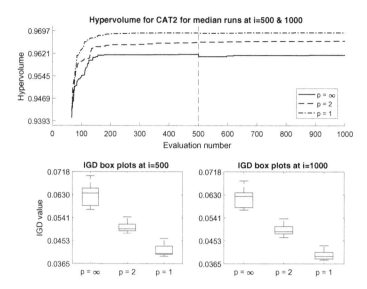

Fig. 4. HV convergence profiles and IGD box plots for static CAT2 tests

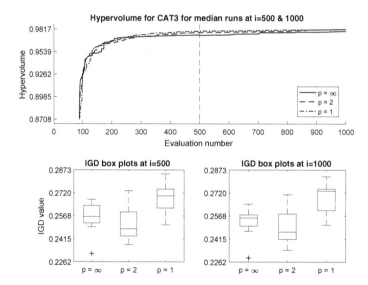

Fig. 5. HV convergence profiles and IGD box plots for static CAT3 tests

run for a total budget of 500 evaluations (exclusive of the budget used for initialisation). The categories of tests remain the same as in the static experiments.

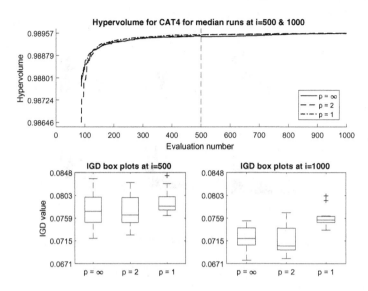

Fig. 6. HV convergence profiles and IGD box plots for static CAT4 tests

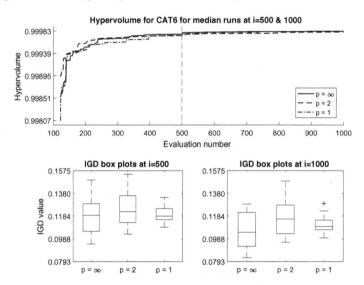

Fig. 7. HV convergence profiles and IGD box plots for static CAT6 tests

Convergence profiles relate to the run with the median HV at $i = 500$. Comparisons are made to the equivalent results for the $p = 1$ and $p = \infty$ norms.

2-objective instances: The results for the dynamic CAT1 test on the bi-objective concave problem are shown in Fig. 9. The ability of the dynamic scheme to improve convergence over the $p = 1$ static scheme is clear, with transitions

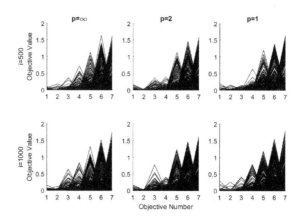

Fig. 8. Parallel coordinates plots relating to median HV runs for static CAT6 tests

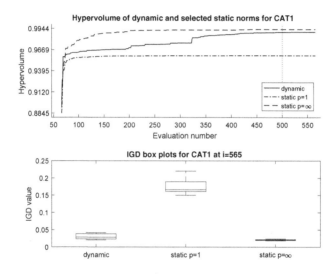

Fig. 9. HV convergence profiles and IGD box plots for dynamic CAT1 tests

to HV values superior to the saturated $p = 1$ results as the order of the norm is increased. The final result at the end of the optimization budget is close to the $p = \infty$ performance. No convergence acceleration advantage is seen during the early stages of the search, reflecting the result already seen for the static CAT1 tests. For the bi-objective convex problem (see Fig. 10), the median HV results suggest some deterioration in convergence speed arising from the progression to higher-order norms in the latter stages of the optimization process; this finding is confirmed by the IGD results, suggesting that there is a penalty in terms of convergence when the higher-order norm is unnecessary.

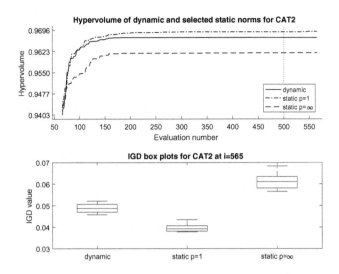

Fig. 10. HV convergence profiles and IGD box plots for dynamic CAT2 tests

4-objective instances: The results for the 4-objective concave problem are shown in Fig. 11. The median HV result indicates that the dynamic scheme has offered a speed-up over the static $p = \infty$ alternative, whilst retaining the benefits of improved coverage of the front; however the IGD results suggest that the performance benefits over $p = \infty$ are inconclusive when considering all 11 runs. In the static CAT4 tests (shown in Fig. 12), whilst it is unsurprising that the dynamic scheme is unable to outperform the $p = 1$ equivalent, the deterioration in search speed shown for the median HV in the early stages of the search is unexpected. This sits alongside the previous unexpected strong performance of $p = \infty$ reported earlier.

7-objective instances: The parallel coordinates plots in Fig. 8 have already indicated the problems encountered by the static schemes in converging to the global front of the 7-objective CAT5 and CAT6 tests. In the concave CAT5 test, the median HV result shows surprisingly poor performance in the early part of the search in comparison to both the $p = 1$ and $p = \infty$ cases; however performance is similar across all three schemes by the end of the budget (Fig. 13). Looking across all 11 runs, the IGD results hint – again, surprisingly – that the dynamic scheme might be faring worse than $p = 1$ on this concave problem. Results for CAT6 are not shown, but follow a similar trend to the CAT5 findings.

The bi-objective experiments offer partial support to hypothesis 3. For the concave geometry, the dynamic scheme showed little deterioration over using a static $p = \infty$, but no speed-up was observed during the early stages of the search. It is possible that this finding may be because DTLZ2 with only 5 variables in the 'g' function is an easy problem to solve. Meanwhile, the dynamic result on the convex problem supported hypothesis 3, offering improved performance over

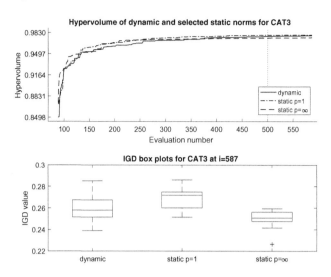

Fig. 11. HV convergence profiles and IGD box plots for dynamic CAT3 tests

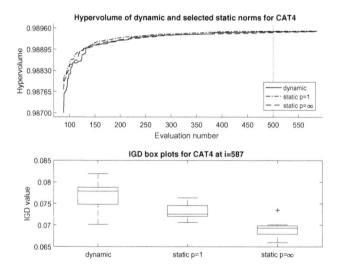

Fig. 12. HV convergence profiles and IGD box plots for dynamic CAT4 tests

$p = \infty$. The 4-objective experiment on the concave problem is similar to the bi-objective result (offering partial support to hypothesis 3), but the result in the convex case is confounded by the curious, rapid convergence of the static $p = \infty$ norm. The 7-objective experiments offer no support to hypothesis 3, with some evidence that the dynamic approach is actually harmful to convergence.

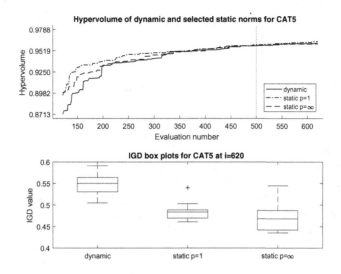

Fig. 13. HV convergence profiles and IGD box plots for dynamic CAT5 tests

6   Conclusion

The paper has investigated two hypotheses that arise from the theory of decomposition based, multi-objective optimization: (1) that the chosen scalarization norm must be of a large enough order to capture the geometry of the front; (2) that higher-order norms converge more slowly than their lower-order equivalents. These issues are particularly important when the evaluation budget for the optimization process is heavily constrained, and so we focused our inquiry on a class of surrogate-based methods for expensive optimization represented by the seminal ParEGO algorithm. Our practical experiments with a new implementation of ParEGO suggest that, for optimization on a limited budget, these hypotheses hold for bi-objective problems but are not supported in many-objective situations. The third hypothesis we considered involved the potential for dynamic adjustment of the norm (from lower-order earlier in the search to higher-order later in the search) to overcome the trade-off implied by hypotheses 1 and 2 for problems with concave geometries. Our experiments showed that, for bi-objective concave problems, the dynamic norm was a more favourable choice than $p = 1$ but did not offer accelerated convergence over $p = \infty$ and, as such, did not represent a 'dominating' option. This favourability was retained in the 4-objective case, but had disappeared in the 7-objective case. Overall, the findings suggest that ParEGO (at least in the new implementation) may not be performing well for many-objective problems.

6.1   Limitations and Future Work

This project was limited by the resource constraints imposed on a 3-month postgraduate project. In a number of cases, convergence trajectories were slow but had not completely saturated within the number of evaluations available and it would have been useful to further understand these convergence trends. The number of replications was limited to 11, which is less than is desirable for robust statistical analysis. Specifically here, it would be interesting to further examine the apparent deterioration of the dynamic scheme in the 7-objective cases. We were also limited in the number of test problems for which we could investigate performance – DTLZ2 and its convex hybrid captured the Pareto front geometries of interest to our hypotheses, but it would have been potentially insightful to explore other (particularly, more challenging) problem formulations. In particular, it would be interesting to consider mixed convex-concave surfaces and asymmetrical problems. The dynamic norm scheme we implemented was quite crude and should be seen as a basis for further exploration. Such further work could involve both sensitivity analysis for schemes with a priori norm scheduling, but also consider adaptive schemes based around on-line convergence metrics.

Acknowledgments. This work was supported by Jaguar Land Rover and the UK-EPSRC grant EP/L025760/1 as part of the jointly funded *Programme for Simulation Innovation*. The authors thank Joshua Knowles for discussions on ParEGO and surrogate-based optimization that helped inspire the research directions in this paper.

References

1. Cristescu, C., Knowles, J.: Surrogate-based multiobjective optimization: ParEGO update and test. In: Workshop on Computational Intelligence (UKCI) (2015)
2. Davins-Valldaura, J., Moussaoui, S., Pita-Gil, G., Plestan, F.: ParEGO extensions for multi-objective optimization of expensive evaluation functions. J. Global Optim. **67**, 1–18 (2016)
3. Deb, K., Thiele, L., Laumanns, M., Zitzler, E.: Scalable test problems for evolutionary multiobjective optimization. In: Abraham, A., Jain, L., Goldberg, R. (eds.) Evolutionary Multiobjective Optimization. Springer, London (2005)
4. Fonseca, C.M., Paquete, L., López-Ibáñez, M.: An improved dimension-sweep algorithm for the hypervolume indicator. In: 2006 IEEE International Conference on Evolutionary Computation, pp. 1157–1163. IEEE (2006)
5. Giagkiozis, I., Fleming, P.J.: Methods for multi-objective optimization: an analysis. Inf. Sci. **293**, 338–350 (2015)
6. Giagkiozis, I., Lygoe, R.J., Fleming, P.J.: Liger: an open source integrated optimization environment. In: Proceedings of the 15th Annual Conference Companion on Genetic and Evolutionary Computation, pp. 1089–1096. ACM (2013)
7. Horn, D., Wagner, T., Biermann, D., Weihs, C., Bischl, B.: Model-based multiobjective optimization: taxonomy, multi-point proposal, toolbox and benchmark. In: Gaspar-Cunha, A., Henggeler Antunes, C., Coello, C.C. (eds.) EMO 2015. LNCS, vol. 9018, pp. 64–78. Springer, Cham (2015). doi:10.1007/978-3-319-15934-8_5
8. Huband, S., Hingston, P., Barone, L., While, L.: A review of multiobjective test problems and a scalable test problem toolkit. IEEE Trans. Evol. Comput. **10**(5), 477–506 (2006)
9. Jeong, S., Obayashi, S.: Efficient global optimization (EGO) for multi-objective problem and data mining. In: 2005 IEEE Congress on Evolutionary Computation, vol. 3, pp. 2138–2145. IEEE (2005)

10. Jones, D.R., Schonlau, M., Welch, W.J.: Efficient global optimization of expensive black-box functions. J. Global Optim. **13**(4), 455–492 (1998)
11. Knowles, J.: ParEGO: a hybrid algorithm with on-line landscape approximation for expensive multiobjective optimization problems. IEEE Trans. Evol. Comput. **10**(1), 50–66 (2006)
12. Mc Ginley, B., Maher, J., O'Riordan, C., Morgan, F.: Maintaining healthy population diversity using adaptive crossover, mutation, and selection. IEEE Trans. Evol. Comput. **15**(5), 692–714 (2011)
13. Miettinen, K., Ruiz, F., Wierzbicki, A.P.: Introduction to multiobjective optimization: interactive approaches. In: Branke, J., Deb, K., Miettinen, K., Słowiński, R. (eds.) Multiobjective Optimization. LNCS, vol. 5252, pp. 27–57. Springer, Heidelberg (2008). doi:10.1007/978-3-540-88908-3_2
14. Ponweiser, W., Wagner, T., Biermann, D., Vincze, M.: Multiobjective optimization on a limited budget of evaluations using model-assisted S-metric selection. In: Rudolph, G., Jansen, T., Beume, N., Lucas, S., Poloni, C. (eds.) PPSN 2008. LNCS, vol. 5199, pp. 784–794. Springer, Heidelberg (2008). doi:10.1007/978-3-540-87700-4_78
15. Scheffé, H.: Experiments with mixtures. J. Roy. Stat. Soc. Ser. B Methodol. **20**, 344–360 (1958)
16. Tabatabaei, M., Hakanen, J., Hartikainen, M., Miettinen, K., Sindhya, K.: A survey on handling computationally expensive multiobjective optimization problems using surrogates: non-nature inspired methods. Struct. Multidiscip. Optim. **52**(1), 1–25 (2015)
17. Van Veldhuizen, D.A., Lamont, G.B.: On measuring multiobjective evolutionary algorithm performance. In: Proceedings of the 2000 Congress on Evolutionary Computation, vol. 1, pp. 204–211. IEEE (2000)
18. Voutchkov, I., Keane, A.: Multi-objective optimization using surrogates. In: Tenne, Y., Goh, C.-K. (eds.) Computational Intelligence in Optimization, pp. 155–175. Springer, Heidelberg (2010)
19. Wagner, T.: Planning and multi-objective optimization of manufacturing processes by means of empirical surrogate models. Ph.D. thesis (2013)
20. Zaefferer, M., Bartz-Beielstein, T., Naujoks, B., Wagner, T., Emmerich, M.: A case study on multi-criteria optimization of an event detection software under limited budgets. In: Purshouse, R.C., Fleming, P.J., Fonseca, C.M., Greco, S., Shaw, J. (eds.) EMO 2013. LNCS, vol. 7811, pp. 756–770. Springer, Heidelberg (2013). doi:10.1007/978-3-642-37140-0_56
21. Zhang, Q., Liu, W., Tsang, E., Virginas, B.: Expensive multiobjective optimization by MOEA/D with Gaussian process model. IEEE Trans. Evol. Comput. **14**(3), 456–474 (2010)

Multi-objective Big Data Optimization
with jMetal and Spark

Cristóbal Barba-Gonzaléz, José García-Nieto,
Antonio J. Nebro$^{(\boxtimes)}$, and José F. Aldana-Montes

Dept. de Lenguajes y Ciencias de la Computación, ETSI Informática,
University of Malaga, Campus de Teatinos, 29071 Malaga, Spain
{cbarba,jnieto,antonio,jfam}@lcc.uma.es

Abstract. Big Data Optimization is the term used to refer to optimization problems which have to manage very large amounts of data. In this paper, we focus on the parallelization of metaheuristics with the Apache Spark cluster computing system for solving multi-objective Big Data Optimization problems. Our purpose is to study the influence of accessing data stored in the Hadoop File System (HDFS) in each evaluation step of a metaheuristic and to provide a software tool to solve these kinds of problems. This tool combines the jMetal multi-objective optimization framework with Apache Spark. We have carried out experiments to measure the performance of the proposed parallel infrastructure in an environment based on virtual machines in a local cluster comprising up to 100 cores. We obtained interesting results for computational effort and propose guidelines to face multi-objective Big Data Optimization problems.

Keywords: Multi-objective optimization · Big Data · jMetal · Spark · Parallel computing

1 Introduction

Over the past few years, Big Data technologies have attracted more and more attention, leading to an upsurge in research, industry and government applications. There are multiple opportunities and challenges in Big Data research. One area in particular where Big Data is promising is Global Optimization [26]. The issue is that Big Data optimization problems may need to access a massive amount of data to be solved, which introduces a new dimension of complexity apart from features such as non-linearity, uncertainty and conflicting objectives.

Focusing on multi-objective optimization, metaheuristic search methods, such as evolutionary algorithms, have been widely applied to a great number of academic and industry optimization problems [6]. Depending on the problem, a metaheuristic may need to perform thousands or even millions of solution evaluations. In the case of complex system optimization, the computational effort, in terms of time consumption and resource requirements, to evaluate the quality

© Springer International Publishing AG 2017
H. Trautmann et al. (Eds.): EMO 2017, LNCS 10173, pp. 16–30, 2017.
DOI: 10.1007/978-3-319-54157-0_2

of solutions can make it impracticable to apply current optimization strategies to such problems. This issue is even harder when dealing with Big Data environments, where a huge volume of data hast to be accurately and quickly managed.

One strategy to cope with these difficulties is to apply parallelism [17]. In the last few years, a number of approaches consisting in adapting metaheuristic techniques to work in parallel on Hadoop, the de facto Big Data software platform, have been proposed. These proposals are related to data mining or data management applications, such as: feature selection [1], data partitioning [12], dimension reduction [23], pattern detection [5], graph inference [16], and task scheduling [22]. Most of these approaches are based on the MapReduce programming model [15].

However, MapReduce entails a series of drawbacks that make it unsuitable to be properly integrated with metaheuristics in particular and with global optimization techniques in general. Chief among them are: high latency queries, non-iterative programming model, and weak real-time processing. Therefore, there is a demand for new challenging approaches to integrate Big Data based technologies with global optimization algorithms in order to cope with all these issues.

In this paper, our approach is to address the parallelization of metaheuristics in Hadoop-based systems with Apache Spark [25], which is defined as a fast and general engine for large-scale data processing. Our proposal is to use the jMetalSP framework[1], which combines Spark with the jMetal multi-objective optimization framework [11][2]. Concretely, we have included support in jMetalSP to parallelize metaheuristics with Spark in an almost transparent way, hence avoiding the intrinsic shortcomings of the usual MapReduce model when applied to global optimization.

We aim to consider two scenarios: first, to use Spark as an engine to evaluate the solutions of a metaheuristic in parallel, and, second to study the influence of accessing a massive amount of data in each evaluation of a metaheuristic algorithm. Instead of focusing on a particular optimization problem, we have defined a generic scenario, in which a benchmark problem is modified to artificially increase its computing time and to read data from the Hadoop file system (HDFS). We have carried out a number of experiments to measure the performance of the proposed method in a parallel virtualization infrastructure of multiple machines in an in-house cluster.

The main contributions of this paper are as follows:

– We provide a software solution for parallelizing multi-objective metaheuristics included in the jMetal framework to take advantage of the high performance cluster computing facilities provided by Spark. This way, developers and practitioners are provided with an attractive tool for Big Data Optimization.
– We perform a thorough experimentation of our proposal from three different viewpoints. First, by measuring the performance of algorithms in terms of computational effort in a Hadoop parallel environment; second, by analyzing

---

[1] In URL https://github.com/jMetal/jMetalSP.
[2] In URL http://jmetal.github.io/jMetal/.

the influence of accessing data stored in HDFS in each evaluation of a meta-heuristic; and third, by combining both approaches. To carry out this study, we define different data access/processing tasks to be done when evaluation a solution, so we can measure the performance of the algorithms according to computational effort and size of data. This allows us to compute the speedups that can be obtained and identify the system's limits, to determine whether or not it is worth using more resources to solve the problem.

The remainder of this article is organized as follows. The next section presents an overview of the related work in the literature. Section 3 details our Big Data optimization approach. In Sect. 4, the experimental framework and parameters settings are described. Section 5 details the experimental results and analyses. Finally, Sect. 6 outlines some concluding remarks and plans for future work.

2   Related Work

Large amounts of data and high dimensionality characterizes many optimization problems in interdisciplinary domains such as, biomedical sciences, engineering, finance, and social sciences. This means that optimization problems handling such spatio-temporal restrictions often deal with tens of thousands of variables or features extracted from documents, images and other objects.

To tackle such challenging problems, a series of proposals have appeared in the last decade, which combine metaheuristics with data mining or data management applications, and adapt them to perform in Hadoop environments. Concretely, in [1] a swarm intelligence approach is adapted to optimize the features that exist in large protein sequences using a two-tier hybrid model by applying both filter and wrapper methods. A similar approach is proposed in [12], where a particle swarm optimization algorithm (PSO) is used to discover clusters in data that are continuously captured from students' learning interactions. In [2], intrusion detection is managed with a MapReduce strategy based on a PSO clustering algorithm.

An interesting method has been reported in [23], in which, by using sensor data to generate a PCA (Principal Component Analysis) model to forecast photovoltaic energy, it is possible to reduce the dimensionality of data with the collaboration of other artificial intelligence techniques, such as: fuzzy interference, neural networks, and genetic algorithms.

In the case of biomedical sciences, the reconstruction of gene regulatory networks is a complex optimization problem that is attracting particularly special from the research community, since it is considered to be a potential Big Data problem in the specialized literature [3,26]. Along the same lines, in the study carried out in [16], the authors proposed a parallel method consisting in a hybrid genetic algorithm with PSO by means of the MapReduce programming model. The resulting approach was tested in different cloud computing environments.

Most of these approaches are based on the MapReduce (MR) programming model [15], which yields a competitive performance in comparison with other parallel (and sequential) models for the specific problems tackled. In addition, other algorithmic adaptations to MapReduce operations can be found in meta-heuristics, such as: PSO [18], Differential Evolution [9] and Ant Colony Optimization [24]. However, as stated, the MapReduce model entails a series of drawbacks that make it unsuitable to be integrated with metaheuristics in particular and with global optimization techniques in general. These are directly related to: high-latency queries, a non-iterative programming model, and weak real-time processing. For instance, the following issues which we aim to cover with our proposal, combining jMetal and Spark:

– MapReduce uses coarse-grained tasks to do its work, which can be too heavy-weight for iterative algorithms, like metaheuristic algorithms. In the proposals analyzed, developers use various MapReduce hacks or alternative tools to overcome these limitations, but this highlights the need for a better computation engine that supports these algorithms directly, while continuing to support more traditional batch processing of large datasets. Our software proposal follows an iterative programming model, which eases the adaptation of algorithms and the integration with software classes managing a multitude of optimization problems.
– Another problem with current optimization algorithms using MapReduce is that they have no awareness of the total pipeline of Map plus Reduce steps, so they cannot cache intermediate data in memory for faster performance. Instead, they flush intermediate data to disk between each step. With Spark the managed data can be cached in memory explicitly, thus improving performance significantly.
– Existing proposals in the literature were not evaluated on well-grounded Big Data environments. Most of them were tested to show their ability to solve a given optimization problem in a parallel infrastructure composed of up to ten machines, and therefore critical aspects such as data volume and variable computational effort remain open issues. In the present study, a thorough experimentation is carried out to measure the performance of algorithms in terms of scalability in a 100-core Hadoop-based cluster, and to analyze the influence of accessing a large amount of data in each evaluation of a multi-objective metaheuristic.

3   Big Data Optimization Approach

To develop Big Data optimization applications it is necessary to have software tools capable of coping with the requirements of such applications. Our contribution in this sense is to propose jMetalSP [4,7], an open-source platform[3] combining the jMetal optimization framework [11] with the Apache Spark cluster computing system [25]. Below we describe the adopted Big Data Optimization scheme in the context of jMetal, Spark, and jMetalSP.

---

[3] URL: https://github.com/jMetal/jMetalSP.

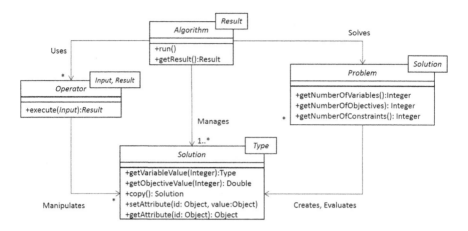

Fig. 1. Class diagram with core classes and interfaces of jMetal 5.

Algorithm 1. Template of a metaheuristic

1: $A(0) \leftarrow$ GenerateInitialSolutions()
2: $t \leftarrow 0$
3: Evaluate($A(0)$)
4: **while not** StoppingCriterion( ) **do**
5:     $S(t) \leftarrow$ Generation($A(t)$)
6:     Evaluate($S(t)$)
7:     $A(t+1) \leftarrow$ Update($A(t)$, $S(t)$)
8:     $t \leftarrow t+1$
9: **end while**

---

jMetal is an algorithmic framework, which includes a number of optimization metaheuristics of the state of the art [11]. It mostly centers in multi-objective optimization, although it also provides single-objective algorithms. In the work presented here, we use jMetal 5 [19], which follows the architecture depicted in Fig. 1. The underlying idea is that an algorithm (metaheuristic) manipulates a number of solutions with some operators to solve an optimization problem.

jMetal 5 provides algorithm templates mimicking the pseudo-code of a generic metaheuristic like that shown in Algorithm 1, where a set A of some initial solutions is iteratively updated by generating a set S of new solutions until a stopping condition is achieved. Another feature of jMetal is that it offers an interface (called `SolutionListEvaluator`) to encapsulate the evaluation of a list of solutions (i.e., a population in the context of evolutionary algorithms):

```
public interface SolutionListEvaluator<S> {
  List<S> evaluate(List<S> solutionList, Problem<S> problem);
  void shutdown();
}
```

The encapsulated behavior of this interface is that the method `evaluate` of the problem (see Fig. 1) is applied to all the solutions in the list, yielding to a new list of evaluated solutions. Metaheuristics using this interface can be empowered with different evaluator implementations, so that the current way of evaluating solutions is transparent to the algorithms. For example, many algorithms incorporate a method similar to this one (which corresponds to step 6 of the template shown in Algorithm 1):

```
protected List<S> evaluatePopulation(List<S> population) {
    population = evaluator.evaluate(population, problem);
    return population;
}
```

In this way, the actual evaluator is instantiated when configuring the settings of the metaheuristic, so no changes in the code are needed. jMetal 5 currently includes two implementations of `SolutionListEvaluator`: sequential and multi-threaded. Our approach has been then to develop an evaluator based on Spark.

Apache Spark [25] is based on the concept of Resilient Distributed Datasets (RDD), which are collections of elements that can be operated in parallel on the nodes of a cluster by using two types of operations: transformations (e.g., map, filter, union, etc.) and actions (e.g., reduce, collect, and count). The Spark based evaluator in jMetal creates an RDD with all the solutions to be evaluated, and a map transformation is used to evaluate each solution. The evaluated solutions are then collected and returned to the algorithm.

It is worth noting that algorithms do not need to be modified to use Spark, although the problem to be solved must fulfill the requirements imposed by this platform, as the algorithms run map processes. For example, the evaluate method of the problem must not modify variables outside the scope of the RDD containing the list of solutions to be evaluated.

Currently, five multi-objective metaheuristics in jMetal 5 use evaluators, so all of them can take advantage of the one based on Spark: NSGA-II [10], SPEA2 [27], SMPSO [20], GDE3 [14] and PESA2 [8]. This scheme can be also used in a number of single-objective algorithms: generational genetic algorithm (gGA), differential evolution (DE), and two PSO algorithms.

jMetalSP is a new project for Big Data Optimization with multi-objective metaheuristics [4] based on jMetal and Spark. It is currently intended to solve dynamic multi-objective optimization problems in Hadoop environments by using the streaming data processing capabilities of Spark, while jMetal provides the optimization infrastructure for implementing the dynamic problems and the dynamic algorithms to solve them. We have extended jMetalSP with the Spark evaluator, so it can be used also to solve non-dynamic optimization problems.

The attractive point of the adopted approach is that a number of single and multi-objective metaheuristics can be executed in parallel without requiring any modification. If we look closely at steps 3 and 6 in Algorithm 1 we can see that the set S can be evaluated in parallel by the Spark-based evaluator, so the resulting parallel model is a heartbeat algorithm: a parallel step is alternated with a sequential one (for the rest of the phases in the main loop of the metaheuristic).

This obviously prevents linear speedups, and this is the price to pay for having a very simple mechanism that uses jMetal's algorithms in a Big Data infrastructure. In this paper, our main interest is to measure the effective time reductions that can be achieved in different contexts.

4   Experimental Framework

To evaluate the performance of the proposed approach, a series of experiments have been conducted from three points of view: (1) computational effort, in terms of which we measure the performance of the parallel model; (2) data management, which is focused on testing the ability to manage a large number of data files; and (3) a combination of (1) and (2). Therefore, for each, we follow a different problem configuration involving: time consuming delays, different data block sizes, and different cluster sizes. We are seeking the maximum limit that a multi-objective algorithm can manage when handling Big Data without a loss in performance.

For this reason, with the aim of guiding us and to simplify this experimentation, we have centered on one optimization problem and one algorithm to solve it, although without emphasizing on the solution quality, as the behavior of the parallel algorithm is the same as its sequential counterpart. Specifically, we have selected the NSGA-II algorithm [10] and the multi-objective optimization problem ZDT1 [13].

From an algorithmic point of view, we have used a common parameter setting of NSGA-II to test our proposal in all the experiments. The variation operators are SBX crossover and Polynomial mutation, with crossover and mutation rates $p_c = 0.9$ and $p_c = 1.0/L$, respectively (L being the number of decision variables of the problem), and both having a distribution index value of 20. The selection

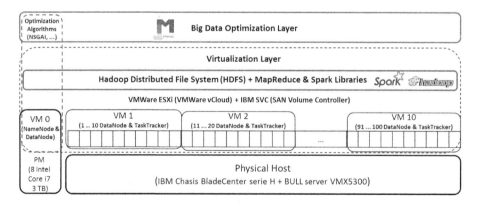

Fig. 2. Computational environment for Big Data Optimization used to test the performance of the proposed jMetal+Spark software solution, under different conditions of computational effort and Big Data management

strategy is binary tournament and the population size has been set to 100 individuals. Finally, the stopping condition is met when the total number of 25,000 candidate solutions have been evaluated. In others words, the NSGA-II performs 250 evolution steps or iterations of the population throughout the running time.

All the experiments have been conducted in a virtualization environment running on a private high-performance cluster computing platform. This infrastructure is located at the *Ada Byron Research Center* at the University of Málaga (Spain), and comprises a number of IBM hosting racks for storage, units of virtualization, server compounds and backup services.

Our virtualization platform is hosted in this computational environment, whose main components are illustrated in Fig. 2. Concretely, this platform is made up of 10 virtual machines (VM1 to VM10), each one with 10 cores, 10 GB RAM and 250 GB virtual storage (adding up to 100 cores, 100 GBs of memory and 2.5 TB HD storage). These virtual machines are used as Slave nodes with the role of TaskTracker (Spark) and DataNode (HDFS) to perform fitness evaluations of algorithmic candidate solutions in parallel. The Master node, which runs the core algorithm (NSGA-II), is hosted in a different machine (VM0) with 8 Intel Core i7 processors at 3.40 GHz, 32 GB RAM and 3 TB storage space. All these nodes are configured with a Linux CentOS 6.6 64-bit distribution. The whole cluster is managed with Apache Ambari 1.6.1 and executes the Apache Hadoop version 2.4.0. This Hadoop distribution integrates HDFS and Apache Spark 1.6.

The jMetalSP framework is then deployed on this infrastructure, providing optimization algorithms with Spark methods to evaluate candidate solutions in parallel, in addition to managing HDFS files.

5   Experiments

This section describes the set of experiments and the analyses carried out to test our approach. We focus on the speedup and efficiency analysis in terms of computational effort and data management.

5.1   Speedup and Efficiency

One of the most widely used indicators for measuring the performance of a parallel algorithm is the *Speedup* (S_N). The standard formula of the speedup is represented in Eq. 1 and calculates the ratio of T_1 over T_N, where T_1 is the running time of the analyzed algorithm in 1 processor and T_N is the running time of the parallelized algorithm on N processing units (processors or cores).

$$S_N = \frac{T_1}{T_N} \tag{1}$$

$$E_N = \frac{S_N}{N} \times 100 \tag{2}$$

Table 1. Experimental results of NSGA-II (jMetalSP) executed on 1, 10, 20, 50, and 100 cores with different time delays in each problem evaluation

Delay	Running time (h)					Speedup				Efficiency			
	T_1	T_{10}	T_{20}	T_{50}	T_{100}	S_{10}	S_{20}	S_{50}	S_{100}	E_{10}	E_{20}	E_{50}	E_{100}
10 s	76.50	10.90	8.10	5.00	3.50	7.02	9.44	15.30	21.85	70.20%	47.22%	30.60%	21.85%
30 s	216.70	34.10	24.40	16.30	10.80	6.35	8.88	13.29	20.06	63.50%	44.41%	26.59%	20.06%

A related measure is the *Efficiency* of a parallel algorithm, which is calculated with the formula of Eq. 2. An algorithm scales linearly (ideal) when it reaches a speedup $S_N = N$ and hence, the parallel efficiency is $E_N = 100\%$.

5.2 Computational Effort

As stated, to measure the parallel computing performance of our approach we have used the NSGA-II algorithm to solve a modified version of the ZDT1 problem. The running time of NSGA-II with those settings in a laptop equipped with an Intel i7 processor is less than a second, so we have artificially increased the computing time of the evaluation functions of ZDT1 (by adding an idle loop) to simulate a real scenario where the total computing time would be in the order of several hours. After a number of preliminary experiments, we set the evaluation time to two values, 10 and 30 s; this way, we estimated the total running time of the sequential NSGA-II algorithm to be around 76.5 and 216.7 h, respectively.

Table 1 shows the running time in hours used by the NSGA-II approach of jMetalSP running on 1, 10, 20, 50, and 100 cores, with regards to the two delays considered, applied in each solution evaluation. This table also contains the corresponding speedup and efficiency values to the resulting times. As mentioned, T_1 (one core) is 76.5 h or 3.19 days in the case of the 10 s delay and 216.7 h or 9.03 days with a delay of 30 s. As expected, these times are reduced in relation to the increase in the number of cores used in the parallel model. The highest reductions in time are obtained when our approach is configured with 100 cores in parallel, for which the running time is reduced to 3.5 h (95.42%) in the case of a delay of 10 s and to 10.8 h (95.02%) using a 30 s delay.

In terms of efficiency, the highest percentage 70.2% is reached with 10 cores and it decreases as the number of resources gets larger, to reach 47.22%, 30.6%, and 21.85% with 20, 50, and 100 cores, respectively.

This behavior was somewhat expected as the parallel model is based on alternating parallel and sequential steps. Considering the results, it is worth mentioning that both problem configurations yield similar speedup and efficiency values, which indicates that the bottleneck is due to both the parallel model and the parallel infrastructure, so increasing the evaluation time does not compensate the synchronization and communication costs.

5.3   Data Block Size

We now turn to a number of experiments to measure the influence of using different data block sizes (file sizes) on the evaluation process of NSGA-II. For this purpose, we have set NSGA-II to manage data files with different sizes in each problem evaluation. This way, we have centered on file sizes of 32 MB, 64 MB, and 128 MB; hence, each complete algorithm's execution manages a total volume of 0.8 TB, 1.6 TB, and 3.2 TB, respectively. We have arranged a pool of files stored in HDFS totalling 1 TB, so the pool size is 1 TB/file size.

Table 2. Experimental results of NSGA-II (jMetalSP) executed on 1, 10, 20, 50, and 100 cores with different block sizes management in each problem evaluation

Block	Running time (h)					Speedup				Efficiency			
	T_1	T_{10}	T_{20}	T_{50}	T_{100}	S_{10}	S_{20}	S_{50}	S_{100}	E_{10}	E_{20}	E_{50}	E_{100}
32 MB	41.47	2.47	1.94	2.18	2.56	16.78	21.37	19.02	16.19	167.79%	106.8%	32.38%	16.19%
64 MB	41.28	4.20	3.72	3.99	4.52	9.82	11.09	10.34	9.13	98.20%	55.45%	20.68%	9.13%
128 MB	48.45	7.89	8.94	9.92	9.47	6.14	5.41	4.88	5.11	61.40%	27.05%	9.76%	5.11%

Once a problem solution has been evaluated, the optimization algorithm randomly (uniformly) selects a file from the pool. Neither additional computing tasks nor delays are performed in the evaluation step for this analysis.

Table 2 shows the running time in hours used by the NSGA-II approach of jMetalSP running on 1, 10, 20, 50, and 100 processors, for the three different configurations of file size. As in the first experiments, the sequential time T_1 is an estimation. In addition, this table contains the speedup and efficiency with regards to the resulting computing times. As we can observe, the running time with one core T_1 took approximately 1.72 days in the case of 32 MB (41.47 h) and 64 MB (41.28 h) file sizes, and it was close to 2 days (48.45 h) in the case of 128 MB. In fact, in the context of one processing unit, for the 32 MB file size the optimization algorithm took longer than for 64 MB, which could be due to the optimal block size in HDFS, which is 64 MB [21] since no extra operations of either wrapping or splitting are required with this file size.

As expected, the running time is reduced in general when using a higher number of nodes in the parallel model. However, the highest reductions in running time are not obtained with 100 cores, but rather with 10 and 20. For the latter, the execution time (T_{20}) is 1.94 h in the case of 32 MB file size and 3.72 h when using 64 MB, which means a reduction of 95.32% and 90.98%, respectively, with regards to the algorithm running time in one processor (T_1). In the case of 128 MB file size, the highest reduction in computing time (83.71%) is reached when running NSGA-II in parallel with 10 nodes (T_{10}). Although this behavior could be counter-intuitive, the fact of focusing only on data management without spending extra computational effort in evaluating the solutions, causes the network throughput in the parallel model to impair the overall performance of the optimization approach, since the greater the number of cores used, the higher the transference of data between nodes.

Table 3. Experimental results of NSGA-II (jMetalSP) executed on 1, 10, 20, 50, and 100 processors with different time delays and 64 MB block size management in each problem evaluation

Delay	Running time (h)					Speedup				Efficiency			
	T_1	T_{10}	T_{20}	T_{50}	T_{100}	S_{10}	S_{20}	S_{50}	S_{100}	E_{10}	E_{20}	E_{50}	E_{100}
10 s	85.90	19.50	11.80	6.80	7.40	4.40	7.28	12.63	11.61	44%	36.40%	25.26%	11.61%
30 s	237.5	39.90	31.70	17.30	12.50	5.95	7.49	13.73	19.01	59.50%	37.45%	27.46%	19.01%

In terms of speedup, a number of 10 and 20 cores in the parallel model reach close to linear (even superlinear) speedup values, although they get worse with a higher number of nodes. In fact, the most efficient configuration is obtained with 10 nodes for the three file sizes. As our computational environment is composed of 10 virtual machines with 10 cores each (see Fig. 2), it is clear that the algorithmic configuration with 10 nodes is able to launch all jobs within the same machine, thereby avoiding having to use the underlying network and reaching the highest efficiency in terms of data management. The achievement of superlinear speedups is related to the application of data cache and file replication techniques in the underlying Hadoop system.

5.4   Computing Data Plus Access Performance

Following the two experiments in the context of parallel numerical performance and parallel data access performance, the next step is to combine both aspects to simulate a real Big Data optimization problem, which involves both heavy data access and data processing. Consequently, we have designed a new experiment in which each evaluation task requires reading a file of 64 MB size and then simulating a computational effort of 10 and 30 s.

The resulting times, as well as speedup and efficiency metrics, are shown in Table 3. We can observe that the running times are different depending on the computational delay in the evaluations. With 10 s, the highest reduction in running time is not obtained with 100 cores, but with 50: the execution time (T_{50}) is 6.8 h which means a reduction of 92.08% with regards to the one processor running time (T_1); the efficiency value is 25.26%. In the case of using a 30 s delay, the best running time is with 100 cores with a reduction of 94.73% with regards to one core. In this case, the efficiency value is 19.01%.

5.5   Discussions

As a general observation, in the scope of our experimental framework, Fig. 3 plots the running times consumed by our jMetalSP approach executed on 1, 10, 20, 50, and 100 cores with all the used configurations of computational delay (10 s and 30 s) and data management (32 MB, 64 MB, and 128 MB), plus the experiment combining computational effort and data access.

In this figure, it is clearly observable that the highest reduction in computing time is reached when the evaluation of solutions in the evolution process of

NSGA-II entails a computational effort (30 s delay), as well as data management (64 MB file size). So, a complete execution of the optimization algorithm requires several days (as much as 10) in a platform environment with fewer than 10 cores, although this computational time is reduced to less than a day when using more than 20 cores in our parallel model. We therefore can suggest that a good trade-off between performance and resource requirement is obtained with a configuration of jMetalSP in the range of 20 to 50 cores, since all jobs are finalized in less than half a day and it is 50% more cost efficient.

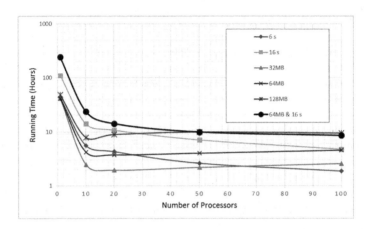

Fig. 3. Running time in hours (logarithmic scale) of jMetalSP version of NSGA-II executed on 1, 10, 20, 50 and 100 cores with all configurations of computational delay and data management

The results we have obtained in the experiments suggest demonstrates the systems limits when dealing with a real-world Big Data optimization problem. The stopping condition of the NSGA-II algorithm used is to compute 25,000 function evaluations, which is a usual value used to optimize the ZDT1 problem. However, in a real scenario the number of evaluations would be most certainly beyond that number.

If we consider the lowest times of the experiments carried out, they range from 1.86 h (computational effort, 10 s delay, 100 cores) to 8.58 h (combined experiment, 100 cores). Assuming that the algorithm would need to compute 100,000 function evaluations, the times would increase to 7.44 and 34.32 h, respectively. While the first time could be acceptable, the second could be at the limit of an acceptable value. Consequently, augmenting the number of evaluations to the order of 1 million or more would go beyond a reasonable time in our Hadoop-based system, and specific algorithms should be designed instead of merely using the standard NSGA-II metaheuristic, e.g. by incorporating local search methods or by designing problem-oriented operators to reduce the number of evaluations to achieve satisfactory solutions.

Finally, it is worth mentioning that a similar experiment has been carried out with SMPSO [20], a swarm-intelligence multi-objective approach, in the scope of the experimental framework used here. SMPSO performs a different learning procedure from NSGA-II, although they both use similar mechanisms in terms of solution evaluation and multi-objective operation, i.e., archiving strategy and crowding density estimator. As expected, SMPSO registered similar speedups to NSGA-II, which leads us to claim that our jMetalSP approach behaves efficiently with different optimizers.

6   Conclusions

In this paper, we have presented a study related to multi-objective Big Data optimization. Our goal has been twofold: first, we have extended the jMetalSP framework (which combines jMetal and Spark) in such a way that jMetal metaheuristics can make use of the Apache Spark distributed computing features almost transparently; second, we have studied the performance of running a metaheuristic based in our parallel scheme on three scenarios: parallel computation, parallel data access, and a combination of both. We have carried out experiments to measure the performance of the proposed parallel infrastructure in an environment based on virtual machines in a local cluster composed of up to 100 cores.

The results lead us to get interesting conclusions about computational effort and to propose guidelines when facing Big Data optimization problems:

- Our approach is able to obtain actual reductions in computing time from more than a week to just half a day, when addressing complex and time consuming optimization tasks. This performance has been obtained in the scope of an in-house (virtualized) computational environment with limited resources.
- In those experiments where we only focused on data management, without spending extra computational effort on evaluating the solutions, the overall performance of the parallel model is usually impaired when using a large number of nodes (more than 50), because of the network overhead and the increasing data transfer between nodes.
- A noteworthy trade-off between performance and resource requirements is obtained with a configuration of jMetalSP in the range of 50 to 100 cores.

As for future work, we plan to evaluate the performance of our proposal in the context of public in-cloud environments, with the aim of studying whether a different behavior in the optimization of Big Data problems is observed in these kinds of environments (compared with virtualized), or not. As a second line of future work, we intend to use the available optimization algorithms in jMetalSP to deal with real-world complex and data intensive optimization problems.

Acknowledgement. This work has been partially funded by Grants TIN2014-58304-R (Spanish Ministry of Education and Science) and P11-TIC-7529 (Innovation, Science and Enterprise Ministry of the regional government of the Junta de Andalucía) and P12-TIC-1519 (Plan Andaluz de Investigación, Desarrollo e Innovación). Cristóbal Barba-González is supported by Grant BES-2015-072209 (Spanish Ministry of Economy and Competitiveness).

References

1. Abdul-Rahman, S., Bakar, A.A., Mohamed-Hussein, Z.-A.: Optimizing big data in bioinformatics with swarm algorithms. In: IEEE 16th International Conference on Computational Science and Engineering (CSE), pp. 1091–1095, December 2013
2. Aljarah, I., Ludwig, S.A.: Mapreduce intrusion detection system based on a particle swarm optimization clustering algorithm. In: IEEE Congress on Evolutionary Computation (CEC 2013), pp. 955–962, June 2013
3. Thomas, S.A., Jin, Y.: Reconstructing biological gene regulatory networks: where optimization meets big data. Evol. Intel. **7**(1), 29–47 (2014)
4. Barba-González, C., Nebro, A.J., Cordero, J.A., García-Nieto, J., Durillo, J.J., Navas-Delgado, I., Aldana-Montes, J.F.: jMetalSP: a framework for dynamic multi-objective big data optimization. Applied Soft Computing (2016, submitted)
5. Cabanas-Abascal, A., García-Machicado, E., Prieto-González, L., de Amescua Seco, A.: An item based geo-recommender system inspired by artificial immune algorithms. J. Univ. Comput. Sci. **19**(13), 2013–2033 (2013)
6. Coello, C., Lamont, G.B., van Veldhuizen, D.A.: Multi-objective Optimization Using Evolutionary Algorithms, 2nd edn. Wiley, New York (2007)
7. Cordero, J.A., Nebro, A.J., Barba-González, C., Durillo, J.J., García-Nieto, J., Navas-Delgado, I., Aldana-Montes, J.F.: Dynamic multi-objective optimization with jmetal and spark: a case study. In: Pardalos, P.M., Conca, P., Giuffrida, G., Nicosia, G. (eds.) MOD 2016. LNCS, vol. 10122, pp. 106–117. Springer, Heidelberg (2016). doi:10.1007/978-3-319-51469-7_9
8. Corne, D.W., Jerram, N.R., Knowles, J.D., Oates, M.J.: PESA-II: region-based selection in evolutionary multi-objective optimization. In: Genetic and Evolutionary Computation Conference (GECCO 2001), pp. 283–290. Morgan Kaufmann (2001)
9. Daoudi, M., Hamena, S., Benmounah, Z., Batouche, M.: Parallel diffrential evolution clustering algorithm based on MapReduce. In: 6th International Conference of Soft Computing and Pattern Recognition (SoCPaR 2014), pp. 337–341 (2014)
10. Deb, K., Pratap, A., Agarwal, S., Meyarivan, T.: A fast and elitist multiobjective genetic algorithm: NSGA-II. IEEE Trans. Evol. Comput. **6**(2), 182–197 (2002)
11. Durillo, J.J., Nebro, A.J.: jMetal: a Java framework for multi-objective optimization. Adv. Eng. Softw. **42**, 760–771 (2011)
12. Govindarajan, K., Somasundaram, T.S., Kumar, V.S., Kinshuk: Continuous clustering in big data learning analytics. In: IEEE Fifth International Conference on Technology for Education (T4E), pp. 61–64, December 2013
13. Kitzler, E., Deb, K., Thiele, L.: Comparasion of multiobjective evolutionary algorithms: empirical results. Evol. Comput. **8**(2), 173–195 (2000)
14. Kukkonen, S., Lampinen, J.: GDE3: the third evolution step of generalized differential evolution. In: IEEE Congress on Evolutionary Computation (CEC 2005), pp. 443–450 (2005)
15. Lammel, R.: Google's MapReduce programming model revisited. Sci. Comput. Program. **70**(1), 1–30 (2008)

16. Lee, W., Hsiao, Y., Hwang, W.: Designing a parallel evolutionary algorithm for inferring gene networks on the cloud computing environment. BMC Syst. Biol. **8**(1) (2014). http://bmcsystbiol.biomedcentral.com/articles/10.1186/1752-0509-8-5
17. Luque, G., Alba, E.: Parallel Genetic Algorithms, 1st edn. Springer, Heidelberg (2011)
18. McNabb, A.W., Monson, C.K., Seppi, K.D.: Parallel PSO using MapReduce. IEEE Cong. Evol. Comput. CEC **2007**, 7–14 (2007)
19. Nebro, A.J., Durillo, J.J., Vergne, M.: Redesigning the jMetal multi-objective optimization framework. In: Genetic and Evolutionary Computation Conference (GECCO 2015) Companion, pp. 1093–1100, July 2015
20. Nebro, A.J., Durillo, J.J., García-Nieto, J., Coello Coello, C.A., Luna, F., Alba, E.: SMPSO: a new PSO-based metaheuristic for multi-objective optimization. In: IEEE Symposium on Computational Intelligence in Multicriteria Decision-Making (MCDM 2009), pp. 66–73. IEEE Press (2009)
21. Shvachko, K., Kuang, H., Radia, S., Chansler R.: The Hadoop distributed file system. In: Proceedings of the 2010 IEEE 26th Symposium on Mass Storage Systems and Technologies (MSST 2010), Washington, DC, USA, pp. 1–10. IEEE Computer Society (2010)
22. Sun, W., Zhang, N., Wang, H., Yin, W., Qiu, T.: PACO: a period ACO based scheduling algorithm in cloud computing. In: International Conference on Cloud Computing and Big Data (CloudCom-Asia), pp. 482–486, December 2013
23. Tannahill, K.B., Jamshidi, M.: System of systems and big data analytics bridging the gap. Comput. Electr. Eng. **40**(1), 2–15 (2014)
24. Wu, B., Wu, G., Yang, M.: A MapReduce based ant colony optimization approach to combinatorial optimization problems. In: 8th International Conference on Natural Computation (ICNC 2012), pp. 728–732, May 2012
25. Zaharia, M., Chowdhury, M., Franklin, M.J., Shenker, S. Stoica, I.: Spark: cluster computing with working sets. In: Proceedings of the 2nd USENIX Conference on Hot Topics in Cloud Computing, HotCloud 2010, Berkeley, CA, USA, p. 10. USENIX Association (2010)
26. Zhou, Z., Chawla, N.V., Jin, Y., Williams, G.J.: Big data opportunities and challenges: Discussions from data analytics perspectives. IEEE Comput. Intell. Mag. **9**(4), 62–74 (2014)
27. Zitzler, E., Laumanns, M., Thiele, L.: SPEA2: improving the strength pareto evolutionary algorithm. In: Evolutionary Methods for Design, Optimization and Control with Applications to Industrial Problems, EUROGEN 2001, Greece, Athens, pp. 95–100 (2002)

An Empirical Assessment of the Properties of Inverted Generational Distance on Multi- and Many-Objective Optimization

Leonardo C.T. Bezerra[1(✉)], Manuel López-Ibáñez[2], and Thomas Stützle[3]

[1] DCC-CI, Universidade Federal da Paraíba (UFPB), João Pessoa, PB, Brazil
leo.tbezerra@ci.ufpb.br
[2] Alliance Manchester Business School, University of Manchester, Manchester, UK
manuel.lopez-ibanez@manchester.ac.uk
[3] IRIDIA, Université Libre de Bruxelles (ULB), Brussels, Belgium
stuetzle@ulb.ac.be

Abstract. The inverted generational distance (*IGD*) is a metric for assessing the quality of approximations to the Pareto front obtained by multi-objective optimization algorithms. The *IGD* has become the most commonly used metric in the context of many-objective problems, i.e., those with more than three objectives. The averaged Hausdorff distance and IGD^+ are variants of the *IGD* proposed in order to overcome its major drawbacks. In particular, the *IGD* is not Pareto compliant and its conclusions may strongly change depending on the size of the reference front. It is also well-known that different metrics assign more importance to various desired features of approximation fronts, and thus, they may disagree when ranking them. However, the precise behavior of the *IGD* variants is not well-understood yet. In particular, IGD^+, the only *IGD* variant that is weakly Pareto-compliant, has received significantly less attention. This paper presents an empirical analysis of the *IGD* variants. Our experiments evaluate how these metrics are affected by the most important factors that intuitively describe the quality of approximation fronts, namely, spread, distribution and convergence. The results presented here already reveal interesting insights. For example, we conclude that, in order to achieve small *IGD* or IGD^+ values, the approximation front size should match the reference front size.

Keywords: Multi-objective optimization · Performance assessment · Inverted generational distance

1 Introduction

Due to the conflicting nature of the multiple objectives to be optimized in a single run, the goal of an EMO algorithm is to find a set of high-quality, trade-off solutions, rather than a single one. Such solution sets are approximations to the Pareto-optimal front. Pareto-optimality only provides a partial ranking between such approximation fronts. Thus, their relative quality is typically evaluated with the aid of *quality metrics*, also known as quality indicators [14], which

© Springer International Publishing AG 2017
H. Trautmann et al. (Eds.): EMO 2017, LNCS 10173, pp. 31–45, 2017.
DOI: 10.1007/978-3-319-54157-0_3

provide a complete ranking. Many quality metrics have been proposed in the literature [9, 11–14], and multiple quality metrics are often used simultaneously, because each metric assigns different importance to various desirable features of approximation fronts, such as convergence, spread and distribution. These differences between metrics may lead to "disagreements", when each metric chooses different approximation fronts as the best ones. Empirical studies have shown that the degree of disagreement strongly depends on features of the approximation fronts, such as convexity, and on the correlation between objectives and their number [10]. Understanding the properties of these metrics is critical for correctly selecting which metrics to use and interpreting their outcome.

One of the most desirable properties of a quality metric is Pareto compliance. A quality metric is Pareto-compliant if, and only if, the ranking it establishes over approximation fronts does not contradict Pareto optimality [14]. In other words, it cannot happen that the metric ranks one front better than another while the latter would always be preferred according only to Pareto optimality. The use of a non-Pareto-compliant metric to evaluate algorithms that attempt to approximate the Pareto front may lead an analyst to prefer an algorithm that returns approximation fronts that are strictly worse in terms of Pareto optimality.

Three widely-used unary quality metrics are the hypervolume (I_H [13]), the (additive or multiplicative) epsilon (I_ϵ^1 [14]) and the *inverted generational distance* (*IGD* [4]). Both I_H and I_ϵ^1 are Pareto-compliant [14],[1] whereas the *IGD* is not [9]. Despite this drawback, the *IGD* has become widely adopted in EMO studies of many-objective optimization problems, i.e., problems with more than three objectives. A possible explanation is that the *IGD* is cheaper to compute than I_H, since the computational cost of I_H grows exponentially with the number of objectives [1]. Moreover, it is commonly assumed that both *IGD* and I_H are able to measure the desired features of approximation sets, that is, convergence, spread and distribution. However, recent empirical studies have shown that the disagreement between I_H and *IGD* increases with the number of objectives [2, 10]. Thus, understanding what is exactly being measured by the *IGD* under various scenarios is of critical importance.

Alternative versions of the *IGD* have been proposed in recent years [9, 11] with hopes of addressing its potential drawbacks. In addition to its lack of Pareto-compliance, the *IGD* is not strictly a distance since under some conditions it does not satisfy the triangle inequality among approximation fronts [11]. Moreover, the *IGD* may evaluate two approximation fronts as almost equal if the main difference between the two is that one of them contains a very poor objective vector (an outlier). Another potential drawback of the *IGD* is that its value is quite sensitive to changes in the size of the reference front, that is, a finer-grained reference front may significantly alter conclusions previously obtained with a smaller reference front, even if both reference fronts only present optimal solutions. The averaged Hausdorff distance (Δ_p [11]) is a variant of *IGD* that attempts to alleviate some of these drawbacks, except for the lack of Pareto-compliance. More recently, the *IGD*$^+$ has been proposed as a Pareto-compliant

---

[1] To be more precise, the ϵ-metric is only weakly Pareto-compliant, but we do not make a distinction between weakly and non-weakly Pareto-compliance in the remainder of this paper.

variant of *IGD* [9], being very similar to the *IGD* and being as robust as Δ_p to different sizes of reference fronts. The properties of the IGD^+ are still poorly understood and so far no investigation has been conducted about the behavior of the IGD^+ with respect to convergence, spread, and distribution.

In this work, we conduct an empirical investigation specifically targeting *IGD* and its Pareto-compliant variant IGD^+. Concretely, we generalize an existing bi-objective benchmarking problem [7] for any number of objectives and we consider scenarios with up to ten objectives. Then, we design a series of experiments where approximation fronts are evolved for increasing convergence, spread, or distribution, and evaluate how the selected metrics respond to these changes. Effectively, our experiments isolate the effects of convergence, spread, and distribution. Furthermore, we design two experiments that simulate practical scenarios that EMO algorithms may run into, namely, when the approximation font has converged to the central region of the Pareto front and when the approximation front has achieved a good convergence and maximum spread but its distribution is poor.

Our results show that the *IGD* and IGD^+ behave exactly the same for all the practical purposes considered in this work. In addition, the factors that affect one variant affect the other in the same degree. For instance, we observe that the most important feature to ensure low *IGD* values is to have an approximation front that matches the size of the reference front adopted. Knowledge of this feature is critical as the default practice in the performance assessment of EMO algorithms is to use very large reference fronts, and here we demonstrate that in this circumstance the *IGD* and IGD^+ values may start worsening even if the spread of the approximation front is improving without worsening any other desirable feature. Another important insight concerns the effect of the parameter meant to regulate the importance of outliers. We observed that a setting often used in the literature leads to a constant *IGD* value despite changes in the distribution of the approximation front.

The remainder of this work is structured as follows. In Sect. 2, we briefly review the most relevant conceptual definitions related to the performance assessment of EMO algorithms, highlighting the desirable features of approximation fronts that we use as factors in our empirical investigation. Next, Sect. 3 presents an overview of *IGD* variants and explains why we focus on *IGD* and IGD^+. Section 4 details our experimental setup, and Sect. 5 reports the empirical investigation we conduct. Finally, Sect. 6 presents our conclusions and discussion of future work.

2   Performance Assessment of EMO Algorithms

In multi-objective optimization (MO), the goal is to simultaneously optimize M objective functions.[2] Therefore, the image of each potential solution is an *objective vector* with M components. The conflicting nature of objectives typically prevents the existence of a single, globally optimal solution that optimizes all objectives at once. In the absence of preference information regarding the importance of each objective, solutions are often compared in terms of Pareto-optimality, where a solution with objective vector a is said to *dominate* another

---

[2] In the following we assume maximization, without loss of generality.

solution with objective vector b iff $a_i \geq b_i$, $\forall i = 1, \ldots, M$ and $\exists j\ a_j > b_j$. Two objective vectors are mutually nondominated if none of them dominates the other. The goal then becomes to find the set of Pareto-optimal solutions, that is, those solutions that are not dominated by any other feasible solutions; or rather the image of this set in the objective space, the Pareto front.

Since finding the Pareto front is often intractable, EMO algorithms attempt to find a high-quality approximation of it, namely, an *approximation front* composed of mutually nondominated objective vectors. Hence, the performance assessment of EMO algorithms requires the evaluation of the relative quality of approximation fronts. Although Pareto-optimality may sometimes be enough to conclude that one approximation front is better than another, the most common case is that fronts are mutually incomparable. Nonetheless, there are features that, in addition to Pareto-optimality, are desirable in high-quality approximations [10]:

1. **Convergence** refers to the (near-)optimality of individual solutions. A front is said to have converged if all of its solutions are Pareto-optimal.
2. **Spread** refers to the extent of the front, more specifically to the distance between the extreme solutions of a front.
3. **Distribution** refers to the evenness of the front, more specifically to the uniformity of the distances between pairs of adjacent solutions.

Instead of directly measuring each individual feature, quality metrics can be found in the literature that attempt to evaluate all features at once [9,11–14]. However, each metric assigns a different importance to each feature and behaves differently depending on the characteristics of the problem and the particular fronts being evaluated, thus it is common that multiple metrics are used for performance assessment. In particular, experiments have shown [10] that the IGD and the I_H consistently disagree in typical scenarios arising in many (more than three) objective problems. Hence, understanding the behavior of quality metrics under various scenarios is crucial for performance assessment. While the I_H and, to some extent, the IGD are fairly well understood nowadays, newer variants such as IGD^+ have received little attention.

3   The Inverted Generation Distance and Its Variants

The predecessor of the IGD, **the generational distance** (GD [12]), was proposed nearly two decades ago. The GD is defined as the distance between each objective vector a in a given approximation front A and the closest objective vector r in a reference front R, which is either the actual Pareto front or a very good approximation to it, averaged over the size of A. Formally,

$$GD(A, R) = \frac{1}{|A|} \left(\sum_{a \in A} \min_{r \in R} d(a, r)^p \right)^{1/p}, \quad d(a, r) = \sqrt{\sum_{k=1}^{M} (a_k - r_k)^2} \quad (1)$$

A value of $p = 2$ was used in the original proposal, but this choice was later superseeded by $p = 1$ for simplicity of interpretation and computation. With $p = 1$, the GD becomes an average of the Euclidean distances between each objective vector in A and its closest objective vector in R. The GD metric is fast to compute and correlates with convergence to the reference set. However, the GD is not Pareto-compliant [14] and it is also sensitive to the size of the approximation front A. Thus, large approximation fronts of poor quality may be ranked highly by GD.

The **inverted generational distance** (IGD [4]) was proposed as an improvement over the GD based on the very simple idea of reversing the order of the fronts considered as input by the GD, i.e., $IGD(A, R) = GD(R, A)$. In other words, the IGD equals the GD metric but computing the distance between each objective vector in the reference front and its closest objective vector in the approximation front, averaged over the size of the reference front. Parameter p plays a similar role as in the GD and often defaults to $p = 1$. The IGD is not sensitive to the size of the approximation fronts and it provides a ranking that intuitively matches more closely the desirable convergence, spread and distribution. Since it is also computationally fast to compute, IGD soon became the most widely used metric to assess many-objective EMO algorithms. Nonetheless, the IGD has been shown recently to lack Pareto-compliance [9].

The **averaged Hausdorff distance** (Δ_p [11]) was proposed as an attempt to address three potential drawbacks of the IGD. First, despite being characterized as a distance metric, the IGD sometimes violates the triangle inequality property. Second, the size of the reference front has a significant effect on the IGD values, to the point that adding additional solutions to the current reference set may change the relative ranking of approximation fronts. Third, if the main difference between two approximation fronts is that one contains a clearly poorer objective vector, the IGD may still regard both fronts as roughly equal, in other words, the IGD is often lenient about outliers. To overcome the first and third drawbacks, Δ_p is defined as an averaged Hausdorff distance metric, regulated by the numerical parameter p. In particular, larger values of p mean stronger penalties for outliers and fewer triangle inequality violations. Concerning the second drawback, Δ_p uses an alternative version of IGD (IGD_p), where the denominator is also affected by the parameter p. The formal definition of Δ_p is given below:

$$\Delta_p(A, R) = \max\left(IGD_p(A, R), IGD_p(R, A)\right) \tag{2}$$

$$IGD_p(A, R) = \left(\frac{1}{|R|} \sum_{r \in R} \min_{a \in A} d(r, a)^p\right)^{1/p} \tag{3}$$

Finally, the **modified inverted generational distance** (IGD^+ [9]) proposes the following modification of the distance function of the original IGD. For objective vectors that are dominated by the reference front, the traditional Euclidean distance is adopted. However, for objective vectors that are nondominated w.r.t. to the reference front, only the dominated objective vector components are used for computing the distance. Formally, given a problem where all M objectives must be maximized, the distance function d in Eq. (3) is replaced by:

$$d^+(r, a) = \sqrt{\sum_{k=1}^{M}(\max\{r_k - a_k, 0\})^2} \tag{4}$$

This modification is enough to make IGD^+ weakly Pareto compliant, similarly to I_ϵ^1. In addition, the definition of IGD^+ includes the denominator $|R|$ under the exponent $1/p$, as in the IGD_p definition proposed for the Δ_p (Eq. 3).

Nowadays, the IGD has effectively superseeded GD, however, Δ_p has not gained the attention of the community as a widely used performance assessment metric. A possible explanation is that Δ_p still lacks Pareto-compliance and the drawbacks addressed by Δ_p only arise in unusual scenarios, such as for very small reference fronts [11]. On the other hand, IGD^+ is a small modification of IGD that adds Pareto-compliance, thus IGD^+ seems a more likely candidate to superseed IGD. Yet, there is little understanding so far about the behavior of IGD^+ [8], with current investigations focused on its Pareto-compliance. In the remainder of this paper, we experimentally compare the behavior of IGD and IGD^+ under various scenarios.

4   Experimental Setup

In the following, we conduct a series of experiments in order to understand how IGD and IGD^+ react to various desirable features of approximation fronts. A first set of experiments is designed to evaluate each feature in isolation. In addition, we design experiments that simulate common scenarios in the context of EMO, where features are not isolated.

Benchmark problem. As a starting point, we consider the bi-objective optimization problem designed by Ishibushi et al. [7] that presents a linear-shaped Pareto front (Fig. 1, left). Generalized to any number of objectives M (Fig. 1, right), we have that a solution s is optimal if $\sum_{i=1}^{M} f_i(s) = 10$, where $f_i(s)$ is the i-th objective value of solution s. In this work, we study $M \in \{2,3,5,10\}$.

Reference fronts. Since generating optimal solutions is trivial, we produce reference fronts of different resolution by using different front sizes. In the case of $M \in \{2,3\}$, each reference front $R_{\langle M,d \rangle}$ is created using a uniform weight vector generation method, parameterized by the number of divisions d, and its size equals $\binom{M-1+d}{d}$. In the case of $M \in \{5,10\}$, a uniform distribution of weights generates fronts with many more solutions on the extremes of the objective space than in the center. Instead, we adopt the two-layer approach proposed by [5]. In more detail, for a given d value, we generate a uniform set of $\lfloor d/2 \rfloor$ weights on the extremes of the objective space where are least one objective function is equal to zero (the outer layer), and a uniform set of $d - \lfloor d/2 \rfloor$ weights in the center of the objective space (the inner layer). We consider three different d values, representing small, moderate and large reference fronts, for each value of M, as shown in Table 1.

Approximation fronts. In general, we generate approximation fronts using the same method explained above for reference fronts, that is, according to the number of objectives M and a parameter d (Table 1).

Effect of p. As explained earlier, most of the literature only considers $p = 1$ while some works use $p = 2$. To understand how different values of p affect the behavior of the quality metrics, we consider $p \in \{1,2,3\}$.

Seeds for random number generation. When sampling is adopted, we repeat each experiment 25 times with a common set of random seeds in order to reduce variance between experiments.

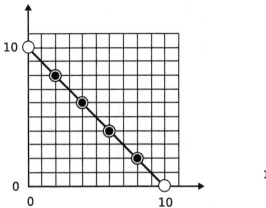 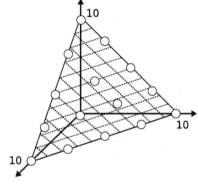

Fig. 1. Linear-shaped maximization problem suggested by [7]. Black dots represent approximation front solutions, whereas white circles represent reference front solutions created with $M = 2$, $d = 3$ (left) and $M = 3$, $d = 5$ (right).

Table 1. Size of the generated reference front ($|R|$) for each number of objectives (M) and value of parameter d. We use the same method to generate approximation fronts.

M	d	$|R|$	M	d	$|R|$	M	d	$|R|$	M	d	$|R|$
2	3	4	3	6	28	5	5	50	10	4	110
	19	20		13	105		8	140		5	275
	99	100		19	210		13	540		7	935

5   Empirical Assessment of *IGD* and *IGD* plus

5.1   Desirable Features of Approximation Fronts

We first explain how we designed experiments to evaluate one desirable feature of approximation fronts at a time. After that, we discuss our conclusions from these experiments.

Convergence. To evaluate convergence, we consider approximation fronts that are obtained by translating a reference front until it intersects with the axes. These approximation fronts are then iteratively "evolved" using a linear interpolation for each pair $\langle a', r' \rangle$, where a' is the solution from the approximation front obtained by translating reference front solution r'. To measure how far the approximation front is from the reference front at a given iteration, we compute the Euclidean distance for the pair $\langle a', r' \rangle$. Each iteration reduces this distance by the same value until it becomes zero, and the step value is calculated in order to perform 100 iterations. Figure 2 illustrates a translated front before being evolved (left) and at a later iteration (right).

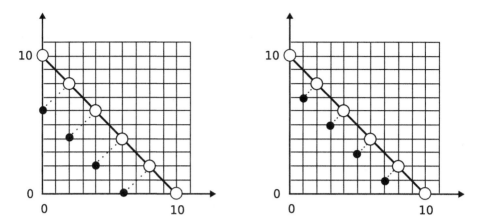

Fig. 2. Illustration of the experiments where convergence is isolated. Left: approximation front in its initial state after translation. Right: approximation front after 50 iterations of linear interpolation. All intermediate states present the same spread and distribution.

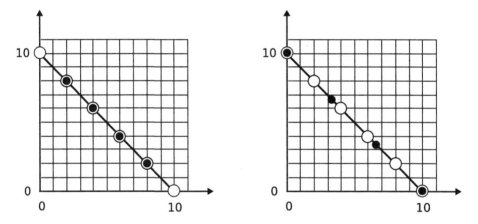

Fig. 3. Illustration of the experiments where spread is isolated. Approximation front with a spread of 6 (left) and after 100 iterations of evolution, with a spread of 10 (right). All iterations present the same distribution and convergence.

Spread. We evaluate spread by initially selecting the desired distance along a single objective between the extreme solutions. Next, we multiply this distance by the uniform weight set obtained for the given M and d, and translate this front so that its solutions become optimal. By increasing the distance between the extreme solutions, we are able to iteratively generate approximation fronts with increasing spread but the same convergence (since all solutions are optimal) and distribution (since solutions are equally distributed between the extremes). Figure 3 illustrates an approximation front for $M = 2$ with spread of 6 (left) and with a spread of 10 (right).

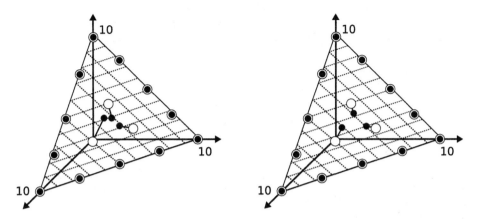

Fig. 4. Illustration of the experiments where distribution is isolated. Left: approximation front with maximum spread and inner layer solutions sampled around the center of the objective space with a Gaussian distribution. Right: intermediate stage of the evolutionary process, where each solution is getting closer to its point of destination.

Distribution. To isolate distribution, we generate approximation fronts in two steps. First, we copy the outer layer (those objective vectors that have at least one zero component) from the reference front of the same size. Next, we sample n_{inner} Pareto-optimal solutions by sampling coordinates using a Gaussian distribution around the center of the Pareto front, where $n_{\text{inner}} = n - n_{\text{outer}}$, n is the size of the corresponding reference front, and n_{outer} is the amount of solutions in the outer layer. To sample solutions that are concentrated around the center of the objective space, we use the algorithm provided in the supplementary material [3]. Figure 4 (left) illustrates a front generated with this method. Once a front is sampled, we evolve it towards an even distribution. More precisely, we first associate each approximation front solution a^r with the point r that is closest to it (in terms of Euclidean distance) in the reference front of the same size. Next, at each iteration, we use linear interpolation to translate each a^r towards its corresponding r, so that the resulting approximation front is more evenly distributed than the previous one. Figure 4 (right) illustrates an intermediary stage of evolution of the front depicted in Fig. 4 (left). Since only solutions from the inner layer are sampled, only those solutions are translated at each intermediary stage. In addition, both convergence and spread remain constant, as all solutions are always optimal and the extreme solutions do not change. To compute how far a front is from a perfect distribution at a given iteration, we define an entropy-like metric defined as $e(A, R) = \frac{1}{|R|} \sum_{r \in R} d(a^r, r)$. The difference between this e metric and the IGD with $p = 1$ is that e does not allow a reference front solution to be associated with multiple approximation front solutions. Effectively, this metric can be seen as a simplification of the *root mean square error*, an average of how far each solution a_i is from where it should be.

We next discuss the most important, high-level insights we observe in the results produced.

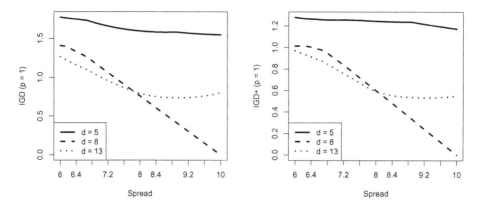

Fig. 5. Effect on IGD (left) and IGD^+ (right) of various approximation front sizes for increasing spread. Reference front is generated with settings $M = 5$, $d = 8$. Approximation fronts are generated with settings $M = 5$ and $d \in \{5, 8, 13\}$. Notice that, despite different in ranges, the shapes of the curves for both IGD and IGD^+ are similar.

Effect of Front Sizes. In the case of varying spread and distribution, we observe that, whichever M considered, the best IGD and IGD^+ values are obtained when the approximation front size matches the reference front size. That is, the IGD metrics will rank an approximation front of the same size as the reference front better than a larger or smaller front, and the difference will increase with higher spread. This result is specially counter-intuitive in the case of fronts larger than the reference front, as one would expect that, everything else being equal, larger approximation fronts are to be preferred. This is illustrated in Fig. 5, depicting the response of IGD and IGD^+ when $M = 5$, in dependence of different spread values (x-axis) and different approximation set sizes ($d \in \{5, 8, 13\}$), using a fixed reference set size ($d = 8$). Both IGD and IGD^+ metrics assign the best quality to the largest size approximation front when spread is small. However, when the spread of the front gets close to the maximum tested, only the quality of the front with the same size as the given reference front ($d = 8$) continues to improve until the minimal possible IGD value is obtained. For the largest approximation front size ($d = 13$), the quality according to IGD and IGD^+ even worsens again, despite neither the convergence or distribution are actually worse.

Similarity between IGD and IGD^+ results. The main difference between IGD and IGD^+ is that only components of objective vectors dominated by the reference front contribute to the computation of the distance. This means that, as observed, the IGD and IGD^+ produce equal results for approximation fronts that are dominated by the reference front, however, one would expect strong differences when the approximation front is mostly nondominated with respect to the reference front, as is the case in most of our experiments. Our experiments show that, under the same conditions, the IGD and IGD^+ differ in range but their behavior with respect to changes in convergence, spread and distribution is very similar. This is observed by comparing the shapes of the curves corresponding to IGD and IGD^+ in most of our plots, for example, in Fig. 5. The only

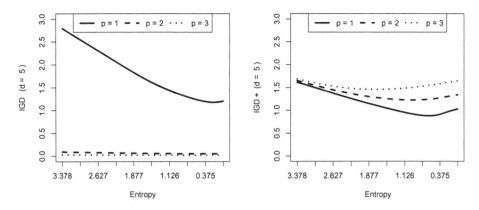

Fig. 6. Illustration of the effects of p. Distribution analysis of IGD (left) and IGD^+ (right) for $M = 10$, with an approximation front generated with $d = 5$ and a reference front generated with $d = 7$.

exception that we observed concerns changes in distribution when $M \in \{5, 10\}$ and the sizes of the approximation and reference fronts are not equal. Figure 6 depicts such a situation, where the shape of the curves when $p \geq 2$ is much smoother for IGD than for IGD^+.

Effect of p. Under the conditions tested in this work, the only observable effect of p is changing the ranges of the IGD variants. However, it is interesting to notice that, while the range of the IGD decreases with larger p, the opposite is observed for the IGD^+. This effect is shown in Fig. 6. Another important effect that is also depicted in this figure concerns the IGD metric only. In particular, a value of $p = 2$, as sometimes used in the literature, makes the IGD insensitive to changes in distribution. This is a potentially dangerous limitation, and we observe that this effect becomes ever stronger with the increase in reference front sizes.

Particularities of Many-Objective Problems. The convergence analysis when $M = 10$ differs from the overall patterns we have so far discussed, as in this case having larger approximation fronts is indeed a winning strategy. More precisely, when the reference fronts created with $d = 4$ or $d = 5$ are used, the approximation front created with $d = 7$ is either the best-performing according to both metrics, or at least very competitive. This situation is illustrated in Fig. 7 for the IGD^+ (the same is observed for the IGD), and may be due to the ratio between solutions in the inner and outer layers. Concretely, when $d \in \{5, 7\}$, the number of solutions in the inner layer is far greater than the number in the outer layer, whereas for $d = 4$ this ratio equals one. Another hypothesis is that the IGD values become better as long as the approximation front does not have more points in the outer layer than the reference front. Understanding this behavior will require further analysis.

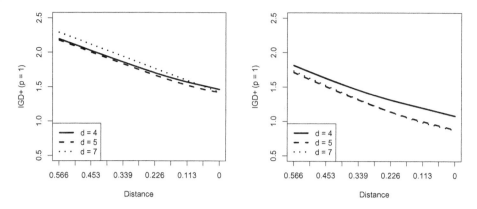

Fig. 7. Illustration of the particularities of many-objective optimization. IGD^+ convergence analysis indicate that, when $M = 10$, having larger approximation sets lead to better results. This effect is observed both when the reference front is created with $d = 4$ (left) and $d = 5$ (right).

5.2 Practical EMO Scenarios

In addition to the experiments above, where we analyzed in isolation the desirable features of approximation fronts, we now design experiments that resemble the evolution of approximation fronts by an EMO algorithm. The first experiment simulates the scenario where an EMO algorithm may converge to the optimal front with maximum spread, but still needs to further improve the distribution of its approximation front. The second experiment simulates a scenario where an EMO algorithm reaches the optimal front but lacks both spread and distribution.

Distribution. In this scenario, the approximation front already has maximum spread and has converged (all of its solutions are Pareto-optimal), but still needs to improve distribution. This happens in practice when high-quality solutions can be found by decomposing the problem into single objective ones. This is the case when using scalarization-based local search to tackle the bi-objective permutation flowshop problem [6]. It may also happen when the EMO algorithm internally uses a quality metric that favors spread over distribution. We generate approximation fronts for this scenario by copying solutions from the corresponding reference front and then applying a small, random uniform perturbation to each non-extreme solution (inner layer solutions). Figure 8 (left) illustrates a front generated in this fashion when $M = 3$ and $d = 5$. We then simulate the evolution of the approximation front by translating the approximation front solutions towards their original position in the reference front by linear interpolation. The entropy-like metric e defined in the previous section is used here to measure the perturbation level of the intermediate fronts. Figure 9 (left) shows results for this scenario, where it can be seen that both IGD variants are barely affected by this kind of perturbation. When investigating the reason for such robustness, it becomes clear that this kind of perturbation introduces very

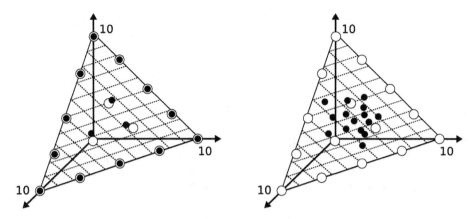

Fig. 8. Illustration of experiments where (left) inner layer solutions have been subject to a perturbation, and; (right) front has converged to the center of the objective space.

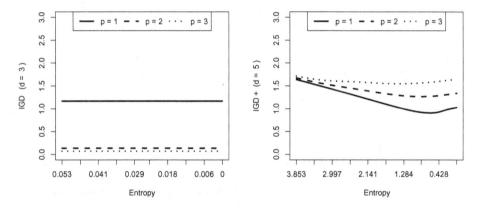

Fig. 9. Results for practical EMO scenarios. Left ($M = 2$): selected IGD variants are robust to perturbations in distribution for a small approximation set ($d = 3$) and a large reference set ($d = 99$). Right ($M = 10$): intermediate size approximation front ($d = 5$) gets worse IGD^+ values when trying to approximate a larger size reference front ($d = 7$), even though its distribution is improving.

little entropy to the approximation front, as evidenced in the x-axis of the plot depicted in Fig. 9 (left).

Distribution and Spread. In this scenario, the algorithm has converged to the center of the objective space, lacking both spread and distribution. This scenario happens often in practice, whenever (i) the algorithm does not preserve extreme solutions, (ii) extreme solutions are hard to find, or (iii) the algorithm focuses on converging to the Pareto front first, and spreading later. To simulate this scenario, we produce fronts using the same solution sampling used in the previous section to generate fronts with uneven distribution. We also use the entropy-like metric presented earlier to assess how far the front is from perfect distribution.

The main difference with respect to those experiments is that, in this case, we do not generate an outer layer that maximizes the spread, but instead we sample additional objective vectors that are likely to be concentrated in the inner layer. Figure 8 (right) illustrates a front generated using this method. Results for this scenario confirm what was observed in the previous set of experiments, where the most important factor was matching the size of the approximation front with the size of the reference front. In fact, the striking similarity between Fig. 6 (right) and Fig. 9 (right) indicates that the outer layer kept to ensure maximum spread from the previous experiment did not play a relevant part in the response from the IGD metrics. This is an important finding, since it could explain the strong disagreements between the IGD and the I_H, that is, fronts with very good spread at the cost of distribution will be favored by I_H whereas IGD will likely do the opposite.

6   Conclusions

In this work, we have conducted a preliminary empirical investigation on the properties of two relevant quality metrics, namely the inverted generational distance (IGD [4]) and a variant that has been proven weakly Pareto-compliant (IGD^+ [9]). In particular, the IGD is perhaps the most adopted performance metric in the context of many-objective optimization due to its low computational cost and its ability to assess several desirable features of approximation fronts, namely, convergence, spread, and distribution. By contrast, IGD^+ is a recent variant that needs further investigation, a task we undertook in this work.

In order to analyze the behavior of the selected metrics, we designed a series of experiments that either (i) isolated the desirable features of approximation fronts, or (ii) simulated real-world situations often faced by EMO algorithms. Perhaps the most important insight is the observation that, under some circumstances, the value of IGD or IGD^+ may become worse if the size of the approximation front grows beyond the size of the reference front, even if convergence, spread and distribution remain equal. In addition, we have shown that, for the scenarios tested in this work, the selected metrics have a similar behavior. The only exception to this pattern concerns IGD, which we have shown to be unable to detect poor distribution depending on the value of the parameter p, which is sometimes not specified in the literature. Finally, another important observation concerns a possible explanation for the previously reported disagreement between I_H and IGD. We have shown that, under some circumstances, outer layer solutions have little impact on the value of the IGD.

Our investigation opens a number of possibilities for future analysis of quality metrics. The first and most straightforward step is to deepen the analysis of IGD variants, both by considering more variants (namely Δ_p) and more examples of problems with different geometries. A second step is to design yet more elaborate experiments to simulate other real EMO scenarios, besides the obvious approach of assessing actual EMO algorithms in practice. Finally, it is imperative that these results be related to other metrics, specifically the hypervolume (I_H) and epsilon (I_ϵ^1) metrics, helping the community further understand their disagreements.

Acknowledgments. The research presented in this paper has received funding from the COMEX project (P7/36) within the IAP Programme of BelSPO. T. Stützle acknowledges support from the Belgian F.R.S.-FNRS, of which he is a senior research associate.

References

1. Beume, N., Fonseca, C.M., López-Ibáñez, M., Paquete, L., Vahrenhold, J.: On the complexity of computing the hypervolume indicator. IEEE Trans. Evol. Comput. **13**(5), 1075–1082 (2009)
2. Bezerra, L.C.T.: A component-wise approach to multi-objective evolutionary algorithms: from flexible frameworks to automatic design. Ph.D. thesis, IRIDIA, École polytechnique, Université Libre de Bruxelles, Belgium (2016)
3. Bezerra, L.C.T., López-Ibáñez, M., Stützle, T.: An empirical assessment of the properties of inverted generational distance indicators on multi- and many-objective optimization: supplementary material (2016). http://iridia.ulb.ac.be/supp/IridiaSupp.2016-006/
4. Coello Coello, C.A., Reyes Sierra, M.: A study of the parallelization of a coevolutionary multi-objective evolutionary algorithm. In: Monroy, R., Arroyo-Figueroa, G., Sucar, L.E., Sossa, H. (eds.) MICAI 2004. LNCS (LNAI), vol. 2972, pp. 688–697. Springer, Heidelberg (2004). doi:10.1007/978-3-540-24694-7_71
5. Deb, K., Jain, S.: An evolutionary many-objective optimization algorithm using reference-point-based nondominated sorting approach, part I: Solving problems with box constraints. IEEE Trans. Evol. Comput. **18**(4), 577–601 (2014)
6. Dubois-Lacoste, J., López-Ibáñez, M., Stützle, T.: Improving the anytime behavior of two-phase local search. Ann. Math. Artif. Intell. **61**(2), 125–154 (2011)
7. Ishibuchi, H., Akedo, N., Nojima, Y.: Behavior of multiobjective evolutionary algorithms on many-objective knapsack problems. IEEE Trans. Evol. Comput. **19**(2), 264–283 (2015)
8. Ishibuchi, H., Masuda, H., Nojima, Y.: A study on performance evaluation ability of a modified inverted generational distance indicator. In: Silva, S. et al. (ed.) GECCO, pp. 695–702. ACM Press (2015)
9. Ishibuchi, H., Masuda, H., Tanigaki, Y., Nojima, Y.: Modified distance calculation in generational distance and inverted generational distance. In: Gaspar-Cunha, A., Henggeler Antunes, C., Coello, C.C. (eds.) EMO 2015. LNCS, vol. 9019, pp. 110–125. Springer, Heidelberg (2015). doi:10.1007/978-3-319-15892-1_8
10. Jiang, S., Ong, Y.S., Zhang, J., Feng, L.: Consistencies and contradictions of performance metrics in multiobjective optimization. IEEE Trans. Cybern. **44**(12), 2391–2404 (2014)
11. Schütze, O., Esquivel, X., Lara, A., Coello, C.A.C.: Using the averaged Hausdorff distance as a performance measure in evolutionary multiobjective optimization. IEEE Trans. Evol. Comput. **16**(4), 504–522 (2012)
12. Van Veldhuizen, D.A., Lamont, G.B.: Multiobjective evolutionary algorithms: analyzing the state-of-the-art. Evol. Comput. **8**(2), 125–147 (2000)
13. Zitzler, E., Thiele, L.: Multiobjective evolutionary algorithms: a comparative case study and the strength Pareto evolutionary algorithm. IEEE Trans. Evol. Comput. **3**(4), 257–271 (1999)
14. Zitzler, E., Thiele, L., Laumanns, M., Fonseca, C.M., da Fonseca, V.G.: Performance assessment of multiobjective optimizers: an analysis and review. IEEE Trans. Evol. Comput. **7**(2), 117–132 (2003)

Solving the Bi-objective Traveling Thief Problem with Multi-objective Evolutionary Algorithms

Julian Blank[1]([✉]), Kalyanmoy Deb[2], and Sanaz Mostaghim[1]

[1] Otto-von-Guericke University, 39106 Magdeburg, Germany
julian.blank@gmx.de, sanaz.mostaghim@ovgu.de
[2] Michigan State University, East Lansing, MI 48824, USA
kdeb@egr.msu.edu
http://www.research-blank.de
http://www.is.ovgu.de/Team/Sanaz+Mostaghim.html
http://www.egr.msu.edu/~kdeb/

Abstract. This publication investigates characteristics of and algorithms for the quite new and complex Bi-Objective Traveling Thief Problem, where the well-known Traveling Salesman Problem and Binary Knapsack Problem interact. The interdependence of these two components builds an interwoven system where solving one subproblem separately does not solve the overall problem. The objective space of the Bi-Objective Traveling Thief Problem has through the interaction of two discrete subproblems some interesting properties which are investigated. We propose different kind of algorithms to solve the Bi-Objective Traveling Thief Problem. The first proposed deterministic algorithm picks items on tours calculated by a Traveling Salesman Problem Solver greedily. As an extension, the greedy strategy is substituted by a Knapsack Problem Solver and the resulting Pareto front is locally optimized. These methods serve as a references for the performance of multi-objective evolutionary algorithms. Additional experiments on evolutionary factory and recombination operators are presented. The obtained results provide insights into principles of an exemplary bi-objective interwoven system and new starting points for ongoing research.

Keywords: Traveling Thief Problem · Traveling Salesman Problem · Knapsack problem · Interwoven systems · Multi-objective optimization · Discrete optimization · Combinatoric problems

1 Introduction

People are facing real-world problems with dependencies and interwovenness every day. Often researchers divide complex problems into various well investigated subproblems to solve them independently. After applying state of the art methods for the subproblems, partial solutions are combined to a solution for the complex problem. However, considering an interwoven problem the optimal solutions for the subproblems do not build an optimal solution for the complex

© Springer International Publishing AG 2017
H. Trautmann et al. (Eds.): EMO 2017, LNCS 10173, pp. 46–60, 2017.
DOI: 10.1007/978-3-319-54157-0_4

problem because of the interaction. In order to provide an interwoven problem with real-world characteristics, this publication presents the Bi-Objective Traveling Thief Problem (TTP), where the well-known Traveling Salesman Problem (TSP) and Binary Knapsack Problem (KNP) interact.

The focus of this publication is to provide a starting point for doing research on the more complex bi-objective problem by showing characteristics of the objective space and proposing deterministic as well as evolutionary multi-objective algorithms. Research on the Bi-Objective Traveling Thief Problem will give insights into how to handle interwoven problems with multiple objectives and therefore how to solve problems with real-world characteristics.

The outline of this work is as follows: Sect. 2 explains the TTP itself and its two interweaving components. Furthermore, the interdependencies are described and an example scenario is provided. Section 3 presents related work, while Sect. 4 describes properties of the input variables and characteristics of the objective space. Moreover, the proposed algorithms are explained in Sect. 5 and the obtained results are presented in Sect. 6. Finally, a conclusion and further considerations are provided in Sect. 7.

2 Problem Definition

2.1 Traveling Thief Problem

The TTP is a combinatorial optimization problem that consists of two interweaving subproblems, namely TSP and KNP [1]. After explaining the components separately, the interaction in the TTP of these two subproblems is shown.

In the TSP [2] a salesman has to visit n cities. The distances are given by a map represented as a distance matrix $D^{n \times n} = (d_{i,j})$ with $i, j \in \{1, .., n\}$ whereby $d_{i,j} = d_{j,i}$. The salesman has to visit each city once and the result is a permutation vector $\pi = (\pi_1, \pi_2, ..., \pi_n) \in \mathbb{P}^n$, where π_i is the i-th city visited by the salesman. Often $\pi_1 = 1$ is required which means the starting city is fixed.

$$min \quad f(\pi) \tag{1}$$

$$f(\pi) = \sum_{i=1}^{n-1} d_{\pi_i, \pi_{i+1}} + d_{\pi_n, \pi_1}$$

$$s.t. \quad \pi \in \mathbb{P}^n \text{ with } \pi_1 = 1 \tag{2}$$

For the KNP [3] a knapsack has to be filled with items by considering the maximal capacity Q. Each item j has a value $b_j \geq 0$ and a weight $w_j \geq 0$ where $j \in \{1, .., m\}$. The binary decision vector $z = (z_1, .., z_m) \in \mathbb{B}^m$ defines, if an item is picked or not. The aim is to maximize the profit $g(z)$:

$$max \quad g(z) \tag{3}$$

$$g(z) = \sum_{j=1}^{m} z_j \, b_j$$

$$\text{s.t.} \quad \sum_{j=1}^{m} z_j \, w_j \leq Q$$

The TTP combines the above defined subproblems and let them interact together. The interdependency relation between the components of the Bi-Objective TTP is visualized in Fig. 1. The tour π is only part of the TSP component. Therefore, the profit $g(z)$ remains the same when π is modified. However, the packing plan z is part of both components and both objectives are influenced by z. Since both components need the packing plan z as a parameter, it is hard to solve the problem. In fact, such problems are called interwoven systems as the solution of one subproblem highly depends on the solution of the other subproblems.

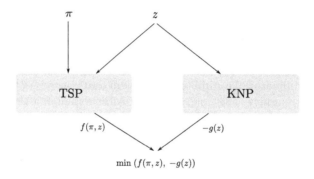

Fig. 1. Interdependence of the Bi-objective Traveling Thief Problem

Items are assigned to each city on the map through the assignment matrix $A^{m \times n} = (a_{i,j})$. Each item is exactly assigned to one city, but one city can have multiple items. The thief can collect items from each city he is visiting. The items are stored in a knapsack with the maximum capacity Q carried by him. As interaction between the subproblems the velocity $v(q)$ depends on the current knapsack weight $q = w(\pi_i)$. When the thief picks an item, the weight of the knapsack increases and the velocity of the thief decreases. The velocity is always in a specific range $v = [v_{min}, v_{max}]$ and could not be negative for a feasible solution. Whenever the knapsack is heavier than the maximum weight Q, the capacity constraint is violated. The variables π and z as well as the item information w and b are defined analogously to TSP and KNP. The two objective functions $f(\pi, z)$ and $-g(z)$ have to be minimized. The whole problem is defined as follows:

Symbol	Description
$\pi \in \mathbb{P}^n$	Tour as permutation vector with length n. π_i represents the i-th city visited by the thief
$z \in \mathbb{B}^m$	Picking plan as binary vector with length m. $z_i = 1$ means the i-th item is picked up
$D^{n \times n} = (d_{i,j})$	Distance matrix: $d_{i,j}$ represents the distance between city i and city j whereby $d_{i,j} = d_{j,i}$
$w \in \mathbb{Z}^m$	Item's weight vector: w_i is the weight of item i
$b \in \mathbb{Z}^m$	Item's value vector: b_i is the value of item i
$v_{min} \in \mathbb{R}^+$	Minimum velocity
$v_{max} \in \mathbb{R}^+$	Maximum velocity
$Q \in \mathbb{Z}^+$	Maximum knapsack capacity
$A^{m \times n} = (a_{i,j})$	Assignment of items to cities where $a_{i,j} \in \{0,1\}$ and $\forall j \in \{1,..,m\} : \sum_{i=1}^{n} a_{j,i} = 1$. (each item is assigned to exactly one city)

$$\min_{\pi,z} F\ (\pi, z) \tag{4}$$

$$F(\pi, z) = (f(\pi, z), -g(z))$$

$$f(\pi, z) = \sum_{i=1}^{n-1} \frac{d_{\pi_i,\pi_{i+1}}}{v(w(\pi_i))} + \frac{d_{\pi_n,\pi_1}}{v(w(\pi_n))}$$

$$g(z) = \sum_{j=1}^{m} z_j b_j$$

$$v(q) = v_{max} - \frac{q}{Q} \cdot (v_{max} - v_{min})$$

$$w(\pi_i) = \sum_{k=1}^{i}\sum_{j=1}^{m} z_j w_j a_{j,\pi_i}$$

$$\text{s.t.} \quad \sum_{j=1}^{m} z_j w_j \leq Q$$

$$\pi \in \mathbb{P}^n \text{ with } \pi_1 = 1$$

Note, that the item value drop effect mentioned in [4] is neglected in our considerations. An example scenario is provided in Sect. 2.2. Given a small map, a tour π and a packing plan z the traveling time and the profit for the thief are calculated and the interdependency effect is shown. Also, the Pareto front for the given example is presented tabularly.

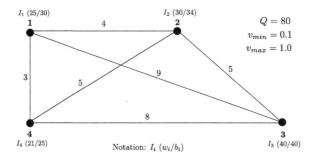

Fig. 2. An example scenario for the Traveling Thief Problem

2.2   Example Scenario

The thief travels on a map represented as a graph shown in Fig. 2. He starts at city 1 and has to visit city $2, 3, 4$ exactly once and return to city 1. In this example each city provides one item and the thief could decide to steal item I_j or not.

A permutation vector, which contains all cities exactly once, and a binary picking vector are needed to calculate the objectives. Even though, this is a very small example with four cities and four items the solution space consists of $(n-1)! \cdot 2^m = 6 \cdot 16 = 96$ combinations.

In order to understand how the objectives are calculated, an example hand calculation for the tour $\pi = [1, 3, 2, 4]$ and the packing plan $z = [0, 1, 0, 1]$ is done as follows: The thief starts with the maximum velocity, because the knapsack is empty. He begins his tour at city *1* and picks no item there.

Table 1. Hand calculations on the example scenario for the tour $\pi = [1, 3, 2, 4]$ and packing plan $z = [0, 1, 0, 1]$

i	π_i	$w(\pi_i)$	$v(w(\pi_i))$	$d_{\pi_i, \pi_{i+1}}$	$t_{\pi_i, \pi_{i+1}}$	
1	1	0	1	9	9	-
2	3	0	1	5	5	9
3	2	30	0.6625	5	7.5472	14
4	4	51	0.42625	3	7.0381	21.547
-	1	51	0.42625	-	-	**28.585**

For an empty knapsack $w(1) = 0$ the velocity is $v(0) = v_{max} = 1.0$. The distance from city 1 to city 3 is 9.0 and the thief needs 9.0 time units. At city 3 the thief will not pick an item and continue to travel to city 2 with $w(3) = 0$ and therefore with v_{max} in additional 5.0 time units. Here he picks item I_2 with $w_2 = 30$ and the current weight becomes $w(2) = 30$, which means the velocity will be reduced to $v(30) = 1.0 - (\frac{30}{80}) \cdot 0.9 = 0.6625$. For traveling from

Table 2. Pareto front of the example scenario.

π	z	$f(\pi, z)$	$-g(z)$
$[1, 2, 3, 4]$	$[0, 0, 0, 0]$	20.0	0.0
$[1, 4, 3, 2]$	$[0, 0, 0, 0]$	20.0	0.0
$[1, 2, 3, 4]$	$[0, 0, 0, 1]$	20.93	-25.0
$[1, 4, 3, 2]$	$[0, 1, 0, 0]$	22.04	-34.0
$[1, 4, 3, 2]$	$[0, 0, 1, 0]$	27.36	-40.0
$\mathbf{[1, 3, 2, 4]}$	$\mathbf{[0, 1, 0, 1]}$	**28.59**	$\mathbf{-59.0}$
$[1, 4, 3, 2]$	$[1, 1, 0, 0]$	32.75	-64.0
$[1, 2, 3, 4]$	$[0, 0, 1, 1]$	33.11	-65.0
$[1, 4, 3, 2]$	$[0, 1, 1, 0]$	38.91	-74.0
$[1, 3, 2, 4]$	$[1, 1, 0, 1]$	53.28	-89.0

city 2 to city 4 the thief needs the distance divided by the current velocity $\frac{5.0}{0.6625} \approx 7.5472$. At city 4 he picks I_4 with $w_4 = 21$ and the current knapsack weight increases to $w(4) = 30 + 21 = 51$. For this reason the velocity decreases to $v(51) = 1.0 - (\frac{51}{80}) \cdot 0.9 = 0.42625$. For returning to city 1 the thief needs according to this current speed $\frac{3.0}{0.42625} \approx 7.0381$ time units. Finally, we sum up the time for traveling from each city to the next $\sum_{k=1}^{n} t_{\pi_k, \pi_{k+1}} = 9 + 5 + 7.5472 + 7.0381 = 28.5853$ to calculate the whole traveling time (c.f Table 1).

The final profit is calculated by summing up the values of all items which is $34 + 25 = 59$. So the variable (π, z) with $\pi = [1, 3, 2, 4]$ and $z = [0, 1, 0, 1]$ is mapped to the point $(28.5853, -59.0)$ in the objective space.

The Pareto front contains 10 solutions and our hand calculation is printed bold in Table 2.

3   Related Work

The TTP was introduced by Bonyadi [4]. He postulated the necessity of research on interwoven problems and provided the mathematical formulation of the TTP. Also, the single and bi-objective version of the problem was introduced. Bonyadi laid the foundation for further research. Afterwards, a new large-scale benchmark was proposed [5]. The benchmark is based on the TSPLIB [6] instances which are available for TSP. Since the TTP also needs the KNP part, items with weight and profit attributes are added to the cities. Three different weight-value correlations and ten different capacity categories are considered. Instances with $1, 3, 5$ and 10 items per city are provided. All this together builds a benchmark set with 9,720 different instances.

Later, most of the publications aimed to solve these single-objective benchmark problems by using different types of algorithms: Heuristic Based [5, 7–10], Local Search [5], Coevolution [8, 11], Evolutionary Algorithm [11–13], Ant Colony Optimization [9].

However, so far less research is done on the Bi-Objective TTP where the solution space has more than one dimension and pareto-optimal solutions have to be considered.

4 Characteristics of the Bi-objective Traveling Thief Problem

In the following characteristics of the TTP are investigated. Since our example scenario considers only a few cities and items, the optimal Pareto front can be determined by solving the problem exhaustively. Therefore, inferences about problem specific characteristics can be drawn and finally generalized. The objective space of our example scenario colored by tours and annotated with characteristics is shown in Fig. 3.

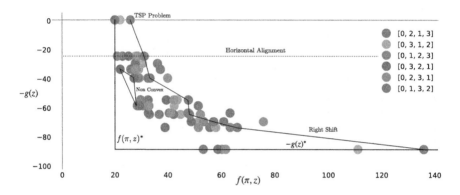

Fig. 3. Objective space of the example scenario colored by tour. (Color figure online)

Symmetry of π

For the TTP it is considered that the thief travels on a symmetric map. Whenever the knapsack is empty $z = z^{empty} = \sum_{i=1}^{m} z_i = 0$, the symmetry of π holds. Let us assume we have the tour π and the symmetric tour $sym(\pi)$, then

$$f(\pi, z^{empty}) = f(sym(\pi), z^{empty}) \tag{5}$$

is guaranteed. Whenever the thief starts to pick up an item this fact does not hold anymore. Whenever TSP Solvers are used this fact has to be kept in mind.

Horizontal Alignments in the Objective Space

The objective space has many horizontal alignments. Solutions on the same horizontal line have the same profit $-g(z)$. Considering the TTP as a problem consisting of two subproblems, every horizontal line represents an embedded TSP. The topmost horizontal line where $g(z) = 0$ represents the TSP without any interaction with the KNP component. In the objective space of the example scenario (c.f. Fig. 3) six different solutions are on each horizontal line. Because of the symmetry of π when $z = z^{empty}$ only three different points exist in the objective space on the topmost TSP line.

Right Shift

For solutions where $g(z) = 0$ the thief does not make any profit and $z = z^{empty}$. For these solutions the velocity during the tour is consistently v_{max}. If the thief starts to pick an item, he gets slower because the weight increase leads to velocity decrease. All solutions on the x-axis with $-g(z) = 0$ are shifted to the right when items are picked (c.f. Fig. 3). These right shifts occur when z is optimized for a fixed π.

5   Algorithms

In the following different algorithms for the Bi-Objective TTP are proposed. The first two algorithms are deterministic and explained by pseudo code. Furthermore, the principles of the multi-objective evolutionary algorithm NSGA-II are described. Different factory and recombination operators are directly evaluated in Sect. 6.

Greedy Algorithm. This algorithm represents a naive deterministic approach. The procedure is shown in Algorithm 1. First, an existing TSP Solver - in our implementation LKH with the Lin-Kernighan heuristic [14] - is used to calculate a tour π^*. A set which will contain only non dominated solutions in the further optimization procedure is initialized with $F(\pi^*, z^{empty})$ and $F(sym(\pi^*), z^{empty})$. As long as the maximum capacity constraint is not violated items are picked. The $getNextItemGreedily$ method returns the next item which is not picked so far and provides the best $\frac{b_i}{w_i}$ rate. Then z combined with π^* and $sym(\pi^*)$ is added to the set. Whereby the add method needs to ensure to keep a set of non dominated solutions. Finally the resulting Pareto front is returned.

Algorithm 1. Greedy Algorithm

Require: $P \leftarrow$ Thief Problem
   $\pi^* \leftarrow$ tspSolver(P)
   $z \leftarrow z^{empty}$
   front $\leftarrow \{F(\pi^*, z), F(sym(\pi^*), z)\}$
   while $\sum_{j=1}^{m} w_j z_j \leq Q$ **do**
      $i \leftarrow$ getNextItemGreedily(P, z)
      $z_i \leftarrow = 1$
      add(front,$F(\pi^*, z)$)
      add(front,$F(sym(\pi^*), z)$)
   end while
   return front

The *Greedy Algorithm* serves as a reference for the other algorithms and does maximally evaluate $O(2m) = O(m)$ times the $F(\pi, z)$ function. There is no maximum number for non dominated solutions in the final front which means for problem instances with many items and a high maximum capacity Q many solutions will be added to the front.

Independent Subproblem Algorithm. Since a lot of research has been done on the knapsack problem, the approach above can be extended by using a KNP Solver instead of solving it greedily. The pseudo code for the algorithm is shown in Fig. 2. After calculating π^*, the maximum weight capacity is divided equally into T intervals. For instance, if $Q = 12$ and $T = 4$ the corresponding maximum capacities are $I = \{0, 4, 8, 12\}$. For each of these maximum capacities the KNP Solver is used to calculate a packing plan z^*. Then - analogous to the Greedy Algorithm - the resulting packing plan is combined with π^* and $sym(\pi^*)$.

Algorithm 2. Independent Subproblem Algorithm

Require: $P \leftarrow$ Thief Problem
Require: $T \leftarrow$ Maximum of Solutions
    $\pi^* \leftarrow \text{tspSolver}(P)$
    $I \leftarrow \text{getLinearMaxWeight}(P, T)$
    $front \leftarrow \{\}$
    for all $i \in I$ **do**
        $z^* \leftarrow \text{knpSolver}(P, i)$
        $\text{add}(front, F(\pi^*, z^*))$
        $\text{add}(front, F(sym(\pi^*), z^*))$
    end for
    if localOptimize **then**
        while hasEvaluations() **do**
            $\text{add}(front, \text{localOptimize}(\text{getRandom}(front)))$
        end while
    end if
    return front

Additionally, a further local optimizing procedure can be applied by selecting randomly one solution of the Pareto front and optimizing it. In our implementation we perform with a probability of $\frac{1}{3}$ a swap tour mutation, of $\frac{1}{3}$ a bitlfip pack mutation and of $\frac{1}{3}$ both mutations. If the new solution is not dominated by any other solution, it is added to the front. The local optimization is done until no function evaluations are left. The standard version is labeled with ISA and with ISA-LOCAL if the front is local optimized.

NSGA-II. As optimization framework the evolutionary multi-objective **N**on-dominated **S**orting **G**enetic **A**lgorithm-II (NSGA-II) [15] is used. As usual for evolutionary algorithms all individuals are factored and added to the population in the beginning. Then each individual gets a rank based on its level of domination and a crowding distance which is used as a density estimation in the objective space. The binary tournament selection compares rank and crowding distance of randomly selected individuals in order to return individuals for the recombination. After executing crossover and mutation operators the offspring is added to the population. In the end of each generation the population is truncated after sorting it by domination rank and crowding distance. All details of the algorithm can be found in [15].

Also, meta-algorithms provide a great possibility to use the same method to solve many problems, different domain specific operators have to be defined. NSGA-II needs to know how to create and recombine individuals. For the experiment we used a population size of 100 individuals and performed the mutation operations in 30% of cases. Moreover, an additional method to remove duplicates each generation is executed in order to prevent diversity loss. The evaluation of different operators is presented in Sect. 6.

6   Results

In the following the proposed algorithms are evaluated and the results are shown. After describing the problem instances, different factory and recombination operators for NSGA-II are compared. Afterwards, the final evaluation takes the best found operator combination for NSGA-II and the other proposed algorithms into account.

All algorithms were executed 30 times on the same problem instance with maximum 100000 function evaluations each run. The median Pareto fronts of these runs were calculated by using the **E**mpirical first-order **A**ttainment **F**unction (EAF) [16].

Problem Instances

We created different problem instances, that vary in the city location generator, number of cities, number of items per city and the knapsack capacity. The number of cities are 10, 20, 50 and 100 with 1, 5 and 10 items per city. The maximum capacity is divided into three different categories, namely small $c = 0.2$, medium $c = 0.5$ and large $c = 0.8$ where $Q = c\sum_{i=1}^{m} w_i$. The velocity range is set to $v_{min} = 0.1$ and $v_{max} = 1$.

The name pattern for problem instances with the random city generator looks as follows:

$$multi - numOfCities - itemsPerCity - knapsackCapacity$$

For all *multi-cluster-XX* problem instances 100 cities are assigned to *XX* different clusters. Only one item is assigned to each city and the knapsack has a medium capacity. This imitates a map with cities where many locations occur in densely populated areas.

NSGAII - Knapsack Component

The factory is an important part of an evolutionary algorithm. Therefore, we made up an experiment in order to test three different packing plan factories.

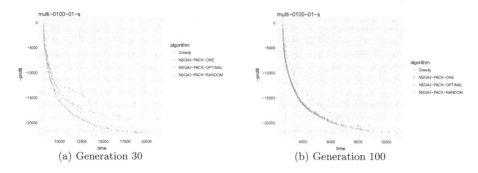

(a) Generation 30   (b) Generation 100

Fig. 4. Objective space after 30 and 100 generations on the problem instance *multi-0100-1-s* with different knapsack factories.

Note, that this experiment showed the same results whatever tour factory and recombination operators we combined it with. Two packing plans are combined by using a single point crossover and mutated by a bitflip mutation.

- PACK-ONE: Produces knapsacks with only one randomly picked item.
- PACK-RANDOM: Completely random knapsacks without violating the maximum capacity constraint.
- PACK-OPTIMAL: Define randomly $0 \le Q' \le Q$ and return the result of a KNP Solver [17] with maximum capacity Q'

The obtained results for the knapsack factories for one problem instance are shown exemplarily in Fig. 4. The factory PACK-OPTIMAL produces better solutions than the naive *Greedy* approach. PACK-ONE needs more time to converge with solutions with higher profit than PACK-RANDOM. All NSGA-II algorithms outperform *Greedy* after 100 generations. The same effect has been observed on all other problem instances.

NSGAII - Tour Component

As it seems to be a good choice to factor packing plans z calculated by a KNP Solver, it should also be considered to return a tour π^* calculated by a TSP Solver. However, in this case only π^* and $sym(\pi^*)$ in the starting population exist and no diversity of different tours is given. Moreover, the 2OPT factory returns two-opt optimal [14] or the RANDOM factory completely random tours. The edge recombination crossover [18] (EDGE) is used to combine tours and a swap mutation to mutate. The recombination of packing plans is done as described above and the optimal packing plan factory is used. In order to see the usefulness of EDGE, we tried another factory FIXED as well which does not perform any crossover on the tours. To sum up, this experiment includes the following four combinations:

- NSGAII-RANDOM-EDGE: Random tours with EDGE
- NSGAII-2OPT-EDGE: Two-opt optimal tours with EDGE

- NSGAII-OPT-EDGE: Tours calculated by TSP Solver and EDGE
- NSGAII-OPT-FIXED: Tours calculated by TSP Solver and no crossover

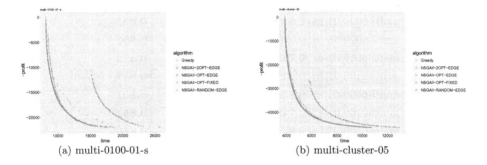

(a) multi-0100-01-s   (b) multi-cluster-05

Fig. 5. Objective space after 1000 generations with different tour factories.

Figure 5 illustrates the results of the different tour factories. The algorithm NSGAII-RANDOM-EDGE shows the worst results because the evolution is not able to find tours that are fast enough. The OPT factory outperforms the 2OPT factory on the multi-0100-01-s instance. OPT-FIXED and OPT-EDGE showed almost the same results which means the edge recombination crossover did not bring much improvement. In fact, as mentioned above only two tours are in the starting population. A crossover between π^* itself or π^* and $sym(\pi^*)$ is difficult to perform. However, when a mutation brings an improvement, a diversity of tours might exist.

Evaluation

After analyzing the evolutionary components of NSGA-II, a complete evaluation including the other proposed algorithm is presented. The hypervolume of the normalized median Pareto fronts is shown in Table 3. The algorithm denoted by NSGAII is the combination of operators with the best performance found above. It uses for the packing plan the OPT factory, the bitlfip mutation and the single point crossover and for the tour the OPT factory, the swap mutation and the edge recombination crossover. The other algorithms are implemented as described in Sect. 5.

Clearly NSGA-II shows the best performance on all problem instances. Since *Greedy* and ISA do not use as many evaluations as the other algorithms, they are more or less a reference that must be outperformed. In fact, using a KNP Solver instead of picking up items greedily improves the overall result. But also note, that ISA-LOCAL is not able to show significantly better results than ISA.

Moreover, further analysis of NSGA-II Pareto fronts showed the following: In average only 3.39 different tours exist in each Pareto front. Also, there were runs where only the two starting tours exist. The maximum diversity was found in *multi-0100-01-s* with 15 different tours. In fact, this shows the challenge to start

Table 3. Hypervolume of normalized median Pareto fronts after 100000 evaluations

	Greedy	ISA	ISA-LOCAL	NSGA-II
multi-0010-01-m	0.562	0.852	**0.874**	**0.874**
multi-0010-05-m	0.79	0.861	0.863	**0.881**
multi-0020-01-m	0.616	0.797	0.805	**0.835**
multi-0020-05-m	0.626	0.816	0.816	**0.853**
multi-0050-01-m	0.701	0.847	0.848	**0.872**
multi-0100-01-l	0.803	0.805	0.806	**0.847**
multi-0100-01-m	0.772	0.777	0.781	**0.851**
multi-0100-01-s	0.741	0.752	0.755	**0.85**
multi-0100-05-l	0.832	0.830	0.830	**0.849**
multi-0100-05-m	0.808	0.806	0.806	**0.838**
multi-0100-05-s	0.801	0.800	0.800	**0.851**
multi-0100-10-l	0.843	0.840	0.840	**0.852**
multi-0100-10-m	0.825	0.822	0.822	**0.84**
multi-0100-10-s	0.808	0.806	0.806	**0.846**
multi-cluster-03	0.827	0.846	0.846	**0.867**
multi-cluster-05	0.813	0.867	0.868	**0.884**
multi-cluster-10	0.804	0.866	0.866	**0.881**

with fast tours in order to have a good initial population, but to preserve diversity during the evolution. Note, that all Pareto fronts seemed to be highly converged for each type of algorithm. The hypervolume of minimum and maximum Pareto fronts showed almost the same results over all runs.

7   Conclusion

After introducing the Bi-Objective Traveling Thief Problem, showing the interaction of its subproblems, and presenting their interdependencies the problem is defined mathematically. Characteristics of the problem itself and the objective space provide basic knowledge in order to propose algorithms for the Bi-Objective Traveling Thief Problem. The evaluation of these algorithms shows the advantages of multi-objective evolutionary methods in comparison to deterministic approaches. However, the diversity of tours in the Pareto front and therefore a better convergence might be improvable with other evolutionary operators. Therefore, operators that consider the interwovenness and are not performing recombinations on tour and packing plan independently might be useful. Furthermore, the performance of these algorithms on problem instances with a large-scale has to be verified. Even though, the Traveling Thief Problem combines two well-known problems, the interdependence is challenging and therefore the complexity is more than just the sum of these two subproblems.

It is more, because the interwovenness does not allow applying state-of-the-art algorithms for TSP or KNP. Therefore, one plus one is more than two and further studies should be done on the overflow part.

Acknowledgment. This work was supported by a fellowship within the FITweltweit programme of the German Academic Exchange Service (DAAD).

References

1. Ishibushi, H., Klamroth, K., Mostaghim, S., Naujoks, B., Poles, S., Purshouse, R., Rudolph, G., Ruzika, S., Sayin, S., Wiecek, M.M., Yao, X.: Multiobjective Optimization for Interwoven Systems. Schloss Dagstuhl-Leibniz-Zentrum fuer Informatik (2015)
2. Applegate, D.L., Bixby, R.E., Chvatal, V., Cook, W.J.: The Traveling Salesman Problem: A Computational Study. Princeton Series in Applied Mathematics. Princeton University Press, Princeton (2007)
3. Lagoudakis, M.G.: The 0–1 Knapsack Problem - An Introductory Survey (1996)
4. Bonyadi, M.R., Michalewicz, Z., Barone, L.: The travelling thief problem: the first step in the transition from theoretical problems to realistic problems. In: IEEE Congress on Evolutionary Computation, pp. 1037–1044. IEEE (2013)
5. Polyakovskiy, S., Bonyadi, M.R., Wagner, M., Michalewicz, Z., Neumann, F.: A comprehensive benchmark set and heuristics for the traveling thief problem. In: Proceedings of the 2014 Annual Conference on Genetic and Evolutionary Computation, ser. GECCO 2014, pp. 477–484. ACM, New York (2014). http://doi.acm.org/10.1145/2576768.2598249
6. Reinelt, G.: TSPLIB - A t.s.p. library. Universität Augsburg, Institut für Mathematik, Augsburg. Technical report 250 (1990)
7. Faulkner, H., Polyakovskiy, S., Schultz, T., Wagner, M.: Approximate approaches to the traveling thief problem. In: Proceedings of the 2015 on Genetic and Evolutionary Computation Conference, ser. GECCO 2015, pp. 385–392. ACM, New York (2015). http://doi.acm.org/10.1145/2739480.2754716
8. Bonyadi, M.R., Michalewicz, Z., Przybylek, M.R., Wierzbicki, A.: Socially inspired algorithms for the travelling thief problem. In: Proceedings of the 2014 Annual Conference on Genetic and Evolutionary Computation, ser. GECCO 2014, pp. 421–428. ACM, New York (2014). http://doi.acm.org/10.1145/2576768.2598367
9. Birkedal, R.: Design, implementation, comparison of randomized search heuristics for the traveling thief problem, Master's thesis. Technical University of Denmark, Department of Applied Mathematics, Computer Science, Richard Petersens Plads, Building 324, DK-2800 Kgs. Lyngby, Denmark, compute@compute.dtu.dk (2015). http://www.compute.dtu.dk/English.aspx
10. Mei, Y., Li, X., Salim, F., Yao, X.: Heuristic evolution with genetic programming for traveling thief problem. In: IEEE Congress on Evolutionary Computation, CEC 2015, Sendai, Japan, 25–28 May 2015, pp. 2753–2760 (2015). http://dx.doi.org/10.1109/CEC.2015.7257230
11. Mei, Y., Li, X., Yao, X.: On investigation of interdependence between sub-problems of the travelling thief problem. Soft Comput., 1–16 (2014). http://dx.doi.org/10.1007/s00500-014-1487-2

12. Mei, Y., Li, X., Yao, X.: Improving efficiency of heuristics for the large scale traveling thief problem. In: Proceedings of the Simulated Evolution, Learning - 10th International Conference, SEAL 2014, Dunedin, New Zealand, 15–18 December 2014, pp. 631–643 (2014). http://dx.doi.org/10.1007/978-3-319-13563-2_53
13. Wachter, C.: Solving the travelling thief problem with an evolutionary algorithm. Diplomarbeit, Technischen Universitt Wien (2015)
14. Applegate, D., Cook, W., Rohe, A.: Chained lin-kernighan for large traveling salesman problems. INFORMS J. Comput. **15**(1), 82–92 (2003). http://dx.doi.org/10.1287/ijoc.15.1.82.15157
15. Deb, K., Pratap, A., Agarwal, S., Meyarivan, T.: A fast elitist multi-objective genetic algorithm: NSGA-II. IEEE Trans. Evol. Comput. **6**, 182–197 (2000)
16. Fonseca, V.G., Fonseca, C.M.: The attainment-function approach to stochastic multiobjective optimizer assessment and comparison. In: Experimental Methods for the Analysis of Optimization Algorithms, pp. 103–130. Springer, Heidelberg (2010). http://dx.doi.org/10.1007/978-3-642-02538-9_5
17. Martello, S., Pisinger, D., Toth, P.: Dynamic programming and strong bounds for the 0–1 knapsack problem. Manage. Sci. **45**(3), 414–424 (1999)
18. Oliver, I.M., Smith, D.J., Holland, J.R.C.: A study of permutation crossover operators on the traveling salesman problem. In: Proceedings of the Second International Conference on Genetic Algorithms on Genetic Algorithms and their Application, pp. 224–230. L.E. Associates Inc., Mahwah (1987)

Automatically Configuring Multi-objective Local Search Using Multi-objective Optimisation

Aymeric Blot[1(✉)], Alexis Pernet[1], Laetitia Jourdan[1],
Marie-Éléonore Kessaci-Marmion[1], and Holger H. Hoos[2,3]

[1] Université de Lille, Inria, CNRS, UMR 9189 – CRIStAL, Lille, France
aymeric.blot@inria.fr, alexis.pernet@etudiant.univ-lille1.fr,
{laetitia.jourdan,me.kessaci}@univ-lille1.fr
[2] University of British Columbia, Vancouver, Canada
hoos@cs.ubc.ca
[3] Universiteit Leiden, Leiden, The Netherlands

Abstract. Automatic algorithm configuration (AAC) is becoming an increasingly crucial component in the design of high-performance solvers for many challenging combinatorial optimisation problems. This raises the question how to most effectively leverage AAC in the context of building or optimising multi-objective optimisation algorithms, and specifically, multi-objective local search procedures. Because the performance of multi-objective optimisation algorithms cannot be fully characterised by a single performance indicator, we believe that AAC for multi-objective local search should make use of multi-objective configuration procedures. We test this belief by using MO-ParamILS to automatically configure a highly parametric iterated local search framework for the classical and widely studied bi-objective permutation flowshop problem. To the best of our knowledge, this is the first time a multi-objective optimisation algorithm is automatically configured in a multi-objective fashion, and our results demonstrate that this approach can produce very good results as well as interesting insights into the efficacy of various strategies and components of a flexible multi-objective local search framework.

Keywords: Algorithm configuration · Multi-objective optimisation · Local search · Permutation flowshop scheduling

1 Introduction

The performance of many single- and multi-objective optimisation methods strongly depends on the setting of their parameters. For most classes of problem instances, to achieve good performance, the values of these parameters must be specifically optimised. Thus, it is becoming increasingly common practice to use automatic algorithm configuration (AAC) procedures, such as irace [14], ParamILS [11] or SMAC [10]. While these configurators optimise a single performance metric only, such as solution quality or running time of a given target algorithm, recently, multi-objective AAC procedures, such as MO-ParamILS [2],

© Springer International Publishing AG 2017
H. Trautmann et al. (Eds.): EMO 2017, LNCS 10173, pp. 61–76, 2017.
DOI: 10.1007/978-3-319-54157-0_5

have become available and shown to be effective for optimising multiple performance metric simultaneously, using a multi-objective approach.

The performance of multi-objective optimisation (MOO) algorithms is generally assessed using multiple performance indicators, in order to characterise several distinct quality properties, such as convergence or diversity. However, so far, the automatic configuration of multi-objective algorithms has used standard, single-objective configurators to optimise either a single performance indicator or an aggregation of several indicators [1].

The hypothesis we investigate here is that automatic configuration of multi-objective optimisation algorithms is best achieved using a multi-objective configuration procedure, such as MO-ParamILS. To study this hypothesis, we consider multi-objective local search algorithms, which are known to be very efficient for a broad class of MOO problems, and specifically, multi-objective iterated local search (MO-ILS), a metaheuristic known to achieve excellent performance if its constituent components are chosen and configured carefully. We introduce a flexible, highly parametric MO-ILS framework for the bi-objective permutation flowshop scheduling problem (PFSP), a classic problem on which multi-objective local search algorithms are known to achieve excellent performance [8]. Our results show that using an MO configuration approach, we achieve better results than obtained from single-objective configuration approaches. Specifically, using the same configuration budget, we obtain a broader range of non-dominated trade-offs between two performance indicators, hypervolume and Δ spread, without significant loss in the quality of the configurations thus obtained. We also report new insights into the components of our local search framework that are effective for solving PFSP instances of different sizes.

2 Preliminaries

In multi-objective optimisation, multiple criteria (or objective functions) characterising the quality of solutions to a given problem are optimised simultaneously. The concept of *Pareto dominance* is used to capture trade-offs between those criteria: solution s_1 is said to dominate solution s_2 if, and only if, (i) s_1 is better than or equal to s_2 according to all criteria, and (ii) there exists at least one criterion according to which s_1 is strictly better than s_2. A set S of solutions in which there are no $s_1, s_2 \in S$ such that s_1 dominates s_2 is called a *Pareto set*, a *Pareto front*, or – in the context of multi-objective local search algorithms – an *archive*.

It is not straightforward to assess or compare the performance of multiple multi-objective algorithms. In the literature, many performance indicators have been proposed [12,17] and classified according to several properties: (i) cardinality, (ii) convergence and (iii) distribution. It has also been shown that it is generally not possible to aggregate such properties into a single indicator. Thus, it is recommended to consider multiple performance indicators, preferably ones that complement each other, in order to assess the efficiency multi-objective optimisation algorithms fairly.

Here, we use two complementary indicators: unary hypervolume [20], a volume-based convergence performance indicator, and Δ spread [4], a distance-based distribution metric. In the following, we assume that all objective values have been normalised to $[0, 1]$ and are to be minimised, meaning that the nadir is at $(1, 1)$ and the ideal at $(0, 0)$. The unary hypervolume indicator [20] measures the hypervolume of the objective space between the solutions of a given Pareto set and the nadir point. Hypervolume is maximised when the Pareto set is reduced to the ideal point. The Δ spread indicator [4] has been proposed to measure the distance-based distribution of set of solutions in a bi-objective context. Given a Pareto set S, ordered regarding the first criterion, we define

$$\Delta := \frac{d_f + d_l + \sum_{i=1}^{|S|-1} |d_i - \bar{d}|}{d_f + d_l + (|S| - 1) \cdot \bar{d}},$$

where d_f and d_l are the Euclidean distances between the extreme positions $(1, 0)$ and $(0, 1)$, respectively, and the boundary solutions of S, and $\bar{d}$ denotes the average over the Euclidean distances d_i for $i \in [1, |S| - 1]$ between adjacent solutions on the ordered set S. This indicator is to be minimised; it takes small values for large Pareto sets with evenly distributed solutions, and values close to 1 for Pareto sets with few or unevenly distributed solutions.

3   Multi-objective Algorithm Configuration

The goal of *automatic algorithm configuration (AAC)* is to automatically determine a configuration (*i.e.*, parameter setting) optimising the performance of a given algorithm for a given class of problem instances. In this context, we call the algorithm whose parameters are being optimised the *target algorithm* and the procedure that configures the target algorithm a *configurator*. The general concept of AAC is illustrated in Fig. 1.

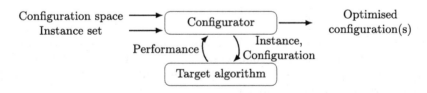

Fig. 1. Automatic configuration of a given, parameterised target algorithm for performance optimised for a given set problem instances.

Most applications of AAC consider a single-objective target algorithm, using a single performance metric – for optimisation algorithms, usually either the running time required to reach a specific solution quality or the solution quality achieved within a given running time. State-of-the-art AAC procedures from the literature include irace [14], SMAC [10] and ParamILS [11]. In principle, these and other single-objective AAC procedures can easily be applied to multi-objective optimisation (MOO) target algorithms, using a single performance indicator or an aggregation of several indicators (*e.g.*, using the hypervolume of normalised indicators [1]). A conceptually attractive alternative is to directly optimise multiple performance indicators, requiring a multi-objective configurator, such as the recently proposed MO-ParamILS [2], an extension of the original, single-objective ParamILS configuration framework. To the best of our knowledge, multi-objective configurators have so far only been applied to single-objective target algorithms – for example, for simultaneously optimising solution quality and running time of single-objective optimisation algorithms. Here, we investigate the efficacy of the state-of-the-art MO configurator MO-ParamILS when applied to MOO algorithms, using multiple multi-objective performance indicators.

In the standard AAC framework shown in Fig. 1, only the performance indicator values of the target algorithm are given to the configurator, rather than the entire target algorithm output. For multi-objective algorithms, this means that we cannot use binary performance indicators or performance indicators that use dynamic reference points. (Of course, this limitation could be overcome in future MO configurator designs.)

Algorithm configuration is a machine learning process that involves three separate phases: *training*, *validation* and *testing*. In the training phase, illustrated in Fig. 1, the configurator is used to optimise the configuration of the target algorithm on a given set of training instances. However, configurators are based on stochastic search procedures that are sensitive to random decisions made during the search process, including the order in which training instances are considered. Thus, the training phase is performed multiple times, independently, each with a different pseudo-random number seed. In the validation phase, the performance of the final configurations thus obtained is measured on a common set of instances (usually a subset of the training set). Dominated configurations are discarded, resulting in a set of non-dominated configurations. In the final testing phase, the set of configurations thus obtained is evaluated on a set of instances that does not contain any of the instances used for training or validation.

Note that from each independent run of a single-objective configuration procedure, a single configuration is obtained, while a run of a multi-objective configurator results in a Pareto set of configurations. In the latter case, the validation phase over n independent configurator runs will result in n Pareto sets, which are then merged, eliminating any configuration that is dominated with respect to the given performance objectives. Likewise, in the testing phase, every configuration is evaluated w.r.t. all given performance objectives, and we do not report configurations that turn out to be dominated on the given test set.

4   Multi-objective Local Search Algorithms

Stochastic local search (SLS) algorithms have been widely used for single-objective optimisation [9], and extensions to multi-objective optimisation are

---

Algorithm 1. Multi-Objective Iterative Improvement (`moii_algo`); as described in detail in Sect. 4.3, 'exploration' makes use of a reference.

Input: Initial archive
Output: Archive of non-dominated solutions
`archive` ← initial archive;
until *termination criterion is met* **do**
    /* Selection */
    `selection` ← select(`archive`);
    /* Exploration */
    `candidates` ← empty list;
    foreach `solution` ∈ `selection` **do**
        `candidates` ← `candidates` ∪ exploration(`solution`);
    /* Archive */
    `archive` ← combine(`archive`, `candidates`);
return `archive`;

---

known to achieve excellent performance. The most popular multi-objective SLS algorithms include Pareto local search (PLS, 2004) [18] and its numerous variants, such as the iterated PLS (2010) [5], stochastic PLS (2012) [6], anytime PLS (2015) [7], and dominance-based multi-objective local search (DMLS, 2012) [13]. In the following, we use a parametrised general local search framework that incorporates most of the strategies used by these algorithms. Our framework can be configured to replicate the behaviour of efficient algorithms of the literature, while making possible numerous intermediate behaviours by combining known building blocks in novel ways. More complex and problem-specific hybrid strategies based on local search strategies have also been studied in the literature. We do not consider these here, as our goal is to automatically configure effective yet widely applicable local search procedures rather than to design a state-of-the-art local search algorithm for a given problem using problem-specific expert knowledge.

Single-objective local search algorithms often get trapped in or around local optima of the search space they are exploring. This can be overcome in many ways, among which iterated local search (ILS) is well-known for its efficacy and versatility; as a general stochastic local search method, ILS provides the basis for state-of-the-art algorithms for many challenging combinatorial optimisation problems [15]. Our highly parametric ILS framework, described in the following, facilitates the design of powerful multi-objective local search algorithms based on an iterative improvement procedure.

4.1 Multi-objective Iterative Improvement

Algorithm 1 outlines a general procedure for iteratively improving an archive (*i.e.*, a Pareto set) until a stopping condition is satisfied. It works in three phases: (i) the *selection* phase, in which solutions are selected from the archive;

Algorithm 2. Multi-Objective Iterated Local Search

Input: Initial archive
Result: Archive of the best solutions
current_archive ← initial archive;
current_archive ← moii_algo(current_archive);
until *termination criterion is met* **do**
     tmp_archive ← perturb(current_archive);
     tmp_archive ← moii_algo(tmp_archive);
     current_archive ← combine(current_archive, tmp_archive);

return current_archive;

(ii) the *exploration* phase, in which the neighbourhood of each selected solution is explored, based on a reference set of solutions, and some neighbours are accepted as candidates; and (iii) the *archive* phase, in which the current archive is updated using the candidate neighbours.

Many strategies exist for these three phases and will be described in Sect. 4.3. We note that structurally, Algorithm 1 resembles DMLS [13]; however, as explained in Sect. 4.3, we have augmented the exploration procedure to make use of a reference point or set in order to replicate the behaviour of PLS algorithms.

4.2 Multi-objective Iterated Local Search

Our multi-objective iterated local search framework is obtained by embedding the iterative improvement procedure from Algorithm 1 into a typical ILS algorithm, as shown in Algorithm 2. After an initial run of the iterative improvement procedure, three search phases are iterated: (i) a perturbation phase, in which a copy of the current archive is perturbed, (ii) an iterative improvement phase, during which the perturbed archive is optimised, (iii) an acceptance phase, during which the current archive is updated based on the previous and newly optimised archives. In our case, the acceptance phase consists of merging the two archives followed by pruning based on the Pareto criterion.

4.3 Parameters and Configuration Space

We now describe the different strategies included in our multi-objective ILS framework, its parameters and their instantiation for simulating the behaviour of prominent local search algorithms from the literature.

Initialisation. We always initialise our iterative improvement procedure with 10 solutions generated uniformly at random. It is known that additional performance improvements can be realised by using an auxiliary optimisation algorithm to generate initial solutions [8], but we decided to not include this type of initialisation mechanism in order to focus our investigation on the core ILS procedure and its configuration.

Selection. In the selection phase, we choose a subset of solutions from the archive on which exploration will be performed. We then distinguish two main selection strategies (`selectStrat`): The first of these selects `all` solutions from the archive for exploration. The second strategy chooses a subset of $k \in \{1, 2, 3\}$ solutions from the archive (parameter `selectSize`), either uniformly at `random`, or the k `newest` or `oldest` solutions. Additionally, if some solutions have already been fully explored, we filter these out before starting the selection process.

Exploration. In this phase, the neighbourhood of each selected solution is explored: a set of neighbours is evaluated, and some of these are then added to the candidate set, which is later merged with the archive. We consider two types of exploration strategies (`explorStrat`), one of which evaluates all neighbours, while the other limits exploration to a subset of the neighbours. If all neighbours are evaluated, we can then either add all non-dominated neighbours to the candidate set (strategy `all`), or only add all dominating neighbours (strategy `all_imp`). Otherwise, we consider two different termination criteria for the exploration. The first of these ends exploration when k dominating neighbours have been evaluated. In that case, we can either only add these dominating neighbours to the candidate set (strategy `imp`), or also include the non-dominated neighbours that have been evaluated so far (strategy `imp_ndom`). The second criterion terminates exploration when k non-dominated neighbours have been evaluated, which are then added to the candidate set (strategy `ndom`). For both termination criteria, the value of k is specified by the parameter `explorSize`.

These strategies are further elaborated using another parameter, `explorRef`, which specifies the reference point for both *dominating* and *non-dominated* neighbours. This reference point can be either set to the current solution being explored (`sol`), or to a set of solutions, in which case we implicitly mean *dominating every solution in the set* and *non-dominated regarding each solution in the set*; we support taking the current archive as reference during exploration (`arch`) as well as taking the subset of the solutions that have been selected (`select`).

Archive. After all selected solutions have been explored, the set of candidates is merged with the current archive, and every dominated solution is removed. Thus, the archive always contains the best non-dominated solutions found during the search process. We note that the size of the archive is unbounded.

Termination Criteria. The termination criterion of the iterative multi-objective improvement procedure (Algorithm 1) is satisfied either when all the solutions in the current archive have been entirely explored, or when a given number of iterations have been performed without any modification of the archive. The termination criterion of the multi-objective iterated local search (Algorithm 2) is simply time-based.

Table 1. Parameters of our MO-ILS framework and their possible values

Phase	Parameter	Parameter values
Initialisation	`initStrat`	`rand`
Initialisation	`initSize`	`10`
Selection	`selectStrat`	$\{$`all, rand, newest, oldest`$\}$
Selection	`selectSize`	$\{1, 2, 3\}$
Exploration	`explorStrat`	$\{$`all, all_imp, imp, imp_ndom, ndom`$\}$
Exploration	`explorRef`	$\{$`sol, select, arch`$\}$
Exploration	`explorSize`	$\{1, 2, 3\}$
Perturbation	`perturbStrat`	$\{$`restart, kick, kick_all`$\}$
Perturbation	`perturbSize`	$\{1, 2, 3\}$
Perturbation	`perturbStrength`	$\{3, 5\}$

Perturbation. In Algorithm 2, the starting point of every subsidiary local search is obtained by applying perturbation to the current archive. The perturbation strategy (`perturbStrat`) can either be `restart`- or `kick`-based. In the first case, we consider 10 new solutions, generated uniformly at random, following the initialisation strategy. Otherwise, we either select $k \in \{1, 2, 3\}$ solutions from the archive (strategy `kick` with parameter `perturbSize`) or take all the solutions of the archive (strategy `kick_all`), before performing a *kick* move on each of them; when a solution is *kicked*, it is replaced by one of its neighbour selected uniformly at random. The parameter `perturbStrength` specifies how many times the selected solutions are kicked.

Overall Configuration Space. Table 1 shows all parameters exposed by our multi-objective iterated local search framework (Algorithms 1 and 2) and their possible values; these jointly give rise to $1 \cdot (1+3 \cdot 3) \cdot (1+3+3 \cdot (3 \cdot 3)) \cdot (1+3 \cdot 2+2) = 2790$ valid configurations.

5   Experiments

In our experiments, we focus on the bi-objective permutation flowshop scheduling problem, for which multi-objective local search algorithms are known to be very efficient [8]. In the following, we first give a brief description of this classical MOO problem, followed by the experimental protocol we used to compare our MO configuration approach with two single-objective approaches.

5.1   The Bi-objective Permutation Flowshop Scheduling Problem

The Permutation Flowshop Scheduling Problem (PFSP) involves scheduling a set of N jobs $\{J_1, \ldots, J_N\}$ on a set of M machines $\{M_1, \ldots, M_M\}$. Each job J_i

is processed sequentially on each of the M machines, with fixed processing times $\{p_{i,1}, \ldots, p_{i,M}\}$, and machines can only process one job at a time. The sequencing of jobs is identical on every machine, so that a solution is represented by a permutation of size N. Here, we consider the bi-objective PFSP, minimising both the makespan and the flowtime of the schedule, two objectives widely investigated in the literature [16], where makespan is the total completion time of the schedule, and flowtime is the sum of the individual completion times of all N jobs.

Classical PFSP neighbourhoods include the exchange neighbourhood, where the positions of two jobs are exchanged, and the insertion neighbourhood, where one job is reinserted at another position in the perturbation. In the following, we consider the union of these two classical neighbourhoods, as this hybrid neighbourhood has been shown to lead to better solutions than either of the two constituting neighbourhoods by itself [7].

5.2 Experimental Design

Benchmark Instances and Configuration Scenarios. We considered two sizes of bi-objective PFSP instances, leading to two configuration scenarios: one with instances of 20 jobs and a configuration budget (training time) of 12 CPU hours, and the other one with instances with 50 jobs and a configuration budget of 24 CPU hours. The classical PFSP instances in the literature are the widely-used Taillard instances [19]. Because the set of training and testing instances need to be completely disjoint and independent, we used the original Taillard instances only in the testing phase and generated new Taillard-like instances for training and validation phases.

In the testing phase, we used 30 classical Taillard instances for both 20 and 50 jobs scenarios, with 5, 10 or 20 machines (10 different instances for each combination). For the training phase, we generated 80 new instances with 5 to 20 machines (18 different instances for 5, 10 and 20 machines, and 2 instances for each intermediate instances sizes) for both 20 and 50 jobs scenarios. We also limited the maximum number of training runs of a given configuration to 240 (*i.e.*, 3 runs on each instance) in order not to spend too much time on too few instances. For the validation phase, a single target algorithm run was performed on each of the 80 training instances. The final test assessment was performed spending 5 runs on each of the 30 Taillard instances for a total of 150 runs.

In all our configuration experiments, we considered the configuration space defined by the parameters and parameter values specified in Table 1 (Sect. 4.3), containing a total of 2790 valid configurations. The maximum running time for all target algorithm runs was dynamically fixed to $n \cdot m/50$ CPU sec, where n and m are the number of jobs and the number of machines of the instance being solved. We note that these running times were chosen to be smaller than those commonly found in the literature ($n \cdot m/10$ in [16] and [7], or from $6 \cdot n \cdot m$ to $0.9 \cdot n \cdot m$ in [13] for instances of similar size), in order to permit the configurator to perform more runs and thus consider a larger number of configurations. We set the maximum number of consecutive iterations of Algorithm 1 without improvement to n. Finally, our algorithm framework and all of its components have been implemented in ParadisEO [3], a white-box, object-oriented framework dedicated to the flexible design of metaheuristics, in order to facilitate fair comparisons between arbitrary instantiations of our MO-ILS framework.

Performance Assessment. We used the hypervolume and Δ spread indicators to assess the performance of each MO-ILS configuration. Let us recall that the hypervolume measures the convergence, and the spread indicator is only used as a complementary indicator that measures the distribution along the Pareto set; we note that Δ spread is meaningless when used alone. To achieve a formulation where both indicators are to be minimised, we used $HV = 1 - \text{hypervolume}$, *i.e.*, the complement between hypervolume and the hypervolume of the ideal point.

AAC Experimental Protocol. We evaluated three configuration approaches: $HV\|\Delta$, a multi-objective approach, in which both hypervolume and Δ spread are minimised, using MO-ParamILS [2]; HV, a single-objective approach, in which only hypervolume is minimised; and $HV+\Delta$, a single-objective approach, in which we minimise a weighted sum of hypervolume and Δ-spread. For both single-objective approaches, we used single-objective ParamILS [11], as implemented in our MO-ParamILS framework. In $HV+\Delta$, we normalised both performance indicators and use weights of 0.75 for the hypervolume and 0.25 for the spread, as we see the latter as a secondary performance indicator. For each of these configuration approaches, we performed 30 independent configurator runs each with a total budget of 12 CPU hours for the 20-job benchmark and 24 CPU hours for the 50-job benchmark. The resulting configurations were evaluated on the entire given training set, and Pareto-dominated configurations were removed.

6   Results and Discussion

Figure 2 shows the performance of the non-dominated configurations obtained in our three configuration scenarios for the 20- and 50-job Taillard benchmark sets.

The single-objective configuration approach optimising hypervolume only (HV) resulted in 3 and 2 final configurations on the 20- and 50-job instance sets, respectively. While the hypervolume values achieved by these configurations were amongst the best obtained by any of our approaches, we observed somewhat average spread values on the 20-job instances, and poor spread on the 50-job set. Two of the three configurations for the 20-job set obtained exactly the same hypervolume and spread values.

The single-objective $HV+\Delta$ configuration approach optimising a weighted combination of hypervolume and spread also produced small numbers of non-dominated configurations (2 on the 20-job and 3 on the 50-job benchmarks). As expected, these configurations achieved different tradeoffs between the two

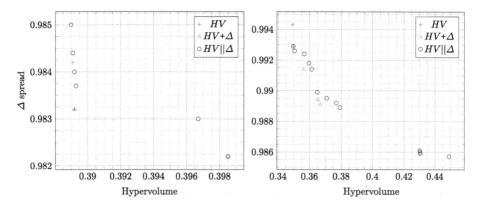

Fig. 2. Performance of automatically determined, non-dominated configurations on test sets of 20-job (left) and 50-job (right) Taillard instances. Each point shows the mean performance of a single configuration over the test set.

performance metrics. While it is clear that a broader range of tradeoffs could be obtained by using different weight vectors, this would require additional, costly configurator runs. Additionally, the $HV+\Delta$ approach requires non-trivial normalisation of both performance metrics, which we achieved based on the configurations obtained by HV and $HV\|\Delta$; this is unproblematic only because the purpose of studying $HV+\Delta$ was to provide a second baseline for our multi-objective configuration approach.

In contrast to HV and $HV+\Delta$, our multi-objective configuration approach, $HV\|\Delta$, resulted in substantially larger numbers of configurations covering a wide range of tradeoffs between hypervolume and spread. (The same phenomenon was observed when evaluating the configurations produced by ParamILS and MO-ParamILS on the respective training sets of instances.) We note that, while on the 20-job benchmark, $HV\|\Delta$ and $HV+\Delta$ achieved very similar extreme values for both performance indicators, on the more challenging 50-job benchmark, the configurations found by $HV\|\Delta$ spanned a wider range of hypervolume and spread values. Overall, for both benchmarks, we obtained two rather well-separated clusters of non-dominated configurations, but for the set of larger instances, the second cluster – characterised by low Δ spread and high hypervolume – was only reached by the multi-objective $HV\|\Delta$ configuration approach, while the single-objective $HV+\Delta$ achieved excellent results only in the direction of the weight vector used for aggregation.

Whereas Fig. 2 shows average performance across each benchmark set, we also examined average performance on subsets of instances with the same number of machines and on individual instances within these sets. This analysis revealed significant variation in both performance metrics with number of machines as well as between instances. For example, while the largest difference in Δ spread for 50-job instances, averaged over the entire set, that we measured for any pair of configurations found by our three approaches is below 0.01 (as seen in Fig. 2),

Table 2. Optimised configurations for Taillard instances with 20 jobs

HV	Δ	Approach	Selection		Exploration			Perturbation		
0.3891	0.9842	HV	rand	2	imp_ndom	select	1	kick	1	5
0.3891	0.9842	HV	oldest	2	imp_ndom	sol	1	restart	.	.
0.3892	0.9832	HV	all	.	imp_ndom	select	1	kick_all	.	5
0.3892	0.9832	$HV+\Delta$	all	.	imp_ndom	select	1	kick_all	.	5
0.3985	0.9822	$HV+\Delta$	all	.	ndom	arch	1	restart	.	.
0.389	0.985	$HV\|\|\Delta$	oldest	1	imp_ndom	sol	1	kick	1	5
0.3891	0.9844	$HV\|\|\Delta$	all	.	imp_ndom	select	1	kick	1	5
0.3892	0.984	$HV\|\|\Delta$	oldest	1	imp_ndom	sol	1	kick	3	3
0.3893	0.9837	$HV\|\|\Delta$	oldest	1	imp_ndom	sol	1	restart	.	.
0.3967	0.983	$HV\|\|\Delta$	all	.	ndom	arch	1	kick	1	3
0.3985	0.9822	$HV\|\|\Delta$	all	.	ndom	arch	1	restart	.	.

we observed differences as large as 0.03 between instances within the set for individual configurations produced by any of our approaches. Furthermore, similar spread values were observed across and within instance subsets with 5, 10 and 20 machines. This indicates that the small differences in mean spread observed between the configurations found by $HV+\Delta$ and the closest configurations produced by $HV\|\|\Delta$ are somewhat insignificant. This finding is further supported by our observation that differences in the spread values measured for pairs of configurations are inconsistent across the instances within each set.

In contrast, the differences in mean hypervolume observed between the two clusters of configurations in Fig. 2 are not only quite large, but also correspond well to the differences in hypervolume for instance subsets with the same, fixed number of machines and on individual instances within these sets. Furthermore, the hypervolume values for instance subsets with 5, 10 and 20 machines fall into intervals with very little overlap, and those intervals differ markedly between configurations from the two clusters seen in Fig. 2 (right). Overall, this indicates that these clusters of configurations are indeed well separated, which highlights the significance of our finding that, at least for the larger, 50-job instances, only the multi-objective $HV\|\|\Delta$ approach finds the second cluster of configurations, characterised by high hypervolume and low Δ spread values.

Finally, we examined the non-dominated configurations found by HV, $HV+\Delta$ and $HV\|\|\Delta$ in detail. Tables 2 and 3 show the performance metrics and parameter settings for each of the configurations shown in Fig. 2; inactive conditional parameters are indicated by dots (*e.g.*, when all the solutions are selected, the selection size parameter is inactive).

For the 20-job scenario (see Table 2), the configurations found by our three approaches are quite varied. The selection strategy is mostly oldest or all, the exploration strategy either imp_ndom or ndom, and other parameters take various values. Overall, this suggests that for these small instances, a broad range

Table 3. Optimised configurations for Taillard instances with 50 jobs

HV	Δ	Approach	Selection		Exploration			Perturbation		
0.3492	0.9943	*HV*	newest	1	ndom	select	2	kick	2	3
0.3496	0.9929	*HV*	newest	1	ndom	sol	1	kick_all	.	3
0.3564	0.9914	*HV*	all	.	ndom	arch	1	kick_all	.	3
0.3655	0.9894	*HV+Δ*	newest	1	ndom	arch	3	kick	2	3
0.3669	0.9891	*HV+Δ*	newest	1	ndom	arch	2	kick_all	.	3
0.3496	0.9929	*HV$\|\Delta$*	newest	1	ndom	sol	2	kick_all	.	3
0.3505	0.9926	*HV$\|\Delta$*	newest	1	ndom	sol	1	kick	1	5
0.3566	0.9924	*HV$\|\Delta$*	newest	3	ndom	arch	2	kick_all	.	3
0.3596	0.9918	*HV$\|\Delta$*	newest	2	ndom	arch	2	kick_all	.	3
0.3614	0.9914	*HV$\|\Delta$*	newest	3	ndom	arch	1	kick_all	.	3
0.3649	0.9899	*HV$\|\Delta$*	newest	1	ndom	arch	3	kick	3	3
0.3669	0.9891	*HV$\|\Delta$*	newest	1	ndom	arch	2	kick_all	.	3
0.379	0.9889	*HV$\|\Delta$*	newest	1	ndom	arch	1	kick	1	3
0.4303	0.9861	*HV$\|\Delta$*	all	.	all		.	kick	3	3
0.4305	0.986	*HV$\|\Delta$*	all	.	all		.	kick	1	5
0.4305	0.986	*HV$\|\Delta$*	all	.	all		.	kick	3	5
0.4306	0.9859	*HV$\|\Delta$*	all	.	all		.	kick	1	3
0.449	0.9857	*HV$\|\Delta$*	all	.	imp_ndom	arch	2	restart	.	.

of design choices within our framework achieves good tradeoffs between hyper-volume and spread. For the larger, 50-job scenario (see Table 3), the situation appears to be markedly different. There are two distinct types of configurations, corresponding to the two clusters seen in Fig. 2. The first of these consistently uses the newest selection strategy and the ndom exploration strategy, whereas the second employs the all selection strategy and (with one exception) the all exploration strategy; almost all configurations make use of the kick or kick_all perturbation strategy. The restart perturbation strategy, which seems to work well on the 20-job instances, appears to be less effective on the more challenging 50-job instances. These findings suggest that, as instance size and difficulty increases, effective combinations of design choices become more constrained, and finding these combinations more challenging. Our multi-objective configuration approach, $HV\|\Delta$, appears to be considerably more effective at exploring this design space than the two single-objective configuration approaches, HV and $HV+\Delta$, as witnessed by the size and diversity of the sets of non-dominated configurations found by each approach. We note that rand is a selection strategy commonly used in the literature; interestingly, it was chosen only once by our configuration approaches, which, in contrast, favoured newest and oldest. The prevalence of the ndom and imp_ndom exploration strategies in the configurations obtained from our approaches highlights the importance of considering the non-dominated neighbours in multi-objective iterated local search for the bi-objective PFSP.

The DMLS recommendations [13] regarding design choices within multi-objective local search algorithms depending on instance size are mostly consistent with the results obtained from our automatic configuration approaches, with some interesting differences. For small instances, the recommendation is to use exploration until the first dominating neighbour is found (i.e., exploration strategy imp_ndom with selectSize = 1), with either the selection of a single random solution from the archive or the selection of the entire archive (i.e., selection strategy all or rand with selectSize = 1). For larger instances, the recommendation is to explore until the first non-dominated neighbour is found (exploration strategy ndom with selectSize = 1). In both cases, the only reference considered is the current solution being explored (explorRef = sol), and the recommendation is to use full restarts. In contrast, our results indicate that the random selection strategy is not a good choice, and that either the entire archive should be selected, or a selection of the oldest or newest solutions from the archive should be considered instead. Furthermore, we found no evidence that the full restart strategy is effective for larger instances; on the other hand, using the current archive as a reference for exploration – something not considered in DMLS – appears to be very effective, especially for larger instances.

7   Conclusions

Based on our empirical investigation, we conclude that there is substantial promise in the use of multi-objective (MO) automated algorithm configuration procedures, such as MO-ParamILS, for optimising the performance of highly heuristic algorithms for multi-objective optimisation (MOO) problems. While standard, single-objective algorithm configurators are effective tools for optimising a single performance metric of an MOO algorithm, such as hypervolume, in many circumstances, it can be important to pay attention to multiple performance indicators – in particular, to ones that assess the overall quality of a set of solutions and its diversity. Conceptually, this turns the configuration problem into a multi-objective optimisation problem, and our results suggest that this problem is solved most effectively using an MO configuration procedure.

For our experiments, we devised a highly parameterised iterated local search framework for the widely studied bi-objective permutation flow-shop problem. Our framework comprises a broad range of building blocks from the literature as well as a few simple novel choices. It is important to emphasise that in this work, our goal was not to improve the state of the art in solving the bi-objective PFSP, but rather to determine whether the automated configuration of flexible and powerful algorithm frameworks for this type of MOO problem should make use of multi-objective algorithm configuration procedures, such as MO-ParamILS. We firmly believe that automated algorithm configuration will greatly facilitate improvements in the state of the art for solving challenging MOO problems, such as the bi-objective PFSP – and this belief is amply supported by prior work on automatically configuring single- and multi-objective optimisation algorithms. Moreover, based on our findings reported in this work, we believe that MOO algorithms should be configured using an MO algorithm configuration procedure for the Pareto-optimisation of multiple performance indicators – an approach which differs from prior work on automated configuration of MOO algorithms.

As a secondary contribution of our work, based on a detailed analysis of the configurations of our multi-objective iterated local search framework, we proposed several revisions to the recommendations from prior work on dominance-based multi-objective local search [13]. In particular, we found evidence that simple alternatives to random selection provide improved overall performance. This type of finding confirms that the use of effective automated algorithm procedures and protocols can yield valuable insights into the efficacy of various algorithmic strategies and components for solving challenging MOO problems.

References

1. Bezerra, L.C.T.: A component-wise approach to multi-objective evolutionary algorithms. Ph.D. thesis, IRIDIA, Université Libre de Bruxelles, Belgium, July 2016
2. Blot, A., Hoos, H.H., Jourdan, L., Kessaci-Marmion, M.É., Trautmann, H.: MO-ParamILS: a multi-objective automatic algorithm configuration framework. In: Festa, P., Sellmann, M., Vanschoren, J. (eds.) LION 2016. LNCS, vol. 10079, pp. 32–47. Springer, Heidelberg (2016). doi:10.1007/978-3-319-50349-3_3
3. Cahon, S., Melab, N., Talbi, E.: Paradiseo: a framework for the reusable design of parallel and distributed metaheuristics. JoH **10**(3), 357–380 (2004)
4. Deb, K., Pratap, A., Agarwal, S., Meyarivan, T.: A fast and elitist multiobjective genetic algorithm: NSGA-II. IEEE TEVC **6**(2), 182–197 (2002)
5. Drugan, M.M., Thierens, D.: Path-guided mutation for stochastic pareto local search algorithms. In: Schaefer, R., Cotta, C., Kołodziej, J., Rudolph, G. (eds.) PPSN 2010. LNCS, vol. 6238, pp. 485–495. Springer, Heidelberg (2010). doi:10.1007/978-3-642-15844-5_49
6. Drugan, M.M., Thierens, D.: Stochastic Pareto local search: Pareto neighbourhood exploration and perturbation strategies. JoH **18**(5), 727–766 (2012)
7. Dubois-Lacoste, J., López-Ibáñez, M., Stützle, T.: Anytime Pareto local search. EJOR **243**(2), 369–385 (2015)
8. Dubois-Lacoste, J., López-Ibáñez, M., Stützle, T.: A hybrid TP+PLS algorithm for bi-objective flow-shop scheduling problems. C&OR **38**(8), 1219–1236 (2011)
9. Hoos, H.H., Stützle, T.: Stochastic Local Search: Foundations and Applications. Elsevier/Morgan Kaufmann, San Francisco (2004)
10. Hutter, F., Hoos, H.H., Leyton-Brown, K.: Sequential model-based optimization for general algorithm configuration. In: Coello, C.A.C. (ed.) LION 2011. LNCS, vol. 6683, pp. 507–523. Springer, Heidelberg (2011). doi:10.1007/978-3-642-25566-3_40
11. Hutter, F., Hoos, H.H., Leyton-Brown, K., Stützle, T.: ParamILS: an automatic algorithm configuration framework. JAIR **36**, 267–306 (2009)
12. Knowles, J., Corne, D.: On metrics for comparing nondominated sets. In: IEEE CEC, vol. 1, pp. 711–716 (2002)
13. Liefooghe, A., Humeau, J., Mesmoudi, S., Jourdan, L., Talbi, E.: On dominance-based multiobjective local search: design, implementation and experimental analysis on scheduling and traveling salesman problems. JoH **18**(2), 317–352 (2012)

14. López-Ibáñez, M., Dubois-Lacoste, J., Stützle, T., Birattari, M.: The irace package, iterated race for automatic algorithm configuration. Technical report TR/IRIDIA/2011-004, IRIDIA, Université Libre de Bruxelles, Belgium (2011)
15. Lourenço, H.R., Martin, O.C., Stützle, T.: Iterated local search. In: Glover, F., Kochenberger, G.A. (eds.) Handbook of Metaheuristics, pp. 320–353. Springer, New York (2003)
16. Minella, G., Ruiz, R., Ciavotta, M.: A review and evaluation of multiobjective algorithms for the flowshop scheduling problem. IJOC **20**(3), 451–471 (2008)
17. Okabe, T., Jin, Y., Sendhoff, B.: A critical survey of performance indices for multi-objective optimisation. In: IEEE CEC, vol. 2, pp. 878–885 (2003)
18. Paquete, L., Chiarandini, M., Stützle, T.: Pareto local optimum sets in the biobjective traveling salesman problem: an experimental study. In: Gandibleux, X., Sevaux, M., Sörensen, K., T'kindt, V. (eds.) Metaheuristics for Multiobjective Optimisation, pp. 177–199. Springer, Heidelberg (2004)
19. Taillard, E.: Benchmarks for basic scheduling problems. EJOC **64**(2), 278–285 (1993)
20. Zitzler, E., Thiele, L.: Multiobjective evolutionary algorithms: a comparative case study and the strength Pareto approach. IEEE TEVC **3**(4), 257–271 (1999)

The Multiobjective Shortest Path Problem Is NP-Hard, or Is It?

Fritz Bökler$^{(\boxtimes)}$

Department of Computer Science, TU Dortmund, Dortmund, Germany
`fritz.boekler@tu-dortmund.de`

Abstract. To show that multiobjective optimization problems like the multiobjective shortest path or assignment problems are hard, we often use the theory of **NP**-hardness. In this paper we rigorously investigate the complexity status of some well-known multiobjective optimization problems and ask the question if these problems really are **NP**-hard. It turns out, that most of them do not seem to be and for one we prove that if it is **NP**-hard then this would imply **P** = **NP** under assumptions from the literature. We also reason why **NP**-hardness might not be well suited for investigating the complexity status of intractable multiobjective optimization problems.

1 Introduction

This paper is concerned with one of the fundamental questions of algorithm design. When do we regard an algorithm as being *efficient*? In the past 60 years, the most important measure of the efficiency of exact algorithms is their running time. We usually regard an algorithm to be efficient, if its worst-case running time can be bounded by a polynomial in the input size. To reason that there is most probably no efficient, i.e., polynomial-time algorithm for a given problem at hand, **NP**-hardness theory has proven to be a very valuable tool especially for (combinatorial) optimization.

In multiobjective optimization the notion of **NP**-hardness has been adopted since the pioneering work by Serafini in 1986 [17]. Many papers cite Serafini to show that the multiobjective version of the shortest path, matching or matroid optimization problem are hard to solve. We will take the multiobjective shortest path problem as an example how **NP**-hardness is usually applied in multiobjective optimization.

The *multiobjective shortest path (MOSP)* problem is defined in the following way: We are given a directed graph G, consisting of a node set V and an arc set A. Formally, $A \subseteq V \times V$, i.e., we write an arc from a node v to a node w as an ordered pair (v, w). And we are given a *source node* s and a *target node* t. A solution is an s-t-path, i.e., a sequence of edges $((v_1, v_2), (v_2, v_3), \ldots, (v_{k-1}, v_k))$ from A, such

---

The author has been supported by the Bundesministerium für Wirtschaft und Energie (BMWi) within the research project "Bewertung und Planung von Stromnetzen" (promotional reference 03ET7505) and by DFG GRK 1855 (DOTS).

© Springer International Publishing AG 2017
H. Trautmann et al. (Eds.): EMO 2017, LNCS 10173, pp. 77–87, 2017.
DOI: 10.1007/978-3-319-54157-0_6

that for each subsequent edges a_1 and a_2, the second component of a_1 is equal to the first component of a_2 and $v_1 = s$ and $v_k = t$. We denote the set of all s-t-paths as $\mathcal{P}_{s,t}$. Moreover, we are given a *multiobjective edge-cost function* $c : A \to \mathbb{N}^d$, where for a path $p \in \mathcal{P}_{s,t}$ the function is extended to $c(p) := \sum_{e \in p} c(e)$. We want to find the minimal elements of $c(\mathcal{P}_{s,t}) := \{c(p) \mid p \in \mathcal{P}_{s,t}\}$ with respect to the component-wise less-or-equal partial order "$\leq$". Usually, we also want to find a representative solution path $p \in \mathcal{P}_{s,t}$ for every minimal element y, such that $c(p) = y$.

One of the first papers which is concerned with the MOSP problem is the work by Hansen [9]. He defines a version of the biobjective shortest path problem where $d = 2$. And he finds the first proof of *intractability*, i.e., he shows that the Pareto-front can be of exponential size in the input size. He also mentions that intractability must not be confused with **NP**-hardness as it is still possible to solve **NP**-hard problems "in polynomial time in the very improbable case of **P** = **NP**" [9].

Warburton [23] and Serafini [17] are the first authors who discuss the following decision problem:

Definition 1 (Canonical Decision Problem (MOSP^{DEC})). *Given a digraph $G = (V, A)$, an edge-cost function $c : A \to \mathbb{N}^d$, a vector $k \in \mathbb{N}^d$ and nodes $s, t \in V$. Decide whether there exists an s-t-path p in G for which $c(p) \leq k$ holds.*

Both authors show that MOSP^{DEC} is **NP**-hard and state this as an informal evidence that MOSP is a hard problem, while they also state that MOSP is actually intractable.

Later, in many papers, e.g., [6,8,10,12,13,15,16,18–22], it is claimed that MOSP is **NP**-hard. Therein, the arguments are based on the paper by Serafini, or the argument is given that since the output is exponential, the problem is **NP**-hard. Of course, the latter argument is not a formal proof of **NP**-hardness. In fact, it is what Hansen warned against, that **NP**-hardness should not be confused with intractability. It is by no means clear, why an output of exponential size implies that there is a polynomial-time reduction from every problem in **NP** to MOSP.

Similar examples can be found for the multiobjective versions of the bipartite matching (or assignment) problem and the spanning tree and matroid optimization problems. In this paper, we will rigorously look into the **NP**-hardness of these problems, focusing on the paper by P. Serafini. First, we define multiobjective combinatorial optimization problems and other concepts which will be needed in this paper in Sect. 2. Then, in Sect. 3, we will give a short introduction to **NP**-hardness theory. In Sect. 4, we will be concerned with the paper by P. Serafini and clarify what he has actually shown in his paper and what not. In Sect. 5, we discuss **NP**-hardness of multiobjective optimization problems in the light of the previous sections and then in Sect. 6 we discuss further means to investigate the complexity of multiobjective optimization problems.

2   Preliminaries

With $[n]$ we denote the set $\{1, \ldots, n\}$. For a set $M \subseteq \mathbb{Z}^d$, the set $\min M$ denotes the minimal elements of M, or $\min M := \{x \in M \mid$ there is no $x' \in M \backslash \{x\}$ with $x' \leq x\}$, where "$\leq$" is the component-wise less-or-equal partial order. For an n-polyhedron P, $\operatorname{vert} P := \{x \in P \mid \exists a \in \mathbb{R}^n : a^T x > a^T x'$ for all $x' \in P\}$, the set of *extreme points* of P. We will be concerned with the following problems.

Definition 2 (Multiobjective Combinatorial Optimization Problem).
A Multiobjective Combinatorial Optimization (MOCO) problem consists of a set of instances $\mathcal{I}$. An instance of a multiobjective combinatorial optimization problem $(\mathcal{S}, C)$ consists of

- *a set of* solutions $\mathcal{S} \subseteq \{0, 1\}^n$,
- *an* objective function matrix $C \in \mathbb{Z}^{d \times n}$.

The goal is to output for a given instance $(\mathcal{S}, C)$ the set $\mathcal{Y}_N := \min C \cdot \mathcal{S} = \min\{Cx \mid x \in \mathcal{S}\}$. The input size is measured in n and the encoding length of C.

For an instance $I = (\mathcal{S}, C)$ of a MOCO problem P, we call the set $\mathcal{Y}_N$ the *Pareto-front* of I. A solution $x \in \mathcal{S}$ is called *efficient* or *Pareto-optimal*, if $Cx \in \mathcal{Y}_N$. Usually, we expect algorithms also to produce an efficient solution $x \in \mathcal{S}$ for every $y \in \mathcal{Y}_N$, such that $Cx = y$, though we do not require this formally. Observe, that this definition also subsumes single objective combinatorial optimization problems.

Generalizing MOSP^{DEC}, we define the canonical decision problem for MOCO problems:

Definition 3 (Canonical Decision Problem (P^{DEC})). *Given a multiobjective combinatorial optimization problem P with instances $\mathcal{I}$. Then P^{DEC} is the following decision problem: For an instance $(\mathcal{S}, C) \in \mathcal{I}$ and a vector $k \in \mathbb{Z}^d$, decide whether there exists a feasible solution $x \in \mathcal{S}$ for which $Cx \leq k$ holds.*

3   Revisiting the Theory of NP-Hardness

Let us first recall formally what **NP**-hardness means. For a deeper understanding of this topic, we refer the reader to the textbook by Arora and Barak [1].

The notion of C-hardness bases on two things: The complexity class C and an often implicit reduction notion $\preceq$. A complexity class can be perceived as a set of computational problems and a reduction is a binary relation on computational problems. For a complexity class C, a problem P is said to be C-hard w.r.t. to a reduction $\preceq$ if for every problem $P' \in C$ we can reduce P' to P or $P' \preceq P$. For **NP**-hardness, we usually implicitly use Karp-reductions, or polynomial-time many-one reductions, when we reduce among decision problems and Cook-reductions, or polynomial-time Turing-reductions, if we reduce to other problems, e.g., optimization problems.

A decision problem P' can be Karp-reduced to a decision problem P, if there exists a polynomial-time transformation f from the inputs x of P' to the inputs of P, such that $x \in P' \Leftrightarrow f(x) \in P$. A computational problem P' can be Cook-reduced to a computational problem P if there exists a polynomial-time algorithm solving P' which may use an oracle solving P which accounts only for a constant in the running time analysis.

So, the first question is what reduction concept to use? Since MOCO problems are certainly not decision problems, only Cook-reduction is feasible. Hence, if we want to prove **NP**-hardness of a MOCO problem P, we are looking for problems which we can solve by an algorithm which has access to an oracle solving P.

As an example, imagine a biobjective version of the travelling salesperson problem (TSP): We are given a complete graph on n nodes. The feasible solutions are all edge sets which describe a cycle, or *tour*, in the graph, containing each node exactly once. Each column of the objective function matrix describes the costs of an edge between two nodes of the graph. Thus, the cost of a feasible tour is the (vector-valued) sum of the costs of the edges in the tour. The single objective version is well-known and **NP**-hard due to [7] and we will review the proof in the next subsection. Now, we can solve the single objective problem by an algorithm which has access to an oracle solving the biobjective problem in the following way: We copy the objective function of the single objective problem twice to get the two objectives of the biobjective problem. This biobjective problem has a Pareto-front size of at most 1. If the problem is feasible (i.e., is a valid instance), we can make this transformation, look at the Pareto-front and at the efficient solution and output it as a solution and value for the single objective problem. Since TSP is **NP**-hard, so is biobjective TSP.

But what if the single objective problem of a given MOCO problem is not **NP**-hard? For example the multiobjective versions of the shortest path, spanning tree or assignment problems? It is well-known that the Pareto-fronts can be exponential in the input size in the worst case. So, in general, looking at the Pareto-front cannot be done in polynomial time in a Cook-reduction, even though computing it accounts only for constant time.

3.1 NP-Hardness of Single Objective Optimization Problems

In single objective optimization, we usually prove **NP**-hardness of an optimization problem by showing the **NP**-hardness of a certain decision problem. Let us consider the TSP again as an example: To show that single-objective TSP is **NP**-hard, we show that TSP^{DEC} is **NP**-hard. Recall, that TSP^{DEC} is the decision problem whether there exists a feasible tour of cost less than a given number $k \in \mathbb{Z}$. The classical reduction is by reduction from the Hamiltonian cycle problem. In the Hamiltonian cycle problem, we are given a graph G and have to decide if there exists a cycle in the graph which contains every node exactly once. We get an instance of the TSP problem by giving each edge the cost 1 and add for each edge which does not exist in G an edge of cost 2. Thus, there exists a Hamiltonian cycle in G iff there exists a feasible tour of costs at most n in the new graph. But why does showing that TSP^{DEC} is **NP**-hard show that TSP is **NP**-hard?

The key is that there is a canonical Cook-reduction from P^{DEC} to P for every (single objective) optimization problem P. Given an instance of P^{DEC}, we can solve the optimization problem and check if k is at least the value of an optimal solution. If it is, then there exists a solution with cost at most k, i.e., each optimal solution. If it is not then since every solution has at least the optimal cost which is already higher than k, there does not exist a solution with cost less than k.

Again, since the Pareto-front can be of exponential size, the same reduction does not seem to generally hold in the context of multiobjective optimization.

4 The Paper by P. Serafini

Serafini is the first author doing a deep complexity analysis of multiobjective combinatorial optimization problems in 1986. In his paper [17], Serafini investigates the computational complexity of the general multiobjective combinatorial optimization problem

$$(\text{S1}) \min \ f(x)$$
$$\text{s.t. } x \in F$$

for $f : \{0,1\}^n \to \mathbb{Z}^d$, $F \subseteq \{0,1\}^n$ and $n, d \in \mathbb{N}$. He leaves the actual solution concept open. Observe that if the goal is to find the Pareto-front, this is a generalization of MOCO problems of Definition 2 to the case of general objective functions. We will identify problem S1 with the problem of finding the Pareto-front to ensure consistency to Definition 2.

In the aforementioned paper, Serafini is generally interested in problems were the set of solutions F is too large to be scanned exhaustively. He also points out that Pareto-front sizes can be exponential in n and thus they cannot be enumerated in polynomial time. So, he restricts the problems he studies to problems with Pareto-fronts of polynomial size by requiring that the differences in solution values are bounded by a polynomial in the input, i.e.,

$$\max_{x,y \in F} ||f(x) - f(y)||_\infty = K \in \mathcal{O}(n^k), \text{for } k > 1. \tag{1}$$

Here, $||x||_\infty$ for a vector $x \in \mathbb{Q}^d$ denotes the *maximum-norm* or ℓ_∞-*norm*: $||x||_\infty := \max_{i \in [d]} |x_i|$. He then shows reductions among several solution concepts of problem S1, e.g., finding the Pareto-front only, finding all Pareto-optimal solutions and S1^{DEC}. His general conclusion is that, assuming (1), finding the Pareto-front of problem S1 is equivalent to S1^{DEC}. He argues that because of this equivalence we can identify MOCO problems with their canonical decision problem. He also puts two other (non mathematical) reasons forward to support this conclusion: The first being that the canonical decision problem is also studied in single objective optimization. The second is that the canonical decision problem often plays a role in methods to solve MOCO problems in a posteriori or interactive approaches. In the following sections, Serafini shows for several well-known MOCO problems that their canonical decision problem is **NP**-hard and, in the light of the above argumentation, states that the MOCO problem itself is **NP**-hard.

He starts by studying S1 with linear objective functions, which are MOCO problems by Definition 2, and for which the feasible set F contains the extreme points of a polyhedron P which again contains F, i.e., vert $P \subseteq F \subseteq P$:

$$(S2) \min \ Cx$$
$$\text{s.t. } x \in F$$

for $C \in \mathbb{Z}^{d \times n}$. He investigates the complexity status of this general problem by looking at the following special case, the biobjective unconstrained combinatorial optimization (BUCO) problem.

$$(BUCO) \min \ (c^1, -c^2)^T x$$
$$x \in \{0,1\}^n$$

Serafini concludes that S2 is **NP**-hard, because BUCO^{DEC} is the binary knapsack problem, which is well-known to be **NP**-hard. He then goes on and investigates the biobjective shortest path and biobjective assignment problems as special cases of S2.

In the last section, Serafini proves **NP**-hardness for the biobjective matroid optimization problem and discusses neighborhoods induced by topological sorting and locally nondominated solutions.

5   Discussion of NP-Hardness in Multiobjective Optimization

Let us first state that Serafini did write that S2, the biobjective shortest path, assignment and matroid optimization problems are **NP**-hard, but he proves it only for the respective decision problems. The reason for this is the identification of a MOCO problem with its canonical decision problem, which was based on the informal reasons stated above and the equivalence of the problems under (1).

Hence, let us revisit the three reasons Serafini offers why a MOCO problem P can be identified with its canonical decision problem P^{DEC}. The third reason being that the hardness of P^{DEC} influences the choice of solution method to use to solve a problem at hand. This is a very important reason why the complexity status of P^{DEC} is interesting. But the incapability of one or a set of solution methods for solving a problem does not determine the complexity status of problem itself.

The second reason is that the problem is also investigated in single-objective optimization. But in single-objective optimization this is well justified as we discussed in Sect. 3.1. And the justification might not transfer to the multiobjective case.

The first reason that the solution concept of finding the Pareto-front and solving the canonical decision problem is equivalent under the bound (1) is solid on the first glance, but only applies if (1) holds. So in the general case, in which it is stated that, e.g., the biobjective shortest path problem is **NP**-hard, the bound does not apply.

Now we can ask the question, whether these problems are indeed **NP**-hard using the solution concept of Definition 2. If the output is exponential there does not seem to be hope. Since we do not impose an ordering on the output, it can be given in any order by an oracle. Finding a specific item in the Pareto-front cannot be done faster than $\Omega(|\mathcal{Y}_N|)$.

Thus, Serafini's bound (1) is a very interesting possibility to restrict the Pareto-front to a size which is searchable in polynomial time. So, we can ask the question if the above problems are **NP**-hard under this restriction.

In the case of the BUCO problem with nonnegative c^1 and c^2, we can prove that if it is **NP**-hard under (1), then $\mathbf{P} = \mathbf{NP}$ despite the fact that BUCO^{DEC} with nonnegative c^1 and c^2 is **NP**-hard in general. This is a strong evidence that the assumption that from **NP**-hardness of P^{DEC} follows **NP**-hardness of P under (1) for MOCO problems P does not hold. We will prove this by showing that the Nemhauser-Ullmann algorithm [14] for the knapsack problem solves BUCO in polynomial time if (1) is assumed.

Let us first discuss the Nemhauser-Ullmann algorithm in some detail. We are given an instance of the knapsack problem, i.e., $c^1, c^2 \in \mathbb{N}^n, W^1, W^2 \in \mathbb{N}$ and we are to decide whether there exists $x \in \{0,1\}^n$ with $c^{1^T} x \geq W_1$ and $c^{2^T} x \leq W_2$. We will use the following interpretation: The vector $(c_i^1, c_i^2)^T$ encodes the gain and loss if we pack item $i \in [n]$, thus we can view this decision problem as a biobjective optimization problem with objective functions $\max c^{1^T} x$ and $\min c^{2^T} x$ and get a BUCO-instance. We observe that if there is a feasible packing, then there is one which is also Pareto-optimal.

The algorithm follows a dynamic programming scheme: Assuming an arbitrary ordering of the items, we look at the Pareto-optimal packings for the first i items. We get the Pareto-optimal packings for the first $i + 1$ items by adding the $(i + 1)$-th item to each of our packings, retaining the old ones. We can now delete all packings which are dominated, since they will never be a subsolution of a Pareto-optimal packing. To solve the knapsack problem, we can delete all solutions for which $c^{2^T} x > W_2$. This algorithm is well known to be pseudo-polynomial and has a running time of $\mathcal{O}(nW_2)$.

Proposition 4. *If the BUCO problem with nonnegative c^1 and c^2 is **NP**-hard assuming (1) then $\mathbf{P} = \mathbf{NP}$.*

Proof. To solve the BUCO problem, we will set $W_2 := c^{2^T} \mathbf{1}$ and thus keep all Pareto-optimal solutions. Observe, that packing nothing, i.e., the solution $x = \mathbf{0}$, is always a Pareto-optimal packing. Using this and because of (1), we have for some $k > 1$

$$\mathcal{O}(n^k) \ni K = \max_{y,z\in F} ||f(y) - f(z)||_\infty$$

$$= \max_{x\in F} ||C\mathbf{0} - Cx||_\infty$$

$$= \max_{x\in F} ||Cx||_\infty$$

$$= \max\{\sum_{i\in[n]} c_i^1, \sum_{i\in[n]} c_i^2\} \geq c_i^j, \text{ for } j \in \{1,2\} \text{ and } i \in [n].$$

And since $nW_2 = n \cdot c^{2^T}\mathbf{1} \in \mathcal{O}(n^{k+1})$, the Nemhauser-Ullmann algorithm runs in polynomial time on BUCO instances. $\qquad\square$

As for the other problems, the MOSP problem, the multiobjective assignment and biobjective matroid optimization problem, it is not obvious what happens if (1) is assumed.

Regarding the MOSP problem, there do exist pseudo-polynomial time algorithms based on the generation of Pareto-optimal paths from s to the other nodes (cf. e.g. [11]). Using the bound (1), we can restrict the size of the labels at the target node t, but we are not able to directly bound the number of candidate paths at intermediate nodes $v \in V\backslash\{s,t\}$. Moreover, the bound we showed in the proof of Proposition 4 does not apply.

In conclusion, while the **NP**-hardness of canonical decision problems is interesting, it does not seem to have (mathematical) implications on the **NP**-hardness of the MOCO problem. Further, we might even want to ask the question what we can draw from the insight of a MOCO problem with exponential Pareto-front size actually being **NP**-hard. On one hand, **NP**-hardness of a problem is a good reason to believe that the problem cannot be solved in polynomial time. On the other hand, intractability proves that there is no polynomial-time algorithm in general.

6   Open Problems

The theory of **NP**-hardness in multiobjective optimization is still very interesting, if we cannot rule out polynomial-time algorithms by other means. The bound by Serafini is one possibility to restrict problems to Pareto-fronts of polynomial size. A more theoretical option is to restrict a problem to all instances which have polynomial Pareto-front size. But similar to Serafini's bound (1), it might be impossible to test this property in polynomial time in general and thus even testing the feasibility of an instance becomes hard.

It could also be interesting to modify MOCO problems by imposing an ordering on the Pareto-front to output. That is, requiring to output the Pareto-front of a MOCO problem in an ordering which is part of the problem, i.e., output the Pareto-front in lexicographic order. Then again, **NP**-hardness can be an interesting tool eventhough the output is exponential in the worst-case.

But there are also other kinds of running time analyses, which we want to discuss in short. These are *smoothed analysis* and *output-sensitive complexity*.

In smoothed analysis, we try to depart from classic worst-case analysis by looking at worst-case inputs which are then perturbed by some source of randomness. In the work by Beier, Röglin and Vöcking from 2007 [2], the following game-theoretic model was introduced: We choose a number $\phi > 1$. Then, an adversary chooses a solution set $S \subseteq \{0,1\}^n$ and the coefficients of the first row of C and fixes for each coefficient of the other rows a probability density function $f_{i,j} : [-1,1] \to \mathbb{R}$. From this distribution we draw our instances. Beier, Röglin and Vöcking showed that the expected Pareto-front size is at most $\mathcal{O}(\phi n^2)$ for biobjective combinatorial problems. This bound was generalized by Brunsch and Röglin who showed in [5] that the expected Pareto-front size is $\mathcal{O}(n^{2d} \phi^d)$ for general multiobjective problems. In this framework, the Nemhauser-Ullmann algorithm has an expected polynomial running time for the knapsack problem and also for solving the BUCO problem [2]. It is also easy to see that the well-known algorithm by Martins [11] has an expected polynomial running time for solving the MOSP problem.

Output-sensitive complexity in multiobjective optimization is a very young field of research. Here, the worst-case perspective is retained. The target is to express the worst-case running time of algorithms for multiobjective optimization problems not only in terms of the input size, but also the output size. An algorithm is said to be output-sensitive, if it runs in time polynomial in $\mathcal{Y}_N + \langle C \rangle$, where $\langle C \rangle$ denotes the encoding lenght of C. In [4], Bökler and Mutzel show that there is an output-sensitive algorithm for the multiobjective linear programming problem and finding extreme points of the Pareto-front of multiobjective combinatorial optimization problems. In contrast to smoothed analysis it has also been shown that certain problems are not solvable in an output-sensitive way: In [3], Bökler, Morris, Ehrgott and Mutzel show that there is no output-sensitive algorithm for the MOSP problem unless $\mathbf{P} = \mathbf{NP}$.

The bound Serafini considers in his paper is interesting also in the context of output-sensitive complexity. It can be seen that if a multiobjective optimization problem is hard under (1), then there does not exist an output-sensitive algorithm unless $\mathbf{P} = \mathbf{NP}$.

In both theories the complexity status of many multiobjective optimization problems is unknown. For smoothed analysis a challenging problem is to find a smoothed polynomial algorithm for the multiobjective spanning tree problem. In output-sensitive complexity even an output-sensitive algorithm for the BUCO problem, or showing that no such algorithm exists, seems to be a challenge.

Acknowledgements. The author especially thanks Paolo Serafini for a very kind and indepth discussion of this topic. Also many thanks to Kathrin Klamroth and Michael Stiglmayr for many fruitful discussions on the complexity of multiobjective combinatorial optimization problems.

References

1. Arora, S., Barak, B.: Computational Complexity - A Modern Approach. Cambridge University Press, Cambridge (2009)
2. Beier, R., Röglin, H., Vöcking, B.: The smoothed number of Pareto optimal solutions in bicriteria integer optimization. In: Fischetti, M., Williamson, D.P. (eds.) IPCO 2007. LNCS, vol. 4513, pp. 53–67. Springer, Heidelberg (2007). doi:10.1007/978-3-540-72792-7_5
3. Bökler, F., Ehrgott, M., Morris, C., Mutzel, P.: Output-sensitive complexity for multiobjective combinatorial optimization. J. Multi-Criteria Decis. Anal. (2017, accepted)
4. Bökler, F., Mutzel, P.: Output-sensitive algorithms for enumerating the extreme nondominated points of multiobjective combinatorial optimization problems. In: Bansal, N., Finocchi, I. (eds.) ESA 2015. LNCS, vol. 9294, pp. 288–299. Springer, Heidelberg (2015). doi:10.1007/978-3-662-48350-3_25
5. Brunsch, T., Röglin, H.: Improved smoothed analysis of multiobjective optimization. In: ACM SToC. pp. 407–426 (2012)
6. Galand, L., Ismaili, A., Perny, P., Spanjaard, O.: Bidirectional preference-based search for multiobjective state space graph problems. In: SoCS 2013 (2013)
7. Garey, M.R., Johnson, D.S.: Computers and Intractability; A Guide to the Theory of NP-Completeness. Series of Books in the Mathematical Sciences, 1st edn. W. H. Freeman & Co., New York (1979)
8. Guerriero, F., Musmanno, R.: Label correcting methods to solve multicriteria shortest path problems. J. Optim. Theory Appl. 111(3), 589–613 (2001)
9. Hansen, P.: Bicriterion path problems. In: Fandel, G., Gal, T. (eds.) Multiple Criteria Decision Making Theory and Application. Lecture Notes in Economics and Mathematical Systems, vol. 177, pp. 109–127. Springer, New York (1979)
10. Horoba, C.: Exploring the runtime of an evolutionary algorithm for the multiobjective shortest path problem. Evol. Comput. 18(3), 357–381 (2010)
11. Martins, E.Q.V.: On a multicriteria shortest path problem. Eur. J. Oper. Res. 16, 236–245 (1984)
12. Mohamed, C., Bassem, J., Taicir, L.: A genetic algorithm to solve the bicriteria shortest path problem. Electr. Notes Discret. Math. 36, 851–858 (2010)
13. Müller-Hannemann, M., Schulz, F., Wagner, D., Zaroliagis, C.: Timetable information: models and algorithms. In: Geraets, F., Kroon, F., Schoebel, A., Wagner, D., Zaroliagis, C.D. (eds.) Algorithmic Methods for Railway Optimization. LNCS, vol. 4359. Springer, Heidelberg (2007). doi:10.1007/978-3-540-74247-0_3
14. Nemhauser, G.L., Ullmann, Z.: Discrete dynamic programming and capital allocation. Manag. Sci. 15(9), 494–505 (1969)
15. Raith, A., Ehrgott, M.: A comparison of solution strategies for biobjective shortest path problems. Comput. Oper. Res. 36(4), 1299–1331 (2009)
16. Sanders, P., Mandow, L.: Parallel label-setting multi-objective shortest path search. In: Parallel & Distributed Processing (IPDPS), pp. 215–224. IEEE (2013)
17. Serafini, P.: Some considerations about computational complexity for multiobjective combinatorial problems. In: Jahn, J., Krabs, W. (eds.) Recent Advances and Historical Development of Vector Optimization. Lecture Notes in Economics and Mathematical Systems, vol. 294, pp. 222–231. Springer, Heidelberg (1986). doi:10.1007/978-3-642-46618-2_15
18. Shekelyan, M., Jossé, G., Schubert, M.: Paretoprep: fast computation of path skyline queries. In: Advances in Spatial and Temporal Databases (2015)

19. Shekelyan, M., Jossé, G., Schubert, M., Kriegel, H.-P.: Linear Path Skyline Computation in Bicriteria Networks. In: Bhowmick, S.S., Dyreson, C.E., Jensen, C.S., Lee, M.L., Muliantara, A., Thalheim, B. (eds.) DASFAA 2014. LNCS, vol. 8421, pp. 173–187. Springer, Heidelberg (2014). doi:10.1007/978-3-319-05810-8_12
20. Skriver, A.J.V.: A classification of bicriterion shortest path algorithms. Asia-Pac. J. Oper. Res. **17**, 199–212 (2000)
21. Tarapata, Z.: Selected multicriteria shortest path problems: An analysis of complexity, models and adaptation of standard algorithms. Int. J. Appl. Math. Comput. Sci. **17**(2), 269–287 (2007)
22. Tsaggouris, G., Zaroliagis, C.D.: Multiobjective optimization: Improved FPTAS for shortest paths and non-linear objectives with applications. Algorithms and Computation **4288**, 389–398 (2006)
23. Warburton, A.: Approximation of Pareto optima in multiple-objective shortest-path problems. Oper. Res. **35**(1), 70–79 (1987)

Angle-Based Preference Models
in Multi-objective Optimization

Marlon Braun$^{(\boxtimes)}$, Pradyumn Shukla, and Hartmut Schmeck

Karlsruhe Institute of Technology, Karlsruhe, Germany
{marlon.braun,pradyumn.shukla,hartmut.schmeck}@kit.edu

Abstract. Solutions that provide a balance between different objective values in multi-objective optimization can be identified by assessing the curvature of the Pareto front. We analyze how methods based on angles have been utilized in the past for this task and propose a new angle-based measure—angle utility—that ranks points of the Pareto front irrespective of its shape or the number of objectives. An algorithm for finding angle utility optima is presented and a computational study shows that this algorithm is successful in identifying angle utility optima.

Keywords: Multi-objective optimization · Preference modeling · Angle utility · Scalarization · Evolutionary algorithm

1 Introduction

Multi-objective optimization deals with problems that possess more than one goal. These goals usually contradict each other. Solving a multi-objective problem yields a set of solutions that can only be improved in one objective by impairing at least one other objective. These solutions are called Pareto optimal or efficient and form a so-called Pareto front in the objective space [7].

In real-world applications, a decision maker (DM) is usually tasked with implementing only a single alternative, requiring him to choose one solution among the set of Pareto optimal solutions. Such a decision provides a difficult challenge as the set of Pareto optimal solutions is usually large. Although most classic optimization approaches approximate the Pareto front by a finite set of points, even choosing an alternative among a small sample may overburden the DM [4].

In order to counteract this issue, many different approaches have been developed in the field of multiple-criteria decision-making (MCDM) for providing guidance in the selection process [4,15]. The scientific community generally distinguishes between two different schools of thought [7]: the French school places its focus on outranking methods [15, Part III], whereas the American school is based on Multi Attribute Utility Theory (MAUT) [15, Part IV]. Approaches within MAUT use a utility or scalarization function that maps the vector of objectives to a single real number signifying its desirability. Such techniques possess the advantage that they allow a total ordering of the objective space

© Springer International Publishing AG 2017
H. Trautmann et al. (Eds.): EMO 2017, LNCS 10173, pp. 88–102, 2017.
DOI: 10.1007/978-3-319-54157-0_7

and thereby invoke a complete ranking between the elements of the Pareto optimal set.

There exists a myriad of approaches to formulate such scalarization functions that all use different sources of inspiration [16,18]. Since the Pareto front of a real-valued optimization problem usually forms a hypersurface in the objective space, it seems intelligible to consider its geometric properties for assessing the desirability of its elements. Those geometric properties can be closely related to trade-offs, which are a good indicator for quantifying human preferences [6,15]. This work, therefore, focuses on the applicability of angles to measure utility in multi-objective optimization. The main contributions may be summarized in the following way:

- a rigorous analysis of existing preference models based on angles that examines their strengths, weaknesses, and applicability to different problem classes,
- presentation of a new angle-based utility measure that counteracts shortcomings of previous angle measures and is applicable to a large range of problem classes,
- a quantitative analysis of how angle-based utility optima can be obtained using different optimization techniques.

The paper is structured in the following way. The next section introduces basic concepts in multi-objective optimization and provides an extensive review of existing angle-based preference models. The subsequent section presents a new approach for defining an angle-based scalarization function—angle utility. Thereafter, we present an algorithm tailored to find solutions that are optimal according to this angle measure. We present a quantitative analysis of different approaches for finding angle utility optima in Sect. 5. The paper is concluded by a summary and an outlook on future work.

2 Preliminaries and Related Work

We consider the minimization of m real-valued objective functions $f = (f_1, \ldots, f_m)$ using decision variables of a subset of an n-dimensional vector space $X \subseteq \mathbb{R}^n$. The image $f(X)$ of the search space X is denoted by $\mathcal{Y}$. We define domination and Pareto optimality directly in the objective space to simplify the notation of our subsequent analysis.

Definition 1 (Domination [17]). *Let* $\mathbf{u}^1, \mathbf{u}^2 \in \mathcal{Y}$ *be given.* $\mathbf{u}^1$ *dominates* $\mathbf{u}^2$, *expressed as* $\mathbf{u}^1 \prec \mathbf{u}^2$, *if* $\forall i \in [m] := \{1, 2, \ldots, m\}$ *it holds that* $u_i^1 \leq u_i^2$ *and* $\exists j \in [m]$ *such that* $u_j^1 < u_j^2$.

Definition 2 (Pareto optimality [17]). *Let* $\mathbf{u}^* \in \mathcal{Y}$ *be given.* $\mathbf{u}^*$ *is called* Pareto optimal *if there exists no* $\mathbf{u} \in \mathcal{Y}$ *such that* $\mathbf{u} \prec \mathbf{u}^*$.

The set of Pareto optimal solutions, i.e. the Pareto front, is denoted by $\mathcal{Y}_p$.

2.1 Reflex Angle

An extensive literature review has revealed three different approaches that explicitly use angles in determining the desirability of a solution. These approaches have been mostly used to identify convex bulges on the Pareto front, so-called knees. Our analysis, on the other hand, focuses on how well each approach is suitable as scalarization function to rank elements of the Pareto front.

The reflex angle was proposed by Branke et al. [3] for two-dimensional optimization problems. The reflex angle of a point $\mathbf{u} \in \mathcal{Y}_p$ is formed by connecting $\mathbf{u}$ by straight lines to its two adjacent neighboring points on the Pareto front. The resulting angle of these three connected points that points towards the direction $(-1, -1)$ is the reflex angle. Branke et al. extend their approach by also considering second-degree neighbors. The maximum of all angles that can be formed with direct and second-degree neighbors constitutes the utility of $\mathbf{u}$. A larger reflex angle implies a higher desirability of the given solution (Fig. 1).

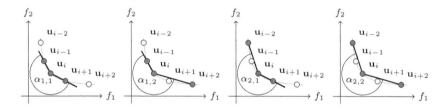

Fig. 1. Illustration of the reflex angle.

Definition 3 (Reflex-Angle [3]). *Let us assume that the elements of the Pareto front are ordered lexicographically in ascending order and let i denote the index of a point $\mathbf{u}^i \in \mathcal{Y}_p$ in said order. The reflex angle $a_r(\mathbf{u}^i)$ of $\mathbf{u}^i$ can then be defined in the following way:*

$$a_r(\mathbf{u}^i) = \max\left(\alpha_{1,1}(\mathbf{u}^i), \alpha_{1,2}(\mathbf{u}^i), \alpha_{2,1}(\mathbf{u}^i), \alpha_{2,2}(\mathbf{u}^i)\right), \tag{1}$$

with

$$\alpha_{1,1}(\mathbf{u}^i) = 2\pi - \angle\mathbf{u}^{i-1}\mathbf{u}^i\mathbf{u}^{i+1}, \quad \alpha_{1,2}(\mathbf{u}^i) = 2\pi - \angle\mathbf{u}^{i-1}\mathbf{u}^i\mathbf{u}^{i+2}, \tag{2a}$$

$$\alpha_{2,1}(\mathbf{u}^i) = 2\pi - \angle\mathbf{u}^{i-2}\mathbf{u}^i\mathbf{u}^{i+1}, \quad \alpha_{2,2}(\mathbf{u}^i) = 2\pi - \angle\mathbf{u}^{i-2}\mathbf{u}^i\mathbf{u}^{i+2}. \tag{2b}$$

Whenever a left or right neighbor does not exist in computing (2), it is replaced by a vertical or horizontal line, respectively.

The reflex angle was originally proposed as a niching technique replacing the crowding distance metric in NSGA-II [11] to identify convex bulges on the Pareto front. The computation of this utility measure is strictly based on local information as only neighboring solutions are considered. This circumstance presents a drawback in utilizing the reflex angle to rank points of a continuous front, since the concept of direct neighbors bears no meaning for curves. The approach is suitable if the Pareto front is discrete or if only a discretized subset of a continuous front is considered. However, the expressiveness of the reflex angle is expected to diminish as the density of points in the objective space increases. The original definition of the reflex angle is limited to two objectives only, since there exists no total ordering in higher dimensions from which left or right neighbors may be identified.

2.2 Bend Angle

The bend angle [9] is a utility measure for convex Pareto fronts of two-dimensional optimization problems. For any point $\mathbf{u} \in \mathcal{Y}_p$, two right-angled triangles are formed *above* the Pareto front with the left and right extreme point of the Pareto front. The bend angle is computed by subtracting the angle at $\mathbf{u}$ of the right triangle from the angle at the left extreme point of the left triangle. A point is considered to be more desirable the larger its bend angle is (Fig. 2).

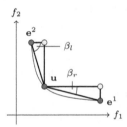

Fig. 2. Illustration of the bend angle.

Definition 4 (Bend Angle [9]). *Let $\mathbf{e}^k = arg\,max_{\mathbf{v} \in \mathcal{Y}_p} v_k$ denote the extreme point of the k-th objective. The bend angle $a_b(\mathbf{u})$ of a point $\mathbf{u} \in \mathcal{Y}_p$ is given by*

$$a_b(\mathbf{u}) = \beta_l - \beta_r \qquad (3)$$

with

$$\beta_l = \arctan\left(\frac{e_2^2 - u_2}{u_1 - e_1^2}\right) \quad and \quad \beta_r = \arctan\left(\frac{u_2 - e_2^1}{e_1^1 - u_1}\right). \qquad (4)$$

The bend angle was proposed as an extension to the reflex angle to identify entire regions of interest on the Pareto front. A threshold is employed specifying the maximum bend angle a DM is willing to accept. Special consideration is given to the point exhibiting the maximum bend angle on the front. This so-called bend angle knee provides an excellent balance between both objectives and therefore presents itself as an interesting choice to the DM.

In contrast to the reflex angle, the bend angle is applicable to both discrete and continuous Pareto fronts. The approach, however, is limited to two objectives, since there exists no natural extension of the bend angle to higher

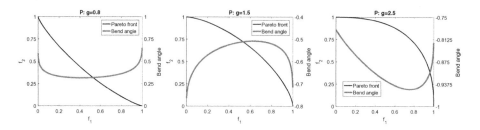

Fig. 3. The figure shows three instances of the parametric Pareto front given by $P = \{f \in [0,1]^2 \mid f_1^g + f_2^g = 1\}$ with $g > 0$ including corresponding bend angle values. Shallow convex curvatures ($g \in [0.6, 1)$) induce smaller bend angles towards the center of the front. The same observation holds for large concave bend angles ($g > 2$). In case of shallow concave curvatures $g \in (1, 2)$, bend angle values become larger towards the center of the front. Bend angles are constant for $g = 1$ or $g = 2$.

dimensions. Deb and Gupta intended the bend angle to be applied to convex Pareto fronts. We have noticed that the magnitude of the curvature is decisive to, whether a smaller or larger bend angle indicates a better trade-off between both objectives. An illustration of this issue is given in Fig. 3. We would like to point out that it is reasonable to apply the bend angle to concave fronts as well. However, the curvature must again be considered to decide, whether smaller or larger bend angles signals higher or lower desirability. These observations make the application of the bend angle difficult, since the curvature of the Pareto front is usually not known prior to any optimization effort.

2.3   Extended Angle Dominance

Sudeng and Wattanapongsakorn [19] proposed the extended angle dominance as a pruning mechanism for identifying regions of interest on the Pareto front that exhibit good trade-off properties. Their approach is applicable to problems featuring two or more objectives. Although extended angle dominance utilizes scalarization in comparing solutions with each other, it may be generally characterized as an outranking method. The approach requires multiple steps.

Firstly, for every pair of points $\mathbf{u}, \mathbf{v} \in \mathcal{Y}_p$ and objective k we compute the extended dominance angle. Since the extended angle dominance approach is a pruning mechanism, we would usually consider only a finite subset of $\mathcal{Y}_p$, e.g., the final population obtained by an evolutionary algorithm. The extended dominance angle $\gamma(\mathbf{u}, \mathbf{v})_k$ is computed for every objective $k \in [m]$ (Fig. 4):

$$\gamma(\mathbf{u}, \mathbf{v})_k = \arctan\left(\frac{\sqrt{\sum_{i=1, i \neq k}^{m} (u_i - v_i)^2}}{|u_k - v_k|}\right). \tag{5}$$

Next, we compute the so-called threshold angle δ_k for each objective, which serves as an estimator for the density of $\mathcal{Y}_p$ along each individual dimension in

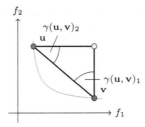

Fig. 4. Illustration of the extended dominance angle.

the objective space. For computing the threshold angle, the following procedure is performed for each objective k: All points of $\mathcal{Y}_p$ are sorted in ascending order according to their k-th objective value. Let IQR_k denote the inter-quartile range of the set of objective values of $\mathcal{Y}_p$ in the k-th dimension and $\tau_k \in [0,1]$ be a so-called bias intensity. The threshold angle is computed as the arctangent of the inter-quartile range divided by the arithmetic mean of the differences in the k-th objective of consecutive solutions. The resulting angle is exponentiated by the bias intensity to account for different user preferences.

$$
\delta_k = \left(\arctan \left(\frac{IQR_k}{\frac{1}{|\mathcal{Y}_p|} \sum_{i=1}^{|\mathcal{Y}_p|-1} \left(u_k^{i+1} - u_k^i \right)} \right) \right)^{\tau_k}
\tag{6}
$$

Finally, extended angle dominance is defined in the following way. For two given points $\mathbf{u}, \mathbf{v} \in \mathcal{Y}_p$, the following computation is performed. For an objective i, we check whether u_i is smaller than v_i. If this is the case, we further check, whether the extended dominance angle for the given objective is smaller than the threshold angle (see Eq. 5). If both conditions are satisfied for at least one objective, $\mathbf{u}$ extended angle dominates $\mathbf{v}$. A formal description is given in Definition 5.

Definition 5 (Extended Angle Dominance [19]). *Let* $\mathbf{u}, \mathbf{v} \in \mathcal{Y}_p$ *be given.* $\mathbf{u}$ *is said to extended angle dominate* $\mathbf{v}$ *if*

$$
\bigvee_{k=1}^{m} (u_k \leq v_k) \wedge (\gamma(\mathbf{u}, \mathbf{v})_k \leq \delta_k).
\tag{7}
$$

Since extended angle dominance was conceived as a binary relation, it cannot be applied straightaway as scalarization function to assign utility values to individual solutions. Instead, it works as a preference model by identifying regions of interest on the Pareto front. Applying extended angle dominance to continuous fronts poses some challenges as well, since the average distance between points and the inter-quartile range of objective values are not defined in this context. Extended angle dominance, however, is applicable irrespective of the shape of the Pareto front.

There exist other approaches in the literature that are implicitly related to angles. Batista et al. [2] change the angle of the Pareto domination cone in

ϵ-domination. Similarly, Emmerich et al. [13] apply said change for hypervolume computation. Both approaches are used to obtain an optimally distributed approximation of the Pareto front.

3 Proposed Approach

Drawing inspiration from the approaches presented in the previous section, we have developed a new angle-based preference model. In our approach, angles are computed in the same manner as the extended dominance angle. However, instead of considering pairwise angles between all points, we only focus on the angles of a point with respect to the extreme points of the Pareto front. The maximum of all said angles then constitutes the utility value of the given point. An illustration is provided in Fig. 5.

Definition 6 (Angle utility). *Let* $\mathbf{e}^k$ *denote the extreme point of the k-th objective. The angle utility $a(\mathbf{u})$ of a points $\mathbf{u} \in \mathcal{Y}_p$ is given by*

$$a(\mathbf{u}) = \max_{k \in [m]} \left(\gamma(\mathbf{u}, \mathbf{e}^k)_k \right). \tag{8}$$

A smaller angle utility implies a larger desirability. We would like to point out that skipping the inverse tangent computation in Eq. 5 does not change the ranking induced by angle utility, as the inverse tangent is a strictly monotonous transformation.

In three and more dimensions, there may exist multiple points that all have the largest objective value among the set of Pareto optimal points. This is for example the case for the well-known problem DTLZ3 [12], whose Pareto front is the unit sphere segment in the positive orthant. For this reason, we define the k-th extreme point as the minimizer of the following achievement scalarization function

$$s(\mathbf{u})_k = \max_{k \in [m]} (w_k u_k), \tag{9}$$

where $\mathbf{u} \in \mathcal{Y}_p$ and $\mathbf{w}$ is the vector of the k-th axis direction, where all entries having the value 0 are replaced by $1/\varepsilon$, where ε is a value close to 0.

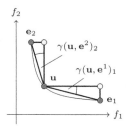

Fig. 5. Illustration of angle utility.

Angle utility is applicable irrespective of the number of objectives and the curvature of the Pareto front, and it can both be used with discrete and continuous fronts. This preference model favors points on convex bulges that usually exhibit excellent trade-off properties. On the other hand, interior points are preferred to extreme points on strictly concave fronts. Both aspects are illustrated in Fig. 6. Since a DM is usually interested in finding a solution that balances all objectives among each other, this is a desirable property.

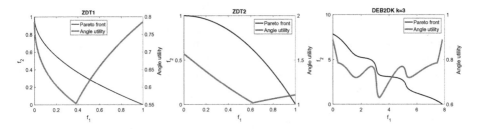

Fig. 6. Example for angle utility landscapes of two-objective optimization problems. The angle optima of convex (ZDT1 [20]) and concave (ZDT2 [20]) Pareto fronts are both located in the interior. In case of multiple convex bulges (DEB2DK [3]), the angle optimum is located at the strongest bulge. Many popular techniques such as the weighted sum or the Chebyshev method only fulfill either property. The weighted sum captures only extreme points on concave fronts and depending on the weights, Chebyshev identifies points on concave bulges as optima on mixed front curvatures, which usually exhibit a bad trade-off.

We conclude this section by a comparison between the different angle-based preference models presented in this work. Table 1 provides a summary of the applicability of these approaches to different problem classes.

4   Algorithmic Approach

Finding the angle utility optimum of an optimization problem poses two main challenges. First of all, computing angle utilities requires the knowledge of the extreme points, which are commonly not known prior to any optimization effort. Secondly, in case an extreme point $\mathbf{e}^k$ dominates the point $\mathbf{u}$, the angle $\gamma(\mathbf{u}, \mathbf{e}^k)_k$ is no longer a suitable indicator for measuring trade-offs. At the same time, however, angle utility satisfies Pareto compliance. Pareto compliance implies that for any $\mathbf{u}, \mathbf{v} \in \mathcal{Y}$ if $\mathbf{u}$ dominates $\mathbf{v}$, then $a(\mathbf{u}) < a(\mathbf{v})$. Of course, this statement is only valid if neither $\mathbf{u}$ nor $\mathbf{v}$ are dominated by any extreme point. This observation is formalized in Proposition 1.

Proposition 1. *Let $\mathbf{u}, \mathbf{v} \in \mathcal{Y}$ be given and let there exist no extreme point $\mathbf{e}^k$ with $\mathbf{e}^k \prec \mathbf{u}$ or $\mathbf{e}^k \prec \mathbf{v}$. If $\mathbf{u} \prec \mathbf{v}$ then it follows that $a(\mathbf{u}) < a(\mathbf{v})$.*

Table 1. Classification of angle-based preference models.

Approach	Problem class	Number of objectives	Curvature	Methodology
Reflex angle	Discrete	Two	Convex and concave	Niching mechanism
Bend angle	Discrete and continuous	Two	Convex and concave with limitations	Scalarization
Extended dominance angle	Discrete	Any	Convex and concave	Outranking
Angle utility	Discrete and continuous	Any	Convex and concave	Scalarization

Proof. Without loss of generality, let us assume that for all $i \in [m]$ we have $v_i = u_i$ with the exception of index $j \in [m]$ for which $v_j = u_j + \varepsilon$, where $\varepsilon > 0$. We compare $\gamma(\mathbf{u}, \mathbf{e}^k)_k$ and $\gamma(\mathbf{v}, \mathbf{e}^k)_k$. Looking at Eq. 5 we can distinguish between the cases $j = k$ or $j \neq k$. If $j = k$, then the denominator of (5) decreases for $\gamma(\mathbf{v}, \mathbf{e}^k)_k$ in comparison to $\gamma(\mathbf{u}, \mathbf{e}^k)_k$. If $j \neq k$ the numerator increases. We conclude that for all $k \in [m]$ that $\gamma(\mathbf{v}, \mathbf{e}^k)_k > \gamma(\mathbf{u}, \mathbf{e}^k)_k$ and thereby $a(\mathbf{u}) < a(\mathbf{v})$.

We propose the following procedure for obtaining an approximation to the angle optimum of a multi-objective problem that addresses the challenges mentioned above and utilizes the Pareto compliance property of angle utility. Let N denote the population size. In each iteration, we generate N new individuals using evolutionary operators. Next, we identify the extreme points of the combined populations of parents and offspring using Eq. 9. For each objective k, we carry λ with $\lambda < N/m$ individuals exhibiting the smallest achievement scalarization values over to the next generation. These points are then removed from the combined population. Subsequently, we identify all those solutions that are not dominated by the extreme points. Angle utility values are computed for these solutions and we fill the next generation by those solutions having the smallest angle utilities. In case the next generation has not reached its maximum size, we repeat the procedure with the remaining combined population. The complete procedure is described in Algorithm 1.

Next, we analyze the computational complexity of our proposed algorithm. Identifying and extracting extreme points requires calculating achievement scalarization values and sorting the combined population. The former can be performed in $O(Nm^2)$ and the latter in $O(N \log(N))$. Extracting extreme points requires $O(\lambda)$. This procedure is performed for each objective, while the population size is reduced in every iteration by λ. Assuming that $\lambda << N$ and $m << N$, $O(Nm^2 + N \log N)$ is the upper bound for identifying and extract-

---

Algorithm 1. Angle utility optimum finding algorithm

```
1  begin
2  |    Generate initial population P_0, t = 0
3  |    repeat
4  |    |    Generate offspring population Q_t
5  |    |    R_t = P_t ∪ Q_t
6  |    |    P_{t+1} = ∅
7  |    |    while |P_{t+1}| < N do
8  |    |    |    E = ∅
9  |    |    |    for ∀k ∈ [m] do                          /* Retain extreme points */
10 |    |    |    |    Compute ∀p ∈ R_t s(p)_k
11 |    |    |    |    Sort R_t according to s(p)_k
12 |    |    |    |    E = E ∪ {R_t(1)}                     /* k-th extreme point */
13 |    |    |    |    P_{t+1} = P_{t+1} ∪ {R_t(1) ∪ ... ∪ R_t(λ)}
14 |    |    |    |    R_t = R_t \ {R_t(1) ∪ ... ∪ R_t(λ)}
15 |    |    |    S = {p ∈ R_t | ∀e ∈ E : e ⊀ p}
16 |    |    |    if |P_{t+1}| + |S| ≤ N then
17 |    |    |    |    P_{t+1} = P_{t+1} ∪ S
18 |    |    |    |    R_t = R_t \ S
19 |    |    |    else                                     /* Selection by angle utility */
20 |    |    |    |    Compute ∀p ∈ R_t a(p) using E
21 |    |    |    |    Sort R_t according to a(p)
22 |    |    |    |    P_{t+1} = P_{t+1} ∪ {R_t(1) ∪ R_t(N − |P_{t+1}|)}
23 |    |    t = t + 1
24 |    until stopping criterion
```

---

ing extreme points. Computing which solutions are dominated by the current extreme points elicits an effort of $O(Nm^2)$. Finally, computing angle utilities and sorting the population once again requires $O(Nm^2)$ and $O(N \log N)$, respectively, while copying the best solutions to the next generation $O(N)$.

In the worst case, all solutions in the remaining population are dominated by extreme points. Assuming that $\lambda m << N$, which implies worst case behavior, the inner loop is executed $\lceil N/(\lambda m) \rceil$ times per iteration and we obtain a total runtime of $O(Nm^2 + N^2 \log N)$ – no angle utilities are computed and achievement scalarization values only need to be computed once. Such a scenario, however, is rather unlikely to occur, as we expect the majority of the population members to lie in between the extreme points. If this is indeed the case, we arrive at an average complexity of $O(Nm^2 + N \log N)$ per iteration.

5   Simulation Results and Discussion

Our computational study addresses two research questions. First of all, we want to assess the effect of the parameter λ on the performance. There exists an inherent trade-off in dividing computational resources between approximating the extreme points and the angle optimum. We aim at finding the angle optimum, however require the extreme points to correctly compute angle utility. We assess two strategies in this context. For the first approach, to which we will refer by *Angle10*, we set $\lambda = 10$. Angle10 acknowledges that finding the angle optimum is the main task, whereas keeping the effort to determining extreme points to a bare minimum. *AngleEQ* uses $\lambda = N(m-1)/m$ and thereby represents the idea that finding the angle optimum and extreme points are two equally important tasks. Evidently, both approaches possess an average complexity of $O(Nm^2 + N \log N)$.

Secondly, both approaches should be superior to a posteriori methods that approximate the entire Pareto front and select the point exhibiting the highest angle utility from the final population as approximation to the angle optimum. We therefore chose NSGA-II [11] as benchmark for two-objective problems and NSGA-III [10] for higher dimensional problems. Both algorithms have shown excellent results across multiple studies, which makes them ideal candidates.

A population size of 100 was used for problems having less than 10 objectives, a size of 200 otherwise. We used SBX crossover [1] and polynomial mutation [8] employing both a distribution index of 20 (30 for NSGA-III), a crossover probability of 1.0 and mutation probability of $1/n$. Both Angle10 and AngleEQ use binary tournament selection with angle utility as winning criterion. Equidistant reference points generated by the method in [10] were used with NSGA-III, since, in general, no assumptions can be made about the location of the angle optimum on the Pareto front. All algorithms were run 100 times on each problem. Our code and problem setup is available online[1].

The focus of our analysis lies on how the approximation of the angle optimum evolves across subsequent generations depending on the algorithm used. We therefore recorded in every iteration the minimum distance of all population members to the true angle optimum of the problem at hand, which we have determined analytically in advance. The optimization horizon was chosen as 100 000 function evaluations.

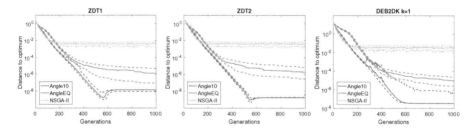

Fig. 7. Median distance to angle optimum (solid line) and IQR (dashed line) at every generation across 100 runs of problems ZDT1, ZDT2 and DEB2DK.

The first part of our analysis concerns the effect of mixed curvatures – Pareto fronts possessing convex and concave regions – on the algorithms' performance.

---

[1] https://sourceforge.net/projects/jmetalbymarlonso/.

The results are displayed in Figs. 7 and 8. We can observe clear performance differences between all three algorithms. The performance gap between Angle10 and AngleEQ, however, becomes smaller if the angle utility landscape exhibits increased multi-modality. The performance of Angle10 sometimes even deteriorates between 200 and 400 generations. We assume that the strong focus on the angle optimum leads the algorithm to get temporarily trapped at local optima, since the extreme points are not sufficiently approximated, yet. Performance decreases of Angle10 in later iterations may be attributed as well to an insufficient approximation of extreme points. Still, Angle10 outperforms the other two algorithms by a large margin.

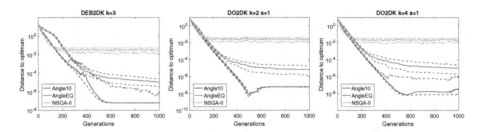

Fig. 8. Median distance to angle optimum (solid line) and IQR (dashed line) at every generation across 100 runs of problems DEB2DK and DO2DK [3].

The effect of curvature is explored in Figs. 9 and 10. Evidently, the performance gap between Angle10 and AngleEQ becomes smaller for strongly curved fronts irrespective of whether they are convex or concave. Performance deterioration can again be observed in early iterations for Angle10, which may be explained by the same line of argument as for the multi-modal case.

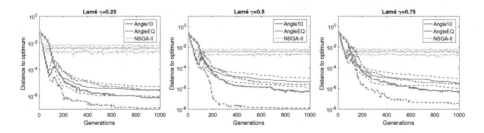

Fig. 9. Median distance to angle optimum (solid line) and IQR (dashed line) at every generation across 100 runs on convex instances of the Lamé hypersphere problem [14]. Smaller values of γ make the front more convex.

The final part of our study focuses on the effect of the number of objectives. Increasing the number of objectives naturally makes finding the angle optimum

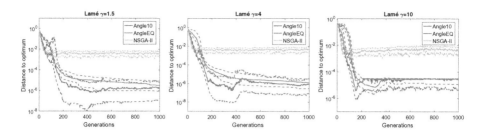

Fig. 10. Median distance to angle optimum (solid line) and IQR (dashed line) at every generation across 100 runs on concave instances of the Lamé hypersphere problem [14]. Larger values of γ make the front more concave.

more difficult, since more extreme points need to be retrieved for correctly computing angle utility. Indeed, we observe in Fig. 11 that convergence of Angle10 and AngleEQ to the angle optimum becomes slower as the number of objectives increases. On concave fronts, both algorithms even failed to approximate the angle optimum altogether. A closer analysis of the results has revealed that this failure is indeed caused by insufficient extreme point approximation. Figure 12 provides a graphical explanation of the problem.

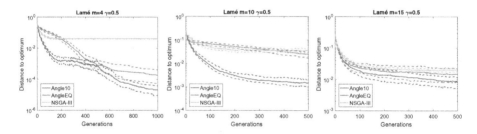

Fig. 11. Median distance to angle optimum (solid line) and IQR (dashed line) at every generation across 100 runs on high-dimensional convex Lamé hypersphere problem instances [14].

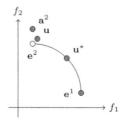

Fig. 12. If $\mathbf{a}^2$ is the currently best known estimate of $\mathbf{e}^2$ we get $a(\mathbf{u}) < a(\mathbf{u}^*)$. The point $\mathbf{u}$ is assumed to be situated at a convex bulge, because its trade-off to $\mathbf{a}^2$ is stronger compared to the trade-off between $\mathbf{a}^2$ and $\mathbf{u}^*$, which is the true angle optimum.

Leaving aside high-dimensional concave Pareto fronts, our study results suggest that our proposed algorithm allows a more precise approximation of the angle optimum compared to the a posteriori approach. NSGA-II and NSGA-III, only outperform AngleEQ in early iterations, which can be attributed to AngleEQ using most of its resources to determine extreme points.

6   Summary and Outlook

We have presented a comparison of angle-based preference models and developed a new scalarization method—angle utility—that utilizes angles to rank Pareto optimal solutions. In contrast to previous approaches, our method can be used with discrete and real-valued Pareto fronts and is applicable irrespective of the curvature or the number of objectives (see Table 1). An algorithm has been designed to find angle utility optima and successfully tested on several benchmark problems. A computational study has revealed that our approach provides a large performance improvement compared to determining the angle optimum from the final population of an algorithm that approximates the entire Pareto front. It has been revealed, however, that our approach struggles to find the angle optimum on high-dimensional concave fronts.

Future research should explore further approaches to determine angle optima on high-dimensional Pareto fronts. Balancing the trade-off between finding the extreme points and the angle optimum plays a crucial part in this context. Angle utility should also be applied to real-world problems to assess its usefulness in a practical optimization context. Problems that feature more than three objectives and are thereby hard to visualize (e.g. [5]) would make an excellent use-case.

References

1. Agrawal, R.B., Deb, K.: Simulated binary crossover for continuous search space. Complex Syst. **9**(3), 1–15 (1994)
2. Batista, L.S., Campelo, F., Guimarães, F.G., Ramírez, J.A.: Pareto cone ϵ-dominance: improving convergence and diversity in multiobjective evolutionary algorithms. In: Takahashi, R.H.C., Deb, K., Wanner, E.F., Greco, S. (eds.) EMO 2011. LNCS, vol. 6576, pp. 76–90. Springer, Heidelberg (2011)
3. Branke, J., Deb, K., Dierolf, H., Osswald, M.: Finding knees in multi-objective optimization. In: Yao, X., et al. (eds.) PPSN 2004. LNCS, vol. 3242, pp. 722–731. Springer, Heidelberg (2004). doi:10.1007/978-3-540-30217-9_73
4. Branke, J., Deb, K., Miettinen, K., Slowinski, R. (eds.): Multiobjective optimization: interactive and evolutionary approaches. Springer, Heidelberg (2008)
5. Braun, M., Dengiz, T., Mauser, I., Schmeck, H.: Comparison of multi-objective evolutionary optimization in smart building scenarios. In: Squillero, G., Burelli, P. (eds.) EvoApplications 2016. LNCS, vol. 9597, pp. 443–458. Springer, Heidelberg (2016). doi:10.1007/978-3-319-31204-0_29
6. Braun, M.A., Shukla, P.K., Schmeck, H.: Preference ranking schemes in multi-objective evolutionary algorithms. In: Takahashi, R.H.C., Deb, K., Wanner, E.F., Greco, S. (eds.) EMO 2011. LNCS, vol. 6576, pp. 226–240. Springer, Heidelberg (2011). doi:10.1007/978-3-642-19893-9_16

7. Coello Coello, C., Lamont, G., Van Veldhuizen, D.: Evolutionary Algorithms for Solving Multi-Objective Problems. Springer, Heidelberg (2007)
8. Deb, K., Goyal, M.: A combined genetic adaptive search (GeneAS) for engineering design. Comput. Sci. Info. **26**, 30–45 (1996)
9. Deb, K., Gupta, S.: Understanding knee points in bicriteria problems and their implications as preferred solution principles. Eng. Optim. **43**(11), 1175–1204 (2011)
10. Deb, K., Jain, H.: An evolutionary many-objective optimization algorithm using reference-point-based nondominated sorting approach, part I: solving problems with box constraints. IEEE Trans. Evol. Comput. **18**(4), 577–601 (2014)
11. Deb, K., Pratap, A., Agarwal, S., Meyarivan, T.: A fast and elitist multiobjective genetic algorithm: NSGA-II. IEEE Trans. Evol. Comput. **6**(2), 182–197 (2002)
12. Deb, K., Thiele, L., Laumanns, M., Zitzler, E.: Scalable test problems for evolutionary multiobjective optimization. In: Jain, L., Wu, X., Abraham, A., Jain, L., Goldberg, R. (eds.) Evol. Multiobjective Optim. Advanced Information and Knowledge Processing, pp. 105–145. Springer, London (2005)
13. Emmerich, M., Deutz, A., Kruisselbrink, J., Shukla, P.K.: Cone-based hypervolume indicators: construction, properties, and efficient computation. In: Purshouse, R.C., Fleming, P.J., Fonseca, C.M., Greco, S., Shaw, J. (eds.) EMO 2013. LNCS, vol. 7811, pp. 111–127. Springer, Heidelberg (2013). doi:10.1007/978-3-642-37140-0_12
14. Emmerich, M.T., Deutz, A.H.: Test problems based on Lamé superspheres. In: Coello, C.A.C., Aguirre, A.H., Zitzler, E. (eds.) EMO 2005. LNCS, vol. 3410. Springer, Heidelberg (2005)
15. Greco, S., Ehrgott, M., Figueira, J. (eds.): Multiple Criteria Decision Analysis: State of the Art Surveys. Springer, New York (2016)
16. Marler, R.T., Arora, J.S.: Survey of multi-objective optimization methods for engineering. Struct. Multidisciplinary Optim. **26**(6), 369–395 (2004)
17. Pareto, V.: Cours d'économie politique. Librairie Droz (1896)
18. Shukla, P.K., Braun, M.A., Schmeck, H.: Theory and algorithms for finding knees. In: Purshouse, R.C., Fleming, P.J., Fonseca, C.M., Greco, S., Shaw, J. (eds.) EMO 2013. LNCS, vol. 7811, pp. 156–170. Springer, Heidelberg (2013). doi:10.1007/978-3-642-37140-0_15
19. Sudeng, S., Wattanapongsakorn, N.: Adaptive geometric angle-based algorithm with independent objective biasing for pruning pareto-optimal solutions. In: 2013 Science and Information Conference (SAI), pp. 514–523 (2013)
20. Zitzler, E., Deb, K., Thiele, L.: Comparison of multiobjective evolutionary algorithms: empirical results. Evol. Comput. **8**(2), 173–195 (2000)

Quantitative Performance Assessment of Multiobjective Optimizers: The Average Runtime Attainment Function

Dimo Brockhoff[1](✉), Anne Auger[1], Nikolaus Hansen[1], and Tea Tušar[2]

[1] Inria Saclay – Ile-de-France and CMAP, UMR CNRS 7641, Ecole Polytechnique,
Palaiseau, France
{dimo.brockhoff,anne.auger,nikolaus.hansen}@inria.fr
[2] Department of Intelligent Systems, Jožef Stefan Institute, Ljubljana, Slovenia
tea.tusar@ijs.si

Abstract. Numerical benchmarking of multiobjective optimization algorithms is an important task needed to understand and recommend algorithms. So far, two main approaches to assessing algorithm performance have been pursued: using set quality indicators, and the (empirical) attainment function and its higher-order moments as a generalization of empirical cumulative distributions of function values. Both approaches have their advantages but rely on the choice of a quality indicator and/or take into account only the location of the resulting solution sets and not *when* certain regions of the objective space are attained. In this paper, we propose the average runtime attainment function as a quantitative measure of the performance of a multiobjective algorithm. It estimates, for any point in the objective space, the expected runtime to find a solution that weakly dominates this point. After defining the average runtime attainment function and detailing the relation to the (empirical) attainment function, we illustrate how the average runtime attainment function plot displays algorithm performance (and differences in performance) for some algorithms that have been previously run on the biobjective bbob-biobj test suite of the COCO platform.

1 Introduction

Performance assessment of black-box algorithms is an important task to *understand* and to *recommend* algorithms in the contexts of practical applications and the design of new algorithms. A lot of progress has been made recently in single-objective optimization on improving standards to assess algorithm performance properly. Particularly, for instance, through the introduction of runtime distributions or data profiles [10,12], performance profiles [4] or software platforms for automated benchmarking, such as COCO [8]. One important aspect of performance assessment, as advocated within the COCO framework, is the need for *quantitative* performance measures. There exist typically two ways of collecting data within single-objective optimization:

© Springer International Publishing AG 2017
H. Trautmann et al. (Eds.): EMO 2017, LNCS 10173, pp. 103–119, 2017.
DOI: 10.1007/978-3-319-54157-0_8

– Record at a given time (budget/function evaluations) the objective function values reached by different runs of an algorithm on a problem. This is referred to as the fixed-budget view (see Fig. 1).
– Collect for a certain function value/target the runtime (typically measured in number of function evaluations) to reach this target. In case the target is not reached by a run, one would record the maximal runtime before stopping the run. This is referred to as the fixed-target scenario (see again Fig. 1).

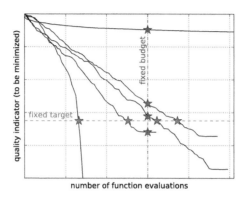

Fig. 1. Fixed-budget versus fixed-target scenarios. Given 5 runs of an algorithm, the fixed budget scenario consists of fixing a cost and recording the objective function values of the 5 runs at this given budget, while the fixed-target scenario consists of fixing a target and recording the number of function evaluations the 5 runs need to reach this target.

While the first scenario is often argued as close to practice where one has a finite budget to solve a problem, it does not allow for a meaningful *quantitative* performance assessment, because the recorded function values can only be interpreted with the scaling of the objective function in mind (reaching with Algorithm A a function value that is two times smaller than the one reached by Algorithm B could mean either that Algorithm A is marginally faster than B or much faster, depending on the objective function to be optimized). On the contrary, the fixed-target view collects runtimes, which allows for direct quantitative comparisons of the type "Algorithm A is two times faster than Algorithm B (to reach a certain target)". Empirical cumulative distributions (ECDFs) of runtimes collected at a given target—compliant with the fixed-target view and originally introduced as *runtime distributions* [10] and *data profiles* [12]—are now standard to assess performance of single-objective algorithms.

For comparing multiobjective algorithms, the fixed-target view has been adopted only recently, in particular in the context of the COCO framework [8], while the less interpretable fixed-budget view is still by far more common. In both cases, most of the time a quality indicator is used to directly exploit single-objective performance assessment techniques such as statistical tests, boxplots, or data/performance profiles. To be more precise, a single, real-valued quality is assigned to each (set) outcome of an optimization algorithm—either as a quality of a single population, or, as for example in the case of COCO, as the quality of all non-dominated solutions found by an algorithm at an arbitrary point in time. This so-called quality indicator approach to performance assessment is simple but relies on the choice of an indicator (or a set of indicators).

The other well-known approach to assessing the performance of multiobjective algorithms has been proposed in the seminal works by Carlos M. Fonseca and his co-authors: the visualization and analysis of the (empirical) attainment function [5,6]. The attainment function is thereby a generalization of ECDFs of the best function value at a given time to the multiobjective case and gives, for each point in the objective space, the probability that this point is *attained* (or in other words weakly dominated) by an algorithm at the end of its run. It is typically approximated as the empirical attainment function (EAF) in practice by estimating the probability to attain a point from a (small) set of independent algorithm runs. The EAF can be plotted to get an idea of *where* in objective space an algorithm produces solutions.

However, when investigating the attainment function for a given algorithm, one looses the information on *when* certain points in objective space have been attained by the algorithm. It is the main goal of this paper to propose, based on the idea of the attainment function, a new performance assessment display which allows to investigate also the runtime—the time an algorithm takes to reach certain points in objective space. We thereby transfer the ideas of expected runtime (ERT) and average runtime (aRT) from the single-objective case [7,9] to the multiobjective case through the so-called average runtime attainment (aRTA) function and its associated plot. Similar to the EAF difference plots from [11], we furthermore introduce the aRTA ratio function to compare the average runtimes of two algorithms graphically. Based on a first preliminary implementation of the aRTA and aRTA ratio plots, the performance of a few algorithms from the BBOB 2016 workshop[1], obtained via the COCO platform, is displayed to showcase the usefulness of the new approach.

The paper is organized as follows. Section 2 gives the background on the attainment function approach as well as on the concepts of expected and average runtime, our aRTA functions are based upon. Section 3 details the new displays while Sect. 4 showcases them for a few data sets, obtained on the `bbob-biobj` test suite of the COCO platform, and gives details on the provided implementation. Finally, Sect. 5 concludes the paper and discusses the limitations of the proposed performance displays.

---

[1] see https://numbbo.github.io/workshops/BBOB-2016/.

2   Preliminaries

Throughout the paper, we consider the minimization of a multiobjective problem with m objective functions defined over a general search space Ω, i.e. we minimize

$$x \in \Omega \mapsto (f_1(x), \ldots, f_m(x)) \in \mathbb{R}^m \tag{1}$$

in which no specific assumption on the search space Ω is made. The search space can actually be discrete, continuous, etc. and in the remainder, we therefore focus our investigations on objective vectors $z \in \mathbb{R}^m$ only and, for simplicity, use the terms *solution* and *objective vector* interchangeably. We denote the coordinates of an objective vector $z \in \mathbb{R}^m$ as $(z_1, \ldots, z_m)$.

The weak dominance relation is defined for two objective vectors y and z of $\mathbb{R}^m$ as y *weakly dominates* z, denoted $y \preceq z$, if and only if $y_i \leq z_i$ for all i. A generalization of the weak dominance relation towards sets of objective vectors is straightforward by defining weak dominance between two sets Y and Z (both subsets of $\mathbb{R}^m$) whenever for each $x \in Z$ there exists a $y \in Y$ such that y weakly dominates z. In this case, we follow the notation of [6] and write $Y \trianglelefteq Z$.

If in a set of objective vectors A, all pairs of objective vectors are mutually non-dominated (in terms of the above weak dominance relation), we call A a set of mutually non-dominated objective vectors, or also a set of non-dominated vectors or even simpler, a non-dominated set.

2.1   Empirical Attainment Function

Given a set of non-dominated vectors $\mathcal{X} = \{X^1, \ldots, X^p\}$ (of random) size p and given a target vector $z \in \mathbb{R}^m$ of the objective space, we say that the target is reached (or attained) by the vector-set $\mathcal{X}$ if $\mathcal{X} \trianglelefteq z$.

Given N such sets of non-dominated vectors $\{\mathcal{X}_1, \ldots, \mathcal{X}_N\}$ (each containing a random number of non-dominated vectors), the empirical attainment function (EAF) introduced in [6] is defined as

$$\alpha(z) = \frac{1}{N} \sum_{i=1}^{N} \mathbf{1}\{\mathcal{X}_i \trianglelefteq z\}. \tag{2}$$

The EAF maps the objective space $\mathbb{R}^m$ to $[0, 1]$. In practice, the EAF is computed for N sets of non-dominated vectors that are the outcome of N independent trials collected at the end of a run or at a fixed budget T. It estimates the probability of an optimizer to find, within the budget T, an objective vector which is at least as good as the target vector z, where "at least as good" is interpreted in the weak dominance sense. Equivalently, the EAF estimates the probability to attain the region

$$\mathcal{A}(z) = \{y \in \mathbb{R}^m | z \preceq y\}. \tag{3}$$

To emphasize the dependence on T, we denote by $\alpha_T(z)$ the empirical attainment function of non-dominated vectors collected at a time T. In the case of a single objective ($m = 1$), $z \to \alpha_T(z)$ is the empirical cumulative distribution of the objective functions distribution reached at T [6]. That is, it is the empirical cumulative distribution of the data collected within the fixed budget scenario introduced in the introduction.

What Is Collected and Exploited? To summarize, the empirical attainment function relies solely on N sets, each composed of a random number of non-dominated objective vectors, which have been collected at some point in time T of an algorithm run. In order to allow for meaningful comparisons of algorithms, the time T shall be the same for all algorithms considered. To see the evolution of the algorithm performances during the search, multiple empirical attainment functions (for varying T) have to be displayed.

2.2 Expected Runtime (ERT) and Average Runtime (aRT)

While the EAF assumes a fixed budget, we remind here the definition of the expected runtime and average runtime that assume a fixed-target scenario. Consider the case where algorithm A either successfully reaches the target value f_{target} or it does not. The ERT [7] corresponds to the expected runtime of a conceptual algorithm that would restart A till obtaining a success, i.e., till f_{target} is reached. Given that algorithm A has a probability of success of p_s, an expected runtime for successful runs of $E[\text{RT}^{\text{s}}]$ and an expected runtime for unsuccessful runs of $E[\text{RT}^{\text{us}}]$, the ERT can be expressed as

$$\text{ERT}(f_{\text{target}}) = \frac{1 - p_s}{p_s} E[\text{RT}^{\text{us}}] + E[\text{RT}^{\text{s}}] \,. \tag{4}$$

It allows to compare in a meaningful and quantitative way algorithms that have a small probability of success, but converge fast when they do, with algorithms with a larger probability of success and a slower convergence rate.

An estimator for ERT is the average runtime (aRT, see also for example [7]). Given N runs of an algorithm with N_{s} successes to reach the target f_{target} and an overall number of function evaluations of $\text{FE}(N) = \sum_{i=1}^{N} T_i$ that includes the number of function evaluations for successful and unsuccessful runs, the aRT equals

$$\text{aRT}(f_{\text{target}}) = \frac{\text{FE}(N)}{N_{\text{s}}} \,. \tag{5}$$

The estimator for ERT when all unsuccessful runs have a number of function evaluations equal to a cutoff number was actually first proposed in [9].

3 Average Runtime Attainment Functions

We introduce in this section the average runtime attainment (aRTA) function that can be seen as a generalization of the attainment function where the information on the runtime to reach a target vector is re-introduced.

Similar to the EAF difference plots for comparing the EAFs of two algorithms in [11], we introduce the aRTA ratio function in addition for an easier comparison of the average runtimes between two algorithms.

3.1   Average Runtime Attainment Function

Compliant with the fixed-target approach, we fix a target vector $z \in \mathbb{R}^m$ and collect the minimal number of function evaluations (runtime) $T(z)$ to obtain a solution that weakly dominates z. If a run was not successful, that is, it did not find a solution that weakly dominates z, we collect the runtime of the run when it stopped. We assume that over N trials of the algorithm, we have collected all N runtimes, $T_1(z), \ldots T_N(z)$, and obtained N_s successes ($\leq N$). Then, the aRTA is the function defined as

$$\mathrm{aRTA}(z) = \frac{\sum_{i=1}^{N} T_i(z)}{N_s} \ . \tag{6}$$

Comparing (5) with (6), we see that aRTA is the natural generalization of the aRT estimator used in the single-objective case where we have adapted the notion of success from reaching a function value below a certain target to reaching a solution that weakly dominates a target vector. The aRTA function maps $\mathbb{R}^m$ to $\mathbb{R}^+$. Note in particular that, like in the single-objective case, the maximum number of function evaluations recorded for an algorithm effects the aRT values which Sect. 4.1 investigates in more detail.

In order to plot $\mathrm{aRTA}(z)$ in practice, the average runtime values of $\mathbb{R}^+$ need to be mapped to a color as we will showcase in the following section. Before, however, let us transfer another known concept around empirical attainment functions.

3.2   Average Runtime Attainment Ratio Function

In order to compare the aRTs of two algorithms more easily, we advocate to display the plots of the so-called aRTA ratio function, similar to the EAF difference plots of [11].[2]

To compare the aRTA functions of algorithms A and B, we can, in principle, plot the ratio of the two aRTA function values for both algorithms and each objective vector directly, i.e. we can plot

$$\mathrm{aRTA_{ratio}}(z) = \frac{\mathrm{aRTA^B}(z)}{\mathrm{aRTA^A}(z)} \tag{7}$$

as long as $\mathrm{aRTA_{ratio}}(z)$ is well defined (it is not well defined as soon as one or both of the aRTA function values are not finite). Since the measured runtimes

---

[2] We opt for displaying ratios here instead of differences as the ratio scale is more natural for statements on runtimes and also has stronger theoretical properties than the interval scale [13].

are comparable on a ratio scale with a non-arbitrary zero, we prefer aRTA *ratios* here over *differences* like in the EAF case [11]. This has the immediate effect that aRTA ratio functions are interpretable without the need to know any absolute values. To have an easier-to-read plot and to also cope with undefined aRTA$_{ratio}$ values, we actually propose to display a slight variant of the above.

If aRTA$_{ratio}(z)$ is well-defined and larger than 1, indicating an advantage for algorithm A, we simply plot aRTA$_{ratio}(z)$. Likewise, if aRTA$_{ratio}(z)$ is well-defined and smaller than 1, indicating an advantage for algorithm B, we plot $\frac{1}{\text{aRTA}_{ratio}(z)}$, color-coded with a different colormap instead—making it possible to easily compare advantages of algorithm A with advantages of algorithm B in the sense of statements like "Algorithm A is X times better than algorithm B in attaining the objective vector z". Undefined values of aRTA$_{ratio}(z)$, where only one algorithm possesses a finite aRT value (because for the other, all runs are unsuccessful), can nevertheless be plotted in a color that indicates the algorithm with the more favorable behavior.

3.3   What Is Collected and Exploited?

In comparison to the empirical attainment function, the aRTA and aRTA ratio functions rely on additional information about algorithm runs. In particular, each solution, which is not dominated by already evaluated solutions, needs to be recorded together with the runtime (in number of function evaluations) when it was evaluated by the algorithm. The input to the aRTA function and the aRTA ratio is therefore a sorted list of (number of function evaluations, objective vector) pairs such that for each algorithm run/problem instance, at each point in time, the current (external) archive of non-dominated solutions found so far can be reconstructed from the data.

Note that in the case of a single recorded run, the aRTA function plot is a visualization of all solutions over time and can be seen as a generalization of a single-objective convergence graph to the multiobjective case, see Fig. 2 for an example.

4   Numerical Examples from COCO

In order to showcase the usefulness of the proposed aRTA plots, we implemented a (preliminary) visualization in Python, which is made available on GitHub[3] and is able to display the algorithm performance from the archive of non-dominated solutions, recorded by the COCO platform [8] on functions from the bbob-biobj test suite [14].

In particular, the provided source code reads in the algorithm data from a COCO archive folder in the form of objective vector and runtime pairs (i.e. the function evaluation counter when the objective vector was produced) for each of 10 problem instances per function and dimension n. The code then computes and

---

[3] https://github.com/numbbo/coco/tree/master/code-postprocessing/aRTAplots.

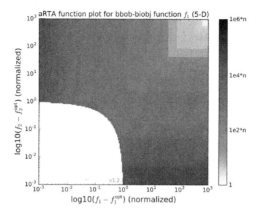

Fig. 2. The aRTA plot for a single algorithm run reduces to a visualization of all recorded non-dominated solutions. The example plot above shows a single run of the algorithm SMS-EMOA with polynomial mutation and SBX crossover on the `bbob-biobj` function 1 (sphere - sphere function) with 5 variables. (Color figure online)

displays the aRTA function values to weakly dominate for the first time a given objective vector z. When displaying the aRTA ratio function, the data of two algorithms is read in and the aRTA ratios are computed after the calculations of the single aRTA values for both algorithms. In both cases, we use a regular grid of objective vectors z for which we compute the aRT values instead of computing the aRTA areas of constant value exactly. Areas in between grid points are then colored according to the aRTA value (and aRTA ratio respectively) of its lower left corner. All objective values are normalized so that the ideal point is at $[0, 0]$ and the nadir point is at $[1, 1]$.[4]

All aRTA function plots shown in this paper are in log-scale and, if not specified differently, use a grid of 200×200 points, chosen equidistant on the log-scale between the ideal point $[0, 0]$ and the point $[10, 10]$.[5] The color-coding of the aRTA values is done in a log-scale as well so that the same color ranges are used, for example, for the first 100 function evaluations ("white to yellow") and the function evaluations $10^4 n, \ldots, 10^6 n$ ("red to black"). Note that the color scheme is absolute to allow for comparisons across figures and all solutions produced beyond the maximal budget of $10^6 n$ function evaluations are not used in the display. Although from all data sets submitted to the BBOB-2016 workshop only the one of HMO-CMA-ES contains solutions beyond this threshold, Sect. 4.1 investigates the influence of this parameter on the aRTA function plots in detail.

---

[4] Note that such a normalization allows for objective values to be larger than 1 and that our plots clips the display to objective values smaller than 10.

[5] Note that with the `logscale` parameter in the provided source code, the log-scale can be easily turned on and off.

In order to cope with the large data sets produced by the COCO platform[6], our implementation removes all but one of the recorded solutions within each grid cell before the computation of the aRT values. Thereby, the solution with the smallest function evaluation count per instance and grid cell is kept to not alter the plots while downsampling. As we will see later on in Sect. 4.2, this *downsampling* significantly reduces the computation time for the aRTA function plots. This section closes with showing a few examples of algorithm comparisons in Sect. 4.3.

All experiments for this paper have been run with COCO, version 1.2.1—more precisely with the code of the feature-branch as of commit 1c22851 on Dec. 31, 2016.

4.1   The Influence of the Maximal Budget of Function Evaluations

As a first investigation of the aRTA function plots we consider the influence of the maximal budget parameter, which specifies a threshold for function evaluations after which no solution is taken into account anymore. While it is typically not needed to change this parameter from its default setting of $10^6 n$ function evaluations to display the available COCO data, Fig. 3 shows the influence of the maximal budget on the aRTA function plots for the Matlab implementation of NSGA-II on the 5-dimensional separable ellipsoid - Rastrigin problem (function f_{16} in the bbob-biobj test suite).

Two observations can be made from Fig. 3. First, we see that a larger maximal budget value (and thus data from longer runs) results in a larger range of aRTA values, which are distributed in a larger area of the objective space. Secondly, increasing the maximal budget can change the color of those areas in the objective space that have not been attained in all runs at the lower budget value. The color can get darker (see the difference for maximal budget set to $10n$ and $10^2 n$) or lighter (see the difference on the upper left part of the plot for maximal budget set to $10^2 n$ and $10^3 n$), depending on which change in the aRTA fraction (the increasing runtimes in the numerator or the increasing success rate in the denominator) has a larger effect. Once an area of the objective space has been attained in all runs, its color cannot change any longer.

4.2   The Influence of Downsampling Data and Different Grid Sizes

To investigate the influence of the possible downsampling on the aRTA plots as well as on the time it takes to produce them[7], we use the data from the Matlab implementation of NSGA-II, as submitted to the BBOB-2016 workshop, on the example of the f_1 function (sphere - sphere). Without downsampling, the 10 instances of the data have 6755 to 9710 non-dominated solutions per instance

---

[6] A single function/dimension combination with 10 instances produces up to 930 MB of data.

[7] All experiments were performed on an Intel Core i7-5600U CPU Windows 7 laptop with 8 GB of RAM.

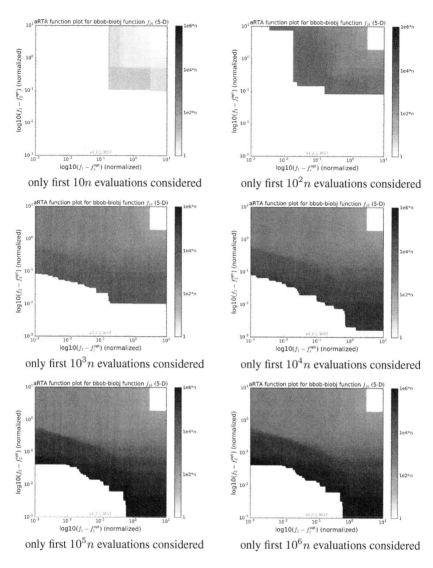

only first $10n$ evaluations considered   only first $10^2 n$ evaluations considered

only first $10^3 n$ evaluations considered   only first $10^4 n$ evaluations considered

only first $10^5 n$ evaluations considered   only first $10^6 n$ evaluations considered

Fig. 3. Influence of the maximal number of function evaluations considered on the average runtime plots for GA-MULTIOBJ-NSGA-II on function f_{16} (separable ellipsoid - Rastrigin) in dimension 5. Shown are, from top left to bottom right, the aRTA function plots when only the first $10n$, $10^2 n$, $10^3 n$, $10^4 n$, $10^5 n$, and $10^6 n$ function evaluations are considered.

(78,318 solutions in total), and it takes about 29 min to produce the single aRTA plot with 200×200 gridpoints, see Fig. 4. The provided source code is certainly not optimized for speed, but a runtime to produce a single aRTA plot of about half an hour is, of course, not acceptable in practice. With downsampling, i.e. taking only into account a single solution per grid cell, the time to produce the

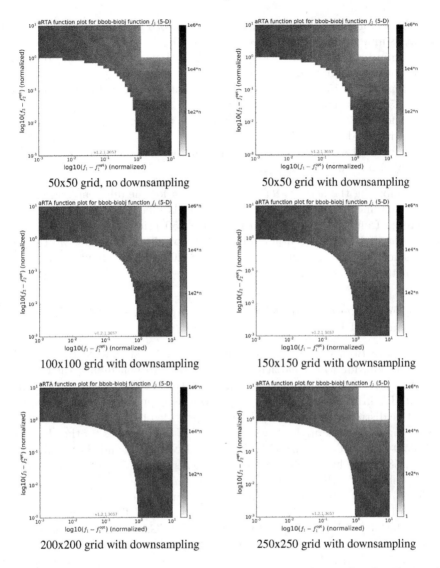

Fig. 4. First row: Downsampling the input data does not change the plot if per grid cell the solution with the lowest function evaluations count is kept. All other plots show the influence of the grid size on the aRTA plots, from a 50×50 grid, which takes about 6 s to produce, up to a 250×250 grid, which takes about 6.5 min to produce.

same plot can be reduced significantly. In order to further decrease the time to produce a single aRTA plot, we can trade the runtime with accuracy and change the grid size. This results in the following runtimes:

- ca. 6.5 min for a 250×250 grid (826–1062 solutions per instance[8]), down from about 49 min without downsampling,
- ca. 3.5 min for a 200×200 grid (685–905 solutions per instance), down from about 29 min without downsampling,
- ca. 1.5 min for a 150×150 grid (523–698 solutions per instance), down from about 15 min without downsampling,
- ca. 32 s for a 100×100 grid (370–483 solutions per instance), down from about 7 min without downsampling, and to
- ca. 6 s for a 50×50 grid (173–244 solutions per instance), down from about 1.5 min without downsampling.

When comparing the actual aRTA plots of Fig. 4 for the different grid sizes, we observe that increasing the number of grid cells increases the accuracy while an increase from the 200×200 to the 250×250 grid is hardly visible. As a good trade-off between accuracy and time to produce the plots, we therefore recommend using downsampling and the 200×200 grid as default, which is used for all plots in the remainder of the paper.

4.3   A Few Examples of Algorithm Comparisons with aRTA Function Plots

In this last section, we investigate some of the data, submitted to the BBOB 2016 workshop, for which participants were asked to benchmark their favorite algorithm on the `bbob-biobj` test suite via the COCO platform. The plots of the aRTA in Figs. 6, 7, and 8 can be seen as supplements to the empirical cumulative distribution functions (ECDFs), provided by the COCO platform by default. Figure 5 for example shows the ECDF of the runtimes for the algorithms RS-5, MO-DIRECT-hv-rank, GA-MULTIOBJ-NSGA-II, and SMS-EMOA-PM to reach 58 target hypervolume indicator values on the 5-dimensional sphere - sphere problem (f_1). RS-5 is thereby a simple random search within the domain $[-5, 5]$ [2], MO-DIRECT-hv-rank is an extension of the DIviding RECTangles approach to multiobjective optimization [15], GA-MULTIOBJ-NSGA-II is the default Matlab implementation of the standard NSGA-II [1], and SMS-EMOA-PM is the standard SMS-EMOA variant with polynomial mutation and SBX crossover [3]. The random search will be used here as a reference algorithm to which the other three algorithms are compared to.

   If we look carefully at Fig. 5, we see that for this particular problem, MO-DIRECT-hv-rank is at all times better than RS-5, SMS-EMOA-PM is worse in the beginning (because it initializes its population in the much larger space $[-100, 100]$) and better in the end compared to RS-5, and finally the NSGA-II is better in the very beginning, then worse, and finally better again than RS-5. Figures 6, 7, and 8 show the corresponding aRTA plots for the single algorithms as well as the corresponding aRTA ratio plots—displaying the same trends of

---

[8] Note that it is not necessarily the case that the instance with the smallest (largest) number of solutions recorded results in the smallest (largest) set of downsampled points.

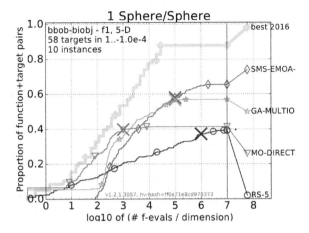

Fig. 5. Empirical cumulative distribution function of the runtimes of four algorithms to reach 58 target values on the `bbob-biobj` function f_1 (sphere - sphere) in dimension 5.

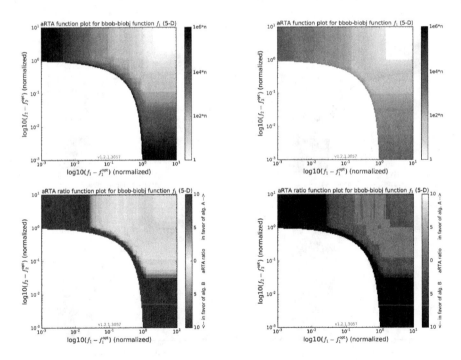

Fig. 6. Average runtime plots for RS-5 (Algorithm A, top left) and MO-DIRECT-hv-rank (Algorithm B, top right) together with the corresponding aRTA ratio plot (bottom row, left: colored, right: optimized for grayscale) on the 5-dimensional `bbob-biobj` function f_1.

when an algorithm is better than another and in addition also *where* in objective space and by *how much*.

A large difference between the shown algorithms lie in particular in their different initialization strategies. While the random search samples always uniformly at random in the set $[-5, 5]^n$ with n being the search space dimension, SMS-EMOA-PM samples its initial population from the much larger space $[-100, 100]^n$. The NSGA-II variant, displayed here, samples all but the first solution in its initial population also from $[-100, 100]^n$ and the first solution according to an isotropic Gaussian distribution around the search space origin. MO-DIRECT-hv-rank, finally, evaluates the search space origin as first solution. These different initialization strategies have a large impact on the algorithm performance during the first evaluations and beyond, which can be seen both in the ECDFs of Fig. 5 and the aRTA function plots. For the larger budgets and therefore areas close to the Pareto front in the aRTA (ratio) function plots, the initialization strategy seems to have no influence anymore and it is the algorithm's ability to approximate the Pareto front well which plays the biggest role in both the ECDFs and the aRTA plots.

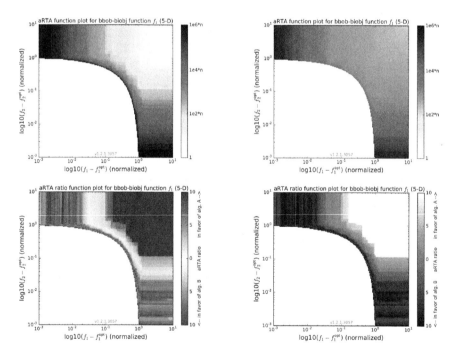

Fig. 7. Average runtime plots for RS-5 (Algorithm A, top left) and SMS-EMOA-PM (Algorithm B, top right) together with the corresponding aRTA ratio plot (bottom row, left: colored, right: optimized for grayscale) on the 5-dimensional `bbob-biobj` function f_1.

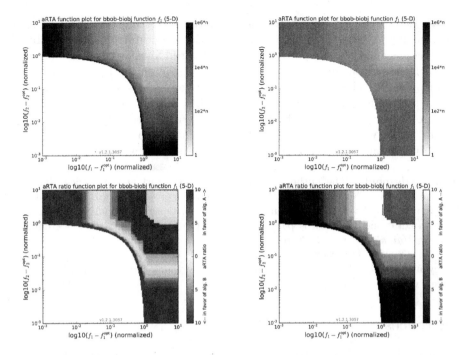

Fig. 8. Average runtime plots for RS-5 (Algorithm A, top left) and GA-MULTIOBJ-NSGA-II (Algorithm B, top right) together with the corresponding aRTA ratio plot (bottom row, left: colored, right: optimized for grayscale) on the 5-dimensional bbob-biobj function f_1.

5   Conclusion

We have proposed the average runtime attainment (aRTA) function as an alternative to the empirical attainment function to evaluate performance of multiobjective optimizers. In contrast to the latter, the aRTA function displays *quantitative* measurements of *when* the region that weakly dominates an objective vector z was reached for the first time. We have illustrated a simple display of the aRTA function (and of the aRTA ratio function for comparing two algorithms) on a grid using some data from the COCO platform.

Two shortcomings of the current implementation must be mentioned. The running time of the code that produces exact plots on an objective space grid is relatively high for practical purposes (in the order of minutes) even when the number of input solutions is downsampled to a single solution per grid cell and per instance. In addition, the displayed data, resulting from the COCO platform, contains results from 10 different instances of the same problem with potential discrepancies in the objective space. Here, the proposed aRTA function would be even more useful (and interpretable) if applied to data from independent runs on the same problem instance.

Last, let us discuss the generalization of the aRTA function displays to a higher number of objectives. While their definition is not restricted to two objective functions and their computation on a similar grid in higher dimension is possible with the same computational complexity per grid point, the aRTA function cannot be practically displayed in the same way as for two-objective problems: already for a three-dimensional grid, any display can only show 2-dimensional cuts through the grid or the surfaces of all points with a certain, predefined aRTA function value. We therefore expect the aRTA function to be less informative in higher dimensions than for two-objective problems as showcased here.

Acknowledgments. The authors acknowledge the support of the French National Research Agency (ANR) within the Modèles Numérique project "NumBBO – Analysis, Improvement and Evaluation of Numerical Blackbox Optimizers" (ANR-12-MONU-0009). In addition, this work is part of a project that has received funding from the *European Union's Horizon 2020 research and innovation program* under grant agreement No. 692286. This work was partially funded also by the Slovenian Research Agency under research program P2-0209. We finally thank the anonymous reviewers for their valuable comments.

References

1. Auger, A., Brockhoff, D., Hansen, N., Tušar, D., Tušar, T., Wagner, T.: Benchmarking MATLAB's Gamultiobj (NSGA-II) on the bi-objective BBOB-2016 test suite. In: GECCO (Companion) Workshop on Black-Box Optimization Benchmarking (BBOB 2016), pp. 1233–1239. ACM (2016)
2. Auger, A., Brockhoff, D., Hansen, N., Tušar, D., Tušar, T., Wagner, T.: Benchmarking the pure random search on the bi-objective BBOB-2016 testbed. In: GECCO (Companion) Workshop on Black-Box Optimization Benchmarking (BBOB 2016), pp. 1217–1223. ACM (2016)
3. Auger, A., Brockhoff, D., Hansen, N., Tušar, D., Tušar, T., Wagner, T.: The impact of variation operators on the performance of SMS-EMOA on the bi-objective BBOB-2016 test suite. In: GECCO (Companion) Workshop on Black-Box Optimization Benchmarking (BBOB 2016), pp. 1225–1232. ACM (2016)
4. Dolan, E.D., Moré, J.J.: Benchmarking optimization software with performance profiles. Math. Program. **91**, 201–213 (2002)
5. Fonseca, C.M., Fleming, P.J.: On the performance assessment and comparison of stochastic multiobjective optimizers. In: Voigt, H.-M., Ebeling, W., Rechenberg, I., Schwefel, H.-P. (eds.) PPSN 1996. LNCS, vol. 1141, pp. 584–593. Springer, Heidelberg (1996). doi:10.1007/3-540-61723-X_1022
6. Grunert da Fonseca, V., Fonseca, C.M., Hall, A.O.: Inferential performance assessment of stochastic optimisers and the attainment function. In: Zitzler, E., Thiele, L., Deb, K., Coello Coello, C.A., Corne, D. (eds.) EMO 2001. LNCS, vol. 1993, pp. 213–225. Springer, Heidelberg (2001). doi:10.1007/3-540-44719-9_15
7. Hansen, N., Auger, A., Brockhoff, D., Tušar, D., Tušar, T.: COCO: performance assessment. CoRR abs/1605.03560 (2016). http://arxiv.org/abs/1605.03560
8. Hansen, N., Auger, A., Mersmann, O., Tušar, T., Brockhoff, D.: COCO: a platform for comparing continuous optimizers in a black-box setting. CoRR abs/1603.08785 (2016). http://arxiv.org/abs/1603.08785

9. Hoos, H., Stützle, T.: Evaluating Las Vegas algorithms: pitfalls and remedies. In: Proceedings of the Fourteenth Conference on Uncertainty in Artificial Intelligence, pp. 238–245. Morgan Kaufmann Publishers Inc. (1998)

10. Hoos, H.H., Stützle, T.: Stochastic Local Search: Foundations and Applications. Elsevier, San Francisco (2004)

11. López-Ibáñez, M., Paquete, L., Stützle, T.: Exploratory analysis of stochastic local search algorithms in biobjective optimization. In: Bartz-Beielstein, T., Chiarandini, M., Paquete, L., Preuss, M. (eds.) Experimental Methods for the Analysis of Optimization Algorithms, pp. 209–222. Springer, Heidelberg (2010). Chap. 9

12. Moré, J., Wild, S.: Benchmarking derivative-free optimization algorithms. SIAM J. Optim. **20**(1), 172–191 (2009). Preprint available as Mathematics and Computer Science Division, Argonne National Laboratory, Preprint ANL/MCS-P1471-1207, May 2008

13. Stevens, S.S.: On the theory of scales of measurement. Science **103**(2684), 677–680 (1946)

14. Tušar, T., Brockhoff, D., Hansen, N., Auger, A.: COCO: the bi-objective black box optimization benchmarking (bbob-biobj) test suite. CoRR abs/1604.00359 (2016). http://arxiv.org/abs/1604.00359

15. Wong, C., Al-Dujaili, A., Sundaram, S.: Hypervolume-based DIRECT for multi-objective optimisation. In: GECCO (Companion) Workshop on Black-Box Optimization Benchmarking (BBOB 2016), pp. 1201–1208. ACM (2016)

A Multiobjective Strategy to Allocate Roadside Units in a Vehicular Network with Guaranteed Levels of Service

Flávio Vinícius Cruzeiro Martins[(✉)], João F.M. Sarubbi, and Elizabeth F. Wanner

Centro Federal de Educação Tecnológica de Minas Gerais,
Belo Horizonte, MG, Brazil
{flaviocruzeiro,joao,efwanner}@decom.cefetmg.br

Abstract. In this work, we propose the Delta-MGA, a specific multi-objective algorithm for solving the allocation of Roadside Units (RSUs) in a Vehicular Network (VANETs). We propose two multiobjective models to solve two different problems. The first one, our objectives are to find the minimum set of RSUs and to maximize the number of covered vehicles. The second one, our objectives are to find the minimum set of RSUs and to maximize the percentage of time that each vehicle remains connected. Our metric is based on Delta Network metric proposed in literature. As far as we concerned, Delta-MGA is the first multiobjective approach to present a deployment strategy for VANETs. We compare our approach with two mono-objective algorithms: (i) Delta-r; (ii) Delta-GA. Our results demonstrate that our approach gets better results when compared with Delta-r algorithm and competitive results when compared with Delta-GA algorithm. Furthermore, the main advantage of Delta-MGA algorithm is that with it is possible to find several different solutions given to the planning authorities diverse alternatives to deploy the RSUs.

Keywords: Vehicular network · Roadside unit deployment · Quality of service

1 Introduction

Vehicular Networks (VANETs) [3,8] are a particular case of Mobile Networks (MANETs) in which each node is represented by a moving car that uses wireless communication to exchange information with others cars and/or with an infrastructure support [4]. The first reason for the development of these networks was the traffic safety [7]. The VANETs can increase security in several ways: (i) the monitoring real-time traffic can solve problems such as traffic jams, avoid congestions and also sending emergency alerts such as accidents; (ii) a vehicle involved in an accident can broadcast warning messages about its position to others vehicles; (iii) surveillance cameras can be installed and transmit real-time

H. Trautmann et al. (Eds.): EMO 2017, LNCS 10173, pp. 120–134, 2017.
DOI: 10.1007/978-3-319-54157-0_9

information; (iv) two vehicles in the potential crash route can be warning and mend their ways;

However, according to [10] there are different VANETs applications such as:

- Internet access: customarily using an infrastructure support acting as a router;
- Digital map downloading: the map of a particular region can be download by the drivers;
- Value-added advertisement: used by service providers, who want to attract customers to their stores, e.g., restaurants, hotels, petrol pumps, etc.;
- Electronic Toll Collection: the toll payment can be made electronically saving time and fuel to the drivers.
- Parking Availability: possibility to show available slots in parking lots.

According to [16], these applications have "the potential to make travel (considerably) more efficient, convenient and pleasant". Figure 1a represents a Vehicular Network.

In VANETs, there are two kinds of communications: (i) Vehicle-to-Vehicle (V2V) [4]; (ii) Vehicle-to-Infrastructure (V2I) [17]. In V2V network, the communication occurs without any support infrastructure (pure ad hoc). However, according to [4,17] the V2V communication can become inefficient in sparse areas, rural zones and low peak hours due to the lack of vehicles. Besides, in V2V communication the signal propagation can be impaired due to some obstacles as trees, building reflexion, hills, other vehicles, etc.

On the other hand, in the V2I network, the communication occurs through connections between vehicles and communication units, called Roadside Units (RSUs). The RSUs are fixed infrastructure positioned along pathways. The main advantage of the V2I network is to serve as a communication layer for vehicular applications, mainly to ones that require a large bandwidth as video and music streaming and online games. Figure 1b represents a V2I network example.

Although a V2I network can improve the general efficiency of a Vehicular Network, the main drawback of this network is the cost to install RSUs, turning the decision of RSUs number and location a challenge to network providers [2,9,11,13,20]. If on the one hand, these planning authorities want to minimize the number of RSUs to reduce cost; on the other hand, they must guarantee a minimum Quality of Service (QoS) to the general population. As the number of RSUs is normally directly proportional to the QoS, usually we have two conflicting objectives: (i) minimize the number of RSUs and, (ii) maximize the QoS.

In this way, to choose the best places to deploy RSUs, we first need a metric to measure the QoS in Vehicular Networks. In this work, we choose the Delta Network metric [17]. In this metric, the QoS is measured using two different perspectives: (i) the individual user (ρ_1) that wants to stay more time connected as possible or, at least, staying connected for a sufficient time to receive the desired information; (ii) the traffic authorities (ρ_2) that want, at least, a fraction of the vehicles receives the sought information. At Delta Network, the individual user perception is entirely dependent upon the application. For traffic and time monitoring, the vehicle can receive information from time to time. On the other

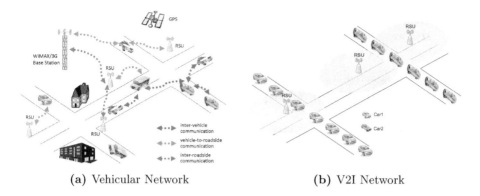

(a) Vehicular Network   (b) V2I Network

Fig. 1. Two different Vehicular Networks. (a) present a Vehicular Network with V2V and V2I communications. (b) presents a V2I Network.

hand, for music and video streaming, the vehicle must get a more "continuous" communication. In other words, Delta Network is based on two measurements: (i) connectivity duration; and, (ii) percentage of vehicles presenting such connectivity duration.

Although Silva and Meira [17] have developed a mono-objective approach, this problem is highly multiobjective. There are, at least, two alternatives to understand the real problem in a bi-objective way: (I) fixing ρ_1, our objectives are to minimize the number of RSUs and also to maximize the number of covered vehicles. A vehicle is covered if and only if it remains connected ρ_1 percent of its entire trip; (II) fixing ρ_2, our objectives are to minimize the number of RSUs and to maximize the percentage of time that each vehicle remains connected - (ρ_1). In both cases, the considered objectives are conflicting giving rise to a set of compromise solutions.

In this work, we propose a multiobjective algorithm, called Delta-MGA, for solving the RSUs deployment. As far as we concerned, it is the first multiobjective approach to solving the deployment of RSUs in a vehicular network. This algorithm can solve two different problems depending on which parameterer, ρ_1 or ρ_2, is given in advance. Delta-MGA is based on the NSGA-II multiobjective algorithm and in a variation of Delta-GA mono-objective algorithm proposed by Sarubbi et al. [14]. The main advantage of a multiobjective approach instead of the mono-objective version is that with the multiobjective algorithm we just run the algorithm once, for a specific parameter (ρ_1 or ρ_2), and we have all combinations of (ρ_2 and ρ_1) given to the planning authorities several alternatives to deploy the RSUs.

Our results demonstrate that:

– All of the solutions found by Delta-MGA algorithm dominate the solutions found by Delta-r algorithm proposed by Sarubbi and Silva [15] and dominate some solutions of Delta-GA algorithm [14];

– The Pareto curve given by Delta-MGA algorithm is close to the mono-objective solution provided by Sarubbi et al. [14] for most of the tested instances;

This work is organized as follows: Sect. 2 explains the used metric. Section 3 presents a selection of related work. Section 4 presents our proposal to represent complex road network. Section 5 formalizes the Multiobjective Deployment Δ^{ρ_1}. Section 6 presents our proposed solution. Section 7 presents our experiments. Section 8 concludes our work.

2 Delta Deployment

The Delta Deployment, proposed by Silva and Meira [17], is a metric for evaluating the quality of deployments. In this metric, the Quality of Service (QoS) is measured by two parameters (ρ_1, ρ_2). The first one, ρ_1, denotes how long each vehicle must stay connected to belong to the solution. The parameter ρ_1 is related to the total travel time of each vehicle. For instance, if the network provider wants each vehicle to remain connected to the RSUs during 30% of its trip, ρ_1 must be set to 0.3. The second one, ρ_2, denotes how many vehicles (from the total number of vehicles) must experience the connectivity defined by ρ_1. Thus, a Deployment $\Delta_{\rho_2}^{\rho_1}$ must guarantee that ρ_2 percent of the vehicles are connected to RSUs during ρ_1 percent of its trip. For instance, a deployment is $\Delta_{0.5}^{0.3}$ if 50% of the vehicles are connected to RSUs during 30% of its journey.

In this way, depending upon the desired application, the network provider can choose different values for the parameters ρ_1 and ρ_2. Thus, the Deployment $\Delta_{\rho_2}^{\rho_1}$ can find the best locations to install the RSUs to achieve the expected QoS minimizing the number of RSUs. We must notice that the Deployment $\Delta_{\rho_2}^{\rho_1}$ is just a metric. It does not specify how the QoS is achieved since it is technology-independent. The metric is independent on the access technology (Wi-Fi, 4G, Bluetooth, etc.) used to perform the communication.

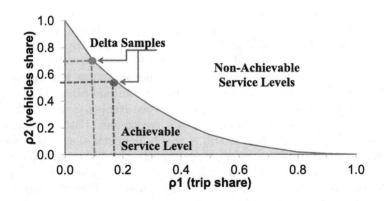

Fig. 2. Delta is a curve that rules the combinations of ρ_1 and ρ_2. [17]

Figure 2 illustrates the Delta metric, $\Delta_{\rho_2}^{\rho_1}$. Differently from typical approaches, the metric is not represented by a single value. Instead, Delta is represented as a curve in a 2D-plan. The x-axis indicates ρ_1, while the y-axis indicates ρ_2. In fact, Delta is the relation between ρ_1 and ρ_2.

3   Related Work

There are several metrics to measure the QoS in Vehicular Networks. Zheng et al. [21] present the evaluation of a deployment strategy considering the contact opportunity. The contact opportunity measures the fraction of distance (or time) a vehicle is in contact with the infrastructure. Authors indicate that such metric is closely related to the quality of data service that a mobile user might experience while driving. Zheng, Sinha, and Kumar [22] present the Alpha Coverage metric that minimizes the number of Roadside Units ensuring that each path of length α from a road network must have at least one RSU.

As far as we concerned, no work deals with a multiobjective approach to deploying RSUs. In VANETs, the multiobjective works deal mainly with QoS Routing. Barolli et al. [1] use Genetic Algorithms (GAs) and multiobjective optimization for QoS Routing in Ad-hoc Networks. Toutouh and Alba [18] applied a multiobjective optimization metaheuristic, to find optimal OLSR parameterizations that improve the QoS of the OLSR RFC. Thus, the optimized configuration obtained significantly reduces OLSR scalability problems keeping competitive packet delivery rates. Perera and Jayalath [12] use realistic vehicular traces to evaluate the performance of their proposed routing algorithms. Fazio et al. [6] implement two parallels multiobjective algorithms based on an evolutionary algorithm and a swarm intelligence for Ad hoc On Demand Vector routing protocol for vehicular networks.

For the original mono-objective Delta Network problem just a few works deal with it. In the Delta Network seminal article, Silva and Meira [17] compare two greedy approaches: (i) DL algorithm proposed in Trullols et al. [19] that iteratively selects the densest urban cell still not having a roadside unit; (ii) Delta-g algorithm that computes the sum of times of all uncovered vehicles that cross each urban cell and iteratively selects the urban cell that presents the highest sum. After, Sarubbi and Silva [15] present a greedy relative contact time approach called Delta-r and compare with DL algorithm [19] and Delta-g algorithm [17]. The authors also present an Integer Linear Formulation, but no exact results were present. As showed in Sarubbi and Silva [15] work, the Delta-r algorithm seems to be a better option when compared with Delta-g and DL algorithms. However, even being a better option then both DL and Delta-g algorithms, Delta-r algorithm is still a greedy heuristic that just searches a restricted part of the solution space. More recently, Sarubbi et al. [14] present a genetic algorithm for this problem with good results. The authors use a variation of Delta-r algorithm to create the initial population. However, they present solutions only for three pairs of ρ_1 and ρ_2.

4   Representing Road Networks

In order to represent a road network of arbitrary topology we use the same strategy used in Sarubbi and Silva [15]. As Sarubbi and Silva [15], we partitioned the urban area into a set of adjacent same size cells (i.e., grid model, Fig. 3) and, once the city or region is partitioned, we abandon the original road network. The main benefits of this strategy are: (i) the possibility to be more/less accuracy just increasing/decreasing the number of grid cells inside the region; (ii) the possibility to reduce the computational efforts reducing the number of possible locations to install RSUs; and, (iii) the complexity of the solution that is not depending upon the flow and works in the same manner for big and small regions.

 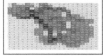  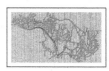

  (a) Road Network.   **(b)** 20×20 grid.   **(c)** 40×40 grid.   **(d)** 80×80 grid.

Fig. 3. Distinct grid setups. (a) shows a real road network (Ouro Branco city, Brazil). (b–d) show how such road network may be modeled by grid setups from 20×20 up to 80×80.

5   Problems Definition

In this section we formally define the two problems that will be solved in this work: (i) Multiobjective Deployment Δ^{ρ_1} and, (ii) Multiobjective Deployment Δ_{ρ_2}.

5.1   Multiobjective Deployment Δ^{ρ_1}

The Multiobjective Deployment is Δ^{ρ_1} whenever the maximal number of vehicles must be connected to roadside units during at least ρ_1 percent of its trip using the minimal possible number of RSUs. Formally, this problem can also be defined by a linear integer formulation (ILP). Suppose the sets:

V: set of vehicles where $V = \{1, 2, ..., k\}$;
U: set of urban cells where is possible to put a roadside unit where $U = \{0, 1, ..., u\}$;

    Our goal is to find the minimum set of points to install RSUs in order to maximize the number of vehicles that keep the connectivity established by the parameter ρ_1.

We have the following set of variables:

$$a_u = \begin{cases} 1 \text{ if the urban cell } u \text{ belongs to the solution} \\ 0 \text{ otherwise.} \end{cases}$$

$$v_k = \begin{cases} 1 \text{ if the vehicle } k \text{ belongs to the solution} \\ 0 \text{ otherwise.} \end{cases}$$

and the following set of parameters:

t_{uk}: time the vehicle k remained in urban cell u

tv_k: total travel time of vehicle k

The mathematical model is represented by:

$$minimize \ f_1 = \sum_{u \in U} a_u \tag{1}$$

$$maximize \ f_2 = \sum_{k \in V} v_k \tag{2}$$

Subject to

$$\sum_{u \in U} (t_{uk}/tv_k)a_u \geq \rho_1 v_k \ \forall k \in V \tag{3}$$

$$a_u \in \{0, 1\} \ \forall u \in U \tag{4}$$

$$v_k \in \{0, 1\} \ \forall k \in V \tag{5}$$

Objective function (1) consists of minimizing the number of roadside units. Objective function (2) consists of maximize the number of covered vehicles. Constraints (3) guarantee that a vehicle is chosen to belongs to the solution only if it is connect $\rho_1\%$ of its travel time. Constraints (4) and (5) are the integrality constraints.

5.2 Multiobjective Deployment Δ_{ρ_2}

The Multiobjective Deployment is Δ_{ρ_2} whenever at least ρ_2 percent of the total number of vehicles must stay connected and the objective is to maximize the percentage of time that each vehicle remains connected using the minimal possible number of RSUs. Formally: This problem can also be defined by a linear integer formulation and the mathematical model is:

$$minimize \ f_1 \tag{6}$$

$$maximize \ f_3 = \sum_{u \in U} (t_{uk}/tv_k)a_u \tag{7}$$

Subject to

$$\sum_{k \in V} v_k \geq \rho_2 |V| \tag{8}$$

$$a_u \in \{0, 1\} \ \forall u \in U \tag{9}$$

$$v_k \in \{0, 1\} \ \forall k \in V \tag{10}$$

Objective function (7) consists of to maximize the percentage of time that each vehicle remains connected. Constraint (8) ensure that a minimum number of the vehicles is chosen to belongs to the solution. Constraints (9) and (10) are the integrality constraints.

6   Proposed Algorithm

In this section, we present our multiobjective genetic algorithm, called Delta-MGA, to solve the Multiobjective Deployment Δ^{ρ_1} and the Multiobjective Deployment Δ_{ρ_2}. Our algorithm is based on the NSGA-II [5]. Our algorithm shares the elitism and selection mechanisms with the original NSGA-II, but it employs problem-based encoding scheme and operators. Sarubbi et al. [14] have introduced these operators in a mono-objective design context.

6.1   Solution Encoding

In our multiobjective algorithm, an individual consists of a list of several possible coordinates to install RSUs. The size of this list defines the number of RSUs present in the solution. The Fig. 4 represents an example of one individual with 8 RSUs.

RSU 1_B		RSU 2_B		RSU 3_B		RSU 4_B		RSU 5_B		RSU 6_B		RSU 7_B		RSU 8_B	
47	43	41	39	40	41	33	42	38	45	41	49	36	44	37	40

Fig. 4. Representation of an individual with 8 RSUs

6.2   Initial Population

Each initial solution is built using a Grasp-like algorithm. This algorithm, proposed in [14], is a randomized greedy technique that provides feasible solutions. Each feasible solution is iteratively constructed, one element at a time. We implemented a variation of the Delta-r algorithm proposed in [15] to generate the initial population. However, instead of always selecting the best solution (the cell with maximum sum of relative contact time) as the Delta-r algorithm, a Restricted Candidate List (RCL) of good elements is built, and one element (not necessarily the top candidate) is randomly selected. An RCL parameter α determines the level of greediness or randomness in the Construction Phase. We use, as a Build_RCL Procedure, a variation of the Delta-r algorithm [15]. However, instead of choosing the urban cell with maximal **relative** contact time, as the original Delta-r algorithm, we create a list of urban cells with the best **relative** contact time and randomly select one element of the list.

A RCL parameter α, that can vary from 0.0 to 1.0, determines the level of greediness or randomness at the Construction Phase. When $\alpha = 0.0$, the Construction Phase becomes the original Delta-r algorithm, a pure greedy algorithm. Otherwise, when $\alpha = 1.0$ the Construction Phase becomes totally random. If the RCL is built with many elements, then many different solutions will be produced, according to chosen α value.

The Build_RCL Procedure works as follows: first, the RCL list is set to empty. Then we compute the Maximal and Minimal relative contact time. For each remaining urban cell, we verify if the relative contact time of this cell is less than $Max - \alpha(Max - Min)$. If the condition is true, we add this urban cell to the RCL list.

In this work, for each instance, we generate 792 initial solutions using the Grasp-like algorithm. For instance, when we want to solve the Multiobjective Deployment Δ_{ρ_2} and $\rho_2 = 0.5$ then we generate 88 solutions for $\rho_1 = 0.1$, 88 solutions for $\rho_1 = 0.2$, 88 solutions for $\rho_1 = 0.3$ and so forth, until $\rho_1 = 0.9$. For each set of 88 solutions, 22 are generated with $\alpha = 0.1$, 22 are generated with $\alpha = 0.2$, 22 are generated with $\alpha = 0.3$ and 22 are generated with $\alpha = 0.4$. From these 792 solutions, we choose randomly 141 solutions, and we add nine greedy solutions from original Delta-r algorithm. Each one of the nine solutions uses a different ρ_1 parameter.

6.3   Evaluation

The Evaluation procedure is responsible for measuring the quality of all individuals of the population. Because the multiobjective nature of problems, the quality of one individual is measured using different *fitness* functions. We implement three *fitness* functions:

$$F_1 = numRSUs \tag{11}$$

wherein $numRSUs$ represents the solution cardinality.

$$F_2 = numVehicles \tag{12}$$

wherein $numVehicles$ represents the number of connected vehicles.

$$F_3 = \rho_1 \tag{13}$$

wherein, for each individual, $crho_1$ is computed in these following steps:

1. Compute ρ_1 for each $v \in V$;
2. Sort the vehicles in decrescent order of ρ_1;
3. Compute Index $= \rho_2 \times |V|$;
4. Get the ρ_1 value of $V[Index]$;

Taking all of these into account, the multiobjective problems can written as shown in (14) and (15).

$$\mathcal{U}^*_{\rho_1} = \arg\min_{\mathcal{U}} \left\{ \begin{array}{l} F_1[\mathcal{U}] \\ -F_2[\mathcal{U}] \end{array} \right. \tag{14}$$

$$\mathcal{U}^*_{\rho_2} = \arg\min_{\mathcal{U}} \left\{ \begin{array}{l} F_1[\mathcal{U}] \\ -F_3[\mathcal{U}] \end{array} \right. \tag{15}$$

For the first problem, multiobjective Deployment Δ^{ρ_1} is computed by two functions: (i) Eq. (11) and (ii) Eq. (12). For the second problem, multiobjective Deployment Δ_{ρ_2}, the *fitness* is also computed by two functions: (i) Eq. (11) and, (ii) Eq. (13).

6.4 Crossover

The crossover operation combines two individuals generating two new individuals. The genetic combination is carried out using a cutoff point chosen randomly in each individual (cut and splice). The new individuals are formed with based on a portion of the first parent code and, the remainder, from the second parent code. The Fig. 5 shows an example of a crossover between individuals A and B generating two new individuals AB and BA. At each generation we randomly create pairs between the individuals using a crossover probability $tCros$.

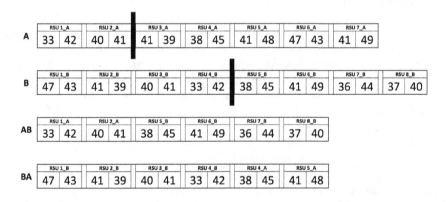

Fig. 5. Crossover

6.5 Mutation

At each generation, every individual has a $tMut$ chance to mutate. If one individual is chosen to mutate, the algorithm chooses randomly the number of mutations between one and three. Note that, at each generation, the same individual can have the same mutation more than once.

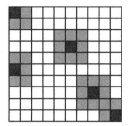

Fig. 6. RSU 2-exchange neighborhood

We implemented four mutations:

1. **insertion**: insertion of a new RSU in a random coordinate;
2. **remotion**: remotion of a RSU in a random coordinate;
3. **random 2-exchange**: insertion of a new RSU in a random coordinate and remotion of a RSU in a random coordinate, i.e., union of mutation one and two;
4. **neighborhood 2-exchange**: we choose a random RSU and for this RSU we make a 2-exchange with a random neighborhood urban cell. The neighborhood is present in Fig. 6.

7   Experiments

Experiments are based on the realistic mobility trace of Cologne, Germany (http://kolntrace.project.citi-lab.fr/). The entire trace is composed of 7,200 s of traffic from 75,515 vehicles. All experiments are performed using the SUMO (Sumo Simulator: http://sumo-sim.org) simulator and a set of tools designed by our team. SUMO runs the Cologne scenario and outputs the location of each vehicle (our mobility trace T) over time. The Partition Program reads the mobility trace, computes the bounding box of the mobility trace, divider Cologne into a grid of $\psi \times \psi$ urban cells, and then translates the mobility trace from Cartesian coordinates to Grid coordinates. For all experiments we use $\psi = 100$, resulting in a covered area of about $260m \times 260m$ for each urban cell. For all experiments we use the same parameters presented in Delta-GA algorithm [14]: (i) Population size = 150; (ii) Number of generations = 500; (iii) Probability of Mutation ($tMut$) = 0.2; (iv) Crossing Probability ($tCros$) = 0.8. For each instance we run the Delta-MGA algorithm 11 times.

In this work, we tested six different instances of this problem. For the Multiobjective Deployment Δ^{ρ_1}, we present solutions for three values of ρ_1: (i) $\rho_1 = 0.1$; (ii)$\rho_1 = 0.5$; and, (iii)$\rho_1 = 0.9$. For the Multiobjective Deployment Δ_{ρ_2}, we present solutions for three values of ρ_2: (i) $\rho_2 = 0.1$; (ii)$\rho_2 = 0.5$; and, (iii)$\rho_2 = 0.9$. For each instance we also present solutions for both Delta-r and Delta-GA algorithms for nine different values. For instance, when we fix $\rho_1 = 0.1$ then we present solutions for $\rho_2 = 0.1$, $\rho_2 = 0.2$, $\rho_2 = 0.3$, $\rho_2 = 0.4$, $\rho_2 = 0.5$,

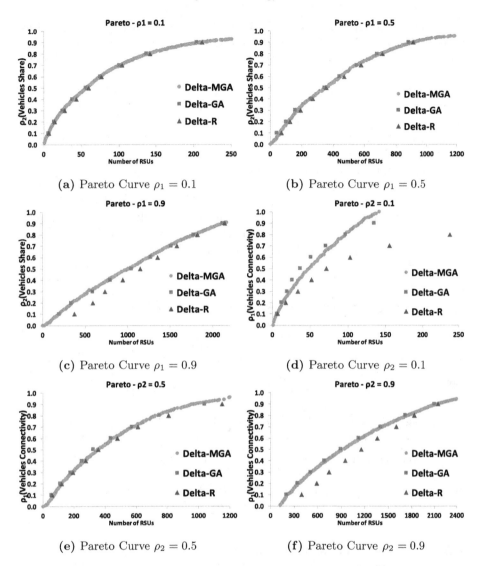

Fig. 7. This graphic represents the combined Pareto Curve after 11 runs for three different values of ρ_1 and ρ_2. For each ρ_1 or ρ_2 we compare the solution of the Delta-MGA algorithm with the solution of Delta-r and Delta-GA algorithms.

$\rho_2 = 0.6$, $\rho_2 = 0.7$, $\rho_2 = 0.8$ and $\rho_2 = 0.9$. These solutions can be used for performing ϵ-constrained searches for the multiobjective solutions. In this case, for a specific ρ_1 and ρ_2, the best solution found after 11 runs of the Delta-GA and the solution from Delta-r represent, for each, one multiobjective solution in the Pareto. Our objective is to compare the solutions given by the mono-objective and the multiobjective approaches. Different from Delta-MGA that just requires the parameter ρ_1 OR ρ_2, Delta-r and Delta-GA require parameters ρ_1 AND ρ_2.

In Fig. 7, for each QoS parameter ρ_1 or ρ_2, we present a combined Pareto curve after 11 runs of the Delta-MGA algorithm. Each curve gives a relation between the number of RSUs and: (i) the percentage of covered vehicles (parameter ρ_2 in the mono-objective version) when we fix parameter ρ_1; (ii) the percentage of time that each vehicle remains connected (parameter ρ_1 in the mono-objective version) when we fix parameter ρ_2; With this curve, the planning authorities can measure the cost-benefit to attend a specific number of vehicles with a given QoS established by ρ_1 or ρ_2.

In Fig. 8 we present the absolute number of RSUs for two instances: (i) when we fixed $\rho_1 = 0.9$ and, (ii) when we fixed $\rho_2 = 0.9$. For each instance we compare the number of RSUs found by each algorithm: (a) Delta-MGA; (b) Delta-r and; (c) Delta-GA. In the first case, we compare the three algorithms for nine different values of ρ_2 and, in the second case, we compare the three algorithms for nine different values of ρ_1.

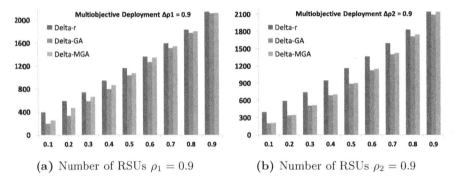

(a) Number of RSUs $\rho_1 = 0.9$ **(b)** Number of RSUs $\rho_2 = 0.9$

Fig. 8. Each figure presents the number of RSUs to each algorithm: (a) Delta-MGA; (b) Delta-r and; (c) Delta-GA. We compare each one for nive different values of ρ_2.

Observing Figs. 7 and 8, our multiobjective algorithm finds better solutions when we compare with the Delta-r [15] algorithm for almost all tested instances. On the other hand, the Delta-GA [14] solutions are, in most of the cases, superior to Delta-MGA, i.e., for the same QoS, Delta-GA finds a solution with fewer RSUs than Delta-MGA. This behavior is expected because we use, in a multiobjective algorithm Delta-MGA, a variation of the mono-objective Delta-GA. However, in some situations, the Delta-MGA algorithm finds less RSUs than Delta-GA, for the same pair (ρ_1, ρ_2). While Delta-MGA tries to find the best solutions for all ρ_1 (from $\rho_1 = [0..1]$) or ρ_2 (from $\rho_2 = [0..1]$) values just running Delta-MGA once, the Delta-GA and Delta-r algorithms work for a specific pair (ρ_1, ρ_2). For instance, when $\rho_1 = 0.1$, just running Delta-MGA eleven times, we have 226 distinct solutions. When $\rho_1 = 0.5$, Delta-MGA finds 326 different solutions and, at last, when $\rho_1 = 0.9$, Delta-MGA finds 286 different solutions. It is also true when we fix parameter ρ_2 (Multiobjective Deployment Δ_{ρ_2}). When $\rho_2 = 0.1$, just running Delta-MGA eleven time, we have 108 distinct solutions. When $\rho_2 = 0.5$, Delta-MGA finds 244 different solutions and, at last, when $\rho_1 = 0.9$, Delta-MGA finds 403 different solutions making possible to give different deployment possibilities to the planning authorities. It is also important to inform that both algorithms, Delta-MGA and Delta-GA, have almost the same execution time.

8   Final Remarks

In this work, we proposed a multiobjective algorithm, Delta-MGA, for solving the allocation of Roadside Units (RSUs) in a Vehicular Network. Delta-MGA can solves two different problems based on Delta Network and for each one, we also present an integer linear programing formulation. In the first problem, our objectives are to find the minimum set of RSUs and to maximize the number of covered vehicles. In the second problem, our objectives are to find the minimum set of RSUs and also to maximize the percentage of time that each vehicle remains connected. Our metrics also takes into account the Quality of Service given by parameter ρ_1, that represents the percentage of the time journey that each vehicle must be connected or by parameter ρ_2 number of vehicle attended by ρ_1.

Our algorithm was, in all cases, a better option the Delta-r algorithm and also presented competitive results when compared with Delta-GA algorithm. In some instances, Delta-MGA found fewer RSUs than Delta-GA to achieve the same QoS. Delta-MGA seemed to be a better alternative to planning authorities that must have to analyze the trade-off choose between cost (RSUs installation) and the number of covered vehicles and the percentage of time that each vehicle remains connected.

Acknowledgments. The authors would like to thank the Brazilian funding agencies, CNPq, CAPES and Fapemig for financial support.

References

1. Barolli, A., Spaho, E., Barolli, L., Xhafa, F., Takizawa, M.: Emerging wireless and mobile technologies. Mob. Inf. Syst. **7**(3), 169–188 (2011)
2. Barrachina, J., Garrido, P., Fogue, M., Martinez, F.J., Cano, J.C., Calafate, C.T., Manzoni, P.: Road side unit deployment: a density-based approach. IEEE Intell. Transp. Syst. Mag. **5**(3), 30–39 (2013)
3. Barrachina, J., Sanguesa, J.A., Fogue, M., Garrido, P., Martinez, F.J., Cano, J.C., Calafate, C.T., Manzoni, P.: V2X-d: a vehicular density estimation system that combines V2V and V2I communications. In: IEEE/IFIP Wireless Days, Valencia, Spain, November 2013
4. Blum, J., Eskandarian, A., Hoffman, L.: Challenges of inter vehicle ad hoc networks. IEEE Trans. Intell. Transp. Syst. **5**(4), 347–351 (2004)
5. Deb, K., Pratap, A., Agarwal, S., Meyarivan, T.: A fast and elitist multiobjective genetic algorithm: NSGA-II. IEEE Trans. Evol. Comput. **6**(2), 182–197 (2002)

6. Fazio, P., Rango, F.D., Sottile, C., Santamaria, A.F.: Routing optimization in vehicular networks: a new approach based on multiobjective metrics and minimum spanning tree. Int. J. Distrib. Sens. Netw. **9**(11), 1–13 (2013)
7. Fogue, M., Garrido, P., Martinez, F.J., Cano, J.C., Calafate, C.T., Manzoni, P., Sanchez, M.: Prototyping an automatic notification scheme for traffic accidents in vehicular networks. In: 4th IFIP Wireless Days (WD), pp. 1–5 (2011)
8. Hartenstein, H., Laberteaux, K.: A tutorial survey on vehicular ad hoc networks. IEEE Commun. Mag. **46**(6), 164–171 (2008)
9. Kchiche, A., Kamoun, F.: Centrality-based access-points deployment for vehicular networks. In: 2010 IEEE 17th International Conference on Telecommunications (ICT), Doha, pp. 700–706. IEEE (2010)
10. Kumar, V., Mishra, S., Chand, N.: Applications of vanets: present & future. Commun. Netw. **5**, 12–15 (2013)
11. Mershad, K., Artail, H., Gerla, M.: ROAMER: roadside units as message routers in vanets. Ad Hoc Netw. **10**(3), 479–496 (2012)
12. Perera, O., Jayalath, D.: Cross layer optimization of VANET routing with multiobjective decision making. In: IEEE Australasian Telecommunication Networks and Applications Conference (ATNAC) (2012). http://eprints.qut.edu.au/55146/
13. Reis, A., Sargento, S., Tonguz, O.: On the performance of sparse vehicular networks with road side units. In: 2011 IEEE 73rd Vehicular Technology Conference (VTC Spring), pp. 1–5, May 2011
14. Sarubbi, J.F.M., Martins, F.V.C., Silva, C.M.: A genetic algorithm for deploying roadside units in VANETS. In: IEEE Congress on Evolutionary Computation (CEC). IEEE, July 2016
15. Sarubbi, J.F.M., Silva, C.M.: Delta-r: a novel and more economic strategy for allocating the roadside infrastructure in vehicular networks with guaranteed levels of performance. In: IEEE/IFIP Network Operations and Management Symposium (NOMS). IEEE, April 2016
16. Sichitiu, M., Kihl, M.: Inter-vehicle communication systems: a survey. IEEE Commun. Surv. Tutorials **10**(2), 88–105 (2008)
17. Silva, C.M., Meira, W.: Evaluating the performance of heterogeneous vehicular networks. In: 2015 IEEE Vehicular Technology Conference (VTC), September 2015
18. Toutouh, J., Alba, E.: Multi-objective OLSR optimization for VANETS. In: IEEE Wireless and Mobile Computing, Networking and Communications (WiMob) (2012)
19. Trullols, O., Fiore, M., Casetti, C., Chiasserini, C., Ordinas, J.B.: Planning roadside infrastructure for information dissemination in intelligent transportation systems. Comput. Commun. **33**(4), 432–442 (2010)
20. Wu, Y., Zhu, Y., Li, B.: Infrastructure-assisted routing in vehicular networks. In: 2012 Proceedings of IEEE INFOCOM, pp. 1485–1493. IEEE (2012)
21. Zheng, Z., Lu, Z., Sinha, P., Kumar, S.: Maximizing the contact opportunity for vehicular internet access. In: 2010 Proceedings IEEE INFOCOM, pp. 1–9, March 2010
22. Zheng, Z., Sinha, P., Kumar, S.: Alpha coverage: bounding the interconnection gap for vehicular internet access. In: INFOCOM 2009, pp. 2831–2835. IEEE, April 2009

An Approach for the Local Exploration of Discrete Many Objective Optimization Problems

Oliver Cuate[1][(✉)], Bilel Derbel[2,3], Arnaud Liefooghe[2,3], El-Ghazali Talbi[2,3], and Oliver Schütze[1]

[1] Computer Science Department, CINVESTAV-IPN,
Av. IPN 2508, Col. San Pedro Zacatenco, 07360 Mexico City, Mexico
`ocuate@computacion.cs.cinvestav.mx`, `schuetze@cinvestav.mx`
[2] Univ. Lille, CNRS, Centrale Lille, UMR 9189 – CRIStAL, 59000 Lille, France
{`bilel.derbel,arnaud.liefooghe,el-ghazali.talbi`}`@univ-lille1.fr`
[3] Inria Lille – Nord Europe, 59650 Villeneuve d'Ascq, France

Abstract. Multi-objective optimization problems with more than three objectives, which are also termed as many objective optimization problems, play an important role in the decision making process. For such problems, it is computationally expensive or even intractable to approximate the entire set of optimal solutions. An alternative is to compute a subset of optimal solutions based on the preferences of the decision maker. Commonly, interactive methods from the literature consider the user preferences at every iteration by means of weight vectors or reference points. Besides the fact that mathematical programming techniques only produce one solution at each iteration, they generally require first or second derivative information, that limits its applicability to certain problems. The approach proposed in this paper allows to steer the search into any direction in the objective space for optimization problems of discrete nature. This provides a more intuitive way to set the preferences, which represents a useful tool to explore the regions of interest of the decision maker. Numerical results on multi-objective multi-dimensional knapsack problem instances show the interest of the proposed approach.

Keywords: Many objective optimization · Multi-criteria decision making · Discrete optimization · Knapsack · Evolutionary computation

1 Introduction

In many real-world applications from engineering or finance, one has to face the issue that several objectives have to be optimized concurrently. If more than three objectives are involved, the resulting problem is often termed a *many objective optimization problem* (MaOP) in the literature. Though the treatment of MaOPs is a relatively young research field (so far, mainly problems with two or three objectives have been studied) it is a very important one as the decision

© Springer International Publishing AG 2017
H. Trautmann et al. (Eds.): EMO 2017, LNCS 10173, pp. 135–150, 2017.
DOI: 10.1007/978-3-319-54157-0_10

making processes are getting more and more important nowadays. One important characteristic of MaOPs is that its solution set, the so-called Pareto set, does typically not consist of one single solution as for 'classical' scalar optimization problems (SOPs, i.e., *one* objective is considered). Instead, the Pareto set of a continuous MaOP typically forms a $(k-1)$-dimensional object, where k is the number of objectives involved in the problem. For discrete MaOPs, as we will consider here, the magnitude of the solution set typically grows exponentially with k [16]. Specialized evolutionary algorithms have caught the interest of many researchers over the last decades; see, e.g., [11,12,32] and references therein. Reasons for this include that these algorithms are applicable to a wide range of problems, are of global nature and hence in principle do not depend on the initial candidate set (i.e., the initial population). Further, due to their set-based approach, they compute a limited-size representation of the entire Pareto set in a single run of the algorithm. Most of these specialized algorithms, called MOEAs (multi-objective evolutionary algorithms) are designed for the treatment of problems with just few objectives (say, 2 to 4). However, more and more algorithms are proposed that deal with many objectives. For instance the use of MAOPs with large population size (e.g. 10,000 individuals) [20], the Dynamical Multi-objective Evolutionary Algorithm (DMOEA) [34] and the Grid-Based Evolutionary Algorithm (GrEA) [31]. However, due to their huge magnitude, the Pareto sets of MaOPs can typically not be computed efficiently, and further on, these sets cannot be visualized adequately. Thus, even with the aid of evolutionary algorithms, if the number of objectives is high, one cannot expect that the solution selected by the *decision maker* (DM) is in fact the preferred solution of the given MaOP with respect to the given setting.

In this paper, we argue that a fine tuning of a selected optimal solution makes sense. More precisely, we propose an approach where, starting from a given initial solution x_0, further solutions x_i, $i = 1, \ldots, N$, are generated such that the sequence of the candidate solutions performs a movement into user-specified directions. Numerical results on different scenarios will show the benefit of the novel approach. We stress that the idea to steer the search along the Pareto set/front into a given user-specified direction is not new. The Directed Search Descent Method [28] and the Pareto Tracer [23] are capable of performing such a search. However, they are both restricted to continuous optimization problems, and they cannot be extended to discrete domains as gradient information is required. Moreover, several specialised MOEAs to solve MaOPs have surged in recent years, for instance in [3] a method which employ the hypervolume for this purpose is proposed and a posteriori method to deal with MaOPs is considered in [7].

The remainder of this paper is organized as follows: in Sect. 2, we will shortly present the required background and will discuss the related work. In Sect. 3, we propose a method for the fine tuning of a given solution from the considered MaOP, and we compare possible realizations of this framework. In Sect. 4, we present some numerical results with a selected approach, and finally, we will conclude and give paths for future research in Sect. 5.

2   Background

2.1   Definitions

From a mathematical point-of-view, a *multi-objective optimization problem* (MOP) can be defined as follows:

$$\max_{x \in \Omega \subset \mathbb{R}^n} F(x), \text{ s.t. } g(x) \leq 0 \text{ and } h(x) = 0, \tag{1}$$

where $F : D \to \mathbb{R}^k$ is a vector function mapping the *feasible set* $D := \{x \in \Omega \mid g(x) \leq 0 \text{ and } h(x) = 0\}$ to its image $Z := \{F(x) = (f_1(x), \dots, f_k(x)) \mid x \in D\} \subseteq \mathbb{R}^k$, where $f_i : \Omega \to \mathbb{R}$ stands for the i-th objective. In the combinatorial case, feasible solutions forms a discrete set D. In a maximization context, an objective vector $z \in Z$ is *dominated* by an objective vector $z' \in Z$, denoted by $z \prec z'$, iff $\forall i \in \{1, 2, \dots, k\}$, $z_i \leqslant z_i'$ and $\exists i \in \{1, 2, \dots, k\}$ such that $z_i < z_i'$. Similarly, a solution $x \in D$ is dominated by a solution $x' \in D$, denoted by $x \prec x'$, iff $F(x) \prec F(x')$. A solution $x^\star \in D$ is said to be *Pareto optimal*, if there does not exist any other solution $x \in D$ such that $x^\star \prec x$. The set of all Pareto optimal solutions is called the *Pareto set* (PS), and its mapping in the objective space is called the *Pareto front* (PF). One of the most challenging task in multi-objective optimization is to identify a minimal complete Pareto set, i.e., one Pareto optimal solution for each point from the Pareto front. In the combinatorial case, generating a complete Pareto set is often infeasible for two main reasons [16]: (*i*) the number of Pareto optimal solutions is typically exponential in the size of the problem instance, and (*ii*) deciding if a feasible solution belongs to the Pareto set may be NP-complete. Therefore, the overall goal is often to identify a good *Pareto set approximation*. For this purpose, heuristics in general, and evolutionary algorithms in particular, have attracted a lot of attention from the optimization community since the late eighties [10,12].

One of the most studied NP-hard problem from combinatorial optimization is the *multi-objective (multi-dimensional) 0–1 knapsack problem* (MOKP). Given a collection of n items and a set of k knapsacks, the MOKP seeks a subset of items subject to capacity constraints based on a *weight function* vector $w : \{0, 1\}^{k \times n}$, while maximizing a *profit function* vector $p : \{0, 1\}^{k \times n}$. More formally, it can be stated as:

$$\begin{array}{ll} \max & \sum_{j=1}^{n} p_{ij} \cdot x_j & i \in \{1, \dots, k\} \\ \text{s.t.} & \sum_{j=1}^{h} w_{ij} \cdot x_j \leqslant c_i & i \in \{1, \dots, k\} \\ & x_j \in \{0, 1\} & j \in \{1, \dots, n\} \end{array}$$

where $p_{ij} \in \mathbb{N}$ is the profit of item j on knapsack i, $w_{ij} \in \mathbb{N}$, is the weight of item j on knapsack i, and $c_i \in \mathbb{N}$ is the capacity of knapsack i. For the MOKP, the cardinality of the Pareto set can grow exponentially with the problem size [16], which makes this problem very appealing to investigate. Notice however, that the proposed approach is not specific to the MOKP and could be applied for other combinatorial problems as well.

2.2 Literature Overview

The success of MOEAs can be essentially attributed to the fact that they do not require special features or properties from the objective functions, by relying on stochastic search procedures that are able to deal with complex MOPs and a large variety of application domains. MOEAs are in fact inspired by the basic principles of the evolutionary process on a population (a set of individuals or solutions), by means of the so-called evolutionary operators. Considering the wide variety of MOEAs, and the different principles guiding their design, e.g., Pareto dominance [14], indicator-based [5], aggregation-based [32] methods, a lot of progress have been made to better harness the complexity of solving MOPs using an evolutionary process and to better understand the main challenges one has to face. In particular, the dimensionality of the objective space is believed to be one of the main difficult challenges one has to address. In fact, several researchers have pointed out different issues on the use of MOEAs to solve problems having 4 or more objectives [17,21,27]. These issues are essentially related to the fact that, as the number of objectives increases, the proportion of non-dominated elements in the population grows. An expression for the portion e in a k-dimensional criteria domain, such that the dominance concept classifies as equivalent solutions is given by $e = \frac{2^k - 2}{2^k}$ [17].

In recent years, approaches for improving the behavior and the performance of EMO algorithms in order to deal with so-called MaOPs have received a growing interest. We can classify them in two groups: *(i)* Methods using alternative preference relations, and *(ii)* Methods transforming the original MaOP into a SOP. In the first group, different preference relations were reported such as, to cite a few, the so-called Preference Ordering [15], a generalization of Pareto optimality which uses two more stringent definitions of optimality, or fuzzy relations [17] based on the number of components which are bigger, smaller or equal between two objective vectors. In the second group, different algorithmic approaches can be found such as those based on indicators (e.g., hypervolume [4]), those based on dimensional reduction techniques (e.g., the so-called Pareto Corner Search Evolutionary Algorithm [29]) or those based on space partitioning (e.g., the so-called ϵ Ranking-Evolutionary Multi-objective Optimizer [1]).

It is important to remark that the rationale behind any MOEA is the computation of a *representative* set of solutions from which the DM can eventually pick one or some. However, and apart from difficulties inherent to the optimization/solving process itself, taking the DM requirements into account is a challenging issue when tackling a MaOP. This is essentially because a MOEA might actually fail in providing high-quality solutions in the regions of interest for the DM, or also to entirely cover the Pareto front due to the high dimensionality of MaOPs. For this purpose, taking the DM preferences into account is a hot issue that is being increasingly addressed in the evolutionary multi-objective optimization and multi-criteria decision making communities; see, e.g., [6,9]. Although reviewing all the literature on the subject is beyond the scope of this paper, we can still comment on three classes that one might encounter [9,24], namely, (i) *a priori* approaches that aim at guiding the evolutionary process through pre-defined regions of interest provided by the decision maker, e.g., [22], (ii) *interactive* approaches [26] where the decision maker preference(s) is iteratively and progressively refined during the evolutionary search procedure until the decision maker is satisfied, and (iii) *a posteriori* approaches, e.g., [18], taking into account the preference of the decision makers solely after the evolutionary process has ended up with a reasonable approximation set.

The work presented in this paper falls at the crossroads of interactive and a priori approaches. In fact, on the one hand, the proposed approach allows the DM to refine a given solution iteratively by providing a direction where to steer the evolutionary search process. On the other hand, we use a reference point approach to transform the direction provided by the decision maker into a reference-point SOP, which is solved using a multi-objective search process. It is to notice that there exist some work in line with our proposal. For instance, Cheng *et al.* [8], and Deb and Jain [13] proposed different methods that allow the decision maker to attain preferred solutions using reference vectors. However, that methods depart from our proposal in several aspects. First, it falls into the a posteriori class of approaches. More importantly, it aims at providing the decision makers with additional preferred solutions that map different preferred objective regions, whereas we aim at allowing the decision maker to navigate through the Pareto front and to locally explore nearby solutions, hence refining his/her preferences and eventually discovering new preferred solutions. This is similar to the so-called NIMBUS system [25], but our approach allows to include the preferences of the DM modeled using a direction in the objective space more explicitly.

3   Fine Tuning Method and Application to Knapsack

3.1   Basic Idea and Motivations

In the following, we introduce the design principles of the proposed method. Basically, the motivations of our proposal is to allow the DM to navigate along the Pareto front by starting from a given initial solution x_0, possibly coming from the output of any available optimization technique. This initial solution x_0 is to be viewed as a departure point in the objective space from where the DM can refine his/her preferences by discovering on-line the vicinity of x_0, and eventually finding new preferred points in the objective space. Consequently, we shall provide the DM with the necessary tools in order to explore a whole path of solutions being as near as possible to the Pareto front and to locally explore the landscape of Pareto optimal solutions in an iterative manner, i.e., from one solution to a nearby one. For this purpose, the DM is required to provide a direction in the objective space, in which the search process shall be steered. Informally speaking, steering the search along the given direction means providing the DM with a sequence of candidate solutions x_i, $i = 1, \ldots, N$, that can be viewed as forming a path in the objective space and such that every solution x_i is improving the previous solution x_{i-1} with respect to the direction given by the DM in the objective space.

In order to effectively set up this idea for combinatorial MOPs, we need to precisely define the role of the direction provided by the DM and the meaning of improving a given solution according to this direction. Before going further, let us comment that specifying a direction with respect to a starting solution can easily be thought by the DM in many different ways and for different purposes. For the sake of illustration, let us consider a MOP with three objectives (f_1, f_2, f_3) and the following scenario: the DM has an optimal solution for this problem that he/she is not fully satisfied with, e.g., he/she would like to minimize the value of f_2 as much as possible. However, a lot of options can be considered for the above example. For instance, the vector $d_1 = (1, -1, 0)$ can refer to a direction (in the objective space) aiming at reducing the second objective, while increasing the first one. Similarly, the direction $d_2 = (0, -1, 1)$ would imply a reduction of the second objective together with a growth on the third objective. In both cases, we could obtain the minimum value for f_2, but following the direction d_1 or d_2 typically produces different paths that can be associated with different DM preferences. By defining a direction, the DM can actually decides which objectives to improve and which ones to "sacrifice" in order to refine his/her preferences. Performing such local movements in the objective space while following a whole path of Pareto optimal solutions with respect to the preferred direction of the DM is the goal of the proposed framework.

For the target framework to work, it is important to keep in mind the notion of Pareto optimality when performing a movement in the objective space. For instance, assuming that the starting solution is a Pareto optimal solution, then it is obviously not possible to improve all the objectives simultaneously. Consequently, the DM can still define a direction which does not involve an optimal movement, because *no prior knowledge* on the shape of the Pareto front is assumed. In the rest of this section, we provide a step-by-step description of the proposed framework and the necessary algorithmic components for its proper realization. This shall also allow us to better highlight the different issues one has to address in such context.

3.2   Framework for the Fine Tuning Method

As mentioned above, the fine tuning method proposed in this paper is based on the assumption that the DM has a preferred direction d in the *objective space*. More formally, given an initial solution x_0, we assume that the DM is interested in a solution x_1 which is in the vicinity of x_0 such that the following holds:

$$F(x_1) \approx F(x_0) + td, \tag{2}$$

where $t > 0$ is a given (typically small) step size.

As it is very unlikely that such a point x_1 exists where the exact equality in Eq. (2) holds (note also that the Pareto front around $F(x_0)$ is not known),

---

Algorithm 1. Fine Tuning Framework

Require: starting point x_0
Ensure: : sequence $\{x_i\}$ of candidate solutions
   for $i = 1, 2, \ldots$ **do**
      Let $d \in \mathbb{R}_+^m$ ▷ Direction for search in objective space
      Let $\delta \in \mathbb{R}_+$ ▷ Step size in objective space
      $Z_i = \text{REFPOINT}(x_{i-1}, d)$ ▷ Reference point in step i
      $\text{SOP}_i = G(Z_i)$ ▷ Next single objective problem to solve
      $x_i = \text{EMO\_OPTIMIZER}(\text{SOP}_i)$ ▷ evolutionary search for the next solution
   end for

---

we propose to consider a 'best approximation' using an 'approximated' *reference point* Z_1 as follows:

$$Z_1 := F(x_0) + \bar{t}d, \tag{3}$$

where $\bar{t} > 0$ is a given, fixed (problem dependent) step size. Then, we propose to compute the 'closest' Pareto optimal solution to the reference point Z_1 in the objective space, which will hence constitute the next point x_1 to be presented to the DM. Notice that we still have to define a metric specifying the closeness of optimal points with respect to the reference point – this will be addressed later. Once x_1 is computed, the DM can consequently change his/her mind or not, by providing a new direction or by keeping the old one. The framework then keeps updating the sequence of reference points and providing the DM with the corresponding closest optimal solutions in an interactive manner. The proposed framework is hence able to provide the DM with a sequence of candidate solutions such that the respective sequence of objective vectors (ideally) performs a movement in the specified direction d.

In Algorithm 1, we summarize the high-level pseudo code of the proposed method. We first remark that the procedure REFPOINT implements the idea of transforming the direction d provided by the DM into a reference point, which we simply define as follows:

$$Z_i = F(x_{i-1}) + \delta d, \tag{4}$$

where x_{i-1} is the previous (starting) solution and δ a parameter specifying the magnitude of the movement in the objective space. Actually, δ can be viewed as the preferred Euclidian distance between two consecutive solutions $\|F(x_{i-1}) - F(x_i)\|$ in the objective space. It hence defines the preferred step size of the required movement, which is kept at the discretion of the DM. Notice also that both the direction d and the step size δ can be changed interactively by the DM, which we do not include explicitly in the framework of Algorithm 1 for the sake of simplicity.

Given this reference point, one has to specify more concretely which solution should be sought for the decision maker. This is modeled by function G, which takes the current reference point into account and output a (single-objective) *scalar optimization problem* (SOP) to be solved. At last, the procedure

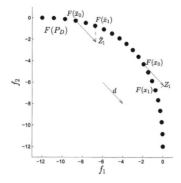

Fig. 1. Illustrative example

EMO_OPTIMIZER refers to the (evolutionary) algorithm that effectively computes the next solution to be presented to the DM. At this stage of the presentation, it is still not fully clear how to define function G and how to effectively implement the evolutionary solving procedure, which is at the core of this paper. Before going into the technical details of these crucially important issues, let us comment on Fig. 1 showing two hypothetical scenarios in the two-objective case, chosen for the sake of a better visualization. For $F(\bar{x}_0)$, the reference point $\overline{Z}_1$ is feasible when choosing the direction $d = (1, -1)^T$. That is, there exists a point $\bar{x}$ such that $F(\bar{x}) = \overline{Z}_1$. We want a function $G(\overline{Z}_1)$ that prevents x to be actually chosen. Instead, the solution $\bar{x}_1$ should be a natural candidate since it is a Pareto optimal solution where $F(\bar{x}_1)$ is the closest element to $\overline{Z}_1$ in the Pareto front. The second scenario is for a given point x_0 such that Z_1 is infeasible. Here it is clear that the solution of $G(Z_1)$ must be a Pareto optimal solution whose image $F(x_1)$ is the closest to the given reference point. Notice that in both cases, the Pareto Front is not known when defining function $G(Z)$.

In the following, we propose a possible answer for the definition of G, as well as some alternative (single- and multi-objective) evolutionary procedures for solving the corresponding SOP.

3.3   Framework Instantiation

Defining the Next Single-Objective Problem to Be Solved. Since the direction provided by the DM could be arbitrary, and given that we do not assume any prior knowledge neither about the Pareto front, nor on the initial solution from where to steer the search, we propose the following modeling of function $G(x|Z)$, defining the next point to be computed by our framework. We rely on the so-called *Wierzbicki's achievement scalarizing function* (WASF) [30]. More precisely, let $\lambda \in \mathbb{R}^m$ be a weighting coefficient vector (which is different from the direction d provided by the DM). Given a reference point Z, the WASF is defined as follows:

$$g(x|Z, \lambda) := \max_{i=1,\ldots,k} \{\lambda_i(f_i(x) - Z_i)\} + \rho \sum_{i=1}^{k} \lambda_i(f_i(x) - Z_i), \qquad (5)$$

where the parameter ρ is the so-called *augmentation coefficient*, that must be set to a small positive value. The motivation of using such a function is that the optimal solution to Problem (5) is a Pareto optimal solution [24], independently of the choice of the reference point. This is an interesting property of the WASF that allows us to deal with reference points that might be defined on the feasible or the infeasible region of the objective space. Notice that this is to contrast to other scalarizing functions, such as the widely-used Chebychev function, that constraint the reference point to be defined beyond the Pareto front.

In our framework, the WASF is intended to capture the DM preferences, expressed by the reference point computed with respect to the DM preferred direction. However, the weighting coefficient vector still has to be specified. It is known that for a given reference point, the solutions generated using different weight vectors are intended to produce a diverse set of solution in the objective space. We here choose to set the weight vector λ as $(1/k, 1/k, \ldots, 1/k)$, which can be viewed as one empirical choice implying a relative fairness among the objectives while approaching the reference point.

The Evolutionary Solving Process. In order to solve the previously defined SOP, we investigate two alternative evolutionary approaches.

The first one consists in using a standard *Genetic Algorithm* (GA). More precisely, and with respect to the experimented knapsack problem, we use the same evolutionary mechanisms and parameters than [2], i.e., a parent selection via a random binary tournament with probability 0.7, an elitist replacement strategy that keeps the best individual, a binary crossover operator with probability 0.5, a single point mutation, and an improve and repair procedure [2] for handling the capacity constraints. However, the initial population of the GA is adapted with respect to the iterations of our proposed framework as follows. Each time the SOP defined by the WASF and the corresponding reference point is updated, we initialize the population with 1/4 of the best individuals from the previous iteration, that we complement with randomly generated individuals. In the first iteration of our framework, the initial population is generated randomly. In our preliminary experiments, this was important in order to obtain a good trade-off between quality and diversity within the evolutionary process. Actually, this observation leads us to consider the following alternative evolutionary algorithm, where population diversity is maintained in a more explicit manner, by using an MOEA for solving the target SOP.

More precisely, our second alternative solving procedure is based on an adaptation of the MOEA/D framework [32]. We recall that MOEA/D is based on the decomposition of a given MOP into multiple subproblems using different weight vectors, which are then solved cooperatively. In contrast to the original algorithm, where the entire Pareto front is approximated using an ideal reference point and a diverse set of weight vectors, typically generated in a uniform way in the objective space; we are here interested in a single solution with respect to the target reference point. Hence, we still consider a set of uniformly-distributed weight vectors, but we use the WASF where the reference point is fixed in order to focus the search process on the region of interest for the DM. At the end of the MOEA/D search process, we output the best-found solution for the weight vector $\lambda = (1/k, 1/k, \ldots, 1/k)$, which precisely corresponds to the target SOP defined with respect to the DM preferred direction. Similarly to the GA, our preliminary experiments revealed that the choice of the initial population for MOEA/D has an important impact on the quality of the target solution. Accordingly, apart from the first iteration where the initial population is generated at random, we choose to systematically initialize MOEA/D with the population obtained with respect to the previous reference point. Due to the explicit diversity of the MOEA/D population, this initialization strategy revealed a reasonable choice in our initial experiments.

Illustrative Scenarios. To exemplify the possible scenarios, we experiment in the following the tuning method on the MOKP with different assumptions. This is in order to highlight the behavior of the framework under some possible representative scenarios and to identify the main raised issues.

We define the exemplary scenarios changing the input values of the fine tuning method, i.e., the initial optimal solution $F(x_0)$, the direction in objective space d and the step size δ. For each investigated scenario, we provide plots rendering the computed reference points, the projection of the selected solutions (x_i) in the objective space, and the final population of each of the two considered evolutionary algorithms, together with the best-known PF approximation. This is reported in Fig. 2 for a bi-objective MOKP instance from [33]. Notice that, since we are interested in the impact of the input parameters, we assume that the initial solution x_0 could be optimal or not, which implies that the first-obtained reference point can also be optimal or not. Thus, by simplicity we omit the first update of the reference point, and we consider that $Z_1 = F(x_0)$. At last, we consider 10 iterations of the proposed method, a population size of 150, and 7, 500 function evaluations when running the solving procedure in each iteration.

In Fig. 2 (left), we consider a Pareto optimal solution as a starting point and a fixed direction vector (provided by the DM) corresponding to the scenario where the second objective is to be refined repeatedly. We can clearly see that running the proposed framework is able to gradually improve the output solutions and to effectively steer the search along the desirable input direction. However, the output solutions are not necessary optimal, which we clearly attribute to the relatively few amount of computational effort used when running the evolutionary solving procedure. Interestingly, using the MOEA/D algorithm as a solving procedure (bottom) for the single-objective reference point target problem appears to work much better than the single-objective GA (top). We clearly attribute this to the diversity issues that the evolutionary process is facing when trying to find Pareto optimal solutions. This is confirmed in our second scenario depicted

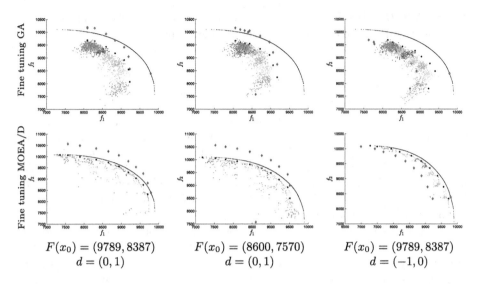

$$F(x_0) = (9789, 8387)$$
$$d = (0, 1)$$

$$F(x_0) = (8600, 7570)$$
$$d = (0, 1)$$

$$F(x_0) = (9789, 8387)$$
$$d = (-1, 0)$$

Fig. 2. Illustrative scenarios on a bi-objectives. The best-known PF approximation is in thin black points. Reference points are shown in red squared points. The output solutions are in shown as circled black point. The population is depicted using a variable color scale. $\delta = 500$. (Color figure online)

in Fig. 2 (middle), where the initial solution is chosen to be a non-optimal one. This second scenario also demonstrates that the proposed approach behaves in a coherent manner even if the solution considered in each iteration is not optimal. Notice that these two scenarios consider the same preferred direction, which is actually pointing to regions where there exist some non-dominated points. In Fig. 2 (right), we instead consider the scenario where a non-optimal direction $d = (-1, 0)$ is provided, that is a direction that points towards a dominated region of the objective space. This leads to the critical situation where the computed reference point might be dominated. Again, we notice that the proposed approach can handle this situation properly and that the MOEA/D-based solving procedure performs better than the GA.

4   Numerical Results

In this section, we present some numerical results of our approach using the modified (fine tuning) MOEA/D as a solver, since it was shown to provide better performance than the fine tuning GA. In order to appreciate the behavior of the proposed method, we consider to compare it against the original MOEA/D algorithm. However, since the original MOEA/D is intended to compute an approximation of the whole PF, a special care has to be taken. First, the proposed method enables to only output a path of solutions based on the computed reference points. Hence, we consider a modification of the Inverted Generational

Distance (IGD_Z) [19], which allows us to work with a set of reference points. More specifically, given the set Z of reference points and a reference set archive A, the distance of Z toward $F(A)$ is measured as follows:

$$IGD_Z(F(A), Z) := \frac{1}{|Z|} \sum_{i=1}^{|Z|} \min_{j=1,\dots,|A|} dist(Z_i^*, F(a_j)), \tag{6}$$

where Z_i^* denotes the point from the PF which is closest to Z_i, i.e., $\|Z_i - Z_i^*\| = dist(Z, F(P_Q))$. Hereby, $dist$ measures the distance between point and set and between two sets as $dist(u, A) = \inf_{v \in A} \|u - v\|$ and $dist(B, A) = \sup_{u \in B} dist(u, A)$, where u and v denote points from sets A and B. Like this, the optimal IGD_Z value is always zero. Notice, however, that the evaluation of the IGD_Z value requires the knowledge of the exact PF. Instead, we use the best-available PF approximations.

The value of IGD_Z can be straightforwardly computed for our approach using the set of reference points computed at each iteration. For the original MOEA/D, we consider to first extract from the archive maintained by MOEA/D the nearest solutions (in the objective space) to the same reference points computed by our approach. Then, these solutions are considered in order to compute an IGD_Z value for the original MOEA/D. By comparing the IGD_Z values for our method as well as for the original MOEA/D, our intent is to highlight the benefits that can be expected when *locally* steering the search along a preferred direction in an interactive way, against computing a *global* approximation set form which we steer the search *a posteriori*. It is however worth-noticing that such a comparison is only conducted for the sake of illustrating the accuracy of our approach and its effective implementation which should not be considered as an alternative to existing (global) multi-objective optimization algorithms.

In the following, we consider some benchmark instances of the considered MOKP[1], as specified in Table 1, which also summarizes the different parameter setting used for the proposed method. MOEA/D was experimented using the same setting than in the original paper [32]. Notice that, overall, the same number of function evaluations are used for both the original MOEA/D and the proposed method. Table 2 shows the obtained results for the consider scenarios over 20 independent runs for each instance and algorithm.

We notice that, for $k = 2$, the original MOEA/D is able to obtain better results than the proposed Fine Tuning method. This is because MOEA/D can generate points close to the entire PF when the number of objectives is limited. However, we can observe that, the larger the number of objectives, the better the Fine Tuning approach. This is because MOEA/D requires more approximation points and function evaluations in order to cover the entire PF as the dimensionality grows, while the Fine Tuning approach is able to naturally focus on certain regions of the PF. We remark that this fact also improves the execution time, because the Fine Tuning approach does not require any external archive, while for MOEA/D to output a high-quality global PF approximation, an archive is actually used for the MOKP.

---

[1] http://www.tik.ee.ethz.ch/sop/download/supplementary/testProblemSuite/.

Table 1. Parameter setting of the fine tuning method for numerical results. Number of knapsacks (KS), items, population size (P), maximal number of functions evaluations for each reference point (ZEvs), number of considered reference points ($|Z|$), step size δ, initial reference point Z_0 and desirable direction d_k.

KS	Items	P	ZEvs	$|Z|$	δ	Z_0	d_k
2	250	150	7500	10	300	$(10000, 8000)$	$(0, 1)$
2	500	200	10000	10	300	$(16000, 19000)$	$(1, 0)$
3	100	351	17600	10	200	$(4056, 3314, 3228)$	$(0, 1, 1)$
3	100	351	17600	10	200	$(4056, 3314, 3228)$	$(0, 0, 1)$
4	500	455	17500	10	300	$(13643, 14224, 16968, 16395)$	$(1, 1, 0, 0)$
4	500	455	17500	10	300	$(16716, 16867, 14178, 13234)$	$(0, 0, 1, 1)$

Table 2. Numerical results. Number of knapsacks (KS), items, population size (P), maximal number of functions evaluations (Ev), and IGD_Z; minimum, average, standard deviation (in small font) and maximum of 20 independent runs are presented for IGD_Z.

KS	Items	P	Ev	IGD_Z							
				Finite Tuning				Original MOEA/D			
2	250	150	75000	69.86	80.73	(7.02)	90.22	**40.34**	**47.51**	(4.73)	**59.01**
2	500	200	100000	241.45	271.04	(15.34)	291.85	**153.12**	**173.79**	(10.89)	**192.08**
3	100	351	176000	35.84	**41.62**	(3.56)	47.93	**34.05**	46.82	(5.13)	57.03
3	100	351	176000	**24.10**	**30.85**	(4.82)	41.29	25.63	32.54	(4.35)	**40.79**
4	500	455	175000	**184.73**	**228.05**	(26.95)	**289.81**	365.29	494.72	(66.79)	577.90
4	500	455	175000	**144.85**	**198.48**	(20.12)	**231.05**	311.34	463.83	(70.58)	589.30

5   Conclusions and Future Work

In this work, we addressed a decision making tool for discrete many objective optimization problems, where we used the multidimensional multi-objective 0–1 knapsack problem as demonstrator. Since the number of non-dominated solutions grows (even exponentially) with the number of objectives k, it becomes difficult or even intractable to compute an approximation of the entire Pareto set for $k \geqslant 4$. Instead, it is likely that each of the chosen solutions obtained by a solver does not represent the most-preferred one from the set of given optimal alternatives. In order to overcome this issue, we proposed a local fine tuning method that allows the search process to be steered from a given solution along the Pareto front in a user-specified direction. More precisely, we presented a framework and two possible realizations of it: one by means of a GA for directly solving the dynamic reference point problem, and another one based on MOEA/D that focuses on a region of the Pareto front delimited by the reference point. Given that only a particular segment of the Pareto front is computed, one retrieves a much more accurate search efficiency compared against the classical method (i.e., aiming to compute *all* Pareto optimal solutions), which we have demonstrated on several benchmark problems. We think that this method can be used as a post-processing step to all existing many objective optimization solvers, and that this will actually help the decision maker to identify his/her most-preferred solution.

Though this work demonstrates as proof-of-principle the application of the novel approach, there is still much to be done. First of all, the tuning of the method is an issue to guarantee in order to get better solutions in lower time, as well as other solvers might be interesting to consider. Next, we think that the tuning method can be extended by other approaches to steer the search along the Pareto front. Finally, it would be interesting to apply the novel method on many objective optimization problems derived from real-worlds applications in order to appreciate its impact on the decision making process.

References

1. Aguirre, H., Tanaka, K.: Many-objective optimization by space partitioning and adaptive ε-ranking on MNK-landscapes. In: Ehrgott, M., Fonseca, C.M., Gandibleux, X., Hao, J.-K., Sevaux, M. (eds.) EMO 2009. LNCS, vol. 5467, pp. 407–422. Springer, Heidelberg (2009). doi:10.1007/978-3-642-01020-0_33
2. Alves, M., Almeida, M.: MOTGA: a multiobjective Tchebycheff based genetic algorithm for the multidimensional knapsack problem. Comput. Oper. Res. **34**(11), 3458–3470 (2007)
3. Auger, A., Bader, J., Brockhoff, D., Zitzler, E.: Articulating user preferences in many-objective problems by sampling the weighted hypervolume. In: Proceedings of the 11th Annual Conference on Genetic and Evolutionary Computation, pp. 555–562. ACM (2009)
4. Bader, J., Zitzler, E.: Hype: an algorithm for fast hypervolume-based many-objective optimization. Evol. Comput. **19**(1), 45–76 (2011)
5. Beume, N., Naujoks, B., Emmerich, M.: SMS-EMOA: multiobjective selection based on dominated hypervolume. EJOR **181**(3), 1653–1669 (2007)
6. Branke, J., Deb, K.: Integrating user preferences into evolutionary multi-objective optimization. In: Jin, Y. (ed.) Knowledge Incorporation in Evolutionary Computation, pp. 461–477. Springer, Heidelberg (2005)
7. Brockhoff, D., Saxena, D.K., Deb, K., Zitzler, E.: On handling a large number of objectives a posteriori and during optimization. In: Knowles, J., Corne, D., Deb, K., Chair, D.R. (eds.) Multiobjective Problem Solving from Nature, pp. 377–403. Springer, Heidelberg (2008)
8. Cheng, R., Jin, Y., Olhofer, M., Sendhoff, B.: A reference vector guided evolutionary algorithm for many-objective optimization. IEEE Trans. Evol. Comput. **20**(5), 773–791 (2016)
9. Coello, C.: Handling preferences in evolutionary multiobjective optimization: a survey. In: Proceedings of the 2000 Congress on Evolutionary Computation, vol. 1, pp. 30–37 (2000)
10. Coello, C., Lamont, G., Van Veldhuizen, D.: Evolutionary Algorithms for Solving Multi-objective Problems, 2nd edn. Springer, Heidelberg (2007)
11. Coello, C., Van Veldhuizen, D., Lamont, G.: Evolutionary Algorithms for Solving Multi-objective Problems, vol. 242. Springer, Heidelberg (2002)

12. Deb, K.: Multi-objective Optimization Using Evolutionary Algorithms, vol. 16. Wiley, Hoboken (2001)

13. Deb, K., Jain, H.: An evolutionary many-objective optimization algorithm using reference-point-based nondominated sorting approach, part I: solving problems with box constraints. IEEE Trans. Evol. Comput. **18**(4), 577–601 (2014)

14. Deb, K., Pratap, A., Agarwal, S., Meyarivan, T.: A fast and elitist multiobjective genetic algorithm: NSGA-II. IEEE Trans. Evol. Comput. **6**(2), 182–197 (2002)

15. Di Pierro, F., Khu, S.T., Savic, D., et al.: An investigation on preference order ranking scheme for multiobjective evolutionary optimization. IEEE Trans. Evol. Comput. **11**(1), 17–45 (2007)

16. Ehrgott, M.: Multicriteria Optimization, 2nd edn. Springer, Heidelberg (2005)

17. Farina, M., Amato, P.: On the optimal solution definition for many-criteria optimization problems. In: Proceedings of the NAFIPS-FLINT International Conference, pp. 233–238 (2002)

18. Giagkiozis, I., Fleming, P.: Pareto front estimation for decision making. Evol. Comput. **22**(4), 651–678 (2014)

19. Hernández Mejía, J.A., Schütze, O., Cuate, O., Lara, A., Deb, K.: RDS-NSGA-II: a memetic algorithm for reference point based multi-objective optimization. Eng. Optim. 1–18 (2016)

20. Ishibuchi, H., Sakane, Y., Tsukamoto, N., Nojima, Y.: Evolutionary many-objective optimization by NSGA-II and MOEA/D with large populations. In: IEEE International Conference on Systems, Man and Cybernetics, SMC 2009, pp. 1758–1763. IEEE (2009)

21. Knowles, J., Corne, D.: Quantifying the effects of objective space dimension in evolutionary multiobjective optimization. In: Obayashi, S., Deb, K., Poloni, C., Hiroyasu, T., Murata, T. (eds.) EMO 2007. LNCS, vol. 4403, pp. 757–771. Springer, Heidelberg (2007). doi:10.1007/978-3-540-70928-2_57

22. Marler, R., Arora, J.: The weighted sum method for multi-objective optimization: new insights. Struct. Multi. Optim. **41**(6), 853–862 (2010)

23. Martín, A., Schütze, O.: A new predictor corrector variant for unconstrained bi-objective optimization problems. In: Tantar, A.-A., et al. (eds.) EVOLVE - A Bridge Between Probability, Set Oriented Numerics, and Evolutionary Computation V. AISC, vol. 288, pp. 165–179. Springer, Heidelberg (2014). doi:10.1007/978-3-319-07494-8_12

24. Miettinen, K.: Nonlinear Multiobjective Optimization. Kluwer, Boston (1999)

25. Miettinen, K., Mäkelä, M.M.: Interactive multiobjective optimization system WWW-NIMBUS on the Internet. Comput. Oper. Res. **27**(7), 709–723 (2000)

26. Miettinen, K., Ruiz, F., Wierzbicki, A.P.: Introduction to multiobjective optimization: interactive approaches. In: Branke, J., Deb, K., Miettinen, K., Słowiński, R. (eds.) Multiobjective Optimization. LNCS, vol. 5252, pp. 27–57. Springer, Heidelberg (2008). doi:10.1007/978-3-540-88908-3_2

27. Purshouse, R., Fleming, P.: On the evolutionary optimization of many conflicting objectives. IEEE Trans. Evol. Comput. **11**(6), 770–784 (2007)

28. Schütze, O., Martin, A., Lara, A., Alvarado, S., Salinas, E., Coello, C.A.: The directed search method for multiobjective memetic algorithms. Comput. Optim. Appl. **63**, 305–332 (2016)

29. Singh, H.K., Isaacs, A., Ray, T.: A Pareto corner search evolutionary algorithm and dimensionality reduction in many-objective optimization problems. IEEE Trans. Evol. Comput. **15**(4), 539–556 (2011)

30. Wierzbicki, A.: The use of reference objectives in multiobjective optimization. In: Fandel, G., Gal, T. (eds.) Multiple Criteria Decision Making Theory and Application, pp. 468–486. Springer, Heidelberg (1980)
31. Yang, S., Li, M., Liu, X., Zheng, J.: A grid-based evolutionary algorithm for many-objective optimization. IEEE Trans. Evol. Comput. **17**(5), 721–736 (2013)
32. Zhang, Q., Li, H.: MOEA/D: a multiobjective evolutionary algorithm based on decomposition. IEEE Trans. Evol. Comput. **11**(6), 712–731 (2007)
33. Zitzler, E., Thiele, L.: Multiobjective evolutionary algorithms: a comparative case study and the strength Pareto approach. IEEE Trans. Evol. Comput. **3**(4), 257–271 (1999)
34. Zou, X., Chen, Y., Liu, M., Kang, L.: A new evolutionary algorithm for solving many-objective optimization problems. IEEE Trans. Syst. Man Cyber. Part B Cybern. **38**(5), 1402–1412 (2008)

A Note on the Detection of Outliers in a Binary Outranking Relation

Yves De Smet$^{(\boxtimes)}$, Jean-Philippe Hubinont, and Jean Rosenfeld

CoDE Department, Ecole Polytechnique de Bruxelles,
Université libre de Bruxelles, Brussels, Belgium
`yves.de.smet@ulb.ac.de`

Abstract. We address the problem of outliers detection in a binary outranking relation. These elements are supposed to be rare, dissimilar to the majority of other elements and are likely to influence the outcomes of the considered method. We propose a model based on the distance introduced by De Smet and Montano and extend it to different samplings of the set of alternatives (which are used as a comparison basis). This leads to study the distribution of distance values. The presence of outliers is detected by the identification of bi-modal distributions. We illustrate this on examples based on the Human Development Index, the Environmental Performance Index (where artificial outliers are added) and the Shanghai Ranking of World Universities.

1 Introduction

Many strategic decision problems involve the simultaneous optimization of several conflicting criteria. For instance, in the design of integrated circuits: a manufacturer will try to simultaneously maximize the performance while minimize the cost of the chip. Also, producing high-end integrated circuits can be subject to more difficulties in terms of thermal dissipation. Finally, a criterion based on ecological standards may have impacts on the cost and the performances [5]. In road design problems [10], one tries to simultaneously minimize investment and operational costs, maximize security (for all types of users), optimize environmental impacts (greenhouse gases emissions, noise, etc.), improve traffic mobility, etc. A huge number of multicriteria applications have been reported in the literature [1,12]. To manage these problems, researchers have developed numerous multicriteria methods (and software). Among them, one usually distinguishes three main families [11]: Multiple Attribute Utility Theory (MAUT) [6], Interactive and Outranking methods [2,7]. In what follows, we will focus ourselves on binary outranking methods. These are based on the pairwise comparison of alternatives leading to state if a given alternative a_i is at least as good as another alternative a_j or not. This will be denoted by $a_i S a_j$. Different approaches have been proposed in order to exploit this binary relation in the context of ranking, choice or sorting [7].

More recently, researchers working on multiple criteria analysis have also investigated issues related to robustness [8]. The origin of these works lies in

© Springer International Publishing AG 2017
H. Trautmann et al. (Eds.): EMO 2017, LNCS 10173, pp. 151–159, 2017.
DOI: 10.1007/978-3-319-54157-0_11

the fact that most multiple criteria methods are based on the instantiation of (preference) parameters that are likely to affect (or not) the results (for instance, a small modification of a given weight value could have an impact on the first ranked alternative). It is thus important to assess if a given recommendation is robust or not.

The study of robustness has already a long history in the field of statistics, leading to the development of robust estimators and the detection of so-called *outliers*. For instance, adding only one arbitrarily high (or low) observation to a given sample is sufficient to completely modify the value of the mean and the standard deviation. Then, computing z-scores (for instance) will not help to properly detect the outlier (as the estimators are themselves influenced by it) [9].

The aim of this work is to investigate how to transpose this notion in the particular context of binary outranking problems. To the best of our knowledge, the notion of *multicriteria outlier* has not been treated yet in the literature. Without being exhaustive, this work could constitute a first step in this new research direction.

At this point, it is worth noting that the definition of outlier is not clearly formalized (even in statistics). We will thus call an outlier an element that is:

- not frequent;
- different from the majority of the remaining elements;
- likely to strongly affect the outcome of the considered method.

As in statistics, the detection of outliers can be based on a distance measure. Of course, it is known that the presence of outliers is likely to influence the estimation of the parameters characterizing the distance (for instance the covariance matrix in the case of the Mahalanobis distance). Therefore, researchers often rely on the computation of robust parameters based on samplings of the set of alternatives. We adopt a similar approach in this contribution. Initially, the distance between two alternatives is computed based on the preferential relations they share with all the other alternatives. In order to limit the effect of outliers, one relies on comparisons based on samplings of the set of alternatives (as outliers are assumed to be not frequent, the probability to take them into account in the comparison is low). This model is developed in Sect. 2 and then illustrated, in Sect. 3, on three simplified examples (based on the Environmental Performance Index, the Human Development Index and the Shanghai Ranking of World Universities).

2   The Model

Let us consider a set of n alternatives, denoted $A = \{a_1, a_2, \ldots, a_n\}$, and a set of q criteria denoted $F = \{f_1, f_2, \ldots, f_q\}$ (without loss of generality we consider that the criteria have to be maximized). One assumes that the decision maker is able to build a binary outranking relation denoted S on the set of alternatives.

This can be done, for instance, by using the ELECTRE I method (see the details in Sect. 3).

When comparing two alternatives, one may work at the *local* or the *global* level. In the first case, one only considers two given alternatives and investigates the preferential statements:

- $a_i P a_j$: a_i is preferred to a_j if $a_i S a_j$ and $a_j \not S a_i$;
- $a_j P a_i$: a_j is preferred to a_i if $a_j S a_i$ and $a_i \not S a_j$;
- $a_i I a_j$: a_i is indifferent to a_j if $a_i S a_j$ and $a_j S a_i$;
- $a_i R a_j$: a_i is incomparable to a_j if $a_i \not S a_j$ and $a_j \not S a_i$;

In the second case, one could assess if two given alternatives behave in the same way with respect to all the other alternatives. This idea has been introduced by De Smet and Montano [4] while developing a procedure for multicriteria clustering. It has then been refined by De Smet and Eppe [3]. Therefore, each alternative has to be characterized with respect to the whole set A. This will be referred to as the *profile* of the alternative.

Definition 1. *The profile $P(a_i)$ of action $a_i \in A$ is defined as being a 4-uple $< R(a_i), P^-(a_i), I(a_i), P^+(a_i) >$ where:*

- $R(a_i) = \{a_j \in A | a_i R a_j\} = P_1(a_i)$
- $P^-(a_i) = \{a_j \in A | a_j P a_i\} = P_2(a_i)$
- $I(a_i) = \{a_j \in A | a_i I a_j\} = P_3(a_i)$
- $P^+(a_i) = \{a_j \in A | a_i P a_j\} = P_4(a_i)$

Given this concept, one can characterize the similarity between two alternatives by assessing if they share common relations with other alternatives. This can be done by using the following indicator.

Definition 2. *Let $P(a_i)$ be the profile of a_i, the distance between two actions $a_i, a_j \in A$ is defined as follows:*

$$d_A(a_i, a_j) = 1 - \frac{\sum_{k=1}^{4} |P_k(a_i) \cap P_k(a_j)|}{n} \tag{1}$$

It has been proven that d_A satisfies the conditions to be called a distance [4]. Of course, it depends on the whole set A which is used as a basis for the comparison. As a consequence, the presence of outliers in A may have an impact on the values of $d_A(a_i, a_j)$ and so lower the discrimination power of this indicator. Such kind of impacts are well-known in statistics. For instance, the presence of an outlier can strongly affect the parameter estimation in a linear regression [9]. In order to lower this effect, estimators are computed on several randomly chosen subsets of A. The underlying idea being that if an outlier is not frequent it is not likely to be integrated in many of these subsets. We adopt a similar approach.

The method consists to randomly select k subsets $B_l \subset A$ (with $l = 1, \ldots, k$ and $|B_l| = m$). For each subset and for each couple of alternatives, one computes $d_{B_l}(a_i, a_j)$. Each alternative is then characterized by the biggest distance compared to any other alternative:

$$\delta(a_i) = \max_{a_j \in A} d_{B_l}(a_i, a_j)$$

Finally, each instance B_l is characterized as follows:

$$\Delta(B_l) = \sum_{i=1}^{m} \delta(a_i)$$

The k repeated tests lead to k different values of $\Delta(B_l)$ that can be represented, for instance, on a histogram. On the one hand if the data set does not encompass any outlier, the selection of a given B_l should not significantly affect the distribution of $\Delta(B_l)$. As a corollary to the central limit theorem, the general shape of the $\Delta(B_l)$ distribution should be close to a normal distribution (at least for high values of n). On the other hand, the presence of outliers in A is likely to affect the distribution of $\Delta(B_l)$ leading, for instance, to bi-modal distributions. This will be illustrated in the next section.

At this point, it is worth noting that the binary S relation used in this section has a clear multicriteria meaning: $a_i S a_j$ iff a_i is at least as good as a_j. This interpretation allows to derive the four main preferential statements $(a_i P a_j, a_j P a_i, a_i I a_j$ or $a_i R a_j)$. Of course, one can imagine other contexts where such a binary relation could be built (without being related to a multicriteria problem). As a consequence, this model is not limited to multicriteria applications.

3   Illustration

There are many ways to build binary outranking relations. In this section, we will use the ELECTRE I method. The decision maker is assumed to provide criteria weights, w_h, a concordance threshold, λ, and veto thresholds v_h. At first the concordance index is computed as follows:

$$C(a_i, a_j) = \sum_{h | f_h(a_i) \geq f_h(a_j)} w_h$$

The higher the concordance index, the higher confidence we have that a_i is at least as good as a_j. Then, the discordance index is computed as follows:

$$V(a_i, a_j) = 1_{\{\exists h | f_h(a_j) - f_h(a_i) > v_h\}}$$

Intuitively, the discordance index is equal to 1 as soon as a_j is much better than a_i on at least one criterion. This indicator plays the role of a veto acting against the fact that a_i could be considered to be at least as good as a_j.

Finally, we will state that $a_i S a_j$ if the two following conditions are satisfied:

$$C(a_i, a_j) > \lambda$$

$$V(a_i, a_j) = 0$$

In order to test the proposed approach, we decided to work with random samples of 10 alternatives coming from the Human Development Index (HDI) and the Environmental Performance Index (EPI). To keep it simple, only two criteria were considered in both cases. For the HDI, the life expectancy at birth (in years - denoted LEB) and the mean years of schooling (in years - denoted MYS) were selected. For the EPI, a global score for environmental health and another one for ecosystem vitality were computed. Then, an eleventh alternative is artificially added. To make sure that this element may be considered as an outlier we proceed as follows: $a_i \in P_k(a_{11})$ iff P_k is the least represented relation for a_i among all the alternatives within A (let us note that this is the opposite procedure than the one developed in [4] where the aim was to build a centroid and so, a representative element). For these first two tests, the subset B is assumed to be constituted by $m = 5$ alternatives and we considered $k = 1000$ (let us note that these values have to be cautiously studied in the future). As a consequence, the probability to select the outlier in a given B_l is nearly equal to 50 %. Therefore, the expected absolute frequency of the high ΔB_l corresponding to the presence of the outlier is slightly lower than 500. Finally, the concordance threshold considered is $\lambda = 0.65$ and the veto thresholds are equal to the third quartile of the differences between each alternative for each criterion separately. For these two first tests, weights were assumed to be equal (let us note that, in this case, the outranking relation is nothing else than the dominance relation enriched by a veto threshold).

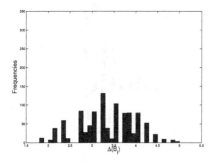

Fig. 1. HDI without outlier

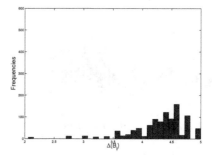

Fig. 2. EPI without outlier

The resulting histograms are shown on Figs. 1, 2, 3 and 4. As expected, when no outlier is added, the histogram seems to follow a normal distribution. Adding

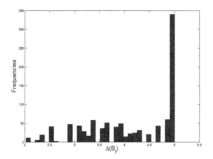

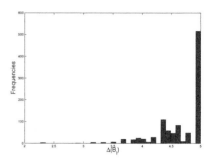

Fig. 3. HDI with an outlier   **Fig. 4.** EPI with an outlier

the outlier leads to modify the histogram by resulting in a bi-modal curve, where the second mode is characterized by the highest ΔB_l values (and is related to the contribution of the outlier during the sampling). One sees that the sum of the frequencies corresponding to the second mode is indeed more or less equal to 500.

Detecting the presence of an outlier is a first step. Now, one has to identify it precisely. A simple way to perform this task is to store each subset B_l and to sort them from the highest ΔB_l values to the lowest. One observes that only the outlier alternative is common to the subsets having the highest ΔB_l values. For all the tests that have been conducted, the method was able to detect the artificial outlier.

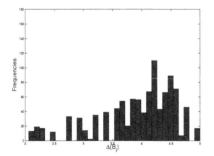

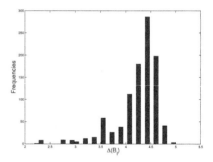

Fig. 5. HDI without outlier   **Fig. 6.** EPI without outlier

Let us note that we added an alternative that can be considered as a "super outlier" (since by construction its profile is in opposition to the profiles of all the other alternatives). In order to verify whether our method is useful in less extreme situations, we considered another construction method (called the *weak outlier construction* method). In this case, the profile of the outlier is built by

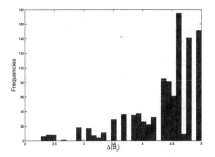

Fig. 7. HDI with a weak outlier

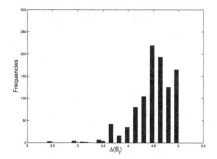

Fig. 8. EPI with a weak outlier

associating (randomly) each alternative among the three preference relations that are the least represented among all the alternatives. The associated results are presented on Figs. 5, 6, 7 and 8. Of course, the impact on the histogram is less pronounced.

At this point, one could question the added value of computing the distribution of $\Delta(A)$ values based on samplings of the set of alternatives. Alternatively, one could simply try to identify outliers based on the maximum values of $\Delta(A)$ (when computed on the entire data set). Tests based on the *weak outlier construction* method have shown that it was not possible to detect the right outlier whereas it was possible with the method based on samples.

Finally, another test was performed on a real problem (without relying on the use of the artificial addition of an outlier) with more alternatives and more criteria. The well-known Shanghai Ranking of World Universities (ARWU) has been selected. Though, only a subset of the alternatives was selected in order to obtain an outlier. Only the best university (Harvard) and the 31th to the 100th best alternatives were considered. Harvard being the outlier in this case, it was

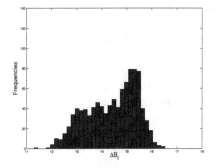

Fig. 9. ARWU without Harvard

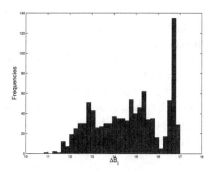

Fig. 10. ARWU with Harvard

withdrawn afterwards to compare the histogram before and after its withdrawal. The subset B has a size equal to 25% of the size of the set A, $\lambda = 0.65$, the veto thresholds are also equal to the 3rd quartile of the differences between each alternative. Finally, the weights are the same as those of the original ranking. The results are shown on Figs. 9 and 10.

One sees that the addition of Harvard in the set of alternatives creates a bi-modal curve on the histogram. The alternatives between the 100th and the 31th are similar while Harvard is by far the best university in the modified data set. When Harvard is in the selected subset, it is, for all the others, the farthest alternative in terms of preference. It explains the very pronounced bimodal form of the Fig. 10.

4 Conclusion

To the best of our knowledge, the concept of *multicriteria outlier* has not yet been treated in the literature. Like in statistics, this concept is strongly related to robustness which is an important issue in decision aid. Starting from this observation, we decided to investigate how we could detect outliers in a binary outranking relation. This is based on the distance defined by De Smet and Montano [4] and sampled over a series of subsets of alternatives (which serves as a basis for the comparison). First experiments have shown that the presence of outliers leads to bi-modal distributions of the $\Delta(B_l)$ indicators. Let us point out that the detection of such bi-modal distributions remains, at this point, purely qualitative. Of course, a number of formalized methods exist to perform this task. Their performances have still to be tested in this specific context.

Of course, there are a number of issues that need to be investigated. First, the tuning phase of the parameters has to be addressed. What are the optimal values for the size of the subsets (m) and the number of repetitions (k)? Then, the model has to be extended to other contexts such as valued outranking relations, MAUT methods, etc.

The formal definition of an outlier is a research topic on its own. For the moment we have considered that these elements are not frequent, different than the majority of other elements and have impacts on the outcomes of the considered methods (in this case the general shape of the Δ distribution). This last point is related to the issue of "rank reversal" (which might be a sub-case of what we call "impacts on the outcomes"). Finally, we have restricted ourselves to the detection of a unique outlier. As expected, the performances of the method could be affected by the presence of multiple outliers (of different kinds). Of course, if they represent a significant portion of the data set, the outlier definition itself can be questionable. This last issue has still to be investigated.

References

1. Behzadian, M., Kazemzadh, R.B., Albadvi, A., Aghdasi, D.: PROMETHEE: a comprehensive literature review on methodologies and applications. Eur. J. Oper. Res. **200**(1), 198–215 (2010)
2. Brans, J.P., De Smet, Y.: Promethee methods. In: Figueira, J., Greco, S., Egrgott, M. (eds.) Multiple Criteria Decision Analysis: State of the Art Surveys, 2nd edn, pp. 187–220. Springer, Boston (2016)
3. De Smet, Y., Eppe, S.: Multicriteria relational clustering: the case of binary outranking matrices. In: Ehrgott, M., Fonseca, C.M., Gandibleux, X., Hao, J.-K., Sevaux, M. (eds.) EMO 2009. LNCS, vol. 5467, pp. 380–392. Springer, Heidelberg (2009). doi:10.1007/978-3-642-01020-0_31
4. De Smet, Y., Montano Guzman, L.: Towards multicriteria clustering: an extension of the k-means algorithm. Eur. J. Oper. Res. **158**(2), 390–398 (2004)
5. Doan, N.A.V., Milojevic, D., Robert, F., De Smet, Y.: A MOO-based methodology for designing 3D-stacked integrated circuits. J. Multi-Criteria Decis. Anal. **21**(1–2), 43–63 (2013)
6. Dyer, J.S.: MAUT multiattribute utility theory. In: Figueira, J., Greco, S., Egrgott, M. (eds.) Multiple Criteria Decision Analysis: State of the Art Surveys, pp. 265–292. Springer, Boston (2005)
7. Figueira, J., Mousseau, V., Roy, B.: ELECTRE methods. In: Figueira, J., Greco, S., Ergott, M. (eds.) Multiple Criteria Decision Analysis: State of the Art Surveys, pp. 133–162. Springer, Boston (2005)
8. Hites, R., De Smet, Y., Risse, N., Salazar-Neumann, M., Vincke, P.: About the applicability of MCDA to some robustness problems. Eur. J. Oper. Res. **174**(1), 322–332 (2006)
9. Rosseeuw, P.J., Leroy, A.L.: Robust Regression and Outlier Detection. Wiley Series in Probability and Mathematical Statistics. Wiley, Hoboken (1987)
10. Sarrazin, R., De Smet, Y.: Design safer and greener road projects by using a multi-objective optimization evolutionary approach. Int. J. Multicriteria Decis. Mak. **61**, 14–33 (2016)
11. Vincke, P.: Multicriteria Decision-Aid. Wiley, New York (1992)
12. Zopounidis, C., Doumpos, M.: Multiple Criteria Decision Making: Applications in Management and Engineering. Springer, Heidelberg (2017, to appear)

Classifying Metamodeling Methods for Evolutionary Multi-objective Optimization: First Results

Kalyanmoy Deb[1]([envelope]), Rayan Hussein[1], Proteek Roy[1], and Gregorio Toscano[2]

[1] Michigan State University, East Lansing, MI 48824, USA
{kdeb,husseinr,royprote}@egr.msu.edu
[2] Information Technology Laboratory, CINVESTAV-Tamaulipas,
87130 Victoria, Tamps, Mexico
gtoscano@cinvestav.mx
http://www.coin-laboratory.com/

Abstract. In many practical optimization problems, evaluation of objectives and constraints often involve computationally expensive procedures. To handle such problems, a metamodel-assisted approach is usually used to complete an optimization run in a reasonable amount of time. A metamodel is an approximate mathematical model of an objective or a constrained function which is constructed with a handful of solutions evaluated exactly. However, when comes to solving multi-objective optimization problems involving numerous constraints, it may be too much to metamodel each and every objective and constrained function independently. The cumulative effect of errors from each metamodel may turn out to be detrimental for the accuracy of the overall optimization procedure. In this paper, we propose a taxonomy of various metamodeling methodologies for multi-objective optimization and provide a comparative study by discussing advantages and disadvantages of each method. The first results presented in this paper are obtained using the well-known Kriging metamodeling approach. Based on our proposed taxonomy and an extensive literature search, we also highlight new and promising methods for multi-objective metamodeling algorithms.

Keywords: Surrogate model · Metamodel · Evolutionary multi-objective optimization · Kriging · Taxonomy

1 Introduction

Many researchers have used metamodels to approximate expensive computer models in an optimization task [1]. The Kriging method is one of the widely used metamodel, which can provide an estimated function value and also simultaneously provide an error estimate of the approximation [15]. The developments in optimization methods have recently led to an increasing interest in metamodeling efforts. Some researchers have made efforts to classify different metamodeling approaches, but only in the realm of single-objective optimization.

© Springer International Publishing AG 2017
H. Trautmann et al. (Eds.): EMO 2017, LNCS 10173, pp. 160–175, 2017.
DOI: 10.1007/978-3-319-54157-0_12

Santana et al. [22] has classified surrogates according to the type of model used (e.g., Kriging, radial basis function, and polynomial regression). Jin [13] proposed a classification based on the way single-objective evolutionary algorithms incorporated surrogate models. Shi and Rasheed [23] classified different metamodeling approaches according to direct or indirect fitness replacement methods. A recent study [11] provided a taxonomy of model-based multi-objective optimization mainly for unconstrained problems. Most metamodeling efforts in multi-objective optimization, so far, seem to have taken a straightforward extension of single-objective metamodeling approaches. First, every objective and constrained function is metamodeled independently. Thereafter, a standard EMO methodology is applied to the metamodels, instead of the original objective and constrained functions, to find the non-dominated front. In some studies, the above metamodeling-EMO combination is repeated in a progressive manner so that refinement of the metamodels can occur with iterations. However, with the possibilities of a combined constrained violation function that can be formulated by combining violations of all constraints in a normalized manner [8] and a combined scalaraized objective function (weighted-sum, achievement variational function or TChebyshev function) [20], different metamodeling methods can certainly be explored. While the straightforward approach requires construction of many metamodels, the suggested metamodels for combined objective and constrained violations will reduce the number of needed metamodels. However, the flip side is that each metamodel of the combined functions is likely to be more complex having discontinuous, non-differentiable, and multi-modal landscapes. Thus, the success of these advanced metamodeling methods is closely tied with the advancements in the metamodeling methods. While these advancements are in progress, in this paper, we outline, for the first time, a number of different and interesting metamodeling methodologies for multi-objective optimization utilizing combined approaches of objectives alone, constraints alone, and objectives and constraints together. Our taxonomy includes one method that requires as many as $(M + J)$ metamodels (where M and J are number of objectives and constraints) to another method that requires only one metamodel. To demonstrate the behavior of each of these metamodeling methods, we present limited results using all six methods and by using Kriging metamodeling approach.

Section 2 describes the proposed taxonomy to classify the different multi-objective metamodeling approaches. Sections 3, 4, 5, and 6 present detail description of each of the six methodologies. Results on unconstrained and constrained test problems are then presented in Sect. 7. Finally, conclusions and extensions of this study are discussed in Sect. 8.

2 Proposed Taxonomy

Multiple and many-objective optimization problems involve a number of (say, M) objective functions ($f_i(\mathbf{x})$) as a function of decision variables $\mathbf{x}$ and a number of (say, J) constrained functions ($g_j(\mathbf{x})$), each as a function of $\mathbf{x}$. For brevity, we do not consider equality constraints in this paper, but with certain modifications, they can be handled in the same way as discussed here.

In this section, for the first time, we propose a taxonomy of various methods of using metamodeling approach in multiple and many-objective optimization algorithms. Our taxonomy finds six different broad methodologies (M1 to M6), as illustrated in Fig. 1. Our approach is based on the cardinality of metamodels for objectives and constraints. In the first method (M1), all objectives and constraints are metamodeled independently, thereby requiring a total of $(M + J)$ metamodels before a multi-objective optimization approach can be applied. This method is a straightforward extension of the single-objective metamodeling studies. Once all such metamodels are constructed, an EMO algorithm can use them to a find one Pareto-optimal solution at a time (like the generative method used in classical optimization literature [20]) and we call this method M1-1, or they can be used to find a number of Pareto-optimal solutions simultaneously (like in evolutionary multi-objective optimization (EMO) literature) and we call this method M1-2.

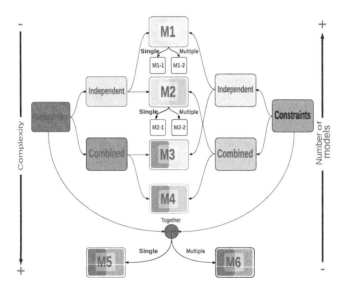

Fig. 1. The proposed taxonomy of six different metamodeling approaches for multiple and many-objective optimization.

The next metamodeling methodology can approximate an overall estimation function of constrained violation, thereby reducing the overall number of metamodels to $(M + 1)$. The well-known normalized, bracket-operator based constrained violation functions [5,8] can be used for this purpose. Like in M1, the constructed metamodels can also be used to find Pareto-optimal solutions as a generative approach (we call it M2-1) or simultaneously (we call it M2-2). The next metamodeling methodology approximates each constrained function independently, but metamodels a combined objective function involving all M objectives. For this purpose, any scaralization approach [20,25] can be used.

Thus, this M3 approach requires $(J + 1)$ metamodels. Since a scaralized formulation finds a single Pareto-optimal solution at a time, the M3 approach, by default, must be applied multiple times, each time finding a single Pareto-optimal solution. Then, our fourth classification (M4) is when only two metamodels are constructed at each iteration, in which one metamodel is for a combined objective function and the second metamodel is made for a combined constrained violation. Due to the use of scalarization function to combine all M objectives, M4 must also be applied multiple times to find a representative set of Pareto-optimal solutions. The methods (M1-1, M2-1, M3, and M4) are ideal for classical point-based optimization algorithms, each requiring multiple applications to find multiple Pareto-optimal solutions. However, methods (M1-2 and M2-2) are ideal for EMO approaches. Methods M3 and M4 can also be followed using other scalaraized EMO approaches as well [20].

A deeper thought will reveal that there could be two more methodologies, in which objectives and constraints are somehow combined to have a single overall *selection* function which when optimized will lead to one or more Pareto-optimal solutions. In M5, the combined selection function has a single optimum coinciding with a specific Pareto-optimal solution and in M6, the combined selection function is multi-modal and makes multiple Pareto-optimal solutions as its optima. Both M5 and M6 methods involve a single metamodel in each iteration, but if K Pareto-optimal solutions are to be found, M5 needs to be applied K times, whereas M6 still involves a single multi-modal metamodel. In EMO algorithms, such as in NSGA-II [7], NSGA-III [9], MOEA/D [28] and others, the combined action of the selection operator involving non-domination and niching operations is an ideal way of visualizing the above-mentioned selection function. In this spirit, we believe that M5 and M6 are uniquely advantageous for EMO approaches.

Thus, it can be observed that our proposed taxonomy classifies metamodeling methods which requires the maximum possible metamodels $(M + J)$ to single metamodel in each iteration. While M6 requires the minimum number of metamodels, this does not come free and it is expected that complexity of the respective metamodels becomes higher from M1 to M6. It then becomes an interesting research task to identify a balance between the number of needed metamodels in each iteration and the reduced complexity of metamodels for particular problem-algorithm combination. In this paper, we do not study the effect of the underlying metamodeling approach, but present results of a particular approach (Kriging) on different problems using all six metamodeling methods.

On a survey of many existing multiple and many-objective metamodeling studies, we have made a classification of them according to our taxonomy. Table 1 shows that the majority of the existing studies use M1 method (in which each objective and constrained function is metamodeled separately) and only a few studies use M2, M3, and M4. Our previous initial proof-of-principle study on M5 is the lone study in this category. There does not exist any past study implementing our M6 approach, which seems to be an interesting and technically challenging proposition. In this paper, we propose one such methodology to spur an interest.

Table 1. Classification of past studies in six proposal methodologies.

Methodology	M1-1	M1-2	M2-1	M2-2	M3	M4	M5	M6
# Studies (of 38)	6	24	0	1	4	2	1	0
Sample citations	[18, 21]	[24, 27]	–	[29]	[2, 17]	[19, 26]	[12]	–

A little thought will prevail that if the multi-objective optimization problem is unconstrained, methodologies M1 and M2 becomes identical and so are M3 and M4. Interestingly, M3, M4 and M5 also become identical to each other. When there is a single objective function, methodologies M1 and M3 are identical and so are M2 and M4. We now provide more description of the adaptations of each of the six methodologies for our initial study here.

3   Methodologies M1 and M2

In these methods, objectives are independently modeled. Thus, any classical generative or any EMO methodology can be applied on the objective metamodels. The difference between M1 and M2 is that in the latter, one constrained violation function is metamodeled. The following constrained violation function $CV(\mathbf{x})$ [8], which accumulates violation of each constrained function $(g_j(\mathbf{x}) \geq 0)$ is used in this study:

$$CV(\mathbf{x}) = \sum_{j=1}^{J} \langle \bar{g}_j(\mathbf{x}) \rangle, \tag{1}$$

where the bracket operator $\langle \alpha \rangle$ is $-\alpha$ if $\alpha < 0$ and zero, otherwise. The function $\bar{g}_j$ is a normalized version of constrained function g_j [4].

In this study, we have used NSGA-II [7] as the EMO methodology, although any other methods could have been used as well. Direct fitness replacement (DFR) [10] has been one of the most straightforward methods to embed surrogate models into MOEAs. DFR assumes that solutions assessed in the surrogate models are comparable to those assessed by the real function (high fidelity function evaluations). DFR is further subdivided into three major model managements [10]: (1) No Evolution control (NEC), which evaluates the MOEA's generated solutions in the surrogate model exclusively (this models trains the surrogate model before the execution of the MOEA), (2) Fixed evolution control (FEC), which only some generations or some individuals are evaluated in the surrogate model while the remaining population is evaluated using the real test function, and (3) Adaptive evolution control (AEC), which avoids any possible poor parameter tuning by the use of an adaptive control that adjusts the number of solutions that will be evaluated in the surrogate model. For the sake of simplicity and performance, we have adopted DFR-FEC model to implement methodologies M1 and M2. Furthermore, the NSGA-II's Pareto dominance and population-based nature is able it to generate multiple solutions in a single run, easing the implementation of methodologies M1-2 and M2-2. Thus, we do not specifically use M1-1 and M2-1 in this study.

The description of M1-2 and M2-2 is as follows. The algorithms start with an archive of initial population created using the Latin hypercube method on the entire search space. Then, metamodels are independently constructed for all M objectives ($f_i(\mathbf{x})$). In M1-2, constraints ($g_j(\mathbf{x})$) are independently metamodeled, but in M2-2, the constrained violation function $CV(\mathbf{x})$ is metamodeled. Then, NSGA-II procedure is run for τ generations with the metamodels. Each non-dominated solution from NSGA-II generations is included in the archive. After τ generations, new metamodels are created again and the process is repeated until termination.

4   Methodologies M3 and M4

In these two methods, we transform the multi-objective optimization problem into a parameterized single-objective optimization problems, similar to ParEGO approach [16]. We use the achievement scalarization function (ASF) using a parameterized reference point $\mathbf{z}$ and a corresponding reference direction $\mathbf{w}$. While the reference direction is an equally-angled direction from each objective axis ($\mathbf{w} = (1, 1, \ldots, 1)^T/\sqrt{M}$), the reference point is initialized as equi-spaced points on a unit hyperplane making equal angle to each objective axis. In this study, we have used Das and Dennis's method [3] to create equi-spaced points on the hyperplane. The objectives are normalized ($\bar{f}_i$) using population maximum and minimum objective values so that reference points on the normalized hyperplane makes sense to normalized objective values of the population members. The ASF formulation is given below:

$$\text{ASF}(\mathbf{x}) = \max_{i=1}^{M} \frac{\bar{f}_i(\mathbf{x}) - z_i}{w_i}. \tag{2}$$

In M3 and M4, we metamodel $\text{ASF}(\mathbf{x})$ instead of each objective function $f_i(\mathbf{x})$ independently.

In M3, each constrained function $g_j(\mathbf{x})$ is modeled separately, but in M4, only the constrained violation function ($CV(\mathbf{x})$), as described in Eq. 1, is metamodeled. Since a parameterized scalarization of multiple objectives are used in both M3 and M4. In both methods, we use a single-objective evolutionary optimization algorithm (real-coded genetic algorithm, RGA [5]). The RGA uses a penalty parameter-less approach [4] to handle constraints. Like in M1 and M2, the algorithms start with an archive of randomly created solutions created using Latin Hypercube method. Each archive member is evaluated exactly and then suitable metamodels are constructed for a particular objective scalarization parameter values. These models are then optimized using RGA and the obtained solutions are included in the archive.

Starting with an extreme scalarization to obtain the respective Pareto-optimal solution (say A), the neighboring scalarization is then optimized to obtain the next Pareto-optimal solution (say B). Since the metamodel construction process for solution B included solutions obtained during the search of solution A, the metamodels constructed for solution B becomes more accurate with the solutions from previous optimizations. This process provides the necessary parallelization and computing advantage of latter scalarizations with prior optimizations, a matter we would like to highlight in this paper. For each scalarization, RGA is started with $\alpha\%$ archive points close to the reference line passing through the specific reference point in the objective space. Then, RGA is applied κ times, to take care of the inherent stochastic of the RGA procedure. The parameters α and κ are settled experimentally and kept fixed for all problems of this study. Again, every RGA solution is included in the archive to make a new metamodel before the new RGA run is performed. After making one pass of consecutive scalarizations involving all reference points, the process is repeated in reverse order to make more refined metamodeling of initial scalarizations.

5   Methodology M5

As discussed, this methodology makes a combined selection function $(S(\mathbf{x}))$ considering all objective functions and all constrained functions, but the final focus is to make the global optimum of S-function as one of the targeted Pareto-optimal solution. This can be achieved with a parameterized scalarization procedure, such as by using the ASF function describe above:

$$S(\mathbf{x}) = \begin{cases} \text{ASF}(\mathbf{x}), & \text{if } \mathbf{x} \text{ is feasible}, \\ \text{ASF}_{\max} + \text{CV}(\mathbf{x}), & \text{otherwise}. \end{cases} \tag{3}$$

Here, the parameter $\text{ASF}_{\max}$ is the worst ASF function value of all feasible solutions of the archive. The ASF formulation is identical to that described in the previous subsection. It is noteworthy that the above S-function is usually discontinuous at the constrained boundaries. Thus, more high fidelity points than those in M1 or M2 are needed to make a better model of the above S-function. However, in all our simulation results here, we use the same number of archive members as in other methods. Clearly, other scalarization approaches, such as weighted-sum function or epsilon-constrained function or Tchebyshev function [20] can be used as well. This methodology is a generative multi-objective optimization procedure in which one Pareto-optimal point is determined to be found at a time using the efficient global optimization approach [12,14]. By changing the reference point $\mathbf{z}$ one at a time and keeping the reference direction $\mathbf{w}$ fixed and by using the sequence of scalarization method, M5 can generate a number Pareto-optimal solutions. Notice that M5 methodology used one metamodel to find one Pareto-optimal solution. If K Pareto-optimal points are needed, the above approach needs to be applied K times, each time constructing a new metamodel using a different reference point.

6   Methodology M6

Finally, M6 methodology suggests an ambitious metamodeling procedure in which only one metamodel is required to find multiple Pareto-optimal solutions.

On the face of it, this may sound an approach which is too good to be true, a little thought will reveal that if by any procedure we are able to construct a multi-modal selection function $M_S(\mathbf{x})$ having K Pareto-optimal solutions as multiple global optima of it, then we can possibly employ a multi-modal, single-objective evolutionary algorithm to find and capture multiple optima in a single simulation run. In this paper, we suggest one such approach based on a recently developed theoretical performance metric for multi-objective optimization [6].

Based on Karush-Kuhn-Tucker (KKT) optimality conditions, the first author and his students have recently developed a performance metric – KKT Proximity Measure (KKTPM) [6] – which makes a monotonous increase in KKTPM values for an increase in domination level of solutions. An interesting aspect of KKTPM is that it's value is zero for all Pareto-optimal solutions and it takes a positive value as a solution gets more and more dominated. Such a property motivates us to use KKTPM as a potential multi-modal function for the purpose of developing an M6 methodology. Also, KKTPM considers all objectives and constrained satisfaction into its computation, thereby making it an ideal M_s-function for our purpose.

First we create a set of sample points from the variable space by using Latin hypercube approach and then compute their objective values by high fidelity computations. We then compute KKTPM value of each sample point. KKTPM function on the entire search space is then metamodeled using the Kriging procedure. Thereafter, a multi-modal real-coded GA (m-RGA) is employed to search metamodeled KKTPM function for new and multiple niched solutions. The procedure is described next. A set of reference directions are initially set up like in MOEA/D [28] or in NSGA-III approaches [9]. Thereafter, each sample point (which is evaluated with high-fidelity computations) is associated with a reference line based on its closeness to the line in the objective space. In each iteration of m-RGA, for each offspring population member, the nearest sample point in the *variable* space is identified. Then, the offspring population member is associated with the same reference line as the nearest sample point. All members associated with a reference line lie on the same cluster. Then, the selection operator is restricted within same cluster solutions in m-RGA. The first parent is chosen at random but care is given to ensure every cluster is considered one at a time. Thereafter, the second parent choice is not random, rather a solution from the same cluster is chosen at random. If no other solution is found in the same cluster, the first parent becomes the winner of the selection process. This restriction of selection operator among similar population members will eventually form multiple niches within the population. At the end of a m-RGA run, the best metamodeled KKTPM value in the cluster of each reference line is sent to the archive. Thus, the m-RGA run sends multiple well-diversified solutions having low KKTPM values for high-fidelity evaluations. This process is continued until the termination of the overall M6 algorithm.

7   Results

In this section, we compare the results obtained by all six proposed approaches. For each methodology, the non-dominated set of all Hi-Fi solutions are presented. We use the same parameter settings for RGA or m-RGA having binary tournament selection operator, simulated binary crossover (SBX), and polynomial mutation, as mentioned below: Population size $= 10n$, where n is a number of variables, number of generations $= 100$, crossover probability $= 0.95$, mutation probability $= 1/n$, Distribution index for SBX operator $= 1$, and distribution index for polynomial mutation operator $= 10$. NSGA-II procedure, wherever used, is also applied with the same parameter values as above. We performed 10 runs of all methodologies on test problems. As mentioned earlier, for unconstrained problems (ZDT) and three-objective problem C2DTLZ2 having a single constrained, the behaviors of both M1 and M2 are identical. In such a case, we only show the results for M1-2. The same situation occurs for M3 and M4 and we only show the results of M3. In order to have a graphical comparison, we show the obtained trade-off frontier for the median IGD run in each case.

7.1   Two-Objective Unconstrained Problems

We performed surrogated methodologies of two-objective unconstrained problems on ZDT1, ZDT2, and ZDT3 with ten variables, 21 reference directions, and with a maximum of only 500 high-fidelity solution evaluations (Hi-Fi SEs). The obtained non-dominated solutions are illustrated in Figs. 2, 3, 4, and 5 respectively, for methodologies M1-2, M3, M5 and M6.

It is clear from the figures that all six methodologies are able to solve ZDT1 and ZDT2 problems fairly well in only 500 Hi-Fi SEs. The obtained points are very close to the true Pareto-optimal front and have a wide distribution of points (with a small IGD value, as shown in Table 2). Most past studies have used tens of thousands of Hi-Fi SEs to have a similar performance, while here using the proposed metamodeling methods, we are able to find a similar set of points in only 500 Hi-Fi SEs. For ZDT3 problem, which has a disconnected Pareto-optimal front, M1-2 and M3 perform the best, followed by M5. However, the

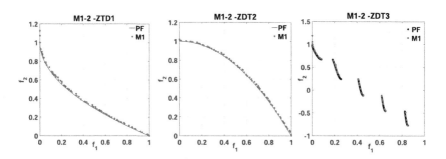

Fig. 2. Obtained non-dominated solutions for ZDT1-ZDT3 using M1-2.

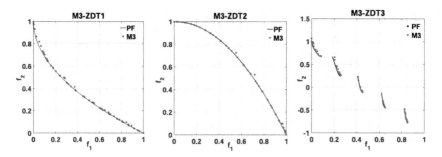

Fig. 3. Obtained non-dominated solutions for ZDT1-ZDT3 using M3.

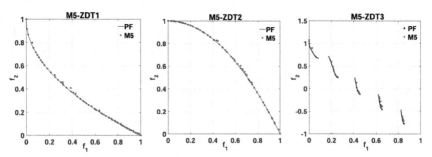

Fig. 4. Obtained non-dominated solutions for ZDT1-ZDT3 using M5.

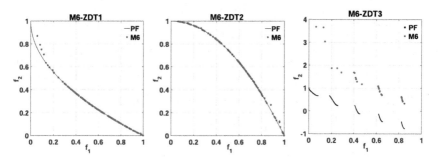

Fig. 5. Obtained non-dominated solutions for ZDT1-ZDT3 using M6

multi-modal KKTPM-based M6 approach is not able to get quite close to the true Pareto-optimal front in 500 Hi-Fi SEs. This is due to the added complexity the multi-modal KKTPM surface offers to the metamodeling approach. Although a single metamodel is progressively modeled, more Hi-Fi SEs are needed to make a better modeling and optimization of this problem.

Table 2 presents the average and standard deviation of the IGD values for an identical set of reference Pareto-optimal solutions for all four methods. Best performing methodologies based p-value in the Wilcoxon signed-ranked test are marked in bold.

7.2   Two-Objective Constrained Problems

Next, we apply our approaches to two-objective constrained problems – BNH, SRN, and TNK with their original problem sizes [5], 21 reference directions, and with a maximum of 800 Hi-Fi SEs. The obtained non-dominated solutions are shown in Figs. 6, 7, 8, 9, 10, and 11, respectively, for M1-2. M2-2, M3, M4. M5 and M6 methodologies.

All six methodologies are able to find a close and well-distributed set of trade-off points to true Pareto-optimal front (shown by a solid line in each case) for BNH and SRN problems. With only 800 Hi-Fi SEs used in our study, the performance of these methods is noteworthy. However, the problem TNK has provided difficulties to all six methodologies, due to discontinuities in its Pareto-optimal front. M3 and M1-2 perform the best on TNK, but as presented in Table 2, followed by M4 and M6. Since constraints are independently modeled in M1-2 and M3 methods, although more metamodels are constructed progressively each

Table 2. IGD values for all six methodologies. Statistically similar methodologies with best performance are marked in bold.

	M1		M2		M3		M4		M5		M6	
	Average	SD	Average	SD	Average	SD	Average	SD	Average	SD	Average	SD
ZDT 1	**0.0115**	0.0006	-	-	0.0185	0.0046	-	-	**0.0103**	0.0028	0.0249	0.0153
	$p = 0.12120$		-		$p = 0.00170$		-		-		$p = 0.02570$	
ZDT 2	0.0105	0.0027	-	-	0.0130	0.0013	-	-	**0.0078**	0.0011	**0.0103**	0.0051
	$p = 0.00170$		-		$p = 0.00019$		-		-		$p = 0.73370$	
ZDT 3	**0.0136**	0.0019	-	-	**0.0254**	0.0154	-	-	0.0246	0.0027	0.7111	0.2054
	-		-		$p = 0.06390$		-		$p = 0.00018$		$p = 0.00018$	
BNH	0.5322	0.2164	0.4605	0.1324	**0.3854**	0.0724	**0.3539**	0.1543	**0.3727**	0.0798	**0.4177**	0.1852
	$p = 0.02070$		$p = 0.00880$		$p = 0.05290$		-		$p = 0.30530$		$p = 0.27120$	
SRN	**0.7452**	0.2473	1.0477	0.1765	1.5564	0.1742	1.9072	0.3237	2.8532	0.5795	1.5824	0.2557
	-		$p = 0.00910$		$p = 0.00024$		$p = 0.00016$		$p = 0.00018$		$p = 0.00025$	
TNK	0.0173	0.0059	0.0748	0.0208	**0.0107**	0.0010	0.0300	0.0075	0.0654	0.0353	0.0388	0.0038
	$p = 0.00033$		$p = 0.00018$		-		$p = 0.00018$		$p = 0.00018$		$p = 0.00018$	
C2DTLZ2	**0.0485**	0.0046	-	-	0.05902	0.0020	-	-	**0.0490**	0.0017	0.1037	0.0268
	-		-		$p = 0.00025$		-		$p = 0.21210$		$p = 0.00018$	

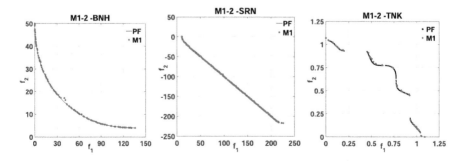

Fig. 6. Obtained non-dominated solutions for BNH, SRN and TNK using M1-2.

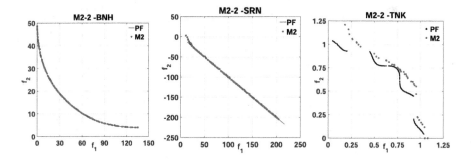

Fig. 7. Obtained non-dominated solutions for BNH, SRN and TNK using M2-2.

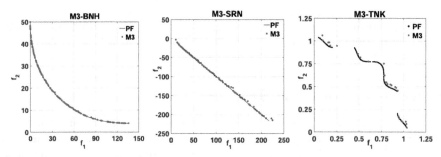

Fig. 8. Obtained non-dominated solutions for BNH, SRN and TNK using M3.

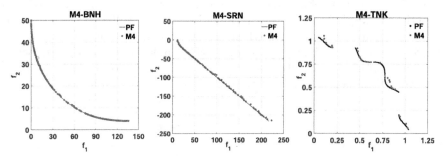

Fig. 9. Obtained non-dominated solutions for BNH, SRN and TNK using M4.

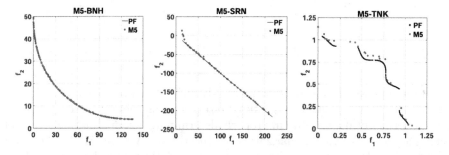

Fig. 10. Obtained non-dominated solutions for BNH, SRN and TNK using M5

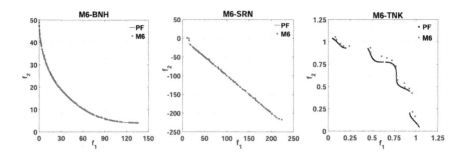

Fig. 11. Obtained non-dominated solutions for BNH, SRN and TNK using M6.

time, the performance of these two methods on this difficult problem is better than other methods in which constraints are metamodeled using a combined function.

7.3   Three-Objective Problems

Next, we apply all six methods to a three-objective constrained problem (C2DTLZ2) having seven variables and 91 reference directions. We fix a total of 1,000 Hi-Fi SEs for this problem. The Pareto-optimal surface of this problem and the obtained non-dominated solutions are shown in Fig. 12 for M1-2, M3, M5 and M6. As noted before, since this problem has a single constraint, M1 and M2 methods will produce identical results, M3 and M4 will also produce identical results. On this problem, methodologies M1-2 and M5 perform the best, followed by M3. Interestingly, methodology M6 is able to find a large number of near Pareto-optimal points from a multi-modal optimization task, but further developments are needed to investigate the full potential of M6.

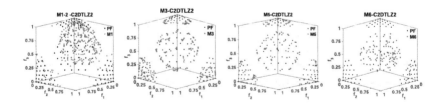

Fig. 12. Results on C2DTLZ2.

As the number of objectives increase, M1 and M2 approaches demand more metamodels (one for each objective function) to be constructed. The error in each metamodel can be detrimental in properly determining non-domination level and niching actions. The ASF scalarization of objectives in M3 and M5 focuses each metamodel for finding a single Pareto-optimal point, thereby making an excellent performance. Method M5 makes a metamodel of the constrained ASF function and hence possesses a more complex metamodeling task. Nevertheless, M5 is able to find multiple widely distributed solutions on the entire Pareto-optimal front. Method M6 gets overwhelmed in modeling the multi-modal KKTPM surface having 91 global optima (desired for three-objective C2DTLZ2 problem). More Hi-Fi SEs are needed to get a better performance by this method, but the use of a single metamodel to solve a multi-objective optimization problem as a multi-modal optimization problem is algorithmically novel and challenging, and further studies are needed to fully exploit its potential.

8   Conclusions

In this paper, we have, for the first time, presented a taxonomy for the use of metamodeling methods in multi-objective optimization. The paper have focused outside the common-sense strategy of extending metamodeling methods for single-objective optimization to multi-objective optimization. The straightforward extension suggests that for every objective function and constrained function, an independent metamodel is needed. Although this is a viable strategy, we have proposed five more potential strategies and classified many existing studies into the proposed six categories. Thereafter, as an initial study, we have implemented each of the six proposed methodologies and applied to a few constrained and unconstrained multi-objective optimization problems. Results are compared against each other using a standard parameter setting and important conclusions about the behavior of each of the methods have been revealed.

Compared to standard EMO studies in which tens of thousands of solution evaluations are used to solve the test problems, here we present results that take only a fraction (a few hundreds to a thousand) of solution evaluations in each case. This is remarkable and holds promise to the successful applications of metamodeling methods in evolutionary multi-objective optimization.

This initial study is motivating for us to pursue a more detail study, which we are currently pursuing. Although identical parameter values are used for each method here, it may very well be true that each methodology performs its best for a different set of parameter values. As a future study, we plan to investigate these aspects and include other metamodeling methods to eventually determine the niche of each of the six possible metamodeling methodologies for solving various types of problems ranging from two to many objectives.

Acknowledgment. Authors acknowledge the Matlab Kriging code provided by Dr. Cem Tutum.

References

1. Cassioli, A., Schoen, F.: Global optimization of expensive black box problems with a known lower bound. J. Glob. Optim. **57**(1), 177–190 (2013)
2. Coelho, R.F., Lebon, J., Bouillard, P.: Hierarchical stochastic metamodels based on moving least squares and polynomial chaos expansion. Struct. Multi. Optim. **43**, 707–729 (2011)

3. Das, I., Dennis, J.E.: Normal-boundary intersection: a new method for generating the Pareto surface in nonlinear multicriteria optimization problems. SIAM J. Optim. **8**(3), 631–657 (1998)
4. Deb, K.: An efficient constraint handling method for genetic algorithms. Comput. Methods Appl. Mech. Eng. **186**(2–4), 311–338 (2000)
5. Deb, K.: Multi-Objective Optimization Using Evolutionary Algorithms. Wiley, Chichester (2001)
6. Deb, K., Abouhawwash, M.: A optimality theory based proximity measure for set based multi-objective optimization. IEEE Trans. Evol. Comput. **20**(4), 515–528 (2016)
7. Deb, K., Agrawal, S., Pratap, A., Meyarivan, T.: A fast and elitist multi-objective genetic algorithm: NSGA-II. IEEE Trans. Evol. Comput. **6**(2), 182–197 (2002)
8. Deb, K., Datta, R.: A fast and accurate solution of constrained optimization problems using a hybrid bi-objective and penalty function approach. In: Proceedings of the IEEE World Congress on Computational Intelligence (WCCI 2010), pp. 165–172 (2010)
9. Deb, K., Jain, H.: An evolutionary many-objective optimization algorithm using reference-point based non-dominated sorting approach, Part I: Solving problems with box constraints. IEEE Trans. Evol. Comput. **18**(4), 577–601 (2014)
10. Díaz-Manríquez, A., Toscano, G., Barron-Zambrano, J.H., Tello-Leal, E.: A review of surrogate assisted multiobjective evolutionary algorithms. Comput. Intell. Neurosci. **2016**, 1–14 (2016)
11. Horn, D., Wagner, T., Biermann, D., et al.: Model-based multi-objective optimization: taxonomy, multi-point proposal, toolbox and benchmark. In: Gaspar-Cunha, A., Henggeler Antunes, C., Coello, C.C. (eds.) EMO 2015. LNCS, vol. 9018, pp. 64–78. Springer, Heidelberg (2015). doi:10.1007/978-3-319-15934-8_5
12. Hussein, R., Deb, K.: A generative kriging surrogate model for constrained and unconstrained multi-objective optimization. In: Proceedings of GECCO 2016. ACM Press (2016)
13. Jin, Y.: A comprehensive survey of fitness approximation in evolutionary computation. Soft Comput. **9**(1), 3–12 (2005)
14. Jones, D., Schonlau, M., Welch, W.: Effcient global optimization of expensive. J. Glob. Optim. **13**, 455–492 (1998)
15. Jones, D.R.: A taxonomy of global optimization methods based on response surfaces. J. Glob. Optim. **21**(4), 345–383 (2001)
16. Knowles, J.: ParEGO: a hybrid algorithm with on-line landscape approximation for expensive multiobjective optimization problems. IEEE Trans. Evol. Comput. **10**(1), 50–66 (2006)
17. Le, M.N., Ong, Y.S., Menzel, S., et al.: Multi co-objective evolutionary optimization: cross surrogate augmentation for computationally expensive problems. In: Proceedings of CEC 2012, pp. 2871–2878. IEEE Press (2012)
18. Martínez, S.Z., Coello, C.A.C.: MOEA/D assisted by RBF networks for expensive multi-objective optimization problems. In: Proceedings of Genetic and Evolutionary Computation Conference. ACM (2013)
19. Martínez-Frutos, J., Pérez, D.H.: Kriging-based infill sampling criterion for constraint handling in multi-objective optimization. J. Glob. Optim. **64**, 97–115 (2016)
20. Miettinen, K.: Nonlinear Multiobjective Optimization. Kluwer, Boston (1999)

21. Namura, N., Shimoyama, K., Obayashi, S.: Kriging surrogate model enhanced by coordinate transformation of design space based on eigenvalue decomposition. In: Gaspar-Cunha, A., Henggeler Antunes, C., Coello, C.C. (eds.) EMO 2015. LNCS, vol. 9018, pp. 321–335. Springer, Heidelberg (2015). doi:10.1007/978-3-319-15934-8_22

22. Quintero, L.V.S., Montano, A.A., Coello, C.A.C.: A Review of techniques for handling expensive functions in evolutionary multi-objective optimization. In: Tenne, Y., Goh, C.-K. (eds.) Computational Intelligence in Expensive Optimization Problems. Springer, Heidelberg (2010)

23. Shi, L., Rasheed, K.: A survey of fitness approximation methods applied in evolutionary algorithms. In: Tenne, Y., Goh, C.-K. (eds.) Computational Intelligence in Expensive Optimization Problems. Springer, Heidelberg (2010)

24. Sreekanth, J., Datta, B.: Multi-objective management of saltwater intrusion in coastal aquifers using genetic programming and modular neural network based surrogate models. J. Hydrol. **393**, 245–256 (2010)

25. Steuer, R.E.: Multiple Criteria Optimization: Theory Computation and Application. Wiley, New York (1986)

26. Tsoukalas, I., Makropoulos, C.: Multiobjective optimisation on a budget: exploring surrogate modelling for robust multi-reservoir rules generation under hydrological uncertainty. Environ. Model. Softw. **69**, 396–413 (2015)

27. Verbeeck, D., Maes, F., De Grave, K., Blockeel, H.: Multi-Objective Optimization with Surrogate Trees. In: Proceedings of GECCO 2013, pp. 679–686. ACM Press (2013)

28. Zhang, Q., Li, H.: MOEA/D: A multiobjective evolutionary algorithm based on decomposition. IEEE Trans. Evol. Comput. **11**(6), 712–731 (2007)

29. Zhang, Y., Hu, S., Wu, J., Zhang, Y., Chen, L.: Multi-objective optimization of double suction centrifugal pump using kriging metamodels. Adv. Eng. Softw. **74**, 16–26 (2014)

Weighted Stress Function Method for Multiobjective Evolutionary Algorithm Based on Decomposition

Roman Denysiuk and António Gaspar-Cunha$^{(\boxtimes)}$

Department of Polymer Engineering, University of Minho,
Campus de Azurém, 4800-058 Guimarães, Portugal
{roman,agc}@dep.uminho.pt

Abstract. Multiobjective evolutionary algorithm based on decomposition (MOEA/D) is a well established state-of-the-art framework. Major concerns that must be addressed when applying MOEA/D are the choice of an appropriate scalarizing function and setting the values of main control parameters. This study suggests a weighted stress function method (WSFM) for fitness assignment in MOEA/D. WSFM establishes analogy between the stress-strain behavior of thermoplastic vulcanizates and scalarization of a multiobjective optimization problem. The experimental results suggest that the proposed approach is able to provide a faster convergence and a better performance of final approximation sets with respect to quality indicators when compared with traditional methods. The validity of the proposed approach is also demonstrated on engineering problems.

1 Introduction

This study considers a multiobjective optimization problem (MOP) of the form

$$
\begin{aligned}
\text{minimize } & (f_1(\boldsymbol{x}), \ldots, f_m(\boldsymbol{x})) \\
\text{subject to } & c_i(\boldsymbol{x}) \leq 0 && i = 1, \ldots, k \\
& l_j \leq x_j \leq u_j && j = 1, \ldots, n
\end{aligned}
\tag{1}
$$

where $\boldsymbol{x} = (x_i, \ldots, x_n)$ is the decision vector, $\boldsymbol{f} = (f_1, \ldots, f_m)$ is the objective vector, $\boldsymbol{c} = (c_1, \ldots, c_k)$ is the vector of inequality constraints, l_j and u_j are the lower and upper bounds of the j-th variable, respectively.

When solving a MOP, multiobjective evolutionary algorithms (MOEAs) attempt to approximate the Pareto set that rises from the conflicting nature of the involved objectives. In this process, MOEAs rely on three major mechanisms such as selection, variation and replacement. Most variation operators are adopted from single objective optimization by inserting existing mechanisms for producing offspring into a framework able to deal with multiple objectives. Parents selection and replacement are based on some fitness assignment scheme intended to emphasize and propagate promising individuals in the presence of multiple objectives. The convergence and diversity are two essential issues that

© Springer International Publishing AG 2017
H. Trautmann et al. (Eds.): EMO 2017, LNCS 10173, pp. 176–190, 2017.
DOI: 10.1007/978-3-319-54157-0_13

must be addressed by the fitness assignment. The convergence refers to guiding the population towards the Pareto set. The diversity implies that a diverse set of solutions is maintained. In current state-of-the-art MOEAs, three major approaches to fitness assignment can be distinguished.

A dominance-based strategy probably is the most frequently used one. It relies on the concept of the Pareto dominance and is usually combined with some diversity preserving mechanism [2]. Its advantage is related to the correspondence with the concept of optimality in multiobjective optimization. Though the performance of such mechanism severely deteriorates when the number of objectives is increased. Also, because the diversity is usually considered as a second sorting criterion, dominance-based MOEAs may face substantial difficulties in solving MOPs due to severe loss of diversity.

Indicator-based MOEAs employ quality indicators to assign fitness to individuals in the population [20]. The development of such MOEAs is based on the idea that it can be beneficial to explicitly optimize a measure that is used for algorithms comparison. Moreover, some quality indicators possess good theoretical properties and are Pareto compliant. The difficulty in their application arises from a high computational cost. In particular, a computational time of the hypervolume grows exponentially with the number of objectives. This significantly limits its applicability. Approximating the hypervolume requires trade-off between accuracy and complexity. The indicator based fitness assignment can also face difficulties in balancing the convergence and diversity.

Decomposition-based approaches aim at decomposing a MOP into a number of subproblems and solving them simultaneously. The decomposition can be based on the aggregation of multiple objectives into a scalarizing function using traditional mathematical techniques [14]. In such MOEAs, convergence is achieved by optimizing the corresponding scalarizing function whereas diversity is ensured by a well distributed set of weight vectors. MOEAs relying on scalarization usually works well on problems with a large number of objectives. The major advantage is their efficiency. Other MOEAs use directional vectors for associating individuals with the corresponding direction so that the diversity of population members is ensured [12,13]. Decomposition-based approaches typically require a set of weights or directional vectors being provided in advance. Generating such set often is not an easy task, especially when the number of dimensions is high. Alternatively, the decomposition can be performed by generating a grid using polar coordinates [5] or by exploring the angles between population members in the objective space [4].

MOEA/D is a representative state-of-the-art approach relying on decomposition by means of scalarization. MOEA/D associates each population member with a subproblem defined by a scalarizing function. In the end, individuals represent solutions to the corresponding subproblems. Neighborhood relations among subproblems are defined using the distances between the weight vectors. These relations are exploited during the search. Since its advent, MOEA/D has been increasingly investigated. This resulted in a number of different variants, including studies of different reproduction operators and improvements of its core framework. The effect of different scalarizing functions was studied in [7]. The results of this study suggest that a proper choice of scalarizing function is an important issue for the performance of MOEA/D. Another issue that heavily influences the search ability of MOEA/D is setting the values of control parameters. Mechanisms involving adaptation of control parameters during the search appeared effective in enhancing the performance of MOEA/D. To balance convergence and diversity, a constrained decomposition approach was proposed in [17]. Recently, it was shown that a better exploration of the search space can be achieved when performing replacement in the neighborhood of the subproblem that best matches offspring [16]. Another important issue in MOEA/D is an efficient allocation of computational resources between different subproblems [19].

This study focuses on the MOEA/D fitness assignment mechanism. A new scalarizing function is suggested to guide the search, which is called a weighted stress function method (WSFM). The use of WSFM is motivated by its promising behavior as a preference articulation method [6]. The inspiration for WSFM stems from mechanics, namely from the stress-strain behavior of thermoplastic vulcanizates [1]. WSFM has particular characteristics that are different from traditional methods for scalarization. As experimental results suggest, MOEA/D with WSFM can provide better results when compared with state-of-the-art scalarization methods.

The remainder of this paper is organized as follows. Section 2 describes the MOEA/D framework. Section 3 introduces the WSFM for MOEA/D. Section 4 discusses the results of the experimental study. Finally, Sect. 5 presents conclusions of the study and outlines some possible future work.

2   Multiobjective Evolutionary Algorithm Based on Decomposition

2.1   Algorithm

The interest from research community and effort in improving the performance led to several variants of MOEA/D. As this study focuses on the fitness assignment, MOEA/D is considered with two different replacement variants. The first uses the mating pool. The second selects the most suitable subproblem to offspring and uses its neighborhood. In the following, this two variants are referred using notation from corresponding papers as MOEA/D [11,18] and MOEA/D-GR [16], respectively. The outline of MOEA/D is given as follows.

Input:

- δ - probability for mating pool;
- T - neighborhood size;
- n_r - maximum number of individuals replaced by offspring;
- μ - population size;
- $maxGen$ - maximum number of generations.

Output:

- $\{x^1, \ldots, x^\mu\}$ - approximation to the Pareto set;
- $\{f(x^1), \ldots, f(x^\mu)\}$ - approximation to the Pareto front.

Step 1 Initialization

Step 1.1 Read input parameters;

Step 1.2 Generate a set of weight vectors $W = \{w^1, \ldots, w^\mu\}$, where $\sum_{j=1}^{m} w_j^i = 1, \forall i \in \{1, \ldots, \mu\}$;

Step 1.3 For each weight vector, select the T closest weight vectors by:
1. Computing the Euclidean distance between any two weight vectors;
2. Setting $B(i) = \{i_1, \ldots, i_T\}, \forall i \in \{1, \ldots, \mu\}$, where $w^{i_1}, \ldots, w^{i_T}$ are the T closest weight vectors to w^i;

Step 1.4 Randomly generate an initial population. Evaluate the population;

Step 1.4 Initialize a reference point, z, by setting $z_j = \min f_j(x^i) \ \forall i \in \{1, \ldots, \mu\}, \forall j \in \{1, \ldots, m\}$.

Step 2 Evolution
For each $i = 1, \ldots \mu$ do:

Step 2.1 Selection
1. Select mating pool:

$$P_m = \begin{cases} B(i) & \text{with probability } \delta \\ \{1, \ldots, \mu\} & \text{otherwise.} \end{cases}$$

2. From P_m, select parents for reproduction;

Step 2.2 Variation
1. Apply evolutionary operators on parents to produce offspring y;
2. Evaluate offspring.

Step 2.3 Update
For each $j = 1, \ldots, m$: $z_j = f_j(y)$ if $f_j(y) < z_j$.

Step 2.4 Replacement
Select pool for replacement P_r, set $c = 0$ and do the following:
1. If $c = n_r$ or $P_r = \emptyset$, go to **Step 3**. Otherwise, pick an index j from P_r;
2. If $fitness(f(y)|w^j) < fitness(f(x^j)|w^j)$, then set $x^j = y$ and $c = c+1$;
3. Remove j from P_r and got to 1.

Step 3 If the stopping criterion is not met, go to **Step 2**. Otherwise, return **Output**:

Depending on the parameter settings, the above depicted algorithm defines one of the three MOEA/D variants considered in the present study. These were originally presented in [11, 16, 18].

2.2 Scalarizing Functions

MOEA/D decomposes a MOP into a set of single-objective subproblems by means of scalarization. Scalarization relies on a scalarizing function to compute a scalar value for the given objectives and weights. The choice of a scalarizing function is an important issue that greatly influences the performance of MOEA/D. Most scalarizing functions are adopted from traditional programming methods for solving MOPs [14]. Owing to the concern of the present study, frequently used scalarizing functions are briefly reviewed in the following.

Weighted Sum (WSUM)

The weighted sum method associates each objective with a weight and minimizes the weighted sum of the objectives. The scalarizing function can be defined as

$$\text{minimize } g(\boldsymbol{f}|\boldsymbol{w}) = \sum_{i=1}^{m} w_i f_i. \tag{2}$$

The advantages of this method are that it does not need a reference point and the resulting scalar optimization problem is convex. The major shortcoming of this method is that it fails to find solutions in nonconvex regions of the Pareto front.

Chebyshev (CHB)

For a reference point $\boldsymbol{z} = (z_1, \ldots, z_m)$, the scalarizing function based on Chebyshev method can be defined as

$$\text{minimize } g(\boldsymbol{f}|\boldsymbol{w}) = \max_{1 \leq i \leq m} w_i |f_i - z_i|. \tag{3}$$

This method belongs to the group of weighted metric methods that seek to minimize the distance between some reference point and the feasible objective region where the weighted L_p metric is used for measuring this distance. The problem in (3) is obtained for $p = \infty$. This method can find solutions in convex and nonconvex regions of the Pareto front. The drawbacks are that it cannot distinguish weakly Pareto optimal solutions and does not provide a uniform distribution of solutions along the Pareto front.

Penalty Boundary Intersection (PBI)

The penalty boundary intersection method was suggested in [18] in order to generate a more uniform approximation to the Pareto front by MOEA/D. The scalarizing function is given by

$$\text{minimize } g(\boldsymbol{f}|\boldsymbol{w}) = d_1 + \theta d_2 \tag{4}$$

where

$$d_1 = \frac{\|(\boldsymbol{f}-\boldsymbol{z})^{\mathrm{T}}\boldsymbol{w}\|}{\|\boldsymbol{w}\|}$$
$$d_2 = \left\| \boldsymbol{f} - \left(\boldsymbol{z} + d_1 \frac{\boldsymbol{w}}{\|\boldsymbol{w}\|}\right) \right\|. \tag{5}$$

The major advantage of this method is that it can provide a reasonably uniform distribution of solutions along the Pareto front. Though it comes at the cost of specifying the value of θ. Different settings of this parameter can heavily affect the performance of MOEA/D.

2.3   Constraint Handling

Real-world problems often involve constraints that must be satisfied. This study considers the constraint handling technique for MOEA/D presented in [9], as it proved effective on a set of challenging constrained problems. This method relies on a penalty function, p, that is dynamically adjusted. For a given individual, the degree of constraint violation, $CV(\boldsymbol{x})$, is estimated as

$$CV(\boldsymbol{x}) = \sum_{j=1}^{k} \max(c_j(\boldsymbol{x}), 0). \tag{6}$$

The value of p is computed on the basis of CV as

$$p(\boldsymbol{x}) = \begin{cases} s_1 CV(\boldsymbol{x})^2 & \text{if } CV(\boldsymbol{x}) < \tau \\ s_1\tau^2 + s_2(CV(\boldsymbol{x}) - \tau), & \text{otherwise.} \end{cases} \tag{7}$$

A threshold value, τ, is defined as

$$\tau = CV_{\min} + 0.3(CV_{\max} - CV_{\min}) \tag{8}$$

where $s_1 = 0.01$ and $s_2 = 20$ are scaling parameters, $CV_{\min}$ and $CV_{\max}$ are the minimum and maximum values of constraint violation in the current population. The penalty function encourages the exploration of both feasible and infeasible regions. The role of parameter τ is to control the amount of penalty. Thus, the fitness of the i-th population member is given as

$$fitness(\boldsymbol{x}|\boldsymbol{w}) = g(\boldsymbol{f}(\boldsymbol{x})|\boldsymbol{w}) + p(\boldsymbol{x}). \tag{9}$$

3   Weighted Stress Function Method

3.1   Analogy with Rubber Elasticity

The weighted stress function method (WSFM) is inspired by the stress-strain behavior of thermoplastic vulcanizates (TPVs) [1]. These materials are a particular group of thermoplastic elastomers possessing high performance elastic and mechanical properties. Stress and strain are two different but closely related concepts. Stress is defined as force per unit area that can cause a change in an object or a physical body. Strain is defined as the amount of deformation experienced due to the application of stress. The relationship between the stress and strain that a particular material exhibits is displayed by the stress-strain curve.

A typical structure of a TPV consists of a very high volume fraction ($0.40 < v_p < 0.9$) of fully cured elastomeric particles surrounded by a continuous thermoplastic matrix. The majority of experimental studies on this type of material show that the elasticity of the material increases when the volume fraction of the elastomeric particle is increased from 0.0 to 1.0. Figure 1 illustrates the stress-strain behavior for different values of v_p. This figure shows that

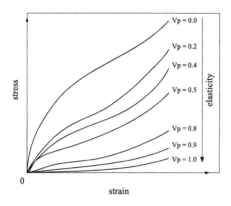
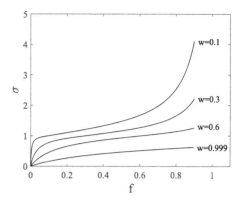

Fig. 1. Stress strain behavior.

Fig. 2. Stress values for different weights

there are zones with different rates of stress and strain, with those having a nonlinear relationship.

WSFM establishes an analogy between the stress-strain behavior of TPV materials and a stress function that takes into account weights defined for each objective. This method assumes that the difference between the ideal point and each solution induces a stress on that solution. The stress depends on the weights associated with the objectives of the given solution. Similarly to the values of v_p, the weights range in the interval [0,1] and resembles the role that v_p plays in increasing or decreasing the stress.

3.2   Scalarizing Function

WSFM transforms each objectives, f_i, into a stress, σ_i, depending on the value of associated weight, w_i. The WSFM problem seeks to minimize the largest stress associated with the given solutions. The scalarizing problem is of the form

$$\text{minimize } g(\boldsymbol{f}|\boldsymbol{w}) = \max_{1 \leq i \leq m} \sigma_i(f_i, w_i). \tag{10}$$

The calculation of the stresses requires the normalization of the objective values so that they are in the range [0,1]. This is done as

$$f_i = \frac{f_i - f_i^{\min}}{f_i^{\max} - f_i^{\min}} \tag{11}$$

where $f_i^{\min}$ and $f_i^{\max}$ are the minimum and maximum values of the i-th objective in the current population.

Assuming minimization of the m objectives $(f_1, \ldots, f_m)$, for the weight vector $(w_1, \ldots, w_m)$, the stress σ_i associated with the i-th objective is calculated as

$$\sigma_i(f_i, w_i) = (1 + \omega_i(f_i, w_i)) \, \xi_i(w_i) \tag{12}$$

where

$$\omega_i(f_i, w_i) = \begin{cases} \dfrac{\tan\left(\frac{\pi}{\psi_i(w_i)}(f_i - w_i)\right)}{\tan\left(\frac{\pi}{\phi_i(w_i)}w_i - \delta_1\right)} \dfrac{\psi_i(w_i)}{\phi_i(w_i)}, & f_i \geq w_i \\[4mm] \dfrac{\tan\left(\frac{\pi}{\phi_i(w_i)}(f_i - w_i)\right)}{\tan\left(\frac{\pi}{\phi_i(w_i)}w_i\right)}, & f_i < w_i \end{cases} \tag{13}$$

$$\psi_i(w_i) = \frac{3}{4}w_i^2 + 2(1 - w_i) + \delta_1 \tag{14}$$

$$\phi_i(w_i) = \frac{3}{4}w_i^2 + 2w_i + \delta_1 \tag{15}$$

$$\xi_i(w_i) = 1 - \frac{\tan\left(\frac{\pi}{2(1+\delta_2)}(2w_i - 1)\right)}{\tan\left(\frac{\pi}{2(1+\delta_2)}\right)}. \tag{16}$$

To ensure that σ_i does not assume infinite values, the extreme values of weights are projected as

$$w_i = \min(\max(w_i, \epsilon), 1 - \epsilon). \tag{17}$$

Figure 2 plots stresses for different weight values. High values of weights correspond to lower stresses. Also, it can be seen that there is a nonlinear relationship.

Figure 3 illustrates contour lines for the different scalarizing functions referred above. In the plots, dashed lines show the directions of search. For CHB and WSFM, these lines correspond to the cases when the terms in the max function in (3) and (12) are equal. For bold dashed lines, the contour lines are depicted by bold solid lines. These lines divide the objective space into region with solutions located under these line being better than those in the other region. The contour line for WSFM is a line that is perpendicular to the direction of search. For the CHB approach, this is a polygonal line with the right angle. A polygonal line is also the contour line for PBI, though the angle depends on the value of θ. A unique characteristic of WSFM can be identified from the corresponding plot. Although the shape of contour lines is identical to CHB, in WSFM there are nonlinear lines that define the directions of search. The curvature existent in the WSFM contour lines allows to get solutions that correspond better to changes made in the set of weigths, *i.e.*, there is a better correspondence between the weigths and the solutions found.

4   Computational Experiments

This section discusses the computational experiments carried out to investigate the performance of MOEA/D when using different scalarizing functions for fitness assignment. The experiments are divided into three different parts according to employed test problems and MOEA/D variants. These include state-of-the-art test suites, problems with complicated Pareto sets and engineering design problems. For each problem, 30 independent runs of each MOEA/D variant are performed. The results are quantitatively assessed using the epsilon and

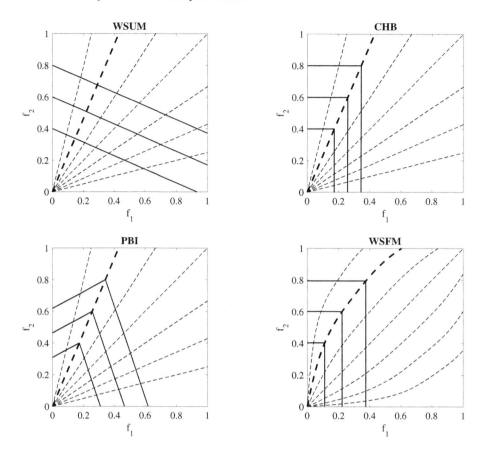

Fig. 3. Contour lines for different scalarizing functions.

hypervolume indicators, which are Pareto compliant quality indicators [10]. The statistical analysis is performed at a significance level of 0.05 using Wilcoxon rank-sum test. In the following tables, symbol † indicates a statistical difference with the best performing variant.

4.1   ZDT and DTLZ Problems

Problems from the ZDT and DTLZ suites are widely used for benchmarking MOEAs. Similarly to the study [18] introducing the MOEA/D framework, this study performs experiments adopting continuous ZDT and three-objective DTLZ1 and DTLZ2 problems. These experiments aim to test a genetic algorithm variant of MOEA/D-WSFM, which uses the SBX crossover and polynomial mutation for reproduction. MOEA/D is run for 300 generations with $\mu = 100$. The other control parameters are $T = n_r = 10$ and $\delta = 1$.

Table 1 summarizes the results with respect to the quality indicators. It is evident that MOEA/D-WSFM is the best performing variant. For the two

Table 1. Results for ZDT and DTLZ problems. The values refer to the median and interquartile range of the epsilon (eps) and hypervolume (hv) indicators.

		WSUM	CHB	PBI	WSFM
		MOEA/D			
ZDT1	eps	2.90e-02 (2.3e-05)†	8.20e-03 (4.7e-05)†	3.43e-01 (3.7e-02)†	5.22e-03 (9.7e-04)
	hv	8.62e-01 (1.1e-06)†	8.72e-01 (2.8e-05)†	7.08e-01 (2.3e-02)†	8.72e-01 (4.9e-05)
ZDT2	eps	3.82e-01 (0.0e+00)†	6.31e-03 (5.0e-05)†	9.98e-01 (2.3e-03)†	5.36e-03 (3.4e-05)
	hv	2.10e-01 (0.0e+00)†	5.37e-01 (2.8e-05)†	1.12e-01 (2.4e-03)†	5.39e-01 (4.2e-04)
ZDT3	eps	2.65e-01 (1.2e-02)†	1.72e-02 (5.0e-05)	2.94e-01 (8.2e-02)†	1.78e-01 (2.0e-01)
	hv	5.56e-01 (2.2e-02)†	7.24e-01 (4.4e-05)	6.09e-01 (6.3e-02)†	7.04e-01 (4.1e-02)
ZDT4	eps	3.15e-02 (1.1e-02)†	1.31e-02 (4.1e-03)	3.33e-01 (4.3e-02)†	1.23e-02 (1.7e-02)
	hv	8.60e-01 (2.5e-03)†	8.65e-01 (5.9e-03)†	7.09e-01 (2.8e-02)†	8.65e-01 (6.4e-02)
ZDT6	eps	4.31e-01 (2.5e-08)†	6.07e-03 (1.4e-05)†	1.13e-01 (3.9e-05)†	5.23e-03 (2.0e-05)
	hv	2.10e-01 (1.9e-08)†	6.11e-01 (3.6e-06)†	5.92e-01 (5.6e-06)†	6.11e-01 (7.7e-07)
DTLZ1	eps	6.29e-01 (3.7e-02)†	9.61e-02 (2.0e-03)†	2.17e-01 (1.0e-01)†	6.29e-01 (2.1e-03)
	hv	3.51e-01 (1.8e-01)†	1.06e+00 (3.8e-03)†	8.33e-01 (6.4e-02)†	1.12e+00 (4.3e-03)
DTLZ2	eps	4.22e-01 (3.3e-08)†	9.83e-02 (5.1e-04)†	7.57e-02 (1.4e-03)†	5.57e-02 (2.9e-04)
	hv	0.00e+00 (0.0e+00)†	6.69e-01 (4.4e-03)†	7.41e-01 (5.6e-05)†	7.46e-01 (2.7e-04)
		MOEA/D-GR			
ZDT1	eps	6.01e-01 (3.6e-01)†	8.21e-03 (3.6e-05)†	3.44e-01 (2.4e-02)†	5.22e-03 (6.4e-04)
	hv	2.15e-01 (9.0e-02)†	8.72e-01 (6.1e-05)†	7.08e-01 (1.4e-02)†	8.72e-01 (3.6e-05)
ZDT2	eps	1.00e+00 (0.0e+00)†	6.35e-03 (1.3e-04)†	9.98e-01 (1.7e-02)†	5.37e-03 (3.3e-05)
	hv	1.10e-01 (0.0e+00)†	5.37e-01 (1.3e-04)†	1.12e-01 (1.7e-02)†	5.38e-01 (3.7e-05)
ZDT3	eps	5.03e-01 (7.6e-03)†	1.72e-02 (5.1e-05)	2.93e-01 (8.9e-02)†	1.78e-01 (1.7e-01)
	hv	1.42e-01 (1.1e-01)†	7.24e-01 (3.3e-05)	6.16e-01 (6.0e-02)†	7.04e-01 (2.1e-02)
ZDT4	eps	2.12e+00 (8.0e-01)†	1.22e-02 (3.8e-03)†	3.48e-01 (4.1e-02)†	1.06e-02 (3.6e-03)
	hv	0.00e+00 (0.0e+00)†	8.65e-01 (4.0e-03)†	7.01e-01 (2.7e-02)†	8.67e-01 (2.9e-03)
ZDT6	eps	4.31e-01 (1.1e-05)†	6.07e-03 (8.6e-06)†	1.13e-01 (4.6e-05)†	5.23e-03 (1.6e-05)
	hv	2.10e-01 (8.5e-06)†	6.11e-01 (5.3e-06)†	5.83e-01 (2.6e-05)†	6.11e-01 (8.0e-06)
DTLZ1	eps	6.76e-01 (9.4e-01)†	9.71e-02 (2.4e-03)†	1.60e-01 (9.8e-02)†	6.13e-02 (4.0e-03)
	hv	0.00e+00 (2.7e-01)†	1.06e+00 (4.0e-03)†	8.83e-01 (7.0e-02)†	1.12e+00 (3.3e-03)
DTLZ2	eps	4.04e-01 (5.2e-02)†	9.85e-02 (4.7e-04)†	8.65e-02 (7.7e-04)†	5.56e-02 (6.2e-04)
	hv	3.44e-02 (7.8e-02)†	6.68e-01 (2.6e-03)†	7.38e-01 (2.5e-04)†	7.46e-01 (3.0e-04)

replacement strategies, WSFM is only outperformed by CHB on the ZDT3 problem, whose Pareto front consists of five disconnected parts. This suggests that WSFM works better for continuous than disconnected shapes of the Pareto front.

Figure 4 depicts the result with the best hypervolume obtained by WSFM for two- and three-objective problems having distinct characteristics. The presented plots show that adequate approximations can be obtained for different Pareto front geometries. Although WSFM loses to CHB on ZDT3 regarding the quality indicators, the Pareto front of this problem is adequately approximated as well.

4.2 Problems with Complicated Pareto Sets

Problems with complicated Pareto sets were introduced in [11] specifically for investigating advantages of MOEA/D. They were designed to resemble properties of real-world problems, such as complex nonlinear interactions between the decision variables. Several studies showed that decomposition based approaches are especially useful for handling such problems. The present study tests a differential evolution variant of MOEA/D-WSFM and its ability to deal with challenging characteristics of these problems. MOEA/D is run for 500

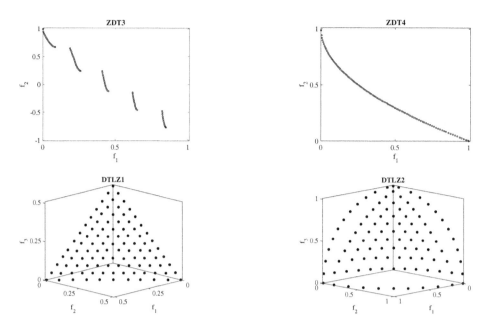

Fig. 4. Pareto front approximations obtained by MOEA/D-WSFM on some ZDT and DTLZ problems.

generations with $\mu = 300$. The other parameter settings are $\delta = 0.9$, $T = 20$ and $n_r = 2$.

Table 2 shows the results in terms of the quality indicators obtained by differ-ent MOEA/D variants. It is apparent that WSFM produces highly competitive performance. When the global replacement strategy is used, WSFM is only out-performed on LZ09_F6. This is a three-objective problem with a spherical Pareto front. The ability to deal with such Pareto front geometry was demonstrated in the previous experiments. Eventually, a larger population size may be needed to successfully handle this problem [11]. Overall, the obtained results further con-firm a trend that was observed so far. Specifically, the performance of WSFM becomes more likely superior to other methods when the replacement is per-formed in the neighborhood of the best matching subproblem to the offspring. The reason is due to competitive characteristics of WSFM for preference articu-lation. As shown in [6], WSFM can identify solutions better reflecting the deci-sion maker preferences expressed by weights when comparing with traditional methods. As results of this study show, this is translated into a better corre-spondence between individuals and respective subproblems during the search, which can improve the MOEA/D performance.

Figure 5 shows the evolution of the epsilon indicator on three challenging problems from the LZ09 suite. The plots in this figure illustrate that the use of WSFM not only allows obtaining approximation sets with better values of the quality indicators but also can provide a faster convergence during the

Table 2. Results for LZ09 problems. The values refer to the median and interquartile range of the epsilon (eps) and hypervolume (hv) indicators.

		WSUM	CHB	PBI	WSFM
		MOEA/D			
LZ09F1	eps	1.08e-02 (8.5e-04)†	2.25e-03 (1.0e-04)†	2.83e-01 (1.5e-02)†	1.80e-03 (6.3e-05)
	hv	8.72e-01 (5.2e-05)†	8.75e-01 (3.7e-05)†	7.41e-01 (7.7e-03)†	8.75e-01 (5.4e-05)
LZ09F2	eps	1.82e-01 (2.3e-01)†	1.18e-02 (9.7e-03)†	3.46e-01 (7.9e-02)†	7.21e-03 (2.3e-03)
	hv	7.66e-01 (1.7e-01)†	8.69e-01 (1.8e-03)†	6.55e-01 (5.6e-02)†	8.71e-01 (1.5e-03)
LZ09F3	eps	1.07e-01 (2.3e-02)†	1.57e-02 (3.2e-02)†	4.25e-01 (4.1e-02)†	9.25e-03 (1.7e-02)
	hv	8.28e-01 (1.4e-02)†	8.70e-01 (4.4e-03)†	6.54e-01 (2.9e-02)†	8.71e-01 (2.1e-03)
LZ09F4	eps	1.35e-01 (3.0e-02)†	4.34e-02 (5.1e-02)†	4.30e-01 (5.2e-02)†	1.04e-02 (1.3e-02)
	hv	8.12e-01 (1.3e-02)†	8.67e-01 (9.9e-03)†	6.51e-01 (3.6e-02)†	8.71e-01 (3.1e-03)
LZ09F5	eps	1.24e-01 (9.6e-03)†	7.46e-02 (3.1e-02)†	3.82e-01 (4.7e-02)†	6.51e-02 (2.6e-02)
	hv	8.12e-01 (9.8e-03)†	8.58e-01 (5.9e-03)†	6.83e-01 (3.0e-02)†	8.60e-01 (4.7e-03)
LZ09F6	eps	3.97e-01 (5.0e-11)†	1.34e-01 (4.6e-02)†	5.02e-02 (7.1e-03)	1.08e-01 (2.1e-02)†
	hv	0.00e+00 (0.0e+00)†	7.07e-01 (1.2e-02)†	7.63e-01 (1.6e-03)	7.34e-01 (7.6e-03)†
LZ09F7	eps	2.12e-01 (5.0e-02)†	2.93e-03 (4.0e-04)†	1.71e-01 (1.8e-02)†	1.91e-03 (1.2e-04)
	hv	7.84e-01 (3.9e-02)†	8.75e-01 (8.5e-05)†	7.89e-01 (7.8e-03)†	8.75e-01 (7.0e-05)
LZ09F8	eps	3.77e-01 (5.4e-02)†	3.98e-03 (1.6e-03)†	3.62e-01 (2.0e-01)†	2.48e-03 (4.5e-04)
	hv	6.10e-01 (7.5e-02)†	8.74e-01 (7.4e-04)†	6.21e-01 (9.5e-02)†	8.74e-01 (2.8e-04)
LZ09F9	eps	3.81e-01 (2.3e-07)†	1.44e-02 (8.9e-03)	1.60e-01 (1.8e-01)†	1.49e-02 (1.4e-02)
	hv	0.00e+00 (0.0e+00)†	5.33e-01 (4.1e-03)	4.89e-01 (8.2e-02)†	5.34e-01 (3.8e-03)
		MOEA/D-GR			
LZ09F1	eps	3.07e-01 (2.9e-02)†	2.21e-03 (6.1e-05)†	2.94e-01 (1.2e-02)†	1.81e-03 (6.0e-05)
	hv	5.39e-01 (1.6e-02)†	8.75e-01 (3.6e-05)†	7.34e-01 (6.9e-03)†	8.75e-01 (2.5e-05)
LZ09F2	eps	5.62e-01 (1.0e-01)†	1.07e-02 (4.3e-03)†	3.31e-01 (8.6e-02)†	6.00e-03 (3.0e-03)
	hv	3.15e-01 (2.3e-01)†	8.70e-01 (1.4e-03)†	6.65e-01 (5.8e-02)†	8.71e-01 (1.1e-03)
LZ09F3	eps	5.33e-01 (4.2e-02)†	1.14e-02 (1.5e-02)†	4.28e-01 (3.6e-02)†	7.16e-03 (5.9e-03)
	hv	2.83e-01 (5.8e-02)†	8.71e-01 (2.2e-03)†	6.51e-01 (2.5e-02)†	8.71e-01 (1.9e-03)
LZ09F4	eps	5.35e-01 (6.3e-02)†	3.07e-02 (2.8e-02)†	4.26e-01 (4.8e-02)†	6.94e-03 (2.4e-03)
	hv	2.94e-01 (1.4e-01)†	8.69e-01 (2.9e-03)†	6.53e-01 (3.4e-02)†	8.72e-01 (8.5e-04)
LZ09F5	eps	5.02e-01 (4.5e-02)†	6.75e-02 (2.9e-02)	3.89e-01 (3.0e-02)†	6.05e-02 (3.0e-02)
	hv	3.27e-01 (7.5e-02)†	8.58e-01 (6.1e-03)†	6.77e-01 (2.0e-02)†	8.61e-01 (4.6e-03)
LZ09F6	eps	3.98e-01 (2.5e-03)†	1.25e-01 (3.6e-02)†	6.70e-02 (5.5e-03)	1.05e-01 (2.3e-02)†
	hv	0.00e+00 (0.0e+00)†	7.10e-01 (1.0e-02)†	7.50e-01 (2.4e-03)	7.39e-01 (9.0e-03)†
LZ09F7	eps	6.05e-01 (1.4e-02)†	3.05e-03 (3.9e-04)†	1.69e-01 (3.1e-01)†	1.90e-03 (1.4e-04)
	hv	2.33e-01 (2.7e-02)†	8.75e-01 (7.5e-05)†	7.88e-01 (1.8e-01)†	8.75e-01 (8.6e-05)
LZ09F8	eps	6.16e-01 (4.0e-03)†	4.06e-03 (8.9e-03)†	5.01e-01 (2.1e-01)†	2.65e-03 (1.0e-03)
	hv	2.13e-01 (8.1e-03)†	8.74e-01 (2.8e-03)	5.96e-01 (1.1e-01)†	8.74e-01 (6.1e-04)
LZ09F9	eps	3.82e-01 (2.5e-02)†	1.56e-02 (1.2e-02)	1.06e-01 (7.8e-02)†	1.43e-02 (1.1e-02)
	hv	0.00e+00 (0.0e+00)†	5.34e-01 (2.2e-03)†	4.94e-01 (4.0e-02)†	5.35e-01 (2.5e-03)

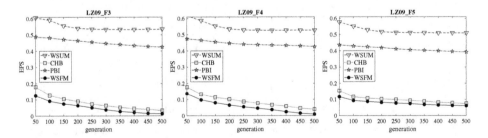

Fig. 5. Evolution of the epsilon indicator. The plots refer to the mean values over 30 runs. The lower the better.

generations. This feature is especially useful when handling real-world problems where function evaluations can be computationally expensive.

4.3 Engineering Problems

Analyzing the performance of different MOEAs on artificially constructed test problems is advantageous, as the Pareto sets and the properties of these problems are known. However, such problems often do not pose difficulties that are encountered in real-world applications, being frequently criticized due to this fact. To illustrate a practical relevance and validity of MOEA/D-WSFM, three engineering problems are selected from the literature for experiments. Two considered problems refer to the design of four bar truss [15] and welded beam [3]. Both problems involve two objectives and four design variables that are restricted by box constraints. In addition, four inequality constraints are associated with the design of welded beam. The third is car side impact problem [8], which involves three objectives and seven variables as well as ten inequality constraints.

Table 3 presents the results of the application of different MOEA/D variants to solving engineering problems. These results suggest that MOEA/D-WSFM performs the best on these problems regarding the quality indicators. This is valid for both replacement strategies. Figure 6 displays the approximations to the Pareto fronts with the best hypervolume values obtained with WSFM. The presented plots show that WSFM can effectively approximate the Pareto fronts for engineering problems. The obtained results demonstrate that WSFM not only exhibits a competitive performance on some test problem but also can be useful in practical applications. Also, WSFM can be successfully combined with a penalty function for handling constraints.

Table 3. Results for engineering problems. The values refer to the median and interquartile range of the epsilon (eps) and hypervolume (hv) indicators.

		WSUM	CHB	PBI	WSFM
		MOEA/D			
		1-objectives			
Four Bar Truss	eps	1.17e-02 (1.0e-03)†	7.20e-03 (3.2e-05)†	2.39e-01 (3.7e-02)†	4.88e-03 (9.8e-05)
	hv	6.71e-01 (2.7e-05)†	6.74e-01 (3.9e-05)†	5.62e-01 (1.8e-02)†	6.74e-01 (1.3e-05)
Welded Beam	eps	4.34e-03 (3.0e-03)	4.91e-03 (2.8e-03)†	4.53e-02 (1.9e-02)†	3.44e-03 (3.5e-03)
	hv	9.93e-01 (2.6e-03)	9.92e-01 (2.6e-03)†	9.49e-01 (1.5e-02)†	9.94e-01 (3.0e-03)
Car Side Impact	eps	5.17e-01 (0.0e+00)†	1.91e-01 (7.3e-02)	9.99e-01 (3.6e-03)†	1.76e-01 (3.0e-02)
	hv	2.68e-01 (3.9e-16)†	4.74e-01 (1.5e-02)†	8.14e-04 (3.6e-03)†	4.85e-01 (8.5e-03)
		MOEA/D-GR			
Four Bar Truss	eps	1.74e-01 (3.6e-02)†	7.19e-03 (2.7e-03)†	2.47e-01 (3.7e-02)†	4.85e-03 (8.2e-05)
	hv	5.02e-01 (3.0e-02)†	6.74e-01 (3.7e-05)†	5.60e-01 (1.4e-02)†	6.74e-01 (2.5e-05)
Welded Beam	eps	4.08e-02 (1.5e-02)†	4.79e-03 (2.8e-03)†	7.60e-02 (3.5e-02)†	2.81e-03 (2.2e-03)
	hv	9.85e-01 (1.5e-02)†	9.93e-01 (1.7e-03)†	9.23e-01 (3.5e-02)†	9.95e-01 (2.8e-03)
CarSide Impact	eps	3.42e-01 (7.6e-02)†	1.74e-01 (4.5e-02)	4.08e-01 (1.8e-01)†	1.74e-01 (1.5e-02)
	hv	1.93e-01 (5.1e-02)†	4.76e-01 (9.1e-03)†	3.33e-01 (1.5e-01)†	4.89e-01 (9.2e-03)

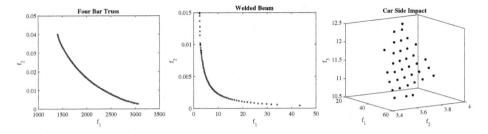

Fig. 6. Pareto fronts obtained by MOEA/D-WSFM for engineering problems.

5   Conclusions

MOEA/D is a popular state-of-the-art framework in the field of evolutionary multiobjective optimization. By means of scalarization, MOEA/D decomposes a MOP into a number of single-objective subproblems. During the run, sub-problems are optimized simultaneously exploiting the neighborhood relations between them. The scalarizing function is an important issue that largely affects the performance of MOEA/D.

This study suggested a new scalarizing function for MOEA/D, named WSFM. It is characterized by the stress function that performs a nonlinear mapping taking into account objective and weight values. This process is inspired by the stress-strain behavior of thermoplastic vulcanizates. The experimental results suggest that MOEA/D with WSFM can achieve a faster convergence and better final results with respect to quality indicators. WSFM works particularly better when the global replacement strategy is used. The validity of the approach is also demonstrated on some engineering problems, highlighting its practical relevance.

As future work, there are plans to investigate the performance of WSFM on other test and real-world problems, including those having a large number of objectives. Also, the development of an adaptive scheme for controlling the parameters in WSFM is a promising research direction.

Acknowledgements. This work has been supported by FCT - Fundação para a Ciência e Tecnologia in the scope of the project: PEst-OE/EEI/UI0319/2014.

References

1. Abdou-Sabet, S., Datta, S.: Thermoplastic Vulcanizates. Polymer Blends: Formulation and Performance. Wiley, New York (2000)
2. Deb, K., Pratap, A., Agarwal, S., Meyarivan, T.: A fast and elitist multiobjective genetic algorithm: NSGA-II. IEEE Trans. Evol. Comput. **6**(2), 182–197 (2002)
3. Deb, K., Pratap, A., Moitra, S.: Mechanical component design for multiple ojectives using elitist non-dominated sorting GA. In: Schoenauer, M., Deb, K., Rudolph, G., Yao, X., Lutton, E., Merelo, J.J., Schwefel, H.P. (eds.) PPSN 2000. LNCS, vol. 1917, pp. 859–868. Springer, Heidelberg (2000). doi:10.1007/3-540-45356-3_84

4. Denysiuk, R., Costa, L., Espírito Santo, I.: MOEA/VAN: multiobjective evolutionary algorithm based on vector angle neighborhood. In: Proceedings of the Conference on Genetic and Evolutionary Computation, pp. 663–670 (2015)
5. Denysiuk, R., Costa, L., Espírito Santo, I., C. Matos, J.: MOEA/PC: multiobjective evolutionary algorithm based on polar coordinates. In: Gaspar-Cunha, A., Henggeler Antunes, C., Coello, C.C. (eds.) EMO 2015. LNCS, vol. 9018, pp. 141–155. Springer, Heidelberg (2015). doi:10.1007/978-3-319-15934-8_10
6. Ferreira, J.C., Fonseca, C.M., Gaspar-Cunha, A.: Methodology to select solutions from the Pareto-optimal set: a comparative study. In: Proceedings of the Conference on Genetic and Evolutionary Computation, pp. 789–796 (2007)
7. Ishibuchi, H., Sakane, Y., Tsukamoto, N., Nojima, Y.: Simultaneous use of different scalarizing functions in MOEA/D. In: Proceedings of the Conference on Genetic and Evolutionary Computation, pp. 519–526 (2010)
8. Jain, H., Deb, K.: An evolutionary many-objective optimization algorithm using reference-point-based nondominated sorting approach, part II: handling constraints and extending to an adaptive approach. IEEE Trans. Evol. Comput. $18(4)$, 577–601 (2002)
9. Jan, M.A., Zhang, Q.: MOEA/D for constrained multiobjective optimization: some preliminary experimental results. In: UK Workshop on Computational Intelligence, pp. 1–6 (2010)
10. Knowles, J., Thiele, L., Zitzler, E.: A tutorial on the performance assessment of stochastic multiobjective optimizers. Technical report 214, TIC, ETH Zurich, Switzerland (2006)
11. Li, H., Zhang, Q.: Multiobjective optimization problems with complicated Pareto sets, MOEA/D and NSGA-II. IEEE Trans. Evol. Comput. $13(2)$, 284–302 (2009)
12. Li, K., Deb, K., Zhang, Q., Kwong, S.: An evolutionary many-objective optimization algorithm based on dominance and decomposition. IEEE Trans. Evol. Comput. $19(5)$, 694–716 (2015)
13. Liu, H.L., Gu, F., Zhang, Q.: Decomposition of a multiobjective optimization problem into a number of simple multiobjective subproblems. IEEE Trans. Evol. Comput. $18(3)$, 450–455 (2014)
14. Miettinen, K.: Nonlinear Multiobjective Optimization. Kluwer Academic Publishers, London (1999)
15. Ray, T., Liew, K.M.: A swarm metaphor for multiobjective design optimization. Eng. Optim. $34(2)$, 141–153 (2002)
16. Wang, Z., Zhang, Q., Zhou, A., Gong, M., Jiao, L.: Adaptive replacement strategies for MOEA/D. IEEE Trans. Cybern. $46(2)$, 474–486 (2016)
17. Wang, L., Zhang, Q., Zhou, A., Gong, M., Jiao, L.: Constrained subproblems in a decomposition-based multiobjective evolutionary algorithm. IEEE Trans. Evol. Comput. $20(3)$, 475–480 (2016)
18. Zhang, Q., Li, H.: MOEA/D: A multiobjective evolutionary algorithm based on decomposition. IEEE Trans. Evol. Comput. $11(6)$, 712–731 (2007)
19. Zhou, A., Zhang, Q.: Are all the subproblems equally important? Resource allocation in decomposition-based multiobjective evolutionary algorithms. IEEE Trans. Evol. Comput. $20(1)$, 52–64 (2016)
20. Zitzler, E., Künzli, S.: Indicator-based selection in multiobjective search. In: Yao, X., et al. (eds.) PPSN 2004. LNCS, vol. 3242, pp. 832–842. Springer, Heidelberg (2004). doi:10.1007/978-3-540-30217-9_84

Timing the Decision Support for Real-World Many-Objective Optimization Problems

João A. Duro[1](✉) and Dhish Kumar Saxena[2]

[1] Automatic Control and Systems Engineering Department,
The University of Sheffield, Sheffield, UK
j.a.duro@sheffield.ac.uk
[2] Department of Mechanical and Industrial Engineering,
Indian Institute of Technology, Roorkee, Roorkee, India
dhishfme@iitr.ac.in

Abstract. Lately, there is growing emphasis on improving the scalability of *multi-objective evolutionary algorithms* (MOEAs) so that many-objective problems (characterized by more than three objectives) can be effectively dealt with. Alternatively, the utility of integrating *decision maker's* (DM's) preferences into the optimization process so as to target some most preferred solutions by the DM (instead of the whole Pareto-optimal front), is also being increasingly recognized. The authors here, have earlier argued that despite the promises in the latter approach, its practical utility may be impaired by the lack of—*objectivity, repeatability, consistency*, and *coherence* in the DM's preferences. To counter this, the authors have also earlier proposed a machine learning based decision support framework to reveal the preference-structure of objectives. Notably, the revealed preference-structure may be sensitive to the timing of application of this framework along an MOEA run. In this paper the authors counter this limitation, by integrating a termination criterion with an MOEA run, towards determining the appropriate timing for application of the machine learning based framework. Results based on three real-world many-objective problems considered in this paper, highlight the utility of the proposed integration towards an *objective, repeatable, consistent*, and *coherent* decision support for many-objective problems.

1 Introduction

Multi-objective evolutionary algorithms (MOEAs) have been well known to approximate the true *Pareto-optimal front* (POF) for a given multi-objective optimization problem, without emphasizing one objective over the other [1]. Notably, unlike the case of two- and three-objective problems, the performance of most existing MOEAs deteriorates as the number of objectives (M) grow beyond three [2]. This perhaps explains as to why optimization problems comprising of four or more objectives are distinctively referred to as *many-objective problems* (MaOPs) and have been receiving a lot of attention recently. The challenges associated with MaOPs relate to: (a) the nature of high-dimensional problems, and (b) the manner in which the existing MOEAs discriminate between

© Springer International Publishing AG 2017
H. Trautmann et al. (Eds.): EMO 2017, LNCS 10173, pp. 191–205, 2017.
DOI: 10.1007/978-3-319-54157-0_14

better and worse solutions (*selection* operation). In terms of the former, visualization of a high-dimensional search space is difficult, and a good approximation (complete convergence and full coverage) of a high-dimensional POF calls for an (impractical) exponential increase in population size with a linear increase in M [3]. In terms of the latter, the *selection* operation either becomes ineffective or computationally too demanding. For instance: (a) the most commonly used *primary selection*, as in NSGA-II [4], is based on Pareto-dominance which fails to induce an effective partial order on the solutions as the number of objectives increases [5], (b) there is huge computational cost involved in dealing with a large number of weight vectors in decomposition based MOEAs, such as MOEA/D [6], and (c) the indicator based MOEAs, such as HypE [7] become impractical since indicators like hypervolume are computationally too demanding.

Acknowledging the poor scalability of most existing MOEAs with the number of objectives, an emergent strategy is to target only a handful of optimal or near-optimal solutions that are most preferred by the *decision maker* (DM). The engagement of a DM with the optimization process has led to what is referred to as *multiple criteria decision making* (MCDM) based MOEAs [8]. The authors in [5] have argued that despite their promise, the utility of MCDM based MOEAs may be impaired by the lack of–*objectivity, repeatability, consistency*, and *coherence* in the DMs' preferences. Towards countering these limitations, the authors have proposed a framework for machine learning based decision support:

1. expressed through *revelation* of the preference–structure of different objectives embedded in the problem model,
2. which can potentially aid the DMs to induce a preference–order over the solutions (guided by the preference–order of the objectives) with *objectivity, repeatability, consistency*, and *coherence*.

This paper distinguishes between: (δ-I) the *capability of a decision support* framework to *reveal* the preference–structure of different objectives in a *given* input solution set (non-dominated solutions obtained from an MOEA), and (δ-II) the *capability to determine the timing of decision support*, implying, the capability to determine the number of generations along an MOEA run at/after which, the corresponding non-dominated solution set could be treated as the most *appropriate* input solution set for application of the machine learning based framework for revelation of the objectives' preference–structure. While the former has been demonstrated in [5], this paper focuses on the latter—a more fundamental aspect of *when to time the decision support*. The criticality of the latter aspect can be gauged from the fact that it is analogous to the fundamental and largely unaddressed question (until [9]) of when to terminate an MOEA in the absence of a termination criterion that is *robust* in terms of its: (a) *generality*, implying that it does not require an a priori knowledge of the POF, and neither depends on MOEA-specific operators, nor on the MOEA-related performance indicators, (b) *on-the-fly* implementation, and (c) *computational efficiency* enabling efficient scalability with the number of objectives. In the absence of [9], the capability of the framework proposed in [5] was demonstrated on input solution sets corresponding to *a priori* fixed number of generations. However, despite its novelty and theoretical contribution, the practical utility of [5] stands implicitly impaired by the fact that if the *a priori* fixed number of generations are not sufficient to ensure that the input solution set corresponds to a *stabilized* MOEA population, the *revealed* objectives' preference–structure and solutions' preference–order can not be treated as accurate, robust and efficient representatives of the problem–structure. In that, if the *a priori* fixed number of generations are:

A. too low: the MOEA population may not stabilize and the *revealed* problem–structure may itself change if the underlying MOEA was allowed to run for more generations, rendering the offered decision support futile,
B. too high: the MOEA population may stabilize far earlier, implying wastage of computational resources despite no variation in *revealed* problem–structure.

In view of the above, this paper aims to bridge the gap between *capability of a decision support* framework (δ-I) and *capability to determine the timing of decision support* (δ-II), and achieves this by integrating the capabilities developed in [5,9]. The utility of this work in terms of the resulting *robust* decision support has been highlighted through three real-world instances of many-objective optimization problems. In that, it is demonstrated: (a) how an ad hoc *a priori* fixation of the timing of application of the framework for decision support [5] may lead to misleading objectives' preference–structure, and (ii) how its integration with an entropy based MOEA–termination criterion can help overcome this challenge, leading up to a *robust* decision support.

The structure of the remaining paper is as follows. The decision-support framework [5] and the MOEA-termination algorithm [9] are briefly described in Sect. 2. This is followed by a brief description of the considered real-world MaOPs in Sect. 3. The results and associated discussions are presented in Sect. 4, while the paper concludes with Sect. 5.

2 Methodology

It has been highlighted above that aiming at an accurate, robust and efficient decision support, this paper integrates a framework for machine learning based decision support [5] and an MOEA-termination algorithm [9]. For completeness, each of these are summarized below.

2.1 Framework for Machine Learning Based Decision Support

The framework for machine learning based decision support proposed in [5], *reveals* the preference–structure of different objectives in terms of:

1. an *essential* objective set, implying a smallest set of conflicting objectives which can generate the same POF as the original problem,

2. ranking of *essential* objectives which further paves the way for revelation of: (i) the smallest objective sets corresponding to pre-specified errors, and (ii) objective sets of pre-specified sizes that correspond to minimum error.

This framework is based on the premise that: (i) most existing MOEAs provide a poor POF-approximation for MaOPs, implying that the dominance relations characterizing the obtained solutions may be different from those characterizing the true POF, and (ii) the dimensionality of the POF need not be the same as the number of objectives M. The problem of *learning* objectives' preference–structure on the true POF, from a solution set that may not represent the true POF, is posed as a machine learning problem. In that, given a set of non-dominated solutions, the objectives' preference–structure is *learnt* by application of PCA (Principal Component Analysis: relying on the correlation matrix R) or MVU (Maximum Variance Unfolding: relying on the kernel matrix K), aided by:

1. an interpretation of objectives as conflicting or non-conflicting based on their relationship along the eigenvectors of R or K,
2. a dynamic interpretation of the correlation between the objectives.

Depending on the choice of R or K, the framework allows for two variants, namely L-PCA based decision support and NL-MVU-PCA based decision support. The two can be distinguished in terms of the former's inability to account for nonlinearity, unlike the latter. While the details of the framework can be found in [5], its building blocks are summarized below.

Computation of Correlation and Kernel Matrices. Let a non-dominated solution set be obtained by running an MOEA with the initial objective set $\mathcal{F}_0 = \{f_1, \ldots, f_M\}$ and a population size of N. Let the objective vector in the non-dominated set, corresponding to the ith objective (f_i) be denoted by $\dot{f}_i \in \mathbb{R}^N$ and its mean and standard deviation by $\mu_{\dot{f}_i}$ and $\sigma_{\dot{f}_i}$, respectively. Furthermore, let $\ddot{f}_i = (\dot{f}_i - \mu_{\dot{f}_i})/\sigma_{\dot{f}_i}$. Then, the input data X and the correlation matrix R can be composed as $X_{M \times N} = [\ddot{f}_1 \ \ddot{f}_2 \ldots \ddot{f}_M]^T$ and $R_{M \times M} = \frac{1}{M} X X^T$. Furthermore, the kernel matrix K can be *learnt* from a semidefinite programming problem, presented in Eq. 1.

$$
\begin{aligned}
&\text{Maximize } trace(K) = \sum_{ij} \frac{(K_{ii} - 2K_{ij} + K_{jj})}{2M} \\
&\text{subject to the following constraints :} \\
&(a) \ K_{ii} - 2K_{ij} + K_{jj} = R_{ii} - 2R_{ij} + R_{jj}, \forall \ \eta_{ij} = 1 \\
&(b) \ \sum_{ij} K_{ij} = 0 \\
&(c) \ K \text{ is positive-semidefinite,} \\
&\text{where: } R_{ij} \text{ is the } (i,j)^{th} \text{ element of the correlation matrix } R \\
&\qquad \eta_{ij} = \begin{cases} 1, \text{ if } \dot{f}_j \text{ is among the } q\text{-nearest neighbor of } \dot{f}_i \\ 0, \text{ otherwise} \end{cases}
\end{aligned}
\tag{1}
$$

Objective Reduction Based on Eigenvalue Analysis. This step aims to:
(i) interpret the objectives as conflicting or non-conflicting based on their relationship along the significant eigenvectors of R or K (depending on the choice
of L-PCA based or NL-MVU-PCA based decision support), and (ii) retain only
the conflicting objectives. Towards it, obtain the eigenvalues of R or K, and
sort these in descending order of magnitude as $\lambda_1 \geq \lambda_2 \geq \ldots \geq \lambda_M$. Let the
corresponding eigenvectors be given by $V_1, V_2, \ldots, V_M$, the contribution of the
i^{th} objective towards V_j be given by f_{ij}, and the normalized eigenvalues be
given by $e_j = \lambda_i / \sum_{j=1}^{M} \lambda_i$ (implying $\sum_{j=1}^{M} e_j = 1$). Subsequently, determine
the number of *significant* eigenvectors as the smallest number (N_v) such that
$\sum_{j=1}^{N_v} e_j \geq \theta$, where the variance threshold $\theta \in [0, 1]$ is an algorithm parameter
prescribed to be set to $\theta = 0.997$. Then, the set of objectives $(\mathcal{F}_e)$ based on
conflict along significant eigenvectors can be composed by picking for each significant V_js: the objective with the highest contribution (in terms of magnitude)
along with all other opposite sign objectives. As an exception, if all objectives
have the same-sign contribution along a particular V_j, then the objectives with
top two contributions by magnitude are selected. Notably $\mathcal{F}_e \subseteq \mathcal{F}_0$.

Objective Reduction Based on Set-Based Correlation. This step aims
to: (i) identify the subsets of correlated objectives within $\mathcal{F}_e$, and (ii) retain from
each such correlated subset, only the most significant objective while discarding
the rest. Towards it, for each $f_i \in \mathcal{F}_e$, constitute a subset $\mathcal{S}_i$ with correlated
objectives $f_{j|j \neq i} \in \mathcal{F}_e$ based on the following:

$$f_j \in \mathcal{S}_i : \begin{cases} sign(R_{ik}) = sign(R_{jk}), \ \forall \ k = 1, 2, \ldots, M, \\ R_{ij} \geq T_{cor} = 1.0 - e_1(1.0 - M_{2\sigma}/M), \end{cases}$$

where T_{cor} is the correlation threshold and $M_{2\sigma}$ is the smallest number, such
that $\sum_{j=1}^{M_{2\sigma}} e_j \geq 0.954$. Subsequently, for each objective in $\mathcal{S}_i$, assign a score sc_i
given by:

$$sc_i = \sum_{j=1}^{N_v} e_j |f_{ij}|$$

following which, the objective with the highest sc_i is retained while the other
objectives in $\mathcal{S}_i$ are eliminated. This step facilitating further objective reduction
leads to an *essential* objective set $\mathcal{F}_s$, where $\mathcal{F}_s \subseteq \mathcal{F}_e \subseteq \mathcal{F}_0$.

Preference-Ranking of All Objectives. The preference-ranking of objectives is guided by the errors to be incurred on elimination of objectives. An error
associated with elimination of a particular objective corresponds to the variance
that is left unaccounted if that objective is eliminated alone, and can be computed as in Eq. 2, depending on whether it belongs to an *essential* or *redundant*
objective set.

$$\left.\begin{array}{l} \mathcal{E}_i = c_i^M \text{ for } f_i \in \mathcal{F}_s \\ \mathcal{E}_i = c_i^M (1.0 - \max_{j \in \mathcal{F}_s}\{\delta_{ij}.R_{ij}\}) \text{ for } f_i \in \mathcal{F}_{redn} \equiv \mathcal{F}_0 \backslash \mathcal{F}_s \\ \text{where:} \\ c_i^M = \sum_{k=1}^{M} e_k f_{ik}^2 \\ \delta_{ij} = \begin{cases} 1, \text{ if } f_i \text{ and } f_j \text{ are correlated} \\ 0, \text{ otherwise} \end{cases} \\ R_{ij} = \text{Correlation between } f_i \text{ and} f_j \end{array}\right\} \tag{2}$$

Finally, the preference-weight for each objective is given by Eq. 3 and the preference-ranking of all objectives is established by the sorted $\mathcal{E}_i$s. For instance, let u and v be two objectives such that $\mathcal{E}_u^n \gg \mathcal{E}_v^n$, implying that the error incurred by eliminating u is far greater than the error incurred if objective v were eliminated. In other words, for higher accuracy, the objective u needs to be preferred over v (or the solutions which are better in u need to be preferred over those which are better in v).

$$w_i = \mathcal{E}_i^n = \mathcal{E}_i / \sum_{j=1}^{M} \mathcal{E}_j \text{ (ensuring that } w_i \geq 0 \text{ and } \sum_{i=1}^{M} w_i = 1) \tag{3}$$

2.2   An Entropy Based MOEA-Termination Algorithm

The MOEA-termination algorithm proposed in [9] is based on the premise that it is prudent to terminate an MOEA if the MOEA population in successive generations does not undergo significant changes. Towards implementation of this principle:

1. a *dissimilarity* measure across two successive MOEA generations, based on information theory concepts of *entropy* and *relative entropy* was proposed in [9],
2. the multidimensional histogram method (Sect. IV in [9]) which relies on partitioning each dimension of the M-dimensional objective space into a fixed number of intervals (n_b), facilitates the computation of a *dissimilarity* measure (Sect. V in [9]),
3. the conformance of the mean and standard deviation of the *dissimilarity* measure up to a pre-specified accuracy level (n_p:number of decimal places), across a pre-specified number of successive MOEA generations (n_s), was treated as a termination criterion (Sect. VI in [9]),
4. just when the termination criterion is satisfied, the MOEA run is terminated and the number of MOEA generations up to that point are denoted by N_{gt}.

The resulting termination algorithm helps identify *on-the-fly* the number of generations beyond which an MOEA *stabilizes*, implying that either a good POF–approximation has been obtained, or that it can not be obtained due to the stagnation of the MOEA in the search space (excluding the POF). The fact that in either case, no further improvement in the POF–approximation can be obtained despite additional computational expense, provides the rationale for MOEA-termination. As highlighted earlier, the hallmark of the proposed algorithm (besides its *on-the-fly* implementation), lies in *generality* and *computational efficiency* enabling efficient scalability with the number of objectives.

3   Real-World Many-Objective Optimization Problems

A brief description of the real-world instances of MaOPs considered is as follows.

3.1   Machining Problem

The authors in [10] have investigated the multi-criteria optimization of machining operations, when applied to a 390 die cast aluminum alloy with VC-3 carbide cutting tools. The optimization problem consists of four objectives, three decision variables and three inequality constraints. The decision variables relate to the cutting speed (x_1), feed rate (x_2), and the depth of cut (x_3). The related constraints and objective functions are, as below.

$$
\left.
\begin{aligned}
&\text{Minimize} \quad f_1(\mathbf{x}) \equiv \text{Surface roughness,} \\
&\text{Maximize} \quad f_2(\mathbf{x}) \equiv \text{Surface integrity,} \\
&\text{Maximize} \quad f_3(\mathbf{x}) \equiv \text{Tool life,} \\
&\text{Maximize} \quad f_4(\mathbf{x}) \equiv \text{Metal removal rate,} \\
&\qquad \text{subject to:} \\
&\qquad g_1(\mathbf{x}) \equiv \text{Upper limit on surface roughness} \\
&\qquad g_2(\mathbf{x}) \equiv \text{Lower limit on surface integrity} \\
&\qquad g_3(\mathbf{x}) \equiv \text{Lower limit on tool life.}
\end{aligned}
\right\} \quad (4)
$$

3.2   Storm Drainage System Problem

Water resources management is a discipline that consists of planning, distributing, developing, and managing the optimum use of water resources. The example here deals with the optimal planning for a storm-drainage system located in a urban area as described in [11]. The decision variables relate to the local detention storage capacity (x_1), maximum treatment rate (x_2), and the maximum allowable overflow rate (x_3). The related constraints and objective functions are, as below.

$$
\left.
\begin{aligned}
&\underset{(x_1,x_2,x_3)}{\text{Minimise:}} \\
&\quad f_1(\mathbf{x}) \equiv \text{Drainage network cost,} \\
&\quad f_2(\mathbf{x}) \equiv \text{Storage facility cost,} \\
&\quad f_3(\mathbf{x}) \equiv \text{Treatment facility cost,} \\
&\quad f_4(\mathbf{x}) \equiv \text{Expected flood damage cost,} \\
&\quad f_5(\mathbf{x}) \equiv \text{Expected economic loss due to flood,} \\
&\text{subject to:} \\
&\quad g_1(\mathbf{x}) \equiv \text{Average no. of floods/year,} \\
&\quad g_2(\mathbf{x}) \equiv \text{Probability of flood depth exceeding 1 basin-inch,} \\
&\quad g_3(\mathbf{x}) \equiv \text{Average no. of pounds/year of suspended solids,} \\
&\quad g_4(\mathbf{x}) \equiv \text{Average no. of pounds/year of settable solids,} \\
&\quad g_5(\mathbf{x}) \equiv \text{Average no. of pounds/year of } BOD, \\
&\quad g_6(\mathbf{x}) \equiv \text{Average no. of pounds/year of } N, \\
&\quad g_7(\mathbf{x}) \equiv \text{Average no. of pounds/year of } P0_4.
\end{aligned}
\right\} \quad (5)
$$

3.3   Work Roll Cooling Design Problem

In metalworking, a rolling operation is the process of shaping a metal strip by reducing its thickness and creating a uniform surface. This is achieved by passing the metal between two rollers that are driven at equal peripheral speed in opposite directions. During this process, heat is transferred from the metal strip to the rolls and in case it becomes too excessive it can lead to the formation of roll cracks or to any other types of damage. The decision variables [12] relate to the roll/stock contact HTC (x_1), stock temperature (x_2), roll/stock contact length (x_3), cooling HTC (x_4), roll speed (x_5), roll temperature (x_6), and delay time (x_7). The related constraints and objective functions are, as below.

$$
\left.
\begin{array}{l}
\text{Minimise:} \\
{\scriptstyle (x_1,\ldots,x_7)} \\
\quad f_1(\mathbf{x}) \equiv \text{Change in temperature at roll surface,} \\
\quad f_2(\mathbf{x}) \equiv \text{Radial stress at the roll surface,} \\
\quad f_3(\mathbf{x}) \equiv \text{Change in temperature at } 9\,\text{mm depth,} \\
\quad f_4(\mathbf{x}) \equiv \text{Radial stress at } 9\,\text{mm depth,} \\
\quad f_5(\mathbf{x}) \equiv \text{Change in temperature at } 15\,\text{mm depth,} \\
\quad f_6(\mathbf{x}) \equiv \text{Radial stress at } 15\,\text{mm depth.}
\end{array}
\right\}
\tag{6}
$$

4   Experimental Results

This section demonstrates the application of the decision support framework to three real-world problems. NSGA-II [4] has been used as the underlying MOEA, and the framework is applied to two populations that are chosen at different generations during the NSGA-II run. The first corresponds to the initial population of solutions ($N_g = 1$), while the second population is chosen by the MOEA-termination algorithm (N_{gt}). The analysis generated by the application of the framework to the two populations are compared and their differences are highlighted. The settings of the parameters involved, are as below:

1. the NSGA-II related parameters include a population size of 200; the probability of crossover and mutation as 0.9 and 0.1, respectively; the distribution index for crossover and mutation as 5 and 20, respectively; and the maximum number of generations as 10000,
2. the chosen parameters for the decision support framework are: $\theta = 0.997$; and the neighborhood size is given by $q = M - 1$ (Eq. 1),
3. the chosen parameters for the MOEA-termination algorithm [9] are $n_b = 20$; $n_s = 20$; and $n_p = 2$.

   Towards analysis of the results, in terms of a comparison between R_{ij} (strength of correlation between objectives f_i and f_j); f_{ij} (contribution of objective f_i along the principal component V_j); or the preference weight w_i for objective f_i, across different N_{gt}, the notion of *relative percentage difference* (RPD) is utilized. In that, if the two quantities being compared (say R_{ij} at two different N_{gt}) are denoted by α_1 and α_2, then RPD is given by Eq. 7.

$$
\text{RPD}(\alpha_1, \alpha_2) = |(\alpha_1 - \alpha_2)| / \max(|\alpha_1|, |\alpha_2|) \times 100,
\tag{7}
$$

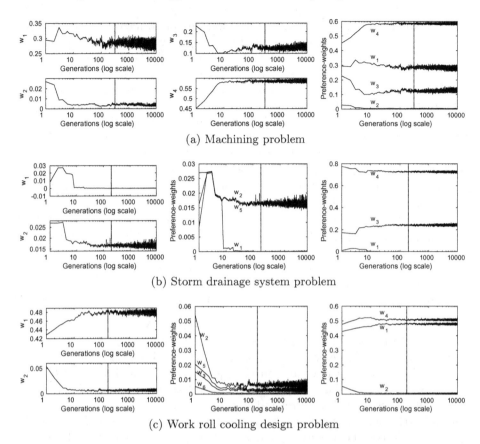

(a) Machining problem

(b) Storm drainage system problem

(c) Work roll cooling design problem

Fig. 1. Preference-weights for the objectives captured along the optimization. The dashed vertical line corresponds to the instant where stability is detected by the termination criterion. The results correspond to one NSGA-II run.

It may be noted that in the tabular results presented ahead, the numbers in the row labeled as RPD(%) denote the RPD(α_1, α_2) values computed by treating the two quantities in the corresponding column above as α_1 and α_2.

Notably, this paper: (i) is based on the premise that the revealed preference-structure may be sensitive to the timing of application of the decision support framework, and (ii) aimed to determine the *appropriate timing* for application of the framework, such that the revealed preference-structure in neither misleading nor comes at an avoidable computational cost. While problem-specific results are discussed in greater details in the following sections, Fig. 1 which presents a snapshot of results for all the problems, not only validates the basic premise of this paper, but also demonstrates that the aim has been realized. In that, regardless of the problem, the following can be observed:

1. there is significant variation in the preference-weight for each objective up to specific number of MOEA generations beyond which it largely stabilizes,

2. in the absence of integration of the decision support framework and MOEA-termination algorithm, when the suitable timing may not be otherwise known, it is highly probable that the *a priori* fixed number of MOEA generations may be such that the preference–structure is revealed either when the preference weights for some/all objectives are undergoing variation or long after they have stabilized. While in the former case, the preference–structure may be misleading, the latter case is marked by avoidable computational wastage,
3. integration of MOEA-termination algorithm has indeed revealed an *appropriate* timing of application of the decision support framework, since the dashed vertical line representing the prescribed timing in each plot is located only just after the preference-weights have stabilized and much before an instance of computational wastage can be cited.

4.1   Machining Problem

For this problem, while the MOEA-termination algorithm infers $N_{gt} = 308$, the corresponding decision support revelations (Table 1) are also compared to those corresponding to *a priori* fixed $N_g = 1$. In that:

1. There is a change in the correlation structure from $N_{gt} = 1$ to $N_{gt} = 308$, since the correlation sign for R_{24} reversed from positive to negative. There are also significant changes to the correlation strengths between some objective pairs. For instance, the change in correlation for R_{13}, and also for R_{23}, is approximately 70%. The analysis of the decision support framework indicates that all objectives are uncorrelated since no two pair of objectives share the same correlation signs with the remaining objectives. One example is f_1 that is positively correlated with f_2 and f_3 (i.e., $R_{12} > 0$ and $R_{13} > 0$), however, since the correlation between f_2 and f_3 is negative (i.e., $R_{23} < 0$), f_1 and f_2 are not interpreted as correlated.
2. There are significant changes to the principal components of the problem. In that: (i) at $N_g = 1$, the first principal component accounts for 74.7079% of the variance (i.e., $e_1 = 0.747079$); and, (ii) at $N_{gt} = 308$, this number rises to 93.5543% (i.e., $e_1 = 0.935543$). It is therefore expected for the preference-weights that are determined at $N_{gt} = 308$, to rely mostly on the first principal component. A high RPD is reported when comparing the elements of the first principal component, that is, almost 200% in some cases. This indicates that the distribution of solutions, across the search space, have experienced a change in direction while approaching stability. The value of T_{cor} suffers a reduction, with an RPD of 15.042%, which is mostly attributed to the increase in e_1.
3. Since no objectives can be interpreted as being correlated due to their correlation signs, the essential objective set is $\mathcal{F}_s = \{f_1, f_2, f_3, f_4\}$. This remains unaltered during the optimization process.
4. There are some significant changes to the preference-weights. In that, a reduction that corresponds to an RPD of 83.569% and 48.278% is reported, respectively, for w_2 and w_3. While an increase that corresponds to an RPD

Table 1. Key highlights for the machining problem.

Gens	Correlation structure						T_{cor}
	R_{12}	R_{13}	R_{14}	R_{23}	R_{24}	R_{34}	
$N_g = 1$	0.512585	0.222373	−0.786626	−0.715817	0.079025	−0.750730	0.626461
$N_{gt} = 308$	0.498700	0.734629	−0.913131	−0.208278	−0.145054	−0.936288	0.532228
RPD(%)	2.7088	69.730	13.854	70.903	154.48	19.818	15.042

Gens	First principal component				Second principal component			
	f_{11}	f_{21}	f_{31}	f_{41}	f_{12}	f_{22}	f_{32}	f_{42}
$N_g = 1$	0.5581	−0.0666	0.2849	−0.7765	−0.4955	−0.2911	0.8178	−0.0312
$N_{gt} = 308$	−0.5404	0.0359	−0.2860	0.7905	−0.5558	−0.1740	0.8091	−0.0793
RPD(%)	196.82	153.84	199.61	198.22	10.839	40.223	1.0666	60.694

Gens	Preference-weights						F_s
	e_1	e_2	w_1	w_2	w_3	w_4	
$N_g = 1$	0.747079	0.249107	0.2946281	0.02694043	0.2272319	0.4511996	$\{f_1, f_2, f_3, f_4\}$
$N_{gt} = 308$	0.935543	0.062604	0.2927858	0.00442904	0.1175300	0.5852552	$\{f_1, f_2, f_3, f_4\}$
RPD(%)	20.145	74.869	0.62530	83.560	48.278	22.905	—

of 22.905% is reported for w_4. The preference-weight w_1 remains almost unaltered. The change in the preference-weights is mostly attributed to the changes experienced by the principal components since all objectives are interpreted as essential. The preference order between the objectives remains unaltered throughout the optimization process and is given by $w_4 > w_1 > w_3 > w_2$.

The framework analysis for the Machining problem has highlighted some important differences when comparing $N_g = 1$ with a population obtained at a stable instant. It is worthy to note the differences in the correlations between the objectives, implying that the correlation structure changes during the optimization process, as solutions approach stability. This could have implications for the problem analysis when validating the relationships between the objectives, in light of their physical meanings. Another important aspect relates to the distribution of the problem variance along the principal components. In that, it has been shown that this can have a significant influence on the accuracy of the preference-weights.

4.2 Storm Drainage System Problem

For this problem, while the MOEA-termination algorithm infers $N_{gt} = 195$, the corresponding decision support revelations (Table 2) are also compared to those corresponding to *a priori* fixed $N_g = 1$. In that:

1. There are changes to the correlation structure since some correlation signs have been reversed. This is the case for R_{12}, R_{23} and R_{24}. There are significant changes to the strength of correlation between some objectives. This is noteworthy for objective f_2 where the correlations R_{12}, R_{23} and R_{24}, all show

an RPD that is higher than 100%. Another significant change is between f_1 and the other objectives where R_{13}, R_{14} and R_{15}, correspond to an RPD of 30.580%, 59.800% and 94.495%, respectively.

2. The eigenvalues analysis shows that there are no significant changes to the principal components. In that, the variance is accounted for mostly by the first principal component (i.e., $e_1 \approx 0.98$), and the RPD is only 0.228%. As a result, T_{cor} is relatively low compared to most correlation strengths between the objectives, which implies that this is a highly redundant problem. The elements of the principal components have a RPD that ranges between 2.6376% and 31.307%. This suggests that the majority of the solutions have kept the same direction in the search space while approaching stability.

3. There are changes to the identified correlated sets. In that: (i) at $N_g = 1$ the sets are $\mathcal{S}_1 = \mathcal{S}_3 = \{f_1, f_3\}$, $\mathcal{S}_2 = \{f_2\}$ and $\mathcal{S}_4 = \mathcal{S}_5 = \{f_4, f_5\}$; and, (ii) at $N_{gt} = 195$ the sets are $\mathcal{S}_1 = \mathcal{S}_3 = \{f_1, f_3\}$, $\mathcal{S}_2 = \{f_2\}$ and $\mathcal{S}_4 = \{f_4\}$ $\mathcal{S}_5 = \{f_5\}$. This is attributed to the changes in the correlation structure, where objectives f_4 and f_5 are no longer interpreted as being correlated at $N_{gt} = 195$. In that, note the changes to the correlation signs where R_{24} is now positive while R_{25} remains negative. This implies that f_4 and f_5 react differently to the presence of f_2, hence, they cannot be interpreted as correlated. This also means that the essential objective set has changed during the optimization process, since $\mathcal{F}_s = \{f_2, f_3, f_4\}$ at $N_g = 1$, while at $N_{gt} = 195$ $\mathcal{F}_s = \{f_2, f_3, f_4, f_5\}$.

4. Despite the changes to the correlated sets, there are no major changes to the preference-weights. In that, the RPD ranges from 3.5827% to 99.226%. The preference order between the objectives is kept the same during the optimization process, which is $w_4 > w_3 > w_2 > w_5 > w_1$.

Table 2. Key highlights for the storm drainage system problem.

Gens	Correlation structure									
	R_{12}	R_{13}	R_{14}	R_{15}	R_{23}	R_{24}	R_{25}	R_{34}	R_{35}	R_{45}
$N_g = 1$	0.0709	0.6916	−0.3657	−0.0234	0.0878	−0.0124	−0.5794	−0.8378	−0.3399	0.3908
$N_{gt} = 195$	−0.3035	0.9963	−0.9096	−0.4246	−0.2863	0.3563	−0.5420	−0.9187	−0.4492	0.4895
RPD(%)	123.36	30.580	59.800	94.495	130.67	103.49	6.4535	8.8141	24.310	20.165

Gens	First principal component					Second principal component				
	f_{11}	f_{21}	f_{31}	f_{41}	f_{51}	f_{12}	f_{22}	f_{32}	f_{42}	f_{52}
$N_g = 1$	0.1586	0.1580	0.4012	−0.8743	0.1566	−0.3238	−0.3319	0.7994	0.1886	−0.3323
$N_{gt} = 195$	0.1254	0.1240	0.4786	−0.8513	0.1233	−0.3408	−0.3448	0.7556	0.2745	−0.3445
RPD(%)	20.945	21.505	16.181	2.6376	21.272	4.9939	3.7450	5.4787	31.307	3.5438

Gens	Preference-weights								$\mathcal{F}_s$	
	T_{cor}	e_1	e_2	w_1	w_2	w_3	w_4	w_5		
$N_g = 1$	0.2117	0.9853	0.0147	0.0084	0.0269	0.1728	0.7757	0.0162	$\{f_2, f_3, f_4\}$	
$N_{gt} = 195$	0.2099	0.9876	0.0124	0.0001	0.0169	0.2373	0.7289	0.0168	$\{f_2, f_3, f_4, f_5\}$	
RPD(%)	0.8520	0.228	15.346	99.226	37.160	27.176	6.0300	3.5827	—	

The above shows a situation where there are changes to the analysis of the decision support framework, first to the correlation structure of the problem, and

Table 3. Key highlights for the work roll cooling design problem.

Gens	Correlation structure										
	R_{12}	R_{13}	R_{14}	R_{15}	R_{16}	R_{23}	R_{24}	R_{25}	R_{26}	R_{34}	R_{35}
$N_g = 1$	−0.7652	0.9566	−0.9529	0.9325	−0.9656	−0.7474	0.8688	−0.6885	0.7832	−0.9737	0.9939
$N_{gt} = 167$	−0.9719	0.9943	−0.9951	0.9899	−0.9954	−0.9656	0.9813	−0.9556	0.9705	−0.9968	0.9989
RPD(%)	21.263	3.8011	4.2367	5.8041	2.9936	22.605	11.461	27.948	19.296	2.3153	0.5015

Gens	Correlation structure				T_{cor}	First principal component					
	R_{36}	R_{45}	R_{46}	R_{56}		f_{11}	f_{21}	f_{31}	f_{41}	f_{51}	f_{61}
$N_g = 1$	−0.9936	−0.9491	0.9862	−0.9833	0.3885	−0.4424	0.3675	−0.4048	0.4647	−0.3728	0.3879
$N_{gt} = 167$	−0.9988	−0.9929	0.9984	−0.9972	0.1752	−0.4564	0.3602	−0.3981	0.4665	−0.3639	0.3917
RPD(%)	0.5201	4.4183	1.2211	1.3912	54.905	3.0682	1.9849	1.6713	0.3909	2.3803	0.9793

Gens	Preference-weights									$\mathcal{F}_s$
	e_1	e_2	e_3	w_1	w_2	w_3	w_4	w_5	w_6	
$N_g = 1$	0.9173	0.0778	0.0042	0.4285	0.0539	0.0154	0.4766	0.0204	0.0052	$\{f_1, f_4\}$
$N_{gt} = 167$	0.9898	0.0095	0.0006	0.4831	0.0059	0.0021	0.5052	0.0031	0.0006	$\{f_1, f_4\}$
RPD(%)	7.3267	87.748	86.528	11.311	89.068	86.541	5.6665	84.896	88.823	—

then to the essential objective set. The fact that one objective is erroneously interpreted as being redundant can have severe consequences for the problem decision making task. This further highlights the importance of detecting stability during an optimization process, before the application of a decision support framework.

4.3 Work Roll Cooling Design Problem

For this problem, while the MOEA-termination algorithm infers $N_{gt} = 167$, the corresponding decision support revelations (Table 3) are also compared to those corresponding to *a priori* fixed $N_g = 1$. In that:

1. There are no changes to the correlation signs, implying that the correlation structure remains unaltered. Most objectives are highly correlated due to the reported high correlation values. The evolution of the correlation strength between pairs of objectives shows a low RPD which ranges between 0.5015% to 27.948%.
2. Most of the problem variance is accounted for by the first principal component with a tendency to increase as the population of solutions approaches stability. In that, e_1 increases from 0.9173 to 0.9898, which corresponds to a RPD of 7.3267%. This increase in e_1 is achieved at the cost of the second and third principal components since their values have decreased with a RPD of 87.748% and 86.528%, respectively. The analysis also shows that this is a highly redundant problem since the T_{cor} value is significantly smaller than the reported correlation values. As the solutions approach stability, the value of T_{cor} decreases from 0.3885 to 0.1752 which corresponds to a RPD of 54.905%.
3. With a low T_{cor} and a relatively high correlation strength values, the correlated sets are $\mathcal{S}_1 = \mathcal{S}_3 = \mathcal{S}_5 = \{f_1, f_3, f_5\}$ and $\mathcal{S}_2 = \mathcal{S}_4 = \mathcal{S}_6 = \{f_2, f_4, f_6\}$. The objectives f_1 and f_4 are identified as the essential objectives of the correlated sets, due to their contribution score towards the principal components.

That is, although only shown for the first principal component it can be seen that, $|f_{11}| > |f_{31}| > |f_{51}|$ and $|f_{41}| > |f_{61}| > |f_{21}|$. This leads to the essential objective set given by $\mathcal{F}_s = \{f_1, f_4\}$.

4. The preference-weights for objectives in $\mathcal{F}_s$ have a low RPD when compared with the other preference-weights. That is, although the RPD for w_1 and w_4 is only 11.331% and 5.6665%, respectively, the other preference-weights show a RPD higher than 84%. This is mostly due to the changes in the correlation strength between redundant and essential objectives, where an increase in the correlation strength leads to a higher preferability for the essential objectives. The preference order between the objectives remains unaltered throughout the optimization process and is given by $w_4 > w_1 > w_2 > w_5 > w_3 > w_6$.

The above situation corresponds to a highly redundant problem where the solutions are mostly spread across objectives f_1 and f_4. The analysis of Fig. 1c shows that the redundant objectives have a reduction in their preference-weights while the essential objectives have a slight increase. There is a high RPD for some preference-weights, such as w_2, w_3, w_5 and w_6, where the RPD is 89.068%, 86.541%, 84.896% and 88.823%, respectively. It is in such cases that the application of an MOEA-termination algorithm becomes significant.

5   Conclusion

This paper presents a framework for an *appropriately timed* and *accurate* decision support in the context of both multi- and many-objective optimization problems. The key features and contributions of this framework based on integration of machine learning based decision support and MOEA-termination algorithm have been highlighted in the context of three real-world problems. This paper is particularly significant in the context of many-objective optimization problems where the user or the decision maker may otherwise be clueless about the number of generations that may be sufficient for the corresponding MOEA population to be treated with some degree of confidence in terms of its conformance or departure from the true POF. Though for the sake of brevity, the *dissimilarity measure* leading up to the prescribed termination has not been shown, reference to it shall also indicate whether or not the *revealed* preference–structure could be treated as representative of the true POF. It is hoped that this integrated framework shall be instrumental in facilitating decision makers' preferences characterized by *objectivity*, *repeatability*, *consistency*, and *coherence* and shall mark a significant contribution towards the practical utility of MCDM based MOEAs.

References

1. Coello, C.A.C., Lamont, G.B., Veldhuizen, D.A.V.: Evolutionary Algorithms for Solving Multi-objective Problems. Springer, New York (2002)
2. Purshouse, R., Fleming, P.: Evolutionary many-objective optimisation: an exploratory analysis. Congr. Evol. Comput. **3**, 2066–2073 (2003)

3. Ishibuchi, H., Tsukamoto, N., Nojima, Y.: Evolutionary many-objective optimization: a short review. In: IEEE Congress on Evolutionary Computation, pp. 2419–2426, June 2008
4. Deb, K., Pratap, A., Agarwal, S., Meyarivan, T.: A fast elitist multi-objective genetic algorithm: NSGA-II. IEEE Trans. Evol. Comput. **6**(2), 182–197 (2000)
5. Duro, J.A., Saxena, D.K., Deb, K., Zhang, Q.: Machine learning based decision support for many-objective optimization problems. Neurocomputing **146**, 30–47 (2014)
6. Zhang, Q., Li, H.: MOEA/D: a multiobjective evolutionary algorithm based on decomposition. IEEE Trans. Evol. Comput. **11**, 712–731 (2007)
7. Bader, J., Zitzler, E.: HypE: an algorithm for fast hypervolume-based many-objective optimization. Evol. Comput. **19**(1), 45–76 (2011)
8. Purshouse, R.C., Deb, K., Mansor, M.M., Mostaghim, S., Wang, R.: A review of hybrid evolutionary multiple criteria decision making methods. In: IEEE Congress on Evolutionary Computation, pp. 1147–1154, July 2014
9. Saxena, D.K., Sinha, A., Duro, J.A., Zhang, Q.: Entropy based termination criterion for multiobjective evolutionary algorithms. IEEE Trans. Evol. Comput. **20**(4), 485–498 (2016)
10. Ghiassi, M., Kijowski, B.A., Devor, R.E., Dessouky, M.I.: An application of multiple criteria decision making principles for planning machining operations. IIE Trans. **16**(2), 106–114 (1984)
11. Musselman, K., Talavage, J.: A tradeoff cut approach to multiple objective optimization. Oper. Res. **28**(6), 1424–1435 (1980)
12. Azene, Y.T.: Work roll system optimisation using thermal analysis and genetic algorithm. Ph.D. thesis, School of Applied Sciences, Cranfield University, May 2011

On the Influence of Altering the Action Set on PROMETHEE II's Relative Ranks

Stefan Eppe and Yves De Smet$^{(\boxtimes)}$

Computer and Decision Engineering Department (CoDE),
École polytechnique de Bruxelles, Université libre de Bruxelles, Brussels, Belgium
`yves.de.smet@ulb.ac.be`

Abstract. For some multi-criteria decision aid methods, the relative ranks of two actions may be inverted when the original set is altered. This phenomenon is known as rank reversal. In this contribution, we formalise rank reversal for the PROMETHEE II method. The aim is not to debate about the legitimacy of such effect but rather to derive the exact conditions for its occurrence when one or more actions are added or removed from/to the original set. These conditions eventually lead us to: (1) assess whether rank reversal between a given pair of actions is, at all, possible, and (2) characterise the evaluations of the actions that have to be added or removed to induce rank reversal. We also propose two metrics that express the "strength" of and the "sensitivity" towards rank reversal. Finally, we show on a toy example how they could be used in a decision making process.

Keywords: Decision making · Rank reversal · PROMETHEE II

1 Introduction

Rank reversal is a topic that is addressed in several disciplines, such as social choice theory, economy [10], decision theory, etc. It has received much attention, because it questions the possible (bounded) rationality of a decision maker. Rank reversal has been reported and studied in numerous multi-criteria decision aid (MCDA) methodologies [12], amongst others in AHP [6,12], TOPSIS [5], the cross-efficiency evaluation method in data envelopment analysis (DEA), ELECTRE [3,13], and PROMETHEE [7,8,11].

Concretely, we address the phenomenon of rank reversal of a pair of actions in a PROMETHEE II ranking, when one action is added to the original set of actions. Anticipating on the illustrative example presented below (Sect. 6), let us consider a set of five actions $A = \{a_1, a_2, a_3, a_4, a_5\}$ that are evaluated on two criteria. Assume that applying PROMETHEE II on A with a given set of preference parameters (relative weight, indifference and preference thresholds for each criterion), results in the following ranking (from the best to the worst action): $(a_5, a_1, a_2, a_3, a_4)$. Adding an action a_6 to the set ($A' = A \cup \{a_6\}$) may, under certain conditions, lead to a new ranking where the relative ranks of two actions are reversed: $(a_5, a_6, a_2, a_1, a_3, a_4)$. In this example, actions a_1 and a_2

© Springer International Publishing AG 2017
H. Trautmann et al. (Eds.): EMO 2017, LNCS 10173, pp. 206–220, 2017.
DOI: 10.1007/978-3-319-54157-0_15

see their relative ranks reversed because of the addition of a_6. With the present contribution, we describe the conditions under which a_6 would provoke the rank reversal of a_1 and a_2.

Although often considered as a phenomenon that should be avoided because it violates the property of *"independence of irrelevant alternatives"*, it has been argued that rank reversal could be legitimate and even desirable [9]. One way of overcoming this debate is to distinguish between the classes of *open* and *closed system* approaches [4]. In a nutshell, adding or removing an action from the considered action set in an *open* system will affect the range of scores for the set of actions. For a *closed* system on the contrary, altering the number of actions will lead to the redistribution of action scores in order not to affect the sum of all scores. As PROMETHEE II is a closed system, rank reversal can be observed and should be considered as intrinsically acceptable. However, our goal is not to judge whether rank reversal is legitimate or not, but rather to acknowledge that this phenomenon may occur with PROMETHEE II and to try getting the best possible grip on it. Practically, the added value of our contribution is to give a quantitative and sound bound on when rank reversal may occur when using PROMETHEE II. We also propose several metrics aimed at the practitioner.

For this outranking method in particular, an upper bound for net flow differences between a pair of actions has been proposed, above which rank reversal cannot occur [7]. However, despite its simplicity, experiments [8] show that this bound is very pessimistic, thereby reducing its practical applicability. In the present paper, we compute the analytical bound for the net flow scores difference of any pair of actions above which rank reversal is not possible. For differences below that bound, rank reversal can always be induced and we provide a characterisation of the actions that lead to it. Based on the preceding, we propose a rank reversal sensitivity measure: the higher this sensitivity, the larger the domain of action evaluations that will lead to rank reversal.

The sequel of the paper is organised as follows: We start by briefly introducing the PROMETHEE II method (Sect. 2). We then delve into rank reversal in two phases: first, we consider the addition of one single action to the original set and we study its impact with respect to rank reversal on a given pair of actions (Sect. 3). We then show how this result can be extended to the cases of several added actions, but also to the removal of existing actions (Sect. 4). These theoretical results lay the ground for defining two related metrics that should help the practitioner get a better grip on that phenomenon (Sect. 5). Finally, we provide an illustrative example to show how rank reversal could be integrated to the analysis of a concrete decision making problem based on PROMETHEE II (Sect. 6).

2   The Promethee II Method

Let $A = \{a_1, \ldots, a_n\}$ be a finite set of n actions. Each action a_i, with $i \in I = \{1, \ldots, n\}$, is characterised by means of an evaluation vector $G(a_i) = \{g_1(a_i), \ldots, g_m(a_i)\} \in \mathcal{G}$, where $g_h(a_i)$ is the evaluation of a_i on criterion $h \in H = \{1, \ldots, m\}$, and $\mathcal{G}$ represents the domain of all possible evaluation vectors. In the following, we assume that this domain is the m-dimensional unit hypercube: $\mathcal{G} = [0, 1]^m$.

To compare any pair of actions $(a_i, a_j) \in A \times A$, a so-called preference function P_h is introduced for each criterion h. It expresses the preference degree, on that criterion, of the first action over the other. Although our approach is not bounded to it, we will consider the widely used "*V-shaped with indifference*" preference function [1] in the following. It uses an indifference and a preference threshold that are, respectively, denoted by q_h and p_h (Fig. 1):

$$
P_h(a_i, a_j) = \begin{cases} 0, & \text{if } \Delta g_h(a_i, a_j) \leq q_h \\ \frac{\Delta g_h(a_i, a_j) - q_h}{p_h - q_h}, & \text{if } q_h < \Delta g_h(a_i, a_j) \leq p_h \\ 1, & \text{if } p_h < \Delta g_h(a_i, a_j) \end{cases} \tag{1}
$$

where $\Delta g_h(a_i, a_j) = g_h(a_i) - g_h(a_j)$ when the criterion has to be maximised, and $\Delta g_h(a_i, a_j) = g_h(a_j) - g_h(a_i)$ otherwise. The pairwise action comparisons are aggregated for each action and provide the unicriterion net flow score $\phi_h(a_i)$ on criterion h:

$$
\phi_h(a_i) = \frac{1}{n-1} \sum_{j \in I} \Delta P_h(a_i, a_j), \tag{2}
$$

where $\Delta P_h(a_i, a_j) = P_h(a_i, a_j) - P_h(a_j, a_i)$. Finally, the unicriterion net flow scores are aggregated once more over all criteria through a weighted sum to yield that action's net flow score:

$$
\phi(a_i) = \sum_{h \in H} w_h \phi_h(a_i), \tag{3}
$$

where $\{w_1, \ldots, w_m\}$ represents the set of the criteria's relative importance, with $w_h \geq 0, \forall h \in H$, and $\sum_{h \in H} w_h = 1$. For a deeper introduction to the PROMETHEE methods, we refer the reader to [1].

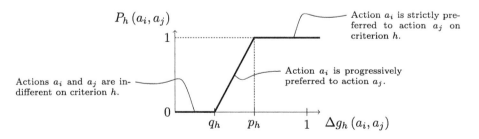

Fig. 1. PROMETHEE's "*V-shaped with indifference*" preference function $P_h(a_i, a_j)$ for criterion h, depending on the evaluation difference $\Delta g_h(a_i, a_j)$ of the action pair (a_i, a_j). It is characterised by an indifference threshold q_h and a preference threshold p_h.

3 Condition for Rank Reversal

Let us now consider the phenomenon of rank reversal in the context of the PROMETHEE II method. Stated in very general terms, we consider that we have rank reversal of two actions a_i and a_j if, after some alteration of the decision making model, their relative ranks are reversed. Alternatively, if a_i was initially better ranked than a_j, then we would say that there is a **rank reversal** of these actions if, after some modification, a_j would be better ranked than a_i.

Practically, in a multi-criteria decision aid context, there are two main types of changes in a decision making model that may lead to rank reversal: *(i)* changes of the considered actions and/or their evaluations or, *(ii)* changes in the preference parameters. In this contribution, we will only address the former ones; we will consider the preference parameters as constants.

Let us consider an added action x and let $A' = A \cup \{x\}$ be the extended set of $n + 1$ actions. Using (2), the net flow score of action a_i with respect to A' becomes:

$$\phi'(a_i) = \frac{1}{n} \sum_{h \in H} w_h \left[\sum_{j \in I} \Delta P_h(a_i, a_j) + \Delta P_h(a_i, x) \right]$$

$$= \frac{1}{n} \left[(n-1) \phi(a_i) + \sum_{h \in H} w_h \Delta P_h(a_i, x) \right]. \tag{4}$$

The relative rank of two actions, a_i and a_j, is given by the sign of their respective net flow scores, i.e., a_i is better ranked than a_j if $\Delta\phi(a_i, a_j) = \phi(a_i) - \phi(a_j) > 0$ and reciprocally. Therefore, rank reversal (RR) under the addition of one action x can be expressed as follows:

$$RR(a_i, a_j; x) \iff \Delta\phi(a_i, a_j) \Delta\phi'(a_i, a_j) < 0.$$

Let us make the unrestrictive assumption that a_i is initially better ranked than a_j. This implies that $\Delta\phi(a_i, a_j) > 0$, and therefore simplifies the preceding condition to find the evaluations of x for which $\Delta\phi'(a_i, a_j) < 0$. Integrating (4), the *extended net flow score difference* can be rewritten as

$$\Delta\phi'(a_i, a_j) = \frac{1}{n} \left[(n-1) \Delta\phi(a_i, a_j) + \sum_{h \in H} w_h z_h(a_i, a_j; x) \right],$$

where

$$z_h(a_i, a_j; x) = \Delta P_h(a_i, x) - \Delta P_h(a_j, x). \tag{5}$$

Hence, we can express the **exact condition for rank reversal**:

$$RR(a_i, a_j; x) \iff \sum_{h \in H} w_h z_h(a_i, a_j, x) < -(n-1) \Delta\phi(a_i, a_j). \tag{6}$$

This general condition is an extension of the work presented in [11]. Let us stress that it does only depend on the preference function differences ΔP_h, but not on a particular preference function type. For the sake of illustration, we express the z_h function explicitly for the "*V-shaped with indifference*" preference function (Fig. 1) in Appendix A and use it in the example of Sect. 6. However, the given example does not reduce the generality of the above condition.

Some Observations

- The inequality in (6) may be interpreted as follows: An additional action x provokes rank reversal of a pair of actions if the weighted sum of the associated z_h functions is lower than a bound that depends on the number of actions and the initial net flow difference between a_i and a_j. The expression also suggests that a net flow difference between two actions may be considered as some kind of *resistance* to their rank reversal: the higher the net flow difference, the more unlikely their rank reversal. Beyond a certain bound, rank reversal becomes impossible.
- According to (6), each criterion h contributes (with its own weight w_h) to the value of $z(a_i, a_j; x) = \sum_{h \in H} w_h z_h(a_i, a_j; x)$. Since we assume that, initially, $\Delta\phi(a_i, a_j) > 0$, the right hand term of (6) will be negative, and, thus, the value of the weighted sum $z(a_i, a_j; x)$ will have to be negative too to induce rank reversal. Looking at the table that describes, as an illustrative example, the possible configurations for the "V-shaped with indifference" preference function, in the Appendix A we notice that for criterion h, $z_h(a_i, a_j; x)$ can be negative only if the evaluation difference $\Delta g_h(a_i, a_j)$ (which is given by the evaluations of the original set of actions) is negative too. Consequently, for a given pair of actions (a_i, a_j) and assuming a maximisation problem, only the criteria with a negative evaluation difference, i.e., $\Delta g(a_i, a_j) < 0$, are capable of contribute to rank reversal. On the contrary, criteria for which $\Delta g(a_i, a_j) > 0$ will tend to reinforce the relative ranking of a_i and a_j. These latter criteria can only, in the best case, not increase the value of $z(a_i, a_j; x)$.
- As has already been pointed out [7], rank reversal of actions a_i and a_j can only occur if they are mutually non-dominating: there must be at least one criterion for which the unicriterion net flow score difference should be negative, despite a globally positive net flow score difference.

4 Extension to the Addition and Removal of Several Actions

Let us generalise the results of the preceding section to the addition of s actions: $S = \{x_1, \ldots, x_s\} \subset \mathcal{A}$ and the removal of r actions: $R = \{y_1, \ldots, y_r\} \subset A$. Note that, while the actions of the first set belong to the set of all possible actions $\mathcal{A}$, the second set, R, only contains actions from the original set A.

Similarly to the case of one single additional action x, rank reversal occurs between a_i and a_j if their relative rank before and after having altered the set of actions is reversed. More concretely, we have rank reversal if, after having added all actions contained in the set S and removed all actions contained in R, the sign of the net flow score difference is reversed. Hence, the condition becomes:

$$RR(a_i, a_j; S, R) \iff \Delta\phi(a_i, a_j)\, \Delta\phi'(a_i, a_j) < 0,$$

where the altered action set is now defined as $A' = (A \cup S) \setminus R$. Again, assuming that action a_i is initially better ranked than a_j, i.e., that $\Delta\phi(a_i, a_j) > 0$, we obtain the general condition for rank reversal:

$$RR\left(a_i, a_j; S, R\right) \iff \sum_{h \in H} w_h\, \mathbf{z}_h\left(a_i, a_j; S, R\right) < -\left(n - 1\right) \Delta\phi\left(a_i, a_j\right),$$

where the unicriterion z-function is extended to

$$\mathbf{z}_h\left(a_i, a_j; S, R\right) = \sum_{x \in S} z_h\left(a_i, a_j; x\right) - \sum_{y \in R} z_h\left(a_i, a_j; y\right).$$

Again, this is a general condition that can be computed for any preference function type. Note that, as the order of adding and/or removing actions does not matter, changing the evaluations of an action from the set can be conveniently modelled by removing that action and adding a new one, that is a modified "copy" of it. We will thus not explicitly consider the change of actions in the sequel.

5 Some Metrics for the Practitioner

In this section we propose some metrics that might be useful to quantitatively evaluate rank reversal related characteristics for a particular decision problem. The illustrative example of Sect. 6 will show how these can be applied in practice.

5.1 How "Strongly" Are the Relative Ranks of Two Actions Reversed?

Depending on the evaluations of x, changing the sign of the net flow score difference may not be very expressive. Indeed, still assuming that $\Delta\phi\left(a_i, a_j\right) > 0$, the above condition would state that there is a rank reversal of two actions, $RR(a_i, a_j)$, even for the smallest negative value of $\Delta\phi'\left(a_i, a_j\right)$. It therefore seems sensible to add a measure that expresses the *rank reversing power* of the added action x:

$$P_{RR}\left(a_i, a_j; x\right) = \frac{\Delta\phi\left(a_i, a_j\right) - \Delta\phi'\left(a_i, a_j\right)}{\Delta\phi\left(a_i, a_j\right)}. \tag{7}$$

Expressing it in relative terms with respect to the initial net flow difference allows it to be interpreted as follows (Fig. 2):

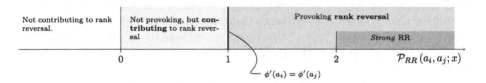

Fig. 2. Value domains of the *rank reversing power* $P_{RR}\left(a_i, a_j; x\right)$ and their respective effect on rank reversal of actions a_i and a_j.

$\mathcal{P}_{RR}\left(a_i, a_j; x\right)$	Effect of action x with respect to rank reversal
<0	**x does *not* contribute to provoking rank reversal** Actually, adding x even contributes to reinforcing the current relative rank of a_i with respect to a_j, as the resulting net flow score difference $\Delta\phi'\left(a_i, a_j\right)$ would be even higher than initially.
$\in]0,1[$	**x contributes but is not sufficient to provoke rank reversal** The effect of adding x to the original action set tends to decrease the net flow score difference with respect to its initial value, but not to change its sign. Eventually, however, it would be possible to provoke rank reversal of a_i and a_j by adding further actions, similar to x, to the set.
1	**provokes the net flows of a_i and a_j to be equal** **$\left(\Delta\phi'\left(a_i, a_j\right) = 0\right)$**
$\in]1,2[$	**x provokes rank reversal of a_i and a_j** The added action x provokes rank reversal, i.e., the sign of the net flow score difference is changed, but the absolute value of that new difference is smaller than the original one. If the score difference is interpreted as expressing (in relative terms) how "strongly" action a_i was better ranked than a_j, then the reversed relative rank is "weaker" than initially.
≥ 2	**x provokes a "strong" rank reversal** The reversed net flow score difference $\Delta\phi'\left(a_i, a_j\right)$ has an absolute value that is greater than the initial one. In terms of "ranking strength", this would mean that the new relative rank of a_i and a_j is reversed and "stronger" than initially, hence the term "strong" rank reversal.

The interpretation of the particular value $\mathcal{P}_{RR}\left(a_i, a_j; x\right) = 2$ used as the limit between "regular" and "strong" rank reversal is related to the amplitude of the initial net flow score difference $\Delta\phi(a_i, a_j)$. When $\mathcal{P}_{RR} \in [1, 2]$, the additional action x changes the sign of the initial score difference. However, this change in sign could be minimal, e.g. $-0.00001 < \frac{\Delta\phi'(a_i,a_j)}{\Delta\phi(a_i,a_j)} < 0$. In such a case, although there is, *stricto sensu* a rank reversal, one would probably interpret it practically as a change that sets actions a_i and a_j to an equal "rank". We have arbitrarily named rank reversals with $\mathcal{P}_{RR} \geq 2$ as "strong rank reversal", as the altered net flow difference would, in absolute terms, be as large as the initial ones, i.e., $|\Delta\phi'(a_i, a_j)| \geq |\Delta\phi(a_i, a_j)|$. Of course, this metric has a relative meaning. It could happen that $|\Delta\phi'(a_i, a_j)|$ and $|\Delta\phi(a_i, a_j)|$ are small numbers (leading to a situation that is close to indifference in both cases) but with $\mathcal{P}_{RR} \geq 2$.

If we compute the average of this power on all actions for the second action parameter, we may express the impact of the added action x on action a_i as follows:

$$P_{RR}\left(a_i; x\right) = \frac{1}{n}\sum_{j \in I}\max\{0, P_{RR}\left(a_i, a_j; x\right)\}.$$

For the derived definition, we only consider the positive values of $P_{RR}(a_i, a_j; x)$ in the aggregation process. The aim is to avoid the compensating effect of action pairs for which adding x would actually reinforce their initial relative ranks. Finally, let us note that other aggregating operators can also be used instead of the average (such as the maximum for instance).

5.2   How Sensitive Are the Ranks of Actions?

For the second metric, we take the point of view of the evaluations. Being able to express the likelihood, for a given pair of actions (a_i, a_j) to experience RR is also of practical interest. We therefore propose the following metric:

$$\rho_{RR}(a_i, a_j) = \int_{x \in \mathcal{A}} \delta_{\Delta \phi'(a_i, a_j) < 0} \, \mathrm{d}x, \qquad (8)$$

where δ_C evaluates to 1 if the condition C is true and to 0 otherwise; $\mathcal{A}$ represents the set of all possible actions. The range of possible values is $\rho_{RR} \in [0, 1]$.

More intuitively, this metric gives the ratio of evaluations (taken from the evaluation space, i.e. $\mathcal{G} = [0, 1]^m$) for an additional action x that lead a given pair of actions (a_i, a_j) to see their ranks reversed. The rationale behind the metric is that the more we have evaluations of x that provoke rank reversal of a_i and a_j, the more the pair (a_i, a_j) is *sensitive* towards rank reversal (Fig. 3). If we arbitrarily assume a uniform distribution of the evaluations on all criteria, this value can be interpreted as the probability to see actions a_i and a_j have their rank reversed when an action is added to the original set (let us point out that the identification of the most appropriate probability distribution is a problem on itself). However, this interpretation also depends on the hypothesis of independence between criteria, which is only rarely met in MCDA problems.

Practically, this metric can be approximated by a sampling, be it by a Monte Carlo approach or by a regular meshing over the evaluation space. The accuracy

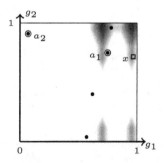

Fig. 3. Areas (shaded) that correspond to evaluations $(g_1(x), g_2(x))$ of the additional action x that provoke rank reversal of actions a_i and a_j for a bi-criterion example. The darker the shade, the "stronger" the observed rank reversal. If x has an evaluation located in a white region, then it does not provoke rank reversal.

of the metric can be modulated depending on the the users needs. Since the evaluation of the term to be integrated in (8) has a constant time complexity, the overall complexity for the sensitivity is linear with respect to the number of samples.

In the next section, we present a concrete example that shows how the proposed metrics can be computed and how the results can be interpreted.

6   Illustrative Example

The conditions for rank reversal described in the previous sections are very general. Here, we provide a toy example that illustrates how rank reversal can be "managed" for a concrete data set. Therefore, we will use the following general workflow (Fig. 4):

1. The z-function is a weighted sum of unicriterion functions: we consider each criterion individually.
2. For each criterion, a type of preference function is chosen by the decision maker. In this contribution, we only consider the "*V-type with indifference*" function.
3. Having chosen a pair of actions (a_i, a_j) for which we want to test their "rank reversibility", a so-called scenario has to be chosen (Appendix A, leftmost column), corresponding to their evaluation difference on that criterion: $\Delta g_h (a_i, a_j) = g_h (a_i) - g_h (a_j)$.
4. Using Appendix A, the characteristic unicriterion features with respect to rank reversal are deduced for the pair of actions (a_i, a_j).

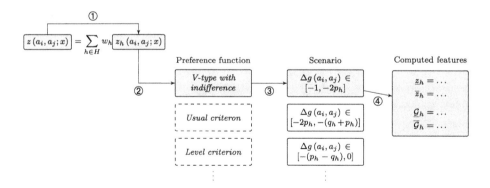

Fig. 4. Flowchart of the presented approach. We decompose the process by considering each criterion h individually; we choose a concrete preference function type to develop the general rank reversal expressions; one specific scenario has to be chosen, depending on the evaluation difference $\Delta g_h (a_i, a_j)$ and the preference parameters; the actual computation of the characteristic rank reversal features is carried out (or retrieved from Appendix A).

This procedure is repeated for each criterion. Once aggregated on all criteria, we can compute the global z-function and finally determine whether a given additional (or removed) action actually leads to rank reversal.

We consider a randomly generated set of actions (using a uniform distribution on all criteria) and a set of preference parameters (Table 1). For the sake of representability, we limit this example to a decision problem with two criteria to which one single action, x, is added. The approach, however, remains the same for higher dimensions ($m > 2$) and for several actions added and/or removed.

Let us explicitly treat the case of the pair of actions (a_1, a_2): The initial net flow score difference $\Delta\phi(a_1, a_2) = 0.05$ (Table 2) is positive, i.e. action a_1 is initially better ranked than a_2. The evaluation difference on criterion 1 and 2 are, respectively, $\Delta g_1(a_1, a_2) = 0.68$ and $\Delta g_2(a_1, a_2) = -0.16$. The latter being negative, it is on criterion 2 that x will contribute to rank reversal, while on criterion 1, it would potentially reinforce the initial relative ranks. Having $q_h = 0.05$ and $p_h = 0.20$ on both criteria, we may deduce the scenario to apply on each. For criterion 1, $\Delta g_1(a_1, a_2) = 0.68 \in [0.40, 1.00] = [2p_h, 1]$: we take the last scenario of Appendix A that yields the minimal value of $z_1(a_1, a_2, x)$: $z_1(a_1, a_2) = \min\{z_1(a_1, a_2; 0); z_1(a_1, a_2; 1)\} = \min\{1; 0\} = 0$ (Fig. 5).

Likewise for criterion 2, $\Delta g_2(a_1, a_2) = -0.16 \in [-0.25, -0.10] = [-(q_2 + p_2), -2q_h]$, and we take the third scenario to find that $z_2(a_1, a_2) = -1$ (Fig. 6).

Slightly modifying condition (6), we can determine whether actions a_1 and a_2 may, at all, experience rank reversal. Rather than exploring all possible evaluations of the added action x, we notice that the "strongest" possible rank reversal (whatever the actual value of $P_{RR}(a_i; x)$) can be computed through the minimum of the aggregated $z(a_i, a_j; x)$ function. As it is a weighted sum, it suffices

Table 1. Evaluations and preference parameters for our toy example.

Criterion h	Action evaluations $g_h(a_i)$					Pref. parameters		
	a_1	a_2	a_3	a_4	a_5	w_h	q_h	p_h
1	0.75	0.07	0.62	0.57	0.78	0.30	0.05	0.20
2	0.75	0.91	0.40	0.04	0.96	0.70	0.05	0.20

Table 2. Net flows for each action, and net flow differences of all pairs of actions.

Sorted net flows		Net flow differences $\Delta\phi(a_i, a_j)$					
Action	$\phi(a_i)$		a_1	a_2	a_3	a_4	a_5
a_5	0.73	a_1	—	0.05	0.60	0.99	-0.50
a_1	0.23	a_2	-0.05	—	0.55	0.94	-0.55
a_2	0.18	a_3	-0.60	-0.55	—	0.39	-1.10
a_3	-0.37	a_4	-0.99	-0.94	-0.39	—	-1.49
a_4	-0.77	a_5	0.50	0.55	1.10	1.49	—

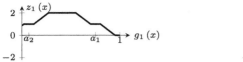

Fig. 5. $h = 1$, $\underline{z}_1 = 0$ **Fig. 6.** $h = 2$, $\underline{z}_2 = -1$

to take, for each criterion h individually, the minimal value $\underline{z}_h(a_i, a_j; x)$ that we have already determined:

$$\min\{\sum_{h \in H} w_h z_h(a_i, a_j, x)\} = \sum_{h \in H} w_h \underline{z}_h \;\; < \;\; -(n-1)\Delta\phi(a_i, a_j)$$

$$0.30 \times 0 + 0.70 \times -1 \;\; < \;\; -4 \times 0.05$$

$$-0.70 \;\; < \;\; -0.20$$

The condition is satisfied; it is hence possible to find at least one additional action x such that a_1 and a_2 have their relative ranks reversed.

Let us now determine the evaluations of x that lead to the "strongest" rank reversal. For the first criterion, we use Appendix A (last scenario) and Fig. 5 to determine that $zArgEvalMin_1(a_1, a_2) = [g_1(a_1) + p_1, 1] = [0.95, 1.00]$. Following the same logic (using the third scenario), $zArgEvalMin_2(a_1, a_2) = [g_2(a_1) - q_2, g_2(a_2) - p_2] \cup [g_2(a_1) + p_2, g_2(a_2) + q_2] = [0.70, 0.71] \cup [0.95, 0.96]$. The strongest rank reversal will thus occur when $g_1(x) \in [0.95, 1.00]$ and $g_2(x) \in [0.70, 0.71] \cup [0.95, 0.96]$.

As an example, let us choose $g_1(x) = 0.95$ and $g_2(x) = 0.70$ and compute the altered net flow scores (Table 3). For this particular choice, $\Delta\phi'(a_1, a_2) = \phi'(a_1) - \phi'(a_2) = 0.12 - 0.22 = -0.10$. Hence the *rank reversing power* of x with respect to the pair (a_1, a_2) is

$$P_{RR}(a_1, a_2; x) = \frac{\Delta\phi(a_1, a_2) - \Delta\phi'(a_1, a_2)}{\Delta\phi(a_1, a_1)} = \frac{0.05 - (-0.10)}{0.05} = 3$$

According to the definition of $P_{RR}(a_1, a_2; x)$ (Fig. 2), the chosen additional action x has a strong rank reversing power. Indeed, $|\Delta\phi'(a_1, a_2)| = 2 \cdot |\Delta\phi(a_1, a_2)|$, i.e., the "new" net flow score difference between a_1 and a_2 not only has changed sign, but its absolute value has become twice as high. If we interpret the net flow score difference as a (relative) indicator about how strongly one action is better ranked than another, then a value of $P_{RR}(a_1, a_2; x) = 3$ indicates that adding x to the original set of actions makes a_2 to be better ranked than a_1 with a "strength" twice as high than previously.

To compute the RR-sensitivity according to (8), we sample the bi-dimensional evaluation space of x. We count the number of sampled evaluations of x that provoke rank reversal of actions a_1 and a_2 and divide this number by the total number of samples. For a sampling granularity of 0.001 on each criterion, i.e. a total of $1000^2 = 10^6$ samples, we obtain $\rho_{RR}(a_i, a_j) \approx 0.13$. As already mentioned, this could be interpreted as a 13% probability that an action added to

Table 3. Net flows for each action for the original and the altered set of actions $A' = A \cup \{x\}$.

Original set A		Altered set A'	
Action	$\phi(a_i)$	Action	$\phi'(a_i)$
a_5	0.73	a_5	0.66
		a_6	0.30
a_1	0.22	a_2	0.22
a_2	0.18	a_1	0.12
a_3	-0.37	a_3	-0.50
a_4	-0.77	a_4	-0.81

the original set provokes a rank reversal of actions a_1 and a_2. Finally, since we compute the RR-sensitivity by sampling the evaluation space, i.e., the domain of possible actions, this information can also be used to graphically represent the area(s) in which an additional action x would provoke the rank reversal of a_i and a_j. The plot corresponding to the present example's dataset is presented in Fig. 3. As stated before, the time complexity for approximating the RR-sensitivity is linear with respect to the number of samples. For the example presented here, our implementation in Octave requires 0.4 s for approximating the sensitivity measure with a granularity of 0.001, i.e. 10^6 samples, for each pair of actions. In order to compute the metric for the $\frac{1}{2}n(n-1) = 10$ possible pairs of actions, the total evaluation time thus represents about $10 \times 0.4 = 4$ s.

7   Conclusion

Rank reversal is a much debated topic in the MCDA community because it may seem counterintuitive that a third action may alter the relative ranks of two other actions. The PROMETHEE II preference model, as others, is a closed system model and, as such, may experience rank reversal. In this paper, we have provided an exact bound that assesses whether a pair of actions may experience this phenomenon. A closer look at the evaluation domain of (added or removed) rank reversing third actions shows that the "intensity" of rank reversal may vary. Therefore, we have defined an easy way to determine "rank reversing domain *kernel*". Besides the theoretical interest of having an exact rank reversal bound, there are also several practical advantages:

- The stability of one (pair of) action(s) with respect to rank reversal can be evaluated. For a given application, this could be integrated to the sensitivity analysis, allowing the decision maker to asses the risk for this phenomenon.
- When increasing the weight of criteria that are less sensitive to rank reversal and decreasing the weights of those that are more sensitive, one could artificially increase the robustness from the point of view of rank reversal. At first sight, this may seem contradictory, because, usually, a decision maker would

not want to change her weights to avoid rank reversal. However, one could use this feature for instance in multi-objective optimisation based preference eliciting techniques [2] as an additional objective function: if a set of weight vectors provides the requested ranking, one would probably favour the weight combinations that increase robustness to rank reversal.

Further investigations could complement the proposed approach. For instance, one could study how the rankings of other actions than the considered pair (a_i, a_j) are affected by a rank reversing action. Would it be meaningful to search for such actions that would exclusively affect the considered pair?

A   Domains of extreme values for the $z_h(x)$ Function

The possible contribution of one additional action x to the rank reversal of two actions a_i and a_j is mainly represented by the corresponding value of the $z_h(x)$ function (defined in Sect. 3). Assuming that a_i is initially better ranked than a_j, i.e., $\Delta\phi(a_i, a_j) = \phi(a_i) - \phi(a_j) > 0$, then the lower the value of $z_h(x)$, the more the additional action x (with its evaluation $g_h(x)$) contributes to a_i and a_j's rank reversal. Depending on the actions' evaluation difference on criterion h, $\Delta g_h(a_i, a_j) = g_h(a_i) - g_h(a_j)$, we observe different scenarios, yielding different characteristic values for the minimum and maximum of $z_h(x)$. The following table presents, for a given criterion h, the main possible scenarios. However, as the $z_h(x)$ function also depends on the preference parameters, q_h and p_h, other "degenerated" cases may also occur. These are not represented here for the sake of readability. Each row of the table corresponds to one scenario, depending on the evaluation difference $z_h(x)$ given in the first column. The second column schematically depicts the shape of the corresponding z_h function as a function of the evaluation $g_h(x)$ of the added action x on criterion h. For the first row, for instance, the plot shows that the addition action evaluation (that contributes the most to rank reversal of a_i and a_j) are located in between $g_h(a_i)$ and $g_h(a_j)$. The third column provides extreme values associated to the scenario:

- $\underline{z}_h = \min_{g_h(x) \in [0,1]} z_h(a_i, a_j; x)$
- $\overline{z}_h = \max_{g_h(x) \in [0,1]} z_h(a_i, a_j; x)$
- $\underline{\mathcal{G}}_h = \arg\min_{g_h(x) \in [0,1]} z_h(a_i, a_j; x)$
- $\overline{\mathcal{G}}_h = \arg\max_{g_h(x) \in [0,1]} z_h(a_i, a_j; x)$

$\underline{z}_h$ and $\overline{z}_h$, respectively, indicate the minimum and maximum value of the function $z_h(a_i, a_j, ; x)$. $\underline{\mathcal{G}}_h$ and $\overline{\mathcal{G}}_h$ represent the corresponding interval(s) of the additional action's evaluation $g_h(x)$ that lead to the minimum and maximum values. While the two first inform the user to what extent the pair of actions (a_i, a_j) may suffer from rank reversal, the two last ones indicate for what values of $g_h(x)$ rank reversal is provoked ($\underline{\mathcal{G}}_h$) or, on the contrary, what values of $g_h(x)$ do "stabilise" most strongly the relative ranking of a_i with respect to a_j.

Scenario	Extreme values for action pair (a_i, a_j)
$\Delta g_h \in [-1, -2p_h]$	$\underline{z}_h = -2$
	$\overline{z}_h = \max\{z_h(0); z_h(1)\}$
	$\underline{\mathcal{G}}_h = [0,1] \cap [g_{hi} + p_h, g_{hj} - p_h]$
	$\overline{\mathcal{G}}_h = [0,1] \setminus [g_{hi} - p_h, g_{hj} + p_h]$
$\Delta g_h \in [-2p_h, -(q_h + p_h)]$	$\underline{z}_h = -2 + \frac{1}{p_h - q_h}(\Delta g_{hij} + 2p_h)$
	$\overline{z}_h = \max\{z_h(0); z_h(1)\}$
	$\underline{\mathcal{G}}_h = [0,1] \cap [g_{hj} - p_h, g_{hi} + p_h]$
	$\overline{\mathcal{G}}_h = [0,1] \setminus [g_{hi} - p_h, g_{hj} + p_h]$
$\Delta g_h \in [-(q_h + p_h), -2q_h]$	$\underline{z}_h = -1$
	$\overline{z}_h = \max\left\{z_h(0); z_h\left(\frac{g_{hi}+g_{hj}}{2}\right); z_h(1)\right\}$
	$\underline{\mathcal{G}}_h = [0,1] \cap ([g_{hi} - q_h, g_{hj} - p_h] \cup [g_{hi} + p_h, g_{hj} + q_h])$
	$\overline{\mathcal{G}}_h = [0,1] \setminus [g_{hi} - p_h, g_{hj} + p_h]$
$\Delta g_h \in [-2q_h, 0]$	$\underline{z}_h = -1 + \frac{\Delta g_{hij} + 2(q_h - p_h)}{p_h - q_h}$
	$\overline{z}_h = \max\left\{z_h(0); z_h\left(\frac{g_{hi}+g_{hj}}{2}\right); z_h(1)\right\}$
	$\underline{\mathcal{G}}_h = [0,1] \cap ([g_{hj} - q_h, g_{hi} - p_h] \cup [g_{hj} + p_h, g_{hi} + q_h])$
	$\overline{\mathcal{G}}_h = [0,1] \setminus ([g_{hj} - p_h, g_{hi} - q_h] \cup [g_{hj} + q_h, g_{hi} + p_h])$
$\Delta g_h \in [0, p_h - q_h]$	$\underline{z}_h = \min\left\{z_h(0); z_h\left(\frac{g_{hi}+g_{hj}}{2}\right); z_h(1)\right\}$
	$\overline{z}_h = 1 - \frac{\Delta g_{hij} + 2(q_h - p_h)}{p_h - q_h}$
	$\underline{\mathcal{G}}_h = [0,1] \setminus ([g_{hj} - p_h, g_{hi} - q_h] \cup [g_{hj} + q_h, g_{hi} + p_h])$
	$\overline{\mathcal{G}}_h = [0,1] \cap ([g_{hj} - q_h, g_{hi} - p_h] \cup [g_{hj} + p_h, g_{hi} + q_h])$
$\Delta g_h \in [p_h - q_h, q_h + p_h]$	$\underline{z}_h = \min\left\{z_h(0); z_h\left(\frac{g_{hi}+g_{hj}}{2}\right); z_h(1)\right\}$
	$\overline{z}_h = 1$
	$\underline{\mathcal{G}}_h = [0,1] \setminus [g_{hj} - p_h, g_{hi} + p_h]$
	$\overline{\mathcal{G}}_h = [0,1] \cap ([g_{hj} - q_h, g_{hi} - p_h] \cup [g_{hj} + p_h, g_{hi} + q_h])$
$\Delta g_h \in [q_h + p_h, 2p_h]$	$\underline{z}_h = \min\{z_h(0); z_h(1)\}$
	$\overline{z}_h = 2 - \frac{1}{p_h - q_h}(\Delta g_{hij} + 2p_h)$
	$\underline{\mathcal{G}}_h = [0,1] \setminus [g_{hj} - p_h, g_{hi} + p_h]$
	$\overline{\mathcal{G}}_h = [0,1] \cap [g_{hi} - p_h, g_{hj} + p_h]$
$\Delta g_h \in [2p_h, 1]$	$\underline{z}_h = \min\{z_h(0); z_h(1)\}$
	$\overline{z}_h = 2$
	$\underline{\mathcal{G}}_h = [0,1] \setminus [g_{hj} - p_h, g_{hi} + p_h]$
	$\overline{\mathcal{G}}_h = [0,1] \cap [g_{hj} + p_h, g_{hi} - p_h]$

References

1. Brans, J.-P., De Smet, Y.: PROMETHEE methods. In: Figueira, J.R., Greco, S., Ehrogott, M. (eds.) Multiple Criteria Decision Analysis, State of the Art Surveys, Chap. 6, 2nd edn, pp. 187–219. Springer, New York (2016)
2. Eppe, S., Smet, Y., Stützle, T.: A bi-objective optimization model to eliciting decision maker's preferences for the PROMETHEE II method. In: Brafman, R.I., Roberts, F.S., Tsoukiàs, A. (eds.) ADT 2011. LNCS, vol. 6992, pp. 56–66. Springer, Heidelberg (2011). doi:10.1007/978-3-642-24873-3_5
3. Figueira, J.R., Roy, B.: A note on the paper, "Ranking irregularities when evaluating alternatives by using some ELECTRE methods", by Wang and Triantaphyllou, Omega (2008). OMEGA **37**(3), 731–733 (2009)
4. Forman, E.H., Selly, M.A.: Decision by Objectives. World Scientific Publishing Co., Singapore (2002)
5. García-Cascales, M.S., Lamata, M.T.: On rank reversal and TOPSIS method. Math. Comput. Model. **56**(5–6), 123–132 (2012)
6. Maleki, H., Zahir, S.: A comprehensive literature review of the rank reversal phenomenon in the analytic hierarchy process. J. Multi-Criteria Decis. Anal. **20**(3–4), 141–155 (2013)
7. Mareschal, B., De Smet, Y., Nemery, P.: Rank reversal in the PROMETHEE II method: some new results. In: IEEE 2008 International Conference on Industrial Engineering and Engineering Management, IEEM 2008, pp. 959–963. IEEE Press, Piscataway (2008)
8. Roland, J., Smet, Y., Verly, C.: Rank reversal as a source of uncertainty and manipulation in the PROMETHEE II ranking: a first investigation. In: Greco, S., Bouchon-Meunier, B., Coletti, G., Fedrizzi, M., Matarazzo, B., Yager, R.R. (eds.) IPMU 2012. CCIS, vol. 300, pp. 338–346. Springer, Heidelberg (2012). doi:10.1007/978-3-642-31724-8_35
9. Saaty, T.L., Sagir, M.: An essay on rank preservation and reversal. Math. Comput. Model. **49**(5–6), 1230–1243 (2009)
10. Sen, A.: Internal consistency of choice. Econometrica **61**(3), 495–521 (1993)
11. Verly, C., De Smet, Y.: Some results about rank reversal instances in the PROMETHEE methods. Int. J. Multicriteria Decis. Making **3**(4), 325–345 (2013)
12. Wang, Y.-M., Luo, Y.: On rank reversal in decision analysis. Math. Comput. Model. **49**(5–6), 1221–1229 (2009)
13. Wang, Y.-M., Triantaphyllou, E.: Ranking irregularities when evaluating alternatives by using some ELECTRE methods. OMEGA **36**(1), 45–63 (2008)

Peek – Shape – Grab:
A Methodology in Three Stages for Approximating the Non-dominated Points of Multiobjective Discrete/Combinatorial Optimization Problems with a Multiobjective Metaheuristic

Xavier Gandibleux[✉]

Faculty of Sciences and Technologies, IRCCyN UMR CNRS 6597,
Université de Nantes, 2, rue de la Houssinière, 44322 Nantes Cedex 03, France
xavier.gandibleux@univ-nantes.fr
http://www.univ-nantes.fr/gandibleux-x

Abstract. The paper discusses on the role of components of multiobjective metaheuristics in the context of multiobjective discrete/combinatorial optimization. It suggests to separate in three stages the design of an operational solver for this class of optimization problems, featuring a methodology in this context. The arguments are illustrated using the knapsack problem as support, along numerical experiments.

Keywords: Multiobjective optimization · Multiobjective metaheuristic · Discrete and combinatorial problems · Algorithmic design methodology

1 Introduction

1.1 Context and Motivations

To solve exactly a Multiobjective Optimization Problem (MOP), without introducing (a priori or interactively) the preferences of a decision maker, means to compute the set of non-dominated points denoted by Y_N. In front of a difficult MOP to solve, a reasonable alternative to exact methods is to derive an approximation method. A Multiple Objective MetaHeuristic (MOMH) is an approximation method in MOP context, which provides a good tradeoff between a high quality approximation of Y_N, and the time and memory requirements to obtain it.

Evolutionary MultiObjective (EMO) algorithms have been originally introduced for approximating Y_N of multiobjective non linear optimization problems with continuous variables (MONLP). A quick look on the pioneer papers ([12,30,31] by example among many others papers), on the standard benchmarks (e.g. ZDT, DTLZ, WFG, all of them can be found online here [20]), and on the

© Springer International Publishing AG 2017
H. Trautmann et al. (Eds.): EMO 2017, LNCS 10173, pp. 221–235, 2017.
DOI: 10.1007/978-3-319-54157-0_16

examples developed into reference books (e.g. [1,2]) confirms this claim. And when the multiobjective discrete/combinatorial optimization (MODCO) side is considered, the knapsack problem is mainly used as benchmark ([19,33], etc.).

It is then not surprising to find specific mechanisms within EMO algorithms for dealing with particularities of MONLP. The filtering of approximated solutions is one illustration of such a mechanism. As the variables are continuous, and EMO algorithms handle discrete values on these variables, the filtering aims to prevent a huge number of discrete values, corresponding to solutions, and moreover, which may over represent a sub set of the non-dominated points searched.

However, such a filtering mechanism is in general not relevant for MODCO where the aims here is to grab the set of non-dominated points or its approximation. A similar observation can be done about the initial population of solutions provided in input of a EMO algorithm. For several reasons, this population is generally composed of randomly generated solutions. However, useful information on the optimization problem to solve, as bounds or bounds sets, may be generally obtained and exploited for MODCO [14].

For situations where the optimization problem to deal with belong to the class of MODCO, EMO algorithms have been coupled with several usual well known techniques in the field of discrete/combinatorial optimization, such as, chronologically Local Search (LS) [24], Pareto Local Search (PLS) [23], 2 Level Local Search (2LLS) [17], etc. The literature presents these blended algorithms under the terminology of "memetic algorithms" (heuristic and local search), "hybrid algorithms" (heuristic mixed with an other exact or heuristic technique), and more recently by "matheuristics" (when exact or heuristics are collaborating within a same solving process).

Different studies have proposed several coupling (such as local search, or again path relinking) and specific mechanisms of local search have been proposed ([4,14,21], etc.). But the introduction of these components has progressively modified the role of the EMO algorithm, itself now viewed as a component of the MOMH.

This paper discusses about the role of each component in the context of MODCO. On basis of contributions that we did to the field of multiobjective metaheuristics since [16], we suggest to separate in three stages the design of a MOMH for MODCO, featuring a methodology in this context. To follow the tradition, the arguments are illustrated using the knapsack problem as support, along numerical experiments.

The rest of the paper is organized as follow. After a brief introduction of notations and definitions in the next subsection, Sect. 2 presents the methodology, first in sketching the main conclusions who led to that proposal, and second, the three stages of the methodology. Section 3 illustrates the methodology with support of numerical examples and mentions two real-life case studies managed with respect to the methodology.

1.2   General Definitions and Notations

Let $\mathbb{R}^n$ and $\mathbb{R}^p$ be Euclidean vector spaces referred to as the decision space and the objective space. $x \in \mathbb{R}^n$ is a decision. Let $X \subset \mathbb{R}^n$ is a feasible set and let f be a vector-valued objective function $f : \mathbb{R}^n \to \mathbb{R}^p$ composed of p real-valued objective functions, $f = (f_1, \ldots, f_p)$, where $f_k : \mathbb{R}^n \to \mathbb{R}$ for $k = 1, \ldots, p$. A multiobjective optimization problem is defined as

$$\min \{ (f_1(x), \ldots, f_p(x)) : x \in X \}. \tag{MOP}$$

This paper addresses MultiObjective Discrete/Combinatorial Optimization problems, which can be formulated as

$$\min \{ Cx : Ax \geqq b, x \in \mathbb{Z}^n \}. \tag{MODCO}$$

Here C is a $p \times n$ objective function matrix, where c^k denotes the k-th row of C. A is an $m \times n$ matrix of constraint coefficients and $b \in \mathbb{R}^m$. Usually the entries of C, A and b are integers. When the feasible set $X = \{ Ax \geqq b, x \in \mathbb{Z}^n \}$ describes a combinatorial structure, we talk about MultiObjective Combinatorial Optimization problems (see [6]), otherwise we talk about MultiObjectve Discrete Optimization problems.

By $Y = f(X) := \{ f(x) : x \in X \} \subset \mathbb{R}^p$ we denote the image of the feasible set in the objective space. We consider optimal solutions of (MODCO) in the sense of efficiency, i.e., a feasible solution $x \in X$ is called efficient if there does not exist $x' \in X$ such that $f_k(x') \leqq f_k(x)$ for all $k = 1, \ldots, p$ and $f_k(x') < f_k(x)$ for some k. In other words, no solution is at least as good as x for all objectives, and strictly better for at least one.

Efficiency refers to solutions x in decision space. In terms of the objective space, with objective vectors $f(x) \in \mathbb{R}^p$ we use the notion of non-dominance. If x is efficient then $f(x) = (f_1(x), \ldots, f_p(x))$ is called non-dominated. The complete set of efficient solutions is X_E, the set of non-dominated points is Y_N. We may also refer to Y_N as the non-dominated frontier. Also, a set of potentially efficient solutions is denoted by X_{PE} and Y_{PN}, respectively in the decision and objective space. More details about definitions and notations are available in [5,8].

2   Methodology in Three Stages

The proposed methodology arises from the major conclusions raised in our previous research since the end of the nineties. The main lines of these conclusions are sketched in this subsection.

2.1   Observations Leading to the Methodology Definition

- **Bound sets on** Y_N [7]. The concept of bound sets has been mentioned in the eighties by Villarreal et al. [32], but it was only developed for the first time in 2001 [7] and fully defined in 2007 [9]. The lower and upper bound sets are useful as criterion for deciding when a solution is promising and then to trigger -e.g.- a local search from that solution (see [17]). An heuristic version of lower bound set has been introduced in [26], as a way to enlarge the area corresponding to promising solutions.

- **Local search and seeding solutions** [22]. The resolution of the biobjective scheduling problem defined by $1 \mid \mid (\sum C_i, T_{max})$ has been considered with a specific multiobjective genetic algorithm (named MGK) coupled with a local search acting in two steps. One main contribution of the paper concerns the "seeding solutions", as principle for accelerating the convergence of the algorithm. The cases (i) without seeding solutions, (ii) with exact seeding solutions and (iii) with an approximation of the seeding solutions are analyzed. A component based on a two steps local search is also integrated. As conclusion of this paper, the positive impact of the seeding solutions is clearly stated.

- **Path-relinking and genetic information** [14]. For the linear assignment problem with two objectives, a path-relinking operator has been embedded for the first time in a population based-method. Seeding solutions, in particular represented by the set of supported solutions which can be easily computed for this combinatorial problem, are intensively used by the method. The conclusion here raises that useful information may be derived from the seeding solutions, and may advantageously be used by the method - notably by the path relinking- for speeding up drastically the convergence of the algorithm.

- **Greedy randomized initial solutions** [3]. Greedy Randomized Adaptive Search Procedure (GRASP) [10,11] is a well-known metaheuristic who has shown its efficiency on several hard combinatorial optimization problems. One proposal developed in the cited paper consists in the use of the first phase of GRASP for computing a subset of initial solutions, presented afterwards in input of a EMO algorithm (a modified "Strength Pareto Evolutionary Algorithm" [34] has been considered in the paper). The proposal has been experimented on set packing problems. In conclusion, in mixing quality and diversity within solutions built, GRASP permits to elaborate very quickly a large population of very good individuals with a control on their distribution into the objective space.

Several of these proposals have been considered in recent papers; some of them are cited here. Haubelt et al. [18] initialize the population in order to guide the search towards the feasible region. They proposed the incorporation of a method called Pareto-Front-Arithmetics for constructing initial populations into existing EMO algorithms. Pasia et al. [26] use a single objective "Ant Colony Optimization (ACO)" algorithm for generating individuals objective by objective for a flowshop problem. [25] proposes a two-phase MOMH. Phase one consists of a weighted sum-based solution method belonging to the "Variable Neighborhood Search (VNS)" metaheuristic. Phase two of is a heuristic module based on Path-relinking. [21] proposes a Two-phase Pareto local search, combining seeding solutions and PLS, for the biobjective traveling salesman problem.

2.2 The Methodology

The proposed methodology is structured in three stages entitled " Peek–Shape–Grab (PSG)". Each stage is characterized by a well stated mission, and is concretized by an algorithmic principle. Of course, each algorithmic principle may be instantiated in different ways. Section 3 presents how the three stages have been instantiated in the experiment reported. It is obvious that an instantiation may be replaced by another one, as long as it complies with the mission of the methodology stage.

1. **Peek**
   The aim of this stage is to probe Y_N in order *to construct a good initial population* (Fig. 1, top). The quality expected by this population is to represent *a good coverage* of the non-dominated frontier Y_N, even if the approximation is sparse due to the local character of the operator applied. It is expected from the resulting population to act as attractor for the approximation mechanism during the next stages. Very high quality solutions are then not specially expected, preferring good but diversified solutions.

   Input: A numerical instance for the considered MODCO problem.
   Output: A non-connected approximation of Y_N which represents of a good initial population, i.e. a representative coverage of Y_N.

2. **Shape**
   The aims of this second stage is to spread the population along Y_N in order *to get a well distributed population* without missing subparts of the non-dominated frontier (Fig. 1, middle). The quality expected is now to deliver a population connected and widespread along Y_N. The generation of a global approximation exploiting genetic information available is typically the strength of genetic algorithms, which have the advantage to explore the search space.

   Input: A constructed population returned at the end of the stage "Peek".
   Output: A population giving a global approximation of Y_N and uniformly distributed.

3. **Grab**
   The last stage aims to improve in an aggressive way the approximation, using all information collected during the search process until now, in order *to deliver a high quality approximation of Y_N* (Fig. 1, bottom). Individuals of the population are locally improved in order to shift the population to "its (local) optimum". Local improvement is a matter of local search procedure bracketed by upper/lower bound sets and guided by components like path-relinking.

   Input: A population uniformly distributed giving the shape of Y_N
   Output: A high quality approximation of Y_N.

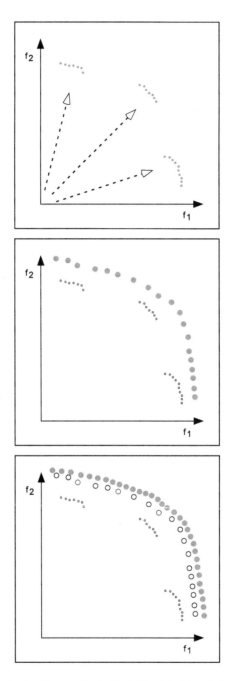

Fig. 1. The figure illustrates the awaited effect (depicted by plain bullets) of *peek* (top), *shape* (middle), and *grab* (bottom) stages. The empty bullets recall the populations presented in input of the preceding stages.

3   Illustrations of the Methodology

3.1   The Version of Knapsack Problem Serving as Support for the Illustration

The MODCO problem is the biobjective ($p = 2$) uni-dimensional ($m = 1$) binary knapsack problem:

$$
\left[
\begin{array}{ll}
\max \sum_{j=1}^{n} c_j^k x_j & k = 1, \ldots, p \\
\text{s.t.} \sum_{j=1}^{n} w_{ij} x_j \leq \omega_i \; i = 1, \ldots, m \\
\quad x_j \in \{0, 1\} & j = 1, \ldots, n
\end{array}
\right]
$$

The computer used for conducting our experiment has these characteristics: intel Core 2 Duo E8500 @ 3.16 GHz; 3.4 GB RAM. The operating system is Ubuntu, and the algorithm is implemented in C++. The instance used is the 1B/2KP500-1A (serie 1B; 2 objectives; 500 variables; random coefficients) available on the MOCOlib[1].

3.2   About the Impact of Seeding Solutions [15]

Figure 2 shows experimentally the impact of seeding solutions on the convergence of the Y_{PN} approximation. In that experiment, the definition of the solution algorithm is fixed and consists in a basic multiobjective genetic algorithm. Its components and parameters (crossover and mutation operators, the probability of crossover and mutation, the size of the population, number of generations) are also freezed. Starting from an initial population composed by the same number of individuals, but built in three different manners, the experiment shows the evolution, in comparison to Y_N (black bullets). That evolution is examined before the first generation (red bullets), after 10 generations (green bullets), and after 50 generations (blue bullets).

- In the first case (top of Fig. 2), the population is only composed of individual randomly generated.

  MOGA algorithm reacts as expected, the population of individuals evolves well towards the non-dominated frontier. However, we remark that the final population returns an approximation that describes only in the central part of non-dominated frontier, and this without reaching Y_N. Therefore, the approximation does not provide a valuable indication of all the non-dominated frontier.
- In the second case (middle of Fig. 2), three seeding solutions are built. They correspond to three solutions computed by application of the following simple greedy constructive algorithm. The convex combination of the two objectives

---

[1] http://xgandibleux.free.fr/MOCOlib/.

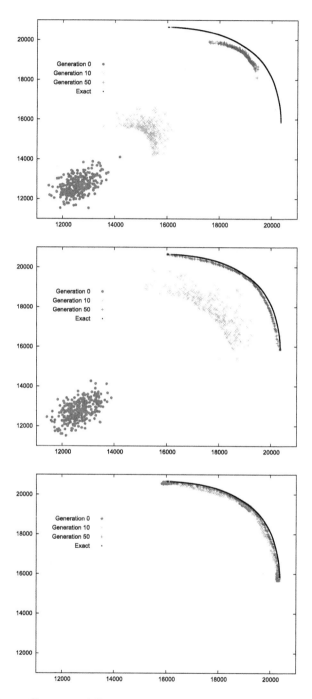

Fig. 2. The figure illustrates different states of the populations along the generations, for a same computing budget, a same tuning of parameters, and three variants of seeding solutions: (top) none seeding solution, (middle) with 3 seeding solutions plus random solutions, (bottom) with 3 sets of greedy solutions. The axes x and y represent respectively f_1 and f_2. (Color figure online)

$f_\lambda = \lambda f_1 + (1 - \lambda) f_2$ is set, giving a parametrized function by λ. The three solutions are obtained by optimizing the resulting single objective problem for the values $\lambda = \{0; 0.5; 1\}$ with a greedy algorithm. Thus, two solutions correspond to the sub-optimal solutions when objectives are optimized separately ($\lambda = 0$ and $\lambda = 1$), the third corresponds to a compromise solution ($\lambda = 0.5$). The rest of the population is only composed by individuals randomly generated.

For the same computing budget given to the algorithm, we observe by comparison with the previous case that the population is closer to Y_N and better spread already after 10 generations. The difference is especially striking at the end of the 50 generations, where the final population gives a good description of the entire non-dominated frontier. Y_{PN} is indicative of Y_N but remains improvable. Nevertheless, the approximation Y_{PN} is able to find several points of Y_N in the neighborhood of the seeding solutions.

- In the third case (bottom of Fig. 2), three sets of seeding solutions are built and constitute exclusively the initial population; here, there is therefore no random solution. These solutions are obtained by application of the stage "peek" presented in Sect. 2, based on the constructive phase of GRASP. The optimized function is the parametric function by defined as in the previous case, with the values $\lambda = 0$, $\lambda = 0.5$, and $\lambda = 1$. All the solutions generated are gathered for constituting the initial population.

For the same computing budget given to the algorithm, the three sets of individuals, which are initially not connected together, are now uniformly distributed along the non-dominated frontier providing only one set of individuals, representative of a good global approximation, already at the tenth generation. At the end of the fiftieth generation, the description of Y_N by Y_{PN} is improved and more points of Y_N are reached, in comparison with the previous case. Y_{PN} is representative of Y_N and "matches" with Y_N.

In preliminary conclusion, we observe that the seeding solutions (1) speed up the convergence of the population of individual toward the non-dominated frontier and (2) incite the individuals in the population to be well distributed along the non-dominated frontier. The saving on the computation budget achieved during the convergence of the population thanks to the seeding solutions allows to relocate the saved time on the others phases of the algorithm. The evolutionary multiobjective component is awaited as component for spreading the individuals of the population, and not any longer as component for generating a high quality approximation. By the way, the construction cost of the population with the proposed principle is insignificant in comparison to the required computing time by the two others phases. That experimental observation validates the suitability of the first phase of the methodology, and the mission assigned to the second phase.

3.3   Experimental Demonstration [15]

1. **Peek** with GRASP

   Principle: The 2 objective functions $f_k, k = 1, 2$ for the considered MODCO problem are agregated using a weight vector λ into a parametric function $f_\lambda = \phi(f_1, f_2, \lambda)$. This function f_λ is then optimized with the construction step of GRASP for a collection of values λ. The utility function defined for ranking the items of the knapsack is the ratio profit/weight.

2. **Shape** with MOGA

   Principle: The biobjective knapsack problem is tackled with a EMO solver. Here, a simple homemade MOGA has been designed. Obviously, all recent advances on EMO based algorithms and all outcomes of recent studies related on solving the knapsack problem by a EMO algorithm are welcome for strengthening this stage.

3. **Grab** with IPLS

   Principle: A selection of solutions are optimized on the objectives f_1, f_2 with a method belonging to the iterated Pareto local search (IPLS). In a short, a niche of size σ allows to select candidates for a local search. A lower bound set is used for deciding if the improvement is triggered. A recursive principle is performed while new potential non-dominated points are grabbed.

The algorithm has been setup with a computing budget of $T = 120\,\text{s}$ dispatched as follow:

$$\text{GRASP}(\approx 0) + \text{MOGA}\left(\frac{T}{2}\right) + \text{IPLS}\left(\frac{T}{2}\right)$$

The resolution has been repeated 10 times. Figure 3 shows the effect of the three stages, in plotting the populations of individuals in the objective space. Again we can notice that the points issued from the first phase (red bullets) are well spread after the second phase (green bullet). The outputs of the second and third phases are visually indistinguishable.

The quality indicator $M1$ is thus computed. It measures the percentage of exact non-dominated points of Y_N detected in Y_{PN}, i.e.

$$M1 = \frac{|Y_{PN} \bigcap Y_N|}{|Y_N|}$$

The values are given in average upon the 10 runs, and also the minimal and maximal values collected. The value of M1 moves from 30.75% after the phase 2, to 77.76% after the phase 3. The positive effect of the last phase of the methodology do not leave any doubt. Moreover, according the minimal and maximal values collected, the algorithm appears robust as the interval is tight.

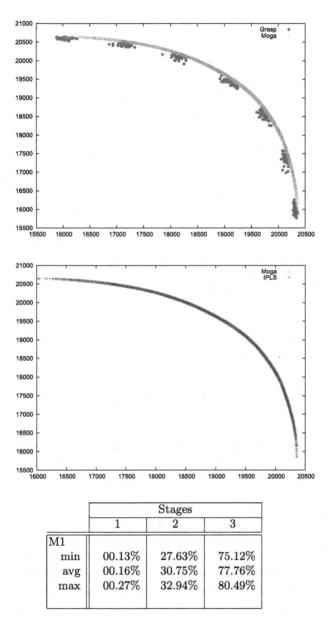

	Stages		
	1	2	3
M1			
min	00.13%	27.63%	75.12%
avg	00.16%	30.75%	77.76%
max	00.27%	32.94%	80.49%

Fig. 3. Dump of Y_{PN} at the end of the first and second stage (top) – Dump of Y_{PN} at the end of the second and third stage (bottom). The axes x and y represent respectively f_1 and f_2. The table report the measure of M1 (percentage min/avg/max of exact non-dominated points detected) at the end of the three stages.

3.4   Two Real Life Problems

- **A biobjective telecommunication network expansion planning problem** [13]. This case study comes from a research collaboration with FranceTelecom. The problem concerns a network expansion planning problem in telecommunication. The mathematical model formalizing this problem belongs to the class of biobjective facility location problems with mixed integer variables. The seeding solutions are obtained with a branch-and-bound algorithm limited in computing time, and embedded into a ϵ-constraint method. The EMO solver is NSGA-II modified (without filtering procedure). A local search and a path-relinking finalize the solver.
- **A multiobjective machine reassignment algorithm for data centers** [27–29]. This work is considered seminal in the data center management community and has led to a collaboration with IBM [27,29]. The problem consists in reassigning virtual machines in large data centers – hence a large and constrained multiobjective problem with mixed variables. We use a GRASP algorithm to get an initial population, followed by NSGA-II and PLS.

4   Conclusion and Open Questions

After having studied and better understood the behavior of MOMH presented since the mid-nineties, identified the strengths and weaknesses, this paper notifies on the importance of a non-monolithic approach to design a resolution method for a MODCO. It therefore presents a methodology of design in three phases, Peek-Shape-Grab. It illustrates its suitability using a very simple algorithmic instantiation but yet effective, when computational resources are limited, and according to the collected observations and measures. In accordance with the definition of a Metaheuristic, this methodology suggests missions and proposes algorithmic directions. For each MODCO problem, the three phases have to be instantiated and to integrate the known results from the literature in order to benefit of last advances and give rise has an operational effective resolution algorithm.

Several questions remain to be addressed, for example on the implementation of the phase 3: local search on all available points or not? How to select a relevant subset of points? How to move forward in the local search without losing information? Also, what questions and answers to bring when the problem rises in the number of objectives to deal with simultaneously? Up to how many objectives it is reasonable to suggest this methodology? So many questions to investigate into future researches.

Acknowledgment. I would like to thank Brice Chevalier, Quentin Delmée, Benjamin Martin, Olga Perederieiva, Sylvain Rosembly, and Jocelyn Willaime, master students in computer science, track "optimization in operations research" from the Université de Nantes (France) who participated to set up the software required for the production of the different cases reported in this paper. I would also thank Takfarinas Saber, doctoral student from University College Dublin (Eire), for the discussions we had on the MOMH that he develops. Last but not least, I would like to thank Dr. Anthony Przybylski, senior lecturer at the Université de Nantes (France) for the fruitful discussions on MOP and MOMH, and the co-supervisions of students we had during these past years.

References

1. Coello, C.A.C., Lamont, G.B. (eds.): Applications of Multi-objective Evolutionary Algorithms. World Scientific, Singapore (2004)
2. Deb, K.: Multi-Objective Optimization using Evolutionary Algorithms. Wiley-Interscience Series in Systems and Optimization. Wiley, Chichester (2001)
3. Delorme, X., Gandibleux, X., Degoutin, F.: Evolutionary, constructive and hybrid procedures for the bi-objective set packing problem. Eur. J. Oper. Res. **204**(2), 206–217 (2010)
4. Dubois-Lacoste, J., López-Ibáñez, M., Stützle, T.: Adaptive "anytime" two-phase local search. In: Blum, C., Battiti, R. (eds.) Learning and Intelligent Optimization. LNCS, vol. 6073, pp. 52–67. Springer, Heidelberg (2010)
5. Ehrgott, M.: Multicriteria Optimization. Springer, New York (2005)
6. Ehrgott, M., Gandibleux, X.: A survey and annotated bibliography of multiobjective combinatorial optimization. OR Spektrum **22**, 425–460 (2000)
7. Ehrgott, M., Gandibleux, X.: Bounds and bound sets for biobjective combinatorial optimization problems. In: Köksalan, M., Zionts, S. (eds.) Multiple Criteria Decision Making in the New Millennium. LNEMS, vol. 507, pp. 241–253. Springer, Heidelberg (2001)
8. Ehrgott, M., Gandibleux, X.: Approximative solution methods for multiobjective combinatorial optimization. Top **12**(1), 1–63 (2004)
9. Ehrgott, M., Gandibleux, X.: Bound sets for biobjective combinatorial optimization problems. Comput. Oper. Res. **34**, 2674–2694 (2007)
10. Féo, T.A., Resende, M.G.C.: A probabilistic heuristic for a computationally difficult set covering problem. Oper. Res. Lett. **8**(2), 67–71 (1989)
11. Féo, T.A., Resende, M.G.C.: Greedy randomized adaptive search procedures. J. Glob. Optim. **6**(2), 109–133 (1995)
12. Fonseca, C., Fleming, P.: Genetic algorithms for multiobjective optimization: formulation, discussion and generalization. In: Forrest, S. (ed.) Proceedings of the Fifth International Conference on Genetic Algorithms, San Mateo, California, 1993. University of Illinois at Urbana-Champaign, pp. 416–423. Morgan Kaufman, San Francisco (1993)
13. Gandibleux, X., Chamayou, C.: Potential efficient solutions of a bi-objective telecommunication network expansion planning problem. In: The Seventh Metaheuristics International Conference, MIC2007, Montreal, Canada, 25–29 June 2007
14. Gandibleux, X., Katoh, N., Morita, H.: Evolutionary operators based on elite solutions for bi-objective combinatorial optimization. In: Coello, C.A.C., Lamont, G.B. (eds.) Applications of Multi-Objective Evolutionary Algorithms, vol. 1, Chap. 23, pp. 555–579. World Scientific (2004)
15. Gandibleux, X., Martin, B., Perederieieva, O., Rosembly, S.: Sur la résolution approchée en trois étapes du sac-à-dos bi-objectif unidimensionnel en variables binaires. In: ROADEF 2011 (12e congrès annuel de la société française de Recherche Opérationnelle et d'Aide à la Décision), Saint-Etienne, France, 2–4 mars 2011

16. Gandibleux, X., Morita, H., Katoh, N.: A genetic algorithm for 0–1 multiobjective knapsack problem. In: Proceedings of the International Conference on Nonlinear Analysis and Convex Analysis (NACA 1998), Niigata, Japan, 28–31 July 1998 (1998)

17. Gandibleux, X., Morita, H., Katoh, N.: Use of a genetic heritage for solving the assignment problem with two objectives. In: Fonseca, C.M., Fleming, P.J., Zitzler, E., Thiele, L., Deb, K. (eds.) EMO 2003. LNCS, vol. 2632, pp. 43–57. Springer, Heidelberg (2003). doi:10.1007/3-540-36970-8_4

18. Haubelt, C., Gamenik, J., Teich, J.: Initial population construction for convergence improvement of MOEAs. In: Coello Coello, C.A., Hernández Aguirre, A., Zitzler, E. (eds.) EMO 2005. LNCS, vol. 3410, pp. 191–205. Springer, Heidelberg (2005). doi:10.1007/978-3-540-31880-4_14

19. Ishibuchi, H., Kaige, S.: Comparison of multiobjective memetic algorithms on 0/1 knapsack problems. In: Proceedings of 2003 Genetic and Evolutionary Computation Conference Workshop Program, Chicago, USA, pp. 222–227 (2003)

20. jMetal: Problems included in jmetal. http://jmetal.sourceforge.net/problems.html. Accessed 28 Sep 2016

21. Lust, T., Teghem, J.: Two-phase pareto local search for the biobjective traveling salesman problem. J. Heuristics 16(3), 475–510 (2010)

22. Morita, H., Gandibleux, X., Katoh, N.: Experimental feedback on biobjective permutation problems solved with a population heuristic. Found. Comput. Decis. Sci. 26(1), 43–50 (2001)

23. Morita, H., Gandibleux, X., Katoh, N.: Experimental feedback on biobjective permutation scheduling problems solved with a population heuristic. Found. Comput. Decis. Sci. J. 26(1), 23–50 (2001)

24. Murata, T., Ishibuchi, H.: MOGA: multi-objective genetic algorithms. In: Proceedings of the 2nd IEEE International Conference on Evolutionary Computing, pp. 289–294 (1995)

25. Parragh, S.N., Doerner, K.F., Hartl, R.F., Gandibleux, X.: A heuristic two-phase solution approach for the multi-objective dial-a-ride problem. Networks 54(4), 227–242 (2009)

26. Pasia, J.M., Gandibleux, X., Doerner, K.F., Hartl, R.F.: Local search guided by path relinking and heuristic bounds. In: Obayashi, S., Deb, K., Poloni, C., Hiroyasu, T., Murata, T. (eds.) EMO 2007. LNCS, vol. 4403, pp. 501–515. Springer, Heidelberg (2007). doi:10.1007/978-3-540-70928-2_39

27. Saber, T., Ventresque, A., Brandic, I., Thorburn, J., Murphy, L.: Towards a multi-objective vm reassignment for large decentralised data centres. In: 2015 IEEE/ACM 8th International Conference on Utility and Cloud Computing (UCC), pp. 65–74. IEEE (2015)

28. Saber, T., Ventresque, A., Gandibleux, X., Murphy, L.: GeNePi: a multi-objective machine reassignment algorithm for data centres. In: Blesa, M.J., Blum, C., Voß, S. (eds.) HM 2014. LNCS, vol. 8457, pp. 115–129. Springer, Heidelberg (2014). doi:10.1007/978-3-319-07644-7_9

29. Saber, T., Ventresque, A., Marques-Silva, J., Thorburn, J., Murphy, L.: MILP for the multi-objective VM reassignment problem. In: 2015 IEEE 27th International Conference on Tools with Artificial Intelligence (ICTAI), pp. 41–48. IEEE (2015)

30. David Schaffer, J.: Multiple objective optimization with vector evaluated genetic algorithms. In: Genetic Algorithms and their Applications: Proceedings of the First International Conference on Genetic Algorithms, pp. 93–100. Lawrence Erlbaum (1985)

31. Srinivas, N., Deb, K.: Multiobjective optimization using nondominated sorting in genetic algorithms. Evol. Comput. **2**(3), 221–248 (1994)
32. Villarreal, B., Karwan, M.H.: Multicriteria integer programming: a (hybrid) dynamic programming recursive approach. Math. Program. **21**(1), 204–223 (1981)
33. Zitzler, E., Laumanns, M.: Test problem suite. http://www.tik.ee.ethz.ch/sop/download/supplementary/testProblemSuite/. Accessed 29 Sept 2016
34. Zitzler, E., Thiele, L.: Multiobjective evolutionary algorithms: a comparative case study and the strength pareto evolutionary algorithm. IEEE Trans. Evol. Comput. **3**(4), 257–271 (1999)

A New Reduced-Length Genetic Representation for Evolutionary Multiobjective Clustering

Mario Garza-Fabre[1]([⊠]), Julia Handl[1], and Joshua Knowles[2]

[1] Decision and Cognitive Sciences Research Centre, University of Manchester,
Manchester M15 6PB, UK
{mario.garza-fabre,julia.handl}@manchester.ac.uk
[2] School of Computer Science, University of Birmingham, Birmingham B15 2TT, UK
j.knowles@cs.bham.ac.uk

Abstract. The last decade has seen a growing body of research illustrating the advantages of the evolutionary multiobjective approach to data clustering. The scalability of such an approach, however, is a topic which merits more attention given the unprecedented volumes of data generated nowadays. This paper proposes a reduced-length representation for evolutionary multiobjective clustering. The new encoding explicitly prunes the solution space and allows the search method to focus on its most promising regions. Moreover, it allows us to precompute information in order to alleviate the computational overhead caused by the processing of candidate individuals during optimisation. We investigate the suitability of this proposal in the context of a representative algorithm from the literature: MOCK. Our results indicate that the new reduced-length representation significantly improves the effectiveness and computational efficiency of MOCK specifically, and can be seen as a further step towards a better scalability of evolutionary multiobjective clustering in general.

Keywords: Data clustering · Evolutionary multiobjective optimisation · Genetic representation · Scalability

1 Introduction

Data clustering is the unsupervised task concerned with classifying a collection of data points, based on some notion of similarity, into a finite number of disjoint subsets called *clusters* [10]. This task is usually modelled as an optimisation problem (tackled *e.g.* by evolutionary algorithms), relying on the use of a clustering quality criterion (or *validity index* [5]) in order to guide the search process [8]. It is often the case, however, that a single criterion is unable to capture all desirable aspects of a clustering solution; this, together with the fact that we generally have little (or poor) information about how to choose k so it corresponds to the number of true natural clusters in the data (or number of classes), renders the conceptual advantages of *evolutionary multiobjective clustering* (EMC) evident [6,13]. EMC approaches are capable of producing, in a single run, a *Pareto front approximation* (PFA) of candidate partitions yielding

© Springer International Publishing AG 2017
H. Trautmann et al. (Eds.): EMO 2017, LNCS 10173, pp. 236–251, 2017.
DOI: 10.1007/978-3-319-54157-0_17

different trade-offs between multiple clustering criteria and potentially covering a wide range of values of k. Intuitively, the EMC methods need to be equipped with appropriate representations and operators which facilitate this outcome.

In the literature, there exists a variety of representations which have been proposed for the data clustering problem. These approaches range from those directly encoding cluster memberships of all data points, or the interaction (shared cluster membership) between points, to the increasingly popular prototype-based strategies encoding cluster centres or representatives. The selection of an appropriate encoding scheme can be seen as a multiobjective problem itself, as usually the approach excelling in one aspect presents weaknesses in some other aspects. The suitability of a representation can be judged in terms of its degree of problematic *non-synonymous redundancy* [16], its capacity to capture arbitrary cluster shapes, or its scalability properties generally related to the length of the genotype (and hence the size of the search pace) which may depend on problem size, dimensionality, and/or number of clusters. Some approaches are also better suited than others for encoding partitions with a non-fixed k, which is essential if this (domain-specific) information is unavailable *a priori*, as discussed above. Finally, some representations (*e.g.* those involving variable-length genotypes) introduce additional complexity and demand a more careful design of the genetic operators. The reader is referred to [8,13] for a review and discussion of problem representations used in evolutionary data clustering.

We introduce a new representation which attempts to provide a better trade-off between the aforementioned aspects in comparison to existing approaches from the literature. The proposed encoding scheme is based on the *locus-based adjacency representation* originally reported in [14]. This graph-based representation (described in Sect. 2.1) is used by one of the most representative methods from the state-of-the-art in EMC: the *multiobjective clustering with automatic k-determination* (MOCK) algorithm [6]. Strengths of this representation include the straightforward definition of meaningful genetic operators, and the ability to *naturally* encode partitions of varying k (removing the need of predefining this parameter). Moreover, this representation has been found to be less non-synonymously redundant than other approaches [7]. Nevertheless, it has also been criticized because its encoding length corresponds to the number of points in the data set. Due to specific strategies adopted during initialisation and genetic variation (discussed later in Sect. 2.3), the resulting size of the search space has not represented a major bottleneck with respect to the scalability of MOCK (as is evident from MOCK's existing performance on data sets with up to ∼4,000 points [6]). The use of the reduced-length representation proposed in this paper, however, can further improve the scalability of MOCK, providing clear benefits in terms of both clustering performance and computational efficiency.

Our new representation preserves the conceptual advantages of the locus-based adjacency representation, but can significantly reduce the length of the genotype and explicitly prune uninteresting regions of the solution space. This is achieved by exploiting relevant information from the *minimum spanning tree* (MST). Furthermore, the new representation enables the incremental processing and evaluation of candidate solutions starting from an initially precomputed state, thus decreasing the computational burden. We implemented this reduced-length representation within the framework of MOCK, which allows us to directly assess the suitability of this proposal with respect to the method's original full-length representation. It is important to remark that such an assessment focuses only on the ability of the reduced-length representation to improve MOCK's performance and efficiency during its *clustering phase*; analysis of the subsequent *model-selection phase* is therefore beyond the scope of this study.[1] Also, the version of MOCK considered here, particularly its optimisation criteria and initialisation scheme, assumes that input data elements are represented by vectors of numerical attributes; scenarios where data can be described by non-numerical attributes, or where only dissimilarity (or similarity) data describing the relationships between elements is available, are not covered by this paper.

This paper proceeds as follows. First, the new reduced-length encoding is described in detail in Sect. 2. Also, this section briefly discusses our implementation of MOCK, which has been adapted in this study to take full advantage of the new representation scheme. Section 3 presents our experimental evaluation and discusses the results obtained. Finally, Sect. 4 summarises the main findings of this study and highlights potential directions for future research.

2   The New Reduced-Length Genetic Representation

This section introduces a new reduced-length representation which seeks to improve the scalability of EMC. The proposed representation is incorporated into the MOCK algorithm in order to investigate its advantages with respect to the method's original (full-length) encoding. Besides equipping MOCK with the new representation, additional changes to the optimisation criteria and search strategy allow us to, respectively, take full advantage of this proposal and of MOCK's specialised initialisation routine.[2] The reduced-length representation and the corresponding adaptation of MOCK are described in Sects. 2.1 and 2.2, respectively. Then, Sect. 2.3 discusses how the new encoding scheme impacts on the size of the solution space which is accessible to the search method.

2.1   Reduced-Length Representation

The new genetic encoding draws from the locus-based adjacency representation originally used by MOCK [14]. As shown in Fig. 1, data points are seen as the

---

[1] Whereas the clustering phase is responsible for generating a PFA comprising high-quality partitions, the model-selection phase is concerned with selecting and reporting one (or more) candidate partition(s) from this PFA as the problem's solution.

[2] It should be noted that changing the optimisation criteria is the only adaptation of MOCK required by the representation scheme proposed in this paper. Such an adaptation, however, is only intended to exploit the advantages that the new representation can provide in terms of computational efficiency; this change is not found to affect MOCK's behaviour and performance as discussed at the end of Sect. 3.2.

Fig. 1. Locus-based adjacency representation. A data set of size $N = 12$ is considered. A gene $x_i = j$ in the genotype denotes a link $i \rightarrow j$ from datum i to another datum j, $i, j \in \{1, \dots, N\}$. Each connected component in the resulting graph is seen as a different cluster. Thus, the genotype of this example encodes a partition with $k = 4$ clusters.

nodes of a graph and the genotype of an individual defines the links between them. These links result in a set of connected components which represents a candidate partition. Despite presenting clear advantages with respect to other existing representations [7], the length of the genotype, given by the size of the problem N, can be seen as the main scalability issue of this approach.

As illustrated in Figs. 2 and 3, the reduced-length representation predefines a potentially large subset of the links based on the MST in order to (explicitly) limit exploration to the most promising regions of the search space. This requires identifying the subsets of relevant and fixed links from the MST, respectively denoted Γ and Δ. Classification of the MST links depends on their *degree of interestingness* (DI) and the setting of parameter δ. As can be seen, the fixed set Δ, consisting of (roughly) the $\delta\%$ less interesting MST links, defines a partial clustering which serves as the basis for the generation of all candidate solutions during the search process. In this way, the optimisation problem is redefined as that of determining only the links not yet defined in the partial clustering solution; such missing pieces of information relate to the relevant links in set Γ and are encoded in the $|\Gamma|$-length genotype of the new representation.[3]

By fixing all non-relevant MST links, only the removal (or replacement) of the relevant MST links is considered. We want to partition the MST links into relevant and non-relevant links by some criterion. We propose here the use of the criterion DI, which was originally introduced in [6] as a means to guide part of the initialisation routine of MOCK (refer to Sect. 2.2 for details). In the ideal case the relevant links would be those whose removal leads to a separation of the MST which is consistent with the inherent cluster structure, but that is not possible to know *a priori*. However, DI seems to do a good job; specifically, the DI approach has been found to be less biased, in comparison to directly using the dissimilarity (distance) between data points, towards classifying as relevant those links connecting outliers. As defined in [6], the DI of a link $i \rightarrow j$ is given

---

[3] Notice that when parameter δ is set to $\delta = 0$, the encoding scheme proposed here is equivalent to the original (full-length) locus-based adjacency representation.

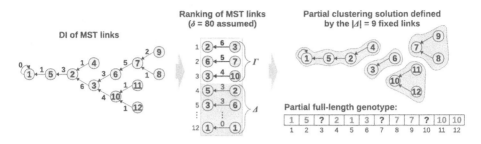

Fig. 2. Classifying MST links based on their degree of interestingness (DI) and the user-defined parameter δ ($0 \le \delta \le 100$). Whereas set Γ is formed by the $|\Gamma| = \lceil \frac{(100-\delta)}{100} N \rceil$ most prominent (highest DI) links, set Δ consists of the $|\Delta| = \lfloor \frac{\delta}{100} N \rfloor$ links with the lowest DI. A value of $\delta = 80$ is used in this example, which produces $|\Gamma| = 3$ and $|\Delta| = 9$. The nine links in Δ (and their corresponding genes in the full-length genotype) are assumed fixed and lead to a partial clustering (with $k = 4$ clusters) which forms the basis for all candidate solutions to be explored during the search process.

by $int(i \rightarrow j) = \min\{nn_i(j), nn_j(i)\}$, where $nn_a(b)$ refers to the ranking position of data point b in the list of nearest neighbours of data point a.

2.2   Adaptation of MOCK

Below, the specifics of MOCK's implementation used during the experiments of this study are described, with particular emphasis on the components which vary with respect to the original version of the algorithm reported in [6].

Optimisation Criteria. MOCK, as reported in [6], optimises two complementary clustering criteria: *overall deviation* (ODV) and *connectivity* (CNN). With the aim of exploiting the benefits that the new encoding can provide in terms of the incremental processing of solutions (see Delta-Evaluation below), ODV is replaced here with a highly correlated criterion: *intracluster variance* (VAR). VAR accounts for cluster compactness (homogeneity) and is given by:

$$var(\mathcal{C}) = \frac{1}{N} \sum_{c \in \mathcal{C}} v(c) \ , \tag{1}$$

where $\mathcal{C}$ is the set of clusters in the candidate partition and $v(c)$ represents the individual contribution of cluster c to this measure: $v(c) = \sum_{i \in c} \sigma(i, \mu_c)^2$. Here, μ_c denotes the centroid of cluster c, and $\sigma(i, \mu_c)$ refers to the dissimilarity between data point i and μ_c (the Euclidean distance is used in this study).

The CNN criterion is preserved as in the original implementation of MOCK. This criterion captures cluster connectedness, reflecting the degree to which neighbouring data points are identified as members of the same cluster:

$$cnn(\mathcal{C}) = \sum_{i=1}^{N} \sum_{l=1}^{L} \rho(i, l) \ , \tag{2}$$

where L specifies the size of the neighbourhood and $\rho(i, l) = \frac{1}{l}$ iff point i and its l-th nearest neighbour are not in the same cluster and $\rho(i, l) = 0$ otherwise.

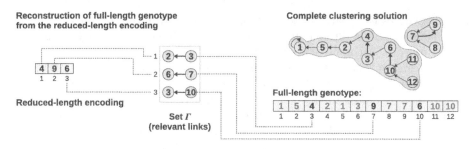

Fig. 3. The new reduced-length genetic representation and the process of reconstructing the full-length genotype which is a first step towards decoding. The new representation operates with genotypes of length $|\Gamma|$, where Γ is the set of relevant MST links (see Fig. 2). Starting from the partial solution given by the set of fixed MST links Δ, the $|\Gamma|$-length encoding is used to define the missing pieces of information in the full-length genotype, which is then decoded into a complete clustering solution.

Both VAR and CNN are to be minimised. While the optimisation of VAR tends to increase k, CNN presents the opposite tendency. Therefore, the simultaneous optimisation of VAR and CNN compensates for these individual biases and produces a PFA of good-quality partitions with a diversity of values for k.

Delta-Evaluation. Given the partial clustering derived from the set of fixed MST links Δ, the decoding and evaluation of a candidate solution only requires processing the non-fixed information encoded in its $|\Gamma|$-length genotype (Figs. 2 and 3). Thus, we can precompute the decoding and evaluation of such a partial solution in order to speed up the processing of individuals during the search.

The decoding of the $|\Gamma|$-length genotype of an individual creates new links which can merge originally separate clusters of the partial solution. Such a change in the phenotype implies an amendment to the initially precomputed values of the VAR and CNN criteria. Adhering to its original definition provided in (1), VAR is recomputed by averaging the individual contributions of all the final clusters to this measure. In this case, however, the contribution of a new joint cluster $c_h = c_i \cup c_j$ to VAR can be more efficiently obtained by leveraging on the original (precomputed) contributions of the clusters being combined [1]:

$$v(c_h) = v(c_i) + v(c_j) + |c_i| \times \sigma(\mu_{c_i}, \mu_{c_h})^2 + |c_j| \times \sigma(\mu_{c_j}, \mu_{c_h})^2 , \qquad (3)$$

where μ_{c_h} denotes the centroid of c_h and is computed as the weighted average of the original centroids μ_{c_i} and μ_{c_j} (this is generalisable to the case where an arbitrary number of clusters are combined). Similarly, adjusting CNN due to the combination of two clusters c_i and c_j requires subtracting all contributions made to this measure as a consequence of the original separation of c_i and c_j.

Search Engine. MOCK's original version is based on the *Pareto envelope-based selection algorithm version 2* (PESA-II) [2]. PESA-II is strongly elitist, and this causes that part of the individuals generated during initialisation (the dominated ones) are filtered and discarded without being considered during the search process. An approach with less selection pressure seems advantageous in this scenario where highly optimised solutions are generated by MOCK's specialised initialisation. We study a different implementation of MOCK based on the *nondominated sorting genetic algorithm version 2* (NSGA-II) [3]. This change in the search engine makes it possible to exploit all of the genetic material introduced during initialisation, as it forms the basis of the initial population.

Initialisation and Genetic Operators. MOCK's implementation studied herein preserves the specialised initialisation routine and the same set of genetic operators as reported in [6]. Initialisation attempts to provide MOCK with a close initial approximation to the Pareto front. An initial population of MST-derived solutions is generated following a two-phase process. The first phase constructs (at most) half of the population. Each individual resulting from this phase encodes a partition created by removing the n highest DI links from the MST (see definition of DI in Sect. 2.1). To obtain a diverse set of initial partitions, n is chosen uniformly and without replacement from the set $\{0, 1, \ldots, \min(k_{user} - 1, I)\}$. Here, k_{user} can be seen as an upper bound on the number of clusters expected; based on preliminary testing this parameter is set to $k_{user} = 2k^*$ in this study, where k^* denotes the real (or estimated) number of clusters in the data set. I is the cardinality of the subset of MST links which are allowed to be removed during this phase; these links, called the *interesting links* in [6], are those links $i \rightarrow j$ such that neither i nor j is one of the L nearest neighbours of the other. Therefore, although this phase attempts to create half of the initial population, the actual number of individuals produced also depends on k_{user} and I. The second phase generates the remainder (at least half) of the population. Every remaining individual is created by first running (a single execution of) k-means [12] for a given target k, and then removing all MST links crossing the cluster boundaries defined by the partition obtained. Each time the target k value used for k-means is drawn uniformly without replacement from the set $\{2, 3, \ldots, k_{user}\}$, which contributes to the diversity of the initial population. Note that when using the reduced-length encoding proposed in this paper, removal of MST links (in both phases defined above) is only permitted for links in Γ (links in set Δ are fixed for all candidate partitions, see Sect. 2.1).

For the genetic operators, we use *uniform crossover* [17] which can produce any possible combination of genetic material from the parent genotypes being recombined. Also, we use the *neighbourhood-biased mutation* [6] which defines individual mutation probabilities for each gene in the genotype based on the specific link it encodes. More precisely, the mutation probability of a gene $x_i = j$, encoding link $i \rightarrow j$, is given by $p_m(x_i) = 1/N + (nn_i(j)/N)^2$. This increases the chances of discarding unfavourable links. Note that the recombination and mutation strategies operate on the $|\Gamma|$-length genotypes of the new encoding.

During initialisation and mutation, any link $i \rightarrow j$ which is removed is replaced either with a self-connecting link $i \rightarrow i$ or with a link from i to one of its L nearest neighbours. The new link is decided uniformly at random from these $L + 1$ choices, but excluding the reintroduction of the original link $i \to j$. It is worth realising that if $i \to j$ is one of the links of the MST, j is not necessarily one of i's L nearest neighbours; every gene x_i in the genotype can thus encode any of (at most) $L + 2$ possible links during the evolutionary process.

2.3   Search Space Reduction

Although the (full-length) locus-based adjacency representation conceptually defines a huge search space of size N^N, MOCK's strategies for the creation of MST-derived solutions and link replacement (discussed at the end of Sect. 2.2) result in a much reduced search space whose size is bounded by $(L + 2)^N$. Moreover, the use of a problem-specific initialisation routine to generate high-quality base partitions contributes to (implicitly) bias exploration in an important manner.[4]

Depending on the setting of parameter δ, the new representation can reduce significantly the length of the genotype and thus explicitly prune the search space further. Specifically, using the reduced-length genetic encoding proposed in this paper the size of the search space is at most $(L + 2)^{|\Gamma|}$.

3   Experiments and Results

This section presents the findings of our experimental study which aims to investigate the suitability of the reduced-length representation proposed in this paper. The new representation is compared with respect to the use of the original full-length representation and is evaluated in terms of its impact on the behaviour and performance of the MOCK algorithm. The data sets, performance assessment measures, and settings adopted for this study are first described in Sect. 3.1. The results of our experiments are then discussed in Sect. 3.2.

3.1   Experimental Setup

A total of 280 data sets are considered for the experiments of this study. As shown in Fig. 4, these data sets present varying sizes and are organised into 28 problem configurations according to their dimensionality and number of clusters. All the data sets were generated using the ellipsoidal generator previously used during the evaluation of MOCK, which is described in detail in [6].

All the experiments of this study consider a total of 21 independent executions of MOCK for each problem instance. Results are evaluated in terms of both the PFAs obtained and clustering performance. PFAs are investigated by visualising the differences between the (first-order) *empirical attainment functions*

---

[4] Since the creation of new solutions relies mainly on recombination (due to the low mutation rates used), the genetic material introduced during initialisation plays a key role in delimiting the extent of the solution space that is reached by the method.

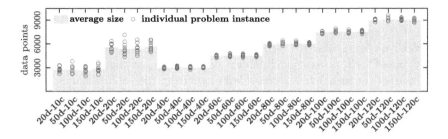

Fig. 4. Randomly generated data sets. 28 problem configurations are considered, each to be referred to as dd-kc, where d is the dimensionality and k is the number of clusters in the data set. 10 random instances were generated for each problem configuration, leading to a total of 280 data sets. The plot shows the size (N) of all individual problem instances, as well as the average size for each given problem configuration.

(EAFs) produced by the two representations [4,11]. This allows us to identify whether, and in which particular regions of the objective space, a representation performs better than the other. Plots of the EAFs were generated using the tools reported in [11]. In all the cases, objective values were normalised to the range $[0, 1]$ and, for visualisation purposes, replaced with their square roots. In addition, we use the *hypervolume indicator* (HV) [18] and the *inverted generational distance indicator with modified distance calculation* (IGD$^+$) [9], both computed after normalising objective values to the range $[0, 1]$. Both HV and IGD$^+$ capture the quality of a PFA with regard to both extent and proximity with respect to the true Pareto front. The reference point for HV was always set to $r = (1.01, 1.01)$ given the normalisation of the objective values. For IGD$^+$, the reference set was constructed in all the cases by merging the PFAs from all runs of all the approaches analysed and then removing the dominated vectors. Whereas HV is to be maximised, IGD$^+$ is to be minimised.

Clustering performance is assessed using the *Adjusted Rand Index* (ARI) measure [15]. ARI is defined in the range $[\sim0, 1]$; the larger the value for ARI, the better the correspondence between the partition obtained and the known cluster structure of the test data set. Each run of MOCK produces a set of candidate partitions. From this set, only the best solution, according to the ARI measure, is selected and considered in the results reported in Sect. 3.2.

Finally, the settings for MOCK adopted in our experiments are as follows. Population size: $P = 100$. Number of generations: $G_{max} = 100$. Recombination probability: $p_r = 1.0$. Mutation probability: defined for each individual link, see Sect. 2.2. Neighbourhood size: $L = 10$. Initialisation parameter: $k_{user} = 2k^*$.

3.2   Results

This section investigates the advantages of the new reduced-length representation from three different perspectives: (i) quality and characteristics of the PFAs obtained; (ii) clustering performance; and (iii) computational efficiency.

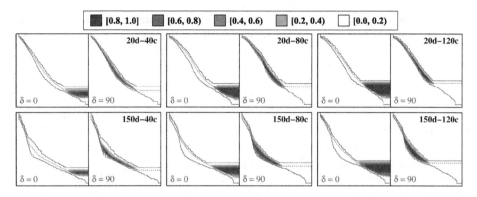

Fig. 5. Differences between the EAFs of the full-length ($\delta = 0$) and reduced-length ($\delta = 90$) representations. Results for a single instance of six problem configurations are shown. In the plots, the x-axis and y-axis denote respectively the CNN and VAR criteria. The magnitudes of the differences in the point attainment probabilities between the two settings are encoded using different intensities of blue; the darker the blue, the larger the difference. Lower and upper solid lines represent the grand best and grand worst attainment surfaces, and the dashed line denotes the median attainment surface.

First, Fig. 5 contrasts the EAFs that were computed from the PFAs obtained when using the full-length representation, *i.e.* adopting a setting of $\delta = 0$, and the reduced-length representation, with $\delta = 90$, which uses only about 10% of the original encoding length. The EAFs in Fig. 5 indicate that the full-length representation performs better with respect to the optimisation of the VAR criterion. This behaviour can be explained by the fact that VAR is relatively easy to optimise; the value of this criterion naturally decreases as the number of clusters (k) in the solution evaluated increases. When considering a large solution space (as that defined by the full-length encoding), it becomes easier for the search method to identify and exploit regions of the space favouring the optimisation of such a criterion. This behaviour is further evidenced by Fig. 6, which highlights that the full-length representation tends to produce PFAs comprising partitions with clearly higher k values (which, again, correlates with lower values of VAR) in comparison to the reduced-length encoding. The figure also illustrates that the full-length encoding obtains k values substantially exceeding the number of clusters in the real cluster structure (k^*) in most cases, including k values which can be considered far beyond practical relevance (in average, though, the k values produced tend to be close to k^*). The differences in the range of k values in the PFAs produced by the two representations are particularly evident (from Fig. 6) for problem configurations with $k^* > 40$ clusters. Consistently, for this specific subset of problem configurations (with $k^* > 40$), the PFAs of the full-length encoding cover a sufficiently larger extent of the objective space (extending widely and deeply into the uninteresting low-VAR, high-CNN regions), leading to significantly higher values for the HV indicator, see Table 1.

The reduced-length representation constrains the number of clusters a partition may involve and the range of values of VAR which can be reached by the method: ~90% of the MST links are fixed for all phenotypes (given the use of $\delta = 90$), and the partial solution defined by such fixed links (see Fig. 2) sets the maximum k and the minimum VAR which can be seen during optimisation.[5] Note, however, that according to Fig. 6 the new representation still produces PFAs covering k values in a wide relevant range around k^*. Moreover, the reduced-length encoding allows the optimisation to be performed within a small promising area of the search space. As can be seen from Fig. 5, this results in an increased convergence ability towards the more challenging central regions of the Pareto front presenting better compromises between the VAR and CNN criteria. This enhanced convergence behaviour is reflected in significantly better scores for the IGD^+ indicator across all 28 problem configurations (Table 1). Similar results (with even more marked differences) to those of IGD^+ have been observed for the GD^+ indicator [9] which focuses exclusively on proximity to the true Pareto front (as represented by a reference set). Such results for the GD^+ indicator have not been included in this paper due to space restrictions.

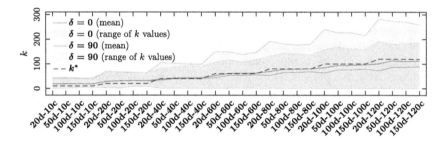

Fig. 6. Number of clusters (k) in the PFAs obtained when using two different settings for parameter δ, namely $\delta \in \{0, 90\}$. In the plot, curves show the arithmetic mean and shaded areas show the full range of k values observed for each problem configuration. A curve indicating the real number of clusters (k^*) is also shown as a reference.

From the perspective of clustering performance, Fig. 7 and Table 1 reveal that the above-discussed improvement to the convergence behaviour of MOCK translates into a better ability to discover partitions of higher quality (as captured by the ARI measure) throughout the search process. For all problem configurations, MOCK consistently reaches higher (significantly different statistically) ARI values at the end of the search process using the reduced-length representation. Two interesting behaviours are worth discussing from Fig. 7. Firstly, the use of the reduced-length encoding causes a drop in the performance of the initial population (generation 0 in the figure). This is due to the fixed nature of a large number (~90%) of the MST links which are not available for removal by

---

[5] During the evolutionary process, the clusters defined by the partial solution are combined, which can only lead to the decrease of k and the increase of VAR.

Table 1. Results for the HV, IGD$^+$, and ARI performance indicators. The table contrasts the performance of two different encoding lengths, resulting from the use of two settings for the new representation: $\delta = 0$ and $\delta = 90$. A mark • indicates that the results of the latter setting are significantly different statistically to those of the former; this is investigated using the *Mann-Whitney U test*, considering a significance level of $\alpha = 0.05$ and *Bonferroni correction* of the p-values. Values for IGD$^+$ have been multiplied by 10^2 in all the cases. For ARI, the average k of the solutions selected for the measure computation (see Sect. 3.1) is indicated in parenthesis, and results for the original implementation of MOCK [6] are included as a reference. For all problem configurations, the best result scored for every performance indicator has been shaded.

	HV		IGD$^+$ $\times 10^2$		ARI (k)		
Problem	$\delta = 0$	$\delta = 90$	$\delta = 0$	$\delta = 90$	$\delta = 0$	$\delta = 90$	MOCK [6]
20d-10c	0.932	**0.942** •	2.408	**1.197** •	0.998 (10)	**0.999 (10)** •	0.996 (10)
50d-10c	0.882	**0.899** •	3.407	**1.359** •	0.999 (10)	**1.000 (10)** •	0.998 (10)
100d-10c	0.863	**0.885** •	4.006	**1.319** •	0.999 (10)	**1.000 (10)** •	0.998 (10)
150d-10c	0.855	**0.873** •	3.838	**1.775** •	0.999 (10)	**1.000 (10)** •	0.998 (10)
20d-20c	0.958	**0.960** •	1.918	**1.257** •	0.992 (20)	**0.994 (20)** •	0.983 (20)
50d-20c	0.921	**0.930** •	3.203	**1.574** •	0.996 (20)	**0.999 (20)** •	0.990 (20)
100d-20c	0.853	**0.871** •	5.001	**2.193** •	0.997 (20)	**0.998 (20)** •	0.993 (20)
150d-20c	0.839	**0.852** •	4.903	**2.357** •	0.997 (20)	**0.999 (20)** •	0.993 (20)
20d-40c	**0.911**	0.908	3.613	**2.216** •	0.865 (47)	**0.910 (45)** •	0.806 (50)
50d-40c	0.936	**0.937**	3.101	**1.830** •	0.932 (44)	**0.960 (42)** •	0.888 (45)
100d-40c	0.941	**0.944** •	3.444	**1.773** •	0.929 (43)	**0.966 (42)** •	0.892 (45)
150d-40c	0.939	**0.943** •	3.359	**1.789** •	0.926 (43)	**0.963 (42)** •	0.885 (45)
20d-60c	**0.896**	0.889 •	4.513	**2.701** •	0.833 (76)	**0.884 (68)** •	0.781 (79)
50d-60c	**0.933**	0.924 •	3.616	**2.678** •	0.912 (68)	**0.941 (65)** •	0.874 (69)
100d-60c	**0.934**	0.926 •	3.768	**2.820** •	0.913 (68)	**0.950 (64)** •	0.867 (70)
150d-60c	**0.939**	0.931 •	3.564	**2.394** •	0.911 (68)	**0.947 (64)** •	0.865 (70)
20d-80c	**0.885**	0.874 •	5.146	**3.151** •	0.806 (108)	**0.850 (93)** •	0.751 (111)
50d-80c	**0.921**	0.910 •	4.204	**2.976** •	0.889 (95)	**0.928 (88)** •	0.847 (97)
100d-80c	**0.926**	0.912 •	4.314	**3.362** •	0.897 (93)	**0.934 (87)** •	0.851 (96)
150d-80c	**0.931**	0.917 •	4.066	**3.444** •	0.904 (91)	**0.939 (86)** •	0.861 (93)
20d-100c	**0.873**	0.864 •	5.745	**3.741** •	0.791 (139)	**0.830 (117)** •	0.743 (141)
50d-100c	**0.913**	0.900 •	4.548	**3.446** •	0.869 (125)	**0.910 (112)** •	0.822 (126)
100d-100c	**0.920**	0.902 •	4.560	**3.918** •	0.880 (120)	**0.919 (110)** •	0.825 (124)
150d-100c	**0.925**	0.907 •	4.322	**3.976** •	0.890 (118)	**0.926 (109)** •	0.834 (120)
20d-120c	**0.866**	0.858 •	6.187	**3.853** •	0.773 (180)	**0.809 (140)** •	0.722 (175)
50d-120c	**0.908**	0.892 •	4.841	**4.204** •	0.863 (152)	**0.904 (134)** •	0.815 (154)
100d-120c	**0.917**	0.898 •	4.773	**3.743** •	0.861 (150)	**0.903 (132)** •	0.806 (151)
150d-120c	**0.922**	0.900 •	4.647	**4.284** •	0.872 (146)	**0.910 (131)** •	0.799 (151)

the specialised initialisation routine (see Sect. 2.2), thus decreasing the diversity of the initially generated set of base partitions. Nevertheless, such a drop in performance is rapidly compensated by conducting a more-focused exploration in the substantially smaller solution space of the new representation. Secondly, the gradual increase in ARI as the search progresses indicates that the simultaneous optimisation of VAR and CNN implicitly and effectively leads to the optimisation of clustering quality. This provides corroborating evidence of the suitability of the clustering criteria considered as objective functions and of the conceptual advantages of the multiobjective approach to data clustering in general.

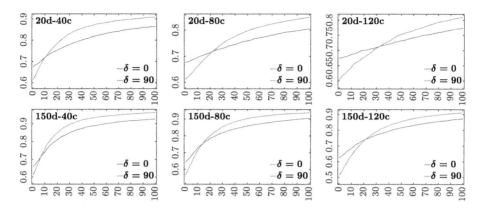

Fig. 7. Highest ARI (y-axis) in the population at every generation (x-axis) of the search process. Plots contrast the convergence behaviour using $\delta = 0$ and $\delta = 90$ for six problem configurations (average results for all instances and repetitions performed).

Figure 8 sustains that the reduced-length representation proposed in this paper delivers clear benefits in terms of computational efficiency. The use of compact (more efficient to handle) genotypes, as well as the strategy implemented for the incremental decoding and evaluation of candidate solutions, leads to a reduction of over 93% (in average) of the execution time derived from the optimisation process. Note that this excludes the computational costs involved with the generation of the initial population and those related to data loading and initial precomputations (*i.e.* distance matrix, nearest neighbours, and minimum spanning tree), since these processes are not affected by the change in representation. Such a decrease of $\sim$93% in the time of optimisation is found to correspond to an average decrease of $\sim$46% in the total execution time of MOCK.

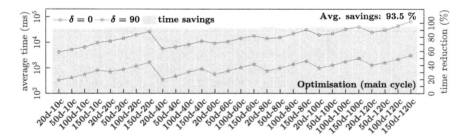

Fig. 8. Execution times scored by MOCK when using $\delta \in \{0, 90\}$. Curves show the average time (left y-axis, which is shown in logarithmic scale) for the main cycle of the optimisation strategy (discarding the time of data loading, initial computations, and generation of the initial population). Bars show the average time savings achieved by the reduced-length encoding with respect to the full-length encoding (right y-axis). The top-right corner indicates the average time savings achieved across all problems.

Finally, and despite that the analysis of other adaptations of MOCK besides the change in encoding is not the focus of this study, it is important to briefly discuss the performance differences observed with respect to the original MOCK reported in [6]. Table 1 indicates that, using a full-length encoding ($\delta = 0$), the new implementation of MOCK based on NSGA-II consistently outperforms the original implementation based on PESA-II in terms of clustering performance. This confirms the relevance of exploiting all the highly optimised solutions generated by MOCK's specialised initialisation, as discussed in Sect. 2.2. No meaningful impact on clustering performance has been observed as a consequence of replacing ODV with the VAR criterion (results not shown); this adaptation, as stated in Sect. 2, does not seek to alter the algorithm's behaviour, but is motivated by the advantages of VAR as it facilitates delta-evaluation.

4   Conclusions and Future Work

This paper studied a new reduced-length genetic representation for evolutionary multiobjective clustering. This representation exploits relevant information from the MST as a means to narrow the extent of exploration by focusing on the most prominent regions of the solution space. When implemented within the MOCK algorithm [6], the new representation scheme was found to offer significant advantages which were analysed from different perspectives.

Owing to its potential to explicitly prune large portions of the search space, the reduced-length representation allowed MOCK to produce Pareto front approximations of much greater quality in comparison to the original full-length representation. These improved convergence capabilities were found to reliably translate into a significantly increased clustering performance on a large number of test data sets. Besides these clear advantages regarding search performance and clustering quality, the new representation reported additional benefits in terms of computational efficiency. By enabling the precomputation of a substantial amount of information with the aim of expediting the processing of candidate solutions during optimisation, the new representation considerably reduced the computational overhead. Together, hence, all these findings indicate that the reduced-length representation proposed in this paper can be seen as an important step towards a better scalability of MOCK (and, in general, of evolutionary approaches to multiobjective clustering), and should impact on the method's ability to deal with larger and more challenging clustering problems.

A single fixed setting ($\delta = 90$) for the new solution representation was considered in this study, which results in the use of only about 10% of the original encoding length. Although promising results have been achieved using this setting, a range of different settings for this approach need to be investigated. More importantly, there remains the question of how far we can go in reducing the length of the encoding, and thus the size of the corresponding solution space, without sacrificing performance. Evidently, this strongly depends on the suitability of the mechanisms through which such a reduction is accomplished. The strategy adopted in this study relies on the ranking of the MST links on the basis of their degree of interestingness (a concept previously exploited with different purposes in [6]); failure to properly discriminate between the MST links could prune and discard key regions of the search space, thus compromising effectiveness. Despite showing promise during the experiments of this paper, therefore, a thorough analysis of this strategy, as well as the exploration of other alternative strategies, is certainly a topic which deserves further investigation and will constitute one of the main directions for our future work.

References

1. Chan, T., Golub, G., LeVeque, R.: Algorithms for computing the sample variance: analysis and recommendations. Am. Stat. **37**(3), 242–247 (1983)
2. Corne, D.W., Jerram, N.R., Knowles, J.D., Oates, M.J.: PESA-II: region-based selection in evolutionary multiobjective optimization. In: Genetic and Evolutionary Computation Conference, pp. 283–290. Morgan Kaufmann Publishers, San Francisco (2001)
3. Deb, K., Pratap, A., Agarwal, S., Meyarivan, T.: A fast and elitist multiobjective genetic algorithm: NSGA-II. IEEE Trans. Evol. Comput. **6**(2), 182–197 (2002)
4. Grunert da Fonseca, V., Fonseca, C.M., Hall, A.O.: Inferential performance assessment of stochastic optimisers and the attainment function. In: Zitzler, E., Thiele, L., Deb, K., Coello Coello, C.A., Corne, D. (eds.) EMO 2001. LNCS, vol. 1993, pp. 213–225. Springer, Heidelberg (2001). doi:10.1007/3-540-44719-9_15
5. Halkidi, M., Batistakis, Y., Vazirgiannis, M.: On clustering validation techniques. J. Intell. Inf. Syst. **17**(2), 107–145 (2001)
6. Handl, J., Knowles, J.: An evolutionary approach to multiobjective clustering. IEEE Trans. Evol. Comput. **11**(1), 56–76 (2007)
7. Handl, J., Knowles, J.: An investigation of representations and operators for evolutionary data clustering with a variable number of clusters. In: Runarsson, T.P., Beyer, H.-G., Burke, E., Merelo-Guervós, J.J., Whitley, L.D., Yao, X. (eds.) PPSN 2006. LNCS, vol. 4193, pp. 839–849. Springer, Heidelberg (2006). doi:10.1007/11844297_85
8. Hruschka, E.R., Campello, R.J.G.B., Freitas, A.A., de Carvalho, A.C.P.L.F.: A survey of evolutionary algorithms for clustering. IEEE Trans. Syst. Man Cybern. Part C (Appl. Rev.) **39**(2), 133–155 (2009)
9. Ishibuchi, H., Masuda, H., Tanigaki, Y., Nojima, Y.: Modified distance calculation in generational distance and inverted generational distance. In: Gaspar-Cunha, A., Henggeler Antunes, C., Coello, C.C. (eds.) EMO 2015. LNCS, vol. 9019, pp. 110–125. Springer, Cham (2015). doi:10.1007/978-3-319-15892-1_8
10. Jain, A.K.: Data clustering: 50 years beyond K-means. Pattern Recogn. Lett. **31**(8), 651–666 (2010)
11. López-Ibáñez, M., Paquete, L., Stützle, T.: Exploratory analysis of stochastic local search algorithms in biobjective optimization. In: Bartz-Beielstein, T., Chiarandini, M., Paquete, L., Preuss, M. (eds.) Experimental Methods for the Analysis of Optimization Algorithms, pp. 209–222. Springer, Heidelberg (2010)
12. MacQueen, J.: Some methods for classification and analysis of multivariate observations. In: Berkeley Symposium on Mathematical Statistics and Probability, vol. 1, pp. 281–297. University of California Press, Berkeley (1967)
13. Mukhopadhyay, A., Maulik, U., Bandyopadhyay, S.: A survey of multiobjective evolutionary clustering. ACM Comput. Surv. **47**(4), 61:1–61:46 (2015)

14. Park, Y.J., Song, M.S.: A genetic algorithm for clustering problems. In: Genetic Programming, pp. 568–575. Morgan Kaufmann, Madison, July 1998
15. Rand, W.M.: Objective criteria for the evaluation of clustering methods. J. Am. Stat. Assoc. **66**(336), 846–850 (1971)
16. Rothlauf, F., Goldberg, D.E.: Redundant representations in evolutionary computation. Evol. Comput. **11**(4), 381–415 (2003)
17. Syswerda, G.: Uniform crossover in genetic algorithms. In: International Conference on Genetic Algorithms, pp. 2–9. Morgan Kaufmann Publishers Inc., San Francisco (1989)
18. Zitzler, E., Thiele, L., Laumanns, M., Fonseca, C.M., Grunert da Fonseca, V.: Performance assessment of multiobjective optimizers: an analysis and review. IEEE Trans. Evol. Comput. **7**(2), 117–132 (2003)

A Fast Incremental BSP Tree Archive
for Non-dominated Points

Tobias Glasmachers[(✉)]

Institute for Neural Computation, Ruhr-University Bochum,
Bochum, Germany
`tobias.glasmachers@ini.rub.de`

Abstract. Maintaining an archive of all non-dominated points is a
standard task in multi-objective optimization. Sometimes it is sufficient
to store all evaluated points and to obtain the non-dominated subset
in a post-processing step. Alternatively the non-dominated set can be
updated on the fly. While keeping track of many non-dominated points
efficiently is easy for two objectives, we propose an efficient algorithm
based on a binary space partitioning (BSP) tree for the general case of
three or more objectives. Our analysis and our empirical results demonstrate the superiority of the method over the brute-force baseline method,
as well as graceful scaling to large numbers of objectives.

1 Introduction

Given $m \geq 2$ objective functions $f_1, \ldots, f_m : X \to \mathbb{R}$, a central theme in multi-
objective optimization is to find the Pareto set of optimal compromises: the set
of points $x \in X$ that cannot be improved in any objective without getting worse
in another one. The cardinality of this set is often huge or even infinite. In a
black-box setting it is best described by the set of mutually non-dominated query
points.

An archive A of non-dominated solutions is a data structure for keeping track
of the known non-dominated points $\{x^{(1)}, \ldots, x^{(n)}\} \subset X$ of a multi-objective
optimization problem. It can result from a single run of an optimization algorithm, or from many runs of potentially different algorithm.

Depending on the application such an archive can serve different purposes.
It can act as a portfolio of solutions accessed by a decision maker, possibly after
going through further post-processing steps. It can also act as input to various
algorithms, e.g., selection operators of evolutionary multi-objective optimization
algorithms (MOEAs), stopping criteria, and performance assessment and monitoring tools. All of these algorithms may involve the computation of set quality
indicators such as dominated hypervolume, for which the computation of the
non-dominated subsets is a pre-processing step.

Some of the above applications require access to the non-dominated set "anytime", i.e., already during the optimization (online case, called dynamic problem
in [16]), while others get along with storing all solutions and extracting the non-
dominated subset after the optimization is finished (offline or static case). The

H. Trautmann et al. (Eds.): EMO 2017, LNCS 10173, pp. 252–266, 2017.
DOI: 10.1007/978-3-319-54157-0_18

storage of millions of intermediate points does not pose a problem on today's computers, and even full non-dominated sorting is feasible on huge collections of objective vectors with suitable algorithms. Hence, we consider the offline case a solved problem, at least for moderate values of m.

In this paper we are interested in archives with the following properties:

- Online updates of the set of non-dominated points shall be efficient. In particular, it shall be feasible to process long sequences of candidate solutions one by one, i.e., as soon as they are proposed and evaluated by an iterative optimization algorithm.
- The archive shall not contain dominated points, even if these points were non-dominated at an earlier stage. I.e., points shall be removed as soon as they are dominated by a newly inserted point.
- The full set of non-dominated points shall be stored, not an approximation set of a-priori bounded cardinality.
- Ideally, the algorithm should scale well not only to large archives, but also to a large number of objectives.

These prerequisites imply the following processing steps. First it is checked whether a candidate point x is dominated by any point in the archive or not. If x is non-dominated then it is added to the archive. In addition, any points in A that happen to be dominated by x are removed. All of these steps should be as efficient as possible. At the very least, they should be faster than the $\mathcal{O}(nm)$ linear "brute force" search through the archive.

In this paper we propose an algorithm achieving this goal. It is based on a binary space partitioning tree (BSP tree). Its performance is demonstrated empirically, for large archives (up to $n = 2^{18}$ non-dominated points) and large numbers of objectives (up to $m = 50$), as well as analytically in the form of asymptotic (lower and upper) runtime bounds. We provide C++ source code for the proposed archive.[1] It is based on the implementation from the (non-public) code base of the black box optimization competition (BBComp)[2], where it is applied for online monitoring of the optimization progress.

In the remainder of this paper we first define the problem and fix our notation. After reviewing related work we present the proposed archiving algorithm. We derive asymptotic lower and upper bounds on its runtime and assess its practical performance empirically on a variety of tasks.

2   Definitions and Notation

Dominance Order. The objectives are collected in the vector-valued objective function $f : X \to \mathbb{R}^m$, $f(x) = \big(f_1(x), \ldots, f_m(x)\big)$. For objective vectors $y, y' \in \mathbb{R}^m$ we define the Pareto dominance relation

$$
\begin{aligned}
y \preceq y' &\quad\Leftrightarrow\quad y_k \leq y'_k \text{ for all } k \in \{1, \ldots, m\} \ , \\
y \prec y' &\quad\Leftrightarrow\quad y \preceq y' \text{ and } y \neq y' \ .
\end{aligned}
$$

---

[1] http://www.ini.rub.de/PEOPLE/glasmtbl/code/ParetoArchive/.
[2] http://bbcomp.ini.rub.de.

This relation defines a partial order on $\mathbb{R}^m$, incomparable values y, y' fulfilling $y \not\preceq y'$ and $y' \not\preceq y$ remain. The relation is pulled back to the search space X by the definition $x \preceq x'$ iff $f(x) \preceq f(x')$.

Pareto Front and Pareto Set. Let $Y = \{f(x) \,|\, x \in X\} = f(X) \subset \mathbb{R}^m$ denote the image of the objective function (also called the attainable objective space). The Pareto front is defined as the set of objective values that are optimal w.r.t. Pareto dominance, i.e., the set of non-dominated objective vectors

$$Y^* = \left\{ y \in Y \,\middle|\, \not\exists y' \in Y : y' \prec y \right\} .$$

The Pareto set is $X^* = f^{-1}(Y^*)$.

Non-dominated Points. In the sequel we will deal with objective vectors, not with actual search points. In this paper we restrict ourselves to maintaining only a single search point for each non-dominated objective vector, although it is possible to observe any number of search points of equal quality. In other words, we aim to approximate the Pareto front, represented by a minimal complete Pareto set.

For our purposes, let $y^{(1)}, \ldots, y^{(n)} \in \mathbb{R}^m$ denote the set of all known non-dominated objective vectors stored in the archive at some point, with n denoting the current cardinality of the archive. The objective vector of the new candidate x is denoted by $y = f(x) \in \mathbb{R}^m$.

3   Related Work

Fixed Memory Approximations. The growth of the set of non-dominated points is in general unbounded. Still, many MOEAs use their (fixed size) population as a rough approximation of the Pareto front. Even if an external archive is employed, computer memory is finite, which sets limits to the approach of archiving an unbounded number of non-dominated points. Therefore limited memory archives (e.g., based on clustering) were subject of intense study [3,11]. Their main disadvantage is the limited precision of their Pareto front representation, which can even lead to oscillatory behavior when used for optimization.

Memory Consumption. Nowadays even commodity PCs are equipped with gigabytes of RAM, enabling the storage of millions of objective vectors in memory.[3] This makes it feasible for MOEAs to maintain all non-dominated points in the population, as proposed by Krause et al. [9]. Although the algorithm is limited to $m = 2$ objectives, it demonstrates successfully that even on standard hardware memory is no longer a limiting factor for archiving of all known non-dominated points.

---

[3] This is usually sufficient for the needs of evolutionary optimization. In data base query problems larger sets must be processed. Hence in some applications memory consumption is still a concern.

Offline Case. Efficient algorithms are known for the offline case of obtaining the Pareto optimal subset from a set of points, i.e., at the end of an optimization run. This problem was first addressed in [10]. Even full non-dominated sorting (delivering not only the Pareto optimal subset, but a partitioning into disjoint fronts) can be achieved in only $\mathcal{O}(n \log(n)^{m-1})$ operations and $\mathcal{O}(nm)$ memory [4,7]. However, computing the full Pareto front from scratch after every update is wasteful, and hence specialized updating algorithms are needed for the online case.

Skylines. A closely related but slightly different problem is found in data base queries: so-called "skyline queries" ask for a skyline of records, which is just a different terminology for the non-dominated (Pareto optimal) subset with respect to a specified subset of fields. While the offline case is the most important one, online algorithms with good anytime performance are of relevance. The requirements differ from the problem under study in a decisive point: in the data base setting the full set of records is available from the start. The process is limited only by computational and memory constraints, but not by the sequential nature of proposing points in an iterative optimizer. Efficient algorithms (e.g., based on nearest neighbor queries) exist for the skyline problem [8,14].

Search Trees. Tree data structures of various sorts are attractive for tackling our problem. They offer a natural problem decomposition, i.e., adding a single point usually affects only a small neighborhood of the current front, possibly represented by a small sub-tree. The dominance decision tree (DDT) data structure was proposed specifically for the problem at hand [16]. It works well if the fraction of non-dominated points and hence the archive is small [16]. Dominated and non-dominated trees [3] represent a different approach. None of these methods achieve a significant reduction of the computational complexity of the problem. Quad trees can reduce the computational effort for large archives [13]. A clear disadvantage of quad trees generalized to $m > 2$ objectives is that partitioning a space cell results in 2^m sub-cells, which limits the approach to small numbers m of objectives. A different approach is based on modeling X^* instead of Y^* [12] with a binary decision diagram. This can work for some problems, however, it cannot be expected to be efficient in general.

4 Algorithms

We first discuss the trivial baseline method and a highly efficient alternative for the special case of $m = 2$ objectives. Then we present our core contribution, an efficient method for the general case of $m \geq 3$ objectives.

4.1 Baseline Method

The naive "brute force" baseline method stores all non-dominated points in a flat linear memory vector (dynamically extensible array) or linked list. The insert operation loops over the archive. It compares the candidate vector y to each stored $y^{(k)}$ w.r.t. dominance. This takes $\mathcal{O}(m)$ operations per point in the archive. If $y^{(k)} \preceq y$ then the point does not need to be archived and the procedure stops. If $y \prec y^{(k)}$ then $y^{(k)}$ is removed. In the list, removal takes $O(1)$ operations. In the vector we overwrite the dominated point with the last point, which is then removed ($\mathcal{O}(m)$ operations). Finally, y is appended to the end of the list or vector (amortized constant time for vectors with a doubling strategy, but even a possible $\mathcal{O}(nm)$ relocation of the memory block does not affect the analysis). Hence, in the worst case the brute force method performs $\Theta(nm)$ operations. It is significantly faster (in expectation over a presumed random order of the archive) only if y is dominated by more than $\mathcal{O}(1)$ points from the archive. Advantages of this algorithm are the trivial implementation and, in case of a linear memory array, optimal use of the processor cache.

4.2   Special Case of Two Objectives

For the bi-objective case the Pareto front is well known to obey a special structure that can be exploited for our purpose: we keep the archive sorted w.r.t. the first objective f_1 in ascending order, which automatically keeps it sorted in descending order w.r.t. to the second objective f_2. Given y we search for the indices

$$\ell = \arg\max \left\{ j \,\middle|\, y_1^{(j)} \leq y_1 \right\} \qquad \text{and} \qquad r = \arg\min \left\{ j \,\middle|\, y_1^{(j)} \geq y_1 \right\}$$

of y's potential "left" and "right" neighbors on the front. If the arg min is undefined because the set is empty then y is non-dominated and we set $r = 0$ for further processing. If y is weakly dominated by any archived point then it is also weakly dominated by $y^{(\ell)}$, which is the case exactly if $y_2^{(\ell)} \leq y_2$. Hence once ℓ is found the check is fast. If y happens to dominate any archived points, then these are of the form $\{y^{(r)}, y^{(r+1)}, \ldots, y^{(r+s)}\}$, for some s. Hence, given r it is easy to identify the dominated points, which are then removed from the archive.

Self-balancing trees such as AVL-trees and red-black-trees are suitable data structures for performing these operators quickly. The search for ℓ and r and the insert operation require $\mathcal{O}(\log(n))$ operations each, and so does the removal of a point. In an amortized analysis there can be at most one removal per insert operation, hence the overall complexity remains as low as $\mathcal{O}(\log(n))$, which is far better than the $\mathcal{O}(n)$ brute-force method (note that here m does not really enter the complexity since it is fixed to the constant $m = 2$). With this archive, the cumulated effort of n iterative updates equals (up to a constant) that of the offline algorithm [10].

4.3   General Case of $m \geq 3$ Objectives

For more than two objectives non-dominated sets obey far less structure than in the bi-objective case. Hence it is unclear which speed-up is achievable. However, it should be possible to surpass the linear complexity of the brute-force baseline. In the following we present an algorithm that achieves this goal.

We propose to store the archived non-dominated points in a binary space partitioning (BSP) tree. Among the different variants a simple k-d tree [1] seem to be best suited. This choice was briefly discussed and quickly dismissed in [3,16]. Nonetheless we propose an archiving algorithm based on a k-d tree, for the following reason. We expect most new non-dominated points to dominate only a local patch of the current front. Therefore it should be possible to limit dominance comparisons to few archived points close to the candidate point. Space partitioning is a natural approach to exploiting this locality.

Let R denote the root node. Each interior node of the tree is a data structure keeping track of its left and right child nodes ℓ and r, an objective index $j \in \{1, \ldots, m\}$, and a threshold $\theta \in \mathbb{R}$. For a node N we refer to these data fields with the dot-notation, i.e., $N.r$ refers to the right child node of N. With $p(N)$ we refer to the parent node, i.e., $p(N.\ell) = N$ and $p(N.r) = N$, while $p(R)$ is undefined. We write $p^2(N) = p(p(N))$, and $p^k(N)$ for the k-th ancestor of N.

The objective space is partitioned as follows. Let $\Delta(N)$ denote the subspace represented by the node N. We have $\Delta(R) = \mathbb{R}^m$. We recursively define $\Delta(N.\ell) = \{y \in \Delta(N) \mid y_{N.j} < N.\theta\}$ and $\Delta(N.r) = \{y \in \Delta(N) \mid y_{N.j} \geq N.\theta\}$ for given j and θ.

Each leaf node holds a set P of objective vectors limited in cardinality by a predefined bucket size $B \in \mathbb{N}$. In addition, each node (interior or leaf) keeps track of the number n of objective vectors in the subspace it represents.

Newly generated objective vectors y are added one by one to the archive with the Process algorithm laid out in Algorithm 1.

Algorithm 1: Process(y)

$s \leftarrow \text{CheckDominance}(R, y, 0, \emptyset)$
if $s \geq 0$ **then**
    $N \leftarrow R$
    while N *is interior node* **do**
        $N.n \leftarrow N.n + 1$
        if $y_{N.j} < N.\theta$ **then** $N \leftarrow N.\ell$ **else** $N \leftarrow N.r$
    end
    $P \leftarrow N.P \cup \{y\}$
    if $|P| \leq B$ **then** $N.P \leftarrow P$; $N.n \leftarrow |P|$
    else
        make N an interior node and create new leaf nodes as children
        select $N.j$ and $N.\theta$ (see text)
        $N.\ell.P \leftarrow \{p \in P \mid p_{N.j} < N.\theta\}$; $N.\ell.n \leftarrow |N.\ell.P|$
        $N.r.P \leftarrow \{p \in P \mid p_{N.j} \geq N.\theta\}$; $N.r.n \leftarrow |N.r.P|$
    end
end

Algorithm 2: CheckDominance(N, y, B, W)

$k \leftarrow 0$
if N *is leaf node* then
 foreach $p \in N.P$ do
  if $p \preceq y$ then return -1
  if $y \prec p$ then
   $N.P \leftarrow N.P \setminus \{p\}$
   $N.n \leftarrow N.n - 1$
   $k \leftarrow k + 1$
  end
 end
end
if N *is interior node* then
 if $y_{N.j} < N.\theta$ then $B' \leftarrow B \cup \{N.j\}$; $W' \leftarrow W$
 else $B' \leftarrow B$; $W' \leftarrow W \cup \{N.j\}$
 if $|W'| = m$ then return -1
 else if $|B'| = m$ then $k \leftarrow k + N.r.n$; $N.r.n \leftarrow 0$
 else
  if $W' = \emptyset \vee B = \emptyset$ then
   $s \leftarrow$ CheckDominance($N.\ell$, y, B, W')
   if $s < 0$ then return -1
   $k \leftarrow k + s$
  end
  if $B' = \emptyset \vee W = \emptyset$ then
   $s \leftarrow$ CheckDominance($N.r$, y, B', W)
   if $s < 0$ then return -1
   $k \leftarrow k + s$
  end
 end
 $N.n \leftarrow N.n - k$
 if $N.\ell.n > 0$ and $N.r.n = 0$ then overwrite N with $N.\ell$
 if $N.\ell.n = 0$ and $N.r.n > 0$ then overwrite N with $N.r$
end
return k

Processing a Candidate Point. Process calls the recursive CheckDominance algorithm (Algorithm 2), which returns the number of points in the archive dominated by y, or -1 if y is dominated by at least one point in the archive. The procedure also removes all points dominated by y. If y is non-dominated (i.e., the return value is non-negative), then y is inserted into the tree by descending into the leaf node N fulfilling $y \in \Delta(N)$. If the insertion would exceed the bucket size ($|P| > B$, with $P = N.P \cup \{y\}$), then the node is split. To this end we need to select an objective index j and a threshold θ. The only hard constraint on the choice of j is that $V_j = \{y_j \mid y \in P\}$ must contain at least two values, so that splitting the space in between yields two leaf nodes holding at most B objective vectors. However, for reasons that will become clear in the next section we prefer to select different objectives if possible when descending the tree. For

$j \in \{1, \ldots, m\}$ we define the distance $d_j = \min\{k \in \mathbb{N} \,|\, p^k(N).j = j\}$ of the node N in its chain of ancestors from the next node splitting at objective j. We set $d_j = \infty$ if the root node is reached before observing objective j. We chose $j \in \arg\max_j \{d_j \,|\, |V_j| > 1\}$. For the selection of θ we have to take into account that multiple objective vectors may agree in their j-th component. We select θ as the midpoint of two values from V_j so that the split balances the leaves as well as possible. For even B (and hence an uneven number of objective vectors) we prefer more points in the left leaf.

Recursive Check of the Dominance Relation. The `CheckDominance` algorithm is the core of our method. It takes a node N, the candidate point y, and two index sets B and W as input. It returns the number of points in the subtree strictly dominated by y, and -1 if y is weakly dominated by any point in the space cell represented by the subtree. The algorithm furthermore removes all dominated points from the subtree.

When faced with a leaf node it operates similar to the brute force algorithm. However, for interior nodes it can do better. To this end, note that the set $\Delta(N) \subset \mathbb{R}^m$ can be written in the form

$$\Delta(N) = \Delta_1(N) \times \cdots \times \Delta_m(N)$$

where $\Delta_j(N) \subset \mathbb{R}$ is the projection of $\Delta(N)$ to the j-th objective. Since the reals are totally ordered, we distinguish the cases $y_j < \Delta_j(N)$, $y_j \in \Delta_j(N)$, and $y_j > \Delta_j(N)$. Note that $y_j < \Delta_j(N)$ for all j implies that y dominates the whole space cell $\Delta(N)$, similarly $y_j > \Delta_j(N)$ implies that y is dominated by any point in $\Delta(N)$. If there exist i, j so that $y_i < \Delta_i(N)$ and $y_j > \Delta_j(N)$ then y is incomparable to all points in $\Delta(N)$.

The algorithm descends the tree by recursively invoking itself on the left and right child nodes, but only if necessary. The recursion is necessary only if the comparisons represented by the recursive calls up the call stack do not determine the dominance relation between y and $\Delta(N)$ yet. At the root node we know that it holds $y_j \in \Delta_j(R)$ for all $j \in \{1, \ldots, m\}$. Hence we have either $y_{R.j} \in \Delta_{R.\ell}$ or $y_{R.j} \in \Delta_{R.r}$, and hence either $y_{R.j} > \Delta_{R.\ell}$ or $y_{R.j} < \Delta_{R.r}$. The sets B and W keep track of the objectives in which y is better or worse than $\Delta(N)$. The two sets are apparently disjoint. If they are non-empty at the same time then the candidate point and the space cell are incomparable, hence the recursion can be stopped. If W equals the full set $\{1, \ldots, m\}$ then y is dominated by the space cell N. The mere existence of the node guarantees that it contains at least one point, so we can conclude that y is dominated. If on the other hand $B = \{1, \ldots, m\}$ then all points in N are dominated are hence removed. If the algorithm finds one of its child nodes empty after the recursion then it recovers a binary tree by replacing the current node with the remaining child. No action is required if both child nodes are empty since this implies $N.n = 0$, and the node will be removed further up in the tree.

Balancing the Tree. In contrast to the bi-objective case it is unclear how to balance the tree at low computational cost. This is not a severe problem for objective vectors drawn i.i.d. from a fixed distribution. This situation is fulfilled in good enough approximation when performing many short optimization runs. However, for a single (potentially long) run of an optimizer we can expect a systematic shift from low-quality early objective vectors towards better and better solutions over time. Hence most points proposed late during the run will tend to end up in the left child of a node, the split point of which was determined early on. We counter this effect by introducing a balancing mechanism as follows. If the quotient $\frac{N.\ell.n}{N.r.n}$ rises above a threshold z or falls below $\frac{1}{z}$ then the smaller child node is removed, the larger one replaces its parent, and the points represented by the smaller node are inserted. Although this process is computationally costly, it can pay off in the long run in case of highly unbalanced trees.

5 Analysis

In this section we analyze the complexity of the BSP tree based archive algorithm. We start with the storage requirements. When storing n non-dominated points, in the worst case there are n distinct leaf nodes, and hence $n - 1$ interior nodes in the tree, requiring $\mathcal{O}(n)$ memory in addition to the unavoidable requirement of $\mathcal{O}(nm)$ for storing the non-dominated objective vectors. Hence the added memory footprint due to the BSP tree is unproblematic. In our implementation the overhead of a tree node is 56 bytes on a 64 bit system, which makes it feasible to store millions of points in RAM.

The analysis of the runtime complexity is more involved. Since archiving small numbers of points is uncritical, we focus on the case of large n, which is well described by an average case amortized analysis. For the analysis we drop the rather heuristic balancing mechanism. For simplicity we set the bucket size to $B = 1$ and assume a perfectly balanced tree of depth $\log_2(n)$. Although optimistic, this assumption is not too unrealistic: note that the depth of a random tree is typically of order $2\log_2(n)$ [15].

The strongest technical assumption we make for the analysis is that objective vectors are sampled i.i.d. from a static distribution. Let P denote a probability distribution on $\mathbb{R}^m$ so that for two random objective vectors $a, b \sim P$ the events $a \preceq b$ and $\exists j \in \{1, \dots, m\} : a_j = b_j$ have probability zero. We consider a BSP archive constructed by inserting n points sampled i.i.d. from P. Then we are interested in bounding the expected runtime T required for processing a candidate point $y \sim P$.

For a node N representing the space cell $\Delta(N)$ we define its order w.r.t. the candidate point y as $k = |\{j \mid y_j \notin \Delta_j(N)\}|$. We call a node comparable to y if its space cell contains at least one comparable point (w.r.t. the Pareto dominance relation). All incomparable cells are skipped by the algorithm, hence the runtime is proportional to the number of comparable nodes. A node is incomparable to y if there exist j_1 and j_2 such that $y_{j_1} < \Delta_{j_1}(N)$ and $y_{j_2} > \Delta_{j_2}(N)$. Furthermore, cells dominating y ($y_j < \Delta_j(N)$ for all j) don't need to be visited, actually, encountering such a cell stops the algorithm immediately. Similarly, nodes dominated by y can be ignored in an amortized analysis since on average only one point can be removed per insertion, and the cost of a removal is as low as $\mathcal{O}(\log(n))$. Hence all nodes of order m can be ignored for the analysis.

In the following we denote the probability for a random node at depth d below the root node (the root has depth 0) to have order $k \in \{0, \ldots, m\}$ with $Q_d(k)$. We have $Q_0(0) = 1$ and $Q_d(k) = 0$ for $k > d$.

The following theorem provides a lower bound on the runtime in the best case, namely when each split of the BSP tree induces the same chance to yield an incomparable child node.

Theorem 1. *Under the conditions stated above, assume that when traversing from root to leaf no two space splits are along the same objective. Then we have $T \in \Omega(n^{\log_2(3/2)})$.*

Proof. The prerequisite implies $m \geq d$ and hence $m \geq \log_2(n)$. Since objective vectors are sampled i.i.d., the probability of the candidate point to be covered by the left or right sub-tree is $1/2$ at each node. This corresponds to a 50% chance to increment the order k when descending an edge of the tree, hence k follows a binomial distribution. We obtain $Q_d(k) = 2^{-d} \cdot \binom{d}{k}$ for $k \leq d$. For given k, the chance of a node to be comparable to the candidate point is $\min\{1, 2^{1-k}\}$, since for this to happen all k decisions (descending left or right in the tree) must coincide. Hence among the 2^d nodes at depth d an expected number of

$$1 + \sum_{k=1}^{d} 2^{1-k} \cdot \binom{d}{k} = 2^{d \cdot \log_2(3/2)+1} - 1$$

nodes is comparable to the candidate point. The statement follows by summing over all depths $d \leq \log_2(n)$. ∎

Under the milder (and actually pessimistic) assumption of random split objectives we obtain a sub-linear upper bound.

Theorem 2. *Under the conditions stated above, assume that each node splits the space along an objective j drawn uniformly at random from $\{1, \ldots, m\}$. Then we have $T \in o(nm)$.*

Proof. In this case Q is described by the following recursive formulas for $d > 0$:

$$Q_d(0) = \frac{1}{2} Q_{d-1}(0)$$

$$Q_d(k) = \frac{1}{2} \left[Q_{d-1}(k-1) \cdot \frac{m-k+1}{m} + Q_{d-1}(k) \cdot \frac{m+k}{m} \right]$$

We obtain $\lim_{d \to \infty} Q_d(m) = 1$, hence only $o(n)$ nodes are of order at most $m-1$, and only a subset of these must be visited. Since processing a leaf node requires $\mathcal{O}(m)$ operations in general, we arrive at $T \in o(nm)$. ∎

6   Experimental Evaluation

All three archives were implemented efficiently in C++. Here we investigate how fast the archives operate in practice, how their runtimes scale to large n and m, and how their practical performance relates to our analysis.

6.1   Analytic Problems

A first series of tests was performed with sequences of objective vectors following controlled, analytic distributions. These tests allow us to clearly disentangle effects caused by different fractions of non-dominated points and non-stationarity due to improvement of solutions over time.

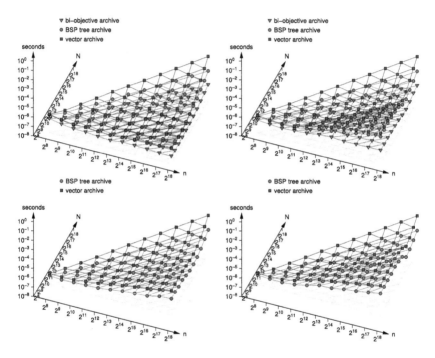

Fig. 1. Processing time per objective vector for $m = 2$ objectives (top) and $m = 3$ objectives (bottom), for a static distribution ($c = 1$, left) and solutions improving over time ($c = 1.1$, right), for systematically varied numbers of overall points ($n = N + D$) and non-dominated points (N) in the range 2^8 to 2^{18}.

We constructed archives from sequences of D dominated ($a > 0$) and N non-dominated ($a = 0$) normally distributed objective vectors according to

$$y^{(k)} \sim \mathcal{N}\left(\frac{a(N+D)}{k}\mathbf{1}, \mathbb{I} - \frac{1}{m}\mathbf{1}\mathbf{1}^T\right) ,$$

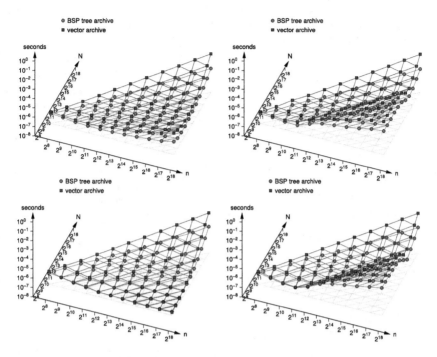

Fig. 2. Processing time per objective vector for $m = 5$ objectives (top) and $m = 10$ objectives (bottom), for a static distribution ($c = 1$, left) and solutions improving over time ($c = 1.1$, right), for systematically varied numbers of overall points ($n = N + D$) and non-dominated points (N) in the range 2^8 to 2^{18}.

where $\mathbb{I} \in \mathbb{R}^{m \times m}$ denotes the identity matrix and $\mathbf{1} = (1, \ldots, 1)^T \in \mathbb{R}^m$ is the vector of all ones. The distribution has unit variance in the subspace orthogonal to the $\mathbf{1}$ vector. The parameter $a \geq 0$ controls the systematic improvement of points over time.[4] At position k of the sequence a dominated point was sampled with probability $c \frac{D'}{N' + D'}$, where D' and N' denote the number of remaining dominated and non-dominated points to be placed into the sequence. Hence for $c > 1$ there is a preference for observing more dominated points early on in the sequence, while for $c = 1$ there is not.

The bucket size B and the tree re-balancing threshold z are tuning parameters of the algorithm. We propose the settings $B = 20$ and $z = 6$ as default values since they gave robust results across problems with varying characteristics in preliminary experiments. The relatively large bucket B size yields well balanced trees. Therefore a high value of the threshold z is affordable, because re-balancing is rarely needed.

Figures 1 and 2 display the average processing times of the different archives over sequences with varying m, $n = N + D$, N, and c. It is no surprise that

---

[4] For $a > 0$ the values $a = 0.1$, $a = 0.2$, $a = 0.5$, and $a = 1$ were used with $m = 2$, $m = 3$, $m = 5$, and $m = 10$ objectives, respectively.

for $m = 2$ the specialized bi-objective archive performs clearly best. For $m \geq 3$ the BSP tree is in all cases superior to the baseline. The vector archive is only competitive for $N \ll n$, i.e., as long as the number of non-dominated points in the archive remains small. Unsurprisingly, this is also the domain where the systematic improvement of points over time has a significant effect on archive performance. The overall effect is similar for all archive types, and in comparison to the static case the BSP archive can even increase its advantage over the linear memory archive.

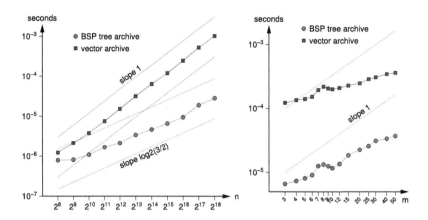

Fig. 3. Empirical scaling w.r.t. n (left) and m (right). The gray lines in the background of the log-log-plots indicate the exponents α and β of the hypothetical scaling laws $T(n) \in \Theta(n^{\alpha})$ and $T(m) \in \Theta(m^{\beta})$, respectively.

Figure 3(left) shows the empirical scaling of the archives for a setting close to the preconditions of the theoretical analysis: $m = 3$, $D = 0$, $c = 1$. The actual scaling is very close to the lower bound of order $n^{\log_2(3/2)} \approx n^{0.585}$ from Theorem 1. In contrast, the vector-based archive scales perfectly linear in n.

Figure 3(right) investigates the scaling to large numbers of objectives. It plots the runtime per processing step for an archive consisting of 2^{15} non-dominated points. In contrast to Theorem 1, in practice the curve is of course not flat. The algorithm still scales gracefully to large numbers of objectives. For the range $3 \leq m \leq 50$ we observe sub-linear scaling. The baseline method (surprisingly) exhibits even slightly better scaling (while taking 10 times more time in absolute terms).

6.2 MOEA Runs

A second series of tests was performed with objective vectors generated by state-of-the-art MOEAs on established benchmark functions. We used two variants of the multi-objective covariance matrix adaptation evolution strategy (MO-CMA-ES), namely with generational and with steady-state selection [5,17]. We applied

Table 1. The table lists the number N of non-dominated points at the end of the optimization run, the average runtimes (in 10^{-6} s) for processing a single point with the BSP tree archive and the vector archive, as well as the speed-up factor of the BSP tree archive over the vector archive.

MOEA	Problem	N	k-d tree	Vector	Speed-up
Generational MO-CMA-ES	DTLZ1	$104,191$	10.70	167.78	15.7
Steady state MO-CMA-ES	DTLZ1	$91,022$	9.58	166.49	17.4
Generational MO-CMA-ES	DTLZ2	$42,153$	8.15	64.71	8.0
Steady state MO-CMA-ES	DTLZ2	$53,814$	10.67	79.58	7.4
Generational MO-CMA-ES	DTLZ3	$7,895$	4.13	32.93	8.0
Steady state MO-CMA-ES	DTLZ3	$61,860$	4.59	72.25	15.7
Generational MO-CMA-ES	DTLZ4	$11,621$	2.75	13.31	4.8
Steady state MO-CMA-ES	DTLZ4	$23,097$	3.86	27.00	7.0

the implementations found in the Shark library,[5] version 3.1 [6]. The population size was set to 100, all parameters were left at their defaults. Sequences of objective vectors were generated by running the optimizers on the scalable benchmark problems DTLZ1 to DTLZ4 [2], with 30 variables, 3 objectives, and a budget of 200,000 function evaluations. The results are summarized in Table 1.

In all cases the BSP tree archive was considerably faster than the baseline. It outperformed the vector-based archive roughly by a factor of 10. The exact speed-up correlates with the number of non-dominated points. This is in line with our analysis, as well with the results for analytically controlled distributions of objective vectors. The results indicate that the tree-based archive also works well for realistic sequences of objective vectors produced by MOEAs. It demonstrates its strengths in particular if the set of non-dominated points grows large.

7 Conclusion

We have presented an algorithm for updating an archive of Pareto optimal objective vectors processed iteratively in a sequence. The data structure is based on a k-d tree for binary space partitioning. We present asymptotic lower and upper bounds on the runtime under a number of technical assumptions. We obtain runtime sub-linear in the archive size n. We demonstrate empirically that the method performs well for medium to large scale archives, where updating performance matters most. The archive is applicable to problems with arbitrary number m of objectives, including the many-objective case $m \gg 3$. Overall, these properties make the proposed algorithm suitable for online processing of non-dominated sets, e.g., in evolutionary multi-objective optimization.

---

[5] http://shark-ml.org.

References

1. De Berg, M., Van Kreveld, M., Overmars, M., Schwarzkopf, O.C.: Computational Geometry. Springer, Heidelberg (2000)
2. Deb, K., Thiele, L., Laumanns, M., Zitzler, E.: Scalable multi-objective optimization test problems. In: Congress on Evolutionary Computation (CEC 2002), pp. 825–830. IEEE (2002)
3. Fieldsend, J.E., Everson, R.M., Singh, S.: Using unconstrained elite archives for multi-objective optimization. IEEE Trans. Evol. Comput. **7**(3), 305–323 (2003)
4. Fortin, F.A., Grenier, S., Parizeau, M.: Generalizing the improved run-time complexity algorithm for non-dominated sorting. In: Proceedings of the Genetic and Evolutionary Computation Conference (GECCO) (2013)
5. Igel, C., Hansen, N., Roth, S.: Covariance matrix adaptation for multi-objective optimization. Evol. Comput. **15**(1), 1–28 (2007)
6. Igel, C., Heidrich-Meisner, V., Glasmachers, T.: Shark. J. Mach. Learn. Res. **9**, 993–996 (2008)
7. Jensen, M.T.: Reducing the run-time complexity of multiobjective EAs: the NSGA-II and other algorithms. IEEE Trans. Evol. Comput. **7**(5), 503–515 (2003)
8. Kossmann, D., Ramsak, F., Rost, S.: Shooting stars in the sky: an online algorithm for skyline queries. In: Proceedings of the 28th International Conference on Very Large Data Bases, VLDB Endowment, pp. 275–286 (2002)
9. Krause, O., Glasmachers, T., Hansen, N., Igel, C.: Unbounded population MO-CMA-ES for the bi-objective BBOB test suite. In: Proceedings of the Genetic and Evolutionary Computation Conference (GECCO) (2016)
10. Kung, H.-T., Luccio, F., Preparata, F.P.: On finding the maxima of a set of vectors. J. ACM (JACM) **22**(4), 469–476 (1975)
11. López-Ibáñez, M., Knowles, J., Laumanns, M.: On sequential online archiving of objective vectors. In: Takahashi, R.H.C., Deb, K., Wanner, E.F., Greco, S. (eds.) EMO 2011. LNCS, vol. 6576, pp. 46–60. Springer, Heidelberg (2011). doi:10.1007/978-3-642-19893-9_4
12. Lukasiewycz, M., Glaß, M., Haubelt, C., Teich, J.: Symbolic archive representation for a fast nondominance test. In: Obayashi, S., Deb, K., Poloni, C., Hiroyasu, T., Murata, T. (eds.) EMO 2007. LNCS, vol. 4403, pp. 111–125. Springer, Heidelberg (2007). doi:10.1007/978-3-540-70928-2_12
13. Mostaghim, S., Teich, J., Tyagi, A.: Comparison of data structures for storing pareto-sets in MOEAs. In: Proceedings of the 2002 Congress on Evolutionary Computation (CEC), vol. 1, pp. 843–848. IEEE (2002)
14. Papadias, D., Tao, Y., Fu, G., Seeger, B.: An optimal and progressive algorithm for skyline queries. In: Proceedings of the 2003 ACM SIGMOD International Conference on Management of Data, pp. 467–478. ACM (2003)
15. Robson, J.M.: The height of binary search trees. Aust. Comput. J. **11**, 151–153 (1979)
16. Schütze, O.: A new data structure for the nondominance problem in multi-objective optimization. In: Fonseca, C.M., Fleming, P.J., Zitzler, E., Thiele, L., Deb, K. (eds.) EMO 2003. LNCS, vol. 2632, pp. 509–518. Springer, Heidelberg (2003). doi:10.1007/3-540-36970-8_36
17. Voß, T., Hansen, N., Igel, C.: Improved step size adaptation for the MO-CMA-ES. In: 12th Annual Conference on Genetic and Evolutionary Computation (GECCO), pp. 487–494. ACM (2010)

Adaptive Operator Selection for Many-Objective Optimization with NSGA-III

Richard A. Gonçalves[1]($\boxtimes$), Lucas M. Pavelski[1], Carolina P. de Almeida[1],
Josiel N. Kuk[1], Sandra M. Venske[1], and Myriam R. Delgado[2]

[1] Computer Science Department - UNICENTRO, Guarapuava, PR, Brazil
{richard,carol,josiel,ssvenske}@unicentro.br, lmpavelski@yahoo.com.br
[2] CPGEI/DAINF, UTFPR, Curitiba, PR, Brazil
myriamdelg@utfpr.edu.br

Abstract. The number of objectives in real-world problems has increased in recent years and better algorithms are needed to deal efficiently with it. One possible improvement to such algorithms is the use of adaptive operator selection mechanisms in many-objective optimization algorithms. In this work, two adaptive operator selection mechanisms, Probability Matching (PM) and Adaptive Pursuit (AP), are incorporated into the NSGA-III framework to autonomously select the most suitable operator while solving a many-objective problem. Our proposed approaches, NSGA-III$_{AP}$ and NSGA-III$_{PM}$, are tested on benchmark instances from the DTLZ and WFG test suits and on instances of the Protein Structure Prediction Problem. Statistical tests are performed to infer the significance of the results. The preliminary results of the proposed approaches are encouraging.

Keywords: Many-objective optimization · Protein structure prediction · Adaptive operator selection · Probability matching · Adaptive pursuit · NSGA-III

1 Introduction

Several real-world problems inherently have multiple objectives. Evolutionary Multi-Objective Algorithms (EMOAs) have been proposed to solve these problems, although their performance greatly deteriorates when the number of objectives increases. This is mostly due to problems like: the lack of selective pressure by the Pareto dominance, the exponential increase in the number of solutions and difficulty of visualization and decision making [12]. Recently, EMOAs like the Non-dominated Sorting Genetic Algorithm III (NSGA-III) [13] have been proposed to deal with these difficulties and scale with the number of objectives.

Crossover and mutation operators used in the evolutionary process of optimization with EMOAs can affect its search ability. The task of choosing the right operators can depend on experience and knowledge about the problem. Adaptive Operator Selection (AOS) strategies aim to automatically select operators from a pool given their performance on previous generations. Another benefit of

© Springer International Publishing AG 2017
H. Trautmann et al. (Eds.): EMO 2017, LNCS 10173, pp. 267–281, 2017.
DOI: 10.1007/978-3-319-54157-0_19

on-line AOS is that the operator choice can adapt to the search landscape and produce better results [16].

This paper investigates the use of two known crossover operators (SBX and Differential Evolution) with two AOS strategies (Probability Matching and Adaptive Pursuit) in NSGA-III framework. The proposed NSGA-III variants are applied on a set of many-objective benchmarks (4 DTLZs [6] and 2 WFGs [11]) and 4 instances of the real-world Protein Structure Prediction (PSP) problem. PSP is the process of determining three-dimensional structures of proteins based on their sequence of amino acids. PSP is of great importance to medicine and biotechnology areas. In this paper, PSP was modeled as a many-objective optimization problem with 7 objective functions to be separately minimized along the evolutionary process.

The remainder of this paper is organized as follows. Section 2 presents an overview of the fundamental concepts related to this work. The proposed approach is detailed in Sect. 3. Experiments and results are presented and discussed in Sect. 4. Section 5 concludes the paper and discusses directions for the future work.

2   Background

This section provides some background of the main concepts used throughout the paper, such as: many-objective optimization, NSGA-III, AOS and PSP problem.

2.1   Many-Objective Optimization and NSGA-III

Multi-objective problems can be formulated as:

$$
\begin{aligned}
&\text{Minimize:} && \mathbf{f} = \{f_1(\mathbf{x}), \ldots, f_M(\mathbf{x})\} \\
&\text{Subject to:} && l_d \le x_d \le u_d && d = 1, \ldots, D
\end{aligned}
\tag{1}
$$

such that, $\mathbf{f}$ is a vector with M conflicting objective functions, $\mathbf{x} = (x_1, \ldots, x_D)$ is the vector of decision variables, $\mathbf{l}$ and $\mathbf{u}$ are the lower and upper bounds, respectively. All feasible decision variables form the decision space and they are mapped by the objective functions to the objective space. In the objective space, a solution is said to dominate another solution if it is equal or better in all objectives and better in at least one. The set of all feasible solutions that are not dominated by any solution in the objective space is called the Pareto-optimal set and its image in the objective space is called Pareto-optimal Front (PF). The goal of many MOEAs (Multi-Objective Evolutionary Algorithms) is to generate a good approximation of the PF by evolving a population of solutions.

When the number of objectives increases to more than three ($M > 3$) the problems are said to have many objectives (MaOPs - Many-Objective Optimization Problems). Some well-known MOEAs face difficulties in finding a good PF approximation for MaOPs. This is mainly due to the fact that most of the solutions in the population become non-dominated. Therefore, the dominance relation is not sufficient to enforce convergence pressure and classic selection approaches rely more on diversity criteria, such as the Crowding Distance.

The NSGA-III improved its predecessor NSGA-II [5] for dealing with MaOPs. The main difference is the substitution of Crowding Distance for a selection based on well-distributed reference points. This helps the maintenance of diversity among population members and, since the reference points are adaptive, NSGA-III also performs well on MaOPs with differently scaled objective values [13].

The NSGA-III was successfully applied to a real-world software engineering problem [19] and has several proposed extensions, such as to integrate alternative domination schemes [26], to solve mono-objective problems [21] and combine different variation operators [27]. More details about NSGA-III algorithm are explained further in Sect. 3 together with our proposed variants.

2.2   Adaptive Operator Selection

Operator selection procedures choose operators for generating new solutions based on the information collected by the credit assignment methods. The operator selection is usually based on probability or Multi-Armed Bandit (MAB) methods. Probability methods use a roulette wheel-like process for selecting an operator while MAB methods use algorithms created to tackle the exploration versus exploitation dilemma [16]. Two examples of probability based operator selection methods are Probability Matching (PM) and Adaptive Pursuit (AP). The Probability Matching method calculates $p_{op}(g+1)$ (the probability of operator op being selected at generation $g+1$) as follows [9,23]:

$$p_{op}(g + 1) = p_{min} + (1 - K * p_{min}) * \frac{q_{op}(g + 1)}{\sum_{s=1}^{K} q_s(g + 1)} \tag{2}$$

where K is the number of operators, p_{min} is the minimal probability of any operator and q_{op} is the quality associated with operator op.

Clearly, $\sum_{s=1}^{K} p_s(g+1) = 1$. From Eq. 2 we note that when only one strategy obtains a reward during a long period of time (with all other strategies receiving no rewards), its selection probability converges to $p_{max} = p_{min} + (1 - K * p_{min})$.

The Adaptive Pursuit method was introduced for adaptive selection operators in the context of Genetic Algorithms [23]. AP calculates $p_s(g+1)$ as follows:

$$p_{s^*}(g + 1) = p_{s^*}(g) + \beta * [p_{max} - p_{s^*}(g)] \tag{3}$$

where $s^* = arg\,max_s(q_s(g + 1))$ and

$$\forall s \neq s^* : p_s(g + 1) = p_s(g) + \beta * [p_{min} - p_s(g)]. \tag{4}$$

This constraint makes sure that if $\sum_{s=1}^{S} p_s(g) = 1$, then $\sum_{s=1}^{S} p_s(g + 1) = 1$ [23]. The AP method has a learning rate $\beta \in (0, 1]$, which controls how greedy the "winner-takes-all" strategy will behave.

Several works propose the integration of AOS to different EMOAs. Some related works include a multi-objective differential evolution used with PM strategy [15], an EMOA based on decomposition (MOEA/D) integrated with PM and AP [25] and ENS-MOEA/D [29] that uses a procedure similar to PM with binary credit assignment in order to adaptively select its operators. Other related approaches use multi-armed bandit AOS with MOEA/D [16] and MOEA/D with adaptive composite operator selection strategy [17]. In this work, the Probability Matching and Adaptive Pursuit methods are combined with the NSGA-III algorithm to deal with MaOPs. This work is an extension of a previous work [8] where PM and AP are also used in NSGA-III to solve DTLZ instances.

2.3 Protein Structure Prediction Problem

For the real-world problem, we consider Protein Structure Prediction (PSP) problem. PSP is an important research topic for bioinformatics. The information that some proteins need to fold properly is encoded in its primary structure (the amino acid or residue sequence) [28]. We consider the *ab initio* approach for PSP, that is a template-free modeling [24]. Also, an off-lattice model is used, representing a protein as a chain of residues or groups of residues that fold in a continuous space [24]. When a protein is in its folded state, its free energy conformation is the lowest one.

The PSP problem analysis considering many objectives is still in its early stages. There are few studies that deal with the problem in this way. In [10], the authors have explored the impact of multiobjectivization on the potential energy functions used in protein structure prediction, minimizing 2 objectives. They conclude that multiobjectivization achieves a reduction in the number of local optima in the landscape whilst simultaneously maintaining some of the important "guidance" (or gradient) that the landscape possesses. The approach proposed in [1] deals with 4 objectives for PSP and presents some initial results.

3   Proposed Approaches: NSGA-III$_{PM}$ and NSGA-III$_{AP}$

This paper proposes the use of adaptive operator selection (Sect. 2.2) with the NSGA-III framework. Since NSGA-III is intended to efficiently solve MaOPs, operator selection strategies could increase its search capability for this kind of problem. The proposed NSGA-III variants, NSGA-III$_{AP}$ and NSGA-III$_{PM}$, adaptively select the operators to be applied. The general procedure of our approaches can be found in Algorithm 1, where the differences to the classical NSGA-III are highlighted.

The general NSGA-III procedure works as follows. Given a MOP, the maximum number of generations ($MaxGen$) and a randomly generated population (P_0) of N individuals, it initially generates N uniformly distributed points across the unit simplex in the objective space (Step 4). For $MaxGen$ generations, new individuals are generated from the current parent population (Step 8) using recombination and mutation operators The generated individuals are merged with the parent population and the next generation is selected from this combined population. The non-dominated sorting selects the non-dominated fronts

Algorithm 1. NSGA-III$_{AP}$ and NSGA-III$_{PM}$ procedure

Require:
 1: **f**: a multi-objective optimization problem;
 2: *MaxGen*: the number of generations;
 3: P_0: the initial population of size N;
Ensure: A PF approximation of **f**
 4: Set Z with N uniformly distributed reference points
 5: $g \leftarrow 0$
 6: **while** $g < MaxGen$ **do**
 7:      $op \leftarrow$ SELECTOPERATOR()
 8:      $Q_g \leftarrow P_g \bigcup$ RECOMBINATIONANDMUTATION(op, P_g)
 9:      $(Front_1, Front_2, \dots) \leftarrow$ NONDOMINATEDSORTING(Q_g)
 10:      $i \leftarrow 1$
 11:      **do**
 12:          $P_{g+1} \leftarrow P_{g+1} \bigcup Front_i$
 13:          $i \leftarrow i + 1$
 14:      **while** $|P_{g+1}| + |Front_i| < N$
 15:      **if** $|P_{g+1}| + |Front_i| > N$ **then**
 16:          Normalize the objectives
 17:          Associate each solution in P_{g+1} with a reference point
 18:          Count the solutions from each cluster
 19:          Fill P_{g+1} with $N - |P_{g+1}|$ solutions from $Front_i$ using niching information
 20:      **end if**
 21:      CALCULATEOPERATORSREWARDS(P_g, P_{g+1})
 22:      UPDATEOPERATORSQUALITIES()
 23:      UPDATEOPERATORSPROBABILITIES()
 24:      $g \leftarrow g + 1$
 25: **end while**
 26: **return** P_g.

partitions in an elitist fashion (Steps 9 to 14). For the last front of non-dominated individuals, NSGA-III normalizes the objectives (Step 16), associates each one with a reference point (Step 18) and iteratively selects individuals in isolated regions (Step 19) [13].

The first difference between our proposed approaches and the classical NSGA-III is the selection of the operator to be applied (Step 7) which occurs based on the probabilities associated with each operator ($p_s(g)$) for both NSGA-III$_{AP}$ and NSGA-III$_{PM}$. As the increase of a probability occurs based on the success of an operator, the most successful operator will be selected more often, theoretically improving the results. The selection mechanisms used in this work are stochastic, this fact is important in the NSGA-III context due to its generational environmental selection: a deterministic approach would select the same operator to be applied during all recombinations (Step 8) of the same generation, which should introduce an undesirable bias to select the operator with the best performance in the initial generations.

Other differences with respect to the classical version of NSGA-III are the calculation of the rewards and the update of the qualities and probabilities associated with each operator (Steps 21, 22 and 23, respectively). The reward calculation is as follows: a reward of 1 is given for each operator that has generated a new solution that is selected to the next generation. Therefore, for each new solution $\mathbf{x}$ selected to be included in the next generation $g + 1$, the reward is:

$$r_s(g) = \begin{cases} 1 & \text{if operator } s \text{ generated } \mathbf{x} \text{ and } \mathbf{x} \notin P_g \text{ and } \mathbf{x} \in P_{g+1} \\ 0 & \text{otherwise.} \end{cases} \tag{5}$$

After calculating the rewards, the *empirical estimate quality* $q_s(g)$ of the sth operator available in the pool at generation g can be updated using Eq. 6 [23].

$$q_s(g + 1) = (1 - \alpha) * q_s(g) + \alpha * r_s(g) \tag{6}$$

where $\alpha \in (0, 1]$ is the adaptation rate. The reward calculation and qualities update are the same for both proposed approaches. Finally, the operator selection probabilities are updated as Eq. 2 for NSGA-III$_{PM}$ variant and Eqs. 3 and 4 for NSGA-III$_{AP}$.

The additional computational effort with the AOS methods is minimal. Calculating the rewards requires only looping through the population once while updating the qualities and probabilities of the operators takes time proportional to the operator pool size.

This work is related to NSGA-III$_{HVO}$ [27] and is an extension of one of our previous works [8]. Similar to [27], we investigate the behavior of different operators and a pool of operators in NSGA-III, but NSGA-III$_{HVO}$ randomly selects the operators of the pool while we use systematic approaches (PM and AP) to do so. In our previous work [8] we also combined NSGA-III with PM and AP, but the pool of operators and the parents selection procedure in the SBX operator (see Sect. 4.1) are different. Furthermore, in our previous work we only tested the approaches on DTLZ instances while in this work we also test the approach in WFG and PSP instances.

4   Results and Discussion

In this section, NSGA-III$_{AP}$ and NSGA-III$_{PM}$ are tested with 6 different benchmark MOPs instances known as DTLZ1 to DTLZ4 [6] and WFG6 and WFG7 [11]. The proposed approach is also used to solve a real-world problem, the protein structure prediction [24].

For DTLZ, 4 instances (DTLZ1-4) with 3, 5, 8, 10 and 15 objectives are used. DTLZ1 has a linear Pareto-optimal front. DTLZ2 is a generic sphere problem. DTLZ3 is a variation of DTLZ2 with multiple local Pareto-optimal fronts. DTLZ4 is a variation of DTLZ2 that is used to evaluate if an algorithm is capable of maintaining a good distribution of solutions. As in [27], the number of variables is set as $D = M + K - 1$, where $K = 5$ for DTLZ1 and $K = 10$ in DTLZ2, DTLZ3 and DTLZ4.

We use 2 WFG instances (WFG6 and WFG7) 3, 5, 8, 10 and 15 objectives. WFG6 and WFG7 are concave, separable and biased. As suggested in [11], the parameters k and l are set to $2 * (M - 1)$ and 20, respectively, therefore, the number of variables is $2 * (M - 1) + 20$. The benchmark instances were chosen in order to cover different problem characteristics.

For PSP we consider 4 real proteins: 1PLW, 1ZDD, 1CRN and 1ROP, with 5, 34, 46 and 56 residues, respectively. In this work we consider PSP problem with 7 objective functions and adopt NSGA-III variants as its optimizer. We also adopt an off-lattice model based on internal coordinates representation with backbone and side-chain torsion angles to model proteins.

The 5 bonded and 2 non-bonded objectives functions considered in this work are defined by Eqs. 7 and 8, respectively.

$$E_b = \left(\sum_{bounds} K_b(b - b_0)^2, \sum_{UB} K_{UB}(S - S_0)^2, \sum_{angles} K_\theta(\theta - \theta_0)^2 \right.$$

$$\left. \sum_{dihedrals} K_\Phi[1 + \cos(n\Phi - \gamma)], \sum_{impropers} K_{imp}(\varphi - \varphi_0)^2 \right) \qquad (7)$$

All terms in Eq. 7 describe a molecule and are represented by the bond length, b; valence angle, θ; distance between atoms separated by two covalent bonds, S; dihedral or torsion angle, Φ; improper angle, φ [24]. Others parameters include the bond force constant and equilibrium distance, K_b and b_0, respectively; the valence angle force constant and equilibrium angle, K_θ, and θ_0; the Urey-Bradley force constant and equilibrium distance, K_{UB} and S_0; the dihedral angle force constant, multiplicity, and phase angle, K_Φ, n, and γ; and the improper force constant and equilibrium improper angle, K_{imp} and φ_0.

$$E_{nb} = \left(\sum_{nb} \varepsilon_{ij} \left[\left(\frac{R_{min_{ij}}}{r_{ij}} \right)^{12} - \left(\frac{R_{min_{ij}}}{r_{ij}} \right)^6 \right], \frac{q_i q_j}{e r_{ij}} \right) \qquad (8)$$

The terms in Eq. 8 describe the interactions between atoms i and j and include the partial atomic charges, q_i; the Lennard Jones well-depth, ε_{ij}; the minimum interaction radius, $R_{min_{ij}}$, used to calculate the van der Waals interactions; and distance between atoms i and j, r_{ij}. In this work we use Tinker (v. 7.1.2) [20] and CHARMM (v. 22) [18] force field parameters to calculate the energy functions of proteins, while NSGA-III and its variants are implemented in Java language using the NSGA-III standard version available in the JMetal framework [7].

4.1   Operators Pool and Parameters

The pool of operators in the proposed variants NSGA-III$_{PM}$ and NSGA-III$_{AP}$ is composed of the well known Simulated Binary Crossover (SBX) [4] and Differential Evolution Crossover (DE) [22]. In NSGA-III, the parents $\mathbf{x}_{p1}$ and $\mathbf{x}_{p2}$

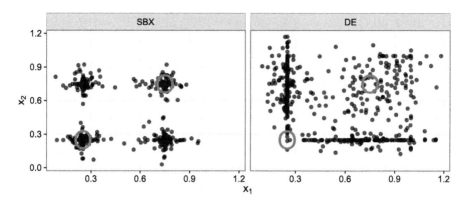

Fig. 1. Comparison between SBX and DE operators. The dots represents 500 solutions generated by both operators for $\mathbf{x} = (0.25, 0.25)$ and $\mathbf{x}_{p1} = (0.75, 0.75)$.

are randomly picked from the population P_t. A small difference in our proposal is that in SBX, the first parent $\mathbf{x}_{p1}$ is equal to the current solution $\mathbf{x}$, as in DE operator. This makes both operators have similar strategies for parent selection and has showed some improvements in earlier tests over selecting both $\mathbf{x}_{p1}$ and $\mathbf{x}_{p2}$ randomly.

As can be seen in Fig. 1, SBX and DE have different capabilities. The SBX operator tends to generate solutions closer to their parents and in orthogonal positions (exploitation) while the DE operator focuses on spreading the new solutions between each parent (exploration). These complementary behaviors of exploration and exploitation can be beneficial for AOS strategies and justify in our opinion the choice of a reduced pool with only these two operators.

The number of evaluations and population size for each problem are show in Table 1. The same values are used in [13], where the population size depends on the number of partitions of each objective axis. For 8, 10 and 15 objectives, the reference points are generated using the layered method described in [14].

Table 1. Number of generations ($MaxGen$) and population size (N) for each benchmark instance.

No. Objs	Number o generations ($MaxGen$)						Pop. size	No.
(M)	DTLZ1	DTLZ2	DTLZ3	DTLZ4	WFG6	WFG7	(N)	Divisions
3	400	250	1000	600	400	400	92	{12}
5	600	350	1000	1000	750	750	212	{6}
8	750	500	1000	1250	1500	1500	156	{3, 2}
10	1000	750	1500	2000	2000	2000	276	{3, 2}
15	1500	1000	2000	3000	3000	3000	136	{2, 1}

For the PSP problem, the population size is $N = 112$ and the number of generations is $MaxGen = 1000 \times D$. As recommended in the literature, the higher $\eta_c = 30$ and lower $CR = 0.1$ values in SBX and DE operators, respectively, generate individuals close to their parents and are preferred for MaOPs [13]. The DE scaling factor $F = 0.5$ and Polynomial Mutation parameters $\eta_m = 20$ and $p_m = 1/D$ are used in this work, as they are frequently used in the literature [27]. The Polynomial Mutation is performed after a SBX or DE recombination. PM minimum probability is $p_{min} = 0.1$, AP learning rate $\beta = 0.8$ and in both AOS strategies, the quality adaptation rate is $\alpha = 0.3$.

For each problem instance and algorithm, 30 independent runs are evaluated with Inverted Generational Distance (IGD) metric [2], given by:

$$IGD(P, R) = \frac{\sum_{\mathbf{x} \in P} dist(\mathbf{x}, R)}{|R|} \tag{9}$$

where P is the non-dominated population in the last generation; R is a reference PF[1], $dist(\mathbf{x}, R)$ is the smallest euclidean distance from $\mathbf{x}$ to another point in R. Smaller IGD values indicate that P has better convergence and coverage of the PF. Aiming to be as fair as possible, in each run, all the comparison approaches perform the same number of fitness computations.

The statistical test used to compare the strategies is the non-parametric pairwise test for multiple comparison of mean rank sums Kruskal-Wallis with Nemenyi post-hoc at 0.05 significance level [3]. The Kruskal-Wallis test is used to detect if there is statistical difference among the IGD found by each algorithm in a group of correlated instances. The results of the Kruskal-Wallis test can be interpreted by the p-values of the corresponding statistical test. When comparing two algorithms, p-values lower than 0.05 indicate that the algorithm with lower IGD is significantly better than the other, while p-values greater than 0.05 indicate lack of statistical confidence in which algorithm is better.

The PSP problem results are evaluated in terms of total energy to be minimized, i.e., the sum of all objective values, and the Root Mean Square Deviation (RMSD). The RMSD metric is used to assess the similarity between the predicted conformation and the native structure [24]. Lower RMSD values are preferable.

4.2 Benchmarks Results

The IGD metric results for the benchmark problems are summarized in Table 2.

The first notable result is that there is a group of instances/problems (DTLZ1, DTLZ2 and DTLZ3, all with higher number of objectives) that favors the DE operator (i.e. NSGA-III$_{DE}$ outperforms or performs comparable to the best approaches). There is another group that is clearly more suitable for SBX (DTLZ4 for every total number of objectives and WFG6 and WFG7 with less than 15 objectives). It seems that NSGA-III$_{DE}$ has difficulties in biased problems

---

[1] The reference Pareto Front R is a set of N uniformly distributed points on each problem. It can be calculated using a generator code available at https://github.com/JerryI00/SamplingPF.

Table 2. Median and standard deviation of IGD for each NSGA-III variants. The best medians are in bold font and grayed cells mean statistically equivalent to the best, accordingly to Kruskal-Wallis test.

Problem	M	NSGA-III$_{SBX}$	NSGA-III$_{DE}$	NSGA-III$_{PM}$	NSGA-III$_{AP}$
DTLZ1	3	$3.243e\text{-}3_{1.28e\text{-}2}$	$7.063e\text{-}4_{1.25e\text{-}4}$	$\mathbf{4.492e\text{-}4}_{7.68e\text{-}5}$	$5.709e\text{-}4_{5.75e\text{-}4}$
DTLZ1	5	$1.861e\text{-}3_{1.16e\text{-}2}$	$3.781e\text{-}4_{7.05e\text{-}5}$	$3.687e\text{-}4_{1.10e\text{-}4}$	$\mathbf{3.428e\text{-}4}_{4.06e\text{-}3}$
DTLZ1	8	$4.533e\text{-}3_{1.93e\text{-}3}$	$\mathbf{2.942e\text{-}3}_{3.62e\text{-}4}$	$4.023e\text{-}3_{1.26e\text{-}2}$	$3.397e\text{-}3_{2.32e\text{-}3}$
DTLZ1	10	$4.163e\text{-}3_{6.44e\text{-}3}$	$\mathbf{2.559e\text{-}3}_{2.80e\text{-}4}$	$3.119e\text{-}3_{7.18e\text{-}4}$	$2.863e\text{-}3_{7.93e\text{-}3}$
DTLZ1	15	$6.482e\text{-}3_{3.17e\text{-}3}$	$\mathbf{1.935e\text{-}3}_{6.83e\text{-}3}$	$2.260e\text{-}3_{1.15e\text{-}2}$	$2.333e\text{-}3_{5.49e\text{-}2}$
DTLZ2	3	$2.218e\text{-}3_{2.35e\text{-}3}$	$4.121e\text{-}3_{3.67e\text{-}4}$	$\mathbf{2.032e\text{-}3}_{2.62e\text{-}4}$	$2.676e\text{-}3_{9.16e\text{-}4}$
DTLZ2	5	$6.105e\text{-}3_{1.18e\text{-}3}$	$5.851e\text{-}3_{4.59e\text{-}4}$	$\mathbf{3.938e\text{-}3}_{4.18e\text{-}4}$	$4.810e\text{-}3_{2.39e\text{-}3}$
DTLZ2	8	$1.724e\text{-}2_{2.93e\text{-}3}$	$1.007e\text{-}2_{8.97e\text{-}4}$	$\mathbf{9.363e\text{-}3}_{6.41e\text{-}4}$	$1.035e\text{-}2_{2.40e\text{-}3}$
DTLZ2	10	$1.745e\text{-}2_{1.18e\text{-}3}$	$8.438e\text{-}3_{5.39e\text{-}4}$	$\mathbf{7.952e\text{-}3}_{4.74e\text{-}4}$	$8.050e\text{-}3_{2.67e\text{-}3}$
DTLZ2	15	$2.037e\text{-}2_{2.51e\text{-}3}$	$\mathbf{6.514e\text{-}3}_{5.98e\text{-}4}$	$6.804e\text{-}3_{5.71e\text{-}4}$	$7.061e\text{-}3_{3.53e\text{-}3}$
DTLZ3	3	$5.249e\text{-}3_{1.52e\text{-}2}$	$2.597e\text{-}3_{2.99e\text{-}4}$	$\mathbf{8.851e\text{-}4}_{1.01e\text{-}4}$	$1.573e\text{-}3_{4.97e\text{-}4}$
DTLZ3	5	$8.387e\text{-}3_{2.75e\text{-}2}$	$4.408e\text{-}3_{4.37e\text{-}4}$	$\mathbf{3.089e\text{-}3}_{3.56e\text{-}4}$	$3.525e\text{-}3_{3.02e\text{-}2}$
DTLZ3	8	$4.097e\text{-}2_{3.20e\text{-}2}$	$\mathbf{1.335e\text{-}2}_{1.02e\text{-}3}$	$1.640e\text{-}2_{4.38e\text{-}3}$	$1.369e\text{-}2_{1.36e\text{-}2}$
DTLZ3	10	$1.768e\text{-}2_{1.67e\text{-}2}$	$8.297e\text{-}3_{4.25e\text{-}4}$	$9.289e\text{-}3_{1.87e\text{-}3}$	$\mathbf{8.095e\text{-}3}_{2.10e\text{-}2}$
DTLZ3	15	$3.240e\text{-}2_{2.16e\text{-}2}$	$\mathbf{5.564e\text{-}3}_{5.89e\text{-}4}$	$6.733e\text{-}3_{9.51e\text{-}2}$	$6.507e\text{-}3_{9.71e\text{-}2}$
DTLZ4	3	$7.292e\text{-}4_{1.73e\text{-}1}$	$3.700e\text{-}3_{3.42e\text{-}4}$	$6.561e\text{-}4_{1.27e\text{-}4}$	$\mathbf{3.845e\text{-}4}_{9.66e\text{-}4}$
DTLZ4	5	$1.282e\text{-}3_{4.31e\text{-}4}$	$6.262e\text{-}3_{5.66e\text{-}4}$	$1.903e\text{-}3_{2.21e\text{-}4}$	$\mathbf{6.134e\text{-}4}_{1.46e\text{-}3}$
DTLZ4	8	$4.528e\text{-}3_{6.60e\text{-}4}$	$1.934e\text{-}2_{1.35e\text{-}3}$	$1.117e\text{-}2_{8.22e\text{-}4}$	$\mathbf{4.173e\text{-}3}_{3.77e\text{-}3}$
DTLZ4	10	$4.709e\text{-}3_{5.17e\text{-}4}$	$1.414e\text{-}2_{9.54e\text{-}4}$	$1.005e\text{-}2_{6.06e\text{-}4}$	$\mathbf{4.583e\text{-}3}_{3.72e\text{-}3}$
DTLZ4	15	$\mathbf{6.843e\text{-}3}_{1.29e\text{-}3}$	$1.056e\text{-}2_{1.04e\text{-}3}$	$8.338e\text{-}3_{7.96e\text{-}4}$	$8.767e\text{-}3_{9.87e\text{-}4}$
WFG6	3	$\mathbf{8.902e\text{-}2}_{1.16e\text{-}2}$	$1.473e\text{-}1_{2.51e\text{-}2}$	$8.905e\text{-}2_{9.16e\text{-}3}$	$9.204e\text{-}2_{2.07e\text{-}2}$
WFG6	5	$1.277e\text{-}1_{1.13e\text{-}2}$	$1.962e\text{-}1_{2.71e\text{-}2}$	$\mathbf{1.231e\text{-}1}_{1.05e\text{-}2}$	$1.281e\text{-}1_{1.41e\text{-}2}$
WFG6	8	$\mathbf{1.655e\text{-}1}_{1.84e\text{-}2}$	$2.127e\text{-}1_{5.45e\text{-}2}$	$1.726e\text{-}1_{2.13e\text{-}2}$	$1.692e\text{-}1_{2.57e\text{-}2}$
WFG6	10	$2.145e\text{-}1_{2.13e\text{-}2}$	$2.398e\text{-}1_{4.24e\text{-}2}$	$2.282e\text{-}1_{3.54e\text{-}2}$	$\mathbf{1.994e\text{-}1}_{2.69e\text{-}2}$
WFG6	15	$5.160e\text{-}1_{5.64e\text{-}2}$	$\mathbf{2.100e\text{-}1}_{6.40e\text{-}2}$	$3.609e\text{-}1_{7.15e\text{-}2}$	$4.694e\text{-}1_{1.61e\text{-}1}$
WFG7	3	$\mathbf{2.933e\text{-}2}_{4.23e\text{-}3}$	$6.329e\text{-}2_{4.99e\text{-}3}$	$4.300e\text{-}2_{4.64e\text{-}3}$	$4.757e\text{-}2_{1.53e\text{-}2}$
WFG7	5	$\mathbf{4.339e\text{-}2}_{5.03e\text{-}3}$	$9.443e\text{-}2_{5.35e\text{-}3}$	$7.336e\text{-}2_{5.13e\text{-}3}$	$6.512e\text{-}2_{1.93e\text{-}2}$
WFG7	8	$\mathbf{6.513e\text{-}2}_{1.19e\text{-}2}$	$1.229e\text{-}1_{1.12e\text{-}2}$	$1.031e\text{-}1_{9.52e\text{-}3}$	$9.241e\text{-}2_{2.06e\text{-}2}$
WFG7	10	$\mathbf{1.066e\text{-}1}_{9.21e\text{-}3}$	$1.522e\text{-}1_{1.04e\text{-}2}$	$1.387e\text{-}1_{8.55e\text{-}3}$	$1.207e\text{-}1_{2.29e\text{-}2}$
WFG7	15	$1.893e\text{-}1_{1.89e\text{-}1}$	$1.714e\text{-}1_{2.10e\text{-}2}$	$\mathbf{1.610e\text{-}1}_{1.97e\text{-}2}$	$1.709e\text{-}1_{1.58e\text{-}1}$

such as DTLZ4 and WFG7. NSGA-III$_{SBX}$ is not affected by the biased problems but it does not perform well on multi-modal problems like DTLZ1 and DTLZ3. So, in case we know that a specific instance has similar characteristics to one group or another, we can chose one among the two operators. On the other hand, if we do not have any prior knowledge, adaptive approaches are advantageous because they have comparable performances to the best operator in most of the cases (WFG7 is an exception).

Another interesting result is that for some subgroup of instances (DTLZ2 and DTLZ4 with less than 15 objectives) the adaptive approaches achieve the best mean values. This could probably indicate that, for these instances, the use of complementary operators during evolution is beneficial for the optimizer performance as a whole. NSGA-III$_{PM}$ seems better for less objectives while the performance of NSGA-III$_{AP}$ improves as the number of objectives increases.

4.3 PSP Results

Table 3 shows the results in terms of best energy values achieved by each NSGA-III variant on PSP problem.

From Table 3 we conclude that, except for 1CRN protein, the adaptive approaches have performances comparable with the best operator (DE in most of cases). From these results we also conclude that, for the addressed PSP problem instances, without prior knowledge of the problem characteristics, adaptive operator selection methods, such as PM and AP, can be used as we avoid testing each available operator individually.

Figures 2 and 3 present the boxplots of best energy and RMSD values comparing NSGA-III variants for each considered protein. In this preliminary work, our contribution to the PSP problem is the optimization approach considering 7 objective functions.

Figure 3 shows the boxplots of RMSD values of the proteins with the best energy values, comparing NSGA-III variants for each considered protein. Regarding the RMSD metric, all variants achieve similar values. For 1PLW protein, NSGA-III$_{AP}$ and NSGA-III$_{SBX}$ have the best results overall. NSGA-III$_{PM}$, NSGA-III$_{DE}$ and NSGA-III$_{SBX}$ yield lower RMSDs in 1ZDD instance. As for 1CRN, the adaptive variants NSGA-III$_{AP}$ and NSGA-III$_{PM}$ are similar to the bests results given by NSGA-III$_{SBX}$. For 1ROP, NSGA-III$_{DE}$ achieves the lowest RMSDs and NSGA-III$_{AP}$ is better in the mean value. In 1ZDD and 1CRN instances, despite having higher energy, the variants with normalization have the best similarity to the original protein. This indicates that some energy types (objectives) might contribute more for the protein conformation (RMSD), even if their energy values are low relative to the total energy estimated (objective sum).

Table 3. Median and standard deviation of the best total energy found for each NSGA-III variants on PSP problem. Bold values emphasize the best results (lowest median) and grayed cells indicate that the variant is statistically equal to the best, accordingly to Kruskal-Wallis test.

Protein	NSGA-III$_{SBX}$	NSGA-III$_{DE}$	NSGA-III$_{PM}$	NSGA-III$_{AP}$
1PLW	$2.19e+02_{6.1e+05}$	$\mathbf{1.56e+02_{3.0e+02}}$	$1.95e+02_{2.0e+06}$	$2.09e+02_{3.4e+06}$
1ZDD	$1.21e+13_{1.4e+28}$	$4.64e+10_{1.4e+20}$	$\mathbf{2.95e+10_{3.2e+20}}$	$7.83e+10_{3.9e+27}$
1CRN	$8.66e+10_{1.7e+18}$	$\mathbf{6.29e+08_{4.6e+16}}$	$6.27e+11_{7.8e+26}$	$2.79e+10_{2.3e+24}$
1ROP	$5.17e+12_{2.5e+91}$	$\mathbf{1.33e+12_{3.6e+34}}$	$8.50e+12_{2.8e+84}$	$1.06e+13_{1.7e+92}$

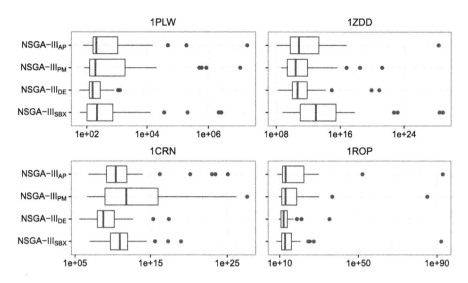

Fig. 2. Boxplots for the best energy values for each algorithm and protein.

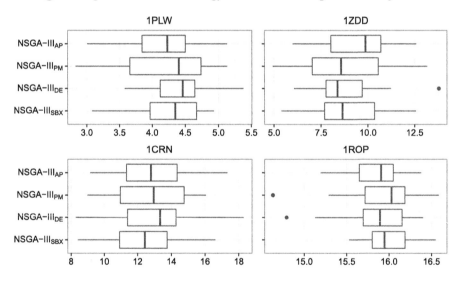

Fig. 3. Boxplots for RMSD values for each algorithm and protein.

5   Conclusions

In this paper we have presented two NSGA-III variants (NSGA-III$_{PM}$ and NSGA-III$_{AP}$) with adaptive operator selection mechanisms to deal with many-objective problems. The proposed approaches have been tested on a set of many-objective benchmarks (DTLZ and WFG), and also extended to a real-world problem (PSP).

Experimental results show that the proposed approaches have comparable performances when compared with best single operator variants (NSGA-III$_{\text{DE}}$ for DTLZ1 to DTLZ3 and NSGA-III$_{\text{SBX}}$ for the remaining ones). A more in-depth analysis demonstrated that the adaptive approaches are particular good to solve DTLZ2 and DTLZ4 instances with less than 15 objectives. When comparing the proposed approaches, NSGA-III$_{\text{PM}}$ is better for less objectives while the performance of NSGA-III$_{\text{AP}}$ improves as the number of objectives increases.

The results for PSP indicate that NSGA-III$_{\text{DE}}$ obtained the best results and that the adaptive approaches have also achieved encouraging performances.

Based on the obtained results, NSGA-III$_{\text{AP}}$ and NSGA-III$_{\text{PM}}$ can be considered as improvements to the NSGA-III framework in the case we have no prior knowledge about the characteristics of each instance as they avoid testing each operator individually.

Future work includes the investigation of different pools of operators and adaptive operation selection mechanisms. Regarding the PSP, we intend to make a more detailed analysis of the impact of each objective.

Acknowledgments. The authors acknowledge CNPq Grants 456179/2014-3, 483974/2013-7, and 311605/2011-7 for the partial financial support.

References

1. Brasil, C.R.S., Delbem, A.C.B., da Silva, F.L.B.: Multiobjective evolutionary algorithm with many tables for purely ab initio protein structure prediction. J. Comput. Chem. **34**(20), 1719–1734 (2013)
2. Coello, C.A.C., Cortés, N.C.: Solving multiobjective optimization problems using an artificial immune system. Genet. Program. Evolvable Mach. **6**(2), 163–190 (2005). http://dx.doi.org/10.1007/s10710-005-6164-x
3. Conover, W.: Practical Nonparametric Statistics, 3rd edn. Wiley, New York (1999)
4. Deb, K., Agrawal, R.B.: Simulated binary crossover for continuous search space. Complex Systems **9**(2), 115–148 (1995)
5. Deb, K., Agrawal, S., Pratap, A., Meyarivan, T.: A fast elitist non-dominated sorting genetic algorithm for multi-objective optimization: NSGA-II. In: Schoenauer, M., Deb, K., Rudolph, G., Yao, X., Lutton, E., Merelo, J.J., Schwefel, H.-P. (eds.) PPSN 2000. LNCS, vol. 1917, pp. 849–858. Springer, Heidelberg (2000). doi:10.1007/3-540-45356-3_83
6. Deb, K., Thiele, L., Laumanns, M., Zitzler, E.: Scalable test problems for evolutionary multiobjective optimization. In: Abraham, A., Jain, L., Goldberg, R. (eds.) Evolutionary Multiobjective Optimization. Advanced Information and Knowledge Processing, pp. 105–145. Springer, London (2005)
7. Durillo, J.J., Nebro, A.J.: Jmetal: a java framework for multi-objective optimization. Adv. Eng. Softw. **42**, 760–771 (2011)
8. Gonçalves, R., Almeida, C., Pavelski, L., Venske, S., Kuk, J., Pozo, A.: Adaptive Operator Selection in NSGA-III. In: To appear in Proceedings of Brazilian Conference on Intelligent Systems, BRACIS 2016 (2016)
9. Gong, W., Fialho, Á., Cai, Z., Li, H.: Adaptive strategy selection in differential evolution for numerical optimization: an empirical study. Inf. Sci. **181**(24), 5364–5386 (2011)

10. Handl, J., Lovell, S.C., Knowles, J.: Investigations into the effect of multiobjectivization in protein structure prediction. In: Rudolph, G., Jansen, T., Beume, N., Lucas, S., Poloni, C. (eds.) PPSN 2008. LNCS, vol. 5199, pp. 702–711. Springer, Heidelberg (2008). doi:10.1007/978-3-540-87700-4_70

11. Huband, S., Hingston, P., Barone, L., While, L.: A review of multiobjective test problems and a scalable test problem toolkit. IEEE Trans. Evol. Comput. **10**(5), 477–506 (2006)

12. Ishibuchi, H., Tsukamoto, N., Nojima, Y.: Evolutionary many-objective optimization: a short review. In: Proceedings of 2008 IEEE Congress on Evolutionary Computation, pp. 2424–2431 (2008)

13. Jain, H., Deb, K.: An evolutionary many-objective optimization algorithm using reference-point based nondominated sorting approach, part ii: handling constraints and extending to an adaptive approach. IEEE Trans. Evol. Comput. **18**(4), 602–622 (2014)

14. Li, K., Deb, K., Zhang, Q., Kwong, S.: An evolutionary many-objective optimization algorithm based on dominance and decomposition. IEEE Trans. Evol. Comput. **19**(5), 694–716 (2015)

15. Li, K., Fialho, Á., Kwong, S.: Multi-objective differential evolution with adaptive control of parameters and operators. In: International Conference on Learning and Intelligent Optimization, pp. 473–487 (2011)

16. Li, K., Fialho, A., Kwong, S., Zhang, Q.: Adaptive operator selection with bandits for a multiobjective evolutionary algorithm based on decomposition. IEEE Trans. Evol. Comput. **18**(1), 114–130 (2014)

17. Lin, Q., Liu, Z., Yan, Q., Du, Z., Coello, C.A.C., Liang, Z., Wang, W., Chen, J.: Adaptive composite operator selection and parameter control for multiobjective evolutionary algorithm. Inf. Sci. **339**, 332–352 (2016)

18. MacKerell, A.D., Banavali, N., Foloppe, N.: Development and current status of the charmm force field for nucleic acids. Biopolymers **56**(4), 257–265 (2000)

19. Mkaouer, W., Kessentini, M., Shaout, A., Koligheu, P., Bechikh, S., Deb, K., Ouni, A.: Many-objective software remodularization using NSGA-III. ACM Trans. Softw. Eng. Methodo. **24**(3), 17:1–17:45 (2015)

20. Ponder, J.W.: Tinker: Software tools for molecular design. Washington University School of Medicine, Saint Louis, MO 3 (2004)

21. Seada, H., Deb, K.: U-NSGA-III: a unified evolutionary optimization procedure for single, multiple, and many objectives: proof-of-principle results. In: Gaspar-Cunha, A., Henggeler Antunes, C., Coello, C.C. (eds.) EMO 2015. LNCS, vol. 9019, pp. 34–49. Springer, Cham (2015). doi:10.1007/978-3-319-15892-1_3

22. Storn, R., Price, K.: Differential evolution - a simple and efficient heuristic for global optimization over continuous spaces. J. Global Optim. **11**(4), 341–359 (1997)

23. Thierens, D.: An adaptive pursuit strategy for allocating operator probabilities. In: Conference on Genetic and Evolutionary Computation, pp. 1539–1546 (2005)

24. Tramontano, A.: Protein Structure Prediction: Concepts and Applications. Wiley, New Year (2006)

25. Venske, S.M., Gonçalves, R.A., Delgado, M.R.: ADEMO/D: multiobjective optimization by an adaptive differential evolution algorithm. Neurocomputing **127**, 65–77 (2014)

26. Yuan, Y., Xu, H., Wang, B.: An improved NSGA-III procedure for evolutionary many-objective optimization. In: Proceedings of the 2014 Annual Conference on Genetic and Evolutionary Computation, pp. 661–668. ACM, New York (2014)

27. Yuan, Y., Xu, H., Wang, B.: An experimental investigation of variation operators in reference-point based many-objective optimization. In: Proceedings of the 2015 Annual Conference on Genetic and Evolutionary Computation, pp. 775–782. ACM, New York (2015)
28. Zaki, M., Bystroff, C. (eds.): Protein Structure Prediction. Methods in Molecular Biology, vol. 413, 2nd edn. Humana Press/Springer, Heidelberg (2008)
29. Zhao, S.Z., Suganthan, P.N., Zhang, Q.: Decomposition-based multiobjective evolutionary algorithm with an ensemble of neighborhood sizes. IEEE Trans. Evol. Comput. **16**(3), 442–446 (2012)

On Using Decision Maker Preferences with ParEGO

Jussi Hakanen[1]([✉]) and Joshua D. Knowles[2]

[1] Faculty of Information Technology, University of Jyvaskyla,
P.O. Box 35 (Agora), FI-40014 University of Jyvaskyla, Finland
`jussi.hakanen@jyu.fi`
[2] School of Computer Science, University of Birmingham,
Edgbaston B15 2TT, Birmingham, UK
`j.knowles@cs.bham.ac.uk`

Abstract. In this paper, an interactive version of the ParEGO algorithm is introduced for identifying most preferred solutions for computationally expensive multiobjective optimization problems. It enables a decision maker to guide the search with her preferences and change them in case new insight is gained about the feasibility of the preferences. At each interaction, the decision maker is shown a subset of non-dominated solutions and she is assumed to provide her preferences in the form of preferred ranges for each objective. Internally, the algorithm samples reference points within the hyperbox defined by the preferred ranges in the objective space and uses a DACE model to approximate an achievement (scalarizing) function as a single objective to scalarize the problem. The resulting solution is then evaluated with the real objective functions and used to improve the DACE model in further iterations. The potential of the proposed algorithm is illustrated via a four-objective optimization problem related to water management with promising results.

Keywords: Surrogate-based optimization · Interactive multiobjective optimization · Preference information · Computational cost · Visualization

1 Introduction

Multiobjective optimization problems have several conflicting objective functions to be optimized simultaneously. Due to the conflict, optimal solutions have inherent trade-offs between the objectives and are called Pareto optimal solutions where none of the objectives can be improved without impairing some other one [15]. Without any additional information, all the Pareto optimal solutions are equally good. On a general level, one can see two approaches for solving multiobjective optimization problems: (1) approximate the whole Pareto front, that is, the set of all Pareto optimal solutions or (2) identify the most preferred Pareto optimal solution(s) utilizing preference information from a human decision maker. In this paper, we concentrate on the latter approach.

H. Trautmann et al. (Eds.): EMO 2017, LNCS 10173, pp. 282–297, 2017.
DOI: 10.1007/978-3-319-54157-0_20

Decision maker preferences are nowadays widely used to find most preferred solutions in both multiple criteria decision making and evolutionary multiobjective optimization communities (see e.g. [2,4,15,21]). Preferences can be used before optimization, after optimization, or iteratively during the optimization process [15]. The methods belonging to the latter approach are called interactive multiobjective optimization methods where the decision maker is iteratively involved in the solution process. The benefits of interactive methods include learning about the interdependencies between the conflicting objectives and ones preferences especially for problems with a high number of objectives. This new knowledge gained during the interactive solution process may necessitate adjusting decision maker's preferences. Interactive methods indeed allow the decision maker to change her preferences which is not possible if the preferences are asked before optimization (see, e.g., [22]). Therefore, they have been found promising for solving real-world problems [17,18].

In computationally expensive multiobjective optimization problems, a single evaluation of the objective and/or constraint functions takes considerable amount of time. To be able to tackle such problems, surrogate-based approaches are commonly used where the idea is to approximate the problem with simpler functions that are faster to evaluate [10,13,24,26]. There are different ways of using surrogates in multiobjective optimization [24], for example, (1) to approximate each objective function separately and apply any suitable multiobjective optimization algorithm by using the approximated functions (the most widely used approach, see e.g. [3,10,24]), (2) to approximate a scalarizing function that converts multiple objectives together with decision maker preferences into a single objective optimization problem [7,12], or (3) to approximate directly the Pareto front [6,9,19]. However from decision making point of view, providing most preferred solutions to the decision maker for computationally expensive multiobjective optimization problems was identified as a future research challenge in a recent survey [24].

The contribution of this paper is to introduce an interactive multiobjective optimization method for computationally expensive multiobjective optimization problems to partly answer the challenge mentioned above. It is based on the ParEGO algorithm [12] developed for approximating the whole Pareto front for computationally expensive multiobjective optimization problems. ParEGO uses a weighted Tchebycheff function to convert multiple objectives into a single objective optimization problem and randomly generates weights at each iteration to be able to approximate the whole Pareto front. Since ParEGO uses a surrogate to approximate a scalarization function, it is convenient to add also decision maker preferences and, thus, try to approximate only preferred parts of the Pareto front. This is done by replacing the weighted Tchebycheff function with an achievement scalarizing function and using reference points instead of weights to include decision maker preferences. *The decision maker guides the search by specifying preferred ranges for each objective function. By specifying ranges, the decision maker deals directly with objective function values (as opposed to e.g. weights that have no clear meaning) and it allows the decision maker to explore*

more if she does not have exact values in mind. When compared to approaches
that use surrogates to approximate directly the whole Pareto front (e.g. [6,9,19]),
the benefit of interactive ParEGO is that it does not need any pre-computed
Pareto optimal solutions as an input. On the other hand, when compared to
the approaches that use surrogates for each objective separately, the number of
surrogates used is always one independently of the number of objectives used.

The rest of the paper is organized as follows. Section 2 presents the problem
formulation along with brief introduction of the basics of the ParEGO algorithm.
In Sect. 3, ideas of incorporating decision maker preferences into ParEGO are discussed and the interactive ParEGO algorithm is introduced. Section 4 illustrates
the potential and ability of interactive ParEGO to utilize changing preference
information through a four-objective example related to water management.
Finally, the paper ends with conclusions and future research ideas presented in
Sect. 5.

2   Basics of ParEGO

We consider a multiobjective optimization problem

$$\text{minimize} \ \{f_1(\boldsymbol{x}), \ldots, f_k(\boldsymbol{x})\}$$

$$\text{subject to } \boldsymbol{x} \in S, \tag{1}$$

where all k objective functions $f_i : \mathbb{R}^n \to \mathbb{R}$, $i = 1, \ldots, k$, are to be minimized.
The feasible region $S \subset \mathbb{R}^n$ denotes the set of feasible decision variable values
$\boldsymbol{x} = (x_1, \ldots, x_n)^T$. Here we assume that the function evaluations are costly in
the sense of taking a long time.

The ParEGO algorithm [12] was developed for computationally expensive
multiobjective optimization problems and it is based on the efficient global optimization (EGO) algorithm [11]. In ParEGO, the multiobjective problem (1) is
converted into a single objective optimization problem by using the augmented
Tchebycheff scalarizing function

$$\text{minimize} \ \max_{i=1,\ldots,k} [w_i f_i(\boldsymbol{x})] + \rho \sum_{i=1}^{k} w_i f_i(\boldsymbol{x})$$

$$\text{subject to } \boldsymbol{x} \in S, \tag{2}$$

where $\boldsymbol{w} = (w_1, \ldots, w_k)^T$, $w_i \geq 0$ and $\sum_{i=1,\ldots,k} w_i = 1$. The term containing $\rho >$
0 is called an augmentation term and it is used to guarantee that the solutions
of problem (2) are Pareto optimal [15]. An important reason for choosing the
Tchebycheff scalarization was that it can find Pareto optimal solutions also in
the non-convex part of the Pareto front. The overall goal in ParEGO is to find
an approximation for the whole Pareto front. Next, the basic idea of ParEGO is
briefly described.

ParEGO starts by using the latin hypercube sampling to find $11n - 1$ points
in the decision space which are then evaluated by the real functions. Those

points are then used to train a Kriging-based design and analysis of computer experiments (DACE) model to approximate the objective function in problem (2). To train the DACE model, Nelder and Mead downhill simplex algorithm is used to maximize the likelihood. Then at each iteration of ParEGO, a new point to be evaluated with the real functions is determined by using a single objective genetic algorithm to maximize the expected improvement. After finding the new point, it is evaluated with the real functions and, then, the DACE model is re-trained by considering also the new point. The updated DACE model is then used to find the next point and this iteration continues until the budget for function evaluations with the real functions is exhausted.

We can make some important observations of ParEGO. Note that at each iteration, one new point is evaluated with the real functions. It is possible to use all the points evaluated so far with the real functions to train the DACE model. However, when the number of the points increases, the more time training takes. It is also possible to use only some subset of the points in training in order to reduce the training time while compromising the accuracy if needed [5]. Further, for each iteration, the weight vector w used in problem (2) is randomly generated in order to finally cover the whole Pareto front. In practice, ParEGO does not set any limitations for the number of objective functions to be considered since it only affects on the generation of the weight vectors. However, the current implementation found in https://github.com/CristinaCristescu/ParEGO_Eigen supports only problems up to four objective functions [5]. More details of ParEGO can be found in [12].

3   Preferences with ParEGO

Next we discuss how to include decision maker preferences into ParEGO. A natural way for this would be to apply the ideas of the interactive weighted Tchebycheff procedure introduced in [23] where the idea is to reduce the space of feasible weight vectors based on preferences from the decision maker. At each iteration, the decision maker sees some Pareto optimal solutions and is asked to select the most preferred one. Based on the selection, the weight vector space is reduced and the new Pareto optimal solutions are generated by using the reduced space of weight vectors. The limitations of the interactive weighted Tchebycheff approach include that one can not go back to the part of the space that has been eliminated (i.e. change preferences). It means that an important property of interactive approaches, a possibility to change one's mind, is not available. For that reason, we do not go towards that path but take another approach utilizing achievement (scalarizing) function (ACH) [25] and preferred ranges for the objective function values as a form of preference information that will enable the decision maker to change her mind during the interactive decision making process. Next, our approach is discussed in more details.

3.1   A Priori Preferences

We start by discussing how to include a priori preferences to ParEGO, that is, how to handle fixed preferences expressed before optimization. That will also serve as a building block for interactive ParEGO described later. As mentioned above, we chose to use preferred ranges for objectives as decision maker preferences since dealing directly with objective function values is found cognitively easy way of expressing preferences for the decision maker [14]. Preferred ranges will result in a k-dimensional hyperbox $H = \{z \in \mathbb{R}^k | a_i \leq z_i \leq b_i\}$ in the objective space. In practice, this hyperbox can be considered as either (1) *optimistic* which means that it does not contain any feasible solutions, (2) *pessimistic* where all the solutions are dominated or (3) *neither optimistic nor pessimistic* which means that it intersects with the Pareto front [7]. Different types of hyperboxes are illustrated in Fig. 1.

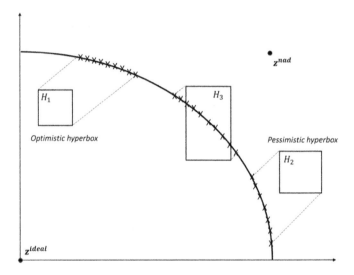

Fig. 1. Examples of different types of hyperboxes. The solutions in the Pareto Front (black curve) that can be reached from each hyperbox by using ACH are marked with Xs. The dotted lines denote the projection direction determined by the weights in ACH.

Further, multiple objectives are scalarized by using ACH instead of the weighted Tchebycheff scalarizing function used in original ParEGO. The mathematical formulation of the augmented ACH used here is

$$\text{minimize} \quad \max_{i=1,\dots,k} w_i(f_i(\boldsymbol{x}) - \bar{z}_i) + \rho \sum_{i=1}^{k} w_i f_i(\boldsymbol{x}) \tag{3}$$

$$\text{subject to } \boldsymbol{x} \in S.$$

The weights typically used with ACH are $w = (z^{nad} - z^{ideal})^{-1}$ where z^{ideal} and z^{nad} denote vectors containing the best and the worst values for the objectives in the Pareto front, respectively. Note that the weights used in the weighted Tchebycheff scalarizing function (2) are used as a preference information to get different Pareto optimal solutions as opposed to the weights in ACH (3) that are used to normalize the scales of the objective function values. In the case of ParEGO, the objective functions are normalized based on the set of evaluated solutions and, therefore, weights $w_i = 1, i = 1, \ldots, k$ are used. For the augmentation term, we here set $\rho = 0.05$ as has been done in original ParEGO. The solution of problem (3) is Pareto optimal [15] and different Pareto optimal solutions can be obtained by varying the reference point $\bar{z} \in \mathbb{R}^k$. Thus, the reference point $\bar{z}$ is the way to include preference information to ACH and, in our approach, to ParEGO. These reference points will be sampled from the hyperbox H determined based on the decision maker preferences in analogy of sampling the weights for the augmented weighted Tchebycheff function in the original ParEGO.

The basic idea here is that decision maker fixes the preferred ranges for the objectives resulting into a hyperbox H in the objective space. Then, ParEGO with ACH is run by using reference points sampled within H. Note that it is not needed to sample reference points from the whole H since it is known that all the reference points along the direction specified by the weight vector w give the same Pareto optimal solution. Therefore, it is enough to sample reference points in H such that they lie in the plane orthogonal to w.

To demonstrate how the modified ParEGO can handle a priori preferences and how our basic building block for the interactive ParEGO works, Figs. 2 and 3 show two example runs for the three-objective DTLZ2 problem used in [12], which has the following formulation:

$$
\begin{aligned}
\text{minimize} \quad & f_1 = (1 + g) \cos(x_1^\alpha \pi/2) \cos(x_2^\alpha \pi/2), \\
\text{minimize} \quad & f_2 = (1 + g) \cos(x_1^\alpha \pi/2) \sin(x_2^\alpha \pi/2), \\
\text{minimize} \quad & f_3 = (1 + g) \sin(x_1^\alpha \pi/2),
\end{aligned}
$$

$$
\begin{aligned}
\text{subject to } g = & \sum_{i \in \{3, \ldots, 8\}} (x_i - 0.5)^2, \\
& x_i \in [0, 1], i \in \{1, \ldots, 8\},
\end{aligned}
\tag{4}
$$

where $\alpha = 1$. Figure 2 illustrates a case where ParEGO was run for 100 real function evaluations by using the preferred ranges $0.4 \le f_1 \le 0.5$, $0.5 \le f_2 \le 0.6$, and $0.4 \le f_3 \le 0.5$. Parallel coordinate plot of the non-dominated points obtained after the initial sampling together with the preferred ranges are shown in Fig. 2 by using blue and red color (together with symbol X), respectively. Similarly, Fig. 3 shows a case where the preferred ranges $0.1 \le f_1 \le 0.3$, $0.7 \le f_2 \le 0.9$, and $0.0 \le f_3 \le 0.2$ were used. In both the figures, one can observe that the obtained solutions follow the given preferences quite well, although not perfectly since the given ranges are not feasible.

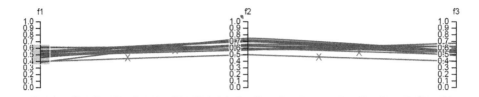

Fig. 2. Non-dominated solutions (in blue) obtained for the DTLZ2 problem by using 100 real function evaluations and the preferred ranges (in red and with symbol X) $0.4 \leq f_1 \leq 0.5$, $0.5 \leq f_2 \leq 0.6$, and $0.4 \leq f_3 \leq 0.5$. (Color figure online)

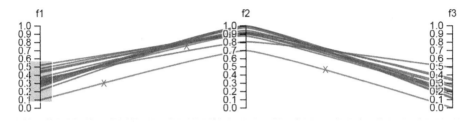

Fig. 3. Non-dominated solutions (in blue) obtained for the DTLZ2 problem by using 100 real function evaluations and the preferred ranges (in red and with symbol X) $0.1 \leq f_1 \leq 0.3$, $0.7 \leq f_2 \leq 0.9$, and $0.0 \leq f_3 \leq 0.2$. (Color figure online)

3.2   Interactive ParEGO

In this paper, we are not satisfied with fixed preferences since our overall goal is to develop an interactive version of ParEGO where the decision maker is able to change her mind if needed to better steer the solution process towards preferred solutions. Therefore, we will use progressive preference articulation that is used in interactive multiobjective optimization approaches [15,17,18]. Algorithm 1 shows the steps of the interactive ParEGO algorithm which has three input parameters. The first is the budget of function evaluations FE_{max} available for the decision making process. The second parameter fixes the frequency of interaction, that is, how many iterations to run ParEGO with given preferences before the next interaction. The last parameter determines how many solutions the decision maker wants to see at each interaction. The output of the algorithm will be the solution most preferred by the DM.

After initialization step 1, $11n - 1$ points are sampled in the decision space by using latin hypercube sampling in the same way than in the original ParEGO algorithm in step 2. The resulting points are evaluated by using the real functions and added to an archive A that will contain all the points evaluated with the real functions. In step 3, non-dominated solutions of A are determined and shown to the decision maker. If there are more than N_S non-dominated solutions in A, the number will be reduced to N_S. This can be done e.g. by using clustering. Otherwise, all the non-dominated solutions are shown to decision maker. The solutions can be visualized to the decision maker with parallel coordinate

Algorithm 1. Interactive ParEGO

Input: FE_{max} = budget for function evaluations; F_{IA} = frequency of interactions; N_S = a number of solutions to be shown to decision maker at each interaction

Output: The solution most preferred by decision maker

1: Initialize the archive of solutions evaluated with real functions $A = \emptyset$, interaction counter $it = 0$, and function evaluation counter $fe = 0$.
2: Create initial population A^0 by latin hypercube sampling of $11n - 1$ points in the decision space, evaluate them by using the original functions, and update $A = A \cup A^0$ $fe = 11n - 1$.
3: Find non-dominated solutions of A and show N_S of them to decision maker.
4: If decision maker wants to stop or $fe + F_{IA} > FE_{max}$, go to step 8.
5: Ask decision maker to indicate preferred ranges for the objectives.
6: Run F_{IA} iterations of ParEGO with given preferences to obtain a set A^{it} of F_{IA} solutions evaluated with real functions.
7: Update $A = A \cup A^{it}$, $fe = fe + F_{IA}$, $it = it + 1$ and go to step 3
8: Ask decision maker to indicate the most preferred solution as the final solution

plots or some other suitable visualization technique depending on the number of objectives [16]. Step 4 is the termination step of the algorithm and it involves two different criteria: stopping if decision maker so desires or if the budget of function evaluations is about to be exceeded. Thus in order to have any interactions, the budget for function evaluations FE_{max} should be more than $11n-1+F_{IA}$. If the decision maker wants to continue and there is budget for function evaluations remaining, then the decision maker is asked to provide preferred ranges for all the objective functions in step 5. The given ranges are then treated as preference information in a way described in Sect. 3.1. Next in step 6, F_{IA} iterations of ParEGO are run with the given preference information resulting into the same amount of new points evaluated with the real functions. Then in step 7, those points are added to the archive A, the function evaluations used and the number of interactions are updated, and the algorithm continues from step 3.

4   Numerical Example

To illustrate the performance and potential of the interactive ParEGO algorithm, we here describe an interactive solution process for the four-objective optimization problem related to water management. We chose this problem because the objective functions in that have some real meaning unlike the objectives in the synthetic test problems often used to evaluate the performance of multiobjective optimization methods. In this way, the interactive solution process becomes more meaningful and easier to follow. In addition, although the example problem is not computationally expensive, it enables faster testing and the limited budget of available function evaluations reflects the challenge with computationally expensive problems. Before describing the water management problem along with the interactive solution process, we illustrate how the interactive ParEGO reacts to changing preferences through the three-objective DTLZ4 problem.

Table 1. Preferred ranges used with DTLZ4.

	f_1	f_2	f_3
1	$[0.0, 0.2]$	$[0.8, 1.0]$	$[0.4, 0.6]$
2	$[0.4, 0.6]$	$[0.0, 0.2]$	$[0.8, 1.0]$
3	$[0.8, 1.0]$	$[0.4, 0.6]$	$[0.0, 0.2]$
4	$[0.4, 0.6]$	$[0.4, 0.6]$	$[0.4, 0.6]$

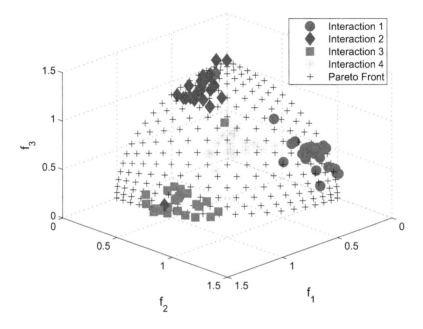

Fig. 4. Illustration of changing preferences for the DTLZ4 problem. (Color figure online)

4.1 DTLZ4

Here, we consider the three-objective DTLZ4 problem which is a modification of the DTLZ2 problem described in Eq. (4) by using $\alpha = 100$. The Pareto front for both DTLZ2 and DTLZ4 is one eighth of a sphere of radius 1, centred on $(0, 0, 0)$ but using $\alpha = 100$ severely biases the density distribution of solutions. To demonstrate the effect of changing preferences, four different preferred ranges (shown in Table 1) were consequently used after initialization.

Parameter values used were $FE_{max} = 200$ and $F_{IA} = 30$. The resulting non-dominated solutions of a single run are shown as a scatter plot in Fig. 4 where different colors/symbols denote solutions obtained after different interactions, i.e., with different ranges. As can be seen, the solutions obtained follow the ranges given despite the limited number of function evaluations used.

4.2   Water Management Problem

The problem deals with water management in the (hypothetical) Fast Water Valley on a stretch of 100 river miles presented in [20]. There is a Fresh Fishery situated near the head of the valley causing pollution to the river. The city of Fortuna having population of 300 000 produces municipal waste pollution and is located 50 miles downstream from the fishery. The measure for water quality is expressed in terms of dissolved oxygen (DO) concentration and both the pollutants are described in pounds of biochemical oxygen demanding material (BOD). There exist primary treatment facilities that reduce the BOD in the gross discharge of the fishery and the city by 30 percent. Additional treatment facilities would increase the tax rate in Fortuna and decrease the return on investment (ROI) from the Fresh Fishery. The decision maker is interested in four objective functions: f_1 = water quality in the fishery, f_2 = water quality in the city, f_3 = ROI from the fishery, and f_4 = addition to the city tax rate resulting in to the following optimization problem

$$
\begin{aligned}
\text{maximize} \quad & f_1 = 4.07 + 2.27x_1, \\
\text{maximize} \quad & f_2 = 2.6 + 0.03x_1 + 0.02x_2 + \frac{0.01}{1.39 - x_1^2} + \frac{0.3}{1.39 - x_2^2}, \\
\text{maximize} \quad & f_3 = 8.21 - \frac{0.71}{1.09 - x_1^2}, \\
\text{minimize} \quad & f_4 = -0.96 + \frac{0.96}{1.09 - x_2^2},
\end{aligned}
\tag{5}
$$

subject to $x_1 :=$ Proportionate amount of removed BOD at fishery,
$\qquad\quad x_2 :=$ Proportionate amount of removed BOD at city,
$\qquad\quad 0.3 \leq x_1, x_2 \leq 1.$

The units for f_1 and f_2 are [mg/L of DO] while the unit for f_3 is [%]. The unit for the addition to the city tax rate f_4 is [1/1000 of $\$$], i.e., $f_4 = \delta$ results in the tax rate of about $\$\delta$ per $\$1000$ assessed value at Fortuna. The decision variables x_1 and x_2 describe the proportionate amount of BOD removed from water discharge at the fishery and at the city, respectively.

4.3   Interactive Solution Process

Parameter values used in this study were $FE_{max} = 100$, $F_{IA} = 20$, and $N_S = 5$. These values were chosen just to illustrate the performance and a study on their effect in the performance of the method is left as a topic for future study. In addition, k-means clustering with squared Euclidean distance was used to reduce the number of produced solutions if it was more than N_S and, from each N_S cluster identified, the solution closest to cluster centroid was chosen. An alternative for k-means would be to use multiobjective clustering techniques where the number of clusters is not fixed but determined by the clustering algorithm [8]. Exploring this option is left for future research.

An important aspect related to interactive multiobjective optimization methods is the graphical user interface that implements the interaction between the decision maker and the method. In this example, we have used the parallel coordinate plot tool developed in Prof. Patrick Reed's group in Cornell [1] to visualize the non-dominated solutions to the decision maker and for her/him to analyze them in order to adjust the preferred ranges if necessary. It is important to note that the figures of parallel coordinate plots used to illustrate the solutions in this paper do not really provide the same possibilities to analyze them as the tool itself. In other words, by using the tool the decision maker can interact with the visualizations and get a better understanding by e.g. filtering the solutions or changing the positions of different objectives to have better understanding of the trade-offs. Developing a more enhanced user interface is one of the future research topics.

To start with, latin hypercube sampling for this problem corresponded to 21 points in the two dimensional decision space and a set of five representative non-dominated solutions among those is shown in Fig. 5. From the figure, the decision maker could also observe the initial ranges for the objectives that help in determining preferred ranges.

Fig. 5. Non-dominated solutions shown to the decision maker after the initial sampling.

Our aim here was to show how changes in preferences during the solution process can be taken into account with the interactive ParEGO and we illustrate this by considering preferences from different stakeholders' points of view. We start from a citizen's perspective who is living in the city of Fortuna. The first goal is to try to maximize the water quality in the city and, thus, the preferred ranges $5.5 \leq f_1 \leq 6.2$, $3.3 \leq f_2 \leq 4.0$, $4.0 \leq f_3 \leq 6.0$, and $3.0 \leq f_4 \leq 5.0$ were used.

Fig. 6. Non-dominated solutions (in blue) shown to the decision maker after the first interaction. The preferred ranges from the first interaction are shown in red with symbol X. (Color figure online)

The resulting non-dominated solutions after clustering are shown in Fig. 6 which also shows the preferred ranges from the first interaction (in red and with symbol X). One can observe that it was not possible to improve water quality in the city for more than around 3.45 which resulted in the city tax rate increase of over 9. This was not found satisfactory and, therefore, the aim was to find satisfactory water quality levels in the city without too large tax rate increase. To continue with, the decision maker gave new ranges $5.5 \leq f_1 \leq 6.2$, $3.0 \leq f_2 \leq 3.5$, $2.0 \leq f_3 \leq 6.0$, and $5.0 \leq f_4 \leq 7.0$.

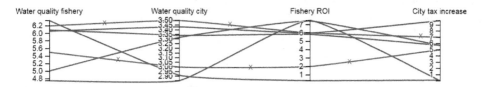

Fig. 7. Non-dominated solutions (in blue) shown to the decision maker after the second interaction. The preferred ranges from the second interaction are shown in red with symbol X. (Color figure online)

The set of five non-dominated solutions together with the preferred ranges are shown in Fig. 7. Of the solutions shown to the decision maker, there were two within the preferred ranges for the city tax increase: $(5.00, 3.32, 7.44, 5.70)$ and $(6.09, 3.34, 5.82, 5.70)$. Both the solutions have similar values for f_2 and f_4 while the values differed for the other objectives. Thus, with these levels for the water quality in the city and the city tax increase, there exists a clear trade-off between the water quality in the fishery and the fishery ROI. For the next interaction, we looked at the problem from the fishery shareholder perspective and the aim was to maximize the ROI from the fishery and see what happens to the other objectives. Thus, the new ranges given were $4.0 \leq f_1 \leq 6.5$, $2.5 \leq f_2 \leq 3.5$, $6.0 \leq f_3 \leq 8.0$, and $0.0 \leq f_4 \leq 10.0$.

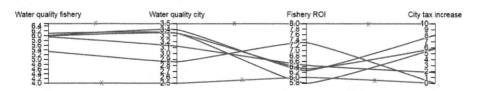

Fig. 8. Non-dominated solutions (in blue) shown to the decision maker after the third interaction. The preferred ranges from the third interaction are shown in red with symbol X. (Color figure online)

The resulting solutions are shown in Fig. 8. One can observe, that in the solution with the best ROI, $(5.34, 2.86, 7.30, 0.00)$, the tax rate in the city does not increase since no additional water treatment is used. On the other hand, the water quality in both the city and the fishery is degraded. Thus, in order to keep the ROI from the fishery at a relatively high level and, simultaneously, not to compromise the water quality too much, the city tax rate should be increased.

Finally, we looked at the problem from perspective of the mayor of Fortuna. From her point of view, all the objectives are important since the quality of the water in the city directly affects to the living conditions, the fishery brings jobs and food to people in the city, and the tax rate should be kept at a relatively low level in order to keep the citizens happy. Therefore, she aims at a balanced solution between all the objectives and the preferred ranges $4.0 \leq f_1 \leq 6.5$, $2.5 \leq f_2 \leq 3.5$, $6.0 \leq f_3 \leq 8.0$, and $0.0 \leq f_4 \leq 10.0$ were taken around the middle of the ranges of each objective obtained among all solutions evaluated so far.

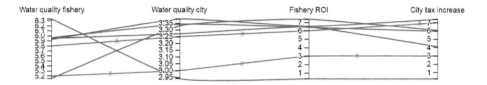

Fig. 9. Non-dominated solutions (in blue) shown to the decision maker after the fourth interaction. The preferred ranges from the fourth interaction are shown in red with symbol X. (Color figure online)

The resulting set of five representative non-dominated solutions along with the preferred ranges are shown in Fig. 9. The first observation is that there is not a single solution that satisfies all the preferred ranges and it might be the case that such a solution does not exist. The closest one has the values $(5.95, 3.27, 6.45, 4.11)$ which seems to balance all the objectives quite nicely although it has quite high ROI from the fishery. However, if one looks at all the non-dominated solutions found during the solution process shown in Fig. 10, there are not many solutions having ROI from the fishery below 5.0. At this point, the interactive solution process ends.

Fig. 10. All the non-dominated solutions found during the interactive solution process.

5   Conclusions

In this paper, an interactive ParEGO algorithm utilizing decision maker preferences in an iterative manner is proposed. As opposed to the original ParEGO algorithm, the aim is not to approximate the whole Pareto front but to help the decision maker to find her most preferred solution. The key modifications to the existing ParEGO algorithm are (1) incorporating decision maker preferences to focus the search, (2) interacting with the decision maker by showing her/him intermediate non-dominated solutions and asking for preferred ranges for the objective functions, and (3) replacing the Tchebycheff scalarization with the achievement scalarizing function enabling the usage of reference points inside the algorithm. The potential of the proposed algorithm was demonstrated through the four objective optimization problem in water management and its ability to consider preference information that changes during the solution process seems promising. The topics for future research include a study on the influence of the three parameters of the algorithm, that is, the budget for real function evaluations, the frequency of interaction, and the number of solutions shown to the decision maker, building a more enhanced graphical user interface for implementing the interaction, using multiobjective clustering to reduce the number of solutions shown to the decision maker, and finally, applying the method to computationally expensive real-world problems.

Acknowledgment. This work was supported by the FiDiPro project DeCoMo funded by TEKES, The Finnish Funding Agency for Innovation. The authors want to thank Mr. Tinkle Chugh, Prof. Yaochu Jin, Prof. Kaisa Miettinen, and Dr. Karthik Sindhya for discussions related to the ideas presented in the paper. In addition, the authors would like to acknowledge Prof. Patrick Reed's group Decision Analytics for Complex Systems (Cornell University) related to their parallel coordinate plotting tool.

References

1. Web-based parallel coordinate plot. https://reed.cee.cornell.edu/parallel-axis-categories/parallel/. Accessed 14 Oct 2016
2. Branke, J., Deb, K., Miettinen, K., Słowiński, R. (eds.): Multiobjective Optimization: Interactive and Evolutionary Approaches. LNCS, vol. 5252. Springer, Heidelberg (2008)
3. Chugh, T., Jin, Y., Miettinen, K., Hakanen, J., Sindhya, K.: A surrogate-assisted reference vector guided evolutionary algorithm for computationally expensive many-objective optimization. IEEE Trans. Evol. Comput. (2017, to appear)
4. Coello, C.: Handling preferences in evolutionary multiobjective optimization: a survey. In: 2000 IEEE Congress on Evolutionary Computation (CEC), pp. 30–37. IEEE (2000)
5. Cristescu, C., Knowles, J.: Surrogate-based multiobjective optimization: ParEGO update and test. Workshop on Computational Intelligence (UKCI) (2015)
6. Eskelinen, P., Miettinen, K., Klamroth, K., Hakanen, J.: Pareto navigator for interactive nonlinear multiobjective optimization. OR Spectr. **32**, 211–227 (2010)

7. Haanpää, T.: Approximation method for computationally expensive nonconvex multiobjective optimization problems. Jyväskylä Studies in Computing 157, PhD thesis, University of Jyväskylä, Jyväskylä (2012). http://urn.fi/URN:ISBN: 978-951-39-4968-6

8. Handl, J., Knowles, J.: An evolutionary approach to multiobjective clustering. IEEE Trans. Evol. Comput. **11**(1), 56–76 (2007)

9. Hartikainen, M., Miettinen, K., Wiecek, M.M.: PAINT: Pareto front interpolation for nonlinear multiobjective optimization. Comput. Optim. Appl. **52**(3), 845–867 (2012)

10. Jin, Y.: Surrogate-assisted evolutionary computation: recent advances and future challenges. Swarm Evol. Comput. **1**(2), 61–70 (2011)

11. Jones, D.R., Schonlau, M., Welch, W.J.: Efficient global optimization of expensive black-box functions. J. Global Optim. **13**(4), 455–492 (1998)

12. Knowles, J.: ParEGO: a hybrid algorithm with on-line landscape approximation for expensive multiobjective optimization problems. IEEE Trans. Evol. Comput. **10**(1), 50–66 (2005)

13. Knowles, J., Nakayama, H.: Meta-Modeling in Multiobjective Optimization. In: Branke, J., Deb, K., Miettinen, K., Słowiński, R. (eds.) Multiobjective Optimization. LNCS, vol. 5252, pp. 245–284. Springer, Heidelberg (2008). doi:10.1007/978-3-540-88908-3_10

14. Larichev, O.: Cognitive validity in design of decision aiding techniques. J. Multi-Criteria Decis. Anal. **1**(3), 127–138 (1992)

15. Miettinen, K.: Nonlinear Multiobjective Optimization. Kluwer Academic Publishers, Boston (1999)

16. Miettinen, K.: Survey of methods to visualize alternatives in multiple criteria decision making problems. OR Spectr. **36**(1), 3–37 (2014)

17. Miettinen, K., Hakanen, J., Podkopaev, D.: Interactive nonlinear multiobjective optimization methods. In: Greco, S., Ehrgott, M., Figueira, J. (eds.) Multiple Criteria Decision Analysis: State of the Art Surveys. International Series in Operations Research & Management Science, vol. 233, 2nd edn, pp. 927–976. Springer, New York (2016). doi:10.1007/978-1-4939-3094-4_22

18. Miettinen, K., Ruiz, F., Wierzbicki, A.P.: Introduction to multiobjective optimization: interactive approaches. In: Branke, J., Deb, K., Miettinen, K., Słowiński, R. (eds.) Multiobjective Optimization. LNCS, vol. 5252, pp. 27–57. Springer, Heidelberg (2008). doi:10.1007/978-3-540-88908-3_2

19. Monz, M., Kufer, K.H., Bortfeld, T.R., Thieke, C.: Pareto navigation - algorithmic foundation of interactive multi-criteria IMRT planning. Phys. Med. Biol. **53**(4), 985–998 (2008)

20. Narula, S.C., Weistroffer, H.R.: A flexible method for nonlinear multicriteria decision-making problems. IEEE Trans. Syst. Man Cybern. **19**(4), 883–887 (1989)

21. Purshouse, R., Deb, K., Mansor, M., Mostaghim, S., Wang, R.: A review of hybrid evolutionary multiple criteria decision making methods. In: 2014 IEEE Congress on Evolutionary Computation (CEC), pp. 1147–1154. IEEE (2014)

22. Sindhya, K., Ojalehto, V., Savolainen, J.: Niemistö, H., Hakanen, J., Miettinen, K.: Coupling dynamic simulation and interactive multiobjective optimization for complex problems: an APROS-NIMBUS case study. Expert Syst. Appl. **41**, 2546–2558 (2014)

23. Steuer, R.E., Choo, E.U.: An interactive weighted Tchebycheff procedure for multiple objective programming. Math. Program. **26**(3), 326–344 (1983)

24. Tabatabaei, M., Hakanen, J., Hartikainen, M., Miettinen, K., Sindhya, K.: A survey on handling computationally expensive multiobjective optimization problems using surrogates: non-nature inspired methods. Struct. Multi. Optim. **52**(1), 1–25 (2015)
25. Wierzbicki, A.P.: Reference point approaches. In: Gal, T., Stewart, T.J., Hanne, T. (eds.) Multicriteria Decision Making: Advances in MCDM Models, Algorithms, Theory, and Applications, pp. 9-1–9-39. Kluwer Academic Publishers, Boston (1999)
26. Zhou, A., Qu, B.Y., Li, H., Zhao, S.Z., Suganthan, P.N., Zhang, Q.: Multiobjective evolutionary algorithms: a survey of the state of the art. Swarm Evol. Comput. **1**(1), 32–49 (2011)

First Investigations on Noisy Model-Based Multi-objective Optimization

Daniel Horn[1]([✉]), Melanie Dagge[2], Xudong Sun[3], and Bernd Bischl[3]

[1] TU Dortmund University, Computational Statistics, 44227 Dortmund, Germany
daniel.horn@tu-dortmund.de
[2] TU Dortmund University,
Statistics with Applications in the Field of Engineering Sciences,
44227 Dortmund, Germany
melanie.dagge@tu-dortmund.de
[3] LMU Munich, Computational Statistics, 80539 Munich, Germany
{xudong.sun,bernd.bischl}@stat.uni-muenchen.de

Abstract. In many real-world applications concerning multi-objective optimization, the true objective functions are not observable. Instead, only noisy observations are available. In recent years, the interest in the effect of such noise in evolutionary multi-objective optimization (EMO) has increased and many specialized algorithms have been proposed. However, evolutionary algorithms are not suitable if the evaluation of the objectives is expensive and only a small budget is available. One popular solution is to use model-based multi-objective optimization (MBMO) techniques. In this paper, we present a first investigation on noisy MBMO. For this purpose we collect several noise handling strategies from the field of EMO and adapt them for MBMO algorithms. We compare the performance of those strategies in two benchmark situations: Firstly, we perform a purely artificial benchmark using homogeneous Gaussian noise. Secondly, we choose a setting from the field of machine learning, where the structure of the underlying noise is unknown.

Keywords: Noisy optimization · Machine learning · Bayesian optimization · Model-based optimization · Multi-objective optimization

1 Introduction

In many practical optimization scenarios it is necessary to consider multiple contradicting objectives. In such multi-objective optimization (MOO) problems several solutions with optimal trade-offs across the objectives are available. Most MOO algorithms assume that repeated evaluations of the same solution will yield the same objective values. However, in many applications this assumption is violated due to the presence of noise.

The field of machine learning offers a lot of different settings in which multiple performance measures can be considered: runtime vs. predictive power, model

© Springer International Publishing AG 2017
H. Trautmann et al. (Eds.): EMO 2017, LNCS 10173, pp. 298–313, 2017.
DOI: 10.1007/978-3-319-54157-0_21

sparsity vs. predictive power or multiple measures from ROC analysis [8]. ROC measures are, in the case of binary classification, derived from the so-called confusion matrix, which tabulates the classes of the prediction versus the classes of the true outcome. Examples are the true-positive and false-positive rate, whose trade-off is usually interesting to consider in imbalanced problems or problems with unknown error cost structure. Machine learning models are usually evaluated by resampling techniques [2]. These split the dataset repeatedly into a training and a corresponding test set – the training sets are used to train the models, their performance is evaluated on the test sets. Naturally, the resulting performance depends on the splits used, which leads to noise in the optimization.

In recent years, many strategies have been proposed to solve noisy MOOs. Most of them are based on EMO algorithms [1,6,9,14,23]. However, evolutionary algorithms (EAs) suffer from some major disadvantages. One drawback is the large number of function evaluations required to achieve a good approximation of the optimal set, however, in many practical settings the budget is limited. For example, the training of a support vector machine (SVM) on a dataset containing about a million samples can take several hours [12]. Hence, it is often not practical to tune the hyperparameters of a SVM on such datasets using EMO techniques. We formerly proposed to use model-based multi-objective optimization (MBMO) to tackle such problems [11]. However, very little work has been published on MBMO algorithms solving noisy MOOs.

Knowles et al. [18] used ParEGO [17], a special MBMO variant to solve noisy MOOs. They were able to show that ParEGO is quite robust to the presence of noise and still achieves reasonable results. However, in settings with noisy observations ParEGO was outperformed by two specialized algorithms [10,28]. One approach of handling noise in MBMO algorithms is to use the mean value of replicated evaluations for each parameter setting [6]. In [19] this approach was compared with a re-interpolating approach.

The main contribution of this paper is a first look into noise handling using MBMO algorithms and the presentation of two naive and two advanced noise handling strategies. All of them can be added to any MBMO algorithm. In two benchmark studies we investigate their effectiveness, however, we are not yet able to give a precise rule which strategy should be used in which situation.

2   Model-Based Multi-objective Optimization

Multi-objective optimization refers to an optimization setting with a vector-valued objective function $\boldsymbol{f} = (f_1, f_2, \ldots, f_m) : \mathcal{X} = \mathcal{X}_1 \times \ldots \times \mathcal{X}_d \to \mathbb{R}^m$ and with the usual convention that each f_i is to be minimized. In general, multiple objectives are contradicting, and we are interested in finding the set of all optimal trade-offs. A solution $\boldsymbol{x} \in \mathcal{X}$ is said to dominate a solution $\boldsymbol{x}' \in \mathcal{X}$ if $\boldsymbol{x}$ is at least as good as $\boldsymbol{x}'$ in all objectives and strictly better in at least one objective. This dominance relation is sufficiently strong for a definition of optimality: a solution $\boldsymbol{x} \in \mathcal{X}$ is called Pareto optimal if and only if it is not dominated by any other solution $\boldsymbol{x}' \in \mathcal{X}$. The set of all non-dominated solutions is called the Pareto set, the set $\{\boldsymbol{f}(\boldsymbol{x}) \mid \boldsymbol{x} \in \mathcal{X} \text{ is Pareto optimal}\}$ is called the Pareto front.

Algorithm 1. General sequential model-based optimization procedure

---
Require: expensive blackbox function $\boldsymbol{f}(\boldsymbol{x}) : \mathcal{X} \to \mathbb{R}^m$, budget n
1: generate initial design $\mathcal{D} = \{\boldsymbol{x}^{(1)}, ..., \boldsymbol{x}^{(n_{init})}\} \subset \mathcal{X}$
2: evaluate $\mathcal{Y} = \boldsymbol{f}(\mathcal{D}) = \{\boldsymbol{f}(\boldsymbol{x}^{(1)}), ..., \boldsymbol{f}(\boldsymbol{x}^{(n_{init})})\}$
3: set $i := n_{init}$
4: **while** $i{+}{+} \leq n$ **do**
5:    fit surrogate model(s) $\hat{\boldsymbol{f}}$ on $\mathcal{Y}$ and $\mathcal{D}$
6:    $\boldsymbol{x}^{(i)} = \arg\max_{\boldsymbol{x}\in\mathcal{X}} \mathrm{Inf}(\boldsymbol{x})$
7:    evaluate $\boldsymbol{y}^{(i)} = \boldsymbol{f}(\boldsymbol{x}^{(i)})$
8:    update $\mathcal{D} \leftarrow \mathcal{D} \cup \{\boldsymbol{x}^{(i)}\}$ and $\mathcal{Y} \leftarrow \mathcal{Y} \cup \{\boldsymbol{y}^{(i)}\}$
9: **end while**
10: **return** Pareto set and front of $\mathcal{D}$ and $\mathcal{Y}$

Many different MBMO algorithms have been proposed within the last years. Most of them are based on ideas of the Efficient Global Optimization (EGO) procedure [15]. At first EGO samples and evaluates an initial design of size n_{init}, typically using Latin Hypercube Sampling. Then a surrogate regression model (typically a Kriging model) is fitted to all design points obtained so far and optimized with regard to a so-called infill criterion (Inf). A new candidate point is proposed and evaluated with the objective function. The procedure iterates until a stopping criterion is reached, which is usually a budget on the number of function evaluations or global runtime.

The general procedure of MBMO algorithms is described in Algorithm 1. Although most steps of the MBMO procedure are close to the EGO algorithm, there are some important differences. Mainly, the model fitting procedure has to be adapted to the multi-objective context. Most MBMO algorithms are adjusted by fitting distinct surrogate models to each objective function. As in EGO, new points are proposed by optimizing an infill criterion. However, this step cannot be adopted directly from EGO, since there might be multiple surrogates.

There is a variety of MBMO algorithms, which are all based on this general procedure, but they differ in the number of fitted models, the infill criterion, the optimization strategy and many more details. An overview of many published MBMO algorithms including their taxonomy is presented in [13].

Although our work on noisy MBMO is compatible with any MBMO algorithm, we decided to focus on one algorithm for simplicity. The SMS-EGO [20] algorithm with its good performance in previous benchmark studies (see [11,13]) appears to be a good choice for this. It works by fitting individual Kriging models to each objective and choosing new design points by maximizing their hypervolume contribution to the current Pareto front approximation.

3   Noisy Multi-objective Optimization

In noisy MOO the true objective functions $\boldsymbol{f}$ cannot be observed. Instead, we have observations $\boldsymbol{y}(\boldsymbol{x}) = \boldsymbol{f}(\boldsymbol{x}) + \boldsymbol{\epsilon}(\boldsymbol{x})$ that are affected by observational noise $\boldsymbol{\epsilon}(\boldsymbol{x}) = \{\epsilon_1(\boldsymbol{x}), ..., \epsilon_m(\boldsymbol{x})\}$. In the general blackbox case we cannot make any assumptions about the characteristics of ϵ. Since this case is hard to handle, most noise handling algorithms do make further assumptions on ϵ. Typical assumptions are: ϵ is stochastically independent of $\boldsymbol{x}$, ϵ is stochastically independent of the point in time of the evaluation, ϵ_j is stochastically independent and identically distributed for $j = 1, ..., m$ or that the expected value of ϵ is $\boldsymbol{0}$ [16].

Noise introduces two main challenges that should be addressed by the specialized optimizer:

(I) The noisy observations provide misleading information. This can induce insufficient results that are not close enough to the true optimal Pareto set.

(II) Noise may lead to overoptimistic assessment of obtained objectives. Thus, the returned Pareto front is very likely to overrate the true Pareto front.

The main idea for handling noise is to perform k (additional) evaluations for each $\boldsymbol{x}$-setting, also called re-evaluations. This reduces the noise's standard deviation by a factor of $\sqrt{k+1}$ [14]. Assuming $E(\epsilon) = \boldsymbol{0}$, the mean value of all re-evaluations of $\boldsymbol{x}$ is an unbiased estimator for $\boldsymbol{f}(\boldsymbol{x})$. On the one hand its quality can be improved by increasing k, on the other hand the optimizer has to spend a part of its budget for each re-evaluation. In particular, in settings with restricted budgets, the number of re-evaluations should be kept as small as possible.

In recent years, many different EMO strategies have been proposed to solve noisy MOOs. Most of them describe how to choose $\boldsymbol{x}$-settings for re-evaluations. In this chapter we will give a brief introduction to some naive and advanced approaches. For an overview please refer to Zhou et al. [27].

3.1   Naive Variants

The following two naive strategies for choosing the re-evaluations serve as a basis for the advanced methods. Due to their simplicity both strategies can be added to most optimizer, in particular to EAs, MBMO and even random search.

The Enlarged Strategy. The simplest idea is to ignore the fact that $\boldsymbol{y}$ is affected by noise and perform optimization as if $\boldsymbol{y}$ was deterministic (i.e. assume that $\boldsymbol{y} = \boldsymbol{f}$). Hence, every setting $\boldsymbol{x}$ is evaluated once, after it was proposed by the algorithm ($k = 0$). Since no budget needs to be spend on re-evaluations, a higher number of different $\boldsymbol{x}$ settings can be evaluated (the space of explored, distinct points is *enlarged*). Since there is no special strategy for handling noise, enlarged algorithms neither address (I) nor (II). (I) may be solved by the brute force of the larger budget, but the returned Pareto front of enlarged algorithms is very likely to be too optimistic.

The Repeated Strategy. The second naive strategy is to perform a constant number of k re-evaluations for every $\boldsymbol{x}$-setting directly (the evaluations are *repeated*). Then the mean value of all $k + 1$ evaluations is used as the new target value for optimization and optimization is performed as if deterministic. This intuitive strategy reduces the strength of noise for each $\boldsymbol{x}$ by a factor of $\sqrt{k+1}$ and has been applied in some practical scenarios (see, e.g., [6]). It is also highly related to resampling strategies from machine learning applications. Performing a k-fold cross-validation is similar to performing k evaluations of the same x-setting. By applying a repeated strategy, the effects of (I) and (II) can be reduced.

3.2 Advanced Strategies

It seems not that meaningful to evaluate a weak noisy, inferior point as often as a very noisy, promising setting. There are several approaches to adapt the number of re-evaluations during optimization. One idea is to increase the number of re-evaluations while optimization proceeds [1]. Other options are to choose the number of re-evaluations according to the observed standard deviation [1] or to use a statistical testing procedure for comparing two solutions and to perform re-evaluations until their difference becomes significant to a given level [23].

In this paper we focus on two different advanced strategies. Firstly, the Rolling Tide EA (RTEA) proposed by Fieldsend [9], which gives a heuristic on how to choose promising settings for re-evaluations. Secondly, we propose a new strategy that combines the ideas of the two described naive approaches.

The Rolling Tide Strategy. The main idea of the RTEA is described in Algorithm 2. It handles noise by re-evaluating only promising settings, while inferior settings are evaluated only once. In each iteration, after a new point has been proposed using evolutionary operators, k already evaluated points are chosen for re-evaluation. The selection of these k points is based on the dominance relation and the number of prior re-evaluations: Sequentially, those Pareto optimal points with the least number of re-evaluations are chosen. New points are only proposed in the first part of the optimization, the second part is used to solely refine the archive. The actual RTEA contains more details on how to perform the evolutionary operations and how to efficiently update the Pareto set. We omit these details since they are not related to noise handling.

Algorithm 2. The Rolling Tide Evolutionary Algorithm (RTEA)

Require: noisy blackbox function $y(x) : \mathcal{X} \to \mathbb{R}^m$, number of re-evaluations k, proportion of evaluations z to solely refine the archive, budget n
1: Generate and evaluate initial design of size n_{init}
2: Set $i := n_{init}$
3: **while** i++ $< n$ **do**
4:     **if** $i < (1 - z) \cdot n$ **then**
5:         Propose a new $x^{(i)}$ using evolutionary operators
6:         Calculate $y(x^{(i)})$ and update archive and Pareto set $\mathcal{A}$
7:     **end if**
8:     **for** k iterations **do**
9:         choose $x^{(j)} \in \mathcal{A}$ with the lowest amount of re-evaluations
10:        Re-evaluate $y(x^{(j)})$ and update both the archive and $\mathcal{A}$
11:    **end for**
12: **end while**
13: **return** Pareto set $\mathcal{A}$ and corresponding Pareto front

---

The Reinforced Strategy. This strategy combines the ideas of the enlarged and the repeated algorithm aiming to eliminate the disadvantages of these naive strategies. Our first results showed that the main drawback of the enlarged strategy are overly optimistic Pareto fronts, while for the repeated strategy it is the high number of unnecessary re-evaluations. Therefore, we propose to start the optimization using the enlarged strategy and only one evaluation per x. In the end k re-evaluations are done for each Pareto optimal point (the final Pareto front is *reinforced*). Thus, the reinforced strategy uses a large budget just like the enlarged strategy, but also returns a reliable estimation of the Pareto front.

3.3   Noise Handling in MBMO Algorithms

To the best of our knowledge, very little work has been published on noise handling in MBMO algorithms up to today. The general capability of MBMO algorithms to solve noisy MOOs was shown in [18]. There are also investigations on solving noisy applications by using the repeated strategy [6,19] and some specialized algorithms have been introduced [10,28]. However, no work on adapting advanced strategies for choosing re-evaluations has been published yet. Since all described strategies solving noisy MOOs can be used within the MBMO framework, we will employ these approaches as a basis for further investigations.

As representative for advanced noise handling strategies we chose RTEA and adapted it toward MBMO by exchanging the evolutionary operations from the RTEA by their MBMO equivalents. This is implemented by inserting lines 8 to 11 of Algorithm 2 into Algorithm 1 after line 9. Moreover, in the RTEA each point is evaluated directly after it was chosen for re-evaluation. We suggest to collect the new proposed point and all k points for re-evaluation in a batch in order to evaluate them in parallel. This procedure comes with the disadvantage that the Pareto front cannot be updated in between the re-evaluations.

Almost all MBMO algorithms use Kriging models as surrogates. But in terms of noisy optimization the interpolating nature of Kriging is problematic: For each $x^{(i)} \in \mathcal{D}$, the model will return the corresponding $y^{(i)} \in \mathcal{Y}$. Consequently, Kriging models cannot be fitted to repeated stochastic evaluations.

We are aware of two solutions for this issue: Firstly, the model can be fitted on the mean values of the repeated evaluations. Thus, information about the strength of the noise is not used. Secondly, by adding a so-called nugget effect to the Kriging model the model loses its interpolating nature. It becomes an approximating model and can be fitted to repeated evaluations. This is achieved by introducing parameter $\lambda \in [0, \infty)$, which controls the strength of smoothing, where $\lambda = 0$ corresponds to use no smoothing, i.e., use interpolating Kriging. Here we focus on the first solution planning to investigate the second one in subsequent studies.

The final Kriging models can also be used to improve the optimization result. Instead of returning the Pareto front of $\mathcal{Y}$ (the *tune front*) the front of the models (the *model front*) can be used. For this the Pareto set of $\mathcal{D}$ is calculated and their outcomes are predicted by using the final models. Naturally, the tune front and the model front are identical, if interpolating Kriging is used. However, if a small nugget effect is added, the model may smooth the affected tune front. Especially for the enlarged strategy we expect the model front to be more realistic.

4   Artificial Experiment

To test and compare the proposed naive and advanced noise handling strategies, we conduct two benchmarks. In this section, we use artificial test functions contaminated with homogeneous Gaussian noise.

4.1   Experimental Setup

For comparison of our four proposed approaches, the noise handling strategies are fused with the SMS-EGO algorithm. Additionally, these algorithms are compared against a baseline comprised of the original RTEA and the two naive strategies combined with a simple random search strategy (rs). This results in a total of seven different algorithms. They are abbreviated using the pattern class_noisehandling, e.g., mbmo_rt refers to the SMS-EGO algorithm using the rolling tide (rt) noise handling strategy. As test functions we use UF1-UF10 from the CEC 09 challenge [26]. For noise we add normal distributed random vectors with expected value $\mathbf{0}$ and covariance matrix $\sigma \cdot \mathbf{I}_m$, $\sigma \in \{0, 0.01, 0.1, 1\}$. This setting mostly matches with an experimental setup proposed by Fieldsend [9].

To simulate an expensive setting, we limit the budget to $40d$ iterations, of which $8d$ are reserved for the initial design and $32d$ for the sequential optimization. The noise handling strategy is allowed to perform k additional evaluations in each sequential iteration, resulting in a total budget of $40d + 32d\,k$ evaluations. Table 1 shows how the different strategies utilize their budget. Since SMS-EGO is an expensive optimization algorithm (m Kriging models have to be fitted in each iteration), we decide to investigate a scenario with $d = 5$ and $k = 4$.

Additional parameters of the SMS-EGO are set as in previous benchmarks [11]. As surrogate we use a Kriging model and add a small nugget effect ($\lambda = 10^{-8}$) for numerical stability. For the RTEA we apply simulated binary crossover ($n = 10, p = 0.7$) and polynomial mutation ($n = 25, p = 0.3$).

The analysis is based on the so-called *test front*, which is generated by evaluating the final Pareto set on the true, deterministic function. Performance

Table 1. Partition of the budget for the individual noise handling strategies. The total budget is $8d + 32d(k + 1)$ for all strategies. The number of iterations for the reinforced variant changes with the size of the final Pareto front.

Algorithm	Token	n_{init}	Evaluations per iteration	Number of iterations
enlarged	enl	$8d$	1	$32d(k + 1)$
repeated	rep	$8d(k + 1)$	$k + 1$	$32d - \lfloor \frac{k}{k+1} 8d \rfloor$
rolling tide	rt	$8d$	$k + 1$	$32d$
reinforced	reinf	$8d$	1	-

of the different approaches is evaluated and compared by two different measures: Firstly, goodness of optimization (see (I)) is measured by the dominated
hypervolume based on the test fronts. For statistical analysis we perform an
ANOVA and post-hoc pairwise one-sided paired t-tests on the normalized hypervolume values [7]. Secondly, to judge the approximation quality of the tune front
(see (II)), we calculate the so-called Average Hausdorff distance [22] (parameter
$p = 1$) between the tune front and the test front.

The experiments are implemented using our own software packages written in
R, including mlrMBO [5], mlr [3, 21] and BatchExperiments [4]. Our experiments
are executed on the LiDOngrL cluster of the TU Dortmund university.

4.2   Results

The resulting hypervolume values for all algorithms, test functions and different
noise strengths are displayed in Fig. 1. First of all, we can state that all SMS-
EGO variants beat the three baselines in nearly all cases.

In the deterministic case ($\sigma = 0$) the results support our expectation. Since
re-evaluating points does not give additional information, the enlarged and reinforced strategies perform best. Surprisingly, this result does also hold for the

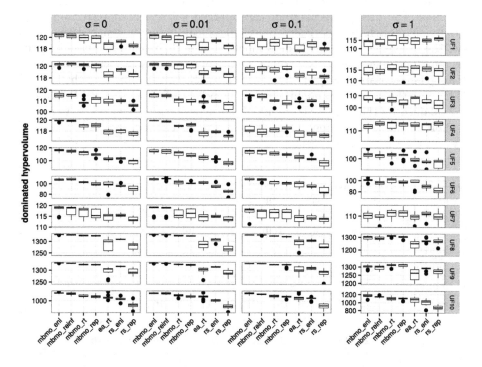

Fig. 1. Dominated hypervolume of the seven algorithms on the ten test functions.
Reference point is (11, 11) for UF1–UF7 and (11, 11, 11) for UF8–UF10.

moderate noise cases $\sigma \in \{0.01, 0.1\}$. The hypervolume values of the individual algorithms are only slightly smaller than the values in the deterministic case, while the ranking of the algorithms stays the same. It seems that ignoring the noise is the best way to handle it. This general picture changes in the presence of strong noise ($\sigma = 1$). Performance of all algorithms deteriorates and there are no clear differences in hypervolume between the algorithms on most of the test functions. On some test functions the performance of SMS-EGO is even equivalent to the random search baselines. We speculate that the combination of the large noise and the low budget is too complicated to be solved much better than using simple random search. In fact, all Pareto fronts of the UF test functions range from $(0, 1)$ to $(1, 0)$, but the noise can range within $[-3, 3]$ [1]. Hence, the noise is even stronger than the actual trade-offs between the objectives.

The results of the ANOVA are presented in Table 2. All parameters have a highly significant influence, though the low p-values are intensified by the rather large number of 2 800 observations. Table 3 shows the results of the post-hoc t-tests, which mainly confirm the impression we got from the figures. The tests give a clear ranking of the algorithms and some groups of strategies can be identified.

Table 2. Results of the ANOVA for the artificial experiment.

Variable	sum Sq	p-value
algorithm	1.61	$<2e-16$
function	10.96	$<2e-16$
σ	0.71	$<2e-16$
replication	0.03	0.0081

The best group contains the enlarged and the reinforced SMS-EGO variants. Although the enlarged variant is significantly better than the reinforced one, the difference between them is much smaller than the differences towards the remaining algorithms. The second best group includes the repeated and the Rolling Tide variant of SMS-EGO, followed by the third group with the original RTEA and the enlarged random search. The poor performance of the RTEA can be explained by the expensive setting of our experiment. Naturally, EAs need a lot more function evaluations to reach good results. The repeated random search is worse than every other variant.

Table 3. Post-hoc pairwise one-sided paired t-tests for the alternative that the method in the row reached a lower hypervolume than the one in the column.

	mbmo_enl	mbmo_reinf	mbmo_rt	mbmo_rep	ea_rt	rs_enl
mbmo_reinf	5.1e−05	-	-	-	-	-
mbmo_rt	$<2e-16$	1.4e−14	-	-	-	-
mbmo_rep	$<2e-16$	$<2e-16$	2.8e−05	-	-	-
ea_rt	$<2e-16$	$<2e-16$	$<2e-16$	$<2e-16$	-	-
rs_enl	$<2e-16$	$<2e-16$	$<2e-16$	$<2e-16$	0.0086	-
rs_rep	$<2e-16$	$<2e-16$	$<2e-16$	$<2e-16$	$<2e-16$	$<2e-16$

---

[1] As per the 3σ rule, 99.7% of normal distributed random numbers are within $\pm 3\sigma$.

Table 4. Average runtimes of the different algorithms.

Algorithm	rs_rep	rs_enl	ea_rt	mbmo_rep	mbmo_rt	mbmo_reinf	mbmo_enl
runtime [sec]	1	3	7	699	8 721	37 327	53 965

Table 4 displays the average runtimes of the seven different algorithms across all test functions. It shows that fitting the Kriging model to a larger number of observations variants results in a notable increase in modeling overhead. However, this can mostly be neglected in the expensive setting.

4.3 Using the Nugget Effect

Although the enlarged strategy reaches the best hypervolume values, it cannot overcome its main disadvantage. Figure 2 shows the Average Hausdorff distances between the tune and the test front for the enlarged and the reinforced variant. Since in the case of $\sigma = 0$ both the tune and the test evaluations return the same values, the distances are 0. For all higher σ–values we observe that the enlarged

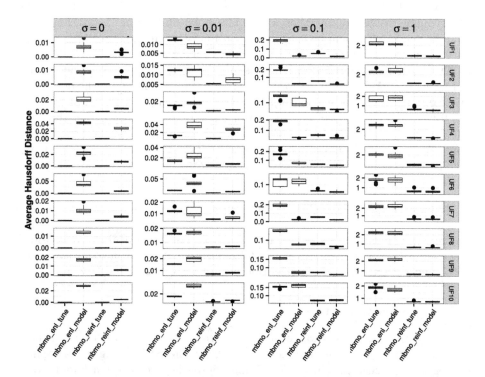

Fig. 2. Average Hausdorff distances for the enlarged and the reinforced SMS-EGO methods. Distances are calculated for the tune fronts with $\lambda = 10^{-8}$ and for the model fronts with $\lambda = 0.01$.

Table 5. Results of the ANOVA for the artificial experiment with added nugget effects (left table). Post-hoc pairwise one-sided paired t-tests for the alternative that the λ in the row reached a higher hypervolume than the one in the column (right table).

variable	sum Sq	p-value
algorithm	0.39	< 2e-16
λ	0.09	< 2e-16
function	11.33	< 2e-16
σ	1.30	< 2e-16
replication	0.07	< 2e-16

λ	1e-08	0.01	0.1
0.01	0.21	-	-
0.1	5.6e-07	7.4e-06	-
1	< 2e-16	< 2e-16	< 2e-16

strategy has high distances and returns overly optimistic Pareto fronts. On the contrary, the reinforced strategy reaches essentially lower distances.

Besides the tune fronts, we can also look at the performance of model fronts. Therefore, we enable the fixed nugget effect in the Kriging model. Subsequently, we rerun the experiments adding different nugget effects $\lambda \in \{0.01, 0.1, 1\}$ to the Kriging models. The results of the corresponding ANOVA are reported in Table 5. The nugget effect has a significant influence on the hypervolume values. The post-hoc tests for the nugget effect show that performance deteriorates while using higher nugget effects. Only the smallest effect ($\lambda = 0.01$) reaches hypervolume values equal to those of the interpolating Kriging model. Thus, the smallest nugget effect can be used during optimization and for calculation of the final model front without deteriorating the hypervolume value.

The boxplots in Fig. 2 additionally show the distances for the model fronts. Naturally, the model fronts perform worse than the tune fronts in the deterministic case ($\sigma = 0$). This relationship flips with increasing noise, until in the case $\sigma = 0.1$ the model front is clearly superior for most test functions. For the reinforced strategy the usage of the model front does not improve the final Pareto front. Furthermore, in most cases the tune fronts of the reinforced strategy are better than the model fronts of the enlarged strategies.

5   Machine Learning Experiment

The second benchmark is a practical setting from machine learning. We consider the bi-objective minimization of the false-negative and the false-positive rate (FNR and FPR) in binary classification using SVMs with radial kernels.

5.1   Experimental Setup

In this setting each function evaluation corresponds to estimating the performance of a given hyperparameter setting for the SVM using a resampling strategy. While classical parameter tuning approaches would use a k-fold cross-validation for each setting, here a single 10% holdout is performed to reflect the noisy optimization. If needed, the noise handling strategy can perform multiple re-evaluations of a parameter setting, i.e., it can perform multiple holdout iterations. This is not equivalent to the usual cross-validation, but especially for the repeated strategies it is highly related.

As a general strategy for balancing FNR and FPR, class weighting can be used. Without loss of generality it is sufficient to adapt the weight ω for the positive class. We optimize the cost parameter C and the inverse kernel each within a region-of-interest of $2^{[-15,15]}$, as well as the class weighting parameter ω within the interval $2^{[-7,7]}$, resulting in an input dimension $d = 3$. All parameters are optimized on a logarithmic scale. We run the experiments on several data sets (see Table 6), all of them are available on the machine learning platform OpenML [24]. For an unbiased comparison of the final Pareto fronts we perform tuning on 50%
of the data points and leave the remaining 50% for calculation of the final test front. The same algorithms and further settings as described in Sect. 4 are used.

Table 6. Description of the used data sets. #obs is the number of observations and #feats is the number of features in the respective data sets.

Name	#obs	#feats
ada_agnostic	4562	48
eeg-eye-state	14980	14
kdd_JapaneseVowels	9961	14
pendigits	10992	16
phoneme	5404	5
spambase	4601	57
wind	6574	14
waveform	5000	40

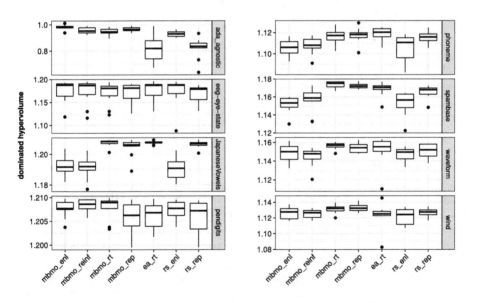

Fig. 3. Dominated hypervolume for the second experiment (reference point (1.1, 1.1)).

5.2 Results

In Fig. 3 the hypervolume values of the resulting test fronts are displayed. In contrast to the artificial experiment, here Rolling Tide and repeated variant of SMS-EGO perform best. Both, the enlarged and reinforced variant reach lower values and perform even worse than the RTEA and the repeated random search. The same effect can be seen comparing the baselines:

The enlarged random search is worse than the repeated one. Both the algorithm and the data set have a significant influence on the performance (Table 7). As in the artificial experiment, we can give a rather clear ranking of the algorithms (Table 8), although the differences are not as explicit as before. However, the ranking is rather different: the best algorithms of the artificial experiment are now on the second and third last ranks.

Table 7. Results of the ANOVA for the artificial experiment.

Variable	sum Sq	p-value
algorithm	2.703e−4	<2e−16
data set	4.088e−4	<2e−16
replication	0.074e−4	0.266

Table 8. Post-hoc pairwise one-sided paired t-tests for the alternative that the method in the row reached a lower hypervolume than the one in the column.

	mbmo_rt	mbmo_rep	ea_rt	rs_rep	mbmo_reinf	mbmo_enl
mbmo_rep	0.00211	-	-	-	-	-
ea_rt	0.00053	0.15213	-	-	-	-
rs_rep	1.5e−07	0.01358	0.07858	-	-	-
mbmo_reinf	1.3e−15	3.2e−14	3.8e−11	1.2e−10	-	-
mbmo_enl	<2e−16	6.5e−16	9.6e−14	3.0e−13	0.01164	-
rs_enl	9.2e−16	6.9e−15	1.7e−13	3.3e−12	0.00087	0.28380

6 Conclusion

In this paper, we took a first look at noise handling within MBMO algorithms. We proposed several variants and evaluated their performance with two experiments. In the first one, we used an artificial test set with homogeneous Gaussian noise. In the second one, from a machine learning context, the noise is unknown and likely does not follow any simple distribution, but is heterogeneous in $\mathcal{X}$.

The results of our two experiments are contradictory in terms of the best variant. In the first experiment, the enlarged and reinforced variant of SMS-EGO performed best. Thus, performing more than one evaluation per x-setting is not necessary to guide the optimization. For a reliable estimation of the Pareto front one should either add a small nugget effect to the underlying Kriging model and use its estimated front, or invest some final evaluations to reinforce the final front. In the second experiment, all strategies using only a single evaluation for each setting performed poorly and even worse than a repeated random search. In this setting, the proposed Rolling Tide MBMO variant reached the best performance.

An intuitive explanation for the different behavior of our methods in the two experiments can be given by looking at the noise effect in the underlying Kriging models. The noise in the artificial experiment is homogeneous. In the Kriging model, the correlation between observations is modeled in the covariance matrix. Thus, the noise effect on individual points can be reduced by evaluating multiple blurry points together due to the smoothing effect of the model. Hence, exploring more points using the enlarged variant gives more information about the global structure of the response. On the contrary, the noise in the machine learning experiment is heterogeneous. Simply exploring more points in the $\mathcal{X}$ space cannot reduce the noise efficiently and fails to gain more information compared to a re-evaluation of the same point for a more reliable estimation of the observation.

In our future, work we will continue to investigate this topic with the aim to understand how the different strategies for choosing the re-evaluations are affected by different types of noise. Furthermore, we think that the smoothing effect of the surrogate model has a great potential in guiding a noisy optimizer. One possibility to improve this effect could be the integration of repeated evaluations into the model, instead of using mean values. But then the interpolating Kriging model cannot be used any longer. A reasonable alternative could be the so-called stochastic Kriging approach [25]. Instead of using a fixed λ, it could also be promising to tune λ during optimization. However, the results of this paper showed that low λ values always give the best results, independent of σ.

References

1. Aizawa, A.N., Wah, B.W.: Scheduling of genetic algorithms in a noisy environment. Evol. Comput. **2**(2), 97–122 (1994)
2. Bischl, B., Mersmann, O., Trautmann, H., Weihs, C.: Resampling methods for meta-model validation with recommendations for evolutionary computation. Evol. Comput. **20**(2), 249–275 (2012)
3. Bischl, B., Lang, M., Kotthoff, L., Schiffner, J., Richter, J., Studerus, E., Casalicchio, G., Jones, Z.M.: mlr: machine learning in R. J. Mach. Learn. Res. **17**(170), 1–5 (2016)
4. Bischl, B., Lang, M., Mersmann, O., Rahnenführer, J., Weihs, C.: BatchJobs and BatchExperiments: abstraction mechanisms for using R in batch environments. J. Stat. Softw. **64**(11), 1–25 (2015)
5. Bischl, B., Richter, J., Bossek, J., Horn, D., Lang, M.: mlrMBO: A Toolbox for Model-Based Optimization of Expensive Black-Box Functions. https://github. com/berndbischl/mlrMBO
6. Breiderhoff, B., Bartz-Beielstein, T., Naujoks, B., Zaefferer, M., Fischbach, A., Flasch, O., Friese, M., Mersmann, O., Stork, J.: Simulation and optimization of cyclone dust separators. In: Proceedings of the 23rd Workshop on Computational Intelligence, pp. 177–195 (2013)
7. Demšar, J.: Statistical comparisons of classifiers over multiple data sets. J. Mach. Learn. Res. **7**, 1–30 (2006)
8. Fawcett, T.: An introduction to ROC analysis. Pattern Recogn. Lett. **27**(8), 861–874 (2006)

9. Fieldsend, J.E., Everson, R.M.: The rolling tide evolutionary algorithm: a multi-objective optimizer for noisy optimization problems. IEEE Trans. Evol. Comput. **19**(1), 103–117 (2015)
10. Hernández-Lobato, D., Hernández-Lobato, J.M., Shah, A., Adams, R.P.: Predictive entropy search for multi-objective bayesian optimization. arXiv preprint arXiv:1511.05467 (2015)
11. Horn, D., Bischl, B.: Multi-objective parameter configuration of machine learning algorithms using model-based optimization. In: IEEE Symposium on Computational Intelligence in Multicriteria Decision-Making (2016, accepted)
12. Horn, D., Demircioğlu, A., Bischl, B., Glasmachers, T., Weihs, C.: A comparative study on large scale kernelized support vector machines. Adv. Data Anal. Classif., 1–17 (2016)
13. Horn, D., Wagner, T., Biermann, D., Weihs, C., Bischl, B.: Model-based multi-objective optimization: taxonomy, multi-point proposal, toolbox and benchmark. In: International Conference on Evolutionary Multi-criterion Optimization, pp. 64–78 (2015)
14. Jin, Y., Branke, J.: Evolutionary optimization in uncertain environments - a survey. IEEE Trans. Evol. Comput. **9**(3), 303–317 (2005)
15. Jones, D.R., Schonlau, M., Welch, W.J.: Efficient global optimization of expensive black-box functions. J. Global Optim. **13**(4), 455–492 (1998)
16. Kleijnen, J.P.C.: White noise assumptions revisited: regression metamodels and experimental designs in practice. In: Proceedings of the 38th Conference on Winter Simulation, pp. 107–117 (2006)
17. Knowles, J.: Parego: a hybrid algorithm with on-line landscape approximation for expensive multiobjective optimization problems. IEEE Trans. Evol. Comput. **10**(1), 50–66 (2006)
18. Knowles, J., Corne, D., Reynolds, A.: Noisy multiobjective optimization on a budget of 250 evaluations. In: International Conference on Evolutionary Multi-criterion Optimization, pp. 36–50 (2009)
19. Koch, P., Wagner, T., Emmerich, M.T., Bck, T., Konen, W.: Efficient multi-criteria optimization on noisy machine learning problems. Appl. Soft Comput. **29**, 357–370 (2015)
20. Ponweiser, W., Wagner, T., Biermann, D., Vincze, M.: Multiobjective optimization on a limited amount of evaluations using model-assisted $\mathcal{S}$-metric selection. In: Proceedings of the 10th International Conference on Parallel Problem Solving from Nature (PPSN), pp. 784–794 (2008)
21. Schiffner, J., Bischl, B., Lang, M., Richter, J., Jones, Z.M., Probst, P., Pfisterer, F., Gallo, M., Kirchhoff, D., Kühn, T., Thomas, J., Kotthoff, L.: mlr tutorial. CoRR abs/1609.06146 (2016)
22. Schutze, O., Esquivel, X., Lara, A., Coello, C.A.C.: Using the averaged hausdorff distance as a performance measure in evolutionary multiobjective optimization. IEEE Trans. Evol. Comput. **16**(4), 504–522 (2012)
23. Syberfeldt, A., Ng, A., John, R.I., Moore, P.: Evolutionary optimisation of noisy multi-objective problems using confidence-based dynamic resampling. Eur. J. Oper. Res. **204**(3), 533–544 (2010)
24. Vanschoren, J., van Rijn, J.N., Bischl, B., Torgo, L.: OpenML: networked science in machine learning. SIGKDD Explor. Newsl. **15**(2), 49–60 (2014)
25. Zhan, J., Ma, Y., Zhu, L.: Multiobjective simulation optimization using stochastic kriging. In: Proceedings of the 22nd International Conference on Industrial Engineering and Engineering Management 2015, pp. 81–91 (2016)

26. Zhang, Q., Zhou, A., Zhaoy, S., Suganthany, P.N., Liu, W., Tiwari, S.: Multiobjective optimization test instances for the cec 2009 special session and competition. University of Essex and Nanyang Technological University, Technical report (2008)
27. Zhou, A., Qu, B.Y., Li, H., Zhao, S.Z., Suganthan, P.N., Zhang, Q.: Multiobjective evolutionary algorithms: a survey of the state of the art. Swarm Evol. Comput. **1**, 32–49 (2011)
28. Zuluaga, M., Krause, A., Püschel, M.: e-pal: an active learning approach to the multi-objective optimization problem. J. Mach. Learn. Res. **17**(104), 1–32 (2016)

Fusion of Many-Objective Non-dominated Solutions Using Reference Points

Amin Ibrahim[1(✉)], Shahryar Rahnamayan[1], Miguel Vargas Martin[2], and Kalyanmoy Deb[3]

[1] Faculty of Electrical Computer and Software Engineering,
University of Ontario Institute of Technology, Oshawa, Canada
{amin.ibrahim,shahryar.rahnamayan}@uoit.ca
[2] Faculty of Business and Information Technology,
University of Ontario Institute of Technology, Oshawa, Canada
miguel.vargasmartin@uoit.ca
[3] Department of Electrical and Computer Engineering,
Michigan State University,
East Lansing, MI, USA
kdeb@egr.msu.edu

Abstract. With recent advancements of multi- or many-objective optimization algorithms, researchers and decision-makers are increasingly faced with the dilemma of choosing the best algorithm to solve their problems. In this paper, we propose a simple hybridization of population-based multi- or many-objective optimization algorithms called fusion of non-dominated fronts using reference points (FNFR) to gain combined benefits of several algorithms. FNFR combines solutions from multiple optimization algorithms during or after several runs and extracts well-distributed solutions from a large set of non-dominated solutions using predefined structured reference points or user-defined reference points. The proposed FNFR is applied to non-dominated solutions obtained by the Generalized Differential Evolution Generation 3 (GDE3), Speed-constrained Multi-objective Particle Swarm Optimization (SMPSO), and the Strength Pareto Evolutionary Algorithm 2 (SPEA2) on seven unconstrained many-objective test problems with three to ten objectives. Experimental results show FNFR is an effective way for combining and extracting (fusion) of well-distributed non-dominated solutions among a large set of solutions. In fact, the proposed method is a solution-level hybridization approach. FNFR showed promising results when selecting well-distributed solutions around a specific region of interest.

1 Introduction

For the last three decades, there has been a number multi- and many-objective algorithms capable of solving complex problems. However, despite recent advancements, researchers and decision-makers are increasingly faced with the difficulty of choosing an appropriate algorithm capable of solving their problem effectively. This is due to a well-established "no-free-lunch" theorem that an algorithm that may have proven to give good performance on a

© Springer International Publishing AG 2017
H. Trautmann et al. (Eds.): EMO 2017, LNCS 10173, pp. 314–328, 2017.
DOI: 10.1007/978-3-319-54157-0_22

particular class of problems may not provide the same level of performance on other classes of problems. As a result many researchers have shifted their focus in developing powerful problem-specific or instance-specific algorithms. One way to accomplish this task is the hybridizations of optimization algorithms where new algorithms are developed by combining one optimization algorithm with another, by combining a standard optimization algorithm with mathematical operators, or incorporating evolutionary operators (selection, mutation, and crossover) into non-evolutionary optimization algorithms [1]. The main hope is that hybridization combines the desirable properties of different approaches so that the hybrid algorithm exhibits improved exploration and exploitation capabilities.

For example, Mirjalili and Hashim [2] proposed a hybrid population-based algorithm called PSOGSA by combining Particle Swarm Optimization (PSO) and Gravitational Search Algorithm (GSA). Their main aim was to integrate the exploitation ability of PSO with the exploration ability of GSA to synthesize both algorithms' strength. Similarly El-hossini et al. [3] proposed three hybrid algorithms based on the PSO and the strength Pareto evolutionary algorithm 2 (SPEA2) to solve multi-objective optimization problems. In all of these algorithms, strength Pareto fitness assignment is used to maintain an external archive; and the three algorithms are developed by alternating the evolutionary and PSO processes in different order. Experimental results showed that the proposed hybrid PSO algorithms have comparable performance to SPEA2. Also, Tang and Wang [4] proposed a novel hybrid multi-objective evolutionary algorithm (HMOEA) for real-valued multi-objective problems by incorporating the concepts of personal best and global best in PSO and evolutionary operators (multiple crossovers) to improve the robustness of evolutionary algorithms to solve variant kinds of optimization problems.

Wang et al. [5] proposed a hybrid evolutionary algorithm based on different crossover and mutation strategies combined with adaptive constrained-handling technique to deal with numerical and engineering constrained optimization problems. Zăvoianu et al. [6] proposed a hybrid and adaptive co-evolutionary optimization method that can efficiently solve a wide range of multi-objective optimization problems. Their approach combines Pareto-based selection for survival, differential evolution's crossover and mutation operators, and decomposition-based strategies. Recently, an ensemble strategy was proposed to benefit from both the availability of diverse approaches and to overcome the difficulty of fine tuning associated parameters. Some of these work include ensemble of ε parameter values and an ensemble of external archives in a multi-objective PSO algorithm [7], ensemble of constraint handling methods to tackle constrained multi-objective optimization problems [8], and ensemble of different neighborhood sizes in multi-objective evolutionary algorithm based on decomposition (MOEA/D) with online self-adaptation [9].

Tan et al. [10] proposed a hybrid multi-objective evolutionary algorithm (HMOEA) featured with specialized genetic operators, variable-length representation and local search heuristic to find the Pareto optimal routing solutions for the truck and trailer vehicle routing problem (TTVRP). Experimental results showed the HMOEA is effective in solving a multi-objective and multi-modal combinatorial optimization problems. Similarly, Xia and Wu [11] proposed a hybrid multi-objective algorithm by combining the PSO algorithm for its explorative power and simulated annealing (SA) for its exploitations to solve flexible job-shop scheduling problem (FJSP). Tavakkoli-Moghaddam et al. [12] also proposed a hybrid multi-objective algorithm based on the features of a biological immune system (IS) and bacterial optimization (BO) to find Pareto optimal solutions for flow shop scheduling problem. Their algorithm uses the clonal selection principle in IS with highest affinity antibodies and criterion distinguishing between antigens and antibodies in BO for Pareto dominance relationship among solutions. Karthikeyan et al. [13] proposed a hybrid discrete firefly algorithm (HDFA) to solve the multi-objective FJSP problem. They have combined the discrete firefly algorithm with local search (LS) method to enhance the searching accuracy and information sharing among fireflies.

In the effort of developing a powerful general-purpose hybridization framework, propose a novel hybridization of population-based multi- or many-objective optimization algorithms called fusion of non-dominated fronts using reference points (FNFR), to gain combined benefit of several algorithms in solution-level. The FNFR method uses well-structured or user supplied reference points to extract targeted solutions from non-dominated solutions gathered during several runs of any multi-objective optimization algorithms (MOOA).

The rest of the paper is organized as follows. Section 2 outlines the FNFR framework in detail. Section 3 presents experimental studies conducted to verify the efficacy the proposed algorithm on 3- to 10-objective benchmark test problems. Concluding remarks are provided in Sect. 4.

2   Proposed Method: Fusion of Non-dominated Solutions Using Reference Points

Generally speaking, hybridization of optimization algorithms can be grouped into two main components. The first includes hybridization of algorithms during the optimization process and the second being the hybridization of the algorithms after the optimization process. Hybridization during the optimization process can also further be grouped into two main categories: algorithms created by combining multiple metaheuristics and algorithms created by combining a metaheuristic algorithm with multiple mathematical operators and/or evolutionary operators.

In this section, we present a novel hybridization technique called Fusion of Non-dominated Fronts using Reference points (FNFR) capable of extracting targeted solutions from a set of non-dominated results collected during several runs of multiple optimization algorithms. The skeleton for the FNFR hybridization procedure is shown as a flowchart in Fig. 1. The FNFR procedure consists of three modules:

Problem Formulation.   In this module, the problem to be optimized is formulated along with the determination of various applicable algorithms that maybe included in the optimization process.

Approximate Front Evaluation.   In this module, the selection process of applicable algorithms are narrowed to the most relevant set for solving the problem. At this point the number of iterations to run for each algorithm is determined by the user, with the foresight of generating enough solutions for a meaningful data set. Although any algorithm can be selected and combined together in this framework, we recommend

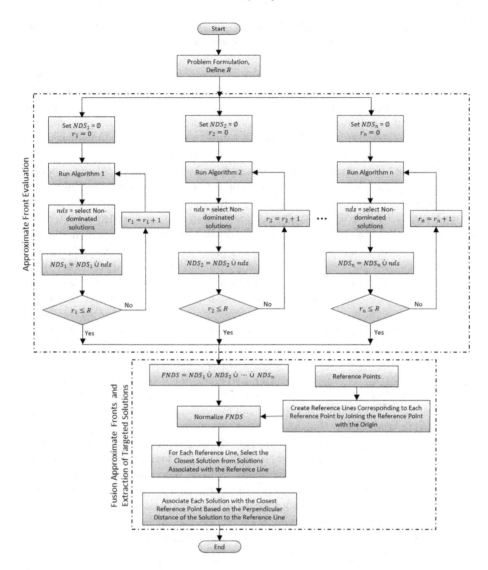

Fig. 1. Flowchart illustrating the FRFR procedure, where ∪̇ indicate the extraction of non-dominated solutions from the union of two non-dominated solution sets.

combining algorithms complimentary to each other, that is, algorithms addressing different constructs to achieve a good representative solutions set on the entire Pareto-optimal front (PF). Note that each algorithm is run independently with the option of running each algorithm in parallel or serial. In every run, the found approximate non-dominated front is combined with previously obtained front to create new non-dominated front by removing dominated solutions from the combined front. This step is a crucial step, because we need a well-represented set of solutions on the entire PF before extracting preferred solutions from this set.

Fusion and Extraction of Targeted Solutions. In this module, targeted solutions are extracted from the union of non-dominated solutions obtained from several runs of multiple algorithms using a procedure similar to reference-point-based non-dominated sorting algorithm (NSGA-III) [14]. Given a union of M-objective non-dominated approximate front obtained from the algorithms, the first step is to combine these results and remove dominated solutions ($FNDS = NDS_1 \dot{\cup} NDS_2 \dot{\cup} \cdots \dot{\cup} NDS_n$ where $\dot{\cup}$ indicate the extraction of non-dominated solutions from the union of two non-dominated solution sets). In the next step, a structured or target predefined set of reference points are selected on a normalized hyper-plane. In the case of structured reference points, one can use the Das and Dennis' [15] procedure to create well-distributed structured reference points. However, if the reference points are targeted points selected by the decision-maker, then M extreme points, one at each objective axis $\{z_1 = (1, 0, 0, \ldots 0), z_2 = (0, 1, 0, \ldots 0), \ldots z_M = (0, 0, \ldots, 1)\}$, are needed to determine the right normalized hyper-plane. Thereafter, we construct an ideal point by finding the minimum value for each objective function from the fusion of non-dominated solutions extracted in the previous step ($FNDS$). Then, we translate each solution in $FNDS$ by the ideal point and find M extreme points from these translated solutions. These extreme points are used to construct M-dimensional linear hyper-plane. Special care is required when finding extreme points so that we construct the proper hyper-plane and be able to extract targeted solutions. We recommend verifying obtained extreme points by finding the individual optimum of each objective function to ensure an appropriate hyper-plane is constructed before moving to the next step. Thereafter, each solution is associated with a reference point by using the shortest perpendicular distance (d) of each solution with a reference line created by joining the reference point with the origin. Finally, for each reference point, we select a solution with the minimum d among solutions associated to each reference line.

The main advantages of the FNFR framework are: (1) we get the full benefit of all algorithms involved in the optimization process, (2) we don't need to investigate how and when to combine merits of different algorithms, (3) many algorithms can be used without the need of extra parameter tuning, and (4) we can run all algorithms in the optimization process in parallel. However, the FNFR framework requires a great amount of time commitment to run all algorithms in the optimization process (i.e., higher time complexity). As result a trade-off occurs from the onset when selecting the number of algorithms to use for the FNFR framework.

3   Experimental Setup and Results

In this section, we describe the optimization algorithms, parameter settings, used test problems, and simulation results of the proposed framework on 3- to 10-objective benchmark test problems.

3.1   Algorithms

In order to assess the capability of the proposed FNFR hybridization framework, we have considered three MOOAs that have considerable differences in their fitness assignment and diversity mechanism to gain combined benefits and achieve a good representative solutions set on the entire PF. These MOOAs are: the Generalized Differential Evolution Generation 3 (GDE3), Speed-constrained Multi-objective Particle Swarm Optimization (SMPSO), and the Strength Pareto Evolutionary Algorithm 2 (SPEA2).

Generalized Differential Evolution Generation 3 (GDE3). GDE3 [16] is an extension of Differential Evolution (DE) for global optimization with an arbitrary number of objectives and constraints (Kukkonen & Lampinen, [16]). GDE3 with a single objective and without constraint is similar to the original DE. GDE3 improves earlier GDE versions in the case of multi-objective problems by giving better distributed solutions. GDE3 uses a growing population and non-dominated sorting with pruning of non-dominated solutions to decrease the population size at the end of each generation. This improves obtained diversity and makes the method more stable for the selection of control parameter values.

Speed-constrained Multi-objective Particle Swarm Optimization (SMPSO). SMPSO [17] is similar to Particle Swarm Optimization (PSO) algorithm which inspired by the social behavior of birds flocking to find food. In PSO, particles move in the search space in a cooperative manner where movements are performed by the velocity operator. The velocity operator is guided by a local and a social behaviour of swarm. SMPSO uses the concept of crowding distance to filter out leader solutions when the leaders archive is full, mutation operator accelerates the convergence of the swarm and it incorporates a mechanism to limit the velocity of the particles which can result an erratic movements of particles towards the upper and lower position limits of particles.

Strength Pareto Evolutionary Algorithm 2 (SPEA2). SPEA2 [18] is an extension SPEA for solving multi-objective optimization problems. Both SPEA and SPEA2 use an external archive to store previously found non-dominated solutions. In SPEA, the external archive maintained based on each individual's strength in which the strength of an individual is measured according to the number of solutions this individual dominates. In every generation the fitness of each member of the current population is computed according to the strengths (closeness to the true PF and distribution of solutions) of all external non-dominated solutions that dominate it. On the other hand, in SPEA2, the external archive is maintained according to each individual's strength not only by the number of individuals that dominate it but also the number of individuals by which it is dominated. Moreover, SPEA2 uses a nearest neighbor density estimation method to guide the search process efficiently and it preserves boundary solutions.

3.2   Test Problems

In order to test the quality of the proposed algorithm, we have used seven many-objective benchmark test problems. The first set of test problems are the DTLZ (DTLZ1 – DTLZ4,

Convex DTLZ2) introduced by Deb et al. [19]. The number of variables are $(M + k − 1)$, where M is the number of objectives and $k = 5$ for DTLZ1, while $k = 10$ for DTLZ2, Convex DTLZ2, DTLZ3, and DTLZ4. The corresponding PFs lie in $f_i \in [0, 0.5]$ for the DTLZ1 problem and in $f_i \in [0, 1]$ for other DTLZ problems. The DTLZ1 problem has a linear PF, Convex DTLZ2 has convex PF, and DTLZ2 to DTLZ4 problems have concave PFs.

The second set of test problems utilized in this study are WFG1 and WFG2 test problems introduced by Huband et al. [20]. The number of position parameters is set to $k = M − 1$, and the number of distance parameters is set to $l = 3$ for all dimensions. The WFG1 has a mixed PF and WFG2 problem has a convex, disconnected PF. The PFs for WFG test problems used in this work lie in $f_i \in [0, 2i]$.

3.3   Parameters Setting

Here, we describe the parameter setting used in the three sample algorithms used in the FNFR method. The GDE3 algorithm has two control parameters: mutation ($F = 0.5$) and crossover ($CR = 0.1$) probabilities. The SMPSO algorithm has three control parameters: external archive size (same as population size), Polynomial mutation ($p_m = 1/n$, where n is the number of variables) and Mutation Distribution Index ($\eta_m = 20$). The SPEA2 algorithm has five control parameters: external archive size (same as population size), SBX probability ($p_c = 0.9$), Polynomial mutation ($p_m = 1/n$, where n is the number of variables), Crossover Distribution Index ($\eta_c = 20$), and Mutation Distribution Index ($\eta_m = 20$). In order to maintain a consistent and fair comparison, the optimal parameter settings reported in [16–18] are used. Table 1 presents the number of reference points (H), the population size (N), and the number of inner and outer divisions used for different dimensions of test problems. In this study we have utilized the Das and Dennis' [15] procedure to create structured reference points used in the FNFR method.

Table 1. Number of reference points and population sizes used in this study.

Number of objectives (M)	Divisions		Reference points (H)	Population size (N)
	Outer	Inner		
3	12	0	91	92
5	6	0	210	212
8	3	2	156	156
10	3	2	275	276

To evaluate the performance of the proposed fusion technique, first we have run each algorithm 20 times independently and the best, the worst, and the median results of each algorithm are recorded. For the performance measure, we have used the inverse generational distance (IGD) metric, which is capable of measuring the convergence and the diversity of the obtained Pareto-optimal solutions. The IGD measure has been predominantly used to evaluate the performance of evolutionary many-objective problems [21, 22]. In this study, the reference Pareto front utilized in the IGD metric is constructed by joining all results from all runs, and then selecting non-dominated solutions from this set.

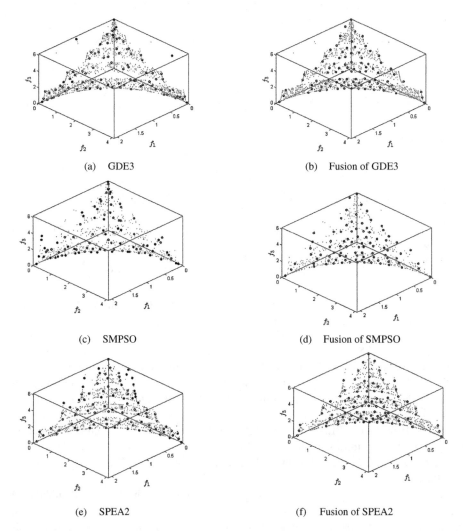

Fig. 2. The trade-off plots of obtained solutions by the GDE3, SMPSO and SPEA2 algorithms for three-objective WFG1 test problem. The grey dotted background indicated the non-dominated solution assembled during 20 runs of the GDE3 ((a) and (b)), SMPSO ((c) and (d)), and SPEA2 ((e) and (f)). The black dots in (a), (c), and (e) indicate the best solution set obtained by each algorithm based on the IGD metric. The black dots in (b), (d), and (f) show well-distributed solutions extracted from the non-dominated solutions assembled in 20 runs of each algorithm.

3.4   Experimental Results and Discussion

In this section, we describe experiments carried out to investigate the effectiveness of the FNFR method. Overall we have conducted three sets of experiments. The first experiment compares the quality of solutions obtained by each algorithm and the FNFR scheme using scatter plots (for three-objective problems) and 3D-RadVis [23] (for many-objective problems, $M > 3$). The second set of experiments involve numerical analysis to evaluate the

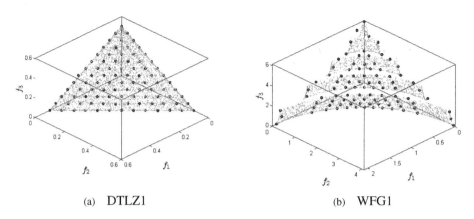

(a) DTLZ1   (b) WFG1

Fig. 3. The trade-off plots of solutions obtained by all algorithms for three-objective DTLZ1 and WFG1 test problems. The grey dotted background indicate non-dominated solutions assembled during 20 runs of the GDE3, SMPSO, and SPEA2 algorithms and the black dots illustrate well-distributed solutions extracted from these solutions using the proposed FNFR scheme.

quality of solutions obtained by each algorithm and the FNFR method. The last set of experiments involve extracting preferred solutions around a specific region.

Visual Analysis of Solutions obtained by each Algorithm against the FNFR Method.
Here we investigate solutions obtained by the three algorithms and solutions extracted by the FNRF method. We run each algorithm 20 times and assembled the non-dominated fronts from each run by removing dominated solution after combining non-dominated solutions from each run. Figure 2 show trade-off plots of solutions obtained by the GDE3, SMPSO, and SPEA2 for 3-objective WFG1 test problem. Figure 3 illustrates trade-off plots of solutions collectively obtained by the GDE3, SMPSO, and SPEA2 and solutions extracted by the FNFR procedure for 3-objective DTLZ1 and WFG1 and test problems.

The black dots in Fig. 2(a), (c), and (e) plots illustrate best solutions (based on the IGD metric) found by the GDE, SMPSO, and SPEA2 in 20 runs for 3-objective WFG1 test problem. From these plots we can see that none of the algorithms are able to find well-converged and well-distributed set of solutions in a single run. However, when we consider all non-dominated solutions collected during 20 runs (grey dots) by each algorithm, we see that they are able to find "satisfactory" solutions on the PF. This analysis indicate that the algorithms used in this experiment cannot consistently find well-distributed and well-converged solutions a single run. This phenomenon is in agreement with the no-free-lunch theorem.

The black dots in Fig. 2(b), (d), and (f) plots show well-distributed and well-converged solution extracted using the FNFR method from the non-dominated solutions gathered by each algorithm during the 20 runs of 3-objective WFG1 test problem. It can be seen that the FNFR method is an effective way of collecting and extracting well-distributed solutions after the optimization process is complete. We also observe that solutions extracted by the FNFR method are better than the best solution set found by any of the algorithms in a single run. Moreover, Fig. 3 shows trade-off plots of non-dominated solutions obtained by

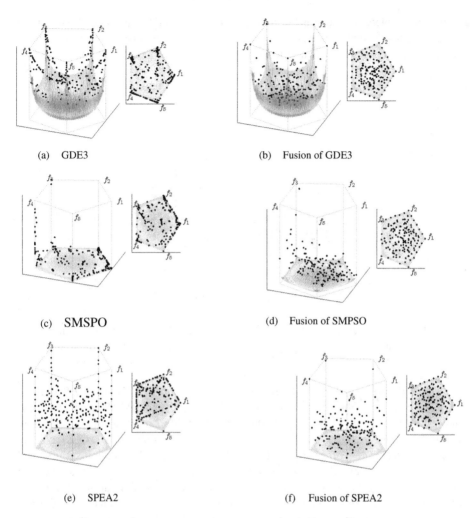

Fig. 4. 3D-RadVis plots of obtained solutions by GDE3, SMPSO and SPEA2 algorithms for five-objective convex DTLZ2 test problem. The surface for all plots are constructed from numerically generated PF. The black dots in (a), (c), and (e) indicate the best solution set obtained by each algorithm based on the IGD metric. The black dots in (b), (d), and (f) show well-distributed solutions extracted from the non-dominated solutions assembled in 20 runs of all algorithms using the FNFR method.

all algorithms for three-objective DTLZ1 and WFG1 test problems. The grey dotted background indicate the non-dominated solutions collected during 20 runs of GDE3, SMPSO, and SPEA2 algorithms and the black dots illustrate well-distributed solutions extracted from these solutions using the FNFR method.

The 3D-RadVis plots in Fig. 4 show the best non-dominated solutions obtained by GDE3, SMPSO, and SPEA2 algorithms (left plots) and solutions extracted using the FNFR method (right plots) for 5-objective convex DTLZ2 test problem. As it can be seen that as the number of objective increases the quality of solutions found by each algorithm start to degrade in terms of solution diversity and accuracy. However, in Figs. 3 and 5, we see that these algorithms collectively are able to find well-distributed solutions on the entire PF and we are able to extract uniformly distributed non-dominated solutions using the FNFR method among solutions gathered during 20 runs of multiple algorithms.

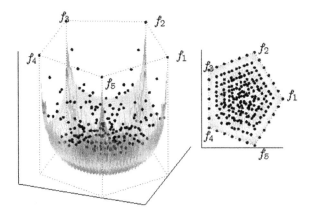

Fig. 5. 3D-RadVis plot of non-dominated solutions extracted using the proposed FNFR method. The black dots illustrate solutions extracted using the proposed scheme from a large set of non-dominated solutions generated by the GDE3, SMPSO, and SPEA2 algorithms in 20 runs.

Numerical Analysis of Solutions obtained by each Algorithm against the FNFR Method.
Here, we investigate the performance of GDE3, SMPSO, and SPEA2 on seven bench-mark test problems with 3-, 5-, 8-, and 10-objectives. Table 2 provides the best, median, worst IGD values for GDE3, SMPSO, and SPEA2 on 3- and 5-objective DTLZ1, convex DTLZ2 and WFG1 test problems in 20 runs. In all instances the solution set obtained through the FNFR method form each algorithm have better IGD values than the best IGD values attained by a single run. Moreover, the IGD value of the solution set extracted by the FNFR method from a large set of solutions collected during each algorithm's run is also better than the three above-mentioned IGD values. The grey shades in Table 2 indicate the IGD values of solutions extracted by the FNFR method. When it comes to higher dimensional test problems, the FNFR scheme further proved its efficacy when comparing the IGD values of solutions extracted by the FNFR method against the best IGD values obtained by each algorithm in a single run. Table 3 shows the best, median, and worst IGD values for GDE3 and SMPSO algorithms against the IGD value of a set solutions extracted by the FNFR method on 8- and 10-objective DTLZ3, convex DTLZ4, and WFG2 test problems.

Table 2. Best, median, and worst IGD values for GDE3, SMPSO, and SPEA2 algorithms against the IGD value of solutions extracted by the FNFR method on 3- and 5-objective DTLZ1, convex DTLZ2, and WFG1 test problems.

	DTLZ1		Convex DTLZ2		WFG1	
M	3	5	3	5	3	5
Max Generation	400	600	400	600	400	600
GDE3 Best	7.43E-04	1.23E-03	7.09E-04	5.96E-04	1.23E-03	8.24E-04
GDE3 Median	8.10E-04	1.27E-03	7.70E-04	6.23E-04	1.40E-03	9.23E-04
GDE3 Worst	8.43E-03	1.31E-03	8.48E-04	7.09E-04	2.36E-03	1.26E-03
GDE3 Fusion	5.68E-04	9.27E-04	6.94E-04	4.96E-04	1.96E-04	7.49E-04
SMPSO Best	9.26E-04	2.21E-03	8.92E-04	1.29E-03	7.45E-03	6.04E-03
SMPSO Median	9.83E-04	2.48E-03	1.00E-03	1.83E-03	7.94E-03	6.44E-03
SMPSO Worst	1.05E-03	2.97E-03	1.15E-03	2.71E-03	8.11E-03	6.74E-03
SMPSO Fusion	6.99E-04	1.75E-03	8.27E-04	1.05E-03	6.41E-03	5.02E-03
SPEA2 Best	6.16E-04	2.23E-01	5.79E-04	2.19E-02	8.13E-04	7.48E-04
SPEA2 Median	6.35E-04	6.38E-01	6.03E-04	3.16E-02	1.04E-03	1.15E-03
SPEA2 Worst	6.90E-04	1.03E+00	6.39E-04	4.02E-02	3.56E-03	1.44E-03
SPEA2 Fusion	5.63E-04	2.23E-01	5.39E-04	2.48E-02	6.05E-04	8.21E-04
Overall Fusion	**4.54E-04**	**8.92E-04**	**7.00E-04**	**4.21E-04**	**1.42E-04**	**7.64E-04**

Table 3. Best, median, and worst IGD values for GDE3, SMPSO, and SPEA2 algorithms against the IGD value of solutions extracted by the FNFR method on 8- and 10-objective DTLZ3, convex DTLZ4, and WFG2 test problems.

	DTLZ3		DTLZ4		WFG2	
M	8	10	8	10	8	10
GDE3 Best	1.26E-03	9.30E-04	2.83E-03	2.35E-03	2.23E-03	1.84E-03
GDE3 Median	2.95E-03	1.24E-03	2.97E-03	2.43E-03	2.64E-03	2.02E-03
GDE3 Worst	7.59E-03	1.42E-03	3.04E-03	2.53E-03	3.26E-03	2.24E-03
GDE3 Fusion	1.08E-03	7.10E-04	3.02E-03	2.40E-03	2.05E-03	1.71E-03
SMPSO Best	3.90E-03	1.11E-03	2.85E-03	2.59E-03	2.73E-03	1.97E-03
SMPSO Median	8.88E-03	2.05E-03	3.95E-03	2.85E-03	3.12E-03	2.18E-03
SMPSO Worst	8.77E-03	1.98E-03	3.67E-03	3.28E-03	2.42E-03	1.93E-03
SMPSO Fusion	3.77E-03	1.83E-03	3.01E-03	2.32E-03	2.55E-03	1.80E-03
Overall Fusion	**3.34E-03**	**1.07E-03**	**1.99E-03**	**2.25E-03**	**2.09E-03**	**1.51E-03**

Visual Analysis of Preferred Solutions Found by the FNFR Method. Here we investigate the effectiveness of the proposed FNFR method when dealing with extracting solutions close to preferred region. In a practical multi- of many-objective optimization problems decision-makers may be interested in visualizing (when possible) the entire PF before selecting the one or more solutions for further investigation. In such scenario, the FNFR scheme is an effective tool to construct the entire PF and select Pareto-optimal points that are close to the supplied reference points. Figure 6 show preferred solutions extracted using the proposed FNFR method. The grey dotted background indicates non-dominated solutions obtained by the GDE3, SMPSO, and SPEA2 algorithms in 20 runs and the black dots illustrate preferred solutions extracted from these solutions using the proposed scheme.

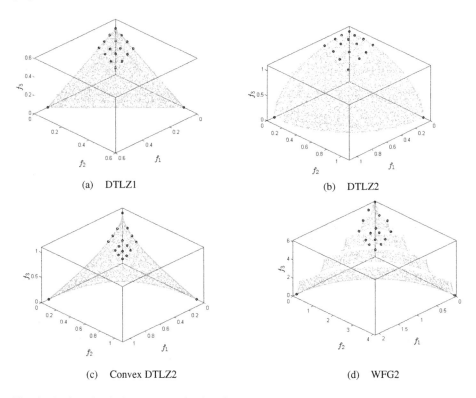

(a)   DTLZ1

(b)   DTLZ2

(c)   Convex DTLZ2

(d)   WFG2

Fig. 6. Preferred solutions extracted using the proposed FNFR method. The grey dotted background indicate non-dominated solutions obtained by the GDE3, SMPSO, and SPEA2 algorithms in 20 runs and the black dots illustrate preferred solutions extracted from these solutions using the proposed FNFR approach.

4   Concluding Remarks

In this paper, we proposed a novel hybridization of population-based metaheuristic algorithms called fusion of non-dominated fronts using reference points (FNFR) to gain combined benefit of several algorithms. The hybridization step in FNFR occurs after the optimization process of multiple runs of several algorithms. In every run, FNFR assembles and constructs the entire Pareto-optimal front from a large set of non-dominated solutions found using multiple algorithms. This step is crucial because no single algorithm is capable of producing well-distributed solutions for all types of problems all the time. Thereafter, FNFR uses predefined structured reference points or user selected reference points to extract preferred solutions from this set.

Experimental results showed the effectiveness of the proposed FNFR scheme with numerical experiments by considering three widely used optimization algorithms, GDE3,

SMPSO, and SPEA2. The FNFR method is able to extract well-distributed solutions in a specific region of interest among thousands of non-dominated solutions collected after every run of multiple algorithms. Therefore the FNFR scheme can effectively be used by decision-makers to select and examine preferred solutions among a large set of solution which can be astronomically difficult to manage. In future, we would like to extend this study further by applying the FNFR scheme during the optimization process so that preferred solutions can be extracted and inserted back to the current population to boost the search process.

References

1. Nwankwor, E., Nagar, A.K., Reid, D.: Hybrid differential evolution and particle swarm optimization for optimal well placement. Comput. Geosci. **17**(2), 249–268 (2013)
2. Mirjalili, S., Hashim, S.Z.M.: A new hybrid PSOGSA algorithm for function optimization. In: 2010 International Conference on Computer and Information Application (ICCIA), pp. 374–377 (2010)
3. Elhossini, A., Areibi, S., Dony, R.: Strength Pareto particle swarm optimization and hybrid EA-PSO for multi-objective optimization. Evol. Comput. **18**, 127–156 (2010)
4. Tang, L., Wang, X.: A hybrid multiobjective evolutionary algorithm for multiobjective optimization problems. IEEE Trans. Evol. Comput. **17**(1), 20–45 (2013)
5. Wang, Y., Cai, Z., Zhou, Y., Fan, Z.: Constrained optimization based on hybrid evolutionary algorithm and adaptive constraint-handling technique. Struct. Multi. Design Optim. **37**, 395–413 (2009)
6. Zăvoianu, A.-C., Lughofer, E., Bramerdorfer, G., Amrhein, W., Klement, E.P.: DECMO2: a robust hybrid and adaptive multi-objective evolutionary algorithm. Soft. Comput. **19**, 3551–3569 (2015)
7. Zhao, S.-Z., Suganthan, P.N.: Multi-objective evolutionary algorithm with ensemble of external archives. Int. J. Innovative Comput. Inf. Control **6**, 1713–1726 (2010)
8. Qu, B.Y., Suganthan, P.N.: Constrained multi-objective optimization algorithm with an ensemble of constraint handling methods. Eng. Optim. **43**(4), 403–416 (2011)
9. Zhao, S.-Z., Suganthan, P.N., Zhang, Q.: Decomposition-based multiobjective evolutionary algorithm with an ensemble of neighborhood sizes. IEEE Trans. Evol. Comput. **16**, 442–446 (2012)
10. Tan, K.C., Chew, Y., Lee, L.H.: A hybrid multi-objective evolutionary algorithm for solving truck and trailer vehicle routing problems. Eur. J. Oper. Res. **172**, 855–885 (2006)
11. Xia, W., Wu, Z.: An effective hybrid optimization approach for multi-objective flexible job-shop scheduling problems. Comput. Ind. Eng. **48**(2), 409–425 (2005)
12. Tavakkoli-Moghaddam, R., Rahimi-Vahed, A., Mirzaei, A.H.: A hybrid multi-objective immune algorithm for a flow shop scheduling problem with bi-objectives: weighted mean completion time and weighted mean tardiness. Inf. Sci. **177**, 5072–5090 (2007)
13. Karthikeyan, S., Asokan, P., Nickolas, S., Page, T.: A hybrid discrete firefly algorithm for solving multi-objective flexible job shop scheduling problems. Int. J. Bio-Inspired Comput. **7**, 386–401 (2015)
14. Deb, K., Jain, H.: An evolutionary many-objective optimization algorithm using reference-point-based nondominated sorting approach, part I: solving problems with box constraints. IEEE Trans. Evol. Comput. **18**, 577–601 (2014)
15. Das, I., Dennis, J.E.: Normal-boundary intersection: A new method for generating the Pareto surface in nonlinear multicriteria optimization problems. SIAM J. Optim. **8**, 631–657 (1998)

16. Kukkonen, S., Lampinen, J.: GDE3: the third evolution step of generalized differential evolution. In: The 2005 IEEE Congress on Evolutionary Computation, pp. 443–450 (2005)
17. Nebro, A.J., Durillo, J.J., Garcia-Nieto, J., Coello, C.C., Luna, F., Alba, E.: Smpso: a new pso-based metaheuristic for multi-objective optimization. In: 2009 IEEE Symposium on Computational Intelligence in Miulti-criteria Decision-Making, pp. 66–73 (2009)
18. Zitzler, E., Laumanns, M., Thiele, L., Zitzler, E., Zitzler, E., Thiele, L. et al.: SPEA2: improving the strength pareto evolutionary algorithm. Ed: Eidgenössische Technische Hochschule Zürich (ETH), Institut für Technische Informatik und Kommunikationsnetze (TIK) (2001)
19. Deb, K., Thiele, L., Laumanns, M., Zitzler, E.: Scalable multi-objective optimization test problems. In: Proceedings of the Congress on Evolutionary Computation (CEC-2002), (Honolulu, USA), pp. 825–830 (2002)
20. Huband, S., Barone, L., While, L., Hingston, P.: A scalable multi-objective test problem toolkit. In: Evolutionary Multi-criterion Optimization, pp. 280–295 (2005)
21. Yang, S., Li, M., Liu, X., Zheng, J.: A grid-based evolutionary algorithm for many-objective optimization. IEEE Trans. Evol. Comput. **17**, 721–736 (2013)
22. Jain, H., Deb, K.: An evolutionary many-objective optimization algorithm using reference-point based nondominated sorting approach, part II: handling constraints and extending to an adaptive approach. IEEE Trans. Evol. Comput. **18**, 602–622 (2014)
23. Ibrahim, A., Rahnamayan, S., Vargas Martin, M., Deb, K.: 3D-RadVis: visualization of pareto front in many-objective optimization. In: 2016 IEEE Congress on Evolutionary Computation (CEC), pp. 736–745 (2016)

An Expedition to Multimodal Multi-objective Optimization Landscapes

Pascal Kerschke$^{(\boxtimes)}$ and Christian Grimme

Westfälische Wilhelms-Universität Münster, Leonardo-Campus 3,
48149 Münster, Germany
{kerschke,christian.grimme}@uni-muenster.de

Abstract. The research in evolutionary multi-objective optimization is largely missing a notion of functional landscapes, which could enable a visual understanding of multimodal multi-objective landscapes and their characteristics by connecting decision and objective space. This consequently leads to the negligence of decision space in most algorithmic approaches and an almost complete lack of Exploratory Landscape Analysis (ELA) tools. This paper dares a first step into this unexplored field based on gradient properties of the multi-objective landscape. For a first time, basins of attraction and superpositions of local optima are visualized and thereby made intuitively accessible. With this work, we hope to highlight the importance of detailed decision space analysis in multi-objective optimization and to stimulate further research in that direction.

Keywords: Multi-objective evolutionary optimization · Multimodality · Fitness landscapes · Basins of attraction · Visualization · Qualitative analysis

1 Introduction

Evolutionary algorithms explore or exploit a given decision space by slightly perturbing and/or recombining existing solutions within the evolutionary loop. For single-objective evolutionary algorithms, the practical advancement of mutation (for exploration) and recombination (for exploitation) operators [13], as well as their theoretical foundation [1], has been driven by understanding their behavior in the decision space. Additionally, understanding and identifying features of the decision space enables the creation of rule sets and finally the educated selection of algorithm parameters or algorithms themselves for (almost) unknown problems [6,7].

Both development of evolutionary operators and *Exploratory Landscape Analysis* (ELA, [11]) are based on a notion of the so-called *function landscape* – also called *fitness landscape* in the evolutionary context – which usually refers to the image of the decision variables under the given objective function. Plotting scalar objective values for a two-dimensional problem is the most common

© Springer International Publishing AG 2017
H. Trautmann et al. (Eds.): EMO 2017, LNCS 10173, pp. 329–343, 2017.
DOI: 10.1007/978-3-319-54157-0_23

approach for visualizing multimodality, identifying basins of attraction of local optima, or finding difficulties for algorithms (e.g., ridges or valleys). However, when dealing with higher dimensional decision spaces or multiple problems at once, such an identification has to be done in an automated fashion. The latter can be done with ELA, which computes (rather general) numerical representations of the landscapes. These numbers can then be used to identify specific structures of the underlying optimization problems.

For multi-objective problems, however, a standard notion of 'function landscapes' is not available, as the image of decision variables is not scalar anymore but comprises several objective values. This severely hinders an informative glance into the functional relationship of decision space and objective space. Not surprisingly, the specific development of evolutionary operators in multi-objective optimization almost not exists. Instead, algorithmic research mainly focuses on tuning selection mechanisms in *multi-objective evolutionary algorithms* (MOEAs). Also, ELA approaches are rare and mostly restricted to the combinatorial case, where knowledge of the problem's phenotype-genotype relation can be exploited in specific measures and features [4].

In this paper, we approach this lack of insight into continuous multimodal multi-objective landscapes by realizing an approach for visualizing such landscapes. Therefore, we adapt definitions for local efficient sets of multi-objective problems [8], which were recently published by a super-set of this paper's authors, and extend the analysis by identifying basins of attraction within bi-objective mixed-sphere problems[1]. The local appearance of a multi-objective landscape is determined via a combined gradient approach and (figuratively speaking) shows how a 'multi-objective ball' behaves in such an environment. Such a ball can be understood as an analog to a gradient descent algorithm and thus, the interpretation of these landscapes is to some extent as intuitive as in the single-objective case.

For a first time, this work allows a glance into continuous multimodal multi-objective landscapes based on well defined locality and paves the ground for future landscape analysis based on characteristics or features of these landscapes. At the same time, we hope that this work also provides some stimulus for algorithm development – focusing more on the decision space of multi-objective search rather than its objective space.

After a brief review of the basic definitions used here and a short discussion of existing multi-objective landscape analysis and visualization approaches in Sect. 2, we detail our visualization approach in Sect. 3, and explore the resulting landscapes in Sect. 4. Section 5 concludes this work.

2   Background

In the following, we briefly summarize some of the existing definitions of multimodality in the multi-objective sense, prior to providing an overview of the

---

[1] Note that we restricted ourselves to bi-objective mixed-sphere problems as we build upon the definitions and problems from [8], but in general our proposed visualization technique can also be applied to problems with more than two objectives.

existing literature, which deals with multi-objective landscapes, as well as their visualizations.

2.1   Multimodal Multi-objective Optimization

As defined in Kerschke et al. [8], we denote $\mathbf{f} : \mathcal{X} \rightarrow \mathbb{R}^m$ as a multi-objective function where $\mathcal{X} \subseteq \mathbb{R}^d$ is the decision space. We denote the component functions of $\mathbf{f}$ by $f_i : \mathcal{X} \rightarrow \mathbb{R}$, $i = 1, \ldots, m$.

A vector $\mathbf{x} \in \mathcal{X}$ is called *Pareto efficient* or *global efficient* iff there does not exist a point $\tilde{\mathbf{x}} \in \mathcal{X}$ which dominates[2] $\mathbf{x}$, i.e., $\nexists \tilde{\mathbf{x}} \in \mathcal{X} : \mathbf{f}(\tilde{\mathbf{x}}) \prec \mathbf{f}(\mathbf{x})$. The subset of $\mathcal{X}$ consisting of all the (Pareto) efficient points of $\mathcal{X}$ is denoted by $\mathcal{X}_E$ and is called the *efficient set* of $\mathcal{X}$. The image of $\mathcal{X}_E$ under $\mathbf{f}$ is called the Pareto front of $\mathbf{f}$. Consequently, the points of $\mathcal{X}_E$ are the multi-objective analogon of the (global) optima within single-objective optimization problems.

The definition of multimodality – of multi-objective problems – is based on the topological notion of connectedness, which can briefly be summarized as follows:

- The set $A \subseteq \mathbb{R}^n$ is called <u>connected</u> if and only if there do not exist two open and *disjoint* subsets U_1 and U_2 of $\mathbb{R}^n$ such that $A \subseteq U_1 \cup U_2$, $U_1 \cap A \neq \emptyset$, and $U_2 \cap A \neq \emptyset$.
- Let $B \subseteq \mathbb{R}^n$. A subset $C \subseteq B$ is a <u>connected component</u> of B iff C is connected and non-empty, and there exists no connected subset of B that is a strict superset of C.

Now, we can state the definitions for locally efficient points in $\mathcal{X}$ and local efficient sets in the multi-objective case:

- A point $\mathbf{x} \in \mathcal{X}$ is called a <u>locally efficient point</u> of $\mathcal{X}$ (or of $\mathbf{f}$) if there is an open set $U \subseteq \mathbb{R}^d$ with $\mathbf{x} \in \mathbf{U}$ such that there is no point $\tilde{\mathbf{x}} \in U \cap \mathcal{X}$ that fulfills $\mathbf{f}(\tilde{\mathbf{x}}) \prec \mathbf{f}(\mathbf{x})$. The subset of all the local efficient points of $\mathcal{X}$ is denoted by $\mathcal{X}_{LE}$.
- A subset $A \subseteq \mathcal{X}$ is a <u>local efficient set</u> of $\mathbf{f}$ if A is a connected component of $\mathcal{X}_{LE}$, where $\mathcal{X}_{LE}$ is the subset of $\mathcal{X}$ which consists of the local efficient points of $\mathcal{X}$.

An important property of local efficient points in the continuous multi-objective case (which we use later on) is stated by the well known Fritz John's [5] necessary condition, which has been extended to a sufficient condition for convex problems by Kuhn and Tucker [10]:

Let the component functions of $\mathbf{f}$ be continuously differentiable in $\mathbb{R}^n$ and $\mathbf{x} \in \mathcal{X}$ be a local efficient point of $\mathcal{X}$. Then, there exists a vector $\nu \in \mathbb{R}^m$ with $0 \leq \nu_i$, $i = 1, \ldots, m$, and $\sum_{i=1}^{m} \nu_i = 1$, such that

$$\sum_{i=1}^{m} \nu_i \nabla f_i(\mathbf{x}) = 0.$$

---

[2] A vector $\mathbf{a} = (a_1, a_2, \ldots, a_n)^T$ dominates another vector $\mathbf{b} = (b_1, b_2, \ldots, b_n)^T$, i.e., $\mathbf{a} \prec \mathbf{b}$, if and only if $\forall i \in \{1, \ldots, n\} : a_i \leq b_i$ and $\exists j \in \{1, \ldots, n\} : a_j < b_j$.

For a local efficient point and a suitable weighting vector, gradients for all objectives cancel each other out. In the special case of a bi-objective problem, gradients become anti-parallel and only differ in length.

This leaves the definition of local and global fronts based on the previously mentioned efficiency for completeness:

- A subset P of the image of $\mathbf{f}$ is a _local Pareto front_ of $\mathbf{f}$, if there exists a local efficient set $E \subseteq \mathcal{X}_{LE}$ such that $P = \mathbf{f}(E)$.
- The (global) _Pareto front_ (PF) of $\mathbf{f}$ is obtained by taking the image under $\mathbf{f}$ of the union of the connected components of the set of global efficient points of $\mathcal{X}$. If $\mathcal{X}_E$ is connected, then the (global) Pareto front of $\mathbf{f}$ is also connected, provided $\mathbf{f}$ is continuous on $\mathcal{X}_E$.

2.2 Multi-objective Landscapes and Visualization Approaches

The generation of function landscapes for single objective problems enables – besides simple visualization – a figurative description of properties directly exposed by the continuous depiction of function values. As such, mountains, valleys, plateaus, or ridges are identified. On the one hand, these geographical notions allow an immediate common understanding of the described structures. On the other hand, they comprise an implicit mathematical interpretation as local optima, modality, areas of similar quality, or discontinuities, respectively. Additionally, this provides some (at least intuitive) notion of problem difficulties posed to algorithms.

Consequently, in some articles, these terms are transferred to the multi-objective domain. As described in [4], families of landscapes can be generated for a given multi-objective problem by successively scalarizing the objective functions using different weighting vectors. The measures for single-objective landscape analysis are then generalized for those families of landscapes. In the works of Knowles and Corne [9], as well as Garrett and Dasgupta [4], problem specific knowledge – e.g., on the quadratic assignment problem or the traveling salesperson problem (TSP) – is used to identify specific measures for combinatorial problems, while Rosenthal and Borschbach [12] estimate modality, ruggedness, correlation and plateaus specifically for multi-objective molecular landscapes for biochemical optimization problems.

All approaches ignore the visualization of landscapes, but directly present abstract geographically labeled features. This is not surprising, as visualizations are mainly restricted to low dimensions and thus not easily generalizable. Additionally, the simultaneous visualization of multiple objective values and combinatorial properties is not trivial. Still, in many cases, visualization is helpful to understand structures and results.

Tušar [14] and later Tušar and Filipič [15] review multiple visualization techniques for multi-objective results and extend these techniques. This provides a good tool set for analyzing algorithm behavior but gives little information on the complete multi-objective landscape of a problem.

To the best knowledge of the authors, the only depiction of a multi-objective landscape is given by Fonseca [3]. In that work, a so-called "cost-landscape" based on population rankings is computed. This gives – at least for the global efficient set's perimeter – some impression of the multi-objective fitness landscape. Note that Fonseca designed his approach to mainly show difficulties, e.g., the influence of constraints of decision maker preferences, for search algorithms around the global solution set. Consequently, it is not designed and used in the more general context of visualizing an entire landscape including locally efficient sets.

3   Analysis and Visualization Approach

As stated within the introduction, we aim at getting a better understanding of multimodal multi-objective continuous landscapes. Therefore, we follow the approach that was introduced in [8], i.e., we focus on bi-objective mixed-sphere problems. That is, we consider two objectives and for each one of them, we create a problem consisting of multiple sphere functions, using the *Multiple Peaks Model 2* generator (MPM2, [16]) as provided within the R package smoof [2] or the python package optproblems [17].

Figure 1 shows the resulting contour plots for two scenarios: a rather simple scenario with one peak for the first objective (indicated by the red solid contour lines) and two peaks for the second objective (blue dotted contour lines) on the left side and a slightly more complex scenario with three optima for the

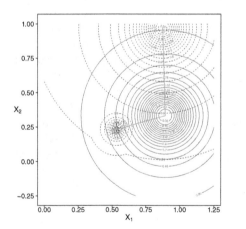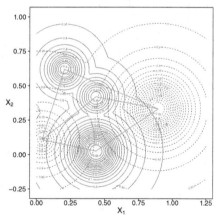

Fig. 1. Examples of two mixed-sphere problems: a simple scenario (left), consisting of one peak for the first (solid red contour lines) and two peaks for the second objective (dotted blue contour lines) and a slightly more complex scenario (right) with three optima for the first objective and two for the second. The connections between all optima from one objective with all optima from the other objective are highlighted by grey lines, indicating the possible area of local efficient sets. (Color figure online)

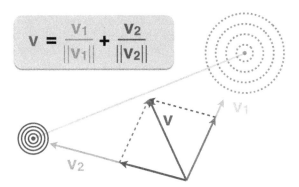

Fig. 2. Schematic view of the combination of the objective-wise (normalized) gradients. The closest optima per objective are indicated by the (dotted) orange and (solid) blue contour lines. The lengths of the two objective-wise gradient vectors v_1 (light orange arrow) and v_2 (light blue arrow) depend on the steepness of the descent towards their respective optimum. The combined bi-objective gradient v (green arrow) is the sum of the objective-wise normalized gradient vectors $v_1/||v_1||$ (orange arrow) and $v_2/||v_2||$ (blue arrow). (Color figure online)

first objective and two for the second on the right side. Note that closer circles of the contour lines indicate a steeper descent towards the optima and as a consequence the attraction towards those optima will be stronger. In addition, these plots contain straight gray lines, connecting each optimum of one objective with each optimum from the other objective, which indicate the areas where local efficient sets could be located [8]. While those contour plots already provide a way of visualizing two objectives within a single plot, the two objectives remain clearly distinguished and it is difficult to grasp the actual interaction of the two. Within this paper, we now want to provide new ways of visualizing two (or even more) objectives (and their interactions) and thereby help to improve the understanding of such landscapes.

In a first step, the search space is divided into a grid of 2 500 by 2 500 points and then the *combined bi-objective gradient* was computed for each of the 6.25 million grid points (see Fig. 2). Given a point from the previously generated grid, the two objective-wise gradients v_1 and v_2, i.e., the corresponding directions of the steepest descent, are computed via

$$v_i = \lim_{\varepsilon \to 0} \frac{f_i(x - \varepsilon) - f_i(x + \varepsilon)}{||\varepsilon||}, \qquad i = 1, 2.$$

The two objective-wise gradients are then normalized to length one (shown as orange and blue arrows within Fig. 2) and afterwards summed up, resulting in a bi-objective gradient v (green arrow). Depending on the location of the grid point w.r.t. its two (closest) objective-wise local optima, the angle between the gradients v_1 and v_2 is somewhere between $0°$ and $180°$. While the former effect occurs when both gradients show into the same direction and thus, double their

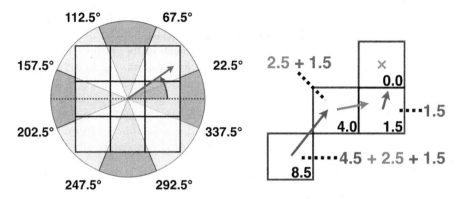

Fig. 3. Finding the path of the gradient vectors towards the next local efficient set. First, one chooses the neighboring cell based on the direction of the gradient vector (green arrow). In this example (left), the resulting neighbor would be the cell on the top right. Following these simple decisions (right), one will ultimately find a path towards a local efficient set. The cumulated length of the gradient vectors along that path then defines the height of the multi-objective landscape. (Color figure online)

effects, the latter effect only occurs, if the grid point is located in a local efficient set – which (in case of mixed-sphere problems) have to lie on the connection of two optima – meaning that the two objective-wise gradients show in opposite directions. Analogously, the length of the resulting gradient vector v is somewhere between 2 (when both normalized objective-wise gradients show into the same direction) and 0 (when the normalized vectors eliminate their effects, by pointing into opposite directions).

As means of visualization of the multi-objective landscape, we plot the cumulated lengths of the gradients along the path towards the local efficient sets for each of the grid points. Hence, for each point, we have a look at the corresponding (combined and normalized) gradient. If its length is below a pre-defined threshold δ (here: $\delta = 10^{-8}$), the point is considered to belong to a local efficient set[3]. Otherwise, we compute the direction into which the gradient points (cf. left image of Fig. 3) and "walk" towards the cell that is next in that direction, i.e., for angles between 22.5° and 67.5° the next cell (on the path towards the local efficient set) would be the top right neighboring cell, for angles between 67.5° and 112.5° it would be the cell located on top of the current cell, and so on.

Based on these decisions, one now follows a path towards a point (in the close vicinity) of the local efficient set. Then, the sum of the lengths of all gradients along that path (cf. right image of Fig. 3) will be used as the "height", i.e., the scalar representation of the multi-objective landscape. Consequently, the initial cell of the exemplary path of Fig. 3, i.e., the cell on the bottom left, receives

---

[3] One can also use more coarse configurations without losing a lot of accuracy. For instance, we tried smaller grids (consisting of 1 000 by 1 000 points) and a smaller threshold ($\delta = 10^{-3}$) while detecting only minor differences in the resulting figures.

a cumulated gradient path length of 8.5, its successor (on the way to the local efficient set) has a cumulated gradient of 4.0 and the last cell before the optimum receives a gradient path length of 1.5, i.e., the length of its gradient itself.

After computing the gradient paths for all grid points, the multi-objective landscapes can be visualized in a two-dimensional heatmap as in Fig. 4 (top: regular cumulated gradient lengths; bottom: cumulated gradient lengths on a $\log_{10}$-scale). In order to simplify analyses, we added contour lines of the objective-wise mixed-sphere problems and the combined gradient vectors (only every 150th value per dimension for better visibility), as well as the (theoretically true) local efficient sets.

4   Landscapes and Observations

For exemplary visualization of multimodal multi-objective landscapes, we select the two instances that were introduced in Sect. 3:

- A simple scenario, comprising a unimodal objective function and a two-peak objective function. This multi-objective problem instance formally consists of three global (shown as green, pink and red lines within Fig. 4) and two local optimal sets (blue and brown lines).
- A comparatively rather complex scenario, comprising a two-peak objective function and a three-peak objective function. This instance exposes two global (shown as dark green and cyan lines within Fig. 5) and seven local efficient sets.

For both instances, different views were generated to capture some first observations and properties of the visualized multi-objective landscapes. Figure 4 shows a projection of the computed aggregated gradient length into the decision space. For each point in the decision space the color denotes the distance along the combined gradients to an (at least) local efficient point. Here, for completeness, the same view is plotted based on the actual length of the cumulated gradient path (upper plot) as well as based on the corresponding $\log_{10}$-transformed values (lower plot). In the following we will only focus on the log-scaled plots, as – for our purposes – specific path lengths are of minor interest and the log-scaled plot exhibits all necessary properties in a clear way.

Basins of Attraction: We observe, that the landscapes provide perspectives on the basins of attraction of locally and globally efficient sets. Essentially, this supports the gradient field perspective in an intuitive way. As for single-objective landscapes, we can easily identify, to which area of attraction an algorithm moves when following the multi-objective gradient direction. This is what we termed the 'multi-objective ball' movement during the introduction. Similar to the single-objective intuition, the area of attraction (i.e., the set of locally efficient points) is – keeping the differences in gradient interpretation in mind – the set of points, where the necessary condition of a zero (single or multi-objective, respectively) gradient holds.

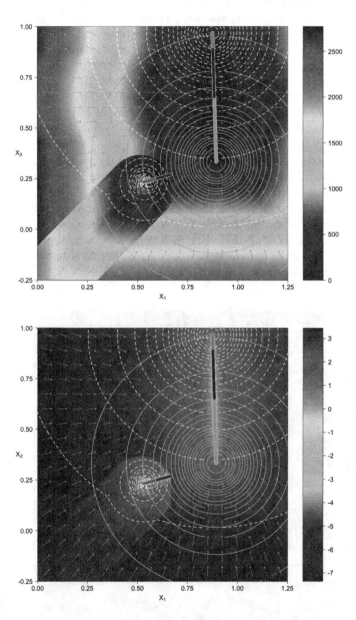

Fig. 4. Plot of the cumulated gradient field landscape for the simple problem. The upper figure shows the landscape according to the actual gradient path length, while the lower figure shows the same situation with log-scaled path length. The two sidebars represent the mapping from the 'height' of the landscape to the used coloring for each of the scenarios. (Color figure online)

Interestingly and different from the single-objective understanding, the area of attraction can comprise disjoint local optimal sets that belong to different domination layers. Even in the very simple instance presented in Fig. 4, the larger area of attraction consists of three distinct efficient sets, two being globally (colored in green and pink) and one intermediate set (colored in brown) being locally efficient. The rational behind this observation is, that this intermediate segment of the attraction set is dominated by the (red colored) efficient solutions produced by the interaction of the unimodal objective function with the second (narrow) peak of the bimodal objective. When focusing on the contour lines of both functions, we detect the same function value for the unimodal objective but a better (i.e., lower) function value in the perimeter of the narrow peak of the bimodal objective. According to the common understanding of dominance (becoming better in one objective and not worsening according to any others), this leads to dominance of the respective efficient set in the second basin of attraction. Still, according to our adapted definitions and the Kuhn-Tucker necessary condition, local efficiency of the dominated set holds.

The inspection of the more complex instance in Fig. 5 reveals multiple basins of attractions and allows the same observations as before.

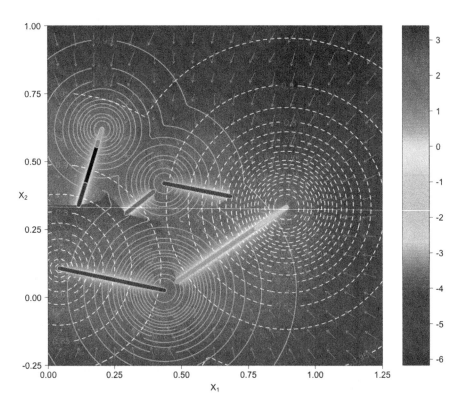

Fig. 5. Depiction of the gradient field landscape on log-scale for the complex problem.

Discontinuities: A most interesting observation in the landscape visualization are the abrupt changes in the landscape. Obviously, basins of attraction are cut through by basins of attraction for different efficient sets. The rational behind this observation is based on the interaction of the combined attraction influence of the considered objectives. As each of the single objectives is a multimodal function itself, gradient influence of different peaks on the combined gradient can switch abruptly. Then, the combined gradient points towards a different efficient set. Consequently, the 'multi-objective ball' following this gradient would change direction and tend for the other efficient set. Therefore, what seems like a discontinuity here, is essentially an abrupt change in attraction resulting in an alternative, vastly longer path.

Expression of Local/Global Fronts and Basins in Objective Space: Based on the colored landscape projections – defined by the cumulated lengths' of the gradient paths towards the attracting local efficient set (see Sect. 3) – we visualized the numerically evaluated decision space elements also in objective space. Figure 6 shows the result for both considered instances. All determined basins of attraction are visible, showing local and global optimal fronts as representatives of (almost) zero length paths. Points in the surrounding of an attraction area are dominated. Increasing distance to efficient sets can be seen in the darker red shading of these points.

Interpretation of 3D Visualization of Multi-objective Landscapes: The 3D visualization of fitness landscapes is usually used to support the visual identification of structures like valleys and ridges and to get a feeling of the landscape's characteristics. Here, plots like those presented in Fig. 7 have to be interpreted carefully. The height plotted in the z-axis must not be interpreted as some kind of multi-objective function value but as the cumulated length of the path of gradients, leading from the current solution to a locally efficient solution in the corresponding basin of attraction. Thus, all basins basically share the same base line. The visualization, however, supports the funnel-like shape of attraction basins and highlights the above described multi-objective discontinuities where attraction changes rapidly.

The Use of Insights into Multi-objective Landscapes: Based on the available landscape visualization, interesting characteristics like the size and geometric form of the areas of attraction as well as their superposition and with it discontinuities in gradient descent behavior can be identified. We also find, that with increasing modality of single objective functions, raggedness of the multi-objective landscapes increases. These observations could lead to advances in landscape analysis as well as in variation operator design. For landscape analysis it may become possible to define features that are inspired by the above observations; e.g. an estimated number of local basins of attraction derived from the raggedness or detected discontinuity in the sample of the decision space. For

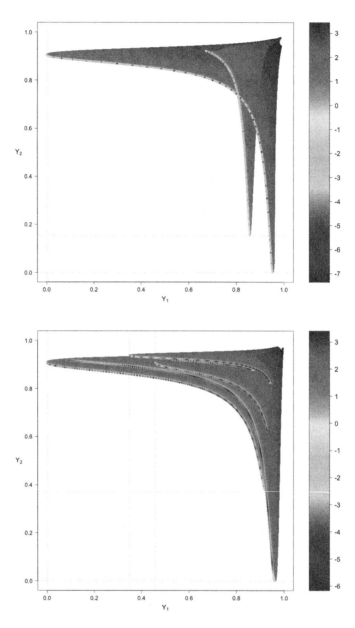

Fig. 6. Mapping of the cumulated gradient field from the decision to the objective space of the simple (top) and complex scenario (bottom). The coloring represents the (log10-scaled) length of the cumulated path towards the local efficient set. (Color figure online)

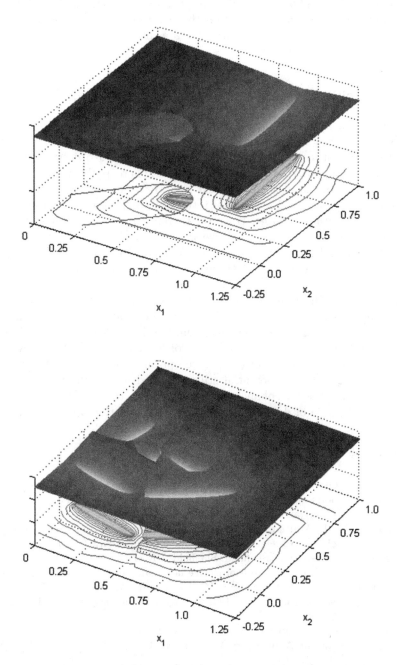

Fig. 7. 3D representation of the cumulated gradient path lengths for both considered problems (simple scenario on top, complex scenario on the bottom).

operator design, the observation of superposition of basins of attraction may be helpful to devise a specific variation that searches for discontinuities. They may be of special interest as they enable the abrupt change into another (maybe better) area of local attraction.

5   Conclusion

In this paper, a scalar visualization of simple continuous multimodal multi-objective landscapes has been realized and qualitatively analyzed. Therefore, a method based on combined gradient information of all objectives has been devised and exemplarily applied to two mixed-sphere problems. To the best knowledge of the authors, this is the first time that multimodal multi-objective landscapes can be explored in a rather intuitive way. It connects decision space information and gradient-based traversal therein to a visually understandable entity. With this technique at hand, we were able to identify basins of attraction for locally and globally efficient sets. Further, we found local and global efficient sets in the same basin of attraction and finally, we saw the interesting appearance of discontinuities, when basins of attraction superpose each other.

Surely, these qualitative findings are only first insights into up to now mostly unknown multi-objective landscapes. Still, we hope, that these first findings will foster the investigation of decision space characteristics, leading to the development of landscape features – i.e., the basis of Exploratory Landscape Analysis (ELA) approaches – in the multi-objective optimization domain. Additionally, those results highlight the potential still concealed within the multi-objective decision space, in order to enhance evolutionary algorithm development on that level, e.g., in the area of dedicated multi-objective variation operator design.

References

1. Beyer, H.G.: The Theory of Evolution Strategies. Springer, Berlin (2001)
2. Bossek, J.: smoof: Single and Multi-Objective Optimization Test Functions (2016). https://github.com/jakobbossek/smoof, r package version 1.4
3. Fonseca, C.M.M.: Multiobjective genetic algorithms with application to control engineering problems. Ph.D. thesis, Department of Automatic Control and Systems Engineering, University of Sheffield, Sheffield, UK (1995)
4. Garrett, D., Dasgupta, D.: Multiobjective landscape analysis and the generalized assignment problem. In: Maniezzo, V., Battiti, R., Watson, J.-P. (eds.) LION 2007. LNCS, vol. 5313, pp. 110–124. Springer, Heidelberg (2008). doi:10.1007/978-3-540-92695-5_9
5. John, F.: Extremum problems with inequalities as subsidiary conditions. In: Studies and Essays. Courant Anniversary Volume, pp. 187–204. Interscience, New York (1948)
6. Kerschke, P., Preuss, M., Wessing, S., Trautmann, H.: Detecting funnel structures by means of exploratory landscape analysis. In: Proceedings of the 17th Annual Conference on Genetic and Evolutionary Computation, pp. 265–272. ACM (2015)

7. Kerschke, P., Preuss, M., Wessing, S., Trautmann, H.: Low-budget exploratory landscape analysis on multiple peaks models. In: Proceedings of the 18th Annual Conference on Genetic and Evolutionary Computation. ACM (2016, accepted)
8. Kerschke, P., Wang, H., Preuss, M., Grimme, C., Deutz, A., Trautmann, H., Emmerich, M.: Towards analyzing multimodality of continuous multiobjective landscapes. In: Handl, J., Hart, E., Lewis, P.R., López-Ibáñez, M., Ochoa, G., Paechter, B. (eds.) PPSN 2016. LNCS, vol. 9921, pp. 962–972. Springer, Heidelberg (2016). doi:10.1007/978-3-319-45823-6_90
9. Knowles, J.D., Corne, D.W.: Towards landscape analyses to inform the design of hybrid local search for the multiobjective quadratic assignment problem. In: HIS, Second International Conference on Hybrid Intelligent Systems (2002)
10. Kuhn, H.W., Tucker, A.W.: Nonlinear programming. In: Proceedings on the Second Berkeley Symposium on Mathematical Statististics and Probability, pp. 481–492. University of California Press (1951)
11. Mersmann, O., Bischl, B., Trautmann, H., Preuss, M., Weihs, C., Rudolph, G.: Exploratory landscape analysis. In: Proceedings of the 13th Annual Conference on Genetic and Evolutionary Computation, GECCO 2011, NY, USA, pp. 829–836. ACM, New York (2011)
12. Rosenthal, S., Borschbach, M.: A concept for real-valued multi-objective landscape analysis characterizing two biochemical optimization problems. In: Mora, A.M., Squillero, G. (eds.) EvoApplications 2015. LNCS, vol. 9028, pp. 897–909. Springer, Cham (2015). doi:10.1007/978-3-319-16549-3_72
13. Schwefel, H.P.: Evolution and Optimum Seeking. Wiley, New York (1995)
14. Tušar, T.: Visualizing Solution Sets in Multiobjective Optimization. Ph.D. thesis, Jožef Stefan International Postgraduate School (2014)
15. Tušar, T., Filipič, B.: Visualization of Pareto front approximations in evolutionary multiobjective optimization: a critical review and the prosection method. IEEE Trans. Evol. Comput. **19**(2), 225–245 (2015)
16. Wessing, S.: Two-stage methods for multimodal optimization. Ph.D. thesis, Technische Universität Dortmund (2015). http://hdl.handle.net/2003/34148
17. Wessing, S.: optproblems: Infrastructure to Define Optimization Problems and Some Test Problems for Black-Box Optimization (2016). https://pypi.python.org/pypi/optproblems/, python package version 0.9

Neutral Neighbors in Bi-objective Optimization: Distribution of the Most Promising for Permutation Problems

Marie-Eléonore Kessaci-Marmion[1]([⊠]), Clarisse Dhaenens[1],
and Jérémie Humeau[2,3]

[1] Univ. Lille, CNRS, Centrale Lille, UMR 9189 - CRIStAL, 59000 Lille, France
{me.kessaci,clarisse.dhaenens}@univ-lille1.fr
[2] Univ. Lille, 59000 Lille, France
[3] Mines Douai, URIA, 90508 Douai, France
jeremie.humeau@mines-douai.fr

Abstract. In multi-objective optimization approaches, considering neutral neighbors during the exploration has already proved its efficiency. The aim of this article is to go further in the comprehensibility of neutrality. In particular, we propose a definition of most promising neutral neighbors and study in details their distribution within neutral neighbors. As the correlation between objectives has an important impact on neighbors distribution, it will be studied. Three permutation problems are used as case studies and conclusions about neutrality encountered in these problems are provided.

Keywords: Neutrality · Multi-objective optimization · Epsilon-indicator · Neighborhood · Fitness landscape

1 Introduction

Multi-objective algorithms evolve archives of Pareto solutions. Among them, multi-objective local search algorithms use neighboring solutions to improve their efficiency iteratively. The neighborhood is explored either exhaustively or partially until finding one or more improving neighbors.

In single-objective optimization, the neutrality concept has been efficiently exploited in both evolutionary and local search algorithms [3,6]. Therefore, we can wonder if neutrality could also be exploited in the multi-objective context. Blot et al. [1] introduced the concept of neutrality in multi-objective optimization as the set of non-dominated neighbors of a solution and showed that archiving neutral neighbors evaluated during the exploration of the neighborhood may improve the performance to solve the permutation flowshop scheduling problem. Recently, the definition of neutral neighbors has been extended [5], using binary indicators of performance. To discuss the interest of exploiting neutral neighbors in multi-objective approaches, we first need to deeper understand their distribution within the neutral part of the neighborhood, in order to identify the most promising ones, *i.e.* the ones that could lead to improve the Pareto front.

© Springer International Publishing AG 2017
H. Trautmann et al. (Eds.): EMO 2017, LNCS 10173, pp. 344–358, 2017.
DOI: 10.1007/978-3-319-54157-0_24

Moreover, when objectives are correlated there is a significant chance for an improvement on one objective to lead to the improvement on the other ones and similarly for a deterioration. On the other hand, when objectives are totally independent, an improvement on one objective does not give any guarantee on what happens on the other ones. Thus, the number and the distribution of neutral neighbors may be influenced by the correlation between the objectives.

In this paper, we study the distribution of neutral neighbors according to the correlation level between the objectives for bi-objective permutation problems. This preliminary study will help to design efficient strategies of exploration of the neighborhood. Therefore, the remainder of this paper is organized as follows. Section 2 introduces neutrality in multi-objective optimization and presents the existing definitions. Section 3 presents a methodology to deeper analyze the distribution of neutral neighbors, especially the most promising ones. Sections 4, 5 and 6 experiment this methodology on three permutation optimization problems. Finally, conclusions and potential future works are discussed.

2   Neutrality in Optimization

2.1   Neutrality in Single-Objective Optimization

In single-objective optimization, neutrality appears when solutions have the same objective value. Formally, let s_1 and s_2 be two different solutions of the search space Ω. s_1 and s_2 are said to be neutral *iff* $f(s_1) = f(s_2)$ where f is the objective function. Both, evolutionary algorithms and local search approaches have been widely used to study neutrality and its effects. It has been shown that local search methods are sensitive to neutrality and other properties of the landscape. Hence, properly exploiting neutrality can improve their performance [6]. As local search algorithms iteratively improve a solution by exploring its neighborhood, the neutral property has been defined between two neighboring solutions. Then, the set of neutral neighbors of a solution s is $\mathcal{N}_n(s) = \{s' \in \mathcal{N}(s)|\ f(s') = f(s)\}$ where $\mathcal{N}(s)$ is the neighborhood of s.

2.2   Neutrality in Multi-objective Optimization

The extension of neutrality to multi-objective optimization is not straightforward and its effects on the dynamics of multi-objective optimization methods are not clearly understood. In the following, a multi-objective optimization problem will be defined by a set of solutions Ω and a set of m ($m \geq 2$) objective functions f_k that associate to each solution s, m values that have to be minimized (without loss of generality). Such a problem is described by:

$$minimize\ F(s) = (f_1(s), f_2(s), ..., f_m(s))$$
$$s \in \Omega$$

According to the Pareto dominance concept, given two solutions s_1 and s_2, either one solution dominates the other ($s_1 \succ s_2$) or ($s_1 \prec s_2$), or the two solutions

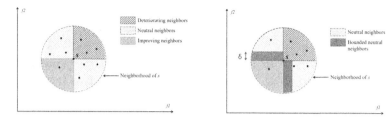

(a) Pareto distribution of neighbors   (b) δ-Bounded subset

Fig. 1. Neutral neighbors (minimization)

are non-dominated with respect to each other ($s_1 \nprec s_2$ and $s_2 \nprec s_1$). These last
are considered as neutral neighbors, $s' \in \mathcal{N}(s)$, as defined in [1] (see Fig. 1(a)).

However, other definitions have been proposed in [5], based on binary multi-
objective quality indicator like the epsilon indicators (additive and multiplicative
versions) or the hypervolume difference indicator. Therefore, different partitions
of the neighborhood can be characterized following the definition considered.
While comparing two solutions, a solution s and a neighboring solution $s' \in
\mathcal{N}(s)$, the epsilon additive indicator (denoted by ε) represents the minimum
value required for s' to dominate s and is computed as:

$$I_\varepsilon(s', s) = max\{max_{\forall k}\{f_k(s') - f_k(s)\}; 0\} \tag{1}$$

If $I_\varepsilon(s', s)$ equals 0, s' already dominates s, otherwise if it is strictly positive, s'
is either a deteriorating neighbor or s' and s are non-dominated each other. I_ε may
also be defined to compare s with s'. Therefore if $I_\varepsilon(s', s) > 0$ and $I_\varepsilon(s, s') > 0$,
s' and s are non-dominated each other. Let us remark that the definition of neu-
trality based on the Pareto-dominance and the symmetric definition based on
epsilon difference indicator lead to the same subset of neutral neighbors. In the
following, we consider as the set of neutral neighbors of a solution s:

$$\mathcal{N}_n(s) = \{s' \in \mathcal{N}(s) \mid s \nprec s' \text{ and } s' \nprec s\} \tag{2}$$

that can be rewritten as follows:

$$\mathcal{N}_n(s) = \{s' \in \mathcal{N}(s) \mid I_\varepsilon(s', s) > 0 \text{ and } I_\varepsilon(s, s') > 0\} \tag{3}$$

As shown, the epsilon indicator gives the same information about neutral
neighbors than the Pareto dominance but in addition, it allows to define several
subsets among neutral neighbors. Indeed, let s and $s' \in \mathcal{N}(s)$ non-dominated
each other, we can define a positive bound δ such as $I_\varepsilon(s', s) < \delta$ that leads to
a particular subset of neutral neighbors, denoted as δ-Bounded (see Fig. 1(b)).

3   Characterization of Promising Neutral Neighbors

3.1   Motivations

Considering neutral neighbors during the search has already been successful in the past [3,6]. However, such neutral neighbors may be very numerous and considering all of them can be very costly (for the archiving task for example). In addition, it is interesting to see if some of these neutral neighbors can be considered as most promising (leading to other parts of the search space, giving more diversity, contribution greater than the storing cost...). Hence, the aim of this study is to analyze the interest of considering neutral neighbors during the search. In particular the following questions will be addressed.

- How to define promising neutral neighbors?
- Are these promising neutral neighbors numerous? Is it dependent on the correlation between objectives?
- Where are they located? What is their distribution in the neighborhood?

3.2   Definition of Promising Neutral Neighbors

Defining promising neighbors seems to be problem-dependent. However, among properties that are important to consider, many will agree to say that a neutral neighbor could be promising if, for example:

- it allows to increase the spread along the Pareto front,
- it improves the convergence,
- it improves the hypervolume measure.

We propose to summarize these properties by defining promising neutral neighbors as solutions that bring to a deterioration on one objective function smaller than the improvement they bring on the other one (see Fig. 2(a)). Let us note $\mathcal{P}_n(s) \subset \mathcal{N}_n(s)$, the set of promising neutral neighbors. It is defined as follows:

$$\mathcal{P}_n(s) = \{s' \in \mathcal{N}_n(s) \mid I_\varepsilon(s', s) < I_\varepsilon(s, s')\} \tag{4}$$

Within those promising neighbors, it is also possible to control the deterioration between s and its neighbors s', represented by the value of $I_\varepsilon(s', s)$ with a bound $\delta \geq 0$ (see Fig. 2(b)). This leads to the definition of $\mathcal{P}_n^\delta(s) \subset \mathcal{P}_n(s)$, the set of promising neutral neighbors depending on δ, that may be defined by:

$$\mathcal{P}_n^\delta(s) = \{s' \in \mathcal{P}_n(s) \mid I_\varepsilon(s', s) \leq \delta\} \tag{5}$$

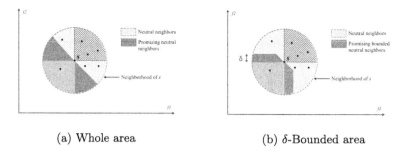

(a) Whole area   (b) δ-Bounded area

Fig. 2. Promising neutral neighbors

3.3 Methodology to Study the Distribution of the Promising Neutral Neighbors

When promising neutral neighbors are numerous in comparison to the number of neutral neighbors and more widely to the whole neighborhood, a deeper study of their location in the search space would help to exploit them into strategies. Therefore, we propose to consider several values for the bound δ to study their distribution among the neutral neighbors. Let us remark that values of objective functions of a solution s and its neighbors s' are first normalized using the whole neighborhood $\mathcal{N}$, so that $I_\varepsilon(s', s) \in [0; 1]$. Let $\Delta = \{\delta_0, \delta_1, \delta_2 \ldots \delta_D\}$ a discretization of $[0; 1]$ where $\delta_0 = 0$ and $\delta_D = 1$.

If we focus on the left-upper side quarter of the neighborhood, different values of δ lead to horizontal layers ($\mathcal{H}$-layer) of the promising neighbors area (similar layers may be found in the other part dealing with neutral neighbors). Such a $\mathcal{H}$-layer allows to study the distribution of neutral neighbors regarding a bounded authorized degradation (see Fig. 3-left). This may be formalized by:

$$\mathcal{H}_n^{\delta_i} = \{s' \in \mathcal{P}_n^{\delta_i}(s) \mid \delta_{i-1} < I_\varepsilon(s', s), \ 1 \le i \le D\} \tag{6}$$

Let us remark that the first $\mathcal{H}$-layer ($\mathcal{H}_n^{\delta_1}$) is the closest to the dominance area and hence neighbors belonging to it represent a great interest.

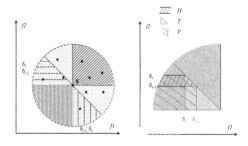

Fig. 3. Space decomposition

In addition, we propose to define two other sub-areas. First, triangles $\mathcal{T}_n^{\delta_i}$, in which both $I_\varepsilon(s, s')$ and $I_\varepsilon(s', s)$ are bounded by δ_i:

$$\mathcal{T}_n^{\delta_i} = \{s' \in \mathcal{P}_n^{\delta_i}(s) \mid I_\varepsilon(s, s') \leq \delta_i, \ 0 \leq i \leq D\} \tag{7}$$

Secondly, vertical layers ($\mathcal{V}$-layer), denoted as $\mathcal{V}_n^{\delta_i}$, to group neighbors further and further from the solution explored s (see Fig. 3-right).

$$\mathcal{V}_n^{\delta_i} = (\sum_{k=\delta_0}^{\delta_i} \mathcal{H}_n^k) \backslash \mathcal{T}_n^{\delta_{i-1}} = \{s' \in \mathcal{P}_n^{\delta_i}(s) \mid I_\varepsilon(s, s') > \delta_{i-1}, \ 1 \leq i \leq D\} \tag{8}$$

The $\mathcal{V}$-layers allow to study solutions that produce a minimal improvement (δ_{i-1}) with a bounded degradation (δ_i). Neighboring solutions belonging to $\mathcal{V}_n^{\delta_D}$ produce an improvement larger than the worst degradation encountered within the whole neighborhood. Let us remark that $\mathcal{V}_n^{\delta_1} == \mathcal{H}_n^{\delta_1} == \mathcal{P}_n^{\delta_1}$.

Focussing on the first $\mathcal{H}$-layer, $\mathcal{H}_n^{\delta_1}$, neighbors belonging to this layer are either in the triangle $\mathcal{T}_n^{\delta_1}$ or in the rest of the layer $\mathcal{P}_n^{\delta_1} \backslash \mathcal{T}_n^{\delta_1}$. The triangle groups solutions very close to the initial solution s; these solutions have a very low interest regarding convergence and diversity. The rest of the layer groups solutions that, on the contrary, produce on one objective an improvement larger than the worst possible degradation for the second objective in this layer (*i.e.* δ_1).

3.4   Experimental Protocol for Permutation Problems

The goal of this paper is to study where are located the promising neighbors in the neighborhood of a bi-objective optimization problem. The correlation between the objectives largely influences the distribution of the whole neighborhood. Hence, we propose to study the distribution of the promising neighbors among the neutral ones and more largely, among the whole neighborhood, according to the correlation between objectives. We focus our study on bi-objective permutation problems since they are widely used in the literature to validate new optimization methods. Therefore, knowing the structure of these permutation problems may help in the design of new methods. The permutation problems considered are: the *permutation flowshop scheduling problem*, the *quadratic assignment problem* and the *travelling salesman problem*. The neighborhood of each one is studied under its best-known neighborhood operator.

The experimental protocol we designed is identical for each bi-objective permutation problem. We study three different problem sizes $N \in \{20, 50, 100\}$ as found in the literature benchmarks, with three levels of correlation between the two objectives: no correlation, mid correlation and high correlation. For each couple size-correlation, we generated 10 instances, and for each instance, the neighborhood of 30 random solutions, denoted as *Rnd*, and 30 Pareto local optimal solutions, denoted as *Opt* are evaluated and analyzed.

As mentioned in Sect. 3.3, the epsilon-indicator values are normalized and then a reference value is needed per objective. Here, the value of each neighbor is studied under the whole neighborhood of s. Then, we propose to define $\mathtt{ref}_k = max(f_k(s'))$ as the maximum value found among the neighbors per objective $k \in \{1, 2\}$, and $\varepsilon_k = \dfrac{f_k(s') - f_k(s)}{\mathtt{ref}_k - f_k(s)}$. Then we compute $I_\varepsilon(s', s) = max_k(\varepsilon_k, 0)$ and $I_\varepsilon(s, s') = max_k(-\varepsilon_k, 0)$ to obtain a symmetric measure for non dominated solutions in bi-objective optimization with a same reference.

The analysis consists in computing different measures for both random and optimal solutions, for each considered instances. Then, an average of these values are applied to get one value per instance of:

- the percentage of neutral neighbors ($\mathcal{N}_n$),
- the percentage of promising ($\mathcal{P}_n$) neighbors among the neutral ones,
- their distribution in the different subsets $\mathcal{P}_n{}^\delta$, $\mathcal{T}_n{}^\delta$, $\mathcal{H}_n{}^\delta$ and $\mathcal{V}_n{}^\delta$ for $\delta \in \Delta = \{0, 0.1, 0.2, 0.3, 0.4, 0.5, 0.6, 1\}$.

4 Case Study 1: Bi-objective Permutation Flow-Shop Scheduling Problem - FSP

4.1 Presentation of the Permutation Bi-objective FSP

The Flow-shop Scheduling Problem (FSP) consists in scheduling a set of N jobs on a set of M machines. Machines are critical resources that can only process one job at a time. A job i is composed of M tasks $\{t_{i,1}, \ldots, t_{i,M}\}$ for the M machines respectively. A processing time $p_{i,j}$ is associated to each task $t_{i,j}$. In the permutation version of the problem, the operating sequence is the same on every machine. Therefore, a schedule may be represented as a permutation of jobs $\pi = \{\pi_1, \ldots, \pi_N\}$.

Several objectives may be defined for the classical multi-objective FSP like the makespan, the total flowtime or the total tardiness (when a due date is associated to each job). But in this work, we need to control the correlation between the objectives. Then, we propose to study a bi-objective Permutation FSP where the objectives are the makespan (sum of completion times) computed from different correlated, or not, matrices of processing times. Let two matrices $(P^k)_{k=\{1,2\}}$ of processing times such that, $p_{i,j}^k$ is the processing time of job i on machine j with $k \in \{1, 2\}$. The completion time of each job i regarding to matrix P^k, denoted as C_i^k, is the time needed for the M tasks of i to be processed on the M machines. Hence, the objectives (to be minimized), f_k ($k \in \{1, 2\}$) are defined as:

$$f_k(\pi) = \sum_{i=1}^{N} C_{\pi_i}^k \tag{9}$$

The neighborhood operator considered in this paper is the insertion operator, in which a job located at position i is inserted at position $j \neq i$ and the jobs located between positions i and j are shifted. The number of neighbors per solution is then $(N-1)^2$.

4.2 Benchmark

Commonly used instances in the literature, are Taillard instances [7], in which processing times are uniformly generated in $[1; 99]$. In this paper, we need to control the correlation. Hence, we generated our own instances following the idea of uniformly generation. First, the processing times of P^1 are generated following $\mathcal{U}([1; 99])$. Then, in the no-correlated version, P^2 is generated independently of P^1 following $\mathcal{U}([1; 99])$. For the two $\rho-$correlated versions, we use the coverage method to generate P^2 from P^1. If $\mathcal{U}([0; 1]) < \rho$, then $p^2_{i,j} = p^1_{i,j}$ otherwise $p^2_{i,j} = \mathcal{U}([1; 99])$. The higher ρ, the higher the correlation between P^1 and P^2. Preliminary experiments showed that mid-correlated FSP instances can be generated with $\rho = 0.6$ and high-correlated FSP instances with $\rho = 0.9$.

4.3 Experimental Results/Analysis

Table 1 presents results obtained for the FSP.

First, Table 1(a), shows that as expected, the number of neutral neighbors $\mathcal{N}_n$ decreases as the correlation between objectives increases. However, the size of the problem has no impact on $\mathcal{N}_n$ for a same level of correlation. For random solutions the proportion of promising neutral neighbors $\mathcal{P}_n$ is around 50% which shows they are well distributed among the neutral ones. This observation is not true for Pareto local optimal solutions, that have much less neutral neighbors and a smaller proportion of promising ones.

Table 1(b) focusses on the analysis of the first $\mathcal{H}$-layer ($\mathcal{H}_n^{0.1}$). For random solutions, $\mathcal{P}_n^{0.1} \backslash \mathcal{T}_n^{0.1}$ represents 2/3 of solutions of the layer for low correlation instances and one half for high correlated ones. On the contrary, for Pareto local optimal solutions, most of the solutions are within the triangle, which means very close to the initial solution s, and hence less interesting to consider for the diversity and the convergence. Therefore, in the following, only details on the distribution of neutral neighbors for random solutions will be given.

Table 1(c) indicates the distribution of promising neutral neighbors over $\mathcal{H}$-layers, according to several values of δ. The first observation is that the number of solutions decreases rapidly as the value of δ increases. Indeed, when a high degradation is allowed on one objective, a high improvement on the other one is required to compensate it.

The aim of Table 1(d) is to study solutions that produce a minimal improvement with a bounded degradation. The distribution among $\mathcal{V}$-layers shows that for low correlated instances, $\mathcal{V}$-layers with a medium δ have still an interesting number of solutions. This is less true for high correlated problems. Let us note that $\mathcal{V}_n^{-1}$ is not empty which means that there exist some neighbors that produce an improvement larger than the worst degradation encountered within the whole neighborhood.

As a conclusion for the FSP, it seems interesting to consider neutral neighbors during the search. Indeed, for random solutions, 50% of these ones are promising,

Table 1. Results for the bi-objective FSP (in percentage).

(a) Percentage of neutral and promising neighbors for both random and optimal solutions.

	Rnd		Opt	
	$\mathcal{N}_n$	$\mathcal{P}_n$	$\mathcal{N}_n$	$\mathcal{P}_n$
$\rho = 0$				
20	49.44	52.89	19.91	14.38
50	49.89	50.17	11.12	6.94
100	50.07	50.56	6.25	4.01
ρ: mid				
20	38.25	52.61	13.3	13.43
50	41.21	50.3	7.33	7.58
100	41.75	50.19	3.94	4.38
ρ: high				
20	23.11	51.3	5.53	14.47
50	23.01	50.72	2.82	9.7
100	25.76	50.91	1.62	6.15

(b) Distribution of promising neutral neighbors in the first horizontal layer.

	Rnd		Opt	
	$\mathcal{T}_n^{0.1}$	$\mathcal{P}_n^{0.1} \setminus \mathcal{T}_n^{0.1}$	$\mathcal{T}_n^{0.1}$	$\mathcal{P}_n^{0.1} \setminus \mathcal{T}_n^{0.1}$
$\rho = 0$				
20	11,2	22,35	7,19	2,62
50	15,64	21,12	4,66	0,74
100	16,68	21,28	3,37	0,27
ρ: mid				
20	13,44	22,78	7,48	2
50	18,13	21,14	6,25	0,52
100	18,94	20,95	4,25	0,05
ρ: high				
20	21,35	20,76	11,94	1,1
50	26,97	17,97	9,06	0,39
100	26,15	18,73	6,01	0,01

(c) Distribution of the promising neutral neighbors in the $\mathcal{H}$-layers (Random solutions).

	$\mathcal{H}_n^{0.1}$	$\mathcal{H}_n^{0.2}$	$\mathcal{H}_n^{0.3}$	$\mathcal{H}_n^{0.4}$	$\mathcal{H}_n^{0.5}$	$\mathcal{H}_n^{0.6}$	$\mathcal{H}_n^{1}$
$\rho = 0$							
20	33.55	10.74	4.28	1.84	1.08	0.56	0.84
50	36.78	9.08	2.71	0.97	0.37	0.17	0.09
100	37.96	9.17	2.48	0.68	0.2	0.05	0.02
ρ: mid							
20	36.22	10.25	3.55	1.45	0.62	0.29	0.23
50	39.28	8.11	1.99	0.6	0.21	0.07	0.04
100	39.9	7.98	1.77	0.42	0.1	0.02	0
ρ: high							
20	42.1	6.78	1.66	0.5	0.2	0.03	0.03
50	44.94	4.78	0.83	0.13	0.03	0.01	0
100	44.89	5.1	0.77	0.13	0.02	0	0

(d) Distribution of the promising neutral neighbors in the $\mathcal{V}$-layers (Random solutions).

	$\mathcal{V}_n^{0.1}$	$\mathcal{V}_n^{0.2}$	$\mathcal{V}_n^{0.3}$	$\mathcal{V}_n^{0.4}$	$\mathcal{V}_n^{0.5}$	$\mathcal{V}_n^{0.6}$	$\mathcal{V}_n^{1}$
$\rho = 0$							
20	33.55	33.09	24.93	18.25	13.56	10.32	8.52
50	36.78	30.21	18.37	10.82	6.52	4.04	2.56
100	37.96	30.45	17.25	8.96	4.58	2.37	1.25
ρ: mid							
20	36.22	33.02	22.82	15.35	10.44	7.24	5.38
50	39.28	29.25	16.03	8.51	4.61	2.65	1.61
100	39.9	28.93	14.79	6.96	3.23	1.53	0.74
ρ: high							
20	42.1	27.53	15.27	8.27	4.76	2.84	1.73
50	44.94	22.75	9.51	4.15	1.87	0.87	0.43
100	44.89	23.84	9.66	3.7	1.41	0.55	0.23

and more than $1/3$ of them belong to $\mathcal{H}_n^{0.1}$, so very close to the dominance part. Moreover, when objectives are not correlated, many promising neighbors produce an improvement much larger than the degradation they produce on the second objective (they belong to $\mathcal{V}_n^{\delta}$ for large values of δ while belonging to $\mathcal{H}_n^{\delta}$ for small values of δ). Regarding Pareto local optimal solutions, very few neutral neighbors are promising, so the interest of using them, except (maybe) to escape from them is not as relevant as during the search.

5   Case Study 2: Bi-objective Quadratic Assignment Problem - QAP

5.1   Presentation of the Bi-objective QAP

The quadratic Assignment Problem (QAP) consists in assigning a set of N facilities to a set of N given locations, minimizing a cost function that depends on the distance between locations and the flow required between associated facilities. This well-known single-objective form has been extended to a multi-objective form by Knowles and Corne [4]. Hence, the bi-objective form may be described as follows: let a distance-matrix D such that, $d_{i,j}$ is the distance between location i and location j, and two flow-matrices $(W^k)_{k=\{1,2\}}$ such that, $w_{i,j}^k$ is the required flow between facility i and facility j regarding the flow-matrix $k \in \{1, 2\}$. A solution is a permutation $\pi = \{\pi_1, \ldots, \pi_N\}$ where π_i denotes the location of the facility i. The objectives (to be minimized), f_k ($k \in \{1, 2\}$), are defined as:

$$f_k(\pi) = \sum_{i=1}^{N} \sum_{j=1}^{N} w_{i,j}^k d_{\pi_i, \pi_j} \tag{10}$$

The neighborhood operator considered in this paper is the exchange operator, where locations of two different facilities of a solution are switched. The number of neighbors per solution is then $N \times (N - 1)/2$.

5.2   Benchmark

Different types of instances have been proposed in the literature [4]. In this paper, we generated our own instances to control the correlation. First, D is a symmetric matrix of size $N \times N$ computed using the Manhattan distance between N points uniformly generated in the interval $[0; 99]$. Then, the values of W^1 are generated following $\mathcal{U}([0; 99])$. In the no-correlated version, W^2 is generated independently of W^1 following $\mathcal{U}([0; 99])$. For the two ρ-correlated versions, we use the coverage method to generate W^2 from W^1. If $\mathcal{U}([0; 1]) < \rho$, then $w_{i,j}^2 = w_{i,j}^1$ otherwise $w_{i,j}^2 = \mathcal{U}([0; 99])$. The higher ρ, the higher the correlation between W^1 and W^2. As for the FSP, the mid-correlated QAP instances are obtained with $\rho = 0.6$ and the high-correlated QAP instances are obtained with $\rho = 0.9$.

5.3   Experimental Results/Analysis

A synthesis of results is presented in Table 2. The first observation is that results are very similar to the FSP ones.

Table 2. Results for the bi-objective QAP (in percentage).

(a) Percentage of neutral and promising neighbors for both random and optimal solutions.

	Rnd		Opt	
	$\mathcal{N}_n$	$\mathcal{P}_n$	$\mathcal{N}_n$	$\mathcal{P}_n$
$\rho = 0$				
20	50,29	52,36	26,17	16,46
50	50,23	51,18	17,94	11,36
100	50,11	50,8	14,19	8,31
ρ: mid				
20	28,87	51,56	7,73	18,99
50	29,27	50,33	5,22	18,16
100	29,16	50,4	3,69	15,19
ρ: high				
20	13,39	50,54	2,05	28,51
50	14,23	50,24	1,41	26,86
100	13,85	50,25	0,94	25,24

(b) Distribution of promising neutral neighbors in the first horizontal layer.

	Rnd		Opt	
	$\mathcal{T}_n^{0.1}$	$\mathcal{P}_n^{0.1} \setminus \mathcal{T}_n^{0.1}$	$\mathcal{T}_n^{0.1}$	$\mathcal{P}_n^{0.1} \setminus \mathcal{T}_n^{0.1}$
$\rho = 0$				
20	5,9	17,97	7,8	2,73
50	8,47	20,14	8,68	1,23
100	10,78	21,1	7,51	0,37
ρ: mid				
20	10,25	23,31	15,29	2,44
50	13,99	23,53	17,45	0,49
100	17,35	23,16	15,01	0,14
ρ: high				
20	22,36	20,89	28,2	0,17
50	28,22	18,46	26,78	0,08
100	33,46	15	25,23	0,01

(c) Distribution of the promising neutral neighbors in the $\mathcal{H}$-layers (Random solutions).

	$\mathcal{H}_n^{0.1}$	$\mathcal{H}_n^{0.2}$	$\mathcal{H}_n^{0.3}$	$\mathcal{H}_n^{0.4}$	$\mathcal{H}_n^{0.5}$	$\mathcal{H}_n^{0.6}$	$\mathcal{H}_n^{1}$
$\rho = 0$							
20	23,88	12,3	7,2	3,94	2,15	1,33	1,56
50	28,61	12,61	5,71	2,49	1,07	0,42	0,27
100	31,87	11,98	4,52	1,62	0,56	0,18	0,07
ρ: mid							
20	33,57	11,2	4,06	1,74	0,57	0,26	0,16
50	37,51	9,65	2,41	0,6	0,13	0,03	0
100	40,51	8,01	1,56	0,27	0,04	0,01	0
ρ: high							
20	43,26	6,01	0,99	0,21	0,05	0,02	0
50	46,67	3,3	0,25	0,02	0	0	0
100	48,47	1,72	0,06	0	0	0	0

(d) Distribution of the promising neutral neighbors in the $\mathcal{V}$-layers (Random solutions).

	$\mathcal{V}_n^{0.1}$	$\mathcal{V}_n^{0.2}$	$\mathcal{V}_n^{0.3}$	$\mathcal{V}_n^{0.4}$	$\mathcal{V}_n^{0.5}$	$\mathcal{V}_n^{0.6}$	$\mathcal{V}_n^{1}$
$\rho = 0$							
20	23,88	30,28	28,48	24,05	19,02	14,91	12,56
50	28,61	32,74	26,67	19,15	12,79	8,13	5,17
100	31,87	33,07	23,79	14,86	8,58	4,69	2,48
ρ: mid							
20	33,57	34,52	25,76	18,06	12,14	7,78	5,37
50	37,51	33,17	20,17	10,87	5,47	2,71	1,31
100	40,51	31,17	16,09	7,26	3,12	1,27	0,52
ρ: high							
20	43,26	26,9	12,77	5,96	3,01	1,64	0,96
50	46,67	21,75	6,9	1,95	0,57	0,2	0,07
100	48,47	16,73	3,43	0,64	0,11	0,02	0

As for the FSP, Table 2(a), shows that the number of neutral neighbors $\mathcal{N}_n$ decreases as the correlation between objectives increases. However, the size of the problem has no impact on $\mathcal{N}_n$ for a same correlation. For random solutions the proportion of promising neutral neighbors $\mathcal{P}_n$ is around 50% which shows that they are well distributed among the neutral ones. This observation is not true for Pareto Pareto local optimal solutions, that have much less neutral neighbors and a smaller proportion of promising ones.

Table 2(b) focusses on the analysis of the first $\mathcal{H}$-layer. As for the FSP, for random solutions, $\mathcal{P}_n^{0.1}\backslash\mathcal{T}_n^{0.1}$ represents around 2/3 of solutions of the layer for low correlation instances and one half for high correlated ones. On the contrary, for Pareto local optimal solutions, most of the solutions are within the triangle, which means very close to the initial solution s, and hence less interesting to consider. Therefore, in the following, only details on the distribution of neutral neighbors for random solutions will be given.

Table 2(c) indicates the distribution of promising neutral neighbors over $\mathcal{H}$-layers, according to several values of δ. The first observation is that the number of solutions decreases rapidly as the value of δ increases. Indeed, when a high degradation is allowed on one objective, a high improvement on the other one is required to compensate it and to belong to the first layer.

Table 2(d) studies solutions that produce a minimal improvement with a bounded degradation. The distribution among $\mathcal{V}$-layers shows that for low correlated instances, $\mathcal{V}$-layers with a medium δ have still an interesting number of solutions. This is less true for high correlated problems. As for the FSP, there are some neighbors belonging to $\mathcal{V}_n{}^1$; such solutions produce an improvement larger than the worst degradation encountered within the whole neighborhood.

As a conclusion on the QAP problem, and for the same reasons than for the FSP, it seems interesting to consider neutral neighbors during the search, as a large part of them are promising ones. The same question stays open as for the consideration of neutral neighbors when the Pareto local optima is close.

6 Case Study 3: Bi-objective Traveling Salesman Problem - TSP

6.1 Presentation of the Bi-objective TSP

The Traveling Salesman Problem (TSP) can be defined on a complete weighted graph, where nodes are a set of cities and edges are the paths between cities. Here we consider the symmetric TSP – the graph is undirected – and weights are defined as the path lengths. A bi-objective version of the problem can be defined as follows: let N cities and two distance-matrices $(D^k)_{k=\{1,2\}}$ such that, $d_{i,j}^k$ is the distance to travel from city i to city j regarding the distance-matrix $k \in \{1,2\}$. A solution is a cyclic permutation $\pi = (\pi_1, \pi_2, ..., \pi_N)$ of the N cities. The objectives (to be minimized), f_k ($k \in \{1,2\}$), are defined as:

$$f_k(\pi) = \sum_{i=1}^{N}\sum_{j=1}^{N} d_{i,j}^k x_{i,j} \tag{11}$$

with $x_{i,j} = 1$ *iff* cities i and j are adjacent in the solution, otherwise $x_{i,j} = 0$.

The neighborhood operator considered in this paper is the two-opt operator, where two non-adjacent edges of a solution are removed and the resulting two parts are connected by two other edges such that a different tour is obtained [2]. The number of neighbors per solution is then $N \times (N-3)/2$.

6.2   Benchmark

To generate controlled correlated benchmarks[1], the coordinates of the N cities are first uniformly generated in $[0; 3163]^2$. Then, D^1 is computed from the coordinates applying the euclidean distance. In the no-correlated version, D^2 is generated independently of D^1 following the same protocol as D^1. For the two ρ−correlated versions, we use the normal distribution $\mathcal{N}(0, \rho)$ to move each city, and then compute the new euclidean distances between them for D^2. The lower ρ, the higher the correlation between D^1 and D^2. The mid-correlated TSP instances are obtained with $\rho = 600$ and the high-correlated TSP instances are obtained with $\rho = 150$.

6.3   Experimental Results/Analysis

A synthesis of results for the TSP is presented in Table 3.

First, Table 3(a), shows that the number of neutral neighbors $\mathcal{N}_n$ decreases faster as the correlation between objectives increases. However, the size of the problem has no impact on $\mathcal{N}_n$ for a same correlation. For random solutions the proportion of promising neutral neighbors $\mathcal{P}_n$ is still around 50% which shows that they are well distributed among the neutral ones. This observation is not true for Pareto local optimal solutions, that have very much less neutral neighbors and a smaller proportion of promising ones.

The analysis of the $\mathcal{H}$-layer (Table 3(b)) shows that for random solutions, $\mathcal{P}_n^{0.1} \backslash \mathcal{T}_n^{0.1}$ represents 4/5 of solutions of the layer for low correlation instances (more than FSP and QAP) and only 1/3 for high correlated ones (less than FSP and QAP). For Pareto local optimal solutions, these ratio are enforced.

The analysis of the distribution over $\mathcal{H}$-layers (Table 3(c)), shows some similarities with the 2 other problems. Indeed, in the same manner, the number of solutions decreases rapidly as the value of δ increases to reach the value 0 for large δ.

Table 3(d) shows that for low correlated instances, $\mathcal{V}$-layers with a medium δ have still an interesting number of solutions. This is less true for high correlated problems, where for large size problems this number may be null.

As a conclusion for the TSP, we can say that those results are slightly different from the FSP and the QAP. If for low correlated instances, the number of promising neutral neighbors encourages to consider neutral neighbors during the search, this is not true anymore for high correlated instances. This analysis contributes to show the important impact of correlation between objectives.

---

[1] The interval has been chosen as the famous instances proposed by the 8th DIMACS challenge (see http://dimacs.rutgers.edu/Challenges/TSP/).

Table 3. Results for the bi-objective TSP (in percentage).

(a) Percentage of neutral and promising neighbors for both random and optimal solutions.

	Rnd		Opt	
	$\mathcal{N}_n$	$\mathcal{P}_n$	$\mathcal{N}_n$	$\mathcal{P}_n$
$\rho = 0$				
20	50.76	51.15	29.9	9.84
50	50.06	49.79	21.12	7.29
100	50.08	50.27	16.2	5.01
ρ: mid				
20	26.53	50.79	10.04	17.46
50	27.42	50.26	8.62	13.66
100	27.84	50.28	6.84	10.14
ρ: high				
20	8.98	49.67	1.63	37.39
50	8.48	49.89	1.53	26.55
100	8.55	49.77	1.23	25.88

(b) Distribution of promising neutral neighbors in the first horizontal layer.

	Rnd		Opt	
	$\mathcal{T}_n^{0.1}$	$\mathcal{P}_n^{0.1} \setminus \mathcal{T}_n^{0.1}$	$\mathcal{T}_n^{0.1}$	$\mathcal{P}_n^{0.1} \setminus \mathcal{T}_n^{0.1}$
$\rho = 0$				
20	4.5	21.08	1.81	2.49
50	6.52	22.84	2.49	2.12
100	7.89	23.99	2.39	1.4
ρ: mid				
20	12.64	23.43	7.36	5.13
50	16.89	23.69	8.25	3.51
100	18.66	23.47	7.26	2.02
ρ: high				
20	37.65	10.66	32.65	3.02
50	43.71	5.97	26.21	0.34
100	44.76	4.93	25.61	0.27

(c) Distribution of the promising neutral neighbors in the $\mathcal{H}$-layers (Random solutions).

	$\mathcal{H}_n^{0.1}$	$\mathcal{H}_n^{0.2}$	$\mathcal{H}_n^{0.3}$	$\mathcal{H}_n^{0.4}$	$\mathcal{H}_n^{0.5}$	$\mathcal{H}_n^{0.6}$	$\mathcal{H}_n^1$
$\rho = 0$							
20	25,58	11,51	6,25	3,61	1,96	1,09	1,15
50	29,37	11,2	5,24	2,4	0,98	0,38	0,22
100	31,88	11,14	4,61	1,77	0,62	0,18	0,07
ρ: mid							
20	36,08	9,6	3,2	1,16	0,48	0,22	0,05
50	40,58	7,69	1,63	0,3	0,05	0,01	0
100	42,13	6,84	1,16	0,14	0,01	0	0
ρ: high							
20	48,31	1,25	0,11	0	0	0	0
50	49,68	0,21	0	0	0	0	0
100	49,69	0,08	0	0	0	0	0

(d) Distribution of the promising neutral neighbors in the $\mathcal{V}$-layers (Random solutions).

	$\mathcal{V}_n^{0.1}$	$\mathcal{V}_n^{0.2}$	$\mathcal{V}_n^{0.3}$	$\mathcal{V}_n^{0.4}$	$\mathcal{V}_n^{0.5}$	$\mathcal{V}_n^{0.6}$	$\mathcal{V}_n^1$
$\rho = 0$							
20	25.58	32.59	32.22	28.39	23.28	18.83	15.05
50	29.37	34.05	29.46	22.41	15.54	9.94	6.03
100	31.88	35.13	28.23	19.65	12.27	6.96	3.68
ρ: mid							
20	36.08	33.04	23.56	15.2	9.6	6.13	3.66
50	40.58	31.38	17.69	8.66	3.97	1.72	0.7
100	42.13	30.31	15.51	6.67	2.49	0.87	0.28
ρ: high							
20	48.31	11.92	2.81	0.76	0.22	0.09	0.02
50	49.68	6.18	0.35	0.03	0	0	0
100	49.69	5.02	0.16	0	0	0	0

7 Conclusion

This article aims at providing a study on the distribution of neutral neighbors and in particular of the most promising ones. Therefore a definition of promising neutral neighbors has been proposed.

The analyses of random solutions of three bi-objective permutation problems show that, the high number of promising neutral neighbors encourages to consider neutral neighbors during the search, as it may give the opportunity to find other good solutions (even if they do not dominate the initial solution s) and to continue the search in different areas of the search space. These analyses also highlight that the correlation between objectives plays an important role. Indeed, when this correlation is high, first, the number of neutral neighbors decreases, and even if almost 50% are promising ones (according to the definition we proposed), they are located very close to the initial solution s and may be not very interested to consider in an archive as they will require a lot of efforts for storage and exploration without providing much more convergence neither diversity. The less correlated the objective functions, the more interesting to be considered the neutral neighbors. The interest of exploiting neutral neighbors had been showed previously on a version of the permutation flowshop scheduling problem. In this paper, we discussed more widely the opportunity to exploit neutral neighbors for bi-objective permutation problems. Indeed, regarding all these analyses, the way to consider neutrality within multi-objective search methods (and in particular local search methods that use the neighborhood concept) is now more clear, and hopefully will lead to design more efficient methods.

References

1. Blot, A., Aguirre, H., Dhaenens, C., Jourdan, L., Marmion, M.-E., Tanaka, K.: Neutral but a winner! How neutrality helps multiobjective local search algorithms. In: Gaspar-Cunha, A., Henggeler Antunes, C., Coello, C.C. (eds.) EMO 2015. LNCS, vol. 9018, pp. 34–47. Springer, Heidelberg (2015). doi:10.1007/978-3-319-15934-8_3
2. Croes, G.A.: A method for solving traveling-salesman problems. Oper. Res. **6**, 791–812 (1958)
3. Galván-López, E., Poli, R., Kattan, A., O'Neill, M., Brabazon, A.: Neutrality in evolutionary algorithms. What do we know? Evol. Syst. **2**(3), 145–163 (2011)
4. Knowles, J., Corne, D.: Instance generators and test suites for the multiobjective quadratic assignment problem. In: Fonseca, C.M., Fleming, P.J., Zitzler, E., Thiele, L., Deb, K. (eds.) EMO 2003. LNCS, vol. 2632, pp. 295–310. Springer, Heidelberg (2003). doi:10.1007/3-540-36970-8_21
5. Marmion, M.-E., Aguirre, H., Dhaenens, C., Jourdan, L., Tanaka, K.: Multi-objective neutral neighbors': what could be the definition (s)? In: GECCO 2016, pp. 349–356. ACM (2016)
6. Marmion, M.-E., Dhaenens, C., Jourdan, L., Liefooghe, A., Verel, S.: Nils: a neutrality-based iterated local search and its application to flowshop scheduling. In: EvoCOP, pp. 191–202 (2011)
7. Taillard, E.: Benchmarks for basic scheduling problems. Eur. J. Oper. Res. **64**(2), 278–285 (1993)

Multi-objective Adaptation of a Parameterized GVGAI Agent Towards Several Games

Ahmed Khalifa[1]([✉]), Mike Preuss[2], and Julian Togelius[1]

[1] Department of Computer Science and Engineering,
New York University, Brooklyn, NY 11201, USA
ahmed.khalifa@nyu.edu, julian@togelius.com
[2] Department of Information Systems,
Westfälische Wilhelms-Universität Münster, Münster, Germany
mike.preuss@uni-muenster.de

Abstract. This paper proposes a benchmark for multi-objective optimization based on video game playing. The challenge is to optimize an agent to perform well on several different games, where each objective score corresponds to the performance on a different game. The benchmark is inspired from the quest for general intelligence in the form of general game playing, and builds on the General Video Game AI (GVGAI) framework. As it is based on game-playing, this benchmark incorporates salient aspects of game-playing problems such as discontinuous feedback and a non-trivial amount of stochasticity. We argue that the proposed benchmark thus provides a different challenge from many other benchmarks for multi-objective optimization algorithms currently available. We also provide initial results on categorizing the space offered by this benchmark and applying a standard multi-objective optimization algorithm to it.

Keywords: Multi-objective optimization · GVGAI · MCTS

1 Introduction

Very many problems more or less naturally lend themselves to a multi-objective formulation, and thus to be solved or understood through multi-objective optimization. This creates a demand for better multi-objective optimization algorithms, which in turn creates a demand for fair, relevant and deep benchmarks to test those algorithms. It would therefore seem important for the furthering of research on multi-objective optimization to create such benchmarks, ideally through building on real, challenging tasks from important application domains.

In the research field of AI and games, there is a long-standing tradition of benchmarking algorithms through their performance on playing various games. For a long time beginning very early on in the history of computer science, Chess was one of the most important AI benchmarks, and advances in adversarial search were validated through their performance on this classic board game. Later on, as classic board games such as Chess, Checkers and eventually Go

© Springer International Publishing AG 2017
H. Trautmann et al. (Eds.): EMO 2017, LNCS 10173, pp. 359–374, 2017.
DOI: 10.1007/978-3-319-54157-0_25

succumbed to the advances of algorithms and computational power, attention has increasingly turned to video games as AI benchmarks. The IEEE Conference on Computational Intelligence and Games now hosts a plethora of game-based AI competitions, based on games such as *Super Mario Bros* [23], StarCraft [14], *Unreal Tournament* [10], and *Angry Birds* [19]. These offer excellent opportunities to benchmark optimization and reinforcement learning algorithms on problems of real relevance.

For many—but not all—games, one can easily identify a single and relatively smooth success criterion, such as score or percent of levels cleared. This makes it possible to use single-objective optimization algorithms, such as evolutionary algorithms, to train agents to play games. For example, when evolving neural agents to drive simulated racing cars, one can simply (and effectively) use the progress around the track in a given time span and as the fitness/evaluation function [21]. However, in many cases the training process can benefit from multi-objectivization, i.e. splitting the single objective into several objectives, or constructing additional objectives in addition to the original one. For example, when using genetic programming to learn a car-driving agent, adding an extra objective that promotes the use of state in the evolved program can improve performance and generalization ability of the agent [1]. In another example, car-driving agents were evolved both to drive well and to mimic the driving style of a human; while these objectives are partially conflicting, progress towards one objective often helps progress towards the other as well [25]. Finally, when using optimization methods for procedural content generation—generating maps, levels, scenarios, items and such game content—the problem is often naturally multi-objective. For example, when evolving balanced maps for *StarCraft*, it is very hard to formulate a single objective that accurately captures the various aspects of map quality we are looking for [22].

So far we have talked about playing individual games. Recently, there has been a trend toward going beyond individual games and addressing the problem of *general game playing*. The idea here is that to be generally intelligent, it is not enough to be good at a single task: you need to have high performance over some distribution of tasks. Therefore, general video game playing is about creating agents that can play not just a single game, but any game adhering to a specific interface. The *General Video Game AI (GVGAI) Competition* was developed in order to provide a benchmark for general video game playing agents. The competition framework features several dozens of video games similar to early 80s arcade games, and competitors submit agents which are then tested on *unseen* games, i.e. games that the competitor did not know about when submitting the agent [15, 16].

In other words, general video game playing agents should be optimized for playing not just a single game well, but a number of different games well. This immediately suggests a multi-objective optimization benchmark: select a number of games, and optimize an agent to perform well on all of them, using the performance on each game as an objective.

This paper describes a multi-objective optimization benchmark constructed on top of the GVGAI framework. We first describe the underlying technologies this builds on, including the GVGAI framework (Sect. 2), the Monte Carlo tree search algorithm (MCTS, Sect. 3), and the specific parameterizable agent representation developed for the benchmark (Sect. 4). We then quantitatively characterize the benchmark through sampling in the space of agent parameters, and by applying a standard multi-objective optimization algorithm on several versions of the benchmark. More precisely, our experimental analysis (Sect. 5) shall answer the following questions:

- Which games enable learning via proper feedback and are different enough from each other so that multi-objective optimization makes sense?
- What is the experienced noise level and how much noise can be tolerated while doing optimization?
- How do we aggregate scores, winning, and set the run length in a meaningful way? What are the properties of the resulting optimization problem?

Finally, we summarize the features of the obtained optimization problem and compare the setting to the one of other recent multi-objective benchmarks/competitions in Sect. 6 before concluding in Sect. 7.

The main contribution of this paper is a new multi-objective optimization benchmark based on general video game playing, which we hope can help the development of better such optimization algorithms. However, we also envision that this tool leads to an improved understanding of the MCTS algorithm, which our parameterizable agent builds on. Employing multi-objective optimization algorithms could further help elucidating the relations between different games based on what agent configurations play them well. A first contribution in this direction is some insight about the conflicts between adapting the agents towards different games, based on an experimental analysis. We assume that very wide and large Pareto front approximations stand for strongly conflicting game requirements (towards the agent), whereas small and narrow Pareto front approximations mean that the gameplay is rather similar and can be accomplished well by only one agent (or that the problem is so difficult that getting near to the real Pareto front is very hard).

2 Extending the General Video Game AI Framework

The General Video Game AI (GVGAI) framework is a Java framework that allows running different games. Most of these games are ports of old Arcade of home computer games or some newer indie games. The GVG-AI framework has been continuously extended and now contains more than 80 implemented games. Figure 1 shows 4 different games implemented in the framework where they range from puzzle to action/arcade games. The framework also enables users to easily design new games by describing them in the Video Game Description Language (VGDL) [7]. VGDL is a declarative description language that enables defining the game itself and the used levels (usually 5 levels are defined for each game) in

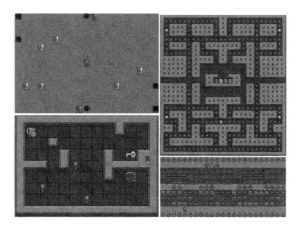

Fig. 1. Different games in the GVGAI framework in order from top left to bottom right: Defem, Pacman, Zelda, and Frogs.

a human-readable form. The game description contains information about game objects, interactions, goals, etc., While the level description contains information about how game objects are spatially arranged.

The GVGAI framework was initially introduced to allow people to implement their own AI agents that can play multiple of unseen games. The best way to encourage people to use a new framework is by organizing competitions. Multiple different competition were organized at different conferences starting at Computational Intelligence and Games (CIG) 2014 [16]. The best agents so far can win only 50% of 50 different games in the framework [3]. After the success of this competition, multiple different tracks were introduced in 2016, such as the level generation track [12] and the 2 player planning track [9]. In the level generation competition, the participants try to design a general level generator that generates game levels for any game provided its description. The two-player planning track is similar to the original track but for games that have two players. Games can be cooperative or competitive based on the game description. In all competition tracks, agents don't know what game they are playing and have to figure it out based on interactions or simulations.

In this work, we are introducing a new track for the GVGAI competitions, intended to attract researchers from a different area, namely multi-objective optimization. The goal for this competition is to optimize a parameterized Monte-Carlo Tree Search (MCTS) algorithm to perform well on different games. This agent and the underlying main algorithm will be described in Sects. 3 and 4.

This new competition track (and its associated benchmark software) aims to obtain a deeper understanding of the induced landscape of different MCTS Upper Confidence Bound equations (explained in Sect. 3). This knowledge will then enable creating better hyper-heuristic agents [13]. Hyper-Heuristic agents are AI agents that can change the current algorithm or heuristic (parameters of the basic equation in our current context), based on the current game. By means

of our approach, we also provide a new benchmark for multi-objective optimization, with a fair number of parameters (14 in our current version), a scalable
number of objectives (we use 3 here but this only depends on the number of
games taken into account), also adding limited noise to problem features, as
could be expected in many real-world scenarios. Thus we obtain a new tool for
measuring the performance of different multi-objective algorithms under near-
realistic conditions, with the ability to explain (after revealing the games and
the exact encoding) the MCTS agent behavior that results from the best found
solutions.

3   Monte-Carlo Tree Search

Monte-Carlo Tree Search (MCTS) [6] is a stochastic game tree search algorithm
that balances between exploiting the best nodes and exploring new nodes. As
usual, every node of the tree is associated with one possible action that may be
played next. Every specific path through the tree resembles the moves of one
game (or at least the first n moves of the game if the game has n levels). MCTS
is divided into four main steps:

1. **Selection:** the algorithm selects a suitable node to investigate further. As
   mentioned before, it tries to balance between exploiting the best nodes and
   exploring new nodes. It uses an equation called Upper Confidence Bound
   (UCB) to achieve this balance.
2. **Expansion:** the algorithm selects which child node to expand at the current
   node. If there is no more nodes to expand, the algorithm returns to step 1.
   In vanilla MCTS, it picks a random node to be expanded.
3. **Simulation:** the algorithm simulates what will happen in future by applying
   a forward model. It plays out random actions until a terminal state is reached
   (win or loss), time is up, or a predefined depth is reached.
4. **Backpropagation:** the algorithm evaluates the simulated state (by means
   of a heuristic if it is not a terminal state), then it updates all parent nodes
   up to the root. This means that for every node, the information stored in
   the children is aggregated. After this step, the algorithm repeats all these
   steps starting from the root, until it finishes (usually after a specified time or
   number of iterations).

   After the whole algorithm has been terminated (there is no natural ending),
we have to choose one move that is actually performed, and usually the one
with the highest win rate is chosen. The nodes that reside directly under the
root aggregate all known information about the consequences of what happens
after the action associated with this node is played next.

   The most well-known UCB equation is UCB1 (1), and it is divided into two
terms. The first term pushes search to exploiting the best node so far while the
second part controls the effort spent on exploring new nodes.

$$UCB1 = \overline{X_j} + C\sqrt{\frac{2 \cdot \ln n}{n_j}} \tag{1}$$

In addition to many other extensions (such as integration of heuristics that represent human expertise) to classic MCTS agents [20], researchers have explored modifying the UCB equation in order to obtain different behaviors of MCTS agents. Some of these modification are done manually, such as using MixMax instead of the average value as the exploitation term [8,11,17]. Other researchers used genetic programming to find alternatives for the current UCB equation. These approaches have been applied to different games such as Go [5] by Cazenave or games in the GVG-AI framework [4] by Bravi et al.

4   A Parameterizable Agent as Optimization Problem

The findings from Bravi's work provide us with multiple different UCB equations for different games. Each equation is evolved over a certain game and tested against all other games in the framework. The evolved equations have terms that explain the nature of the game. For example; if the game goal is to collect all the moving butterflies that move randomly everywhere. The evolved equation will have a term to explore different map locations to find all of them. In this work, we combined features from all of the evolved equations from Bravi's work into one big, parameterized equation, depicted in (2). This equations can be seen as a very general formula for node selection, which can be specified into a very large number of different strategies depending on its parameterization.

$$UCB_{opt} = P_0\overline{X_j} + P_1 max(X_j) + P_2(\frac{\ln n}{n_j})^{P_3} + P_4 E_{xy}^{P_5} + P_6 min(D_{npc})^{P_7}$$
$$+ P_8 min(D_{port})^{P_9} + P_{10} min(D_{mov})^{P_{11}} + P_{12} min(D_{res})^{P_{13}} \qquad (2)$$

Here, X_j is the total reward of the current node, n is the number of visits for the parent node, n_j is the number of visits for the current node, E_{xy} is the number of times a certain tile is visited providing its x,y position, D_{npc} is the distance to a NPC[1] object in the game, D_{port} is the distance to a portal object in the game, D_{mov} is the distance to a movable object in the game, and D_{res} is the distance to a resource object in the game.

The first two terms are the most frequently ones representing exploitation in the evolved equations. If their coefficients equal q and $1-q$, we have the MixMax term. The rest of the terms stand for different ways of exploration, either based on the tree information (3rd term), space information (4th term), or different game sprite information (the rest of the terms).

To summarize, we have 14 different parameters to set (or optimize), where each parameter can take values between -1 to 1. Each term in Eq. 2 has a different meaning, based on the sign of the parameter. However, we can expect that the first two terms (exploitation terms) most likely will have positive values. The other terms (exploration terms) have different meanings based on the sign. Zero value means this term doesn't have any effect on the agent behavior. Negative values mean that this node/tile/sprite should be avoided while playing the game,

---

[1] Non-player character, usually but not necessarily adversarial.

positive values mean that getting closer to a sprite or accessing the same visited tile/node will be rewarded.

In the remainder of this section, we provide some information concerning implementation and interface, being well aware that the actual competition setting may differ from this in details.

We wanted our interface to be as accessible as possible and the problem be treated as any other (noisy) black box optimization problem. Therefore, we designed the interface to be language independent. A solution (14 numbers) is evaluated by providing it as set of command line parameters to an executable Java archive (JAR) file. After running the parameterized MCTS on all three games a predefined number of times, the average objective value per game is written to a file, together with a control flag that indicates if the execution ended correctly or not. The objective values have been encoded according to (3), with *win* being 1 if the game is won, and 0 else. This weights winning the game much higher than achieving a high score, and for most GVGAI games the difference is huge. However, there is no defined maximum score that can be obtained for one game, so that the ratio can vary.

$$f_{\text{game}} = win + \frac{score}{1000} \tag{3}$$

Whenever we perform test runs with a multi-objective optimization method (we employ the SMS-EMOA [2] as it is one of the most popular ones, but it could also be another one) in the following, we switch to minimization by means of the transformation $f_{\text{sms}} = 3 - f_{\text{game}}$, resulting in a Nadir point at $(3, 3, 3)$. We take this also as reference point, such that the maximal (theoretical) hypervolume between the Pareto front and the origin is 27. However, this would mean that we would win every single game with a score of 2000, which is not possible in most games. The real maximum is therefore unknown, but considerably smaller. E.g., if the achieved Pareto front approximation contains points that realiably win every one of the 3 games (but do not get a high score), the hypervolume would be in the range of $3^3 - 2^3 = 19$.

5   Experimental Analysis

By means of our experimental analysis, we generally investigate the suitability of the previously described setup as benchmark problem. We do this in several steps. At first, we select appropriate games by looking at their adaptability value [18]. We then consider the noise level in order to find out how stable single measurements are and how they must be aggregated. Finally, we perform some test optimization runs, relying on the SMS-EMOA [2] in order to see how competition results would look like.

5.1   Game Selection

From the currently available 80 games, we picked 12 that seemed to be interesting and suitable for tackling them with the parametrizable MCTS-based agent described in Sect. 4. These games were selected from different clusters based on Bontrager's work [3]. They were selected equally from the top 3 hardest clusters (all clusters except the easiest one). Additionally, we take into consideration that the games shall either be solvable easily (i.e. Frogs) or provide feedback often (i.e. Pacman). We avoided long rewarding puzzle games because optimization methods would not get any improvement feedback except from *win* and *loose*. This is the list of the 12 considered games (Fig. 1 shows some screenshots):

- **Defem:** a port of an indie-game with the same name. The goal of the game is to survive. The player should avoid/kill enemies that are spawning in the level. The player shoots automatically in any random direction.
- **Eggomania:** a port of a mini-game in Pokemon stadium 2. The goal is to catch all the falling eggs before they reach the ground.
- **Frogs:** a port of Frogger, a famous Atari game. The goal is to cross the street and the lake to the other side without getting hit by cars or drowning in water.
- **Modality:** a puzzle game about two different dimensions. The goal is to push a tree to its correct spot. The level is divided into two colors, each color represent a separate dimension. The player/tree can only pass from one dimension to the other through certain spots in the map.
- **Pacman:** a port of Pacman game. The goal is to eat all the pills without being caught by the chasing ghosts.
- **Painter:** a puzzle game where the player plays as a painter. The goal is to change all the level tiles to be colorful. The tile color toggles between plain and colorful when the player pass over it.
- **Plants:** a port of the indie-game Plants vs Zombies. The goal is to prevent waves of zombie from reaching your house. The player can grow plants that shoot the zombies to kill them and the zombies attack back when they are close enough.
- **Zelda:** a port of the dungeon system in The Legend of Zelda. The goal is to get the key and go to the exit without getting killed. The player can kill the enemies using his sword.
- **Boulderdash:** a port of Boulder Dash a famous Atari game. The goal is to collect 10 gems and go towards the exit without dying. The player should avoid falling boulders and monsters while searching for the gems.
- **Sokoban:** a variant of Sokoban, a famous Japanese puzzle game. The goal is to destroy all the boxes in the level by pushing them into holes.
- **Solarfox:** a port of Solarfox a famous Atari game. The goal is to collect all the jewels and avoid hitting the borders of the screen or getting hit by enemy missiles.
- **Roguelike:** an action game. The goal is to collect as much treasure as you can and reach the exist. The path toward the goal is filled with moving monsters and doors. The player needs to get a sword to protect itself from the monsters, and collect keys to open the locked doors in its way.

We now analyze how well suited to optimization the different games are by performing a well distributed sample over the search space $[0, 1]^{14}$, do repeated measuring and then look at the empirical tuning potential (adaptability, Eq. (4)) value as introduced in [18]. The ETP was originally considered to measure how easy it is to obtain a better algorithm configuration by tuning, but the situation here is quite similar (repeated runs due to noise, unknown peak performance, vast parameter space). It takes the performance of the best $(\mathbf{y}_p)$ and an average configuration $(\mathbf{y}_a)$ and the semi-quartile range (non-parametric alternative to standard deviation) at these points into account. The higher the computed value, the easier it is to improve the performance, starting with an average configuration. In case one semi-quartile range is zero, we set the ETP value to zero as well.

$$\mathrm{ETP}(\mathbf{y}_p, \mathbf{y}_a) := \frac{\mathrm{median}(\mathbf{y}_a) - \mathrm{median}(\mathbf{y}_p)}{\mathrm{sq}(\mathbf{y}_a)} \cdot \frac{\mathrm{median}(\mathbf{y}_a) - \mathrm{median}(\mathbf{y}_p)}{\mathrm{sq}(\mathbf{y}_p)} \quad (4)$$

We perform the following experiment in order to remove unsuitable games - games with very low ETP values - from this set.

Experiment: which games are well suited for a multi-objective benchmark?

Pre-experimental planning. After some preliminary tests, we found that 50 samples per game with 10 repeats each is a good compromise between runtime and result quality. This combination requires about 30 h of computation time for the 12 games[2].

Task. What ETP value do we require in order to keep a game? As concrete values are difficult to estimate, we will order the games according to their ETP and then look for a gap between "near zero" and "clearly larger than zero".

Setup. We generate a common sample of 50 well distributed 14 parameter controller configurations by means of the MaxiMin reconstruction (MMR) method of Wessing [26], and run the agent 10 times for each sample point and each game, recording the objective values.

Results/Visualization. Table 1 holds the measured ETP values for each game. Note that the value gets zero as soon as one of the semi-quartile ranges get zero.

Table 1. Expected tuning potential (ETP) for the 12 considered games.

Game	Defem	Sokoban	Roguelike	Frogs	Zelda	Boulderdash
ETP	0.72	0.0	0.0	0.0	144.65	0.0
Game	Modality	Pacman	Eggomania	Painter	Solarfox	Plants
ETP	0.0	16.49	0.0	891.37	46.25	11.54

Observations. 6 of the games have a zero value, whereas the non-zero values for the other games vary quit a lot. It seems that Painter is the game with the best adaptation/optimization potential, and Defem the one with the worst.

---

[2] Using a single core of an Intel i7-3770 CPU @ 3.4 GHz.

Discussion. Although there is some variance in the observed point samples for Frogs, Eggomania, and Modality, the ETP values clearly recommend only using Defem, Zelda, Pacman, Painter, Solarfox, and Plants. However, Solarfox is a special case: one cannot win, not even have a positive score, but only loose with different negative scores. In consequence, the ETP is positive, but nearly all objective function values for the game would lead to a point beyond the Nadir point (such that the gradual differences in the signal would be cut out). Frogs and Modality provide basically binary feedback (won or lost), and for Eggomania more than half of the samples end up with 10 losses without any point in a row. It seems reasonable that these are not really well suited for optimization; if they are used, an additional difficulty is added (fitness cliffs). We thus end up with a game selection of 8 games: Defem, Zelda, Pacman, Painter, Plants, Frogs, Eggomania, and Modality.

We can also state that the ETP value provides valuable information, but the necessities of the optimization process may lead to a positive ETP value when the game is still unsuitable (in the case of Solarfox). Adding 3 more games with (almost or completely) binary signals may make the problem more interesting and shall be considered.

5.2 Analyzing Variance and Noise

In this section, we analyze the parameterized agent and check its response to noise and variance. We ran two experiments, the first one is running each game using random parameters for 2000 times (without repetitions). Table 2 shows the result of this experiment. As you can see, the mean is very low in all games except for Modality and Painter. The reason is that both these games have a very small map with a size of 4×4 tiles which make it easy for an agent to win it. Also, we can notice there is no agent has won pacman. The main reason is that the pacman map is very huge which takes a lot of time to finish it. Only Modality, Painter, and Frogs have a very high standard deviation compared to the other games which shows that these games are more sensitive to the parameter change than others (or, as described above, have few very extreme attained objective values).

Table 2. Average statistics over 2000 random configurations.

Game	Mean	Std	Min	25%	50%	75%	Max
Defem	0.033	0.099	0.0	0.001	0.032	0.046	1.05
Frogs	0.105	0.306	0.0	0.0	0.0	0.0	1.001
Modality	0.583	0.494	0.0	0.0	1.001	1.001	1.001
Painter	0.561	0.442	0.039	0.118	0.395	1.042	1.482
Zelda	0.011	0.102	−0.001	−0.001	0.0	0.002	1.008
Pacman	0.025	0.036	0.0	0.002	0.009	0.03	0.244
Eggomania	0.01	0.095	0.0	0.0	0.0	0.001	1.116
Plants	0.045	0.154	0.003	0.016	0.021	0.026	1.109

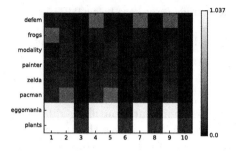 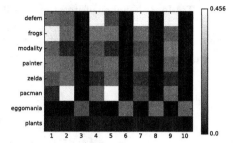

Fig. 2. Average values for 1000 runs over all the games vs. 10 different random parameter configurations. (Color figure online)

Fig. 3. Standard deviation values for 1000 runs over all the games vs. 10 different random parameter configurations. (Color figure online)

In the second experiment, we picked 10 random configurations for the parameters, then we ran the configured agent on each game repeatedly 1000 times. Figure 2 shows a heat map representing the average values of the 1000 runs for all 8 games and all 10 configurations. Darker colors mean the agent neither wins nor obtains a descent score, brighter colors otherwise. Figure 3 shows the standard deviation of the previous experiment. Brighter colors shows more variance in their results than darker colors. From analyzing the heatmaps, we can notice that agent configuration numbers 3, 6, 8, and 10 loose almost all games and therefore have small variance. Also, we can notice that both on Eggomania and Plants repeated measurements of single agent configurations have a very low standard deviation. This means that these games behave more deterministic than the others. On the other hand, some configurations result in very high standard deviations on Defem. One could argue that this is the noisiest game in the set. Pacman is somewhere in between, most likely due to the huge amount of pills and the presence of the chasing ghosts, factors that surely increase variance. Eggomania and Plants are very sensitive to the agent configuration as almost 50% of all the configurations lose the game.

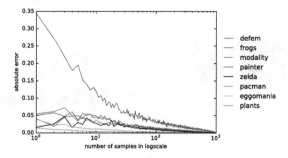

Fig. 4. Absolute error of a point averaged from repeated samples (resampled 50 times).

Table 3. Number of iterations until the average error goes below 0.01.

Game	Defem	Frogs	Modality	Painter	Zelda	Pacman	Eggomania	Plants
Iterations	181	586	148	181	114	15	1	2

Figure 4 shows the absolute error of the first configuration using different number of samples for all 8 games. From the graph, we can see that all games have error values around 0.05 from the start except for Frogs. Frogs is a very noisy game, it takes 150 resamples to obtain an error of less than 0.05. Table 3 shows the number of resamples necessary to get an error of less than 0.01. We notice that Pacman, Eggomania, and Plants need quite few resamples to reach an error of less than 0.01 (at most 15 iterations). Therefore, these games are more suitable for the optimization setting than others.

5.3   Test Optimization Runs

We perform some example runs with a standard multi-objective optimization algorithm in order to find out how well the established optimization problems are suited to be used as basis for a competition.

Experiment: Is the problem difficulty adequate? Do we obtain interesting fronts?

Pre-experimental planning. The run length of 500 evaluations (x3) with 2 repeats each (averaged) is designed to add up to a wallclock time of around 12–15 hours per run (single core).

Task. Our expectation is rather fuzzy: we strive for some variance in the runs, but also some similarity (the problem should be difficult enough but not too difficult). Also, we expect to find large fronts (more than 20 is not possible in this setup), especially not a single dominating optimum.

Setup. We employ the SMS-EMOA [2] with a population size of 20 individuals and otherwise standard parametrization. We choose the games Painter, Pacman, and Plants as objectives, based on our previous findings (others are possible).

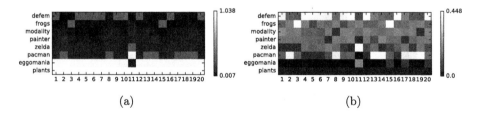

Fig. 5. (a) shows the average fitness of 500 repetitions of the 20 solutions of the best Pareto front approximation of one run on {Painter, Pacman, Plants}, using the SMS-EMOA. (b) shows the standard deviations of the values measured.

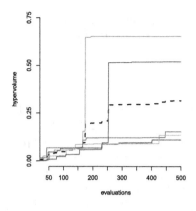

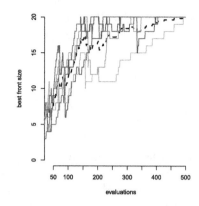

Fig. 6. Best dominated hypervolume of 5 SMS-EMOA runs on Painter, Pacman, and Plants. Average is dashed.

Fig. 7. Best front sizes over the same 5 runs. Average best front size is dashed. Population size is 20.

Results/Visualization. Figures 6 and 7 show the dominated hypervolume and number of solutions in the best front over 5 runs, respectively. Additionally, we take the resulting front of the first optimization run and measure how good the solutions do on all 8 games (average and standard deviation shown in Fig. 5).

Observations. We see gradual improvements in the hypervolume as well as huge jumps. There is a common tendency to improve, but with different speeds. The number of solutions in the best front starts around 5 and reaches the maximum of 20 around 250 evaluations (with one exception).

Discussion. Obviously, the optimization problem is neither trivial nor unfeasible, which corresponds well to our expectations. The amount of solutions in the best front supports that multi-objective optimization actually makes sense here: we have conflicts between the objectives. In future, this shall be investigated further, also broadening the scope to more games.

6   Optimization Problem Characterization

As performance measure for comparing different algorithms on this problem we want to employ the hypervolume dominated by the best Pareto front approximation. This is similar to what is used in the other recent multi-objective competitions/benchmarks, the Bi-objective BBOB [24] and the Black Box Optimization Competition BBComp[3] that features 2 and 3-objective tracks. However, unlike the other recent multi-objective competitions/benchmarks, we have noisy function evaluations here, with 2 types of noise in different strengths, depending on the actual game. If an agent can win often but not always, we have an almost binary noise, only the aggregation of several measurements gets us nearer to the

---

[3] http://bbcomp.ini.rub.de.

real function value. The other type of noise is more gradual and stems from the different rewards the agents get, this is especially visible for Pacman.

7 Conclusions

In this paper, we introduced a multi-objective optimization benchmark based on general video game playing, specifically the GVGAI framework. The task is to optimize the parameters of a general video game playing agent so that it plays several rather different games as well as possible. A parameterizable agent was developed for this purpose, where the parameters are coefficients in a complex formula for node selection in a Monte Carlo Tree Search Algorithm.

Testing was conducted to find a set of games on which the agent would have a large performance variance depending on its parameterization, and which would require so different playing styles that it would be likely that optimizing the agent's performance on them would yield partially conflicting objectives. A set of such games were found, and testing of random parameters as well as multiple samples of the same parameters showed that the variance in performance value was not only high between different configurations, but also relatively high between different samples of the same configuration. Attempts to optimize agents using the SMS-EMOA algorithm showed that it is indeed possible to find good solutions, that the objectives as expected are partially conflicting, and that there is relatively high noise in the fitness evaluation. We believe that these characteristics reflect many real-world problems, and that a benchmark with such characteristics provides a useful complement to existing multi-objective ones.

Finally, it should be noted that the framework and experiments described here are useful not only from the perspective of benchmarking multi-objective optimization algorithms, but also from the perspective of exploring the design space of games and generating new game-playing agents. It is for example an interesting idea to sample several agents from different positions on a Pareto front and then use hyper-heuristic methods to select the most appropriate agent parameterization for a particular game [3,13].

References

1. Agapitos, A., Togelius, J., Lucas, S.M.: Multiobjective techniques for the use of state in genetic programming applied to simulated car racing. In: IEEE Congress on Evolutionary Computation, IEEE 2007, pp. 1562–1569 (2007)
2. Beume, N., Naujoks, B., Emmerich, M.: SMS-EMOA: multiobjective selection based on dominated hypervolume. Eur. J. Oper. Res. **181**(3), 1653–1669 (2007). issn: 0377–2217
3. Bontrager, P., Khalifa, A., Mendes, A., Togelius, J.: Matching games, algorithms for general video game playing. In: AIIIDE (2016)
4. Bravi, I., Khalifa, A., Holmgård, C., Togelius, J.: Evolving UCT alternatives for general video game playing. In: The IJCAI-16 Workshop on General Game Playing, p. 63

5. Cazenave, T.: Evolving Monte Carlo tree search algorithms. Dept. Inf., Univ. Paris 8 (2007)
6. Chaslot, G.: Monte-Carlo tree search. Universiteit Maastricht (2010)
7. Ebner, M., Levine, J., Lucas, S.M., Schaul, T., Thompson, T., Togelius, J.: Towards a video game description language (2013)
8. Frydenberg, F., Andersen, K.R., Risi, S., Togelius, J.: Investigating MCTS modifications in general video game playing. In: Computational Intelligence and Games, pp. 107–113. IEEE (2015)
9. Gaina, R.D., Pérez-Liébana, D., Lucas, S.M.: General video game for 2 players: framework and competition. In: Proceedings of the IEEE Computer Science and Electronic Engineering Conference (CEEC) (2016)
10. Hingston, P.: Believable Bots. Springer, Heidelberg (2012)
11. Jacobsen, E.J., Greve, R., Togelius, J.: Monte Mario: platforming with MCTS. In: Proceedings of the Annual Conference on Genetic and Evolutionary Computation, pp. 293–300. ACM (2014)
12. Khalifa, A., Perez-Liebana, D., Lucas, S.M., Togelius, J.: General video game level generation
13. Mendes, A., Nealen, A., Togelius, J.: Hyperheuristic general video game playing. In: IEEE Computational Intelligence and Games (2016)
14. Ontanón, S., Synnaeve, G., Uriarte, A., Richoux, F., Churchill, D., Preuss, M.: A survey of real-time strategy game AI research and competition in starcraft. IEEE Trans. Comput. Intell. AI Games **5**(4), 293–311 (2013)
15. Perez, D., Samothrakis, S., Togelius, J., Schaul, T., Lucas, S., Couëtoux, A., Lee, J., Lim, C.-U., Thompson, T.: The 2014 general video game playing competition (2015)
16. Perez-Liebana, D., Samothrakis, S., Togelius, J., Lucas, S.M., Schaul, T.: General video game AI: competition, challenges and opportunities. In: Thirtieth AAAI Conference on Artificial Intelligence (2016)
17. Pettit, J., Helmbold, D.: Evolutionary learning of policies for MCTS simulations. In: Proceedings of the International Conference on the Foundations of Digital Games, pp. 212–219. ACM (2012)
18. Preuss, M.: Adaptability of algorithms for real-valued optimization. In: Giacobini, M., et al. (eds.) Applications of Evolutionary Computing, pp. 665–674. Springer, Heidelberg (2009). doi:10.1007/978-3-642-01129-0
19. Renz, J.: AIBIRDS: the angry birds artificial intelligence competition. In: AAAI, pp. 4326–4327 (2015)
20. Soemers, D., Sironi, C.F., Schuster, T., Winands, M.H.M.: Enhancements for real-time Monte-Carlo tree search in general video game playing. In: Proceedings of the IEEE Conference on Computational Intelligence and Games (2016)
21. Togelius, J., Lucas, S.M., De Nardi, R.: Computational intelligence in racing games. In: Baba, N., Jain, L.C., Handa, H. (eds.) Advanced Intelligent Paradigms in Computer Games. Studies in Computational Intelligence, pp. 39–69. Springer, Heidelberg (2007). doi:10.1007/978-3-540-72705-7_3
22. Togelius, J., Preuss, M., Beume, N., Wessing, S., Hagelbäck, J., Yannakakis, G.N., Grappiolo, C.: Controllable procedural map generation via multiobjective evolution. Genet. Program Evolvable Mach. **14**(2), 245–277 (2013)
23. Togelius, J., Shaker, N., Karakovskiy, S., Yannakakis, G.N.: The Mario AI championship 2009–2012. AI Mag. **34**(3), 89–92 (2013)

24. Tusar, T., Brockho, D., Hansen, N., Auger, A.: COCO: the bi-objective black box optimization benchmarking (bbob-biobj) test suite. In: CoRR abs/1604.00359 (2016)
25. Van Hoorn, N., Togelius, J., Wierstra, D., Schmidhuber, J.: Robust player imitation using multiobjective evolution. In: CEC, pp. 652–659. IEEE (2009)
26. Wessing, S.: Two-stage methods for multimodal optimization. Ph.D. thesis, TU Dortmund (2015).http://dx.doi.org/10.17877/DE290R-7804

Towards Standardized and Seamless Integration of Expert Knowledge into Multi-objective Evolutionary Optimization Algorithms

Magdalena A.K. Lang[1] and Christian Grimme[2(✉)]

[1] RWTH Aachen University, Kackertstr. 7, 52072 Aachen, Germany
magdalena.lang@ada.rwth-aachen.de
[2] Westfälische Wilhelms-Universität Münster, Leonardo-Campus 3,
48149 Münster, Germany
christian.grimme@uni-muenster.de

Abstract. Evolutionary algorithms allow for solving a wide range of multi-objective optimization problems. Nevertheless for complex practical problems, including domain knowledge is imperative to achieve good results. In many domains, single-objective expert knowledge is available, but its integration into modern multi-objective evolutionary algorithms (MOEAs) is often not trivial and infeasible for practitioners. In addition to the need of modifying algorithm architectures, the challenge of combining single-objective knowledge to multi-objective rules arises. This contribution takes a step towards a multi-objective optimization framework with defined interfaces for expert knowledge integration. Therefore, multi-objective mutation and local search operators are integrated into the two MOEAs MOEA/D and R-NSGAII. Results from experiments on exemplary machine scheduling problems prove the potential of such a concept and motivate further research in this direction.

Keywords: Multi-objective evolutionary algorithm · Expert knowledge integration · MOEA/D · R-NSGAII · Scheduling

1 Introduction

Multi-Objective Evolutionary Algorithms (MOEAs) aim to obtain useful solutions for real-world optimization problems in a reasonable amount of time. Different quasi-standard approaches have been introduced in research, for instance, the Non-dominated Sorting Genetic Algorithm II (NSGAII) [3] and the Multi-Objective Evolutionary Algorithm Based on Decomposition (MOEA/D) [20].

However, to increase MOEA acceptance in practice, these metaheuristics have to improve regarding reliability and computational efficiency for complex problems. It is widely acknowledged that including problem specific heuristics supports the search and allows faster convergence towards optimal solutions [5, 12, 16]. For instance, in application fields such as scheduling, a broad theoretically founded knowledge base is available which can help to tackle subproblems of multi-objective optimization tasks [5, 6].

© Springer International Publishing AG 2017
H. Trautmann et al. (Eds.): EMO 2017, LNCS 10173, pp. 375–389, 2017.
DOI: 10.1007/978-3-319-54157-0_26

In single-objective optimization, expert knowledge integration is from an algorithmic point of view rather simple and commonly applied via special search operators [2]. However, for multi-objective problems, domain experts face the challenge to effectively introduce their mostly subproblem-oriented knowledge. In particular, the question arises of *when* to apply *which problem-specific operators* on *which individuals* in case only heuristics for single objectives are available. Thus, to utilize the full potential of MOEAs in practice, we suggest a reusable and easily deployable optimization framework that allows the integration of problem-specific expert knowledge by simple interfaces without requiring to redesign the underlying algorithm. An internal logic needs to effectively apply the problem specific heuristics in order to leverage the potential of both the metaheuristic and the problem specific heuristics. In the following, we call such a framework *extended or adapted MOEA*.

Several successful applications of MOEAs integrating expert knowledge already exist. Nevertheless, authors like Peng and Zhang [14] or Konstantinidis and Yang [10] apply specific algorithm architectures that are only designed for the given problems. Grimme et al. [5] follow the idea of an extended MOEA and successfully implement a modular integration of single-objective knowledge into a multi-objective predator-prey model via variation operators. Until now, the more popular, dominance-based MOEAs are regarded as rather unsuitable for such a general adaption, due to their monolithic structure and focus on the selection mechanism [5,18]. Nevertheless, more recent research advances, such as introducing reference points [4] or decomposition-based algorithmic structures [20], motivate to check their potential and openness for a respective framework.

This contribution takes a step towards an extended MOEA by proposing to integrate multi-objective mutation and local search modules as reusable interfaces for subproblem-specific heuristics. The modules are integrated into two MOEAs, MOEA/D and R-NSGAII, which especially recommend themselves for a practical framework. That is, they allow both a-priori and a-posteriori optimization. An experimental approach to multiple a-posteriori parallel machine scheduling problems allows to examine and compare the performance of the adapted frameworks in contrast to plain general-purpose MOEAs. The performance indicators consider the quality and reliability of provided solutions as well the as speed of convergence.

2   Scheduling Problems and Dispatching Rules

Scheduling is '*an optimization process by which limited resources are allocated over time among parallel and sequential activities*' [1, p. 1]. In the following, we focus on machine scheduling because most theoretical models and concepts refer to this class and commonly accepted optimization criteria exist [16]. Additionally, we restrict ourselves to offline, non-preemptive, and non-delay settings, where jobs cannot be paused and unforced idleness of machines is not allowed.

2.1 Notation

A schedule comprises both the assignment of jobs to machines and their processing order [15]. A job j is a task mainly characterized by its processing time $p_{i,j}$ on machine i, due date d_j, and release date r_j. For a given schedule, a job has a start time S_j and a completion time $C_j = S_j + p_{i,j}$.

A scheduling problem is stated via the popular three-field notation $\alpha|\beta|\gamma$, with α being the machine environment of the problem, β describing problem constraints, and γ enumerating the considered objectives. This study focuses on the P_m machine environment, i.e. identical parallel machines, for which a job consists of a single operation and can be processed on any of the available machines with processing time p_j [15]. As objectives, this study considers the minimization of four regular criteria, which are among the most popular in literature and important in practice. The *makespan* C_{max} denotes the total duration of a schedule, i.e. $C_{max} = \max_{j=1,...,n}(C_j)$ with n as the number of jobs. It aims at balancing the load on the identical parallel machines in P_m [15]. In contrast, the *total completion time objective* $C_{sum} = \sum_{j=1}^{n} C_j$ strives for completing all jobs as fast as possible. The due date-related *maximum lateness* objective $L_{max} = \max_{j=1,...,n}(L_j)$ with $L_j = C_j - d_j$ minimizes the largest due date violation. When considering $T_{sum} = \sum_{j=1}^{n} T_j$, the *total tardiness* with $T_j = \max(0, L_j)$ needs to be minimized. It demands that no delay of any job is unacceptably long [15].

Scheduling problems usually strike for combinatorial solutions. For multi-objective problems with regular criteria without preemption, the number of optimal points is finite and rather small in contrast to continuous Pareto fronts [1,7]. Most scheduling problems are (weakly or strongly) NP-hard or open [7]. References such as [15] provide details about complexity hierarchies.

2.2 Dispatching Rules

In practice, dispatching jobs to machines often follows specific rules, which require little computational time and exactly or approximately solve basic scheduling problems. These rules usually rely on job attributes or machines conditions. The *Shortest Processing Time* (SPT) rule sorts the jobs in ascending order of processing time p_j, leading to an optimal solution for the $1||C_{sum}$ problem [17]. In conjunction with the *First Available Machine* (FAM) rule, which assigns the next job to the first available machine in a parallel machine environment, also $P_m||C_{sum}$ is solved to optimality [15]. The reverse of SPT, the *Longest Processing Time* (LPT) rule, heuristically balances the load on parallel machines under the FAM assignment [15]. It is easy to show that the upper bound for this heuristic is $4/3 - 1/(3m)$ times the optimal value of C_{max}.

The *Earliest Due Date* (EDD) rule and *Minimal Slack* (MS) rule aim on solving due date related criteria. The EDD rule sorts jobs in ascending order of due dates and optimally solves the $1||L_{max}$ problem [15]. In contrast to the static rules mentioned before, the MS rule is a so-called dynamic rule [15]. Each time a job needs to be assigned, the job j with the minimum $slack_j = max(0, d_j - p_j)$ is chosen next.

If release dates are present, the *Earliest Release Date* (ERD) rule sorts the jobs according to increasing release dates. So-called *domination results* can help to reduce the search space of a scheduling problem. When considering $1||T_{sum}$, we know that *if $p_j \leq p_k$ and $d_j \leq d_k$, then there exists an optimal sequence in which job j is scheduled before job k* [15, p. 51]. This is further on called DR rule.

For multi-objective scheduling problems, only few efficient algorithms exist. Wassenhove and Gelders [19] introduced an approach to optimally solve the $1||C_{sum}, L_{max}$ problem with $\mathcal{O}(n^3 log(n))$ complexity [7]. The approach leverages respective single-objective dispatching rules to acquire the Pareto front.

3   Methodology

The extended MOEA approach aims to find a better approximation of the Pareto front, i.e. the *approximated front*, for the minimization or maximization of $F(x) = (f_1(x), f_2(x), ..., f_k(x))$ with k objectives, objective functions $f_i(x)$, $i = 1, ..., k$, and x as feasible solution. Thereby, the common quality criteria *convergence to the Pareto front* as well as *diversity* of the approximated front, including the number of solutions, are considered. Secondary criteria are the *speed of convergence* towards good solutions as well as the *reliability* of an approach regarding robustness of the result quality.

3.1   Multi-objective Modules

Similar as for existing successful approaches to integrate multi-objective expert knowledge [6, 10, 14], subproblem-specific expertise is combined by a successive application of alternate subproblem dependent heuristics. They manipulate parts of an individual's genotype. The proposed *multi-objective modules* (MOM) extend the approach by [10].[1] For each objective, the module provides adapters for registering subproblem-specific operators. Figure 1 illustrates the concept for a mutation module (MOM-M). The domain expert can provide single-objective mutation operators to facilitate the local improvement of a temporary solution to the respective objective by small changes in the genotype of the considered individual. If an operator fits multiple objectives, it can be registered on several adapters. Thus, this approach has the advantage that it allows to integrate both single-objective knowledge as well as problem specific improvement mechanisms supporting combinations of objectives. Thereby, constraint-related rules as the ERD rule, concerning multiple objectives, can easily be added.

Furthermore, the module has an interface to the base MOEA. It receives an individual of the population as well as the information needed for the so-called *assignment strategy*. The latter is a logic within the module that decides

---

[1] Note, however, that we focus on the more general context of applicability in MOEAs. Thus, the proposed methodology is not restricted to specific applications but strive for identifying additional integration points of expertise.

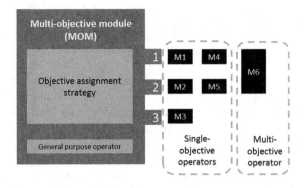

Fig. 1. Schema of the Multi-objective Mutation Module (MOM)

which objective to support for the current individual. Subsequently, the module randomly chooses one operator of the respective queue. The most simple strategy randomly selects one objective. Alternatively, if the individual belongs to a specific weighting of objectives as in a decomposition algorithm like MOEA/D, the module can set selection probabilities according to the weighting of considered objectives. Table 1 lists the assignment strategies considered here. Strategy applicability depends on the underlying algorithm.

Moreover, the MOM can allow a certain probability to apply general purpose operators like swap mutation to facilitate diversity. These do not explicitly bias towards specific objectives but apply undirected random adaption. Similar to [8], this probability is an external parameter.

Table 1. Objective assignment strategies of a MOM

Assignment strategy	Description	MOEA
Random assignment (`random`)	Random selection of the objective	MOEA/D
		R-NSGAII
Probabilistic assignment based on objective weights (`weights prob`)	Weighting of objectives is transferred to selection probabilities for the respective objectives	MOEA/D
Deterministic assignment based on objective weights (`weights det`)	Objective with highest weight is selected (random draw on ties)	MOEA/D
Worst objective assignment (`worst obj`)	The currently worst performing objective is selected, either by looking at	
	1. The worst weighted objective value	MOEA/D
	2. The largest distance to the assigned reference point	R-NSGAII

Such a modular interface combines low maintenance effort and flexible recon-figuration of individual modules, without a need to adapt the overall algorithm structure. Therefore, different implementations of modules using various assign-ment strategies can be tested. A MOM can be integrated at different parts of the MOEA structure to serve as mediator between the metaheuristic and the problem specific heuristics provided by the user. We evaluate such a MOM as mutation operator (MOM-M) as well as for an additional local search (MOM-L) starting from individuals of the population.

A local search module facilitates intensification. Next to choosing problem specific techniques, it applies an internal procedural logic to decide for further actions after producing a new candidate solution. This study uses a greedy strat-egy [8,9]. Therein, the local search module receives a solution from the popula-tion as a starting point. A neighbor of this solution is produced via a problem specific heuristic, which biases the neighborhood structure of a solution. There-after, the module compares fitness values of the old and the new solution based on a fitness function it receives from the underlying algorithm. If the new solu-tion is better or as good as the original one, it serves as new starting point for the next local search step. The overall number of steps as well as the allowed number of steps without improvement are external termination parameters. Our preliminary experiments detected an advantage for accepting new solutions as intermediate step towards better individuals, if they have the same fitness. Note that the problem specific heuristic is selected only once at the beginning of each local search process and pursued for all following steps.

3.2   Adapting MOEA/D

MOEA/D [20] was already successfully extended by single-objective expert knowledge (e.g. in [10,14,18]). It is a decentralized MOEA dividing a MOP into scalar sub-problems each explicitly addressing a specific weighting of the objec-tives. Per generation, the sub-problems are subsequently handled in a single-objective manner and interact in terms of crossover and selection mechanism with their neighboring subproblems to benefit from their findings.

The weights of the scalar functions can be interpreted as relative prioriti-zation of the respective objectives and determine the search direction [18]. A method like the weighted sum approach (see [11]) or the Tchebycheff approach (see [20]) can serve to determine the fitness of an individual. The latter uses a reference point like the ideal vector, consisting of the best values of the individual objectives, as a goal and aims on minimizing the maximum weighted distance between the objectives of a solution and this point. As in the original MOEA/D, we use the Tchebycheff method as scalarizing function, especially because it provides more connecting points for the assignment of expert knowledge.

The MOMs are integrated as follows. For MOM-M, the algorithm ini-tiates the respective module as mutation operator within its reproduction cycle. Still, the module needs to know about the sub-problem under considera-tion to decide which single-objective operator to choose. Different possibilities are considered here. First, the module assigns operators randomly (random).

Second, the algorithm informs the module about the respective weight vector and the module decides either based on the highest weight (`weights det`) or randomly (`weights prob`) for a sub-problem specific operator. Last, the algorithm passes on the worst objective based on the Tchebycheff rule (`worst obj`), see Table 1.

For better exploitation, local search is periodically initialized after a complete generation was determined. All sub-problems are traversed and each individual is used as possible start solution. If the local search module returns a better performing individual for the subproblem, the old solution is replaced. Therefore, the algorithm provides a comparator which evaluates the fitness of the solutions based on the respective weight vector. The assignment strategies are analog to the mutation module.

3.3   Adapting R-NSGAII

From a practical point of view, the reference point based adaption R-NSGAII [4] of NSGAII [3] plays an important role. It allows to integrate preferences in form of one or more reference points guiding search within the objective space.

Two assignment strategies are most obvious for R-NSGAII. First, the random assignment (`random`) and second, analogously to the worst objective assignment in MOEA/D, an allocation based on the objective which performs worst in comparison to the reference point assigned to an individual (`worst obj`). As the reference points do not need to be optimal, an individual can be better in one or more objectives than the respective reference point $\bar{z}$. Therefore, the objective solving $max_i(f_i(x) - \bar{z}_i)$ for $i = 1, ..., k$ is targeted. As main difference to MOEA/D, the individuals are not constantly assigned to one reference point. Instead, the allocation can change over time. Moreover, several individuals can be allocated to the same point and the selection mechanisms do not only compare individuals in similar search directions but the population as a whole.

As MOEA/D, R-NSGAII applies local search periodically to all individuals after a predefined number of generations. The base population and the new individuals are collected in one combined set and the usual generational selection mechanism is applied to determine the population for the next recreation loop. The algorithm provides the module with a comparator which compares individuals based on non-domination and distances to the respective reference points.

3.4   Subproblem-Specific Operators

The following expert knowledge based operators help to test the performance of the multi-objective procedure. Note that they are not themselves in focus of the current analysis. In principle, they build on generic mechanisms allowing to integrate order based rules into a permutation encoding of an individual.

First, the permutation based mutation operator proposed by Grimme et al. [6] allows to integrate scheduling knowledge into the MOM-M. The operator randomly selects a position within the permutation and sorts all jobs within a range δ around this position according to an assigned dispatching rule. Thereby, an overall amount of $2\delta + 1$ positions is in focus during one mutation operation. The concrete value of δ is determined randomly based on a normal distribution, with $\mu = 0$ and σ as external parameter.

For local search, the neighborhood is biased towards individuals which are likely to improve in the respective objective by considering the appropriate dispatching rule. Thereby, a random position j within the permutation is selected and used as starting point to traverse the following job indices for finding a job which can be inserted at position j such that all following jobs are shifted backwards by one position. For instance, the SPT-biased neighborhood approach looks for a job with a lower processing time than the one at random position j. The number of checks can be implemented as predefined parameter which influences the probability to swap jobs and thus, the possible impact of the operator similar to the σ parameter in the mutation operator. While such a general approach is used for the SPT, MS, DR, and ERD rules, more effective, deterministic methods are adopted for the C_{max} and the L_{max} objectives because their respective values are determined by single jobs. Therefore, the phenotype of the individual is used to identify the job causing the value of the objective in focus. The phenotype is the schedule with assignment of jobs to machines, obtained by applying the FAM rule to the permutation. The approach tries to reposition this job by backward traversal, searching for a job with lower processing time or due date, respectively, to prepend it.

The new individual adopting the resulting job order serves as neighbor of the individual in focus. It is evaluated by the local search module to decide on its acceptance as basis for another local search round or as replacement of the start individual in the population.

4   Experimental Analysis

Computational experiments on machine scheduling problems help to evaluate extended MOEA/Ds and R-NSGAIIs.

4.1   Experimental Setup

For each scheduling problem, 40 test instances serve as the basis for a statistical analysis of MOEA performances. Half of the instances consist of 50 jobs and the other half of 100 jobs. The processing times are randomly drawn realizations of a uniform distribution, i.e. $p_j \sim [\mathcal{U}(1, 49)]$. Half of both 50 job and 100 job instances consist of jobs with a due date offset d_j^o between 0 and 100 ($d_j = r_j + d_j^o$ with $d_j^o \sim [\mathcal{U}(0, 100)]$). The other half is less restrictive with $d_j^o \sim [\mathcal{U}(0, 200)]$). In addition, a special 50 job instance results from due date offsets between 0 and 999 to obtain a wider and more populated Pareto front for the $1||C_{sum}, L_{max}$ problem. Individual genotypes are permutation encoded. Each element of the permutation resembles a job. Jobs are assigned in a first available machine (FAM) manner to parallel machines during evaluation.

Table 2. Experimental settings and algorithmic parameters.

Attribute	Final setting	Attribute	Final setting
Individual representation	Permutation encoding	Variation operators	Order crossover and swap mutation
Initial population	Random permutations	Crossover probability	0.9
Population size μ	100 (2 objectives) or 300 (3 objectives)	Mutation probability	0.8
Termination criterion	25,000 (2 objectives) and 75,000 (3 objectives) individuals	Specific parameters	Neighborhood size: 20 (MOEA/D); Epsilon: 0.001 (R-NSGAII)
Selection operators	Binary tournament (R-NSGAII)	Maximum number of checks in local search	10 % of the number of jobs
Number of instances	20 instances with 50 jobs, 20 instances with 100 jobs	Number of runs per experiment	30

Table 2 summarizes relevant settings like variation operators and population size. Most settings result from a preliminary parameter tuning, suggesting same best settings for both algorithms. They are used throughout all experiments unless otherwise explicitly stated.

4.2   Performance Evaluation

The *Hypervolume Indicator (HV)* [21], the *Unary Additive Epsilon Indicator (Epsilon)* [21], and the *number of non-dominated solutions (Count)* serve as performance indicators for the quality of the approximated front. Additionally, an indicator *Endpoint* measures the number of generations after which no significant improvement in HV ($<2\%$) exists anymore, compared to the mean final HV of all runs. The indicator *Comparison Reached Point (CRP)* states when the mean final HV of the respective base MOEA was reached, while the *Comparison Reached Rate (CRR)* measures the quota of runs reaching this HV value. The latter indicators mainly measure the speed of convergence of algorithms.

Every algorithm was applied to each test instance 30 times to account for stochastic effects. The average value of each indicator over the 30 runs serves as result per instance. We look at mean and standard deviation over the 40 test instances to present the overall performance of the algorithm. Thereby, we present the indicators as quota over the respective base algorithm outcome, with values above one indicating a larger indicator value. As the Epsilon and CRR indicators can be zero, we present differences instead of quotas.

We conduct non-parametric tests to check the statistical significance of differences over the 40 instances (significance level of $p \leq 0.05$). The *Friedman test* and the paired *Wilcoxon Rank Sum test (Wilcoxon test)* with *Bonferroni correction* check for significant differences in mean, the *Fligner-Killeen test (Fligner test)* helps to identify whether two algorithm outcomes differ significantly in variance.

Normalization balances the influence of incommensurable objectives on quality indicator values. Per instance, the best and worst objective values of all approximated fronts serve as boundary values. Furthermore, the combined best approximated front serves as reference front for the Epsilon indicator and the worst objective values form the reference point for the HV indicator. Note that this approach circumvents the comparability of results across experiments.

For the implementation of algorithms, we adopted the Java-based jMetal framework [13] (Java 8 Update 65). Data analysis approaches as significance tests and visualizations were mainly conducted via R (version 3.2.2).

4.3   Results

We present representative results with main focus on MOEA/D that sufficiently provide relevant insights for both algorithms with and without multi-objective modules applied. For illustration purposes, the computational easy $1||C_{sum}, L_{max}$ is used as for this problem approximated solutions can be compared to the true Pareto front. Then, the analysis is supported by considering an NP-hard parallel machine problem $5||C_{max}, C_{sum}$ and the also NP-hard three objective problem $1||C_{sum}, L_{max}, T_{sum}$.

Multi-objective Mutation Module (MOM-M). For $1||C_{sum}, L_{max}$, SPT and EDD rule based single-objective mutation operators are used to sort jobs within a randomly selected range of the permutation. Figure 2 illustrates the best and worst approximated fronts over 30 runs for MOEA/D and MOEA/D-MOM-M (random) on a 50 jobs instance. The results look similar for the other assignment strategies. While MOEA/D focuses on the region that primarily supports the C_{sum}-objective, the extended algorithms find solutions spread over the whole front but can have some problems to converge to optimal extreme solutions regarding L_{max} values. The HV development over the runtime is exemplarily visualized in Fig. 3(a). Apparently, the different assignment strategies are

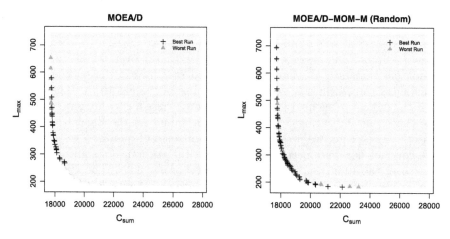

Fig. 2. Pareto front and approximated fronts for $1||C_{sum}, L_{max}$ with 50 jobs, solved by MOEA/D and MOEA/D-MOM-M (random) ($\sigma = 10$, general purpose operators: 0%)

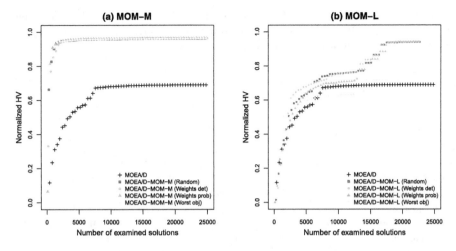

Fig. 3. HV development over the runtime of algorithms for $1||C_{sum}, L_{max}$ problem with 50 jobs for median performing runs of MOEA/D-MOM-M ($\sigma = 10$, general purpose operators: 0%) and MOEA/D-MOM-L

similar in their convergence speed, increasing much faster than the unadapted MOEA/D and ending with higher HV values.

The extended versions outperform MOEA/D in all considered indicators according to Friedman and Wilcoxon tests. The descriptive statistics in Table 3 state, for instance, that HV increases by around 7% and that at most 14% of generations are needed to achieve comparable results to MOEA/D. In general, the assignment strategies do not provide significantly differing results, only `weights det` is outperformed for the Epsilon indicator and converges last. `Worst obj` improves faster than the other adaptions, reaching the mean final HV of MOEA/D and its own endpoint significantly earlier. The Fligner test reveals that, except from HV, the adapted versions perform more robust across different problem instances than MOEA/D.

The R-NSGAII approach applied here uses evenly spread, normalized reference points to allow a search for the whole Pareto front. The analysis over 40 test instances reveals significantly better solutions of the extended R-NSGAII over the base algorithm in all indicators. Table 3 states that the HV values increase by more than 45%. `Worst obj` achieves a significantly faster convergence towards the mean final HV of R-NSGAII and its own endpoint than `random`. Furthermore, both alternatives are more robust than the unadapted R-NSGAII in Epsilon and the time related indicators.

For $5||C_{max}, C_{sum}$, two reverse dispatching rules are needed, i.e. the LPT rule for the first objective and the SPT rule for the second. Because the LPT-rule based mutation is not optimal for the C_{max} objective and the reverse mutations are probably very disruptive, the integration of partly swap mutation instead of problem-specific mutation was advantageous in this case. Table 4 presents the descriptive statistics of the two experiments on the 40 test instances.

Table 3. Descriptive statistics of performance indicators for $1||C_{sum}, L_{max}$ problems (40 instances, $\sigma = 10$, general purpose operators: 0%). Note that MOEA/D and R-NSGAII experiment indicators relate to the respective experiment only and must not be compared.

| | Experiment 1: MOEA/D MOEA/D-MOM-M | | | | Experiment 2: R-NSGAII R-NSGAII-MOM-M | |
	Random	Weights det	Weights prob	Worst obj	Random	Worst obj
HV quota mean	1.07028	1.06613	1.06961	1.06999	1.45713	1.45760
HV quota std.	0.11922	0.12237	0.11980	0.11927	0.59446	0.59442
Epsilon diff. mean	−0.05762	−0.05313	−0.05642	−0.05719	−0.21435	−0.21542
Epsilon diff. std.	0.08227	0.08491	0.08262	0.08212	0.23810	0.23786
Count quota mean	4.41897	3.35208	3.80007	4.55546	11.48547	11.74812
Count quota std.	3.57277	1.99379	2.62486	3.55111	12.72062	13.19304
CRR quota mean	1.13997	1.13498	1.14041	1.14398	1.33191	1.33370
CRR quota std.	0.18320	0.18397	0.17829	0.17114	0.36050	0.35854
CRP quota mean	0.10867	0.13612	0.11501	0.10122	0.11892	0.11141
CRP quota std.	0.06440	0.10302	0.07710	0.07895	0.07812	0.06627
Endpoint quota mean	0.12565	0.16001	0.13705	0.11804	0.14360	0.13400
Endpoint quota std.	0.06242	0.10423	0.08664	0.08458	0.07944	0.06991

Table 4. Descriptive statistics of performance indicators for $5||C_{max}, C_{sum}$ problems (40 instances, $\sigma = 10$, general purpose operators: 20%). Note that MOEA/D and R-NSGAII experiment indicators relate to the respective experiment only and must not be compared.

| | Experiment 1: MOEA/D MOEA/D-MOM-M | | | | Experiment 2: R-NSGAII R-NSGAII-MOM-M | |
	Random	Weights det	Weights prob	Worst obj	Random	Worst obj
HV quota mean	1.05078	1.04538	1.04865	1.05182	1.06638	1.06968
HV quota std.	0.08386	0.08486	0.08377	0.08346	0.07934	0.07875
Epsilon diff. mean	−0.04173	−0.02734	−0.03511	−0.04529	−0.06258	−0.06761
Epsilon diff. std.	0.06339	0.06364	0.06231	0.06215	0.06241	0.06220
Count quota mean	1.99265	1.86450	1.85238	1.98604	3.35462	3.91708
Count quota std.	1.26462	1.16177	1.10300	1.20368	2.60105	3.06321
CRR quota mean	1.15593	1.10869	1.13627	1.16476	1.20999	1.22278
CRR quota std.	0.16691	0.20075	0.18253	0.16551	0.19966	0.19590
CRP quota mean	0.35655	0.45816	0.41022	0.32179	0.35559	0.34094
CRP quota std.	0.13671	0.22237	0.18336	0.10197	0.11080	0.09015
Endpoint quota mean	0.42133	0.52316	0.48718	0.38758	0.44246	0.42077
Endpoint quota std.	0.07969	0.14092	0.12174	0.07047	0.06842	0.07056

The significance tests do not only suggest improvements in all indicators for the extended algorithms, but also a ranking between the different assignment strategies. For MOEA/D, `worst obj` results in best HV and EPSILON values. `Random` is second best and reaches the mean final HV of MOEA/D together with `worst obj` first. The comparatively poor performance of `weights det` possibly results from the missing support of the lower weighted objective. Additionally, all adapted algorithms are more robust in their end HV and Epsilon indicators than MOEA/D.

Multi-objective Local Search Module (MOM-L). We apply local search only after half of the maximum evaluations termination criterion is reached, allowing the MOEA to find promising regions beforehand. Afterwards, local search is activated every ten generations to facilitate intensification. The local search step size is five, while the maximum number of subsequent non-improving individuals is two.[2] To allow a fair comparison between approaches with and without local search, the number of intermediate individuals created during the procedure adds to the overall number of individuals of the algorithm. Figure 3(b) shows the HV development of the median runs for $1||C_{sum}, L_{max}$ and the effects of local search. While MOEA/D-MOM-L `worst obj` falls behind the other extended MOEA/Ds, all extended versions show improvement and perform better than simple MOEA/D.

The Friedman test finds significant improvements in all indicators for the MOEA/D and R-NSGAII experiments. For MOEA/D, no significant differences between the extended algorithms exists according to the Wilcoxon test. All adapted MOEA/Ds are more robust in their comparison reached rate than MOEA/D across problem instances. Additionally, `random` and `worst obj` are less variable in the Epsilon indicator.

For R-NSGAII, both extended algorithms significantly outperform R-NSGAII in all indicators and are more robust across problem instances.

For $5||C_{max}, C_{sum}$, MOEA/D-MOM-L `random` and `weights det/prob` significantly outperform MOEA/D in all indicators. Only in CRR, `worst obj` can outperform the original MOEA/D. According to the Fligner test, the weight-based assignment strategies are more robust in the CRR than MOEA/D and MOEA/D-MOM-L `worst obj`. For R-NSGAII, there are small but significant improvements in all indicators for applying MOM-L. Thereby, the two assignment strategies do not significantly differ.

Three-Objective Problems. Finally, the up to now best performing versions of MOEA/D and R-NSGAII, i.e. MOEA/D-MOM-M `worst obj` and R-NSGAII-MOM-M `worst obj`, are used to analyze their performance on three-objective problems. $1||C_{sum}, L_{max}, T_{sum}$ helps to analyze how two objectives with optimal scheduling rules interact with one objective which is NP-hard to solve in any machine setting. In addition, the domination rule for T_{sum} can be beneficial for the other objectives and is therefore assigned to all objectives. The EDD and the MS rule are applied to both due date related objectives.

For both algorithms, the extended variants can significantly improve in the quality of the approximated front and the speed towards the final end HV of the base algorithm. Additionally, they are more robust in their results.

---

[2] This configuration was applied without fine tuning to demonstrate the methodology. Future rigorous investigation may consider systematically generated configurations.

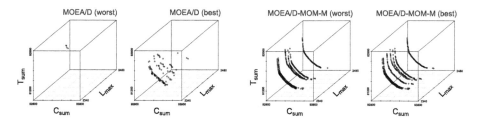

Fig. 4. Approximated fronts of MOEA/D and MOEA/D-MOM-M (worst obj) for $1||C_{sum}, L_{max}, T_{sum}$ problem with 100 jobs

A closer look at one of the instances for the MOEA/D experiment provides an insight into the resulting approximated fronts, see Fig. 4. Apparently, the results of the adapted MOEA/D are far more robust whereas MOEA/D performs very poor in the worst run. Additionally, the approximated fronts of the extended MOEA/D are more diverse, here best visible for the combination of C_{sum} and T_{sum} values.

5 Conclusion

The presented idea of a multi-objective algorithmic framework for ad-hoc integration of expert knowledge uses multi-objective modules that apply subproblem-specific operators based on different assignment strategies. They provide standardized interfaces to allow a seamless integration of domain knowledge by practitioners. Experimental results suggest that the modules can significantly improve the performance of dominance based R-NSGAII and MOEA/D regarding the quality of the approximated front, the convergence speed towards good solutions, as well as the reliability of results. Overall, we prove the potential of such a framework with respect to *usability* and *performance*.

Still, open research issues exist that need to be approached in future work. For instance, the question arises whether the successful integration of expert knowledge is transferable to other scheduling problems or even other application domains of multi-objective optimization. More advanced approaches like adaptive operators can be tested as well. Thereby, other base algorithms and combinations of different integration points may come into play. Furthermore, the underlying forces between the effects seen in the presented experiments should be further investigated to improve the understanding of the problem domain.

References

1. Bagchi, T.P.: Multiobjective Scheduling by Genetic Algorithms. Springer, New York (1999)
2. Burke, E.K., Gendreau, M., Hyde, M.R., Kendall, G., Ochoa, G., Özcan, E., Qu, R.: Hyper-heuristics: a survey of the state of the art. JORS **64**(12), 1695–1724 (2013)
3. Deb, K., Pratap, A., Agarwal, S., Meyarivan, T.: A fast and elitist multiobjective genetic algorithm: NSGA-II. IEEE Trans. Evol. Comput. **6**(2), 182–197 (2002)

4. Deb, K., Sundar, J., Udaya Bhaskara Rao, N., Chaudhuri, S.: Reference point based multi-objective optimization using evolutionary algorithms. Int. J. Comput. Intell. Res. **2**(3), 273–286 (2006)
5. Grimme, C., Kemmerling, M., Lepping, J.: On the integration of theoretical single-objective scheduling results for multi-objective problems. In: Tantar, E., Tantar, A.A., Bouvry, P., Del Moral, P., Legrand, P., Coello Coello, C., Schuetze, O. (eds.) EVOLVE- A Bridge between Probability, Set Oriented Numerics and Evolutionary Computation, Studies in Computational Intelligence, vol. 447, pp. 333–363. Springer, Heidelberg (2013)
6. Grimme, C., Lepping, J., Schwiegelshohn, U.: Multi-criteria scheduling: an agent-based approach for expert knowledge integration. J. Sched. **16**(4), 369–383 (2013)
7. Hoogeveen, H.: Multicriteria scheduling. Eur. J. Oper. Res. **167**(3), 592–623 (2005)
8. Ishibuchi, H., Hitotsuyanagi, Y., Tsukamoto, N., Nojima, Y.: Use of heuristic local search for single-objective optimization in multiobjective memetic algorithms. In: Rudolph, G., Jansen, T., Beume, N., Lucas, S., Poloni, C. (eds.) PPSN 2008. LNCS, vol. 5199, pp. 743–752. Springer, Heidelberg (2008). doi:10.1007/978-3-540-87700-4_74
9. Ishibuchi, H., Hitotsuyanagi, Y., Tsukamoto, N., Nojima, Y.: Use of biased neighborhood structures in multiobjective memetic algorithms. Soft Comput. **13**(8), 795–810 (2009)
10. Konstantinidis, A., Yang, K.: Multi-objective energy-efficient dense deployment in wireless sensor networks using a hybrid problem-specific MOEA/D. Appl. Soft Comput. **11**(6), 4117–4134 (2011)
11. Miettinen, K.: Nonlinear Multiobjective Optimization. International Series in Operations Research and Management Science, vol. 12. Springer, New York (1998)
12. Nagar, A., Haddock, J., Heragu, S.: Multiple and bicriteria scheduling: a literature survey. Eur. J. Oper. Res. **81**(1), 88–104 (1995)
13. Nebro, A.J., Durillo, J.J.: jMetal 4.5 user manual (21 Jan 2014)
14. Peng, W., Zhang, Q.: Network topology planning using MOEA/D with objective-guided operators. In: Coello, C.A.C., Cutello, V., Deb, K., Forrest, S., Nicosia, G., Pavone, M. (eds.) PPSN 2012. LNCS, vol. 7492, pp. 62–71. Springer, Heidelberg (2012). doi:10.1007/978-3-642-32964-7_7
15. Pinedo, M.L.: Scheduling: Theory, Algorithms, and Systems, 4th edn. Springer, New York (2012)
16. Silva, J.D.L., Burke, E.K., Petrovic, S.: An introduction to multiobjective meta-heuristics for scheduling and timetabling. In: Gandibleux, X., Sevaux, M., Soerensen, K., T'kindt, V. (eds.) Metaheuristics for Multiobjective Optimisation - Part I, Lecture Notes in Economics and Mathematical Systems, vol. 535, pp. 91–129. Springer, Heidelberg (2004)
17. Smith, W.E.: Various optimizers for single-stage production. Nav. Res. Logistics Q. **3**(1), 59–66 (1956)
18. Wang, J., Cai, Y.: Multiobjective evolutionary algorithm for frequency assignment problem in satellite communications. Soft Comput. **19**(5), 1229–1253 (2015)
19. Wassenhove, L.N.V., Gelders, F.: Solving a bicriterion scheduling problem. Eur. J. Oper. Res. **4**(1), 42–48 (1980)
20. Zhang, Q., Li, H.: MOEA/D: a multiobjective evolutionary algorithm based on decomposition. IEEE Trans. Evol. Comput. **11**(6), 712–731 (2007)
21. Zitzler, E., Thiele, L., Laumanns, M., Fonseca, C.M., Da Fonseca, V.G.: Performance assessment of multiobjective optimizers: an analysis and review. IEEE Trans. Evol. Comput. **7**(2), 117–132 (2003)

Empirical Investigations of Reference Point Based Methods When Facing a Massively Large Number of Objectives: First Results

Ke Li[1,2]($\boxtimes$), Kalyanmoy Deb[3], Tolga Altinoz[4], and Xin Yao[2]

[1] University of Exeter, North Park Road, Exeter EX4 4QF, UK
k.li@exeter.ac.uk
[2] University of Birmingham, Edgbaston, Birmingham B15 2TT, UK
x.yao@cs.bham.ac.uk
[3] Michigan State University, East Lansing, MI 48864, USA
kdeb@egr.msu.edu
[4] Ankara University, Ankara, Turkey
taltinoz@ankara.edu.tr

Abstract. Multi-objective optimization with more than three objectives has become one of the most active topics in evolutionary multi-objective optimization (EMO). However, most existing studies limit their experiments up to 15 or 20 objectives, although they claimed to be capable of handling as many objectives as possible. To broaden the insights in the behavior of EMO methods when facing a massively large number of objectives, this paper presents some preliminary empirical investigations on several established scalable benchmark problems with 25, 50, 75 and 100 objectives. In particular, this paper focuses on the behavior of the currently pervasive reference point based EMO methods, although other methods can also be used. The experimental results demonstrate that the reference point based EMO method can be viable for problems with a massively large number of objectives, given an appropriate choice of the distance measure. In addition, sufficient population diversity should be given on each weight vector or a local niche, in order to provide enough selection pressure. To the best of our knowledge, this is the first time an EMO methodology has been considered to solve a massively large number of conflicting objectives.

1   Introduction

During the past three decades, a large number of evolutionary multi-objective optimization (EMO) algorithms have been developed for solving multi-objective optimization problems (MOPs) with two and three objectives [8]. With the development of science and technology, real-life applications nowadays consider more complicated problems with four or more objectives, as known as many-objective optimization problems. Unfortunately, the curse of dimensionality has always been the Achilles's heel of optimization algorithms. It is widely accepted that the performance of Pareto dominance-based EMO algorithms such as NSGA-II [7] severely degrade with the increase of the number of objectives [19].

© Springer International Publishing AG 2017
H. Trautmann et al. (Eds.): EMO 2017, LNCS 10173, pp. 390–405, 2017.
DOI: 10.1007/978-3-319-54157-0_27

Although some indicator-based EMO algorithms such as SMS-EMOA [4] claim to be scalable to any number of objectives in theory, their computational over-heads in practice increase exponentially with the number of objectives [2]. In the past three years or so, many-objective optimization has become one of the most active topics within the EMO community and numerous studies have been conducted [13], e.g., remedy strategies for the canonical Pareto domi-nance to improve the convergence property [20], diversity management mech-anisms to reimburse the loss of selection pressure [1], techniques for speeding up the calculation of some computationally expensive performance metrics in the indicator-based methods [3,15]. Recently, the reference point based method[1], e.g., MOEA/D [22] and NSGA-III [9], has shown very competitive performance for handling problems with many objectives. Generally speaking, its basic idea is using a set of pre-defined weight vectors to guide the search process. In par-ticular, the weight vectors can be used to decompose the original MOP into a set of subproblems, either as scalarizing functions or simplified MOPs. As a consequence, the convergence is guaranteed by the optimization of each subprob-lem whereas the diversity is implicitly controlled by the uniform distribution of weight vectors.

Although the existing many-objective optimizers claim to be able to handle problems scalable to any number of objectives, most, if not all, existing stud-ies limit their experiments to problems with the number of objectives up to 15 or 20 [9,14,21]. In view of the competitive performance reported in recent studies [12], this paper empirically investigates the performance of three selected reference point based EMO algorithms, i.e., MOEA/D, NSGA-III and R-NSGA-II [10], on MOPs with a massively large number of objectives. From the experi-mental results, we surprisingly find that R-NSGA-II, which was originally used to approximate the preferred Pareto-optimal solutions rather than the entire Pareto-optimal front (PF), shows the most competitive performance. In con-trast, the performance of MOEA/D and NSGA-III deteriorate significantly with the massively growing number of objectives.

The remainder of this paper is started by a detailed description of the exper-imental settings of our comparative study in Sect. 2. Then, Sect. 3 presents the discussions of the experimental results. At last, Sect. 4 summarizes the main findings and provide an outlook of possible future directions.

2   Experimental Settings

2.1   Benchmark Problems

As a first study, here we use DTLZ1 to DTLZ4 from the widely used DTLZ suite [11] to form the benchmark set. For each benchmark problem, we consider the 25-, 50-, 75- and 100-objective cases separately. As for DTLZ1, the number of decision variables is set as $n = m + 4$; and for DTLZ2 to DTLZ4, it is set as $n = m + 9$, where m is the number of objectives.

---

[1] Also known as decomposition-based method, but here we use the terminology refer-ence point based method without loss of generality.

2.2   Performance Metrics

To have a quantitative comparison of different algorithms, we use the following two indicators as the performance metrics.

- **Convergence Measure (CM):** As discussed in [11], the objective functions of a Pareto-optimal solution $\mathbf{x}^*$ satisfy: $\sum_{i=1}^m f_i(\mathbf{x}^*) = 0.5$ for DTLZ1 and $\sum_{i=1}^m f_i^2(\mathbf{x}^*) = 1.0$ for DTLZ2 to DTLZ4, respectively. Let S be the set of solutions obtained by an EMO algorithm, and the CM is calculated as:

$$
CM(S) = \begin{cases} \frac{\sum_{\mathbf{x} \in S} |\sum_{i=1}^m f_i(\mathbf{x}) - 0.5|}{|S|}, & \text{DTLZ1} \\ \frac{\sum_{\mathbf{x} \in S} |\sum_{i=1}^m f_i^2(\mathbf{x}) - 1.0|}{|S|}, & \text{DTLZ2 to DTLZ4} \end{cases} \tag{1}
$$

   where $|S|$ is the cardinality of S.
- **Inverted Generational Distance (IGD)** [5]: Let P^* be a set of points uniformly sampled along the PF, and the IGD value of S is calculated as:

$$
\text{IGD}(S, P^*) = \frac{\sum_{\mathbf{x}^* \in P^*} dist(\mathbf{x}^*, S)}{|P^*|} \tag{2}
$$

   where $dist(\mathbf{x}^*, S)$ is the Euclidean distance between the point $\mathbf{x}^* \in P^*$ and its nearest neighbor of S in the objective space. We use the method developed in [14] to get the Pareto-optimal samples that form P^*.

   The CM metric only evaluates the convergence property while the IGD metric can evaluate both the convergence and diversity simultaneously. The lower are the CM and IGD metric values, the better is the quality of a solution set for approximating the PF. Each algorithm is independently run 31 times, and we use the Wilcoxon's rank sum test at a 5% significance level to validate the significance of a better result.

2.3   Examined Reference Point Based EMO Algorithms

This paper considers the following three algorithms as the representatives.

- **MOEA/D** [22]: Its basic idea is to decompose a MOP into several scalarizing subproblems and optimizes them in a collaborative manner. To leverage the similarity information among neighboring subproblems, the mating selection and population update have a large chance to take place within the neighborhood. The offspring solution can replace its parent only when it has a better scalarizing function value for the corresponding subproblem. In particular, we consider the penalty-based boundary intersection (PBI) as the scalarizing function [22] in view of its reported superior performance for many-objective optimization [9]. Formally, the PBI function is defined as:

$$
\begin{aligned} \text{minimize } & g^{PBI}(\mathbf{x}|\mathbf{w}, \mathbf{z}^*) = d_1 + \theta d_2, \\ \text{subject to } & \mathbf{x} \in \Omega, \end{aligned} \tag{3}
$$

where Ω is the feasible region, $\mathbf{w}$ is a priori defined weight vector, $\mathbf{z}^*$ is the ideal objective vector, $\theta > 0$ is a user-defined penalty parameter (here we set $\theta = 5.0$), $d_1 = \frac{\|(\mathbf{F}(\mathbf{x}) - \mathbf{z}^*)^T \mathbf{w}\|}{\|\mathbf{w}\|}$, $d_2 = \|\mathbf{F}(\mathbf{x}) - (\mathbf{z}^* + d_1 \mathbf{w})\|$ and $\|\cdot\|$ is the ℓ^2-norm.

- **NSGA-III** [9]: Its subproblem is defined to achieve the local optimum, in terms of convergence and diversity, of a subregion specified by the corresponding weight vector. More specifically, each solution is associated with its closest weight vector according to the perpendicular distance. Afterwards, the solutions associated with a particular weight vector form a niche. The survival of a solution is determined by the Pareto dominance relationship and the crowdedness of a niche. In particular, the ones associated with a less crowded niche have a larger chance to survive to the next generation.

- **R-NSGA-II** [10]: It was originally proposed to consider the decision maker's (DM's) preference information into the search process and to approximate the region of interest (ROI) rather than the entire PF. Specifically, the DM elicits his/her preference information as one or multiple aspiration level vectors which can be regarded as a discrete approximation of the PF. During the selection procedure, solutions closer to the given aspiration level vector have a larger chance to survive to the next generation. To maintain the population diversity, R-NSGA-II employs a ϵ-clearing idea to control the spread of those selected solutions within the ROI. If we set the aspiration level vectors the same as the weight vectors used in MOEA/D and NSGA-III, we can expect R-NSGA-II to approximate the entire PF as well.

2.4   Parameter Settings

All EMO algorithms use the simulated binary crossover (SBX) and polynomial mutation [8] for offspring reproduction. As suggested in [9], the distribution indices for SBX and polynomial mutation are set to $\mu_c = 30$ and $\mu_m = 30$, respectively. The crossover probability is set to $p_c = 0.9$ and the mutation probability is set to $p_m = \frac{1}{n}$. As suggested in [22], the neighborhood size is set as 20 in MOEA/D, and the probability to select within the neighborhood is set as 0.9. In addition, the maximum number of replacement of an offspring is set as 2. As for R-NSGA-II, the ϵ-clearing parameter is set as $\epsilon = 0.01$ as suggested in [10]. The stopping criterion for each algorithm is the predefined number of function evaluations (FEs). The settings of population size and the maximum number of FEs for different benchmark problems are given in Table 1. Note that the population size of R-NSGA-II is set twice as that of MOEA/D and NSGA-III. This is because R-NSGA-II was originally proposed to approximate the ROI where each aspiration level vector is supposed to hold more than one solution.

Note that the selection mechanism of a reference point based method mainly relies on a predefined set of weight vectors. They not only guide the search direction, and their distribution also determines the population diversity to a certain extent. The Das and Dennis's method [6], which samples $N = \binom{H+m-1}{m-1}$ uniformly distributed weight vectors with a uniform space $\Delta = \frac{1}{H}$ ($H > 0$ is the number of divisions considered along each objective coordinate) from a canonical simplex, is the most popular one for weight vector generation. Although

Table 1. Settings of population size and maximum number of FEs

m	MOEA/D	NSGA-III	R-NSGA-II	Problem	♯ of FEs	Problem	♯ of FEs
25	125	128	256	DTLZ1,3,4	250,000	DTLZ2	200,000
50	250	252	500		750,000		600,000
75	375	376	752		1,500,000		1,125,000
100	500	500	1,000		2,500,000		2,000,000

this method works well in the two- or three-objective case, it becomes impractical when the number of objectives becomes large. As discussed in [14], in
order to have intermediate weight vectors within each edge of the simplex, we
should set $H \geq m$. However, even for a 7-objective case, $H = 7$ will results in
$\binom{7+7-1}{7-1} = 1,716$ weight vectors. Such large amount of weight vectors obviously
aggregates the computational burden of an EMO algorithm. On the other hand,
if we simply set $H < m$, all weight vectors will sparsely distribute along the
boundary of the simplex. This is apparently harmful to the population diversity.
To attack this issue, especially when encountering a massively large number of
objectives, the two-layer weight vector generation method suggested in [14] is
a viable resolution. Its basic idea is to make the boundary weight vectors con-
tract inward the simplex by a coordinate transformation. Though it is originally
suggested for generating two layers of weight vectors, it can be generalized to a
multiple-layer case. Suppose that we need to generate σ ($\sigma \geq 2$) layers of weight
vectors. First of all, σ layers of weight vectors (denoted as $W^k = \{\mathbf{w}_1^k, \cdots, \mathbf{w}_{\ell_k}^k\}$,
where $k \in \{1, \cdots, \sigma\}$ and ℓ_k is the number of weight vectors in the k-th layer)
are initialized according to the Das and Dennis's method, with appropriate H
settings. Afterwards, the coordinates of weight vectors in the inside layer (i.e.,
except the first layer) are shrunk by a coordinate transformation. Specifically,
as for a weight vector in the k-th layer, denoted as $\mathbf{w}^k = (w_1^k, \cdots, w_m^k)$, where
$k \in \{2, \cdots, t\}$, its j-th component is re-evaluated as:

$$w_j^k = \frac{1-\tau}{m} + \tau_k \times w_j^k \tag{4}$$

where $j \in \{1, \cdots, m\}$ and $\tau_k \in [0, 1]$ is the shrinkage factor for the k-th layer.
All σ layers of weight vectors combine to form the final weight vector set W.

Note that one of the major assumptions here is that the uniform distribution
of weight vectors can result in a uniform distribution of the obtained solutions.
But unfortunately, this might not always be guaranteed in a high-dimensional
space, especially when encountering a massively large number of objectives. Let
us consider a 25-objective case, where weight vectors are generated by a five-
layer weight vector generation method as shown in Fig. 1(a). In particular, the
shrinkage factors are $\tau = \{1.0, 0.75, 0.5, 0.25, 0.1\}$ for different layers, respectively. And for each layer, we set $H = 1$. Correspondingly, we calculate the
Pareto-optimal samples on the PFs of DTLZ1 and DTLZ2 by using the method
introduced in [14]. From the results shown in Fig. 1(b) and (c) we find that,

although different layers of weight vectors have a uniform scale on the simplex, only the Pareto-optimal samples of DTLZ1, which has a linear PF shape, have a uniform spread over the PF; as for DTLZ2, which has a non-linear PF shape, the Pareto-optimal samples on the first three layers crowd around the boundary of the PF. This means that a linear setting of shrinkage factors, which results in a uniformly scaled weight vector layers, can lead an algorithm to search for a set of solutions having a biased distribution at the end. Obviously, this is harmful to the population diversity. Even worse, it aggravates with the increase of the number of objectives. To alleviate this side effect, here we suggest setting the shrinkage factors in a non-linear manner, according to Proposition 1.

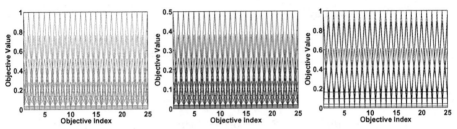

(a) Distribution of weight vectors.   (b) Pareto-optimal samples on DTLZ1.   (c) Pareto-optimal samples on DTLZ2.

Fig. 1. Linear τ setting, i.e., $\tau = \{1.0, 0.75, 0.5, 0.25, 0.1\}$, and their corresponding Pareto-optimal samples on PFs of 25-objective DTLZ1 (linear PF) and DTLZ2 (non-linear PF), respectively.

Proposition 1. *For DTLZ2 to DTLZ4, suppose the expected objective value is f, the appropriate shrinkage factor should be set as the positive τ:*

$$\tau = \frac{-b \pm \sqrt{b^2 - 4ac}}{2a} \tag{5}$$

where $a = (f^2 - 1)m^2 - (f^2 - 2)m - 3$, $b = -2(m - 1)$ and $c = f^2 m - 1$.

Proof. Let us use a simple example to prove this proposition. Suppose an extreme weight vector $\mathbf{w} = (w_1, \cdots, w_m)^T$ is set as: $w_i = 1.0$ and $w_j = 0.0$ where $j \in \{1, \cdots, m\}$ and $j \neq i$. By using the weight vector transformation method introduced in Eq. (4), we can have the corresponding transformed weight vector as:

$$w_i = \frac{1 - \tau}{m} + \tau \times w_i, \quad w_j = \frac{1 - \tau}{m}, \forall j \in \{1, \cdots, m\}, j \neq i \tag{6}$$

According to [14], the i-th objective value on the PF of DTLZ2 is calculated as:

$$
\begin{aligned}
f_i(\mathbf{x}) &= \frac{w_i}{\sqrt{\sum_{k=1}^{m} w_k^2}} \\
&= \frac{\frac{1-\tau}{m} + \tau \times w_i}{\sqrt{\sum_{k=1}^{m-1}(\frac{1-\tau}{m})^2 + (\frac{1-\tau}{m} + \tau \times w_i)^2}} \\
&= \frac{\frac{1-\tau}{m} + \tau \times w_i}{\sqrt{(m-1)(\frac{1-\tau}{m})^2 + (\frac{1-\tau}{m} + \tau \times w_i)^2}}
\end{aligned}
\tag{7}
$$

Let $t = f_i(\mathbf{x})^2$, we can have:

$$
\begin{aligned}
\frac{(\frac{1-\tau}{m} + \tau \times w_i)^2}{\sqrt{(m-1)(\frac{1-\tau}{m})^2 + (\frac{1-\tau}{m} + \tau \times w_i)^2}} &= t \\
\implies (t-1)(\frac{1-\tau}{m} + \tau)^2 + t(m-1)(\frac{1-\tau}{m})^2 &= 0 \\
\implies [(t-1)m^2 - (t-2)m - 3]\tau^2 - 2(m-1)\tau + (tm-1) &= 0
\end{aligned}
\tag{8}
$$

Let $a = [(t-1)m^2 - (t-2)m - 3], b = 2(m-1), c = (tm-1)$, the above equation thus can be treated as a quadratic equation with τ as the unknown. Accordingly, this equation can be solved as:

$$
\tau = \frac{-b \pm \sqrt{b^2 - 4ac}}{2a}
\tag{9}
$$

Obviously, the appropriate τ for the weight vector transformation purpose should be a positive number, i.e., the positive solution of Eq. (9).

Based on Proposition 1, we change the shrinkage factor setting of the example shown in Fig. 1 as $\tau = \{1.0, 0.2, 0.125, 0.08, 0.03\}$. As shown in Fig. 2, by using this τ setting, we can have a better spread of Pareto-optimal samples than using the linearly scaled τ setting. In our experiments, we use the 5-layer weight vector generation method suggested in this subsection to generate initial weight vectors. In particular, we set $\tau = \{1.0, 0.75, 0.5, 0.25, 0.1\}$ for the DTLZ1 problem and $\tau = \{1.0, 0.2, 0.125, 0.08, 0.03\}$ for the DTLZ2 to DTLZ4 problems.

3   Experimental Results

3.1   Comparison Results of MOEA/D, NSGA-III and R-NSGA-II

The CM and IGD metric values obtained by different algorithms are presented in Table 2. The best metric value is highlighted in bold face with a gray background. In Fig. 3, we plot the trajectories of the mean metric values obtained by different algorithms versus different number of objectives. To have a visual comparison,

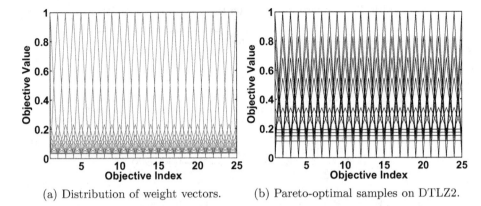

(a) Distribution of weight vectors.   (b) Pareto-optimal samples on DTLZ2.

Fig. 2. Non-linear τ setting, i.e., $\tau = \{1.0, 0.2, 0.125, 0.08, 0.03\}$, and their corresponding Pareto-optimal samples on the PF of 25-objective DTLZ2.

we show the parallel coordinate plot of the population holding the best IGD value for each algorithm[2].

From the experimental results, in terms of the IGD metric values and the parallel coordinate plots shown in the supplementary file, we find that R-NSGA-II is the best candidate on all benchmark problem instances. In particular, the solutions found by R-NSGA-II almost converge to the PF, and they well approximate the expected points on the PF. In contrast, the performance of NSGA-III is not satisfactory. Its performance deteriorates significantly with the growing number of objectives. For NSGA-III, the most direct effect from the curse of dimensionality is the convergence where NSGA-III can hardly drive the solutions fully converge to the PF on problems with the multi-modal property, e.g., DTLZ1 and DTLZ3. Although DTLZ4 does not have local PFs in its search space, its parametric mapping, which causes a biased density of solutions, impairs the selection pressure of NSGA-III and thus makes the solutions be drifted. As for MOEA/D, we observe a very decent performance on the CM metric. However, we also notice that the IGD metric values obtained by MOEA/D are not very competitive. From the parallel coordinate plots shown in the supplementary file, we find that although the solutions obtained by MOEA/D converge to the PF, they crowd on a narrow region. This means that MOEA/D fails to approximate the entire PF and thus explains its poor IGD metric values. It is interesting to note that the CM metric values obtained by MOEA/D are improved with the growth of the dimensionality, while their IGD metric values deteriorate accordingly. This might imply that diversity preservation becomes more different by increasing the number of objectives. In this case, most of the computational budgets in MOEA/D have been devoted to several selected weight vectors, thus results in a biased distribution in a high-dimensional space.

---

[2] Due to the page limit, the parallel coordinate plots are put in the supplementary file, which can be found in https://coda-group.github.io/publications/suppEMO.pdf.

Table 2. Comparison results on the CM and IGD metric

Problem	m	MOEA/D	NSGA-III	R-NSGA-II	Generative Method	s
			CM			
DTLZ1	25	**4.634E-3(2.01E-3)**	8.775E-2(1.52E-1)	2.768E-2(9.88E-3)	3.807E+1(1.70E+0)	†
	50	**3.877E-3(1.82E-3)**	2.594E+0(5.01E+0)	3.781E-2(4.36E-3)	3.871E+1(1.03E+0)	†
	75	**1.848E-3(5.23E-4)**	2.529E+1(9.89E+0)	6.762E-2(1.86E-2)	3.891E+1(1.68E+0)	†
	100	**1.612E-3(8.15E-4)**	2.985E+1(8.50E+0)	1.259E-1(1.74E-2)	3.907E+1(1.12E+0)	†
DTLZ2	25	4.425E-3(9.51E-4)	5.719E-2(3.65E-2)	1.262E-3(3.26E-4)	**4.613E-5(4.63E-5)**	†
	50	3.046E-3(3.30E-4)	5.497E-2(1.38E-2)	7.431E-4(1.34E-4)	**1.599E-3(1.19E-3)**	†
	75	2.722E-3(5.68E-4)	6.992E-2(2.71E-2)	5.566E-4(6.01E-5)	**2.058E-3(1.49E-3)**	†
	100	1.548E-3(6.60E-4)	1.191E-1(3.55E-2)	3.918E-4(6.01E-5)	**9.883E-4(3.56E-4)**	†
DTLZ3	25	2.478E-2(2.32E-2)	1.573E+3(1.79E+3)	**3.068E-3(2.96E-3)**	1.065E+4(5.06E+2)	††
	50	1.775E-2(1.89E-2)	7.793E+3(1.09E+3)	**4.667E-4(3.71E-4)**	1.038E+4(2.53E+2)	†
	75	4.263E-3(8.83E-3)	2.162E+4(1.40E+4)	**1.577E-4(5.21E-5)**	1.051E+4(1.29E+2)	†
	100	**6.794E-5(3.67E-5)**	3.292E+4(1.43E+4)	8.730E-5(2.78E-5)	1.059E+4(2.08E+2)	†
DTLZ4	25	2.974E-3(4.26E-3)	7.520E+1(1.18E+2)	1.068E-3(1.77E-4)	**1.074E-5(2.28E-5)**	†
	50	**9.323E-4(8.35E-4)**	9.869E+2(6.54E+2)	6.911E-4(7.58E-5)	1.011E-3(1.07E-3)	†
	75	**2.909E-4(8.20E-5)**	3.919E+3(2.52E+3)	5.144E-4(7.93E-5)	6.689E-4(4.05E-4)	†
	100	**2.617E-4(1.36E-4)**	2.798E+3(1.23E+3)	3.317E-4(7.02E-5)	6.086E-4(2.41E-4)	†
			IGD			
DTLZ1	25	1.830E-1(9.51E-4)	1.514E-1(2.33E-2)	**1.059E-1(4.41E-3)**	2.460E-1(8.35E-3)	†
	50	1.848E-1(1.72E-3)	2.335E-1(6.07E-2)	**8.388E-2(1.00E-3)**	2.588E-1(3.26E-3)	†
	75	1.848E-1(8.79E-4)	2.839E-1(9.27E-3)	**7.981E-2(3.22E-3)**	2.708E-1(1.50E-3)	†
	100	1.846E-1(4.72E-4)	2.643E-1(1.20E-2)	**7.585E-2(7.28E-3)**	2.731E-1(2.76E-3)	†
DTLZ2	25	2.110E-1(7.58E-3)	2.643E-1(1.20E-2)	1.391E-1(3.04E-3)	**2.368E-2(5.22E-3)**	†
	50	2.557E-1(4.47E-3)	2.207E-1(4.94E-2)	**1.769E-1(1.61E-2)**	4.934E-1(1.55E-2)	†
	75	2.654E-1(2.67E-2)	2.247E-1(2.57E-2)	**1.986E-1(2.24E-3)**	8.294E-1(1.80E-2)	†
	100	3.993E-1(2.95E-1)	2.374E-1(2.97E-2)	**2.108E-1(1.21E-2)**	9.565E-1(1.29E-2)	†
DTLZ3	25	6.798E-1(5.27E-1)	2.974E+0(1.30E+0)	**1.405E-1(2.74E-3)**	6.612E-1(9.61E-2)	†
	50	7.958E-1(5.60E-1)	1.695E+0(1.01E+0)	**1.795E-1(5.74E-3)**	8.573E-1(4.81E-2)	†
	75	1.136E+0(4.65E-1)	1.159E+0(6.99E-2)	**1.978E-1(3.57E-3)**	1.048E+0(2.54E-2)	†
	100	1.349E+0(3.74E-2)	1.134E+0(5.06E-2)	**2.118E-1(5.77E-3)**	1.139E+0(1.04E-2)	†
DTLZ4	25	5.966E-1(8.10E-2)	7.935E-1(3.13E-2)	**1.358E-1(2.37E-3)**	2.013E-1(1.20E-2)	†
	50	6.047E-1(7.11E-2)	8.596E-1(2.58E-2)	**1.731E-1(2.13E-3)**	3.911E-1(7.80E-3)	†
	75	6.058E-1(7.41E-2)	8.775E-1(1.91E-2)	**1.926E-1(3.70E-3)**	7.414E-1(1.30E-2)	†
	100	6.081E-1(6.99E-2)	8.744E-1(1.24E-2)	**2.061E-1(3.54E-3)**	8.871E-1(1.42E-2)	†

Wilcoxon's rank sum test at a 5% significance level is performed between the best metric value and others. † denotes the best mean metric value is significantly better than all other peers.

3.2 Further Discussions

In principle, the ultimate goal of these reference point based EMO algorithms is to find the appropriate solutions for each weight vector. Putting the algorithmic implementations aside, the key differences among them lie in their distance measures. To facilitate the illustration, let us consider the examples shown in Fig. 4. In particular, we connect a weight vector and the origin to form a reference line.

- MOEA/D uses an aggregated distance measure which adds the perpendicular distance between a solution and the reference line to the distance between the ideal point and the projection of a solution onto the reference line.
- NSGA-III only uses the perpendicular distance between a solution and the reference line as the criterion in its niching process.
- R-NSGA-II uses the direct Euclidean distance between a solution and the DM specified aspiration level vector, i.e., the weight vector, as the secondary selection criterion additional to the Pareto dominance.

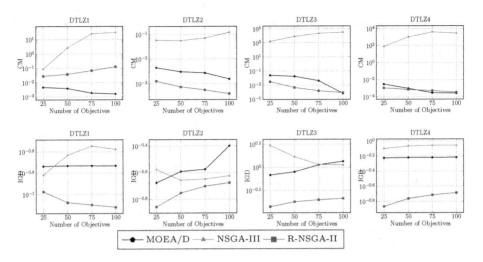

Fig. 3. Trajectories of CM and IGD values on different problems.

The experimental results discussed in Sect. 3.1 provides us an impression that the reference point based methods can be viable for handling MOPs with a massively large number of objectives given the use of a proper distance measure. The superior performance obtained by R-NSGA-II demonstrate that the direct Euclidean distance is the best choice, which might be unexpected at the first glance. The incorporation of the perpendicular distance between a solution and a reference line may help maintain the population diversity, but it may also impair the selection pressure towards the PF.

From another perspective, each weight vector used in R-NSGA-II plays as the DM supplied preference information relating to a particular region of the PF, i.e., the ROI. Accordingly, the use of a set of weight vectors not only implies the prior assumption of the geometrical characteristics of the PF, but also can be

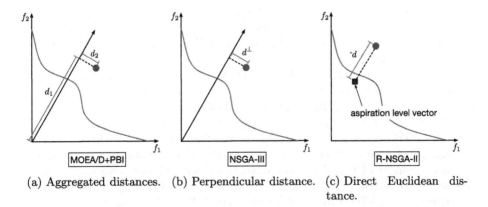

(a) Aggregated distances.   (b) Perpendicular distance.   (c) Direct Euclidean distance.

Fig. 4. Comparisons of different distance measures.

regarded as its discrete approximation. To a certain extent, the approximation of each ROI plays as a reduction of the originally huge objective space; while R-NSGA-II uses a population-based technique to approximate a set of a priori defined ROIs in a parallel and collaborative manner.

3.3   Comparison with the Classical Generative Method

From the experimental results discussed in Sect. 3.1, we find that the reference point based EMO methods, in particular R-NSGA-II, are capable of handling MOPs with a massively large number of objectives. As discussed in Sect. 3.2, the reference point based EMO methods use a set of predefined weight vectors, either supplied by the DM or systematically constructed, to guide the search process and to approximate a set of Pareto-optimal solutions in a single simulation run. However, the classical generative multi-criterion optimization methods [18] repeatedly solve a parameterized single-objective problem might achieve the same effect. Due to lack of parallelism, some studies [9] have reported that the classical generative methods might not be as efficient as the EMO methods. One may ponder whether a classical generative method can have a similar or even better performance than the EMO methods when tackling MOPs with a massively large number of objectives? For this purpose, we formulate a scalarized single-objective optimization problem for each weight vector in the following form:

$$\text{minimize ASF}(\mathbf{x}|\mathbf{z}^*, \mathbf{w}) = \max_{1 \le j \le m} \left(\frac{f_i(\mathbf{x}) - z_i^*}{w_i} \right),$$
$$\text{subject to } \mathbf{x} \in \Omega, \tag{10}$$

where $\mathbf{z}^*$ is an ideal objective vector. In this paper, we set $\mathbf{z}^*$ to be the origin for simplicity. As discussed in some recent studies [9,16,17], the optimal solution of the above optimization problem is the intersecting point of the reference direction of the corresponding weight vector and the PF. To make a fair comparison, for each weight vector, we allocate a maximum of $T = FE_{\max}/N$ (where $FE_{\max}$ is the total number of FEs suggested in Table 1 and N is the population size used by MOEA/D as suggested in Table 1, i.e., the number of weight vectors) FEs to its corresponding optimization procedure. In particular, each optimization procedure is run by the fmincon routine of MATLAB, where a random solution is used for initialization.

The classical generative method is run 31 independent times with different random seeds and the comparative results of the CM and IGD metric values are presented in Table 2 as well. In the supplementary file, we also plot the parallel coordinate plots of solutions which obtain the best IGD value in the corresponding benchmark problem instance. Due to the existence of many local PFs, the classical generative method cannot find any Pareto-optimal solution within the allocated number of FEs. Its poor convergence performance is also reflected by the large CM metric values shown in Table 2. It is also interesting to note that the differences of IGD values obtained by NSGA-III and the classical generative method become small with the growth of dimensionality. Comparing the parallel coordinate plots of NSGA-III and the classical generative method, we find that both methods cannot find any meaningful solution on 50-, 75- and 100-objective DTLZ1 instances. As for the DTLZ2 problem, the classical generative method obtains the best CM metric value on all 25- to 100-objective scenarios. As shown in the parallel coordinate plots, the classical generative method almost obtains a perfect PF approximation on the 25-objective DTLZ2 problem instance. Accordingly, its IGD value is the best comparing to the other three reference point based EMO algorithms. However, the performance of the classical generative method deteriorates significantly with the growth of dimensionality. Although the solutions obtained by the classical generative method almost converge to the PF on the 50-, 75- and 100-objective scenarios, they all bias towards a particular region. This is reflected by its poor IGD metric values. Similar to the DTLZ1 problem, DTLZ3 is also featured by multi-modality. Solutions found by the classical generative method are obviously far away from the PF. However, they seem to have a better spread than those obtained by NSGA-III. Accordingly, the IGD values obtained by the classical generative method are better than NSGA-III. DTLZ4 is featured by its biased distribution. The classical generative method seems to have a relatively acceptable approximation to the PF on the 25- and 50-objective scenarios. This is reflected by its second best IGD metric values. However, with the growth of dimensionality, solutions found by the classical generative method have a obvious bias towards certain objectives.

From the experimental results discussed in this section, we find that the classical generative method is able to have a comparable or even better performance than the reference point based EMO methods when tackling MOPs with a massively large number of objectives without many local optima. However, with the growth of dimensionality and when the problem becomes complicated, the classical generative method can hardly find any meaningful solution as well.

3.4 Multiple-Solution Strategy

All these reference point based EMO algorithms give a high survival rate to the solutions close to the corresponding weight vector. As discussed in Sect. 3.2, one of their major differences is the method for measuring the distance between a solution and a weight vector. Besides, in R-NSGA-II, each weight vector can hold two or more solutions at once, whereas in MOEA/D and NSGA-III, each weight vector is only allowed to accommodate one solution. One may ponder whether the performance of MOEA/D and NSGA-III can be improved in case each weight vector is allowed to hold more than one solution? Bearing this consideration in mind, we make some modifications on the selection mechanisms of MOEA/D and NSGA-III. More specifically, for MOEA/D, each weight vector is assigned two solutions that hold the current best aggregation function values. In addition, the assigned solutions of a particular weight vector have an equal opportunity to participate the mating selection and the update procedure. For NSGA-III, the niche preservation operation is performed in case the niche count of a particular weight vector is greater than two.

Table 3. Comparison results on the CM and IGD metrics

		CM					
Problem	m	MOEA/D△	MOEA/D	NSGA-III△	NSGA-III	R-NSGA-II	s
DTLZ1	25	**2.675E-3(1.18E-3)**	4.634E-3(2.01E-3)	3.568E-3(1.81E-3)	8.775E-2(1.52E-1)	2.768E-2(9.88E-3)	†
	50	**8.160E-4(4.99E-4)**	3.877E-3(1.82E-3)	1.699E-2(2.73E-2)	2.594E+0(5.01E+0)	3.781E-2(4.36E-3)	†
	75	**6.541E-4(4.07E-4)**	1.848E-3(5.23E-4)	6.985E-2(3.75E-2)	2.529E+1(9.89E+0)	6.762E-2(1.86E-2)	†
	100	**5.044E-4(4.44E-4)**	1.612E-3(8.15E-4)	6.985E-2(3.75E-2)	2.985E+1(8.50E+0)	1.259E-1(1.74E-2)	†
DTLZ2	25	2.952E-3(7.58E-4)	4.425E-3(9.51E-4)	2.509E-2(1.33E-3)	5.719E-2(3.65E-2)	**1.262E-3(3.26E-4)**	†
	50	2.133E-3(4.23E-4)	3.046E-3(3.30E-4)	1.557E-2(9.26E-3)	5.497E-2(1.38E-2)	**7.431E-4(1.34E-4)**	†
	75	1.932E-3(3.59E-4)	2.722E-3(5.68E-4)	1.154E-2(2.92E-3)	6.992E-2(2.71E-2)	**5.566E-4(6.01E-5)**	†
	100	1.527E-3(2.95E-4)	1.548E-3(6.60E-4)	3.858E-2(2.93E-3)	1.191E-1(3.55E-2)	**3.918E-4(6.01E-5)**	†
DTLZ3	25	2.335E-2(8.86E-3)	2.478E-2(2.32E-2)	4.796E-1(2.84E-1)	1.573E+3(1.79E+3)	**3.068E-3(2.96E-3)**	†
	50	1.511E-2(1.71E-2)	1.775E-2(1.89E-2)	3.412E-1(1.10E-1)	7.793E+3(1.09E+3)	**4.667E-4(3.71E-4)**	†
	75	1.790E-3(5.62E-3)	4.263E-3(8.83E-3)	3.831E-1(1.10E-1)	2.162E+4(1.40E+4)	**1.577E-4(5.21E-5)**	†
	100	**3.508E-6(4.20E-6)**	6.794E-5(3.67E-5)	3.998E-1(1.95E-1)	3.292E+4(1.43E+4)	8.730E-5(2.78E-5)	†
DTLZ4	25	**6.654E-6(1.97E-5)**	2.974E-3(4.26E-3)	1.46E-2(4.11E-3)	7.520E+1(1.18E+2)	1.068E-3(1.77E-4)	†
	50	**9.553E-7(6.82E-7)**	9.323E-4(8.35E-4)	1.776E-2(4.10E-3)	9.869E+2(6.54E+2)	6.911E-4(7.58E-5)	†
	75	**7.167E-7(4.48E-7)**	2.909E-4(8.20E-5)	1.802E-2(4.02E-3)	3.919E+3(2.52E+3)	5.144E-4(7.93E-5)	†
	100	**5.007E-7(4.50E-7)**	2.617E-4(1.36E-4)	1.867E-2(3.95E-3)	2.798E+3(1.23E+3)	3.317E-4(7.02E-5)	†
		IGD					
DTLZ1	25	1.166E-1(6.96E-4)	1.830E-1(9.51E-4)	1.204E-1(5.45E-3)	1.514E-1(2.33E-2)	**1.059E-1(4.41E-3)**	†
	50	1.208E-1(6.94E-5)	1.848E-1(1.72E-3)	1.502E-1(1.14E-2)	2.335E-1(6.07E-2)	**8.388E-2(1.00E-3)**	†
	75	1.308E-1(7.51E-4)	1.848E-1(8.79E-4)	2.056E-1(1.47E-2)	2.839E-1(9.27E-3)	**7.981E-2(3.22E-3)**	†
	100	1.414E-1(8.55E-5)	1.846E-1(4.72E-4)	2.643E-1(1.20E-2)	2.643E-1(1.20E-2)	**7.585E-2(7.28E-3)**	†
DTLZ2	25	3.442E-1(6.22E-4)	2.110E-1(7.58E-3)	1.026E+0(3.90E-2)	2.643E-1(1.20E-2)	**1.391E-1(3.04E-3)**	†
	50	4.313E-1(1.62E-4)	2.557E-1(4.47E-3)	1.178E+0(2.30E-2)	2.207E-1(4.94E-2)	**1.769E-1(1.61E-3)**	†
	75	4.611E-1(1.27E-3)	2.654E-1(2.67E-2)	1.225E+0(1.88E-2)	2.247E-1(2.57E-2)	**1.986E-1(2.24E-3)**	†
	100	4.926E-1(2.73E-3)	3.993E-1(2.95E-1)	1.276E+0(1.29E-2)	2.374E-1(2.97E-2)	**2.108E-1(1.21E-3)**	†
DTLZ3	25	3.492E-1(8.21E-4)	6.798E-1(5.27E-1)	8.748E-1(1.45E-1)	2.974E+0(1.30E+0)	**1.405E-1(2.74E-3)**	†
	50	8.674E-1(4.57E-1)	7.958E-1(5.60E-1)	8.528E-1(4.45E-2)	1.695E+0(1.01E+0)	**1.795E-1(5.74E-3)**	†
	75	1.070E+0(4.14E-1)	1.136E+0(4.65E-1)	8.610E-1(3.96E-2)	1.159E+0(6.99E-2)	**1.978E-1(3.57E-3)**	†
	100	1.255E+0(2.63E-1)	1.349E+0(3.74E-2)	8.844E-1(5.03E-2)	1.134E+0(5.06E-2)	**2.118E-1(5.77E-3)**	†
DTLZ4	25	6.909E-1(4.57E-2)	5.966E-1(8.10E-2)	3.327E-1(9.74E-4)	7.935E-1(3.13E-2)	**1.358E-1(2.37E-3)**	†
	50	6.917E-1(2.83E-2)	6.047E-1(7.11E-2)	4.221E-1(1.52E-3)	8.596E-1(2.58E-2)	**1.731E-1(2.13E-3)**	†
	75	6.703E-1(3.32E-2)	6.058E-1(7.41E-2)	4.678E-1(2.32E-3)	8.775E-1(1.91E-2)	**1.926E-1(3.70E-3)**	†
	100	6.850E-1(3.60E-2)	6.081E-1(6.99E-2)	5.398E-1(2.63E-3)	8.744E-1(1.24E-2)	**2.061E-1(3.54E-3)**	†

Wilcoxon's rank sum test at a 5% significance level is performed between the best metric value and others. † denotes the best mean metric value is significantly better than all other peers.

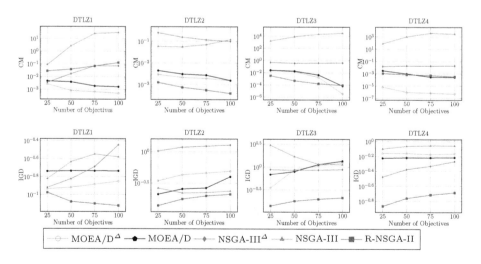

Fig. 5. Trajectories of CM and IGD values on different problems.

The parameters of MOEA/D and NSGA-III are set the same as described in Sect. 2.4, except the population size and the maximum number of FEs. In particular, the population size is doubled, while the maximum number of FEs is reduced by half accordingly. The CM and IGD metric values are presented in Table 3 and Fig. 5. In particular, MOEA/D and NSGA-III with a doubled population size are denoted as MOEA/D$^\Delta$ and NSGA-III$^\Delta$ respectively. From the experimental results, we clearly observe the performance improvement, in terms of convergence and diversity, after doubling the population size. Since each weight vector is able to hold at least two candidates simultaneously, we can expect an improvement on the population diversity. In the meanwhile, due to the existence of more than one candidate associated with a weight vector, we can also expect an enhanced selection pressure towards the PF within a local niche. From the parallel coordinate plots shown in the supplementary file, we find that the solutions obtained by MOEA/D$^\Delta$ and NSGA-III$^\Delta$ have a better convergence than those obtained by MOEA/D and NSGA-III. Nevertheless, we also observe that the IGD metric values obtained by R-NSGA-II are still better than MOEA/D$^\Delta$ and NSGA-III$^\Delta$. This further implies the importance of the use of an appropriate distance measure.

4 Conclusions and Future Directions

In this paper, we conduct a series of experiments that investigate the performance of three selected reference point based EMO methods on MOPs with a massively large number of objectives. From the experimental studies, we find that R-NSGA-II, originally proposed to search for the ROIs, shows very competitive and robust performance when handling problems with a massively large number of objectives. We attribute the success of R-NSGA-II to two aspects: (1) since each weight vector is able to hold more than one solution, a stronger selection pressure can be expected in the local niche; (2) the appropriate distance measure plays an important role in the selection criterion design of the reference point based EMO methods.

We admit that the results observed in this paper might not be conclusive, since we only focus on the reference point based EMO methods. Future research will deepen the insights in the behavior of more types of EMO methods, i.e., the Pareto- and indicator-based EMO methods. However, with a massively large number of objectives, the computational overheads for calculating the Hypervolume metric, which is the most popular metric used in the indicator-based EMO method, might become practically infeasible for the scalability studies. As for the Pareto-based method, we suspect that they might have some troubles to drive the population in a massively high dimensional search space without any direction information provided by the weight vectors. Furthermore, the benchmark problems considered in this paper are relatively less challenging. In our follow-up work, we will test the performance on a wider range of and more complicated benchmark problems. In addition, statistically guided parameter studies will be performed to figure out the suitable parameterizations for different EMO algorithms.

Acknowledgement. This work was partially supported by EPSRC (Grant No. EP/J017515/1).

References

1. Adra, S.F., Fleming, P.J.: Diversity management in evolutionary many-objective optimization. IEEE Trans. Evol. Comput. **15**(2), 183–195 (2011)
2. Auger, A., Bader, J., Brockhoff, D., Zitzler, E.: Hypervolume-based multiobjective optimization: theoretical foundations and practical implications. Theor. Comput. Sci. **425**, 75–103 (2012)
3. Bader, J., Zitzler, E.: HypE: an algorithm for fast hypervolume-based many-objective optimization. Evol. Comput. **19**(1), 45–76 (2011)
4. Beume, N., Naujoks, B., Emmerich, M.T.M.: SMS-EMOA: multiobjective selection based on dominated hypervolume. Eur. J. Oper. Res. **181**(3), 1653–1669 (2007)
5. Coello, C.A.C., Cortés, N.C.: Solving multiobjective optimization problems using an artificial immune system. Genet. Program. Evolvable Mach. **6**(2), 163–190 (2005)
6. Das, I., Dennis, J.E.: Normal-boundary intersection: a new method for generating the pareto surface in nonlinear multicriteria optimization problems. SIAM J. Optim. **8**, 631–657 (1998)
7. Deb, K., Pratap, A., Agarwal, S., Meyarivan, T.: A fast and elitist multiobjective genetic algorithm: NSGA-II. IEEE Trans. Evol. Comput. **6**(2), 182–197 (2002)
8. Deb, K.: Multi-objective Optimization Using Evolutionary Algorithms. John Wiley & Sons Inc., New York (2001)
9. Deb, K., Jain, H.: An evolutionary many-objective optimization algorithm using reference-point-based nondominated sorting approach, part I: solving problems with box constraints. IEEE Trans. Evol. Comput. **18**(4), 577–601 (2014)
10. Deb, K., Sundar, J., Bhaskara, U., Chaudhuri, S.: Reference point based multi-objective optimization using evolutionary algorithms. Int. J. Comput. Intell. Res. **2**(3), 273–286 (2006)
11. Deb, K., Thiele, L., Laumanns, M., Zitzler, E.: Scalable test problems for evolutionary multiobjective optimization. In: Abraham, A., Jain, L., Goldberg, R. (eds.) Evol. Multiobjective Optim. Advanced Information and Knowledge Processing, pp. 105–145. Springer, London (2005)
12. Ishibuchi, H., Setoguchi, Y., Masuda, H., Nojima, Y.: Performance of decomposition-based many-objective algorithms strongly depends on Pareto front shapes. IEEE Trans. Evol. Comput. **PP**(99), 1 (2016)
13. Li, B., Li, J., Tang, K., Yao, X.: Many-objective evolutionary algorithms: a survey. ACM Comput. Surv. **48**(1), 13 (2015)
14. Li, K., Deb, K., Zhang, Q., Kwong, S.: An evolutionary many-objective optimization algorithm based on dominance and decomposition. IEEE Trans. Evol. Comput. **19**(5), 694–716 (2015)
15. Li, K., Kwong, S., Cao, J., Li, M., Zheng, J., Shen, R.: Achieving balance between proximity and diversity in multi-objective evolutionary algorithm. Inf. Sci. **182**(1), 220–242 (2012)
16. Li, K., Kwong, S., Zhang, Q., Deb, K.: Interrelationship-based selection for decomposition multiobjective optimization. IEEE Trans. Cybern. **45**(10), 2076–2088 (2015)

17. Li, K., Zhang, Q., Kwong, S., Li, M., Wang, R.: Stable matching based selection in evolutionary multiobjective optimization. IEEE Trans. Evol. Comput. **18**(6), 909–923 (2014)
18. Miettinen, K.: Nonlinear Multiobjective Optimization, vol. 12. Kluwer Academic Publishers, Boston (1999)
19. Purshouse, R.C., Fleming, P.J.: On the evolutionary optimization of many conflicting objectives. IEEE Trans. Evol. Comput. **11**(6), 770–784 (2007)
20. Sato, H., Aguirre, H.E., Tanaka, K.: Self-controlling dominance area of solutions in evolutionary many-objective optimization. In: Deb, K., Bhattacharya, A., Chakraborti, N., Chakroborty, P., Das, S., Dutta, J., Gupta, S.K., Jain, A., Aggarwal, V., Branke, J., Louis, S.J., Tan, K.C. (eds.) SEAL 2010. LNCS, vol. 6457, pp. 455–465. Springer, Heidelberg (2010). doi:10.1007/978-3-642-17298-4_49
21. Wang, H., Jiao, L., Yao, X.: Two_arch2: An improved two-archive algorithm for many-objective optimization. IEEE Trans. Evolutionary Computation **19**(4), 524–541 (2015)
22. Zhang, Q., Li, H.: MOEA/D: A multiobjective evolutionary algorithm based on decomposition. IEEE Trans. Evolutionary Computation **11**, 712–731 (2007)

Building and Using an Ontology of Preference-Based Multiobjective Evolutionary Algorithms

Longmei Li[1,2]($\boxtimes$), Iryna Yevseyeva[3], Vitor Basto-Fernandes[4,5],
Heike Trautmann[6], Ning Jing[1], and Michael Emmerich[2]

[1] School of Electronic Science and Engineering,
National University of Defense Technology, Changsha 410073, Hunan, China
{longmeili,ningjing}@nudt.edu.cn
[2] Leiden Institute of Advanced Computer Science,
Leiden University, 2333CA Leiden, The Netherlands
{l.li,m.t.m.emmerich}@liacs.leidenuniv.nl
[3] Faculty of Technology, School of Computer Science and Informatics,
De Montfort University, Leicester LE1 9BH, UK
iryna.yevseyeva@dmu.ac.uk
[4] Instituto Universitário de Lisboa (ISCTE-IUL), University Institute of Lisbon,
ISTAR-IUL, Av. das Forças Armadas, 1649-026 Lisbon, Portugal
vitor.basto.fernandes@iscte.pt
[5] School of Technology and Management, Computer Science and Communications
Research Centre, Polytechnic Institute of Leiria, 2411-901 Leiria, Portugal
[6] Department of Information Systems,
University of Münster, 48149 Münster, Germany
trautmann@wi.uni-muenster.de

Abstract. Integrating user preferences in Evolutionary Multiobjective Optimization (EMO) is currently a prevalent research topic. There is a large variety of preference handling methods (originated from Multicriteria decision making, MCDM) and EMO methods, which have been combined in various ways. This paper proposes a Web Ontology Language (OWL) ontology to model and systematize the knowledge of preference-based multiobjective evolutionary algorithms (PMOEAs). Detailed procedure is given on how to build and use the ontology with the help of Protégé. Different use-cases, including training new learners, querying and reasoning are exemplified and show remarkable benefit for both EMO and MCDM communities.

Keywords: Preference · Evolutionary Multiobjective Optimization · Multicriteria decision making · OWL ontology · Protégé

1 Introduction

Evolutionary Multiobjective Optimization (EMO) [13,14] and Multiple Criteria Decision Making (MCDM) [34,50] are two research areas dealing with multiobjective optimization. While traditional MCDM aims at finding one most preferred

H. Trautmann et al. (Eds.): EMO 2017, LNCS 10173, pp. 406–421, 2017.
DOI: 10.1007/978-3-319-54157-0_28

solution based on a preference model using (mainly) mathematical programming, EMO, on the other hand, uses population-based evolutionary algorithms to obtain the whole set of Pareto optimal solutions. User selection is the second step after (an approximation of) the Pareto front (PF) has been achieved. Generally, such methods can be considered as a-posteriori methods in MCDM [34].

Since 2004 when EMO and MCDM researchers met at the Dagstuhl seminar, great attention has been drawn to a collaboration of the two fields. MCDM can help to relieve the selection burden of the Decision Maker (DM) and plays an important role in many-objective optimization (when the number of objectives is no less than three). EMO can tackle difficult problems (such as discontinuous, nondifferentiable, nonconvexity, etc.) that MCDM fails to handle.

Preference-based multiobjective evolutionary algorithms (PMOEAs) are a collaboration of EMO and MCDM. They use preference information provided by the DM to guide the search towards preferred parts of the PF, instead of approximating the whole PF. A variety of PMOEAs have been proposed, they utilize diverse kinds of preferences (reference point, desirability functions, outranking relation, etc.) at different interaction moments (a-priori, interactive, a-posteriori), integrated with several categories of MOEAs (Pareto dominance-based (e.g. PAES, NSGA-II, SPEA2), indicator-based (e.g. SMS-EMOA, HypE, POSEA [53]), decomposition-based (e.g. MOEA/D, NSGA-III), etc.), by means of various integration methods (change of objectives, constraints, or components of algorithms).

There are good review papers along the development of PMOEAs [4,7,12,41], the core value of reviews is to gain knowledge, figure out what has already been done and what are the relations between different works. Ontologies are currently the most suitable and widely used method to formally describe knowledge, by means of comprehensive notations and graphical representations, to help understand concepts and relationships in complex knowledge domains [46].

In this paper we propose an OWL ontology of PMOEAs to model and systematize the results in this domain. With the help of this ontology, researchers can easily understand, access, and analyze methods, or identify future research topics. The PMOEA ontology is made public and is extensible, it can also be reused for building ontologies for EMO and MCDM knowledge domains.

The rest of this paper is structured as follows: In Sect. 2 we give a brief introduction of PMOEAs and OWL ontology. Section 3 provides a detailed procedure of building and extending the PMOEA ontology. Section 4 exemplifies the usage of the proposed ontology. Finally, Sect. 5 concludes the work with a summarizing discussion and the outlook.

2 Background

2.1 PMOEAs

Compared to MOEAs, PMOEAs attracted wide attention in the last decade owing to some advantages. Firstly, the ultimate goal of multiobjective optimization is to assist the DM with finding a best solution. Since visualization and inspection of the whole PF is not a trivial task, a PMOEA can be used to shrink the searching space and to alleviate the selection burden of the DM. Secondly, it is challenging for MOEAs to handle many-objective optimization because most of the solutions are non-dominated. Extending the Pareto order by additional preference information is seen as a great help, if not necessity, in this context. Thirdly, focusing the search within preferred parts will reduce computational efforts spent on developing solutions in less preferred regions.

Because of these benefits, a large number of publications have appeared recently in the PMOEA area. Preference information is an essential element of PMOEAs. In the following, we classify PMOEAs by preference information and list the representative works in each category.

Reference point [52] is probably the most popular approach to embed preferences in PMOEAs. A reference point (or goal vector) is a user-defined point in the objective space indicating the DM's aspiration level for each objective. The closer a solution is to a reference point, the more it is preferred. It was used to change Pareto dominance relation (e.g. g-dominance [36], r-dominance [6], Chebyshev preference relation [26]), modify crowding distance (e.g. R-NSGA-II [18]), alter set quality indicator (e.g. PBEA [47], R2-EMOA [49]) and integrate with swarm-based algorithms (e.g. RPSO-SS [2]).

Reference direction [31] and *light beam search* [27] are extensions of reference point method in MCDM. They were also integrated with EMO (e.g. RD-NSGA-II [16], LBS-EMO [15]).

Preference region is a region in the objective space that is of interest to the DM. Weight functions are widely used as reflection of the DM's degree of satisfaction on objective values (within a user-defined region), such as weighted Hypervolume indicator [11], weighted NSGA-II [21]. Desirability functions are another popular means of preference articulation, which was embedded into NSGA-II [48], SMS-EMOA [51] and MOPSO [37] on both benchmark and real-world applications.

Trade-off is a threshold of how much the DM can sacrifice in the value of one objective in order to improve the other objective(s). It was incorporated with Pareto dominance relation (e.g. in G-MOEA [10]), front sorting (e.g. in pNSGA-II [44]) and Cone-based Hypervolume indicator (CHI) [19,43].

Knee point is the solution with maximal trade-off, often regarded as the most preferred when no explicit preference is given. It was introduced in EMO to guide the search to knee point, e.g. in TKR-NSGA-II [5].

Objective comparison is preference on objectives. For instance objectives can be qualitatively ordered as "most important, important, less important", or quantitatively assigned with different weights. Representative works include development of fuzzy preference relation and dynamic weights [28], relative importance of objectives [42], selection based on fuzzy measure and fuzzy integral [30].

Solution comparison is often used in interactive methods when the DM is asked to compare two alternatives, or select the best (or/and worst) among a sample set during the process of optimization. Preference elicitation is of great significance in this category. Main approaches include value function fitting-based methods (e.g. NEMO [8,9], PI-EMO-VF [17]), machine learning-based methods (e.g. Brain-Computer EMOA [3], Neural Network TDEA [40]), polyhedral cone-based methods (e.g. PI-EMO-PC [45]), rules deduction-based methods(e.g. DRSA-EMO [22]).

Outranking is a binary relation which expresses the degree of truth on the predicate "x is at least as good as y". It was combined with outranking-based dominance in NOSGA [20].

Performance metrics of PMOEAs have also attracted attention in that classical EMO metrics are no longer feasible, because they measure convergence and diversity considering the whole PF, without consideration of the preference information. A new performance metric for reference point-based MOEAs was proposed in [35], a testing framework to compare different reference point-based interactive methods was also introduced in [39], which provide some suggestions on this issue.

Due to the variety of approaches and contexts where preference modelling is combined, to obtain a comprehensive view of PMOEAs becomes increasingly complex. Ontologies are currently an appropriate method to formally describe knowledge in a standard way, to help understand concepts and relationships in complex knowledge domains.

2.2   OWL Ontology

Ontologies are content theories about the sorts of objects, properties of objects and relations between objects that are possible in a specified domain of interest. The most widely accepted definition of ontology in this context is given by Gruber [23]: "An ontology is a formal explicit specification of a shared conceptualization for a domain of interest." It is formal and logic-based, which makes reasoning possible; it has explicit specification, which makes it easy for new learners of this domain; it is a shared conceptualization, which defines a common vocabulary for researchers who need to share information in this domain.

Web Ontology Language (OWL) [25] was approved by World Wide Web Consortium (W3C) to be one of the key Semantic Web technologies in 2004. It employs the eXtensible Markup Language (XML) for the definition of text-based documents syntax/structure, capable of reasoning (OWL-DL [33]) such as consistency and subsumption checking. OWL ontologies can be published online and may refer to or be referred from other OWL ontologies.

Reasons to develop the PMOEA ontology are the following [38]:

(1) To share a common understanding of the structure of information in the PMOEA domain. Concepts in PMOEA domain include preference information, multiobjective evolutionary algorithms and their components (selection operator, variation operator, etc.), multiobjective problems (academic and real-world), preference elicitation methods and so on. It is of high value to represent this information with machine-interpretable vocabulary and analyze it with knowledge discovery tools.

(2) To enable reuse of domain knowledge. Since PMOEA is the connection of EMO and MCDM domains, there are relevant common concepts and relations of concepts in these three domains. Many concepts and relations can be reused in EMO (also Evolutionary Computation) and MCDM ontologies.

(3) To analyze domain knowledge of PMOEA. Building the PMOEA ontology is seen as a step towards the harmonization and systematization of knowledge management in the PMOEA domain. To some extent building an ontology is like defining a set of data and their structure, after collecting information compatible with this structure, analysis can be done to provide more useful information. By analyzing the PMOEA ontology, we can find out what kind of preference information has been integrated in what kind of MOEA, through what kind of integration. We can also query for algorithms that can deal with a specific kind of problem, or find potential combination of MCDM and EMO for future research.

Ontologies are widely used in a variety of research fields, such as knowledge management, recommendation systems, e-Learning, e-Commerce, semantic web, bioinformatics and so on. As far as we know, they have not been applied in the PMOEA field, but an Evolutionary Computation Ontology for e-Learning [29] has been proposed, revealing the feasibility and suitability of using ontologies for optimization algorithms design domain.

Although complete reuse of the e-Learning ontology was not possible for the PMOEA ontology, common concepts and vocabulary were adopted when possible, in order to contribute for knowledge representation harmonization in this domain.

3   Building the PMOEA Ontology

Protégé is a free, open-source platform which provides a growing community with a suite of tools to construct domain models and knowledge-based applications with ontologies [1]. It was developed and maintained by Stanford Center for Biomedical Informatics Research (BMIR) and now has more than 300 thousand registered users. People from different background can publish, view, download, (collaboratively) edit OWL ontologies for research freely. The PMOEA ontology was built with the help of *Protégé Desktop* and made public in *WebProtégé*.

An OWL ontology comprises *Classes*, *Properties* and *Individuals*. A class describes a group of concepts with the same properties and may have necessary and sufficient conditions an individual must verify to belong to that class. A class can have subclasses that represent concepts more specific than the superclass. The hierarchy of classes, which can be represented as a tree structure that relates classes by is-a relation, defines the taxonomy adopted in the ontology. For example, **PMOEA** is the class of preference-based multiobjective evolutionary algorithms. **ReferencePoint_based** is subclass of *PMOEA* which must have reference point as preference information. There are two kinds of properties: object properties and data properties. An object property is a binary relation to relate classes or individuals. For instance, `hasPreferenceInformation` is an object property that can relate **PMOEA** and **PreferenceInformation-FromDM**, which indicates the preference information provided by the DM. A data property relates classes or individuals with a designed primitive data-type (e.g. integer, string, boolean). For example, hasDevelopingYear is a data property of **PMOEA** with datatype **integer**. Individuals represent class instances in the domain of interest, e.g. *r-NSGA-II* [6] is an individual of **PMOEA**.

The general process of building ontologies is given by Noy [38]: (1) determine the scope and domain of the ontology; (2) consider reusing existing ontologies; (3) enumerate important terms in the domain; (4) define the class hierarchy; (5) define object properties; (6) define data properties; (7) create individuals; (8) publish.

Among the above steps (4)–(7) are of core importance for the ontology design, we will give a detailed description of these steps next.

3.1 Class Hierarchy

A tree view of the class hierarchy is shown in Fig. 1.

MetaHeuristic reveals the searching method to find optimal solutions. **MOEA** (multiobjective evolutionary algorithms) and **SOEA** (single-objective evolutionary algorithms) are subclasses of **MetaHeuristic** class, **MOEA** contains **DiversityVSConvergence_based** (individuals *NSGA-II*, *SPEA2*, etc.), **Indicator_based** (individuals *SMS-EMOA*, *HypE*, etc.), **Decomposition_based** (individuals *NSGA-III*, *MOEA/D*, etc.), **Swarm_based** (individuals *MOPSO*, etc.), **Memetic** (individuals *PMA*, etc.), **Coevolution_based** (individuals *CCEA*, etc.)[1]. **PMOEA** is also a subclass of **MOEA**, whose individuals will be introduced in Sect. 3.4. **ObjectiveSpaceTransformation_based** and **ReferencePoint_based** are inferred subclasses of **PMOEA**, which will be introduced in Sect. 4.

ImplementationLibrary is the library or framework used by metaheuristics for implementation. It includes *jMetal*, *KanGAL*, *PISA*, *MOEAFramework*, etc. as individuals.

InteractionTime indicates the moment when the DM interacts with the optimization process: *a-priori*, *a-posteriori* and *progressive* are individuals of this class.

LearningMethod refers to the learning or preference elicitation methods used by some PMOEAs (usually interactive approaches) to mimic the DM's preferences. Subclasses include **OrdinalRegression**, **LinearProgramming**, **QuadraticProgramming**, **SupportVectorMachine**, **NeuralNetwork**.

MOP is the class of multiobjective optimization problems, which has **Academic_Problem** and **Realworld_Problem** as subclasses. **Academic_Problem** has subclasses **DTLZ**, **Knapsack**, **WFG**, **ZDT**.

PreferenceInformationFromDM indicates what the DM should provide to express his/her preferences. Subclasses include **BudgetofDMcalls**,

---

[1] Strictly speaking, not all of these algorithms are evolutionary algorithms, we consider to change MOEA to MOMH (Multiobjective MetaHeuristic) in the future version.

Fig. 1. Class hierarchy

DesirabilityFunction, GroupDecisionMaking, Indicator, Objective-Comparison, OutrankingParameters, PreferenceRegion, ReferenceDirection, ReferencePoint, SolutionComparison (where **PairwiseComparison, SampleRanks** and **SampleSorts** are subclasses), **Trade-off**.

PreferenceIntegration defines how the preference information is integrated in the search method, i.e. what is modified in the optimization process to incorporate preferences. Subclasses include the following: **ASF** (achievement scalarizing function), **Constraints, CrowdingDistance, DominanceRelation, Fitness, FrontSorting, Initialization, Objectives, ParticleUpdate, SelectionCriterion, SetQualityIndicator, TerminationCriterion, TerritorySize.**

PreferenceModel specifies the preference model applied in the PMOEA. It is strongly related to **PreferenceInformationFromDM**, but it focuses on the internal model utilized by the algorithm, about which the DM does not care or know. Its subclasses include **AchievementScalarizingFunction, DecisionRules, FuzzyLogic, KneePoint, LightBeamSearch, ObjectiveRelativeImportance, OutrankingRelation, PolyhedralConeBased, PreferencePoint, PreferenceRegion, UtilityFunction** (which has **Linear, AdditivePiecewiseLinear, GeneralAdditive, Quasiconcave, ChoquetIntegral, Polynomial, SetPreferenceRelation, DesirabilityFunction** as subclasses).

ProgrammingLanguage refers to the language used for implementation of Metaheuristics.

Researcher is the class of researchers in this domain, who are authors of academic papers.

ResultInfluence shows the type of the result, which can be classified as **OneSolution** and **SetOfSolutions**. **BiasedDistribution** and **BoundedRegion** are subclasses of **SetOfSolutions**.

3.2 Object Properties

Object properties are binary relations on individuals, they may be functional, transitive, symmetric and reflexive. Object properties may have a domain and a range specified. For example, *R-NSGA-II* canSolve *ZDT1*, canSolve is an object property whose domain is *MetaHeuristic* and range is *MOP*. The main

Table 1. Object properties

Object property	Domain	Range
hasResultInfluence	MetaHeuristic	ResultInfluence
hasPreferenceModel	PMOEA	PreferenceModel
canSolve	MetaHeuristic	MOP
hasSearchAlgorithm	PMOEA	MetaHeuristic
hasInteractionTime	PMOEA	InteractionTime
hasAuthor	MetaHeuristic	Researcher
hasPreferenceInformationFromDM	PMOEA	PreferenceInformationFromDM
hasPreferenceIntegration	PMOEA	PreferenceIntegration
hasLearningMethod	PMOEA	LearningMethod
isInteractiveVersionOf	PMOEA	PMOEA
hasInteractiveVersion	PMOEA	PMOEA
hasComparison	MetaHeuristic	MetaHeuristic
isExtensionOf	MetaHeuristic	MetaHeuristic
hasExtension	MetaHeuristic	MetaHeuristic
useLibrary	MetaHeuristic	ImplementationLibrary
useLanguage	MetaHeuristic	ProgrammingLanguage

object properties in our ontology are listed in Table 1. Note that one individual can be related to several individuals with the same object property, such as *R-NSGA-II* `hasPreferenceInformationFromDM` *ReferencePoint* and *R-NSGA-II* `hasPreferenceInformationFromDM` *weights* hold at the same time.

`isInteractiveVersionOf` and `hasInteractiveVersion` are inverse of each other, which means "A `isInteractiveVersionOf` B" infers "B `hasInteractiveVersion` A" and vise versa. `isExtensionOf` and `hasExtension` are also inverse of each other. They are both transitive, which means "A `isExtensionOf` B and B `isExtensionOf` C" infers "A `isExtensionOf` C".

3.3   Data Properties

Data properties relate an individual to an XML Schema Datatype value. For example, `hasDevelopingYear` is a data property of *R-NSGA-II* with datatype **integer**, which specifies the year when *R-NSGA-II* was proposed. The main data properties defined in our ontology are listed in Table 2.

Table 2. Data properties

Data property	Domain	Range
isContinuousProblem	**MOP**	boolean
isDiscreteProblem	**MOP**	boolean
isMixedIntegerProblem	**MOP**	boolean
isManyObjectiveProblem	**MOP**	boolean
isMultimodalProblem	**MOP**	boolean
isNoisyProblem	**MOP**	boolean
hasExpensiveEvaluation	**MOP**	boolean
hasNumberOfObjectives	**MOP**	interger
hasDevelopingYear	**MetaHeuristic**	interger
hasReference	**MetaHeuristic, MOP**	string
hasMultipleRegionOfInterest	**PMOEA**	boolean
hasSpreadControl	**PMOEA**	boolean
preservesParetoDominance	**PMOEA**	boolean

Three data properties present the characteristics of PMOEAs. hasMultipleRegionOfInterest describes whether the algorithm can obtain more than one Region of Interest (ROI) in one run. hasSpreadControl indicates whether it allows the DM to control the spread of the obtained ROI, preservesParetoDominance shows whether it preserves the order induced by Pareto dominance. These are important properties of PMOEAs which were also examined by Bechikh [4].

3.4   Creating Individuals

Here we focus on the individuals of **PMOEA** class and use DF-SMS-EMOA [51] as an example, as shown in Fig. 2.

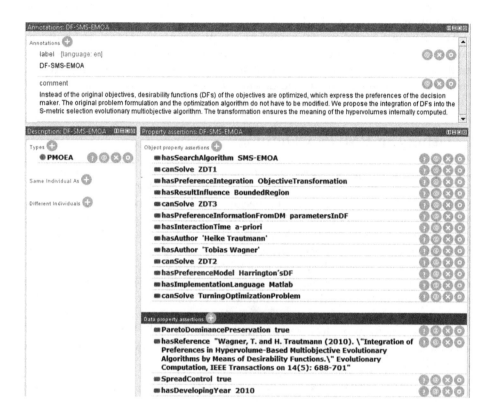

Fig. 2. Individual of DF-SMA-EMOA

Firstly, we create an individual of **PMOEA** named "DF-SMS-EMOA" (*label*) and give a brief introduction as *comment*.

Secondly, basic information of the paper where this algorithm was proposed (authors, reference and publish year) is used to create properties `hasAuthor`, hasReference, hasDevelopingYear.

Thirdly, we consider the interaction time, preference information, preference model, integration method, searching algorithm and result influence of this method, and create properties relating to the corresponding individuals (if there is no such individual, we should create it first).

Then, in the experiment part of the DF-SMS-EMOA paper we set the test problems using property `canSolve`. We should also create **MOP** individuals and fill their properties (isContinuousProblem, isDiscreteProblem, isMixedIntegerProblem, isManyObjectiveProblem, isMultimodalProblem, hasExpensiveEvaluation, hasNumberOfObjectives). Implementation language and library mentioned in the paper are set with `useLibrary` and `useLanguage` properties if they are given.

At last, we create hasMultipleRegionOfInterest, hasSpreadControl and preservesParetoDominance by checking these characteristics. If we are not sure about the property, we can leave it out. Not all the properties are required for an individual to belong to **PMOEA** class.

The PMOEA ontology is published on *WebProtégé* (http://webprotege. stanford.edu/#Edit:projectId=79e2fcdc-b58a-443f-8447-67ce5a388f84). Now it has 92 classes and 507 individuals (including 75 PMOEA individuals). Guests can access, visualize, comment and download the ontology, it also allows collaborative edit after sharing with specified users.

4   Using the PMOEA Ontology

Building the PMOEA ontology is the first step towards its beneficial uses. By checking the information given in the PMOEA Ontology, new learners can quickly get familiar with the domain, because concepts and relations are explicitly defined. Researchers from EMO and MCDM as well as experts and domain engineers can also look for the information they need and comment on the controversial information with the help of *WebProtégé*.

DL Query is a powerful Protégé Desktop plugin for querying and extracting knowledge. It can search for classes or individuals that satisfy certain conditions. The query language is based on the Manchester OWL syntax [24], a user-friendly syntax for OWL-DL.

Next we briefly provide a set of query examples, which shall serve as an overview of possible analysis of the ontology and show how query statements and syntax look like:

- Find all PMOEAs that use a specified kind of preference information, e.g., reference point:
  (`hasPreferenceInformationFromDM value ReferencePoint`)
- Find all PMOEAs that use a specified kind of preference integration method, e.g., objective space transformation:
  (`hasPreferenceIntegration value ObjectiveTransformation`)
- Find all PMOEAs that have been used to solve problems that belong to a certain class, e.g., many objective optimization:
  (`canSolve some (isManyObjectiveProblem value true)`)
- Find all PMOEAs that have been applied to a certain engineering problem (e.g. airfoil optimization), and use a specific implementation language (e.g. Java):
  (`canSolve some AirfoilOptimization and hasImplementationLanguage value Java`)
- Find all PMOEAs by authors or developing year, e.g. Juergen Branke and after 2012:
  (`hasAuthor value 'Juergen Branke'     and     hasDevelopingYear some integer[> "2012"^^integer])`)

- Find all PMOEAs that are extensions of a certain algorithm, e.g., R-NSGA-II:
  (`isExtensionOf value R-NSGA-II`)
  Note that *reasoning* supported by a Protégé Desktop plugin *Reasoner* (e.g. Pellet, HermiT, etc.) is required here for description logics inference. For example, *KR-NSGA-II* is extension of *R-NSGA-II*, *TKR-NSGA-II* is extension of *KR-NSGA-II*, are explicitly defined (asserted) in the ontology by the designer. "*TKR-NSGA-II* `isExtensionOf` *R-NSGA-II*" will be inferred by the *Reasoner* because `isExtensionOf` is a transitive property.
- Find all PMOEAs that use a certain algorithm in comparative experiments, e.g., r-NSGA-II:
  (`hasComparison value r-NSGA-II`)
- Find all PMOEAs that use a specified type of benchmark, so that new algorithms can compare with them, e.g., ZDT1:
  (`canSolve value ZDT1`)

Another important feature of OWL ontology modeling is the use of *Reasoner*. The result of DL Query can be defined as new (inferred) classes. For instance, the result of the first DL Query can be defined as a new class **ReferencePoint-based**, which is a subclass of **PMOEA**. Any PMOEA to be added in the future that has reference point as preference information will be automatically sorted into this subclass. This is systematically guaranteed by the inference engine implemented by the *Reasoner*.

The PMOEA Ontology defines the structure of this domain and provides further information by various queries. One promising advantage of the ontology is its extensibility. New individuals of **PMOEA** can be added following the procedure in Sect. 3.4, as well as new individuals of **MOP**. The value of PMOEA Ontology will increase with the progressive accumulation of information. Moreover, the classes and properties can also be extended and improved by means of *WebProtégé*, which offers a shared Ontology edit platform for collaborative editing. Further use cases of the PMOEA Ontology, including visualization tool, finding appropriate method for an application, discovering opportunities for new research, can be found in [32].

5   Conclusions and Outlook

This paper presents an Ontology of Preference-based Multiobjective Evolutionary Algorithms (PMOEA). Background information of PMOEA and Ontology is given, after which a detailed structure of the ontology is discussed. Then How to build/extend the PMOEA ontology is introduced, simple and practical examples for various use cases of the proposed ontology are described and explained. It demonstrates the utility of web semantic technologies within the PMOEA (EMO and MCDM) communities. It is believed that the more useful and comprehensive data PMOEA Ontology contains, the more valuable information it will provide, for both the research community in EMO and MCDM fields, as well as practitioners who want to use PMOEAs. In the future, on the one hand, new individuals of PMOEA, MOEA and MCDM methods will be added to the ontology, on the other hand, additional properties will be created such as hasBenchmark of real-world problems and hasSourceCode of MetaHeuristic individuals. Moreover, the PMOEA ontology might be reused and serve as inspiration for building more detailed ontologies for larger domains, such as the EMO domain and the MCDM domain.

Acknowledgement. Longmei Li acknowledges financial support from China Scholarship Council (CSC). Heike Trautmann and Michael Emmerich acknowledge support by the European Research Center for Information Systems (ERCIS).

References

1. Protege webpage (2016). http://protege.stanford.edu/
2. Allmendinger, R., Li, X., Branke, J.: Reference point-based particle swarm optimization using a steady-state approach. In: Li, X., et al. (eds.) SEAL 2008. LNCS, vol. 5361, pp. 200–209. Springer, Heidelberg (2008). doi:10.1007/978-3-540-89694-4_21
3. Battiti, R., Passerini, A.: Brain-computer evolutionary multiobjective optimization: a genetic algorithm adapting to the decision maker. IEEE Trans. Evol. Comput. **14**(5), 671–687 (2010)
4. Bechikh, S., Kessentini, M., Said, L.B., Ghédira, K.: Chapter four-preference incorporation in evolutionary multiobjective optimization: a survey of the state-of-the-art. Adv. Comput. **98**, 141–207 (2015)
5. Bechikh, S., Said, L.B., Ghédira, K.: Searching for knee regions of the Pareto front using mobile reference points. Soft Comput. **15**(9), 1807–1823 (2011)
6. Ben Said, L., Bechikh, S., Ghédira, K.: The r-dominance: a new dominance relation for interactive evolutionary multicriteria decision making. IEEE Trans. Evol. Comput. **14**(5), 801–818 (2010)
7. Branke, J.: MCDA and multiobjective evolutionary algorithms. In: Greco, S., Ehrgott, M., Figueira, J.R. (eds.) Multiple Criteria Decision Analysis, vol. 233, pp. 977–1008. Springer, New York (2016)
8. Branke, J., Corrente, S., Greco, S., Słowiński, R., Zielniewicz, P.: Using Choquet integral as preference model in interactive evolutionary multiobjective optimization. Eur. J. Oper. Res. **250**(3), 884–901 (2016)
9. Branke, J., Greco, S., Slowinski, R., Zielniewicz, P.: Learning value functions in interactive evolutionary multiobjective optimization. IEEE Trans. Evol. Comput. **19**(1), 88–102 (2015)
10. Branke, J., Kaußler, T., Schmeck, H.: Guidance in evolutionary multi-objective optimization. Adv. Eng. Softw. **32**(6), 499–507 (2001)
11. Brockhoff, D., Bader, J., Thiele, L., Zitzler, E.: Directed multiobjective optimization based on the weighted hypervolume indicator. J. Multi-Criteria Dec. Anal. **20**(5–6), 291–317 (2013)
12. Coello, C.A.C.: Handling preferences in evolutionary multiobjective optimization: a survey. In: Proceedings of the 2000 Congress on Evolutionary Computation, vol. 1, pp. 30–37. IEEE (2000)
13. Coello, C.C., Lamont, G.B., Van Veldhuizen, D.A.: Evolutionary Algorithms for Solving Multi-objective Problems. Springer, New York (2007)
14. Deb, K.: Multi-objective Optimization Using Evolutionary Algorithms, vol. 16. Wiley, New York (2001)

15. Deb, K., Kumar, A.: Light beam search based multi-objective optimization using evolutionary algorithms. In: IEEE Congress on Evolutionary Computation, 2007. CEC 2007, pp. 2125–2132. IEEE (2007)

16. Deb, K., Kumar, A.: Interactive evolutionary multi-objective optimization and decision-making using reference direction method. In: Proceedings of the 9th Annual Conference on Genetic and Evolutionary Computation, pp. 781–788. ACM (2007)

17. Deb, K., Sinha, A., Korhonen, P.J., Wallenius, J.: An interactive evolutionary multiobjective optimization method based on progressively approximated value functions. IEEE Trans. Evol. Comput. **14**(5), 723–739 (2010)

18. Deb, K., Sundar, J.: Reference point based multi-objective optimization using evolutionary algorithms. In: Proceedings of the 8th Annual Conference on Genetic and Evolutionary Computation, pp. 635–642. ACM (2006)

19. Emmerich, M., Deutz, A., Kruisselbrink, J., Shukla, P.K.: Cone-based hypervolume indicators: construction, properties, and efficient computation. In: Purshouse, R.C., Fleming, P.J., Fonseca, C.M., Greco, S., Shaw, J. (eds.) EMO 2013. LNCS, vol. 7811, pp. 111–127. Springer, Heidelberg (2013). doi:10.1007/978-3-642-37140-0_12

20. Fernandez, E., Lopez, E., Lopez, F., Coello, C.A.C.: Increasing selective pressure towards the best compromise in evolutionary multiobjective optimization: the extended NOSGA method. Inf. Sci. **181**(1), 44–56 (2011)

21. Friedrich, T., Kroeger, T., Neumann, F.: Weighted preferences in evolutionary multi-objective optimization. In: Wang, D., Reynolds, M. (eds.) AI 2011. LNCS (LNAI), vol. 7106, pp. 291–300. Springer, Heidelberg (2011). doi:10.1007/978-3-642-25832-9_30

22. Greco, S., Matarazzo, B., Slowinski, R.: Interactive evolutionary multiobjective optimization using dominance-based rough set approach. In: IEEE Congress on Evolutionary Computation (CEC), pp. 1–8. IEEE (2010)

23. Gruber, T.R.: A translation approach to portable ontology specifications. Knowl. Acquis. **5**(2), 199–220 (1993)

24. Horridge, M., Drummond, N., Goodwin, J., Rector, A.L., Stevens, R., Wang, H.: The Manchester OWL syntax. In: OWLed, vol. 216 (2006)

25. Horridge, M., Jupp, S., Moulton, G., Rector, A., Stevens, R., Wroe, C.: A practical guide to building OWL ontologies using Protégé 4 and CO-ODE tools edition 1.2. The University of Manchester (2009)

26. Jaimes, A.L., Montano, A.A., Coello, C.A.C.: Preference incorporation to solve many-objective airfoil design problems. In: 2011 IEEE Congress on Evolutionary Computation (CEC), pp. 1605–1612. IEEE (2011)

27. Jaszkiewicz, A., Słowiński, R.: The light beam search approach-an overview of methodology applications. Eur. J. Oper. Res. **113**(2), 300–314 (1999)

28. Jin, Y., Sendhoff, B.: Incorporation of fuzzy preferences into evolutionary multi-objective optimization. In: GECCO, p. 683 (2002)

29. Kaur, G., Chaudhary, D.: Evolutionary computation ontology: E-learning system. In: 4th International Conference on Reliability, Infocom Technologies and Optimization (ICRITO) (Trends and Future Directions), pp. 1–6. IEEE (2015)

30. Kim, J.H., Han, J.H., Kim, Y.H., Choi, S.H., Kim, E.S.: Preference-based solution selection algorithm for evolutionary multiobjective optimization. IEEE Trans. Evol. Comput. **16**(1), 20–34 (2012)

31. Korhonen, P.J., Laakso, J.: A visual interactive method for solving the multiple criteria problem. Eur. J. Oper. Res. **24**(2), 277–287 (1986)

32. Li, L., Yevseyeva, I., Basto-Fernandes, V., Trautmann, H., Jing, N., Emmerich, M.: An ontology of preference-based multiobjective evolutionary algorithms. arXiv:1609.08082, submitted to IEEE Transactions on Evolutionary Computation
33. McGuinness, D.L., Van Harmelen, F., et al.: Owl web ontology language overview. W3C Recomm. **10**(10), 77–213 (2004)
34. Miettinen, K.: Nonlinear Multiobjective Optimization, vol. 12. Springer, New York (2012)
35. Mohammadi, A., Omidvar, M.N., Li, X.: A new performance metric for user-preference based multi-objective evolutionary algorithms. In: 2013 IEEE Congress on Evolutionary Computation, pp. 2825–2832. IEEE (2013)
36. Molina, J., Santana, L.V., Hernández-Díaz, A.G., Coello, C.A.C., Caballero, R.: g-dominance: reference point based dominance for multiobjective metaheuristics. Eur. J. Oper. Res. **197**(2), 685–692 (2009)
37. Mostaghim, S., Trautmann, H., Mersmann, O.: Preference-based multi-objective particle swarm optimization using desirabilities. In: Schaefer, R., Cotta, C., Kołodziej, J., Rudolph, G. (eds.) PPSN 2010. LNCS, vol. 6239, pp. 101–110. Springer, Heidelberg (2010). doi:10.1007/978-3-642-15871-1_11
38. Noy, N.F., McGuinness, D.L., et al.: Ontology development 101: a guide to creating your first ontology (2001)
39. Ojalehto, V., Podkopaev, D., Miettinen, K.: Towards automatic testing of reference point based interactive methods. In: Handl, J., Hart, E., Lewis, P.R., López-Ibáñez, M., Ochoa, G., Paechter, B. (eds.) PPSN 2016. LNCS, vol. 9921, pp. 483–492. Springer, Cham (2016). doi:10.1007/978-3-319-45823-6_45
40. Pedro, L.R., Takahashi, R.H.C.: Decision-maker preference modeling in interactive multiobjective optimization. In: Purshouse, R.C., Fleming, P.J., Fonseca, C.M., Greco, S., Shaw, J. (eds.) EMO 2013. LNCS, vol. 7811, pp. 811–824. Springer, Heidelberg (2013). doi:10.1007/978-3-642-37140-0_60
41. Rachmawati, L., Srinivasan, D.: Preference incorporation in multi-objective evolutionary algorithms: a survey. In: IEEE Congress on Evolutionary Computation, 2006. CEC 2006, pp. 962–968. IEEE (2006)
42. Rachmawati, L., Srinivasan, D.: Incorporating the notion of relative importance of objectives in evolutionary multiobjective optimization. IEEE Trans. Evol. Comput. **14**(4), 530–546 (2010)
43. Shukla, P.K., Emmerich, M., Deutz, A.: A theoretical analysis of curvature based preference models. In: Purshouse, R.C., Fleming, P.J., Fonseca, C.M., Greco, S., Shaw, J. (eds.) EMO 2013. LNCS, vol. 7811, pp. 367–382. Springer, Heidelberg (2013). doi:10.1007/978-3-642-37140-0_29
44. Shukla, P.K., Hirsch, C., Schmeck, H.: A framework for incorporating trade-off information using multi-objective evolutionary algorithms. In: Schaefer, R., Cotta, C., Kołodziej, J., Rudolph, G. (eds.) PPSN 2010. LNCS, vol. 6239, pp. 131–140. Springer, Heidelberg (2010). doi:10.1007/978-3-642-15871-1_14
45. Sinha, A., Korhonen, P., Wallenius, J., Deb, K.: An Interactive Evolutionary Multi-objective Optimization Method Based on Polyhedral Cones. In: Blum, C., Battiti, R. (eds.) LION 2010. LNCS, vol. 6073, pp. 318–332. Springer, Heidelberg (2010). doi:10.1007/978-3-642-13800-3_33
46. Staab, S., Studer, R.: Handbook on Ontologies. Springer, Heidelberg (2013)
47. Thiele, L., Miettinen, K., Korhonen, P.J., Molina, J.: A preference-based evolutionary algorithm for multi-objective optimization. Evol. Comput. **17**(3), 411–436 (2009)
48. Trautmann, H., Mehnen, J.: Preference-based Pareto optimization in certain and noisy environments. Eng. Optim. **41**(1), 23–38 (2009)

49. Trautmann, H., Wagner, T., Brockhoff, D.: R2-EMOA: focused multiobjective search using R2-indicator-based selection. In: Nicosia, G., Pardalos, P. (eds.) LION 2013. LNCS, vol. 7997, pp. 70–74. Springer, Heidelberg (2013). doi:10.1007/978-3-642-44973-4_8

50. Vira, C., Haimes, Y.Y.: Multiobjective decision making: theory and methodology. No. 8, North-Holland (1983)

51. Wagner, T., Trautmann, H.: Integration of preferences in hypervolume-based multiobjective evolutionary algorithms by means of desirability functions. IEEE Trans. Evol. Comput. **14**(5), 688–701 (2010)

52. Wierzbicki, A.P.: The use of reference objectives in multiobjective optimization. In: Fandel, G., Gal, T. (eds.) Multiple Criteria Decision Making Theory and Application. LNE, vol. 177, pp. 468–486. Springer, Heidelberg (1980)

53. Yevseyeva, I., Guerreiro, A.P., Emmerich, M.T.M., Fonseca, C.M.: A portfolio optimization approach to selection in multiobjective evolutionary algorithms. In: Bartz-Beielstein, T., Branke, J., Filipič, B., Smith, J. (eds.) PPSN 2014. LNCS, vol. 8672, pp. 672–681. Springer, Cham (2014). doi:10.1007/978-3-319-10762-2_66

A Fitness Landscape Analysis of Pareto Local Search on Bi-objective Permutation Flowshop Scheduling Problems

Arnaud Liefooghe[1,2](✉), Bilel Derbel[1,2], Sébastien Verel[3],
Hernán Aguirre[4], and Kiyoshi Tanaka[4]

[1] Univ. Lille, CNRS, Centrale Lille, UMR 9189 CRIStAL, 59000 Lille, France
`arnaud.liefooghe@univ-lille1.fr`
[2] Inria Lille – Nord Europe, 59650 Villeneuve d'Ascq, France
[3] Univ. Littoral Côte d'Opale, LISIC, 62100 Calais, France
[4] Faculty of Engineering, Shinshu University, Nagano, Japan

Abstract. We study the difficulty of solving different bi-objective formulations of the permutation flowshop scheduling problem by adopting a fitness landscape analysis perspective. Our main goal is to shed the light on how different problem features can impact the performance of Pareto local search algorithms. Specifically, we conduct an empirical analysis addressing the challenging question of quantifying the individual effect and the joint impact of different problem features on the success rate of the considered approaches. Our findings support that multi-objective fitness landscapes enable to devise sound general-purpose features for assessing the expected difficulty in solving permutation flowshop scheduling problems, hence pushing a step towards a better understanding of the challenges that multi-objective randomized search heuristics have to face.

1 Introduction

The multi-objective optimization community has spent a lot of efforts in the design of general-purpose techniques allowing to tackle hard optimization problems. Despite the number of available algorithms, their skillful design and their flexibility when applied to a large spectrum of problems, a key ingredient to make them efficient and effective lies in the choice of their components in order to be specifically adapted to the multi-objective optimization problem (MOP) being tackled. This might even depend on the intrinsic properties of the problem instance being considered. In this respect, there is evidence that new tools dedicated to the understanding of the behavior of existing algorithms in light of the properties of the MOP(s) under study are needed. We in fact argue that it is timely to set up new methodological tools allowing to systematically investigate the properties of multi-objective heuristics and to accordingly study and relate their performance and their effectiveness to a given MOP. Such tools exist in the single-objective optimization literature, where a number of paradigms from the so-called fitness landscape analysis [14] have proved to be extremely helpful in

© Springer International Publishing AG 2017
H. Trautmann et al. (Eds.): EMO 2017, LNCS 10173, pp. 422–437, 2017.
DOI: 10.1007/978-3-319-54157-0_29

attaining such a goal. Motivated by their success and their accuracy, there was recently several studies leveraging the single-objective case and pushing fitness landscape analysis a step toward the development of new statistical methodologies and the identification of general-purpose characteristics and features that fit the multi-objective nature of a given optimization problem; see e.g. [1,5,17]. However, still a relatively huge gap remains in bridging fitness landscape analysis with the design of multi-objective randomized search heuristics.

In this paper, we continue the efforts from the community in this direction by conducting a comprehensive fitness landscape analysis with respect to the co-called permutation flowshop scheduling problem (PFSP). Our interest in this problem class stems from the fact that scheduling problems in general, and the PFSP in particular, are ubiquitous to countless real-world applications and constitute one of the most important and challenging problem class from combinatorial optimization [10,16]. Nonetheless, in the multi-objective setting and from a fitness landscape perspective, this problem class was only studied to a small extent. Given that a number of different objectives (e.g., makespan, tardiness) can be considered, one can naturally consider several multi-objective formulations, which makes the PFSP particularly appealing and challenging to investigate. In particular, we already know that optimizing each objective separately leads to single-objective problems with different degrees of difficulty [16]. However, extending such a knowledge to the multi-objective setting is far from being straightforward, given that different objective combinations can be considered. It is to notice that previous studies on the subject exist [2,8]. We actually extend them in different perspectives, as will be described in the following. In particular, we restrict ourselves to the bi-objective PFSP. In fact, multiple proposals considering bi-objective PFSP formulations and instances exist, but it is not clear what are their differences and similarities, neither is clear their impact on the performance of multi-objective algorithms. We are here specifically interested in measuring such an impact and eliciting in an explicit and comprehensive manner the characteristics that makes a problem formulation and the characteristics of problem instances more difficult to solve than another for Pareto local search algorithms. Our contributions can be summarized as follows:

- By conducting an empirical analysis, we are able to study and characterize small-size bi-objective PFSP instances in terms of the correlation between the objective values, the number of Pareto optimal solutions, the number of ranks in non-dominated sorting, and the number of Pareto local optimal solutions for two widely-used neighborhood operators (namely exchange and insert).
- By plugging these operators into the framework of the so-called Pareto Local Search (PLS) algorithm [13], we conduct a comprehensive study on the behavior of the so-obtained variants of PLS when applied to different bi-objective PFSP formulations and instance types. It is to notice that PLS is reported to be one of the state-of-the-art approach for bi-objective PFSP, especially for small-size instances [12].
- As a byproduct, we are able to measure the individual effect of problem features on the performance of PLS variants in terms of success rate, that is the

probability to identify the whole Pareto set. More importantly, we report the joint impact of these features by considering two scenarios: (i) predicting the degree of difficulty of each instance for each algorithm variant, (ii) predicting which algorithm variant performs better for a given instance. For each scenario, a machine learning classification model based on random forests [4] is considered and analyzed in order to measure and hence to quantify how important problem features are when solving a particular instance. Our findings reveal important knowledge on the impact of multimodality, in terms of the proportion of Pareto local optimal solutions, in order to explain the difficulty faced by PLS when solving bi-objective PFSP instances.

In the remainder, we first describe in Sect. 2 the considered PFSP, the corresponding pairs of objectives that we shall study, and the different types of instances. In Sect. 3, we provide a comprehensive analysis and correlation study for the considered problem characteristics. In Sect. 4, the individual effect and the joint impact of problem features on algorithm performance are analyzed. In Sect. 5, we conclude the paper and discuss some open questions.

2 Bi-objective Permutation Flowshop Scheduling Problems

The Permutation Flowshop Scheduling Problem (PFSP) is one of the most popular optimization problem from scheduling [16]. Most research typically deals with a single-objective formulation aiming at minimizing the *makespan*, i.e. the final completion time. However, many other criteria can be formalized, and multi-objective PFSP formulations have also been investigated in the literature; see e.g. [16] for on overview. This section introduces some necessary definitions and presents different benchmark instances taken from the specialized literature.

Problem Formulation and Objectives. The PFSP consists in scheduling N jobs $\{J_1, J_2, \ldots, J_N\}$ on M machines $\{M_1, M_2, \ldots, M_M\}$. Machines are critical resources, i.e. two jobs cannot be assigned to the same machine at the same time. A job J_i is composed of M consecutive tasks $\{t_{i1}, t_{i2}, \ldots, t_{iM}\}$, where t_{ij} is the jth task of the job J_i, requiring the machine M_j. A processing time p_{ij} is associated with each task t_{ij}, and a due date d_i is assigned to each job J_i. We here focus on the permutation PFSP, where the operating sequences of the jobs are identical and unidirectional on every machine. A candidate solution can be represented by a permutation of size N. Let X denote the set of all feasible solutions (permutations). For an instance of N jobs, there exists $|X| = N!$ feasible solutions. To evaluate the schedule of a PFSP instance, many criteria may be defined. In this study, we focus on minimizing two objectives (f_1, f_2) simultaneously from a panel of five criteria as listed in Table 1, and selected among the most widely investigated ones from the literature [16]. Following the notations from [16], this problem class is actually denoted as $F/prmu, d_i/\#(f_1, f_2)$. Given that the single-objective PFSP (minimizing each of these objectives separately)

Table 1. PFSP objectives ($C_i(x)$ is the completion time of Job J_i in schedule x).

Name	Formulation	Description
$C_{\max}(x) =$	$\max_{i \in [\![1,N]\!]} C_i(x)$	Maximum completion time (makespan)
$C_{\text{sum}}(x) =$	$\sum_{i \in [\![1,N]\!]} C_i(x)$	Sum of completion times
$T_{\max}(x) =$	$\max_{i \in [\![1,N]\!]} \max(0, C_i(x) - d_i)$	Maximum tardiness
$T_{\text{sum}}(x) =$	$\sum_{i \in [\![1,N]\!]} \max(0, C_i(x) - d_i)$	Sum of tardiness
$T_{\text{card}}(x) =$	$\left| \{ i \mid C_i(x) > d_i, i \in [\![1,N]\!] \} \right|$	Number of late jobs

is known to be NP-hard [16] for instances with more than two machines ($M > 2$), bi-objective instances can typically not be solved to optimality.

Definitions. Let $Z \subseteq \mathbb{R}^2$ denote the image of feasible solutions in the *objective space* when using the function vector $f = (f_1, f_2)$ such that $Z = f(X)$. The Pareto dominance relation is defined as follows. A solution $x \in X$ is dominated by a solution $x' \in X$, denoted as $x \prec x'$, if $f_k(x') \leqslant f_k(x)$ for all $k \in \{1, 2\}$, with at least one strict inequality. A solution $x \in X$ is *a Pareto optimal solution* (POS) if there does not exist any other solution $x' \in X$ such that $x \prec x'$. The set of all POS is the *Pareto set*, and its mapping in the objective space is the *Pareto front*. One of the most challenging issues in multi-objective optimization is to identify a minimal complete Pareto set, i.e. one Pareto optimal solution mapping to each point from the Pareto front. Since the PFSP is NP-hard [16], heuristics appear to be well suited to identify a *Pareto set approximation*.

Problem Instances. The most common and challenging set of single-objective PFSP benchmark instances is due to Taillard [15]. Each instance file provides a processing time p_{ij} for each task t_{ij}. Each processing time is generated following a discrete uniform distribution: $p_{ij} \sim [\![1, 99]\!]$. These instances are restricted to the objectives dealing with the completion time (i.e. $C_{\max}, C_{\text{sum}}$). Indeed, given that no due date is provided, tardiness-related objectives (i.e. $T_{\max}, T_{\text{sum}}, T_{\text{card}}$) cannot be considered. For this reason, multiple researchers have been interested in extending these single-objective instances by adding a due date d_i for every job J_i. Three due date generation techniques are described below:

1. Inspired by Basseur et al. [3], we propose to generate the due dates following the discrete uniform distribution: $d_i \sim [\![50 \cdot M, \texttt{lb\_cmax}]\!]$, where $\texttt{lb\_cmax}$ is a lower bound of the optimal makespan for the instance under consideration, as defined in [15]. In [3], the authors used an upper bound instead of a lower bound. Using a lower bound avoids us to use any prior external knowledge (about the optimal solution) and likely results in tighter due dates. Hence, a due date d_i roughly lies between the average completion date of the first scheduled job and an optimistic estimate of the best completion time.
2. Liefooghe et al. [11] propose to generate the due dates as follows: $d_i \sim [\![\bar{p} \cdot M, \bar{p} \cdot (N + M - 1)]\!]$, such that $\bar{p}$ is the average-value of processing times.

As a consequence, if all processing times are the same, a due date d_i would lie between the completion date of the first and of the last scheduled job.

3. Unlike the previous approaches, Minella et al. [12] generate the due dates as: $d_i \sim [\![\, p_i^\star \,,\, p_i^\star \cdot 4]\!]$, such that $p_i^\star = \sum_{j \in [\![1, M]\!]} p_{ij}$ is the sum of the processing times over all machines for job J_i. The authors argue that this method generates due dates that range from very tight to relatively tight values.

We intuitively hypothesize that the way the due dates are generated, together with the pair of objectives to be minimized, have a high impact on the bi-objective PFSP instance to be solved. For this reason, we investigate the impact of the choice of these parameters on the fitness landscape characteristics and difficulty of small-size instances. More particularly, we consider the following parameter setting. The problem size is set to a small value of $N \in \{5, 6, 7, 8\}$ in order to enumerate the solution space exhaustively, from $|X| = 120$ for $N = 5$ up to $|X| = 40\,320$ for $N = 8$. The number of machines is $M \in \{5, 6, 7, 8\}$. For each value of type, N and M, 10 independently generated PFSP instances are considered. We investigate the following objective pairs: $(f_1, f_2) \in \{(C_{\max}, T_{\max}), (C_{\max}, T_{\text{sum}}), (C_{\max}, T_{\text{card}}), (C_{\text{sum}}, T_{\max}), (C_{\text{sum}}, T_{\text{sum}}), (C_{\text{sum}}, T_{\text{card}})\}$.

3   Problem Features

In this section, we explore the characteristics of bi-objective PFSP instances in terms of visualization of the objective space, correlation between the objective values, number of Pareto optimal solutions, number of non-dominated fronts, and number of Pareto local optimal solutions. The section ends with a correlation analysis between each pair of those features and of the instance parameters.

Objective Space. Following [2], in order to give a clear intuition of how the different objectives relate one to the other, we show in Fig. 1 a simple inspection of the objective space for one instance with $N = 8$ and $M = 8$. We can basically see that different objective pairs results in different objective space and Pareto front shapes. The objective space actually looks similar for the instances of type basseur and liefooghe, which is clearly to contrast to minella. This is clearly attributed to the differences of due date generation policies. Let us also observe that the pair of objectives $(C_{\max}, \star)$ seems to provide the same objective space shape than its counterpart $(C_{\text{sum}}, \star)$ However, it is not clear that the convexity of the Pareto front is also similar for different pairs of objectives. Notice also that the degree of conflict between the two pairs of objectives $(C_{\max}, T_{\text{sum}})$ and $(C_{\text{sum}}, T_{\max})$ for both instance types liefooghe and basseur is less perceivable than for the two pairs $(C_{\max}, T_{\max})$ and $(C_{\text{sum}}, T_{\text{sum}})$, for which even the size of the Pareto front seems to be relatively smaller. This informal observation is to be extended in the next two paragraphs where the objective correlation and the Pareto front cardinality are quantified and analyzed more explicitly.

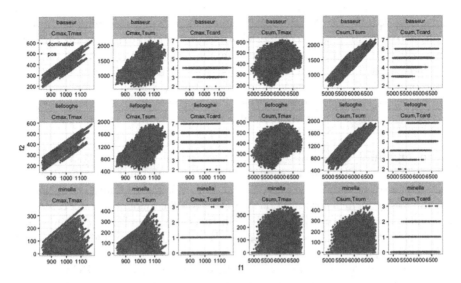

Fig. 1. Objective space ($N = 8$, $M = 8$, one instance).

Objective Correlation. In Fig. 2, we summarize the degree of conflict between the objectives by showing the box-plots obtained when computing the nonparametric Spearman rank correlation coefficient between the objective function values over the whole solution space and for all the instances of each class. Recall that the extreme value of 1 indicates a perfect correlation between the objective values, whereas -1 indicates that they are highly conflicting. A value around 0 indicates that both objectives are uncorrelated. The first notable observation is that the instances of type `minella` have a significantly smaller objective correlation than the two other types, except for $(C_{\max}, T_{\mathrm{card}})$. The second notable observation is that all pairs of objectives are positively correlated, which suggests that by improving f_1, one has a high chance to also improve f_2. This is not surprising for PFSP because lowering the completion time naturally tends to lower the value of other objectives, but this rather uncommon when compared

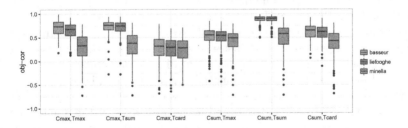

Fig. 2. Spearman rank correlation between the objective function values over the whole solution space for all the instances.

against other MOPs [7]. Notice that there exists a stronger correlation between the objectives dealing with completion time and the sum/maximum tardiness compared to the objective dealing with the number of late jobs, which might be attributed to the high number of plateaus for this function. The lowest correlation is for $(C_{\max}, T_{\text{card}})$ which is validated by Fig. 1, where the convex envelope of the objective space resembles a circle. The relatively high correlation between the objectives suggests that the Pareto set cardinality shall be limited.

Pareto Optimal Solutions. In Fig. 3, we report the proportion of Pareto optimal solutions (POS) in the solution space, computed and aggregated over all the instances of the same type for $N = 8$ and $M = 8$. This is related to the notion of intractability, which arises when the Pareto set cannot be enumerated in a polynomial amount of time [7]. For those instances, the number of POS goes from 0.001% up to 0.1% of the solution space. Besides observing that the objective pair has an impact on the proportion of POS, the most notable observation is that the instances of type `minella` have the highest proportion of POS for $C_{\max}$. This is more mitigated when analyzing the proportion of POS involving C_{sum}. At last, the instances of type `liefooghe` comes in second position, and typically exhibit more POS than `basseur`, which again confirms that the distribution of due date plays an important role in characterizing PFSP instances.

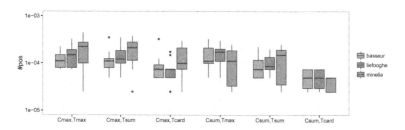

Fig. 3. Proportion of POS in the solution space ($N = 8$, $M = 8$).

Non-dominated Fronts. The previous feature provides a descriptive statistic about the *optimal* Pareto set. In this section, we examine another measure about how *non*-optimal solutions are structured. We use the same measure as [1] by looking at the impact of PFSP instance parameters on the number of non-dominated fronts. This is done by ranking all feasible solutions and layering them into fronts based on non-dominated sorting [9]. In Fig. 4, we report the proportion of non-dominated fronts, with respect to the size of the solution space, computed and aggregated over all the instances of the same type, for size $N = 8$ and $M = 8$, and for each objective pair. This measure can be interpreted as follows. When it is close to 1, the landscape has many fronts which contain few solutions. As this measure goes down to 0, all solutions tend to gather into the same front. From 1 to 0, we then except fewer, but denser, non-dominated fronts. For example, for $N = 8$, when the value is 0.01, we have around 400 fronts, each one containing 100 solutions (i.e. 0.25% of the solution space) in average. When the value is 0.05, we have around 2000 fronts, each one containing 20 solutions (i.e. 0.5% of the solution space) in average. Roughly speaking, when an instance has less POS, it tends to have more fronts. This means that what happens for POS tend to uniformly generalize to the whole solution space. We can observe that the objective pairs $(C_{\max}, \star)$ implies less fronts than the objective pairs $(C_{\text{sum}}, \star)$. We also can notice that the instance type does not have a significant impact on the number of fronts, except for the instances of type `minella` which have less fronts than the instances of type `basseur` and `liefooghe` when considering the objective pairs $(\star, T_{\text{sum}})$.

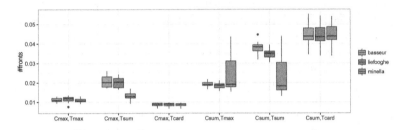

Fig. 4. Proportion of non-dominated fronts in the solution space ($N = 8$, $M = 8$).

Pareto Local Optimal Solutions. The previous considered features are algorithm-independent in the sense that they do not rely on any particular optimization technique or operator. However, we are also interested in the multi-modality, in terms of the number of local optima, of the multi-objective landscape induced by a given PFSP instance and a given neighborhood operator. Hence, we consider a feature that shall directly relate to the search process implied by a local search algorithm for solving the PFSP. Since we are dealing with permutations, we consider the following conventional neighborhood operators [10]:

- The *exchange* neighborhood, for which two solutions are neighbors iff one can be obtained from the other by exchanging two jobs at arbitrary positions.
- The *insert* neighborhood, for which two solutions are neighbors iff one can be obtained from the other by removing a job at a given position and inserting it at another position, hence shifting all other jobs in between.

For both operators, the neighborhood size is quadratic in N, although the insert neighborhood is larger by a factor of about two. Considering these two neighborhood operators, we investigate the number of Pareto local optimal solutions (PLOS) with respect to the different formulations and problem instances. A PLOS is simply a solution which is not dominated by any of its neighbors [13]. In fact, POS are also PLOS, and they are taken into account in our analysis. In Fig. 5, we show a statistic on the proportion of PLOS, computed over all the instances of the same type, for size $N = 8$ and $M = 8$. To be more precise, we report the proportion of POS divided by the proportion of PLOS, which relates global and local optima and which can be interpreted as follows. When this measure is 1, every local optima is a global optima. On the contrary, as this measure goes down to 0, the proportion of PLOS increase substantially, which eventually can be interpreted as having a problem instance which is more challenging to solve for a local search procedure. In fact, when the number of PLOS increases, the local search procedure is more likely to be trapped into local optima, and it is then eventually more difficult for the algorithm to find a new POS. The most notable observation from Fig. 5 is that, when comparing the two neighborhood operators, it appears that the exchange operator induces more difficult landscapes, whatever the instance type and the objective pair. In fact, the insert neighborhood, which is slightly larger, induces less PLOS. This means that we expect the induced landscape to be easier to search. We can also remark that, overall, the objective pairs $(C_{max}, \star)$ induce more PLOS than the objective pairs $(C_{sum}, \star)$, especially for instances of type `minella`. This indicates that the instances corresponding to the former shall be more difficult to solve than the ones corresponding to the latter. In particular, for the objective pairs $(C_{max}, \star)$, instances of type `minella` have typically less PLOS, whereas there is not much difference between the instance types for $(C_{sum}, \star)$. Finally, we can observe that, apart from a few exceptions, no significant difference in the range of the number of PLOS can be reported when comparing the two other instance types.

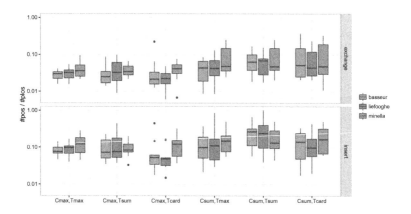

Fig. 5. Proportion of POS *per* PLOS ($N = 8$, $M = 8$).

A Correlation Analysis on Problem Features. To conclude this section, we provide in Fig. 6 a correlation analysis between the different considered features. The goal here is to study and to quantify the association between features. Notice that we consider six numerical features (N, M, objective correlation, number of POS, of fronts, of PLOS) and three categorical ones (instance types, neighborhood structures, and objective pairs). When comparing two numerical features, a Spearman nonparametric correlation coefficient is computed and

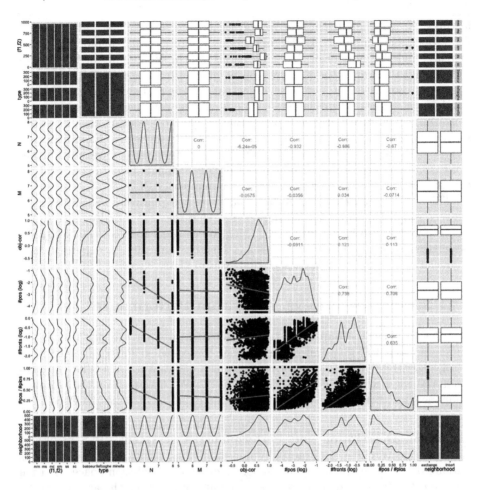

Fig. 6. Association matrix between each pair of problem features. For the objective pair (f_1, f_2), the notation mm (resp. ms, mc, sm, ss and sc) stands for (C_{max}, T_{max}) (resp. $(C_{max}, T_{sum}), (C_{max}, T_{card}), (C_{sum}, T_{max}), (C_{sum}, T_{sum})$ and (C_{sum}, T_{card})).

reported together with a scatter plot and a simple linear regression. When the association between a numerical feature and a categorical one is considered, the density of feature values in every category is shown, as well as box-plots. Notice that, depending on the pair of features considered, the instances are mixed accordingly in Fig. 6 in order to provide a big picture summarizing and generalizing all our findings so far, of course in a more coarse-grained fashion.

Let us start with the problem size N (3rd line, 3rd column). We observe that it has no impact on the correlation between the objectives. However, when N grows, the proportional number of POS decreases, the proportional number of fronts decreases, and the proportional number of PLOS increases. More surprisingly, M (4th line, 4th column) does not show any correlation with the other features. Next, when the objective correlation grows (5th line, 5th column), the proportional number of POS decrease, the proportional number of fronts increase, and the number of PLOS decreases. When the number of POS grows (6th line, 6th column), the proportional number of fronts grows. Although this might seem surprising at first sight, we actually argue that this is an artifact due of mixing instances with different $N-$values, as one can guess by observing the clustering effect in the scatter plot. Actually, when considering each $N-$value separately, the proportional number of fronts slightly decreases. In addition, we can see that the larger the number of POS, the proportionally fewer the number of PLOS. As for the proportional number of non-dominated fronts (7th line, 7th column), it is negatively correlated with the number of PLOS.

This correlation analysis indicates that there is a relatively high interaction between some of the features. We hypothesize that they are actually impactful in terms of instance difficulty. In the following, we consider to effectively study the association between problem features and instance difficulty by experimentally appreciating the ability of a dedicated algorithm class to solve PFSP instances.

4   Problem Features vs. Algorithm Performance

In order to elicit the impact of the previous features on the performance of solving a given PFSP instance, we consider the so-called Pareto Local Search (PLS) algorithm [13]. PLS has proved extremely relevant to solve different combinatorial MOPS, including the PFSP [6,8,12]. We first recall its algorithmic components and the way it is instantiated and experimented for the PFSP.

Algorithm Description and Performance Assessment. PLS maintains an unbounded archive A of mutually non-dominated solutions. This archive is initialized with one (random) initial solution from the solution space. At each iteration, one (unvisited) solution is selected at random from the archive $x \in A$. All neighboring solutions $\mathcal{N}(x)$ from x are then exhaustively evaluated. This of course assumes given a neighborhood relation. The non-dominated solutions from $A \cup \mathcal{N}(x)$ are stored in the archive, and the current solution x is then flagged as visited to avoid a useless revaluation of its neighborhood. The algorithm naturally stops once all solutions from the archive are flagged as visited. PLS always terminates and returns a maximal Pareto local optimum set [13]. In this work, we experiment PLS using the two previously-defined neighborhood operators, namely *exchange* and *insert*, hence ending with two PLS variants. To experiment the so-obtained variants and their relative performance, we consider to run them on the considered instances (30 independent runs for each instance), and to compute the success rate, that is the proportion of runs where PLS is able to identify the Pareto set. The success rate can be interpreted as the empirical probability of solving a given instance to optimality. Notice that, since the considered instances are of small size, this measure is fully accurate for the purpose of our study by avoiding the biases that other performance indicators and their corresponding parameters could introduce. Before studying the impact of problem features, we start by summarizing the main empirical observations with respect to the success rate of PLS on the different instance classes.

Exploratory Analysis. In Fig. 7, we report the success rate of both PLS variants for instances of size $N = 8$ and $M = 8$. The first notable observation is that, independently of the instance type or neighborhood, the success rate is lower for the objective pairs $(C_{max}, \star)$ than for $(C_{sum}, \star)$. This is in accordance with our observations on the number of PLOS, where the former objective pairs were intuited as more difficult to solve. Similarly, we can observe that the considered instances are overall less difficult to solve for PLS when using the insert neighborhood than when using the exchange neighborhood, which we can directly relate to our observation about PLOS. Actually, the objective pairs $(C_{sum}, \star)$ for the instances of type `minella` are relatively easy to solve when using the insert neighborhood, with a median success rate above 50%. On the contrary, the instances of type `basseur` seem to provide the overall lowest success rate, which indicates that they are the most challenging for PLS. Over the objective pairs $(C_{max}, \star)$, the instances of type `liefooghe` are found to be relatively easier to solve than `minella`, whereas this is not the case for $(C_{sum}, \star)$.

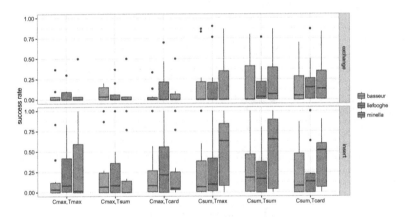

Fig. 7. Success rate of PLS ($N = 8$, $M = 8$).

Correlation Analysis. In the following, we want to understand the relation between the problem features and the success rate of PLS over all the instances. We here focus on this association *across* instance classes. In order to measure this association *within* instance classes, we rather suggest to follow a mixed model analysis such as [5]. In Fig. 8, we can first appreciate the distribution of success rates for both PLS variants over all the instances. Basically, we can observe many failures when using the exchange neighborhood, and relatively more success when using the insert neighborhood. In Figs. 9 and 10, we report the correlation between the success rate and the considered features. Although the two operators have different behavior

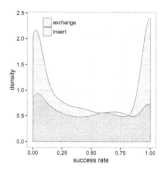

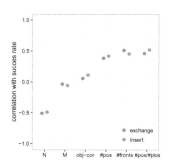

Fig. 8. Density of success rate values over all the instances.

Fig. 9. Non-parametric correlation between problem features and success rate.

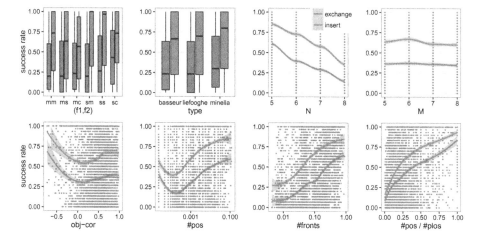

Fig. 10. Scatterplot and locally-fitted polynomial function (LOESS) between problem features and success rate.

in terms of the distribution of their success rate (Fig. 8), we can now observe that their performance is correlated in the same manner with the considered features. In particular, we found a relatively high correlation of success rate with the problem size N, as well as with the number of fronts, of POS, and of PLOS. Interestingly, the correlation with the pair of objectives is rather low, which indicates that the performance of PLS is more impacted by the underlying structure of the PFSP fitness landscape, independently of the nature of the objectives to be optimized. Similarly, the success rate is not impacted by the instance type, which could seem surprising at a first sight. We attribute this to the relatively high variance of success rate across all the instances of same type. Unsurprisingly, the instance difficulty increase with N, since we can see that the success rate decreases with N for both operators. However, the difficulty stays constant whatever the value of M, which is to be related to the very low correlation that was found previously between M and the other considered features. When examining how the performance relates with the objective correlation, we can see that the expected success rate estimation follows a more complicated trend. This is also the reason why the correlation between objective correlation and success rate is low in Fig. 9, i.e. the local regression curve does not follow a monotonic function but rather has a U-shape. Notice also that PFSP instances tend to be easier when the objectives are conflicting (although we only observed few instances in such cases, so that the confidence interval is large), or on the contrary highly correlated. However, when the objective correlation is close to zero, the success rate is smaller for both PLS variants. As for the correlation with the proportion of POS, the general trend of the local regression curve clearly indicates that the success rate increases as the number of POS grows (this also holds when the proportion of POS is very low, that is when the absolute number of POS is 1). This might be explained as follows. When the number of non-dominated solutions is large, the archive maintained by PLS is typically large. Hence, the probability to get trapped in a Pareto local optimum set [13] decreases, so that PLS is able to run longer and to find better approximation sets. This means that it is easier for PLS to find the whole Pareto front when it is larger. This is to be mitigated by the runtime required to effectively find all optimal solutions, which we do not report in this paper due to space restriction. Actually, we observed that the runtime of PLS is much larger when the number of POS is large. At last, as expected, when the proportional number of PLOS increases, PFSP instances are typically more difficult, and then the success rate of PLS decreases, independently of the neighborhood operator.

Relative Importance. The correlation analysis provided in the previous section allowed to us to highlight the individual impact and effect of each feature on the performance of PLS. However, this does not tell us the combined effect of those features. In the following, we investigate this issue by considering a machine learning perspective for classification. More specifically, we consider the two following complementary scenarios: (i) predicting the degree of difficulty of an instance for each algorithm variant, and (ii) predicting which algorithm variant (PLS with insert or PLS with exchange) performs better for a given instance. For each scenario, a classification model based on random forests (RF) is first constructed using the different considered features as input variables. RF is a state-of-the-art ensemble learning method based on the construction of multiple decision trees that outputs the mode of the classes of the individual trees [4]. In the first scenario, the output of the RF model is the class of difficulty of a given instance. Given the distribution of success rates depicted in Fig. 8, we divided PFSP instances into three classes: easy instances for which the success rate is 1 (265 for exchange, 1134 for insert), hard instances for which the success rate is 0 (657 for exchange, 243 for insert), moderate instances for which the success rate is in-between 0 and 1 (1958 for exchange, 1503 for insert). The estimate of error rate of the corresponding RF model is 22.33% for exchange, and 26.56% for insert. In the second scenario, the output class of the RF model is simply whether PLS with insert has a better success rate than PLS with exchange for a

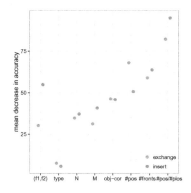

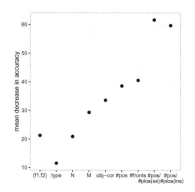

Fig. 11. Variable importance from RF classification model for scenario #1.

Fig. 12. Variable importance from RF classification model for scenario #2.

given instance. In case both variants have the same success rate, the one which evaluated the smallest number of solutions in average is said to be better. PFSP instances are then divided into two classes: 808 instances for which PLS-exchange is better, and 2072 instances for which PLS-insert is better. The estimate of error rate of the corresponding RF model is 23.40%.

Interestingly, RF has the ability to render the relative importance of each input variable (here, features), given in terms of the mean decrease in accuracy [4]. This measure is depicted in Fig. 11 for the first scenario and in Fig. 12 for the second one. We clearly see that the most important feature relates to the proportional number of PLOS. This emphasizes the high impact of multimodality on the difficulty of PFSP instances and on the performance of multi-objective local search. It is also interesting to remark that the next important features are those of general-purpose nature which were specifically investigated in the paper. In fact, the number of POS, the number of fronts and the objective correlation are all found to be more important than the other features that relates directly to the benchmark parameters, namely N, M, the pair of objectives (with the exception of scenario 1 for the insert neighborhood), and the instance type.

5   Conclusions

The multi-objective fitness landscape analysis conducted in this paper allowed us to comprehensively relate the characteristics of bi-objective PFSP instances with the sources of difficulty faced by Pareto local search. From the methodological point of view, we argue that the investigated features and the statistical tools involved all along the paper revealed to be highly valuable in order to accurately harness the complexity of a multi-objective search process, in particular by emphasizing the crucial importance of multimodality. It is our hope that the approach adopted in this paper, as well as our findings regarding the PFSP, will enable to tackle other challenging issues, both from the algorithm design perspective and the more practical problem solving perspective. In fact, leveraging our analysis to other multi-objective scheduling problems and algorithms is still an open challenge in terms of scalability (e.g., large-size instances with more than two objectives, low-cost feature computation or estimation). Tightly related to this issue, a particularly challenging research path would be to set up a sound and complete methodology for algorithm performance prediction, with the ultimate goal of selecting the most appropriate ones for a given instance.

References

1. Aguirre, H., Tanaka, K.: Working principles, behavior, and performance of MOEAs on MNK-landscapes. Eur. J. Oper. Res. **181**(3), 1670–1690 (2007)
2. Basseur, M., Liefooghe, A.: Metaheuristics for biobjective flow shop scheduling. In: Metaheuristics for Production Scheduling, Chap. 9, pp. 225–252. Wiley (2013)
3. Basseur, M., Seynhaeve, F., Talbi, E.G.: Design of multi-objective evolutionary algorithms: application to the flow shop scheduling problem. In: CEC 2002, NJ, USA, pp. 1151–1156 (2002)
4. Breiman, L.: Random forests. Mach. Learn. **45**(1), 5–32 (2001)
5. Daolio, F., Liefooghe, A., Verel, S., Aguirre, H., Tanaka, K.: Problem features vs. algorithm performance on rugged multi-objective combinatorial fitness landscapes. Evol. Comput. (to appear)
6. Dubois-Lacoste, J., López-Ibáñez, M., Stützle, T.: A hybrid TP+PLS algorithm for bi-objective flow-shop scheduling problems. Comput. Oper. Res. **38**(8), 1219–1236 (2011)
7. Ehrgott, M.: Multicriteria Optimization. Springer, Heidelberg (2005)
8. Geiger, M.: On operators and search space topology in multi-objective flow shop scheduling. Eur. J. Oper. Res. **181**(1), 195–206 (2007)
9. Goldberg, D.E.: Genetic Algorithms in Search, Optimization and Machine Learning. Addison-Wesley, Boston (1989)
10. Hoos, H., Stützle, T.: Stochastic Local Search: Foundations and Applications. Morgan Kaufmann, San Francisco (2004)
11. Liefooghe, A., Basseur, M., Jourdan, L., Talbi, E.-G.: Combinatorial optimization of stochastic multi-objective problems: an application to the flow-shop scheduling problem. In: Obayashi, S., Deb, K., Poloni, C., Hiroyasu, T., Murata, T. (eds.) EMO 2007. LNCS, vol. 4403, pp. 457–471. Springer, Heidelberg (2007)
12. Minella, G., Ruiz, R., Ciavotta, M.: A review and evaluation of multiobjective algorithms for the flowshop scheduling problem. INFORMS J. Comput. **20**(3), 451–471 (2008)
13. Paquete, L., Schiavinotto, T., Stützle, T.: On local optima in multiobjective combinatorial optimization problems. Ann. Oper. Res. **156**(1), 83–97 (2007)
14. Richter, H., Engelbrecht, A. (eds.): Recent Advances in the Theory and Application of Fitness Landscapes. Springer, Heidelberg (2014)
15. Taillard, E.D.: Benchmarks for basic scheduling problems. Eur. J. Oper. Res. **64**, 278–285 (1993)
16. T'Kindt, V., Billaut, J.C.: Multicriteria Scheduling: Theory, Models and Algorithms. Springer, Heidelberg (2005)
17. Verel, S., Liefooghe, A., Jourdan, L., Dhaenens, C.: On the structure of multiobjective combinatorial search space: MNK-landscapes with correlated objectives. Eur. J. Oper. Res. **227**(2), 331–342 (2013)

Dimensionality Reduction Approach for Many-Objective Vehicle Routing Problem with Demand Responsive Transport

Renan Mendes[1]([✉]), Elizabeth Wanner[2], Flávio Martins[2], and João Sarubbi[2]

[1] Programa de Pós-Graduação em Modelagem Matemática
e Computacional CEFET, Belo Horizonte, MG, Brazil
renansantosmendes@gmail.com
[2] Departamento de Computação, CEFET, Belo Horizonte, MG, Brazil
{efwanner,flaviocruzeiro,joao}@decom.cefetmg.br

Abstract. Demand Responsive Transport (DRT) systems emanate as a substitute to face the problem of volatile, or even inconstant, demand, occurring in popular urban transport systems. This paper is focused in the Vehicle Routing Problem with Demand Responsive Transport (VRP-DRT), a type of transport which enables passengers to be taken to their destination, as a shared service, trying to minimize the company costs and offer a quality service taking passengers on their needs. A many-objective approach is applied in VRPDRT in which seven different objective functions are used. To solve the problem through traditional multi-objective algorithms, the work proposes the usage of cluster analysis to perform the dimensionaly reduction task. The seven functions are then aggregated resulting in a bi-objective formulation and the algorithms NSGA-II and SPEA 2 are used to solve the problem. The results show that the algorithms achieve statistically different results and NSGA-II reaches a greater number of non-dominated solutions when compared to SPEA 2. Furthermore, the results are compared to an approach proposed in literature that uses another way to reduce the dimensionality of the problem in a two-objective formulation and the cluster analysis procedure is proven to be a competitive methodology in that problem. It is possbile to say that the behavior of the algorithm is modified by the way the dimensionality reduction of the problem is made.

Keywords: Vehicle routing problem · Demand responsive transport · Cluster analysis · Dimensinality reduction · Many-objective optimization

1 Introduction

Flexible services of transport, also named in literature as Demand Responsive Transport Systems (DRT), deal with an extensive range of transport systems that provide services for passengers or non-conventional load transport [28]. The DRT tries to address the problem of volatile and inconstant demand using flexible routes and schedules that may vary according to observed demand. Based on

© Springer International Publishing AG 2017
H. Trautmann et al. (Eds.): EMO 2017, LNCS 10173, pp. 438–452, 2017.
DOI: 10.1007/978-3-319-54157-0_30

this flexibility, transport companies can offer a more efficient service with the inclusion of routes planned to be executed and thus increasing vehicle occupancy rate and providing a better service to the user [12].

Questions involving practical applications and theoretical studies of the DRT systems could be seen in recent papers, such as the assessment of the operation and performance of DRT systems in England and Wales [13]; the study of the main barriers to implementing a flexible transport [21] and the project evaluation and performance of DRT program in Great Britain [7]. According to Gomes [10], design and operation of transport systems based on DRT are similar to vehicle routing problem (VRP) found in Dantzig and Ramser [6].

This work focuses in the Vehicle Routing Problem with Demand Responsive Transport (VRPDRT). That problem appeared for the first time in [11] and deals with the features found in the classical routing problems and management of demand responsive transport systems. DRT systems provide a hybrid approach between a taxi and a bus using shared service. In this kind of system, the vehicles must follow routes and timetables aiming to match the requests and to take as many users as possible in the same vehicle while keeping the quality of the service in terms of pickup and delivery times and tour duration as higher as possible. In that case, the passenger transport is made on demand, trying to meet the passengers needs through flexible routes and schedules.

A many-objective approach to VRPDRT is used in this work, in which seven different objective functions are used. The objective functions represent the company and the human perspectives in the transport system. Since dealing with a many-objective problem is not an easy task, a Cluster analysis-based objective reduction technique, also called dimensionality reduction, is applied in order to transform the many-objective problem into a bi-objective formulation. The resulting bi-objective problem can be solved via two state-of-art multiobjective algorithms, the Non-dominated Sorting Genetic Algorithm II (NSGA-II) and the Strength Pareto Evolutionary Algorithm 2 (SPEA 2). Comparing the results in the situations tested, the NSGA-II algorithm achieves best sets of non-dominated solutions when compared to the SPEA 2 algorithm. The proposed approach is also compared to another objective reduction technique based on Aggregation Tree found in literature and the proposed approach has proven to be very competitive.

The remainder of this work is organized as follows: Sect. 2 presents the related work. Section 3 presents the problem description and solution representation. Section 4 describes the main concepts of multiobjective optimization and presents the NSGA-II and SPEA 2 algorithms. Section 5 defines the cluster analysis technique. Section 6 presents a many-objective approach for the problem. Results are provided in Sect. 7 and then discussed. Finally, Sect. 8 concludes the work.

2   Literature Review

According to [6], DRT services corresponds to the vehicle assignment to transportation requests and the corresponding vehicle routing and is closely related to the Vehicle Routing Problem (VRP), and, in particular, the Dial-A-Ride (DARP) [3]. The VRPDRT extends the classical VRP in many ways and is, at least, as complex than the latter [4]. The VRPDRT can be seen as a VRP with time-windows, a limited number of vehicles and variable travel times and can be said to be NP-complete [11]. The VRPDRT is multi-objective, multi-criteria problem that requires good solutions to be obtained in useful time. For that purpose, heuristic and metaheuristic approaches are well suited to solve the problem at hand and, in the last years, have been proposed to deal with the VRPDRT.

Two variations of GRASP algorithm: GRASP-like and GRASP-reactive, are used to solve the VRPDRT in [11]. In GRASP-like algorithm, only the construction phase is performed while the GRASP-reactive uses both phases. Besides, in the GRASP-reactive, when a better result is found, the algorithm recalibrates the list of parameters. Although the work addresses the problem using a multiobjective formulation, the objective functions are aggregated turning the problem into a mono-objective approach.

A multiobjective evolutionary approach to VRPDRT is proposed in [2]. The goals are three-fold: to minimize the number of vehicles, the journey duration and the total delay. The work uses three evolutionary algorithms: NSGA-II, SPEA 2 and IBEA to solve the problem. A hybrid approach of these algorithms, using the Iterated Local Search (ILS) as a mutation operator, is also used. The insertion of the ILS algorithm improves the average quality of solutions in all algorithms and IBEA achieves the best performance.

A multiobjective evolutionary approach to DRT could be seen in [25]. The mathematical model presents the minimization of three objectives: (i) the number of vehicles; (ii) the total travel time, and (iii) passenger pickup delay. The work uses the NSGA-II algorithm with specific genetic operators to guarantee the feasibility of the solutions. Computational experiments are performed on benchmark instances and results show that the insertion of these operators in NSGA-II improves the algorithm performance.

A comparison between the NSGA-II and SPEA 2 algorithms for VRPDRT with a reduction of five objectives is made in [16]. The dimensionality reduction uses an aggregation tree-based procedure resulting in a bi-objective formulation. The first objective function represents the human perspective and the second represents the company perspective. S-Metric is used to assess the quality of the outcomes of both algorithms and results show the existence of a statistical difference between them favoring NSGA-II over SPEA 2.

Following the aggregation tree-based approach, in [17] two new objective functions are added to the formulation in [16]. The dimensionality reduction technique transforms the VRPDRT containing seven objective functions into a bi-objective formulation. However, results show there is no clear division between the human and the company perspectives since the technique aggregates one of the objectives in a different perspective. Trying to overcome this difficulty, that objective function with different behavior becomes an additional objective function in [18]. The NSGA-II is applied to solve the three objective problem and the results show the algorithm can improve the non-dominated solutions in both reduced and original dimensions.

3   Problem Description

In the VRPDRT, origins and destinations are specified by passengers within a set of predefined stop points which can be added to the route. The pickup time window and a possible delivery time window is also specified by the passenger. A fleet of capacitated vehicles serves the demand for transport. All stopping points, within the route, can be a pickup, delivery or pickup and delivery simultaneously, excluding the depot. Different destinations and delivery time windows can be defined for passengers who board in the same stop point.

The vehicles transport many users at the same time according to the capacity. Every route begins and ends in the depot. Considering that each stopping point contains different time windows and users have different demands of transport, it is possible the overlapping occurrence of these windows. The vehicle can return to a point, already visited, in different periods of time, since the visit is not consecutive.

This work deals with the VRPDRT based in [11]. In short, the main features of the problem is the following: vehicles begin and end the route in a single depot; pickup and delivery time windows are specified by users; at each stopping point, the vehicle may be loading and unloading; a fleet of vehicles with the same capacity (homogeneous fleet); multiple time windows at each stopping point; transport requests can be from anywhere to anywhere (given the stop points set); pickup time windows must be respected and delivery time windows can be violated, however, being penalized.

Using the notation used in [11] for the VRPDRT, the passenger transport requests is represented by the set $P = \{1, 2, \cdots, p\}$ and p is the number of requests. Each passenger corresponds to only one request. A graph with a set of vertices and arcs can model the problem. The set V is called stop vertex set and the pickup and delivery vertices belong to this set. The number 0 represents the depot. A cost c_{ij} and travel time d_{ij} are associated with each arch a_{ij} connecting the vertex i to j in which $i \neq j$ and $i, j \in V$.

The sets P_{in_v} and P_{out_v} store, respectively, the pickup and delivery requests at v for every $v \in V - \{0\}$; and m_V is the capacity index, which is calculated as the number of pickup requests minus the number of delivery requests in a vertex v, $m_i = |P_{in_i}| - |P_{out_i}|$ with $i = 1, 2, \cdots, p$. In the depot, passengers can not board in the vehicle, so $P_{in_0} = P_{out_0} = \varnothing$.

The following information is found in every request p_i with $i = 1, 2, \cdots, p$: identification (id); pickup vertex (p^+); delivery vertex (p^-); pickup time window $([Pe_i, Pl_i])$; delivery time window $([De_j, Dl_j])$; the time in which a passenger (request) i boards in the vehicle k (Pt_i^k) and the time in which a passenger (request) i is delivered by the vehicle k (Dt_i^k).

A homogeneous fleet is represented by a set of vehicles $K = \{1, 2, \cdots, k\}$ corresponding to k vehicles with a capacity of Q seats. One vertex can be visited by the vehicle more than once if the visit is not consecutive. The problem is called static since the demand of the system is known before the routing process; otherwise, it is named as dynamic. In this case, requests come in real time and routes are not built beforehand.

3.1   Solution Representation

The representation of the solution used in this work is the same found in [19], in which a subset of n routes R_i composes a solution to the VRPDRT, for example, $S = \{r_1, \cdots, R_n\}, n \leq k$ in which k is the number of vehicles available in the fleet K. Each route is a list of vertices that the vehicle i must cover, $R_i = < v_0, v_1, \ldots, v_m, v_{m+1} >$.

A route may contain vertices that have already been visited so that further requests can be serviced by a single vehicle, provided that the visit is not consecutive. The representation of the solution consists of a list of routes in which each vehicle in the fleet is allocated to a route. To represent non-attended requests, an auxiliary data structure is used to store such information, in this case a set is used.

A concatenated array of routes can be created for this solution and a movement can be performed on such array (Fig. 1). The movement can perform exchanges inter or intra-routes.

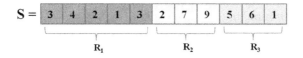

Fig. 1. Representation of concatenated routes vector.

In Fig. 2 is shown a example of inter-route exchange in which vertices 1 and 5 are swapped.

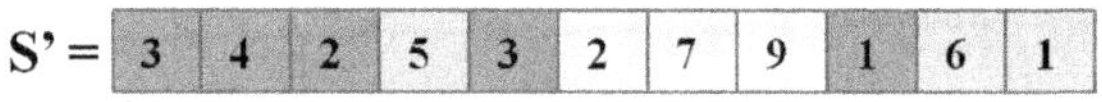

$$S' = \boxed{3 \mid 4 \mid 2 \mid 5 \mid 3 \mid 2 \mid 7 \mid 9 \mid 1 \mid 6 \mid 1}$$

Fig. 2. Concatenated vector after some movement.

After some movement on the vector of concatenated routes, it is necessary to rebuild the solution. The reconstruction is done looking for requests that have become feasible and unfeasible after the movement, and then inserting each feasible vertex from vector of concatenated routes in the solution.

4   Multiobjective Optimization

Real world optimization problems represent, quite often, situations in which conflicting multiple objectives must be optimized simultaneously. In this case, the problem cannot be solved in order to obtain the best solution with regard to the objectives. One problem with multiple objectives originates a set of non-dominated solutions called Pareto-optimal set. In this set, each objective component can be refined by deterioration of other objective components. Any one of the solutions in the Pareto-optimal set can be used since there is no solution that is totally better than the others in the set [5].

A multiobjective optimization problem can be formulated as follows:

$$\mathbf{x}^* = \min_x \mathbf{f}(x)$$

$$\text{subjected to: } \begin{cases} g_i(x) \leq 0; \ i = 1, 2, \cdots, r \\ h_j(x) = 0; \ j = 1, 2, \cdots, p \end{cases} \quad (1)$$

in which $x \in \mathbb{R}^n$, $\mathbf{f}(.) : \mathbb{R}^n \to \mathbb{R}^m$, $\mathbf{g}(.) : \mathbb{R}^n \to \mathbb{R}^r$ and $\mathbf{h}(.) : \mathbb{R}^n \to \mathbb{R}^p$. The functions g_i and h_j are, respectively, inequality and equality constraint functions. The vectors $x \in \mathbb{R}^n$ are called *decision variable vectors* of multiobjective problem and form the *parameters space*. The vectors $\mathbf{f}(x) \in \mathbb{R}^m$ are in a vector space called *objective space*.

Mathematically, in a minimization problem, if $x \in \mathbb{R}^n$ denotes the decision variable vector of the problem and $D \subset \mathbb{R}^n$ is the set of all feasible x, the set of nondominated solutions $X \subset D$ is characterized by:

$$X = \{\bar{x} \in D; \neg \exists \ x \in D : f_i(x) \leq f_i(\bar{x}),$$
$$\forall i = 1, \ldots, m; \mathbf{f}(x) \neq \mathbf{f}(\bar{x})\} \quad (2)$$

4.1 Multiobjective Evolutionary Algorithms

By evolving a population of solutions, multiobjective evolutionary algorithms are able to approximate the Pareto optimal set in a single run. In [22], the first implementation of one multiobjective evolutionary algorithm was done. This implementation was called Vector Evaluated Genetic Algorithm (VEGA) and the fundamental difference in relation to the genetic algorithm is the selection operator. Subpopulations from each objective function, are combined to form the new population.

The Strength Pareto Evolutionary Algorithm 2 (SPEA 2), found in [27], it is an updated version of the SPEA, in which it introduces a revised strategy of fitness assignment. The algorithm uses an internal file with fixed-size that is used during the selection stage. The fitness assignment uses the strength concept that is the number of solutions that one individual dominates.

A Non-dominated Sorting Genetic Algorithm II (NSGA-II) was proposed in [8]. The main idea of this algorithm is to sort the population by non-dominated fronts consisting in ordering the individuals in terms of non-dominated fronts in the objective space and assigning a fitness value for each front.

A variety of multiobjective evolutionary algorithms have been proposed and applied to many scientific and engineering applications. Several other algorithms can be found in literature, for example, Set Preference Algorithm for Multiobjective Optimization (SPAM), Epsilon-Constraint Evolutionary Algorithm

(ECEA), Sampling-based Hyper Volume-oriented algorithm (SHV) and Multiple Single Objective Pareto Sampling (MSOPS). However, in this work, only NSGA-II and SPEA 2 algorithms have been used.

5   Cluster Analysis

Motivated by the fact that the multiobjective approach deals with many conflicting objectives simultaneously, it is necessary to understand the relationship between them, so that decisions are taken with a reasonable understanding of the relations between the possible choices [9]. Multiobjective evolutionary algorithms are able to solve generic problems with two or three objectives functions. However, when the number of functions increases, the performance of these algorithms deteriorates [14].

The technique known as cluster analysis allows decomposing a dataset into significant groups (called clusters). If the goal is to obtain significant groups, the cluster must be able to retain the natural structure of the data [24]. There are several applications of the cluster analysis in different areas, such as market research, education, data mining, life insurance and machine learning.

Cluster analysis uses the principle that objects belonging to the same group have greater similarity and lower values in relation to other groups. For members of a group, the greater its similarity, the greater the difference between the groups [24]. The similarity is a measure of correspondence or affinity among the objects to be grouped. One of possible measures used to calculate the similarity is the distance, for example, the Manhattan, Euclidian, Mahalanobis, Chebyshev or Minkowski distances [23]. Measures of correlation and association are also other ways to calculate the similarity between objects [1].

Many types of clustering, such as partial, full, hierarchical, overlapping and fuzzy, can be found in literature. This work uses the agglomerative hierarchical clustering methods, which consist of creating a hierarchy of relationships between objects and working by creating sets from isolated elements. Besides, they do not have any theoretical justification based on statistics or information theory [1].

There are different ways to measure the distance between two clusters such as: (i) single linkage, (ii) complete linkage and (iii) average linkage [24]. This work adopted the complete linkage method, in which it groups the elements with greater similarity.

5.1   Sample Correlation Matrix

This section presents some necessary definitions to understand cluster analysis technique. Additional details can be found in [20]. Let X be a vector containing p components, where each component is a random variable, in other words, X_i is a random variable, $\forall i = 1, 2, ..., p$. Then X is called a random vector.

Definition 1 (Correlation). *The correlation coefficient between the i-th and j-th variables, ρ_{ij}, of the vector X is defined as:*

$$\rho_{ij} = \frac{\sigma_{ij}}{\sqrt{\sigma_{ii}\sigma_{jj}}} = \frac{\sigma_{ij}}{\sigma_i\sigma_j} \tag{3}$$

in which σ_{ij} is the covariance between the values of the i-th and j-th vector variables X, σ_i and σ_{ii} represent, respectively, the standard deviation and variance of the i-th vector component X.

Definition 2 (Correlation Matrix). *The correlation matrix, $P_{p \times p}$, of random vector X is denoted by:*

$$P_{p \times p} = \begin{bmatrix} 1 & \rho_{12} & \rho_{13} & \cdots & \rho_{1p} \\ \rho_{21} & 1 & \rho_{13} & \cdots & \rho_{2p} \\ \rho_{31} & \rho_{32} & 1 & \cdots & \rho_{3p} \\ \vdots & \vdots & \vdots & \ddots & \vdots \\ \rho_{p1} & \rho_{p2} & \rho_{p3} & \cdots & 1 \end{bmatrix} \tag{4}$$

in which ρ_{ij} denotes the correlation between the i-th and j-th component of vector X. The correlation matrix will be used in cluster analysis as a similarity measure.

6   Many-Objective Approach for VRPDRT

This section presents the objectives used in VRPDRT approach. The functions can be found at [17] and are defined as:

$$Minimize \quad f_1 = \sum_{i=1}^{n} C(R_i) \tag{5}$$

$$Minimize \quad f_2 = \sum_{k=1}^{|K|}\sum_{j=1}^{p} Max\left(Dt_j^k - Dl_j, 0\right) \tag{6}$$

$$Minimize \quad f_3 = |R_{MAX}| - |R_{MIN}| \tag{7}$$

$$Minimize \quad f_4 = |U| \tag{8}$$

$$Minimize \quad f_5 = |K| \tag{9}$$

$$Minimize \quad f_6 = \sum_{k=1}^{|K|}\sum_{j=1}^{p} Max\left(Dt_j^k - Pt_j^k, 0\right) \tag{10}$$

$$Minimize \quad f_7 = \sum_{k=1}^{|K|}\sum_{j=1}^{p} Max\left(Pt_j^k - Pe_j, 0\right) \tag{11}$$

where f_1 minimizes the total route cost, f_2 minimizes the delay in the passengers delivery, f_3 minimizes the difference between the highest route and the lower route, f_4 minimizes the number of non-attended requests, f_5 minimizes the number of vehicles used, f_6 minimizes the total travel time and f_7 minimizes the waiting time.

7   Experimental Results

The Cluster analysis procedure will be used to reduce the size of VRPTRD in the same way the Aggregation Tree (AT) methodology was used in [16]. However, instead of using distance (or rank position in the case of AT), we used the correlation between objective functions. For calculating the sample correlation, 2000 random solutions are used and the following matrix R is obtained:

$$R_{ij} = \begin{bmatrix} 1.00 & 0.05 & -0.22 & -0.04 & 0.63 & 0.04 & -0.16 \\ 0.05 & 1.00 & 0.12 & 0.14 & -0.20 & 0.77 & 0.15 \\ -0.22 & 0.12 & 1.00 & -0.11 & -0.18 & 0.15 & 0.40 \\ -0.04 & 0.14 & -0.11 & 1.00 & -0.13 & -0.04 & -0.02 \\ 0.63 & -0.20 & -0.18 & -0.13 & 1.00 & -0.15 & -0.27 \\ 0.04 & 0.77 & 0.15 & -0.04 & -0.15 & 1.00 & 0.08 \\ -0.16 & 0.15 & 0.40 & -0.02 & -0.27 & 0.08 & 1.00 \end{bmatrix}$$

The matrix R shows the sample correlation between the functions. However, to turn R in a distance matrix is necessary to override the values found on the main diagonal. The following transformation is made:

$$D = R - I_n \tag{12}$$

in which I_n is the identity matrix of order n.

Applying the cluster analysis based on the complete linkage method, the dendrogram, shown in Fig. 3, is obtained. The R language was used to apply cluster analysis and construct the dendrogram. Firstly, functions f_5 and f_7 are grouped on the left side of the dendrogram. After that, functions f_1 and f_3 are grouped. Latter, f_4 and f_6 are grouped to functions f_5 and f_7. Functions f_1 and f_3 are grouped to function f_2. At the end, all functions are grouped in a cluster. Based on the aggregation of objectives, the following bi-objective optimization problem can be formulated:

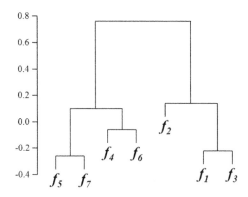

Fig. 3. Dendrogram based on sample correlation for VRPDRT.

$$\begin{aligned} \min \quad & F_1 = \lambda_4 f_4 + \lambda_5 f_5 + \lambda_6 f_6 + \lambda_7 f_7 \\ \min \quad & F_2 = \lambda_1 f_1 + \lambda_2 f_2 + \lambda_3 f_3 \end{aligned} \tag{13}$$

Analyzing the above formulation, it is clear that the aggregated objective functions (F_1 and F_2) represent two global visions of VRPDRT. The objective functions f_1 and f_5 represents the perspective of the company, f_3 represent the driver perspective, and f_2 and f_7 represent the passenger perspective. f_4 can represent the perspective of the company and the passenger, as the non-served passenger means a certain inability of service in relation to quality and may also consider the company view since a non-served passenger may have to appeal to other ways of transport resulting in a reduction of customers using the service. The objective function f_6 it is another example that can be considered from two perspectives. In this case, both the passenger and driver prefer a route having the smallest travel time.

The parameters used in both algorithms were chosen empirically and are given by: Crossing Probability (P_C) = 0.8; Probability of Mutation (P_M) = 0.2; Number of generations = 1000; Population size = 100; Vehicle capability (Q) = 10; $\lambda_1 = 1$, $\lambda_2 = 1$, $\lambda_3 = 20$, $\lambda_4 = 500$, $\lambda_5 = 800$, $\lambda_6 = 1$ and $\lambda_7 = 20$.

The operators used here in both algorithms are the same used in [16] and [17] in which statistical tests were made using the values obtained for S-Metric after thirty executions. The selection method used for both algorithms is roulette. For the NSGA-II algorithm, we used the two-point crossover and the mutation operator, called 2-Shuffle, in which two non-consecutive vertices are selected randomly, and then the vertices between then are shuffled. SPEA 2 uses the single point crossover and the 2-Opt mutation operator that two non-consecutive vertices are selected randomly, and the sequence of vertices between the two vertices selected before are reversed. Since the main goal of this work is to propose the application of cluster analysis as a dimensionality reduction technique, the study of choice of the best operators for each algorithm in the bi-objective formulation used in this paper is left for future work.

The two performance measures to calculate the algorithm performance in this work are Set Coverage Metric and S-Metric. The Set Coverage Metric ($\mathcal{C}(\mathbf{A}, \mathbf{B})$), also called set covering metric, calculates the ratio of solutions of a set $\mathbf{B}$ that are dominated by the solutions of a set $\mathbf{A}$ [26]. The metric S-Metric, found in [26], calculates the hypervolume delimited by a set and a reference point in the space of standard objectives. The point used as reference to calculate the hypervolume for each set is:

$$x_N = [5000 \quad 5000 \quad 100 \quad 110 \quad 50 \quad 10000 \quad 20]$$

The point x_N is evaluated according to the Eq. (13) and then, the point is reduced to:

$$x_N^R = [127000 \quad 12000]$$

Figure 4 presents the mean convergence curve for the NSGA-II and SPEA 2 algorithms using the mean value of S-Metric throughout generations. Observe

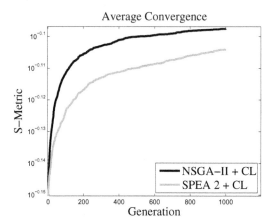

Fig. 4. Average convergence curve of S-Metric of the NSGA-II and SPEA 2.

that there is a stabilization of S-Metric values over generations for both algorithms. Apparently, the NSGA-II algorithm can achieve a better value of S-Metric when compared to SPEA 2.

The combined Pareto set obtained after all runs are shown in Fig. 5. It can be seen that few solutions from combined Pareto set obtained from SPEA 2 dominates some solutions resulting from NSGA-II (lowest values F_2 and highest F_1).

A random test (found in [15]) is used to compare the obtained results statistically. The statistical analysis is carried out using: Mean(NSGA-II) − Mean(SPEA 2) and, in this way, positive differences (which means that the observed difference is located to the right of the histogram) favors NSGA-II compared to SPEA 2. It is important to highlight that the comparison is made

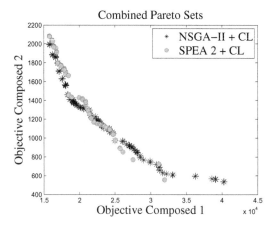

Fig. 5. Pareto combined set of the NSGA-II and SPEA 2.

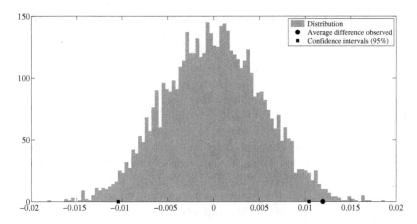

Fig. 6. Random test for NSGA-II and SPEA 2. The result is statistically significant since the black filled circle is outside the confidence interval.

in $\mathbb{R}^2$. Figure 6 shows a histogram of the values of randomized objects with the observed value highlighted.

To assess whether the algorithms show the same behavior within the original objectives (in $\mathbb{R}^7$), set coverage metric is used. Let the set **A** be the combined Pareto set from the NSGA-II algorithm and **B** the combined Pareto set from SPEA 2. Comparing the solutions in $\mathbb{R}^7$, $\mathcal{C}(\mathbf{A}, \mathbf{B}) = 0.0536$ and $\mathcal{C}(\mathbf{B}, \mathbf{A}) = 0.0$. It means that, in $\mathbb{R}^7$, approximately 5.36% solutions in the Pareto combined set obtained from SPEA 2 are dominated by solutions in the Pareto combined set obtained from NSGA-II and no solution from NSGA-II are dominated by any solution of SPEA 2. Since $\mathcal{C}(\mathbf{A}, \mathbf{B}) > \mathcal{C}(\mathbf{B}, \mathbf{A})$, it is possible to say that, in $\mathbb{R}^7$, **A** is better than **B**. In other words, the solutions obtained after applying the NSGA-II algorithm are better than those obtained by the SPEA 2 algorithm [26].

The proposed cluster analysis-based dimensionality reduction is compared to the Aggregation Tree-based dimensionality reduction proposed in [17]. In both cases, the original problem has seven objective functions. However, the resulting bi-objective formulations are different. Having this in mind, the comparison is made using the seven-objective original problem and the set coverage metric is also used. Table 1 summarizes the comparison between the two reduction techniques. **AT** represents the dimensionality reduction using Aggregation Tree and **CL** represents the dimensionality reduction using Cluster Analysis.

Table 1. Set coverage metric values to algorithms using different reduction techniques.

	NSGA-II	SPEA 2
$\mathcal{C}(\mathbf{AT}, \mathbf{CL})$	0.0000	0.0000
$\mathcal{C}(\mathbf{CL}, \mathbf{AT})$	0.0465	0.0217

Table 1 shows that, comparing the solutions in $\mathbb{R}^7$ found by NSGA-II, $\mathcal{C}(\mathbf{AT}, \mathbf{CL}) = 0.0$ and $\mathcal{C}(\mathbf{CL}, \mathbf{AT}) = 0.0465$. It means that, in $\mathbb{R}^7$, approximately 4.65% solutions in the Pareto combined set obtained using aggregation tree are dominated by solutions in the Pareto combined set obtained using cluster analysis and no solution obtained using cluster analysis are dominated by any solution obtained using aggregation tree. Similarly, comparing the solutions in $\mathbb{R}^7$ found by SPEA 2, $\mathcal{C}(\mathbf{AT}, \mathbf{CL}) = 0.0$ and $\mathcal{C}(\mathbf{CL}, \mathbf{AT}) = 0.0217$. It means that, in $\mathbb{R}^7$, approximately 2.17% solutions in the Pareto combined set obtained using aggregation tree are dominated by solutions in the Pareto combined set obtained using cluster analysis and no solution obtained using cluster analysis are dominated by any solution obtained using aggregation tree. According to the results, it can be seen that the usage of cluster analysis in order to reduce the size of the problem is able to obtain approximations for the Pareto front that dominate the solutions found via aggregation tree.

8   Conclusion and Outlook

This work addressed the Vehicle Routing Problem with Demand Responsive Transport (VRPDRT). In this problem, the transport is shared and is done taking passengers from their origin to their destination in order to minimize the company costs and offer a quality service taking passengers on their needs.

A many-objective approach with seven different objective functions was used to solve the VRPDRT and they were added to new functions, generating a bi-objective formulation of the problem. Cluster Analysis technique was used to reduce the number of objectives taking into consideration the relationship among the functions. The bi-objective problem was solved using two state-of-art multiobjective algorithms, NSGA-II and SPEA 2.

Two performance measures, Set Coverage Metric and S-Metric, were used to assess the algorithm performances. When comparing the results, it could be seen that the NSGA-II algorithm was able to find the best sets of non-dominated solutions for both spaces, the reduced space, $\mathbb{R}^2$, and the original objective space, $\mathbb{R}^7$, when compared with SPEA 2.

The results also show that the new way to aggregate the objective functions achieved more non-dominated solutions in relation to the aggregation tree approach found in literature. It could be seen that the behavior of the algorithm is modified by the way the dimensionality reduction of the problem is made.

Acknowledgments. The authors would like to thank the Brazilian funding agencies, CNPq, CAPES and Fapemig for financial support.

References

1. Balcan, M.F., Blum, A., Vempala, S.: A discriminative framework for clustering via similarity functions. In: Proceedings of the 40th ACM Symposium on Theory of Computing (2008)

2. Chevrier, R., Liefooghe, A., Jourdan, L., Dhaenens, C.: Solving a dial-a-ride problem with a hybrid evolutionary multi-objective approach: application to demand responsive transport. Appl. Soft Comput. **12**, 1247–1258 (2012)
3. Cordeau, J.F.: A branch-and-cut algorithm for the dial-a-ride problem. Oper. Res. **54**, 573–586 (2003)
4. Cordeau, J.F., Laport, G.: The dial-a-ride problem: models and algorithms. Ann. Oper. Res. **153**(1), 29–46 (2007)
5. Cruz, A.R., Cardoso, R.T.N., Wanner, E.F., Takahashi, R.H.C.: A multiobjective non-linear dynamic programming approach for optimal biological control in soy farming via NSGA-II. In: IEEE Congress on Evolutionary Computation (2007)
6. Dantzig, G.B., Ramser, J.H.: The truck dispatching problem. Manage. Sci. **6**(1), 80–91 (1959)
7. Davison, L., Enoch, M., Ryley, T., Quddus, M., Wang, C.: A survey of demand responsive transport in great britain. Transp. Policy **31**, 47–54 (2014)
8. Deb, K., Agrawal, S., Pratap, A., Meyarian, T.: A fast and elitist multiobjective genetic algorithm: NSGA-II. IEEE Trans. Evolut. Comput. **6**(2), 182–197 (2002)
9. de Freitas, A.R.R., Fleming, P.J., Guimarães, F.G.: Aggregation trees for visualization and dimension reduction in many-objective optimization. Inf. Sci. **298**(298), 288–314 (2015)
10. Gomes, J.R.: Dynamic vehicle routing for demand responsive transportation systems. Doctoral thesis, Universidade do Porto, Porto, Portugal (2012)
11. Gomes, R.J., Souza, J.P., Dias, T.G.: A new heuristic approach for demand responsive transportation systems. In: XLII Simpósio Brasileiro de Pesquisa Operacional (XLII SBPO), pp. 1839–1850 (2010)
12. Josselin, D., Lang, C., Murilleau, N.: Modelling dynamic demand responsive transport using an agent based spatial representation. Technical report, Universit d' Avignon, Avignon (Março 2009). http://lifc.univ-fcomte.fr/~publis/hal/jlm09:ip. pdf
13. Laws, R., Enoch, M., Ison, S., Potter, S.: Demand responsive transport: a review of schemes in england and wales. J. Public Transp. **12**(1), 19–37 (2009)
14. Lucken, C.V., Barán, B., Brizuela, C.: A survey on multi-objective evolutionary algorithms for many-objective problems. Comput. Optim. Appl. **58**, 707–756 (2014)
15. Manly, B.F.: Randomization, Bootstrap and Monte Carlo Methods in Biology, vol. 70. CRC Press, Boca Raton (2006)
16. Mendes, R.S., Miranda, D.S., Wanner, E.F., Sarubbi, J.F.M., Martins, F.V.C.: Multiobjective approach to the vehicle routing problem with demand responsive transport. In: Proceedings of IEEE Congress on Evolutionary Computation (2016)
17. Mendes, R.S., Wanner, E.F., Martins, F.V.C.: árvore de agregação aplicada ao problema de roteamento de veículos com transporte reativo a demanda. In: XLVIII Simpósio Brasileiro de Pesquisa Operacional (XLVI SBPO) (2016)
18. Mendes, R.S., Wanner, E.F., Sarubbi, J.F.M., Martins, F.V.C.: Optimization of the vehicle routing problem with demand responsive transport using the NSGA-II algorithm. In: Proceedings of IEEE Intelligent Transportation Systems Society Conference Management System (2016)
19. Miranda, D.S.: Aplicação de metaheurísticas para o problema de roteamento de veículos dinâmico para o transporte reativo a demanda. Master's thesis, Universidade Federal de Viçosa, Viçosa (2012)
20. Montgomery, D.C., Runger, G.C., Hubele, N.F.: Engineering Statistics, 4th edn. Wiley, New York (2011)

21. Mulley, C., Nelson, J., Teal, R., Wright, S., Daniels, R.: Barriers to implementing flexible transport services: an international comparison of the experiences in Australia, Europe and USA. Res. Transp. Bus. Manag. **3**, 3–11 (2012)
22. Schaffer, J.D.: Some experiments in machine learning using vector evaluated genetic algorithms. Tese de doutorado, Vanderbilt Unervisty, Nashville, TN (1984)
23. Soler, J., Tencé, F., Gaubert, L., Buche, C.: Data clustering and similarity. In: Proceedings of the Twenty-Sixth International Florida Artificial Intelligence Research Society Conference, pp. 492–495 (2013)
24. Tan, P.N., Steinbach, M., Kumar, V.: Introduction to Data Mining. Pearson, Boston (2005)
25. Viana, R.J.S., Santos, A.G., Arroyo, J.E.C.: Multi-objective evolutionary approach for optimizing a demand responsive transports. In: XLVII Simpósio Brasileiro de Pesquisa Operacional (XLII SBPO) (2015)
26. Zitzler, E., Deb, K., Thiele, L.: Comparison of multiobjective evolutionary algorithms: empirical results. Evol. Comput. **8**(2), 173–195 (2000)
27. Zitzler, E., Laumanns, M., Thiele, L.: SPEA 2: Improving the strength pareto evolutionary algorithm. Technical report, Zurich, Switzerland (2002)
28. Zografos, K.G., Androutsopoulos, K.N., Sihvola, T.: A methodological approach for developing and assessing business models for flexible transport systems. Transportation **35**, 777–795 (2008)

Heterogeneous Evolutionary Swarms with Partial Redundancy Solving Multi-objective Tasks

Ruby L.V. Moritz[(✉)] and Sanaz Mostaghim

Otto-von-Guericke-University Magdeburg, Magdeburg, Germany
{ruby.moritz,sanaz.mostaghim}@ovgu.de

Abstract. Consider a self-organized system of heterogeneous reconfigurable agents solving a multi-objective task. In this paper we analyze an evolutionary approach to make such a system adaptable. In principle, this system is comparable to a multi-objective genetic algorithm, however, requires asynchronous generations and decentralized evaluation- and selection processes. The primary objective of this paper is to introduce the proposed system, to provide several interesting theoretic properties and a primary experimental analysis. The heritable material (genes) compromises a parameter set that encodes an agents configuration and can be communicated between agents. We introduce partial redundancy into the system by supplying a certain number of agents with two parameter sets instead of one. These agents are denoted as redundant and are free to chose which of their two parameter sets is applied. A special focus lies on two strategies for the agents to derive a fitness value based on their property set(s) and the respective objective functions of the multi-objective task suitable for decentralized systems. A slightly more sophisticated approach with weights for each of the objectives performs just as good as a simple method where agents pick the best or respectively worst objective value. The results show that systems with low redundancy tend to lose a lot of diversity, however, redundant systems are slower in their adaptive process.

1 Introduction

The concept of evolution is most prominently represented in computational intelligence by Evolutionary Algorithms (EAs), but also emerged in Swarm Intelligence in the form of *evolutionary swarms*. These are a subcategory of *population based adaptive systems* (PAS) [6]. In general, an evolutionary swarm or evolutionary multi-agent system is a population of autonomous agents or entities that apply evolutionary mechanisms in order to solve problems, e.g. adapt to their environment [7].

Various studies contributed to this research field, e.g. in evolutionary robotics [1] or in multi-agent systems [19,20,22,25]. Most of these studies either focus on homogeneous systems or on systems where each agent or robot adapts on its own, e.g. by using a neural network or a genetic algorithm. We are interested

© Springer International Publishing AG 2017
H. Trautmann et al. (Eds.): EMO 2017, LNCS 10173, pp. 453–468, 2017.
DOI: 10.1007/978-3-319-54157-0_31

in decentralized heterogenous systems with cooperative evolution, i.e. with a transfer of genetic material between agents that is limited by communication expenses.

EAs are typically based on Darwinian evolution, which certainly is adaptive but leads to convergence. Convergence is a desirable feature for EAs, however, is also a sign of complete loss of diversity. This loss of diversity is detrimental for PAS that have to remain adaptive when confronted with dynamic tasks or environments and should avoid stagnation.

Darwinian evolution is based on the notion that each individual has a single set of values (alleles) that defines its properties (genes) encoded in their DNA. When two individuals produce offspring their sets of values are recombined and merged in the child individual. This model is too simple to explain the diversity we observe in nature. However, if extended by the implications to be made from Mendel's research, i.e. *diploidy*, the model allows for sufficiently high diversity. The concept of diploidy describes organisms that carry two sets of values for their properties instead of one. Thus, they carry redundant information – and a lot of it they never apply. However, this redundancy increases the amount of information in the system and its diversity, stability, flexibility, and adaptability.

Fisher, Haldane, and Wright showed in theoretical and empirical studies how important genetic diversity is for adaptive processes. While mutations are rare, and seldom beneficial, recombination of genetic material by sexual reproduction gives diploid more complex organisms an evolutionary advantage over microbes that are missing this information redundancy [12,23].

The concept of diploidy has been introduced into the field of evolutionary algorithms in the late 1980s [5,27]. It is a valid option to avoid premature convergence [8,11,13,21]. In these studies only one of the available sets of values is actually applied, while the other one stays inactive. Both sets of values remain in the population if their carrier has been selected to stay. This allows for solutions of lower quality to remain in the system for an extended period of time by 'catching a ride' with a better solution. Diploid systems have a lower average but a higher peak fitness in fixed environments [4]. [2,14] also studied the benefit of diploidy and the resulting diversity in dynamic multi-agent systems.

In summary, without redundancy the evolutionary system converges rapidly to good solutions, while redundancy allows it to maintain a higher diversity and adaptability. In nature, evolution has created systems who combine both of these features by introducing *haplodiploidy*. Here, males have only one copy of genetic material while females have two. This type of heredity model evolved multiple times independently from each other in many taxonomic clades [17,18].

This concept of asymmetry or heterogeneity in redundancy poses an interesting concept for evolutionary swarms. It was first introduced in [15], where swarms adapting to dynamic environments were studied.

In this study we extend the model from [15] for agents solving a bi-objective task. We analyze how different intensities of redundancy influences the systems adaptability and diversity in face of a multi-objective problem. This includes a comparison of two decentralized methods that allow the agents to derive a single fitness value from multiple objective values without performing any comparisons with other agents fitness values. In the present study we do not analyze dynamic scenarios but focus on the initial adaptive process.

The system contains two types of agents: Redundant and non-redundant agents, i.e. agents with two property sets and agents with one property set. The heterogeneity in information redundancy provides the system with an additional level of diversity beyond allele frequency.

In Sect. 2 we introduce the model that this study is based on followed by a description of the performed experiments and their results in Sect. 3. Finally, we provide a conclusion of this study in Sect. 4.

2   Model

[15] introduced a haplodiploid model where agents have to adapt to various types of dynamic environments. In the present study we extend this model for non-dynamic multi-objective scenarios. For convenience we describe the complete model in this section with a special focus on the multi-objective fitness evaluation to highlight the originality of this study.

The proposed multi-agent model consists of n agents, $A = \{a_1 \ldots, a_n\}$, that have to process a bi-objective task. The agents are reconfigurable and adaptable, such that they can change their capabilities or properties encoded by their configuration in order to better precess the task, i.e. increase their fitness. However, as the task is bi-objective there exists one optimal configuration for each objective but typically no configuration that is optimal for both objectives of the task. The agents configuration is abstractly denoted by a bit-vector. An agent a_i can either contain one or two bit-vectors that are denoted by a_{i1} and, if available, a_{i2}. Redundant agents apply only one of their bit-vectors at any given point in time.

A distributed evolutionary mechanism is employed to adapt the agents configuration to the task. We assume that communication between agents is expensive and should be kept minimal. Otherwise, the communication network is defined by a complete graph where all agents can communicate with each other. This assumption makes any spatial information irrelevant. Thus the agents are not located within any specific space. They have neither positions nor distances to each other or the task, which is omnipresent. Further, there is no centralized control or any instance with global knowledge about the system. In this model we require asynchronous generations because the global comparison of all available agent configurations is too expensive or generally too complex with the assumed communication costs.

2.1   Correlated Bi-objective Problem

The minimal information required from the task is the fitness $\vec{f} = (f_1, f_2)$ that an agent can achieve with its current configuration. Without loss of generality,

$\vec{f}$ has to be maximized. Generally, the fitness of a bit-vector a_{ij} of an agent a_i, $j \in \{1, 2\}$, is defined by

$$f(a_{ij}) = \left(1 - \frac{||V_1 - a_{ij}||_1}{l}, 1 - \frac{||V_2 - a_{ij}||_1}{l}\right) \tag{1}$$

$$= \left(1 - \sum_{k=1}^{l} \frac{|V_{1k} - a_{ijk}|}{l}, 1 - \sum_{k=1}^{l} \frac{|V_{2k} - a_{ijk}|}{l}\right)$$

where $V_1, V_2 \in \{0, 1\}^l$ denote the optimal configuration for the respective objectives. $||V_1 - a_{ij}||_1$ is the 1-norm or Hamming distance between the two vectors. See Fig. 1 for an exemplary case. Each agent $a_i, i \in \{1, \ldots, n\}$ has to configure its bit-vectors, such that they resemble V_1 and V_2 as close as possible, i.e. has as many common bits as possible. The Pareto-front of this problem is defined by

$$\{u \mid \forall i \in \{1, \ldots l\} : V_{1i} = V_{2i} \rightarrow u_i = V_{1i}, u \in \{0, 1\}^l\} \tag{2}$$

Let l' be the number of identical bits in V_1 and V_2, then the Pareto-front contains $2^{l-l'}$ vectors. Thus, the Pareto front size depends on the number of bits where V_1 equals V_2.

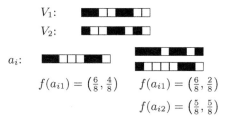

Fig. 1. An agent a_i with one (left) and two (right) bit-vectors; the bit-vector length is $l = 8$.

An important characteristic of multi-objective problems is the correlation between their objectives. It influences the fitness landscape of the problem [26], correlations between Pareto-optimal solutions [10], and generally the size and diversity of the Pareto-front [9,24]. In order to fully understand a method that solves multi-objective problems, it has to be tested on problem instances with varying correlations. Thus, we generate instances for the present problem type by creating bit-vectors V_1 and V_2 with specific correlations.

Assuming we know the distribution of the fitness values of V_1, we can easily design V_2 applying the method from [16]. This allows us to create two bit-vectors with a specific correlation coefficient ρ (Pearson correlation coefficient) where the expected mean value and variance are identical for both vectors. Let V_1 and V_2 be the bit-vectors defining objectives f_1 and f_2, respectively. Then we have

$$V_2 = (|\rho| - \rho)\bar{v} + \rho V_1 + (1 - |\rho|)R_1 + \sqrt{2|\rho|(1 - |\rho|)}(R_2 - \bar{v}), \tag{3}$$

where R_1 and R_2 are vectors with random numbers from a uniform distribution between 0 and 1 and $\overline{v}$ is the mean value of that distribution ($\overline{v} = \frac{1}{2}$). All values of V_2 are rounded such that $v_{2i} = 0$ if the floating point number created by Eq. 3 is smaller than 0.5. Otherwise, $v_{2i} = 1$. This creates a binary vector. The method can be easily extended to problem instances with less discrete properties. In this paper we chose to set the complexity of the problem with the vector size l instead of the complexity of the value types of their individual positions.

Generally, two bit-vectors are more positively correlated the more identical bits they share. This higher similarity reduces the size of the Pareto-front (consider Eq. 2). Indeed, with $\rho = -1$, V_1 would be the inverse of V_2 and as a result any possible bit-vector of length l is in the Pareto-front. When $\rho = 1$, on the other hand, there is only one vector in the Pareto-front, because $V_1 = V_2$, which reduces the bi-objective to a single objective problem with one optimal solution.

2.2   Simulation Processes

The time discrete simulation is a concatenation of T time steps in which the agents calculate their fitness and can communicate their bit-vectors to other agents. For a graphical overview of the procedure see Fig. 2.

Initially n agents are created. The system contains nk agents with two bit-vectors and $(1 - k)n$ agents with one bit-vector. The redundancy constant $k \in [0, 1]$ defines how many of these agents contain two bit-vectors, i.e. are redundant. Thus, there are $(k + 1)n$ bit-vectors in the system.

Each simulation consists of T time steps and each time step is split into two phases. During the first phase all agents calculate their respective fitness in dependance to V_1 and V_2. They add this value to their respective accumulated fitness value, denoted by f_i in Fig. 2. In the follow up step each agent checks

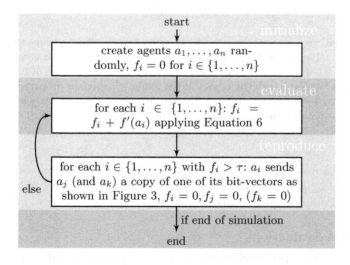

Fig. 2. Flow diagram of the simulation that ends after a specific number of time steps.

whether its accumulated fitness is larger than a certain threshold τ and, if so, reproduces. The threshold τ is the key parameter for the asynchronous selection process of this evolutionary system. Assuming, that a lower and upper bound for the fitness values are known, τ can be used to set the average number of time steps between generations. With an expected fitness of $E[f'(a_i)]$, we have an expected number of $\tau/E[f'(a_i)]$ time steps between generations. Further, by adapting the operation used to accumulate the fitness values in f_i' one can control the influence of differently scaled fitness values, e.g. by changing the sum to a product, adding weights, or exponents.

When agent a_i reproduces, because $f_i > \tau$, it selects for each of its bit-vectors a_{ik}, $k \in \{1, 2\}$, a random other agent a_j and initiates communication with that agent. Agent a_i sends its bit-vector a_{ik} to a_j. Agent a_j receives the bit-vector and selects one of its own bit-vectors, a_{jl}, $l \in \{1, 2\}$. Each bit of vector a_{ik} mutates with a probability of $\mu \in [0, 1]$ and performs a uniform crossover with a_{jl} with a probability of $\chi \in [0, 1]$. Finally, this changed version of a_{ik} replaces a_{jl}. Figure 3 shows this procedure. Both agents a_i and a_j reset their accumulated fitness value to zero.

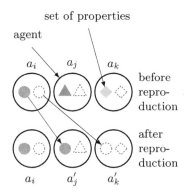

Fig. 3. Schematic overview of the reproductive process. Agent a_i initiates the reproduction and copies all of its property sets (small circles) and mutates those copies. It then chooses for each copy a target agent and recombines and replaces one of their property sets (chosen at random) with the copy. All the while a_i's original property sets remain unchanged. Arrows symbolize unidirectional communication between agents. Dashed property sets are only contained by agents with two property sets. If a_i has only one set of properties, agent a_k remains unchanged.

This mechanism enables the asynchronous selection procedure of this evolutionary model by giving agents with a higher fitness a higher reproduction rate, i.e. their f_i is more frequently above τ.

2.3 Fitness Evaluation

In this study we compare two methods for the system to deal with the multi-objective property of the problem. While the actual fitness evaluation for each individual objective is very simple, there are several ways to handle multiple objectives in order to converge towards bit-vectors close to the Pareto-front. In the following we distinguish between a *weighted* method and a *min-max* method. In this model any approach using the Pareto-dominance relation is unsuitable, because comparisons of sufficiently many configurations is impossible. Such a comparison is a prerequisite for a useful application of the Pareto-dominance relation as it allows to find non-dominated solution among a set of solutions. The following methods provide each agent a_i with a single fitness value $f'(a_i)$ derived from the fitness values of the multiple objectives.

Weights: Each agent a_i has a set of random weights $w_i = (w_{i1}, w_{i2}) \in [0,1]^2$ with $\sum_{w_{ij} \in w_i} w_{ij} = 1$. The weights are chosen from a uniform distribution $U[0,1]$. All agents have individual weights which they apply to calculate their fitness value.

$$f'(a_{ij}) = w_{i1} f_1(a_{ij}) + w_{i2} f_2(a_{ij}) \tag{4}$$

This method is based on a concept from the *Tchebicheff* approach [3] which exploits that for each Pareto-optimal solution x there is at least one set of weights which makes x the optimal solution for Eq. 4.

Min-Max: Depending on whether an agent has two property sets or only one, they either take the minimum or maximum value of the objective fitness values, such that

$$f'(a_{ij}) = \begin{cases} \max(f_1(a_{ij}), f_2(a_{ij})) & \text{if } a_i \text{ contains one property set} \\ \min(f_1(a_{ij}), f_2(a_{ij})) & \text{otherwise.} \end{cases} \tag{5}$$

This additional heterogeneity introduces an interesting aspect to the system. The non-redundant agents prioritize bit-vectors specialized towards one specific objective, while the redundant agents counteract this behavior. They prioritize bit-vectors that do not neglect any of the available objectives. We expect them to spread the solutions more broadly and optimize similar to agents applying the weighted method with $w_{ij} = 0.5$.

Note, that the above methods can be easily extended to more than two objectives. An agent with two property sets applies the one that provides the higher fitness:

$$f'(a_i) = \max_j f'(a_{ij}) \tag{6}$$

$$= \begin{cases} f'(a_{i1}) & \text{if } a_i \text{ contains one property set} \\ \max(f'(a_{i1}), f'(a_{i2})) & \text{otherwise} \end{cases}$$

Considering the min-max method we can make the following observations for scenarios with uncorrelated objectives where the values of V_1, V_2, and the bit-vectors of the concerned agents are uniformly distributed.

An agent a_i with one bit-vector a_{i1} has an expected fitness of

$$E[\max(f_1(a_{i1}), f_2(a_{i1}))] = \frac{2}{3}.$$

For d objectives we have

$$E[\max(f_1(a_{i1}), \ldots, f_d(a_{i1}))] = \frac{d}{d+1}.$$

An agent a_i with two bit-vectors a_{i1} and a_{i2} has an expected fitness of

$$E[\max(\min(f_1(a_{i1}), f_2(a_{i1})), \min(f_1(a_{i2}), f_2(a_{i2})))] = \frac{7}{15}.$$

For d objectives we have

$$E[\max(\min(f_1(a_{i1}), \ldots, f_d(a_{i1})), \min(f_1(a_{i2}), \ldots, f_d(a_{i2})))] = \frac{3d+1}{(d+1)(2d+1)}.$$

This implies that agents with one bit-vector have a higher expected fitness than agents with two bit-vectors and a higher impact on the evolutionary processes of the system. In the following we show how these two behaviors influence the adaptation process.

3   Experiments and Results

A series of experiments has been performed with the parameter value summarized in Table 1. A first set of experiments is used to substantiate the selected values for the mutation and crossover rates. A second set of experiments shows how agents applying the weighted or min-max method perform with various distributions of (non-)redundant agents processing tasks with differently correlated objectives. The system applying the weighted scheme assigns random weights from a uniform distribution to the agents.

As a quality measurement for the systems we apply the *Hypervolume indicator* (see Fig. 4), which denotes the size of the objective space dominated by a

Table 1. Parameter values applied in experiments

Parameter		Standard	Tested values
Time steps	T	15000	
No. of runs		100	
No. of bit-vectors	$(k+1)n$	1000	
No. of objectives	d	2	
Vector length	l	64	
Redundancy factor	k	0, 1	$0.05, 0.5, 0.95$
Reprod. threshold	τ	10	
Mutation rate	μ	0.001	$0.1, 0.01, 0.0001, 0$
Crossover rate	χ	0.1	$1, 0.1, 0.01, 0.0001$
Correlation	ρ	$1, 0, -1$	$-0.5, 0.5$

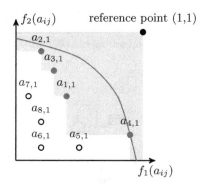

Fig. 4. Agent bit-vectors in the two-dimensional objective space and the actual Pareto front (red line). Bit-vectors $a_{1,1}$ through $a_{4,1}$ are non-dominated with respect to the Pareto dominance relation. The grey area gives the hypervolume of the non-dominated front (red points) with respect to the reference point (black). (Color figure online)

reference point and not dominated by the current fitness values of the agents. The agents have two objectives f_1 and f_2 which they need to maximize. The objective values of both objectives are in $[0, 1]$, thus we use the point $(1, 1)$ as reference point. It dominates the whole objective space which is of size 1. Using the objective values of all available bit-vectors in the system we can derive a non-dominated front and calculate the difference between the space dominated by the reference point and the space dominated by the non-dominated front. The hypervolume indicator is a suitable measurement to show the quality and distribution of many solutions for multi-objective problems. Note, that the system aims at minimizing the hypervolume we measure.

3.1   Influence of Evolutionary Operators

The evolutionary operators, mutation and crossover, are the only opportunity for the agents to change a bit-vector which doubles as genetic material or DNA in this system.

Mutation Rate. The mutation rate μ controls the diversity of specific bits in the bit-vector. For example, a higher mutation rate increases the diversity of the values in the first bit of the vector by pushing the distribution towards an uniform distribution. Figure 5 shows the final hypervolume (time step 15000) of homogeneous systems ($k \in \{0, 1\}$) applying the min-max method and five different mutation rates. The optimal mutation rate depends on the correlation between the objectives. If $\rho = -1$ the system benefits from higher mutation rates. Note, that with $\rho = -1$ all possible bit-vectors are in the Pareto-front and to get lower hypervolumes the system requires very diverse solutions. This type of problem is hard for the system to solve, because there is no "right" direction for the evolutionary drive to go to. Generally, in this case it holds that the more unique bit-vectors there are in the system the lower is the measured

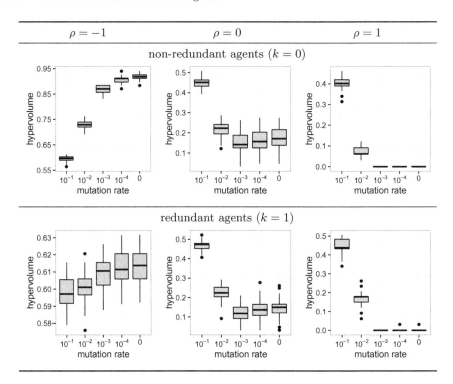

Fig. 5. Performance with different mutation rates. Hypervolume measured after 15000 time steps (the smaller the value the better). (top) Systems with 1000 agents, each containing one bit-vector. (bottom) Systems with 500 agents, each containing two bit-vectors. From left to right: (1) $\rho = -1$, the vectors V_1 and V_2 are inverse to each other, (2) $\rho = 0$, the vectors V_1 and V_2 are expected to be equal in half their bits, (3) $\rho = 1$, $V_1 = V_2$.

hypervolume. If the correlation between the objectives is higher, systems with mutation rates of $\mu = 0.001$ perform best.

Crossover Rate. The crossover rate χ controls the diversity of the system on the bit-vector level. If, for example, one considers the bit-vectors to be binary numbers with l digits, a higher χ increases the diversity of these numbers towards a uniform distribution. With one exception, $\chi = 1$ deals the most promising results (refer to Fig. 6), i.e. in every copy process the bit-vector of the donor agent is recombined with the bit-vector of the receiving agent. Unless $V_1 = V_2$, recombining the bit-vectors supresses the genetic drift towards one of the objectives. This is obviously beneficial when measuring the quality of the system with the hypervolume indicator. Only with $\rho = -1$, the whole system drifts towards one of the objectives and gradually reduces its diversity which inevitably leads to a hypervolume very close to 1. The higher the crossover rate, the faster ends up in one objective, completely neglecting the other one. Thus, the more efficient adaptive process ultimately reduces the number of bit-vectors in the non-dominated front and decreases the hypervolume. This exception is not to be taken as reason to decrease the crossover rate below 1.

3.2   Influence of Distribution and Fitness Evaluation Technique

In the second set of experiments we analyze the proposed systems concerning the distribution of agents with one and two bit-vectors. The final quality measurement is based on the total number of bit-vectors in the system and not the total number of agents. Thus, we use the same number of bit-vectors for all systems to derive comparable results from our experiments. The number of agents with two vectors reduce the total number of agents in the system. Refer to Fig. 7 for a list of the distributions and the number of (non-)redundant agents applied in the experiments. Figure 8 shows all available results for a convenient comparison.

The systems applying the weighted scheme behave expectedly and adapt rapidly, if possible, to the common bits of V_1 and V_2. The non-redundant agents speed up the adaptive process. In almost all cases, we can observe a gradual loss of diversity in the systems and an increase of the hypervolume indicator. This phenomenon can be best seen with $\rho = -1$ where no adaptation is possible and the loss of diversity by random (genetic) drift is the sole force remaining in the system.

The min-max method results are considerably different from the results derived from the weighted method. Although, the systems dominated by agents with two bit-vectors show indistinguishable behavior to the respective weighted system. Apparently, the diversity generated by using a multitude of different weight vectors throughout the system still leads to behavior observed in systems with homogeneous uniform weights. The elitist behavior of the non-redundant agents introduces a change in the systems adaptive behavior.

Note, that with $\rho = 1$ both methods are identical in their behavior because the fitness evaluation is identical. Otherwise, the most striking difference between the two systems is the loss of diversity promoted by agents with only one bit-vector and their greedy optimization. Although, a majority of agents with one bit-vector leeds to a fast adaptive process and promising hypervolume measurements at first, the simultaneous loss of diversity swerves the vectors in the system to resemble either V_1 or V_2 and reduces the non-dominated solution space. This effect is best observable with correlations close to -1. The lower or more negative the correlation is, the sooner this loss of diversity sets in, such that the system quality decreases before it reaches the good values the other systems achieve later on.

With $k = 0.5$ this effect has less influence. The higher number of redundant agents keeps the diversity of the system high enough for it to dominate a large part of the solution space before the diversity reduction sets in.

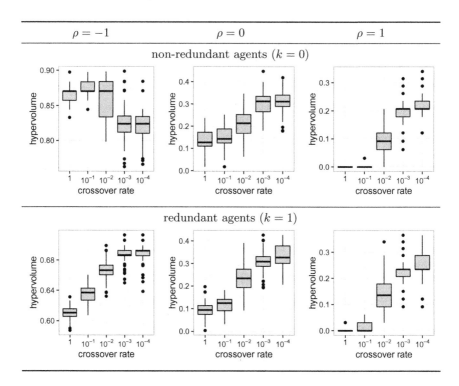

Fig. 6. Performance with different crossover rates. Hypervolume measured after 15000 time steps (the smaller the value the better). (top) Systems with 1000 agents, each containing one bit-vector. (bottom) Systems with 500 agents, each containing two bit-vectors. From left to right: (1) $\rho = -1$, the vectors V_1 and V_2 are inverse to each other, (2) $\rho = 0$, the vectors V_1 and V_2 are expected to be equal in half their bits, (3) $\rho = 1$, $V_1 = V_2$.

k	Agents with one vector	two vectors	n
0.00	1000	0	1000
0.05	950	25	975
0.50	500	250	750
0.95	50	475	525
1.00	0	500	500

Fig. 7. Population sizes for systems with specified redundancy.

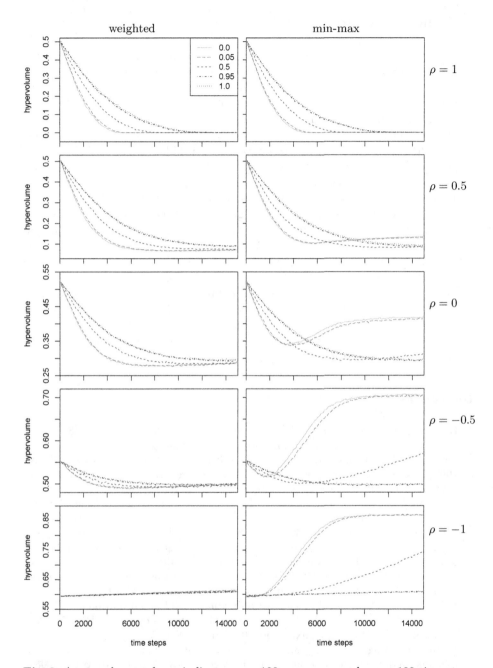

Fig. 8. Average hypervolume indicator over 100 runs measured every 100 time steps. Each column shows the results for one fitness evaluation method (weighted or min-max), each row shows the results for a specific correlation in the problem instance (on the right hand side). Each plot contains five graphs, one for each distribution of (non)-redundant agents, where the number gives the frequency of redundant agents.

4 Conclusion and Outlook

We presented a multi-agent model processing a bi-objective task and compared two methods that allows for asynchronous generation selection. The model contains two types of agents, which either contain one or two sets of values for the parameters that define their fitness concerning the two objectives. The systems perform best with a very high crossover rate of $\chi = 1$, which results in permanent recombination of the parameter sets. Since there are no interdependencies between the parameters, this is most probably a problem specific phenomenon – a question that calls for more research.

The first method is based on weights. The agents denote individual weights to the objectives to derive a suitable fitness value. This approach ruled the same results obtained with a method where agents solely used the worst objective value of all their objective values as their fitness. The second method, though simpler in the design, performed just as good as the weighted method. If the system contains agents that assume the best of their objective values as fitness, they reduce the diversity within the system drastically.

Generally, we can summarize that redundancy in the system can maintain high diversity as it reduces the speed of convergence. However, it will be interesting to see how this type of system performs with many-objective tasks and partially or fully dynamic multi-objective tasks and a theoretic analysis shows that they have a higher influence on the evolutionary process.

References

1. Bongard, J.C.: Evolutionary robotics. Commun. ACM **56**(8), 74–83 (2013)
2. Bowers, R.I., Sevinç, E.: Preserving variability in sexual multi-agent systems with diploidy and dominance. In: Dikenelli, O., Gleizes, M.-P., Ricci, A. (eds.) ESAW 2005. LNCS (LNAI), vol. 3963, pp. 184–202. Springer, Heidelberg (2006). doi:10.1007/11759683_12
3. Bowman, V.J.: On the relationship of the Tchebycheff norm and the efficient frontier of multiple-criteria objectives. In: Thiriez, H., Zionts, S. (eds.) Multiple Criteria Decision Making. LNE, vol. 130, pp. 76–86. Springer, Heidelberg (1976)
4. Calabretta, R., Galbiati, R., Nolfi, S., Parisi, D.: Two is better than one: a diploid genotype for neural networks. Neural Process. Lett. **4**(3), 149–155 (1996)
5. Goldberg, D.E., Smith, R.E.: Nonstationary function optimization using genetic algorithms with dominance and diploidy. In: ICGA, pp. 59–68 (1987)
6. Haasdijk, E., Eiben, A., Winfield, A.: Individual, social and evolutionary adaptation in collective systems. In: Handbook of Collective Robotics, pp. 413–471. Pan Stanford, May 2013. http://dx.doi.org/10.1201/b14908-15
7. Hanna, L., Cagan, J.: Evolutionary multi-agent systems: an adaptive and dynamic approach to optimization. J. Mech. Des. **131**(1), 011010 (2008). http://dx.doi.org/10.1115/1.3013847
8. Herrera, F., Lozano, M.: Adaptation of genetic algorithm parameters based on fuzzy logic controllers. Genet. Algorithms Soft Comput. **8**, 95–125 (1996)
9. Jaszkiewicz, A.: Genetic local search for multi-objective combinatorial optimization. Eur. J. Oper. Res. **137**(1), 50–71 (2002)

10. Knowles, J.D., Corne, D.: Towards landscape analyses to inform the design of hybrid local search for the multiobjective quadratic assignment problem. HIS **87**, 271–279 (2002)
11. Leung, K.S., Duan, Q.H., Xu, Z.B., Wong, C.: A new model of simulated evolutionary computation-convergence analysis and specifications. IEEE Trans. Evol. Comput. **5**(1), 3–16 (2001)
12. Lumley, A.J., Michalczyk, L., Kitson, J.J.N., Spurgin, L.G., Morrison, C.A., Godwin, J.L., Dickinson, M.E., Martin, O.Y., Emerson, B.C., Chapman, T., Gage, M.J.G.: Sexual selection protects against extinction. Nature **522**(7557), 470–473 (2015). http://dx.doi.org/10.1038/nature14419
13. Mauldin, M.L.: Maintaining diversity in genetic search. In: AAAI, pp. 247–250 (1984)
14. Moritz, R., Middendorf, M.: Evolutionary inheritance mechanisms for multi-criteriadecision making in multi-agent systems. In: Proceedings of the 2015 on Genetic and Evolutionary Computation Conference, GECCO 2015, NY, USA, pp. 65–72. ACM, New York (2015)
15. Moritz, R., Mostaghim, S.: The influence of heredity models on adaptability in evolutionary swarms. In: Proceedings of the Conference on Genetic and Evolutionary Computation, GECCO 2016, NY, USA. ACM, New York (2016, to appear)
16. Moritz, R.L.V., Reich, E., Bernt, M., Middendorf, M.: A property preserving method for extending a single-objective problem instance to multiple objectives with specific correlations. In: Chicano, F., Hu, B., García-Sánchez, P. (eds.) EvoCOP 2016. LNCS, vol. 9595, pp. 18–33. Springer International Publishing, Cham (2016). doi:10.1007/978-3-319-30698-8_2
17. Normark, B.B.: The evolution of alternative genetic systems in insects. Annu. Rev. Entomol. **48**(1), 397–423 (2003)
18. Normark, B.B.: Perspective: maternal kin groups and the origins of asymmetric genetic systems-genomic imprinting, haplodiploidy, and parthenogenesis. Evol. Int. J. Org. Evol. **60**, 631–642 (2006)
19. Panait, L., Luke, S.: Cooperative multi-agent learning: the state of the art. Auton. Agent. Multi-Agent Syst. **11**(3), 387–434 (2005)
20. 't Hoen, P.J., Jong, E.D.: Evolutionary multi-agent systems. In: Yao, X., et al. (eds.) PPSN 2004. LNCS, vol. 3242, pp. 872–881. Springer, Heidelberg (2004). doi:10.1007/978-3-540-30217-9_88
21. Potts, J.C., Giddens, T.D., Yadav, S.B.: The development and evaluation of an improved genetic algorithm based on migration and artificial selection. IEEE Trans. Syst. Man Cybern. **24**(1), 73–86 (1994)
22. Shibata, T., Fukuda, T.: Coordinative behavior by genetic algorithm and fuzzy in evolutionary multi-agent system. In: 1993 Proceedings of the IEEE International Conference on Robotics and Automation, pp. 760–765. IEEE (1993)
23. Smith, J.M.: The Evolution of Sex, vol. 32. Cambridge University Press, Cambridge (1978)
24. Verel, S., Liefooghe, A., Jourdan, L., Dhaenens, C.: Analyzing the effect of objective correlation on the efficient set of MNK-landscapes. In: Coello, C.A.C. (ed.) LION 2011. LNCS, vol. 6683, pp. 116–130. Springer, Heidelberg (2011). doi:10.1007/978-3-642-25566-3_9
25. Whitacre, J.M., Rohlfshagen, P., Bender, A., Yao, X.: The role of degenerate robustness in the evolvability of multi-agent systems in dynamic environments. In: Schaefer, R., Cotta, C., Kołodziej, J., Rudolph, G. (eds.) PPSN 2010. LNCS, vol. 6238, pp. 284–293. Springer, Heidelberg (2010). doi:10.1007/978-3-642-15844-5_29

26. Xu, Y., Qu, R., Li, R.: A simulated annealing based genetic local search algorithm for multi-objective multicast routing problems. Ann. Oper. Res. **206**(1), 527–555 (2013)
27. Yukiko, Y., Nobue, A.: A diploid genetic algorithm for preserving population diversity — Pseudo-Meiosis GA. In: Davidor, Y., Schwefel, H.-P., Männer, R. (eds.) PPSN 1994. LNCS, vol. 866, pp. 36–45. Springer, Heidelberg (1994). doi:10.1007/3-540-58484-6_248

Multiple Metamodels for Robustness Estimation in Multi-objective Robust Optimization

Pramudita Satria Palar$^{(\boxtimes)}$ and Koji Shimoyama

Institute of Fluid Science, Tohoku University,
Sendai, Miyagi Prefecture 980-8577, Japan
palar@edge.ifs.tohoku.ac.jp

Abstract. Due to the excessive cost of Monte Carlo simulation, metamodel is now frequently used to accelerate the process of robustness estimation. In this paper, we explore the use of multiple metamodels for robustness evaluation in multi-objective evolutionary robust optimization under parametric uncertainty. The concept is to build several different metamodel types, and employ cross-validation to pick the best metamodel or to create an ensemble of metamodels. Three types of metamodel were investigated: sparse polynomial chaos expansion (PCE), Kriging, and 2nd order polynomial regression (PR). Numerical study on robust optimization of two test problems was performed. The result shows that the ensemble approach works well when all the constituent metamodel is sufficiently accurate, while the best scheme is more favored when there is a constituent metamodel with poor quality. Moreover, besides the accuracy, we found that it is also important to preserve the trend and smoothness of the decision variables-robustness relationship. PR, which is the less accurate metamodel from all, can found a better representation of the Pareto front than the sparse PCE.

Keywords: Multiple metamodels · Robustness estimation · Multi-objective robust optimization · Evolutionary algorithm

1 Introduction

Robust optimization is the framework of choice when one wants to optimize a certain system under the presence of uncertainty. In engineering design, uncertainties are inevitable and have to be incorporated to some degree into the optimization process. In this paper, we have the interest to perform optimization under uncertainty subjected to the uncertain environmental parameters.

It is a common practice to nest the random variables inside the decision variables (double loop approach). This means that the calculation of robustness is performed by varying the random variables with fixed value of decision variables. Although this non-nested relationship can be broken down, this non-nested relationship is more prone to the curse of dimensionality since the total number of variables is now the sum of both the decision and random variables; We then use the standard practice of double loop approach in this paper. Specifically,

© Springer International Publishing AG 2017
H. Trautmann et al. (Eds.): EMO 2017, LNCS 10173, pp. 469–483, 2017.
DOI: 10.1007/978-3-319-54157-0_32

our interest is to solve robust optimization using the evolutionary algorithm, thus we refer the overall optimization framework in this paper as 'evolutionary robust optimization' (ERO).

In metaheuristics literature, one of the early work is the robust evolutionary algorithm of Tsutsui and Ghosh [1]. Another attempt to tackle robust optimization was performed by Deb and Gupta who introduced robustness into NSGA-II by employing Monte Carlo simulation (MCS) [2]. A comprehensive review of robust optimization itself can be found in [3,4], which explain robust optimization from metaheuristics and general perspective, respectively. Optimizing mean and standard deviation as the metric of robustness is a common approach used in the engineering community. We, therefore, employ statistical moments as the robustness metric to be optimized in this paper.

In contrary to the conventional deterministic optimization, the objective function (which is the robustness) in ERO is typically the average from many runs of deterministic simulation, usually perceived from the probabilistic perspective. Metamodel (or surrogate model), which approximates the relationship between the input and output with analytical expression, is a common and useful practice to accelerate this process of robustness estimation/uncertainty quantification (UQ). Some metamodel types that were already successfully used to various degree for UQ and robust optimization are Kriging [5], radial basis function (RBF) [6], polynomial regression (PR) [7], and polynomial chaos expansion (PCE) [8]. It has to be noted that there is no single metamodel that strictly outperforms the others. It is worth to be considered that the field of UQ itself is also progressing. PCE is now widely used in engineering community as a UQ method. Nonetheless, PCE does not always perform better than other metamodels such as Kriging. Within ERO framework, it is important to accurately estimate the robustness value to guide the optimizer into discovering the true optimum solution/s.

The motivation of this paper is to study the effect of the use of various metamodels to estimate the robustness in aiding multi-objective ERO. Besides the individual metamodel, the leave-one-out cross-validation (LOOCV) information of each individual metamodel is also exploited to create the ensemble of metamodels [9], or to select the most potential metamodel in each robustness estimation. In this paper, we studied three different types of metamodels: least-angle-regression based (LARS) sparse PCE [8], Kriging, and 2nd order PR. PCE and Kriging were selected due to their popularity in UQ community. On the other hand, PR is chosen due to its simplicity and it is also more well known in the metaheuristics community. Please note that the work explained in [10], although looks similar, deals with uncertain design variables and not the environmental parameters. Moreover, in this paper, we studied the efficacy of utilizing various multiple metamodels to aid the process of multi-objective ERO. This paper also differs to [11] in a sense that this paper uses metamodel to calculate the robustness, and not to approximate the responses surfaces of mean and standard deviation as a function of design variables.

2   Metamodel for Robustness Estimation

We are interested in optimizing a function $f(\boldsymbol{x}, \boldsymbol{\xi})$, where $\boldsymbol{x} = \{x_1, \ldots, x_{n_d}\}$ and $\boldsymbol{\xi} = \{\xi_1, \ldots, \xi_{n_r}\}$ are the vector of decision and random variables, respectively. Here, n_d and n_r are the number of decision and random variables, respectively. We want to solve robust optimization case where the function is subjected to the uncertainties in environmental parameters, which are modeled as the random variables. Thus, for a solution $\boldsymbol{x}^{(i)}$, our goal is to obtain the mean $\mu[f(\boldsymbol{x}^{(i)}, \boldsymbol{\xi})]$ and standard deviation $\sigma[f(\boldsymbol{x}^{(i)}, \boldsymbol{\xi})]$ of $\boldsymbol{x}^{(i)}$ subjected to the random variables $\boldsymbol{\xi}$. Computation of statistical moments for a solution $\boldsymbol{x}^{(i)}$ can then be done simply by performing MCS.

As for the terminology that we used throughout this paper, we refer to the relationship between the random variables and random output as 'stochastic function' and the space where the random variables live is called the 'random space'. Furthermore, the surface of the stochastic function itself is referred as a 'stochastic response surface'. Moreover, solely 'response surface' means the relationship surface between the decision variables and the objective function (which is the robustness in this paper).

2.1   Robust Optimization Using Evolutionary Algorithm

The problem is expressed as a multi-objective robust optimization problem defined as:

$$\min : \mu[f(\boldsymbol{x}, \boldsymbol{\xi})], \sigma[f(\boldsymbol{x}, \boldsymbol{\xi})] \tag{1}$$

where $\mu[f(\boldsymbol{x}, \boldsymbol{\xi})]$ and $\sigma[f(\boldsymbol{x}, \boldsymbol{\xi})]$ are the mean and standard deviation of a function $f(\boldsymbol{x}, \boldsymbol{\xi})$ subjected to the uncertainties in environmental parameters. For simplicity, we dropped the notation $f(\boldsymbol{x}, \boldsymbol{\xi})$ from $\mu[f(\boldsymbol{x}, \boldsymbol{\xi})]$ and $\sigma[f(\boldsymbol{x}, \boldsymbol{\xi})]$ on the rest of this paper. Without loss of generality, we considered no constraint in this paper. Although in this paper, we used the NSGA-II to solve the robust optimization problem, one is free to use any more efficient method for truly expensive problem.

2.2   Metamodels for Fitness Estimation

Instead of performing MCS on the real function f, the MCS is executed on the metamodel $\hat{f}$. To calculate the robustness of a solution $\boldsymbol{x}^{(i)}$, the first step is to collect the set of experimental design (ED) in the random space $\mathcal{X} = \{(\boldsymbol{x}^{(i)}, \boldsymbol{\xi}^{(1)}), \ldots, (\boldsymbol{x}^{(i)}, \boldsymbol{\xi}^{(k)})\}$ and the corresponding output $\boldsymbol{y} = \{y^{(1)}, \ldots, y^{(k)}\}$. Here, $\boldsymbol{y}$ is obtained by solving k deterministic simulations by varying only the random variables, so $\boldsymbol{y} = \{f(\boldsymbol{x}^{(i)}, \boldsymbol{\xi}^{(1)}), \ldots, f(\boldsymbol{x}^{(i)}, \boldsymbol{\xi}^{(k)})\}$. In this paper $\hat{f}$ is built using three kinds of metamodel explained below:

Kriging. Kriging works by approximating the true function with a combination of the basis functions of:

$$\psi^{(i)} = \exp\left(-\sum_{j=1}^{n_r}\theta_j|\xi_j^{(i)} - \xi_j|^{p_j}\right) \tag{2}$$

The basis of Kriging model is a vector $\boldsymbol{\theta} = \{\theta_1, \theta_2, ..., \theta_{n_r}\}^{\mathrm{T}}$. In addition, the exponent $\boldsymbol{p} = \{p_1, p_2, ..., p_{n_r}\}^{\mathrm{T}}$ is also tunable, although for simplicity it can be set to a constant value $p = 2$.

After the values of the hyperparameters $\boldsymbol{\theta}$ and $\boldsymbol{p}$ are optimized, the Kriging predictor is as follows:

$$\hat{f}_{KRG}(\boldsymbol{x}, \boldsymbol{\xi}) = \hat{\mu}_K + \boldsymbol{\psi}^{\mathrm{T}}\boldsymbol{\Psi}^{-1}(y - 1\hat{\mu}_K) \tag{3}$$

where $\hat{\mu}_K$ and $\boldsymbol{\Psi}$ are the mean of the Kriging approximation and the Kriging correlation matrix, respectively. For details of Kriging implementation and hyperparameters training, readers are suggested to refer to [12].

Polynomial Regression. PR works by approximating the multi-dimensional function with the sum of multi-dimensional polynomial bases $\boldsymbol{\Phi} = \{\Phi_0, \ldots, \Phi_P\}$. The expression for PR is as follows:

$$\hat{f}_{PR}(\boldsymbol{x}, \boldsymbol{\xi}) = \sum_{i=0}^{P}\alpha_i\Phi_i(\boldsymbol{x}, \boldsymbol{\xi}) \tag{4}$$

Sparse Polynomial Chaos Expansion. PCE is the form of polynomial regression with the set of orthogonal polynomials $\boldsymbol{\Theta} = \{\Theta_0, \ldots, \Theta_P\}$ is used as the basis. Legendre polynomials are used here since we assume that all random variables are uniformly distributed.

PCE works by approximating $f(\boldsymbol{x}, \boldsymbol{\xi})$ with:

$$\hat{f}_{PC}(\boldsymbol{x}, \boldsymbol{\xi}) = \sum_{i=0}^{P}\alpha_i\Theta_i(\boldsymbol{x}, \boldsymbol{\xi}) \tag{5}$$

The polynomial bases for the standard PR and PCE can be expanded by using total-order expansion similar to the standard PR. Ordinary least squares technique is then used to compute the coefficients $\boldsymbol{\alpha}$.

In this paper, we use the sparse PCE representation using LARS [8] to optimize the efficiency of PCE with the maximum polynomial order of 10. Note that a metamodel that combines PCE and Kriging in the form of universal Kriging is also exists [13]. However, it is important to note that the overall training process of a PCE-Kriging method is highly expensive [13]. For the PR, a 2nd order polynomial term is used.

In this paper, we use the LOOCV error to evaluate and compare the quality of Kriging, sparse PCE, and PR. LOOCV is performed by taking one point $\boldsymbol{\xi}^{(i)}$

out from the ED and building the metamodel $\hat{f}^{(-i)}$ using $\mathcal{X}\backslash\boldsymbol{\xi}^{(i)}$. One can then calculate the prediction error at $\boldsymbol{\xi}^{(i)}$:

$$\Delta^{(i)} = y^{(i)} - \hat{f}^{(-i)}(\boldsymbol{\xi}^{(i)}) \tag{6}$$

The next step is to calculate $\Delta^{(i)}$ for all $\boldsymbol{\xi}^{(i)}$ in the ED and the leave-one-out error within mean absolute error framework can be calculated as follows:

$$\varepsilon_{LOO} = \frac{1}{k} \sum_{i=1}^{k} |\Delta^{(i)}| \tag{7}$$

Within the context of PCE and PR, the analytical expression for the LOOCV can be used directly [8]. Similarly, such analytical expression also exists for the Kriging model (see [14] for more details).

2.3   Multiple Metamodels Approach

In this paper, we explore the information of LOOCV error to choose the best metamodel for each robustness estimation and to combine the capability of each metamodel into the ensemble of metamodels. For a different value of $\boldsymbol{x}$, the best metamodel with the lowest error might be different. The first multiple meta-models approach that we used is to choose the most potential metamodel with the lowest LOOCV error among M metamodels for each robustness estimation (we call it the BST scheme). On the other hand, the ensemble of metamodels [9] (denoted as ENS) is another technique that might avoid the pitfall of the wrong selection of metamodel. The metamodels ensemble approach works by combining individual metamodels through the following formulation:

$$\hat{f}_{ens} = \sum_{i=1}^{M} w_i \hat{f}_i(\boldsymbol{x}, \boldsymbol{\xi}) \tag{8}$$

To calculate the individual weight w_i of each metamodel, we use the optimal weighted metamodels approach as detailed in Viana, Haftka, and Steffen [15]. Here, we have an interest in how the use of multiple metamodels can aid the process of ERO to find a good representative of the Pareto front. It is important to note that the robustness computed from metamodel is an estimation of the true robustness. Hence, analyzing whether the use of individual and multiple metamodels can properly guide the process of ERO to the true PF is our main interest.

3   Computational Study

We investigate the performance of the individual, BST, and ENS schemes on two robust optimization test functions with $n_r = 2$ and $n_d = 2$. This low-dimensionality made it much easier to do some useful analysis techniques such as visualization. In all examples, we used Sobol sampling to generate the samples in the random space. For both problems, NSGA-II with population size and maximum generation of 200 and 500, respectively, was used to optimize the problem and to ensure that the Pareto front was found.

3.1   Augmented Branin Function

As the first test function, we constructed a normalized Branin function augmented with the modified two-dimensional O'Hagan function [16], reads as follows:

$$f_1(\boldsymbol{x}) = \left(\bar{x}_2 - (5.1/4\pi^2)\bar{x}_1^2 + (5/\pi)\bar{x}_1 - 6 \right)^2 + 10 \left(1 - (1/8\pi) \right) \cos\left(\bar{x}_1 \right) + 10$$

$$+ 20 \left(5 + x_1 + x_2 + 2\cos\left(\xi_1 x_1 \right) + 2\sin\left(\xi_2 x_2 \right) \right) \quad (9)$$

where $\bar{x}_1 = 15x_1 - 5, \bar{x}_2 = 15x_2$, with $\boldsymbol{x} = \{x_1, x_2\}$ is in the range of $[0,1]^2$. Here, ξ_1 and ξ_2 are the random variables which are uniformly distributed in the range of $[0,5]^2$. For each value of $\boldsymbol{x}$, the shape of the stochastic response surface can be linear or even highly non-linear, thus provide a challenge if a single type metamodel is employed for UQ.

Assessment of the quality of metamodel produced by the individual (PCE, KRG, PR), BST, and ENS is done first by calculating the error over 1000 stochastic response surfaces created by randomly varying the value of decision variables. For each stochastic response surface, 10000 validation samples on the random space (n_v) was used to calculate the actual error. The mean absolute error (MAE) over the random space given a sample $\boldsymbol{x}^{(i)}$, is:

$$MAE(\boldsymbol{x}^{(i)}) = \frac{1}{n_v} \sum_{j=1}^{n_v} \left| f(\boldsymbol{x}^{(i)}, \boldsymbol{\xi}^{(j)}) - \hat{f}(\boldsymbol{x}^{(i)}, \boldsymbol{\xi}^{(j)}) \right| \quad (10)$$

The error of the estimated mean $\hat{\mu}$ relative to the true one (computed using MCS with 10000 samples in the random space for the Branin function) is computed using root-mean-square relative error (RMSRE) as follows:

$$RMSRE = \sqrt{\frac{1}{n_{vd}} \sum_{j=1}^{n_{vd}} \left(\frac{\mu^{(j)} - \hat{\mu}^{(j)}}{\mu^{(j)}} \right)^2} \quad (11)$$

where n_{vd} is the number of validation samples in the decision space (the RMSRE for the σ is calculated through the same way). Note that RSMRE is used as the error metric for the statistical moments, since, based on our observation, it can better capture the spiky and spurious behavior of the response surface (explained later) than the other error metrics.

The MAE result is presented in the form of boxplot as shown in Fig. 1. From the viewpoint of actual error, PCE was able to reach a very low MAE on some samples in the decision space but the dispersion is very high. If we compare the performance of individual metamodel, for a low value of k ($k = 10, 15$) Kriging is more robust than PCE and PR as it can be seen from its lower dispersion. On the other hand, PR can also reach a very low MAE on some stochastic response surfaces but its median error performance is worse than PCE and Kriging. Although PCE and PR can approximate some stochastic response surfaces

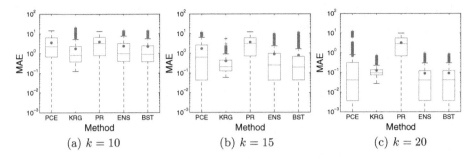

Fig. 1. Boxplots of the collection of MAE obtained from 1000 replicates with different value of k (red filled dots shows the mean of the observations). (Color figure online)

with linear or slightly non-linear behavior very well, they cannot sufficiently capture the behavior of stochastic functions with a highly non-linear trend. This is why their dispersion and median value are higher than the Kriging. It is obvious to see that the characteristics of multiple metamodels methods (BST and ENS) somehow looks like the combination of PCE and Kriging method. Here, we can see that the use of BST and ENS is more reliable than using pure PCE although the performance of the multiple metamodel approaches is comparable to Kriging if seen from the median value in this case. One can see that using either BST or ENS is a safer approach than using a pure single-metamodel approach from the viewpoint of reducing the actual error.

We then analyze the result with k values of 20. The median MAE performance of the PCE surpasses Kriging on $k = 20$, however, one can see that it still has poor performance on some stochastic functions, as indicated by the higher value of the upper quartile and mean. Here, the BST and ENS approaches clearly outperformed all individual metamodel approach. Both BST and ENS have the median value similar to the PCE but with a lower value of the upper quartile. Based on this experiment, we can see that the success degree of BST and ENS increased together with the increment of k. Although the BST scheme outperformed ENS, the difference itself is slight and not significant.

We then performed multi-objective robust optimization of augmented Branin function with $k = 10, 15, 20$ until it finds the non-dominated solutions that are very close to the PF of a given problem with a given metamodel. The robustness of the final non-dominated solutions found by each scheme was recomputed again with MCS simulation of 10000 samples to be compared with the 'true' PF found by NSGA-II + MCS. This was done to assess the accuracy of final non-dominated solutions found by using the metamodel relative to the 'true' values of the statistical moments. The assessment was done by using the hypervolume (HV) ratio (HVR) metric, which is defined as the ratio of the current HV to the HV of the true 'PF'. Here, higher HVR means that the re-estimated robustness value of the non-dominated solutions are closer to the 'true' PF. Calculation of hypervolume was done by normalizing the objective space using the best and worst values of the objectives, with the 1.2× worst value serve as the reference

Table 1. HVR of the final non-dominated solutions found by all metamodeling schemes re-estimated by MCS on Branin function.

k	Method				
	PCE	KRG	PR	ENS	BST
10	0.9592	0.9963	0.98101	0.9647	0.9643
15	0.9634	0.9983	0.9818	0.9738	0.9959
20	0.9237	0.9998	0.9967	0.9982	0.9982

point. For better depiction and analysis of the result, we depicted the attainment surface of the final non-dominated solutions [17].

Table 1 shows that among the three metamodels, Kriging is the most reliable and the most consistent metamodel as it can found the non-dominated solutions closest to the true PF (after re-estimation). It is quite surprising that PR performs better than PCE despite its lower accuracy. On low sample size $k = 10$, the performance of BST and ENS are closer to the PCE, which is the worst performer among the three metamodels. The performance of the multiple metamodels becomes better with $k = 15$, with the BST scheme outperformed the ENS one, while KRG is still the best performer of all. Figure 2 shows the final re-estimated non-dominated solutions found by all schemes with $k = 10$ and $k = 20$ as representatives. We can see further the lower quality of the PF estimated by the PCE, with KRG is closer to the true PF than both PR and PCE. The non-dominated solutions found by PR can better cover a wide section of the PF if compared to the PCE. This is quite unintuitive since the statistical moments calculated by PR were the most inaccurate among the three types of metamodels. On the other hand, we can see visually that the BST and ENS schemes are more accurate when k is high. We can also see the worsening quality of the PCE with $k = 20$, which is quite surprising.

In order to further investigate the previous findings, Fig. 3 shows the true (MCS) and estimated response surface of σ as a function of decision variables with $k = 10$. The response surface of σ with $k = 15, 20$ is not depicted since the behavior is roughly similar with the exception that the quality of multiple metamodels becomes better when k increases. The response surface of μ is highly accurate for all schemes, thus, is not a particular interest. We can see that the response surface made by PCE is highly spiky than either Kriging and PR, where this spiky behavior does not exist in the true surface. This, in turn, contributes to the inaccurate estimation of the true PF when PCE is used as the UQ method. Kriging produced smoother response surface although slight jaggedness can be observed on the upper right side of Fig. 3b. On the other hand, the response surface comes from PR is the smoothest in spite of its inaccuracy if compared to the true response surface. In spite of its less accuracy, it is clear that the response surface trend of the PR is roughly similar to the true one. This smoothness and similarity in trend result in a better estimation of the true PF when PR is used as the UQ method instead of the PCE. One can see that preserving the trend and smoothness of the decision variables and objective functions (which are μ and σ in this case), and not just the actual error, is also important to ensure the convergence to the true PF. In this regards, Kriging is the best choice that can preserve this trend while having high, but not substantially very high, accuracy to approximate the stochastic response surfaces. Figure 4 shows the convergence of RMSRE of the estimated μ and σ relative to the statistical moments estimated using MCS. One can see that Kriging is the most accurate, while the performance of BST and ENS are better than PCE and PR but worse than KRG. The BST and ENS schemes perform better than all individual schemes on $k = 20$, which indicate the higher degree of success of the multiple metamodels schemes when k is increased, although the improvement itself is minimal for this problem. Note that although the error to the actual statistical moments is similar for both the ENS and BST, BST is still (slightly) better than ENS on approximating the stochastic response surfaces (see Fig. 1 again). Here, BST is also better than ENS on approximating the 'true' PF, most notably on $k = 15$. The explanation for this is that the presence of inaccurate surrogate might lead to the loss of accuracy [15]. In this case, the non-linear part is difficult to be modeled by the PCE accurately.

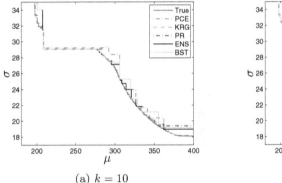 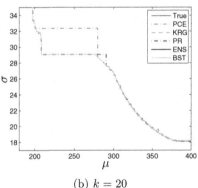

(a) $k = 10$ (b) $k = 20$

Fig. 2. Attainment surface of the final non-dominated solutions from all schemes re-estimated using MCS on Branin problem.

3.2   Airfoil in Euler Flow

The second problem is the robust optimization of a transonic airfoil in Euler flow. The airfoil was parameterized with the 9 variables PARSEC parameters [18]. The random output of interest is the drag coefficient C_d without any lift constraint C_l, where computational fluid dynamics (CFD) code was employed to solve the Euler equation. The random variables are the Mach number and Angle of Attack (AoA) with the distribution of Uniform $[0.78 - 0.82]$ and Uniform $[1.5^0 - 2.5^0]$, respectively. The design variables are illustrated and listed in Fig. 5 and Table 2, respectively. We performed sensitivity analysis on the nominal design condition

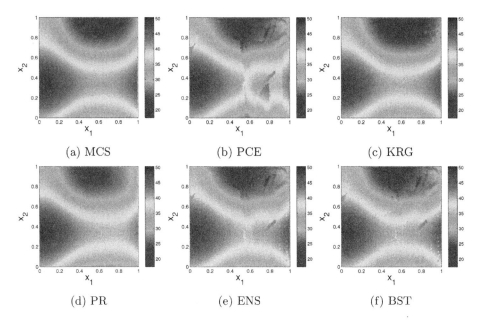

Fig. 3. Response surface of σ with $k = 10$ and various metamodeling schemes on Branin function.

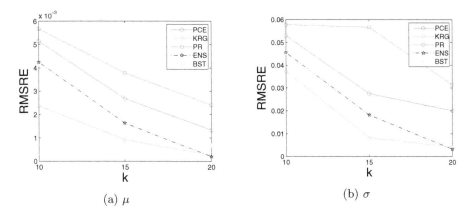

Fig. 4. RMSRE convergence to the 'true' statistical moments on Branin function.

$(M = 0.8, AoA = 2^0)$ to select the most important variables. Based on this analysis, X_{up} (x_1) and y_{up} (x_2) were selected as the decision variables while the rest was kept fixed.

Since the calculation of the true statistical moments is difficult due to the expensive computation of MCS, we used the statistical moments computed from the BST scheme with $k = 20$ as the reference values. We then varied the value of k from $8, 12$ and 16 with the error of statistical moments were computed

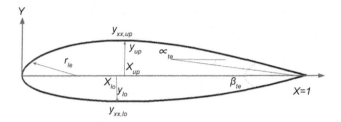

Fig. 5. Definition of PARSEC airfoil parameterization.

Table 2. Design variables of the robust airfoil optimization case.

	r_{le}	X_{up} (x_1)	y_{up} (x_2)	$y_{xx,up}$	X_{low}	y_{low}	$y_{xx,low}$	α_{te}	β_{te}
Fixed	0.0078	–	–	–0.4245	0.3618	–0.0589	0.7069	–0.1126	0.1646
Lower bound	–	0.3465	0.0503	–	–	–	–	–	–
Upper bound	–	0.5198	0.0754	–	–	–	–	–	–

toward this 'true' statistical moments. A grid of 15×15 in the decision space was used to generate the RMSRE convergence results shown in Fig. 6. From Fig. 6, one can see that, surprisingly, PR was the best and the most consistent metamodel from the viewpoint of error convergence to the 'truth'. Moreover, the ENS scheme clearly outperformed the BST scheme and performed comparably to PR as indicated by its better error convergence. An interesting fact is that the BST scheme mostly chose PCE and KRG as the metamodel of choice and not the PR (the mean of the ensemble weights for the 225 samples with all values of k are 0.1912, 0.8085, and 0.0003, for PCE, KRG, and PR, respectively). This means that the ensemble of PCE and KRG also create a more accurate approximation than the combination of the two. There is no obvious difference between the error trend of the PCE and KRG, with the BST scheme closely mirroring the trend of the KRG. Figure 7 depicts a sample of the various response surfaces of σ from the airfoil problem generated using $k = 8$ (note that the design variables were normalized). As one can see from this figure, there is no highly spiky characteristic such as in the Branin problem was observed even with the low value of k ($k = 8$), which is due to the linear and almost linear behavior of the stochastic functions in this problem. However, the response surface generated by all schemes, although preserved the trend of the reference response surface, is still not accurate enough. Here, the response surfaces from the PR and the BST scheme are more similar to the reference response surface than PCE, KRG, and ENS, especially on the lower x_2 area. Since no metamodel performs highly poorly for this problem, all metamodels can give a sufficiently accurate approximation of the statistical moments to a various degree. In fact, this might be the cause of the better performance of ENS when compared to the BST scheme.

Finally, multi-objective optimization procedure was performed by applying NSGA-II. However, instead of calling the true function, we performed the optimization using the interpolated value which comes from the 15×15 grid mentioned

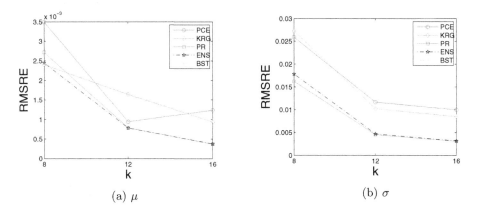

(a) μ (b) σ

Fig. 6. RMSRE convergence to the true statistical moments on airfoil problem.

earlier. This is because it is very difficult to obtain the exact representation of the true PF given the expensive cost of the robust optimization on this problem, in spite of the use of the Euler CFD code. Given the only slightly non-linear behavior of the response surface and a high number of samples in the decision space (225 samples), the interpolated value between this grid should be a highly accurate representation of the true values. The non-dominated solutions found by NSGA-II were then re-estimated again using the interpolated value from the reference response surface (BST scheme with $k = 20$), which we treated as the 'true' values of the statistical moments. Comparison of the individual metamodel (see Table 3) shows that PR is the best metamodel among the three, with the ENS and BST scheme performed roughly the same. Figure 8 shows that using $k = 8$ is not sufficient enough to closely approximate the true PF. However, although it is slightly difficult to observe, one can see that the non-dominated solutions generated by PR and ENS are closer to the true PF than the others.

Table 3. HVR of the final re-estimated non-dominated solutions found by all meta-model schemes on airfoil problem.

k	Method				
	PCE	KRG	PR	ENS	BST
8	0.9797	0.9794	0.9808	0.9799	0.9794
12	0.9994	0.9993	0.9997	0.9993	0.9995
16	0.9994	0.9992	0.9999	0.9997	0.9997

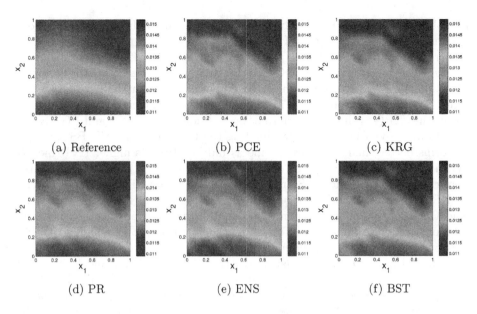

Fig. 7. Response surface of σ on airfoil problem with $k = 8$ and various metamodeling schemes.

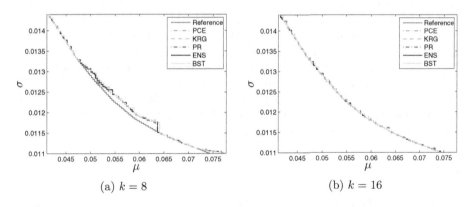

Fig. 8. Attainment surface of the final non-dominated solutions from all schemes re-estimated using MCS on airfoil problem.

4   Conclusion

Computational study on the use of multiple metamodels for robustness estimation within the framework of ERO under parametric uncertainty is presented in this paper. We are motivated by the need to achieve accurate values of statistical moments as the robustness measure to guide the evolutionary optimizer toward to the true PF. The main goal of our study in this paper is not to specifically accelerate

482   P.S. Palar and K. Shimoyama

the robust optimization process, but to accurately estimate the objective functions and guide the discovery of the PF.

Kriging, sparse PCE, and 2nd order PR were considered as the metamodel of choice in this paper, with the LOOCV error was employed to create the ensemble of metamodels and to choose the most potential one with the lowest LOOCV error. Result comparison of individual metamodel on augmented Branin function shows that KRG is the best performer overall, followed by PR, and PCE, when seen from the viewpoint of approximating the true PF. Here, the BST scheme performs better than the ENS scheme as indicated by its higher HVR metric. When seen from the viewpoint of error to the actual statistical moments, the performance of BST and ENS are better than PCE and PR, but is still inferior to KRG. We also found that besides the accuracy, preserving the trend and smoothness of the robustness metric in the objective functions-decision variables relationship is also important. Test on transonic airfoil problem shows that the best performers from the viewpoint of error to the actual statistical moments are the PR and ENS. On approximating the true PF, there is not much difference between the performance of the individual and multiple metamodels scheme except on low number of samples in random space where PR performed as the best followed by the ENS scheme. This is quite the opposite if compared to the Branin function, where BST performed better than ENS. The cause of this is that ENS will perform better on a problem where the performance difference of the individual metamodel is not disparately high (such as in airfoil problem). Overall, when one is not sure which metamodel is to pick for calculating the robustness, the multiple metamodels method is a safer approach to be employed.

As for the future works, we would like to investigate other metamodels such as RBF and support vector regression to be utilized in the multiple metamodels scheme for ERO purpose. Study on higher dimensional problem is also the subject of the future work. The effect of various cross validation method to properly select the metamodel will also be studied. Additionally, application to cases with non-uniform probability distribution should also be investigated.

References

1. Tsutsui, S., Ghosh, A.: Genetic algorithms with a robust solution searching scheme. IEEE Trans. Evol. Comput. **1**(3), 201–208 (1997)
2. Deb, K., Gupta, H.: Introducing robustness in multi-objective optimization. Evol. Comput. **14**(4), 463–494 (2006)
3. Jin, Y., Branke, J.: Evolutionary optimization in uncertain environments-a survey. IEEE Trans. Evol. Comput. **9**(3), 303–317 (2005)
4. Beyer, H.G., Sendhoff, B.: Robust optimization-a comprehensive survey. Comput. Methods Appl. Mech. Eng. **196**(33), 3190–3218 (2007)
5. Rumpfkeil, M.P.: Optimizations under uncertainty using gradients, hessians, and surrogate models. AIAA J. **51**(2), 444–451 (2012)
6. Loeven, G., Witteveen, J., Bijl, H.: A probabilistic radial basis function approach for uncertainty quantification. In: Proceedings of the NATO RTO-MP-AVT-147 Computational Uncertainty in Military Vehicle Design Symposium (2007)

7. Allen, M.S., Camberos, J.A.: Comparison of uncertainty propagation/response surface techniques for two aeroelastic systems. In: 50th AIAA Structures, Structural Dynamics, and Materials Conference, Palm Springs (2009)
8. Blatman, G., Sudret, B.: Adaptive sparse polynomial chaos expansion based on least angle regression. J. Comput. Phys. **230**(6), 2345–2367 (2011)
9. Goel, T., Haftka, R.T., Shyy, W., Queipo, N.V.: Ensemble of surrogates. Struct. Multidiscip. Optim. **33**(3), 199–216 (2007)
10. Paenke, I., Branke, J., Jin, Y.: Efficient search for robust solutions by means of evolutionary algorithms and fitness approximation. IEEE Trans. Evol. Comput. **10**(4), 405–420 (2006)
11. Zhou, X., Ma, Y., Tu, Y., Feng, Y.: Ensemble of surrogates for dual response surface modeling in robust parameter design. Qual. Reliab. Eng. Int. **29**(2), 173–197 (2013)
12. Toal, D.J., Bressloff, N.W., Keane, A.J.: Kriging hyperparameter tuning strategies. AIAA J. **46**(5), 1240–1252 (2008)
13. Schobi, R., Sudret, B., Wiart, J.: Polynomial-chaos-based kriging. Int. J. Uncertain. Quantif. **5**(2) (2015)
14. Dubrule, O.: Cross validation of kriging in a unique neighborhood. J. Int. Assoc. Math. Geol. **15**(6), 687–699 (1983)
15. Viana, F.A., Haftka, R.T., Steffen Jr., V.: Multiple surrogates: how cross-validation errors can help us to obtain the best predictor. Struct. Multidiscip. Optim. **39**(4), 439–457 (2009)
16. Oakley, J., O'Hagan, A.: Bayesian inference for the uncertainty distribution of computer model outputs. Biometrika **89**(4), 769–784 (2002)
17. Knowles, J.: A summary-attainment-surface plotting method for visualizing the performance of stochastic multiobjective optimizers. In: 5th International Conference on Intelligent Systems Design and Applications (ISDA 2005), pp. 552–557. IEEE (2005)
18. Sobieczky, H.: Parametric airfoils and wings. In: Fujii, K., Dulikravich, G.S. (eds.) Recent Development of Aerodynamic Design Methodologies, pp. 71–87. Springer, Braunschweig (1999)

Predator-Prey Techniques for Solving Multiobjective Scheduling Problems for Unrelated Parallel Machines

Ana Amélia S. Pereira[1], Helio J.C. Barbosa[1,2], and Heder S. Bernardino[1(✉)]

[1] Universidade Federal de Juiz de Fora (UFJF), Juiz de Fora, MG, Brazil
aamelia.mg@gmail.com, heder@ice.ufjf.br
[2] Laboratório Nacional de Computação Científica (LNCC), Petrópolis, RJ, Brazil
hcbm@lncc.br
http://www.lncc.br/~hcbm

Abstract. The multiobjective scheduling problem on unrelated parallel machines is tackled by predator-prey techniques. Several objectives adopted in the literature were considered and the corresponding best aligned dispatch rules were associated with the predators. Also, we suggest and analyse modifications in the movement operator, in the number of predators, and in the influence of the objectives in the selection/replacement procedures. Numerical comparisons with the popular NSGA-II were performed and good results were obtained by the predator-prey techniques.

Keywords: Predator-prey · Multiobjective · Scheduling · Unrelated parallel machines

1 Introduction

Scheduling problems are decision making processes, commonly arising in industry, where tasks/elements must be assigned to the available resources within a given period of time [22]. Tasks are operations executed during the productive process, while resources are machines which perform those tasks. The decision making process in a production behaviour aims at optimizing one or more objectives, such as production time and stocking time.

Scheduling problems can be classified according to the machine environment and the production flow [18]. The simplest problem involves a single machine. Also, parallel machines –identical, uniform, or unrelated– can be used in the production process. Finally, another possibility is to separate the production steps in stages which can contain parallel machines (flexible variation of the problems). The scheduling problem is classified as *flow shop* when the same route is used to process every job, while the problem is defined as *job shop* when different stages are allowed to the jobs.

Unrelated Parallel Machines Scheduling Problems (UPMSP) are considered here, where n tasks should be executed in m parallel machines. Additionally, the

© Springer International Publishing AG 2017
H. Trautmann et al. (Eds.): EMO 2017, LNCS 10173, pp. 484–498, 2017.
DOI: 10.1007/978-3-319-54157-0_33

model contains the release date of the tasks and the setup time of the machines, which depend on the sequence of the tasks. The formulation of the scheduling problem considered here can be found in [8]. It is important to highlight the practical significance of this problem as it represents many situations in manufacturing, such as the decision on machine replacement or improvement, and the use of machines with different ages or from different manufacturers.

In view of the broad industrial applications and the computational complexity for solving this problem using exact approaches, it is common to adopt heuristics and meta-heuristics to solve UPMSP [16]. A common algorithmic tool for solving scheduling problems is the incorporation of the so-called dispatch rules, which are greedy algorithms used for deciding which job to run next based on the jobs and the time but neither using the history of jobs running, nor computing tentative schedules.

When solving a scheduling problem, one can think of several objectives, such as reducing the delay in the production, minimizing the total production time, and maximizing profit. At the end of the 1980s the efforts to solve multiobjective scheduling problems started [10,11], as it is common to observe multiple and conflicting objectives in practical industrial situations. Several papers can be found in the literature in which heuristics or meta-heuristics are used to solve multiobjective UPMSP [15,21].

In [26], a heuristic method is proposed to solve a bi-objective optimization problem on unrelated parallel machines in a factory of glass bottles. The objectives are maximizing the total production margin and minimizing the difference in machine workload.

A formulation of a UPMSP with release and due dates (the same values for any task) is adopted in [2]. Constructive and iterative heuristics were used for solving the scheduling problem in which both the total weighted earliness and tardiness are to be minimized.

In [12], the scheduling problem involves the minimization of the sum of the makespan with the total earliness and tardiness. A heuristic based on earliness and tardiness criteria named Initial Sequence (ISETP) is proposed to solve the problem. The results obtained by the heuristic ISETP were compared to those found by an exact method, indicating that the proposed method is able to achieve solutions within an acceptable processing time.

The minimization of the weighted sum of delay and cost of energy consumption are the objectives in [9]. A model was formulated, an integer linear programming problem was solved using Cplex, and the obtained results were compared to those found by two heuristics, which are based on the earliest due date (EDD) and the total weighted tardiness (WSPT) rules. Also, a Particle Swarm Optimization (PSO) technique is proposed and its performance is compared to these three methods, indicating that PSO provides good solutions for that problem. A UPMSP with machine and task dependent setup times and downtime was tackled in [4]. The objectives were minimizing both total earliness and tardiness. Three techniques were compared, namely: GRASP; GRASP with Path Relinking; and a hybrid method in which GRASP, ILS and Path Relinking are combined. The hybrid methods obtained better solutions than those achieved by GRASP. In [25], a Tabu Search (TS) is proposed to find the solution of a UPMSP, where maximum tardiness and maximum completion time (makespan) must be minimized. It was concluded that TS can achieve good solutions using a shorter processing time when solving that problem.

A UPMSP with machine and task dependent setup times, downtime, release time, and adopting penalties for tardiness and earliness as objectives was solved in [20]. A-priori as well as a-posteriori approaches using Genetic Algorithms were evaluated, and competitive results were obtained by the *a-posteriori* approach.

A multiobjective PSO (MOPSO) is proposed in [27] and applied to a UPMSP in which the weighted sum of the flow times and the weighted sum of the lateness are to be minimized. The model also considers machine-dependent setup times, and secondary resource constraints for jobs. The solutions found by MOPSO were compared to those obtained by a standard multiobjective PSO, showing that MOPSO is able to find more diverse solutions and with better quality.

In [10], a multiobjective prey-predator (PP) technique is proposed and used to solve scheduling problems other than those (UPMSP) attacked here, namely, in single-machine and identical parallel machines environments. In that method, a dispatch rule for each target is used in order to improve multiple objectives. When the environment is formed by a single machine, the meta-heuristic was applied to minimize the completion time and the maximum lateness. In the cases containing identical parallel machines, the problem aims at minimizing three objectives: the makespan, the total completion time, and the number of tardy jobs. The proposed PP was compared to the Non-dominated Sorting Genetic Algorithm II (NSGA-II) and the results found in the experiments show that PP performed better than NSGA-II.

This paper discusses PP variants to solve multiobjective scheduling problems on unrelated parallel machines. The meta-heuristic is modified here to increase the diversity in the predators and includes more dispatch rules than those adopted in [10]. Sixteen pairs of objectives are considered in the computational experiments executed in order to access the relative performance of the PP variants.

2   Scheduling Problems on Unrelated Parallel Machines

The scheduling of n tasks to be executed in m unrelated parallel machines is described in this section. The solution is represented as a permutation of the tasks, which indicates the sequence used to assign each task to a machine. The main features of this scheduling problem are:

- a task represents a single procedure that must not be interrupted;
- each task j must be processed alone in a given machine;
- each task j has a processing time p_{ij} which depends on the machine i where it will be performed
- each task j has a due date d_j;
- each task j is available to be processed according to a release date r_j; and

– the setup time S_{ijl} is dependent on the sequence (to execute the task l after the task j) and on the machine i.

Beyond the machine environment, the formulation of the scheduling problem also contains the objectives to be reached. Normally, these objectives involve elements of tardiness and completion time. Here, the following cases are adopted:

– Tardiness:
  - T_{max} is the maximum tardiness
  - $\sum T_j$ is the total tardiness
  - $\sum w_j T_j$ is the weighted total tardiness
  - $\sum U_j$ is the number of tardy jobs
– Completion time:
  - $\sum C_j$ is the sum of the completion times
  - C_{max} is the makespan
  - $\sum c_j C_j$ is the total weighted completion time

where completion time C_j is the instant of conclusion of the j-th task, $T_j = \max\{C_j - d_j, 0\}$, and $U_j = 1$ if $C_j \geq d_j$ and $U_j = 0$, otherwise.

In practice, some of the previously presented objectives must be handled simultaneously in a scheduling problem. Often, the goals of the optimization problem are conflicting. Multiobjective optimization corresponds to the situation where there are more than one conflicting (and often non-commensurable) objectives to be simultaneously optimized. A multiobjective optimization problem has no single solution that optimizes all objectives, but the so-called Pareto set of efficient solutions where no one is better than any other in all objectives can in fact be found. The concept of dominance is used in multiobjective optimization, where a candidate solution a_1 dominates a_2 if $f_i(a_1) \leq f_i(a_2), i = 1, \ldots, M$, and $\exists \ y \in 1, \ldots, M$ such that $f_y(a_1) < f_y(a_2)$. Thus, a_1 dominates a_2 when the evaluation of a_1 is better (lesser) than a_2 in at least one of the M objectives, and it is no worse in the remaining ones. Given two candidate solutions x_1 and x_2, three alternatives can occur regarding dominance: x_1 dominates x_2 $(x_1 \prec x_2)$, x_2 dominates x_1 $(x_1 \succ x_2)$, or x_1 and x_2 are non-dominated $(x_1 \sim x_2)$.

3 The Predator-Prey Method

Before defining the modifications suggested here, the candidate solution representation must be presented.

3.1 Solution Representation

An illustrative example of a scheduling problem with $n = 6$ tasks and $m = 2$ machines is presented next. Table 1 shows the processing times p_{ij} required to execute the tasks $j = 1, \ldots, 6$ in the machines M1 and M2. The setup times S_{ijl} are presented in Tables 2 and 3, respectively, for the machines M1 and M2. For instance, one can see in Table 2 that $S_{154} = 1$. The due dates for this instance are $d_j = \{10, 21, 23, 15, 7, 15\}$ and the release dates are $r_j = \{2, 1, 3, 2, 1, 1\}$.

Table 1. Processing times p_{ij} for the illustrative problem.

n	M1	M2
1	9	10
2	10	8
3	6	10
4	5	8
5	7	10
6	8	6

Table 2. Setup times for the illustrative problem when using machine M1.

n	1	2	3	4	5	6
1	0	7	2	5	7	1
2	6	0	1	2	7	8
3	8	6	0	4	9	1
4	5	6	3	0	4	3
5	3	4	7	1	0	2
6	6	2	5	6	4	0

Table 3. Setup times for the illustrative problem when using machine M2.

n	1	2	3	4	5	6
1	0	2	9	3	1	2
2	9	0	2	2	2	1
3	3	2	0	1	7	9
4	1	2	8	0	9	1
5	2	3	4	8	0	7
6	1	7	1	9	3	0

A candidate solution to a UPMSP is given by a permutation of the tasks. A possible solution for the illustrative problem is $\pi = \{5, 1, 4, 2, 3, 6\}$. A Gantt's chart can be shown in Fig. 1, where the tasks in the sequence π are allocated in the first available machine. Thus, M1 executes the sequence of tasks 5, 4 and 3, while M2 processes the tasks 1, 2 and 6. According to this schedule, the tasks 5, 1, 4, 2, 3, and 6 are concluded, respectively, at times 8, 12, 14, 22, 23, and 29. Thus, one can see that the sum of the completion time is $\sum C_j = 108$. Also, $\sum T_j = 18$, as the tasks $j = 1, 2, 5, 6$ are concluded after the due data d_j.

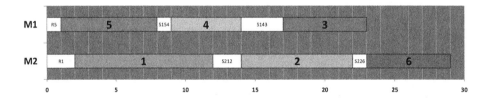

Fig. 1. Gantt's chart of a solution for the illustrative scheduling problem.

3.2   The Predator-Prey Inspiration

In 1998 Laumanns *et al.* [13] presented a novel evolutionary approach for multi-objective optimization problems inspired by the predator-prey model from ecology. The preys, that represent possible solutions to the problem, are placed at the vertices of a lattice, remain stationary, and are chased by predators. A predator take local steps in its neighborhood and chase/improve preys according to its own particular objective function. As there are several predators with different objectives, the idea is that "efficient" prey individuals will be able to produce more descendants and tend to converge to Pareto-optimal solutions. Later, Deb [5] proposed the introduction of a different vector of weights for each predator in order to improve diversity in the original algorithm. Although applications of the PP idea to standard multiobjective optimization in $\mathbb{R}^n$ can be found in the literature, Grimme and co-workers were the first to apply the PP technique to solve scheduling problems. In [10], the mutation of a given prey is performed by a dispatch rule chosen in order to improve the objective function associated with the predator. This is a way to introduce domain knowledge in the move operator definition. Shortest processing time (SPT) and earliest due dates first (EDD) are the dispatch rules used in [10]; also, the heuristic SBC-3 (from Süer, Báez, and Czajkiewicz) is employed in the mutation operator when solving the problems involving parallel identical machines. Figure 2(a) presents the structure of the predators.

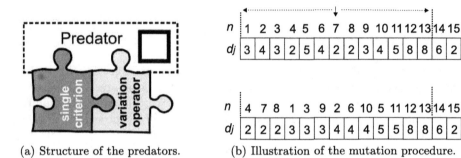

(a) Structure of the predators.   (b) Illustration of the mutation procedure.

Fig. 2. The structure of a predator (a) (from [10]), and an example of mutation (b).

The mutation of a given prey is applied as proposed in [10] using the following two steps: (i) a position $i \in \{1, \ldots, n\}$ in the candidate solution is randomly selected, and (ii) the sequence composed by $2\delta + 1$ design variables and centred by the i-th variable is sorted according to the dispatch rule of the predator. Thus, part of the solution is preserved. The value of δ is obtained from a normal distribution with an externally adjustable step size of α (an user defined parameter) and the value of δ is not modified during the search.

Figure 3 illustrates three stages of the PP method, where the grid is formed by 25 preys (a 5×5 toroidal grid) and contains a single predator. The predator is initially in the cell $(2, 2)$ and concludes its two-step random walk in the cell $(3, 3)$, as shown in Fig. 3(a). Figure 3(b) presents the selection of the preys by the predator in its neighborhood. Finally, the replacement of a prey is shown in Fig. 3(c), where (i) the best neighbor of the prey is selected and it is mutated using the dispatch rule, and (ii) the new generated candidate solution replaces the worst neighbor of the prey when it is not worse than the latter.

It is important to note that neither agglomeration nor diversity preservation mechanisms are explicitly adopted by the PP technique. The variation in the population of preys is produced by predators that use different dispatch rules and objectives when guiding the search. Thus, there is an implicit cooperation among the predators, generated by their different setups, aiming at solving a given multiobjective optimization problem.

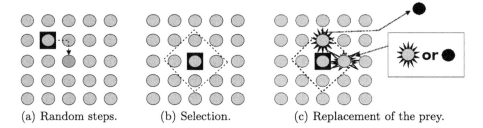

(a) Random steps.   (b) Selection.   (c) Replacement of the prey.

Fig. 3. Illustration of the three main steps of the prey-predator technique. The random steps of a predator are shown in (a), the selection of the prey in its neighborhood is presented in (b), and the replacement procedure is shown in (c). Adapted from [10].

4   The Proposed Variants

The method considered here to solve UPMSP is based on [10], but extends the ideas presented there by (i) increasing the diversification generated by the mutation operator; (ii) analysing and adopting other dispatch rules; and (iii) modifying the number of predators and varying the influence of the objectives in their selection/replacement procedures.

The mutation was modified in order to increase the variation of the move operator. In [10], the value of δ is fixed during the search while here we propose to randomly select S every time the mutation is applied. Also, the value β is picked here uniformly at random. The proposed mutation is presented in Algorithm 1 and an illustrative example can be found in Fig. 2(b). The due dates (d_j) are presented for each one of the $n = 15$ tasks, $porc = 40\%$ (leading to $S = 6$), and EDD is the dispatch rule of the predator; also, $\beta = 7$ was randomly obtained.

Algorithm 1. Mutation operator of a predator-prey method.

```
 1  begin
 2  |   S = n x porc%;
 3  |   β = random(1, n);
 4  |   ini = β − S;
 5  |   fin = β + S;
 6  |   if ini <= 0 then
 7  |   |   ini = 1;
 8  |   end if
 9  |   if fin > n then
10  |   |   fin = n;
11  |   end if
12  |   for task := ini, . . . , fin do
13  |   |   Mutate(task);
14  |   end for
15  end
```

As previously mentioned, the influence of the predator is determined by a dispatch rule, which is used in the move operator (`Mutate` call in Algorithm 1), and by the objective(s) used in the selection/replacement decision of the candidate solutions. Here, an extended list of dispatch rules was used in order to handle the increased number of objective functions considered in the experiments. An analysis was performed in [19] with the purpose of identifying dispatch rules better aligned with the objectives at hand. The results of this study as well as the dispatch rules which are adopted here in each case are presented in Sect. 5.2.

The third proposed variation for PP is a modification in the choices made by the predators based on the objective functions values. While in most previous PP techniques the preference of a given predator is directed by its particular objective, the use of a weighted sum of the objective values is studied here. Thus, the quality F_p of a prey x can be evaluated by a predator p as

$$F_p(x) = \sum_{m=1}^{M} w_{p,m} f_m(x),$$

where $w_{p,m}$ denotes the weight assigned by the predator p to the m-th objective, with $\sum_{m=1}^{M} w_{p,m} = 1$. Thus, the single criterion within the structure of the predators as presented in Fig. 2(a) is replaced by a set of weights which indicate the influence of each objective. One can notice that the previously proposed PPs are recovered when $w_{p,i} = 1$ for a given objective i (which varies between the predators) and $w_{p,j} = 0$, $\forall j \in \{1, \ldots, M\}$ and $j \neq i$.

Tables 4 and 5 illustrate, respectively, the cases with 4 and 10 predators. SPT and minimum slack time (MST) dispatch rules conclude the components of the predators $P1, \ldots, P_{np}$, where np is the number of predators. Both cases –with 4 and 10 predators– are analysed in the computational experiments. It is

Table 4. Example of the proposed structure for the predators with $np = 4$.

Predator	Rule	$w_{p,1}$	$w_{p,2}$
P1	MST	1	0
P2	SPT	1	0
P3	MST	0	1
P4	SPT	0	1

Table 5. Example of the proposed structure for the predators with $np = 10$.

Predator	Rule	$w_{p,1}$	$w_{p,2}$
P1	MST	1.00	0.00
P2	SPT	1.00	0.00
P3	MST	0.75	0.25
P4	SPT	0.75	0.25
P5	MST	0.50	0.50
P6	SPT	0.50	0.50
P7	MST	0.25	0.75
P8	SPT	0.25	0.75
P9	MST	0.00	1.00
P10	SPT	0.00	1.00

important to note that a uniform variation in the values of the weights is not required, allowing for added bias to the search.

It should be noted that this meta-heuristic was validated comparing the exact Pareto front of the single machine scheduling problem presented in [10] with those obtained by PP. The solutions found are almost equal to the optimum solutions. This analysis is detailed in [19].

5   Computational Experiments

Computational experiments were performed in order to evaluate the proposed ideas for solving UMPSPs. The results obtained by the PP variants are compared to those found by NSGA-II [6], perhaps the meta-heuristic most widely applied to scheduling problems. NSGA-II is adapted here to solve UMPSPs problems by using the SJOX [23] crossover and a *shift* mutation operator, as suggested by [28]. Also, a variant of this NSGA-II version is considered here, where a local search based on an iterated greedy algorithm is incorporated [24]. Several experiments were performed using the two variants of both PP and NSGA-II, and the notation adopted is as follows:

- PP-4: PP with 4 predators;
- PP-10: PP with 10 predators;
- NSGA-II: NSGA-II; and
- NSGA-II+LS: NSGA-II with local search.

The stop criterion is 6000 objective function evaluations, as in [10], and the initial population (preys and individuals) is randomly generated in both techniques. A parameter analysis was made in [19] and the better values are used here for both PP and NSGA-II. The PP environment is defined by a 10×10 bi-dimensional toroidal grid populated by 100 preys plus the predators considered in each case, $porc = 30\%$, and the predators are allowed to walk 1 step at a time. For the NSGA-II the population size was set to 100, crossover probability equal to 100%, and mutation probability equal to 50% where one element of the chromosome is shifted.

5.1   Problem Instances

Problem instances with $n \in \{50, 100\}$ tasks and $m \in \{8, 12\}$ machines were solved. For each (n, m) combination 50 instances were considered, leading to a total of 200 instances which were generated according to [1,14]. The processing times p_j were taken randomly in the interval $[50, 100]$. The setup time S_{ijl} for preparing the i-th machine to process the l-th task after processing the j-th task were randomly generated in the interval $[2/3\eta\bar{p}, 4/3\eta\bar{p}]$, where $\bar{p}$ is the average of the processing times and $\eta = 0.25$. Penalties w_j due to lateness as well as the completion time c_j are randomly generated in $[1, 100]$.

The due dates d_j are generated randomly in the interval $[(1 - R)\bar{d}, \bar{d}]$ with probability τ, and along the interval $[\bar{d}, R(C_{max} - \bar{d}) + \bar{d}]$ with probability $(1 - \tau)$, where $\bar{d} = C_{max}(1 - \tau)$ is the median of the due dates and C_{max} is computed according to $C_{max} = n/m(\bar{p} + \bar{S}(0.4 + 10m^2/n^2 - \eta/7))$ where $\bar{S}$ is the average of the setup times. The parameters $\tau = 0.3$ and $R = 0.25$ are the lateness factor and the due date dispersion, respectively. Release times are generated in the interval $[0, (0.1/m) \times \sum p_{ji}/m]$ [17].

5.2   Objectives and Dispatch Rules

A study on the objective functions commonly adopted in the scheduling literature is presented in [19]. That study also includes an analysis of the relationship between pairs of those objectives and indicates the more conflicting ones, that is, those that should form multiobjective scheduling optimization problems. Correlation plots, parallel coordinates graphs, and aggregation trees were used in the study presented in [19]. Figure 4 displays a plot where the colored cells represent the correlation values between the pairs of objectives from -1 to 1. In that plot, blue, white and brown correspond, respectively, to the correlation values 1, 0, and -1; other values are combination of these colors. The 16 pairs of objective functions used here in the computational experiments are marked by an "X" and stand for (i) bi-objective situations usually tackled in the literature, or (ii) cases in which the correlation is negative or close to zero.

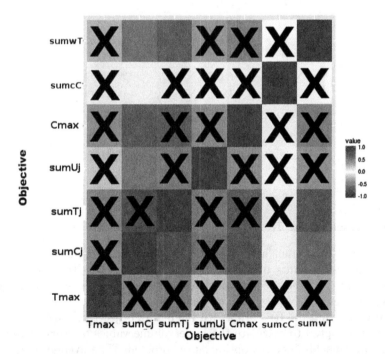

Fig. 4. Correlation between objective function values obtained in a environment formed by unrelated parallel machines with $n = 100$ and $m = 8$. (Color figure online)

The PP techniques assume that the predators are to be equipped with dispatch rules. As the current computational experiments comprise many more objectives than those used in [10], then additional rules are necessary here. The following dispatch rules were analysed: SPT, longest processing time (LPT), EDD, earliest release date first (ERD), MST, WSPT, and weighted longest processing time (WLPT). For each pair of objectives, the dispatch rule that produced the best results was identified. This analysis can be found in [19] and Table 6 presents a summary of the chosen rules.

Table 6. Dispatch rules chosen here as adequate to each pair of objectives (1 and 2).

Objective 1	Rule	Objective 2	Rule
T_{max}	MST	$\sum C_j$	ERD
T_{max}	MST	$\sum T_j$	EDD
T_{max}	MST	$\sum U_j$	EDD
T_{max}	MST	C_{max}	ERD
T_{max}	MST	$\sum c_j C_j$	ERD
T_{max}	MST	$\sum w_j T_j$	$WSPT$
$\sum C_j$	ERD	$\sum T_j$	EDD
$\sum T_j$	EDD	$\sum U_j$	MST
$\sum T_j$	EDD	C_{max}	ERD
$\sum T_j$	EDD	$\sum c_j C_j$	ERD
$\sum U_j$	EDD	C_{max}	ERD
$\sum U_j$	EDD	$\sum c_j C_j$	ERD
$\sum U_j$	EDD	$\sum w_j T_j$	$WSPT$
C_{max}	ERD	$\sum c_j C_j$	SPT
C_{max}	ERD	$\sum w_j T_j$	$WSPT$
$\sum c_j C_j$	ERD	$\sum w_j T_j$	$WSPT$

5.3 Analyses of the Results

Hypervolume [29] is used here in order to compare the quality of the results obtained by the variants of the proposed PP and NSGA-II techniques. The reference point (RP) required to calculate the hypervolume was determined from the worst values achieved by each objective value in all the independent runs. One can notice that the RP varies with the pair of objectives considered but the same RP is adopted to calculate the hypervolume values of every Pareto front found in a given bi-objective optimization problem. The average value of the hypervolumes obtained in the independent runs is used in order to compare the performance of the techniques.

As the current computational experiment is composed by 16 bi-objective optimization problems, performance profiles [7] are adopted to present the relative performance of the compared methods. Performance profiles are used to facilitate the comparative analyses and visualization when one has a large amount of data. We use the average of the hypervolume values as the performance measure and $\rho_a(\tau)$ represents the percentage of the problems which are solved by the algorithm a with a relative performance up to τ times the best performance. Thus, it is possible to verify that [3]: (i) the technique with the largest value of $\rho_a(1)$ achieves the best results in more problems, (ii) the method with the smallest value of τ in which $\rho_a(\tau) = 1$ is the most robust one, and (iii) the normalized area under the performance profiles curves (NAUC) is an indicator of the overall performance.

Figure 5(b) presents the performance profiles for $\tau \in [1; 1.05]$, where one can observe that the variant PP-4 obtained the largest $\rho(1)$ value, indicating that it is the technique achieving the top performance in most problems. This variant achieved the best result in about 60% of the problems. Also, it is possible to verify that the variant PP-10 obtained the best results in about 40% of the problems, while both NSGA-II variants found the best results in less than 10% of the problems.

The robustness of the techniques can be analyzed via Fig. 5(a), where the performance profiles are presented for $\tau \in [1, \tau_{max}]$ and τ_{max} corresponds to the worst relative performance observed. The variant PP-4 is also the most robust one, achieving the smallest value of τ such that $\rho(\tau) = 1$. In addition, one can see that the variant PP-10 has the second best value of τ, and that the PP variants are much more robust that the NSGA-II ones.

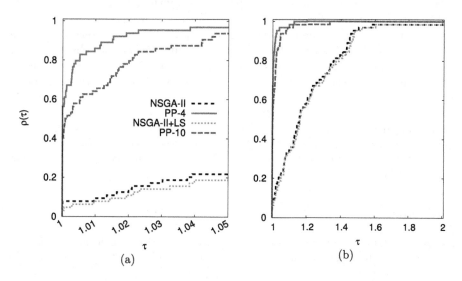

Fig. 5. Performance profiles, where $\tau \in [1; 1.05]$ in (a), and $\tau \in [1; \tau_{max}]$ in (b).

Finally, the normalized areas under the performance profiles curves (NAUC) are presented in Table 7, reinforcing that the proposed PP variants performed better than the NSGA-II ones. In addition, the best overall performance is obtained by the variant PP-4, namely, the proposed PP but with 4 predators, which is the same number of predators used in [10].

Table 7. Normalized areas under the performance profiles curves.

Method	PP-4	PP-10	NSGA-II	NSGA-II+LS
NAUC	**1.0000**	0.9890	0.7997	0.7903

The objectives $\sum T_j$ and C_{max} are often studied. For those objectives, PP-4 presented the best average in all cases. With respect to the objectives T_{max} and C_{max}, the variant PP-10 produced the best results in all cases.

6   Concluding Remarks and Future Work

Scheduling problems are very important in practical situations and the class of unrelated parallel machines is studied here. Sixteen bi-objective optimization cases are considered and a predator-prey (PP) technique is proposed to tackle them. In order to handle this large variety of objectives, additional dispatch rules were studied and the more adequate ones were incorporated into PP. Also, the new ideas include a modified mutation operator and a more general formulation for the selection/replacement of preys performed by the predators.

The results obtained by PP are better than those achieved by NSGA-II with respect to the average of the hypervolume values. Also, it was observed that the proposed PP with 4 predators found the best results in most of the problems, obtained the most robust solutions and reached the best overall performance.

As future work, the use of tools, such as aggregation trees, to allow for a principled consideration of a large number of objectives seems to be promising.

Acknowledgments. The authors would like to thank the reviewers for their comments, which helped to improve the quality of the final version, and the support provided by CNPq (grant 310778/2013-1), FAPEMIG (grants APQ-03414-15), and PPGMC/UFJF.

References

1. Arroyo, J.E.C., de Souza Pereira, A.A.: A GRASP heuristic for the multi-objective permutation flowshop scheduling problem. Intl. J. Adv. Manufact. Technol. **55**(5–8), 741–753 (2011)
2. Bank, J., Werner, F.: Heuristic algorithms for unrelated parallel machine scheduling with a common due date, release dates, and linear earliness and tardiness penalties. Math. Comput. Model. **33**(4), 363–383 (2001)

3. Barbosa, H.J.C., Bernardino, H.S., Barreto, A.M.S.: Using performance profiles to analyze the results of the 2006 CEC constrained optimization competition. In: Proceedings of the IEEE World Congress on Computational Intelligence (2010)
4. de CM Nogueira, J.P., Arroyo, J.E.C., Villadiego, H.M.M., Gonçalves, L.B.: Hybrid grasp heuristics to solve an unrelated parallel machine scheduling problem with earliness and tardiness penalties. Electron. Notes Theor. Comput. Sci. **302**, 53–72 (2014)
5. Deb, K.: Multi-objective optimization using evolutionary algorithms. Wiley- Interscience Series in Systems and Optimization. Wiley, New York (2001)
6. Deb, K., Agrawal, S., Pratap, A., Meyarivan, T.: A fast elitist non-dominated sorting genetic algorithm for multi-objective optimization: NSGA-II. In: Schoenauer, M., Deb, K., Rudolph, G., Yao, X., Lutton, E., Merelo, J.J., Schwefel, H.-P. (eds.) PPSN 2000. LNCS, vol. 1917, pp. 849–858. Springer, Heidelberg (2000). doi:10. 1007/3-540-45356-3_83
7. Dolan, E.D., Moré, J.J.: Benchmarking optimization software with performance profiles. Math. Program. **91**(2), 201–213 (2002)
8. Etcheverry, G.V., Anzanello, M.A.: Sequenciamento em máquinas paralelas com tempos de *setup* dependentes da sequência. In: XXXIII Encontro Nacional de Engenharia de Produção (2013)
9. Fang, K.T., Lin, B.M.: Parallel-machine scheduling to minimize tardiness penalty and power cost. Comput. Ind. Eng. **64**(1), 224–234 (2013)
10. Grimme, C., Lepping, J., Schwiegelshohn, U.: Multi-criteria scheduling: an agent-based approach for expert knowledge integration. J. Sched. **16**(4), 369–383 (2013)
11. Hoogeveen, H.: Multicriteria scheduling. Eur. J. Oper. Res. **167**(3), 592–623 (2005)
12. Kayvanfar, V., Aalaei, A., Hosseininia, M., Rajabi, M.: Unrelated parallel machine scheduling problem with sequence dependent setup times. In: International Conference on Industrial Engineering and Operations Management, Bali, pp. 7–9 (2014)
13. Laumanns, M., Rudolph, G., Schwefel, H.-P.: A spatial predator-prey approach to multi-objective optimization: a preliminary study. In: Eiben, A.E., Bäck, T., Schoenauer, M., Schwefel, H.-P. (eds.) PPSN 1998. LNCS, vol. 1498, pp. 241–249. Springer, Heidelberg (1998). doi:10.1007/BFb0056867
14. Lee, Y.H., Pinedo, M.: Scheduling jobs on parallel machines with sequencedependent setup times. Eur. J. Oper. Res. **100**(3), 464–474 (1997)
15. Lei, D.: Multi-objective production scheduling: a survey. Intl. J. Adv. Manuf. Technol. **43**(9–10), 926–938 (2009)
16. Lin, B., Wu, J.: Bicriteria scheduling in a two-machine permutation flowshop. Intl. J. Prod. Res. **44**(12), 2299–2312 (2006)
17. Lin, Y.K.: Particle swarm optimization algorithm for unrelated parallel machine scheduling with release dates. Mathematical Problems in Engineering (2013)
18. MacCarthy, B., Liu, J.: Addressing the gap in scheduling research: a review of optimization and heuristic methods in production scheduling. Intl. J. Prod. Res. **31**(1), 59–79 (1993)
19. de Souza Pereira, A.A.: Análise de objetivos e meta-heurísticas para problemas multiobjetivo de sequenciamento da Produção. Ph.D. thesis. Universidade Federal de Juiz de Fora, Juiz de Fora, Brasil (2016)
20. de Souza Pereira, A.A., Bernardino, H.S., Barbosa, H.J.C.: Comparação entre abordagens multiobjetivo e mono-objetivo para o problema de programação em máquinas paralelas não relacionadas com tempos de preparação dependentes da sequência e datas de liberação. In: XXI Simpósio de Engenharia de Produção (2014)

21. Pfund, M., Fowler, J.W., Gupta, J.N.: A survey of algorithms for single and multiobjective unrelated parallel-machine deterministic scheduling problems. J. Chin. Inst. Ind. Eng. **21**(3), 230–241 (2004)
22. Pinedo, M.L.: Scheduling: Theory, Algorithms, and Systems, 4th edn. Springer Publishing Company, Incorporated, Heidelberg (2012)
23. Ruiz, R., Şerifoğlu, F.S., Urlings, T.: Modeling realistic hybrid flexible flowshop scheduling problems. Comput. Oper. Res. **35**(4), 1151–1175 (2008)
24. Ruiz, R., Stützle, T.: A simple and effective iterated greedy algorithm for the permutation flowshop scheduling problem. Eur. J. Oper. Res. **177**(3), 2033–2049 (2007)
25. Suresh, V., Chaudhuri, D.: Bicriteria scheduling problem for unrelated parallel machines. Comput. Ind. Eng. **30**(1), 77–82 (1996)
26. T'kindt, V., Billaut, J.C., Proust, C.: Solving a bicriteria scheduling problem on unrelated parallel machines occurring in the glass bottle industry. Eur. J. Oper. Res. **135**(1), 42–49 (2001)
27. Torabi, S.A., Sahebjamnia, N., Mansouri, S.A., Bajestani, M.A.: A particle swarm optimization for a fuzzy multi-objective unrelated parallel machines scheduling problem. Appl. Soft Comput. **13**(12), 4750–4762 (2013)
28. Vallada, E., Ruiz, R.: A genetic algorithm for the unrelated parallel machine scheduling problem with sequence dependent setup times. Eur. J. Oper. Res. **211**(3), 612–622 (2011)
29. Zitzler, E., Thiele, L.: Multiobjective evolutionary algorithms: a comparative case study and the strength pareto approach. IEEE Trans. Evol. Comput. **3**(4), 257–271 (1999)

An Overview of Weighted and Unconstrained Scalarizing Functions

Miriam Pescador-Rojas[1], Raquel Hernández Gómez[2(✉)], Elizabeth Montero[3], Nicolás Rojas-Morales[3], María-Cristina Riff[3], and Carlos A. Coello Coello[2]

[1] ESCOM, Instituto Politécnico Nacional, 07738 Ciudad de México, Mexico
[2] Computer Science Department, CINVESTAV-IPN (Evolutionary Computation Group), 07360 Ciudad de México, Mexico
rhernandez@computacion.cs.cinvestav.mx
[3] Universidad Técnica Federico Santa María, Avenida España, 1680 Valparaíso, Chile

Abstract. Scalarizing functions play a crucial role in multi-objective evolutionary algorithms (MOEAs) based on decomposition and the R2 indicator, since they guide the population towards nearly optimal solutions, assigning a fitness value to an individual according to a predefined target direction in objective space. This paper presents a general review of weighted scalarizing functions without constraints, which have been proposed not only within evolutionary multi-objective optimization but also in the mathematical programming literature. We also investigate their scalability up to 10 objectives, using the test problems of Lamé Superspheres on the MOEA/D and MOMBI-II frameworks. For this purpose, the best suited scalarizing functions and their model parameters are determined through the evolutionary calibrator EVOCA. Our experimental results reveal that some of these scalarizing functions are quite robust and suitable for handling many-objective optimization problems.

Keywords: Scalarizing function · Many-objective optimization · Evolutionary Algorithms · Tuning process

1 Introduction

Multi-Objective Evolutionary Algorithms (MOEAs) mostly rely on three methods for tackling Multi-objective Optimization Problems (MOPs): *Pareto dominance*[1], aggregation, and performance indicators. MOEAs based on Pareto dominance compare individuals preferring those that are less dominated by other members in the population. When a tie occurs, a secondary selection criterion is applied to improve *diversity*, i.e., the uniform distribution of solutions covering all regions in objective space. Aggregation-based MOEAs decompose MOPs into

---

[1] A solution $\mathbf{x} \in \mathcal{S}$ dominates a solution $\mathbf{y} \in \mathcal{S}$ ($\mathbf{x} \prec \mathbf{y}$), if and only if $\forall i \in \{1, \ldots, m\}$, $f_i(\mathbf{x}) \leq f_i(\mathbf{y})$ and $\exists j \in \{1, \ldots, m\}$, $f_j(\mathbf{x}) < f_j(\mathbf{y})$.

© Springer International Publishing AG 2017
H. Trautmann et al. (Eds.): EMO 2017, LNCS 10173, pp. 499–513, 2017.
DOI: 10.1007/978-3-319-54157-0_34

several single-objective subproblems, each one associated with a different target direction or *weight vector*. The best individuals for every subproblem have a better chance to survive. Indicator-based MOEAs favor solutions that highly contribute to a performance indicator, which reflects a quality aspect regarding convergence and diversity of the current population. In this case, *convergence* measures how close the solutions are to the *Pareto Optimal Front (POF)*[2]. The hypervolume [1] and the unary $R2$ [2] indicators are examples of quality measures that incorporate both aspects, and they are also Pareto compliant.

One of the main concerns is that Pareto-based MOEAs are ineffective in MOPs having four or more objective functions, the so-called many-objective optimization problems (MaOPs) [3]. Owing to the fact that at early generations, most individuals will normally become non-dominated, and the search will then be guided solely by the secondary selection criterion. In consequence, individuals that enhance diversity are kept although they may deteriorate convergence [4]. The hypervolume indicator does not suffer from this selection pressure issue since its maximization is equivalent to reaching the Pareto Optimal Set (POS) [5]. However, its computational cost increases exponentially with the number of objectives [6], making it unaffordable for MaOPs.

On the other hand, the methods based on aggregation and the $R2$ indicator can scale to any number of objectives while having a low computational cost. Nevertheless, two key points should be considered when using such approaches: the setting of the scalarizing function and the weight vectors. A *scalarizing function*, also known as utility function or aggregation function, transforms the original MOP into a real value using a predefined weight vector in objective space. Regarding the choice of the weight vectors, several attempts have been proposed for adapting them in order to get a high-quality approximation of the POF (see for example [7,8]). However, not much is known about the proper selection of the scalarizing function and its effect on MOEAs. In fact, only three aggregation functions have been exhaustively researched so far [9–11], neglecting other approaches that may be able to handle MaOPs.

In this paper, we present a general review of fifteen weighted unconstrained scalarizing functions. Even though there are excellent reviews on this topic [12–15], the most comprehensive study dates from 1998 [13], while the most recent ones only consider a small set of utility functions. Furthermore, none of these studies analyzes their ability to scale to any number of objectives.

This paper is organized as follows. In Sect. 2 we present a review of unconstrained scalarizing functions. In Sect. 3, the best scalarizing functions and their model parameters are determined through the Evolutionary Calibrator (EVOCA) [16], using both the Multiobjective Evolutionary Algorithm Based on Decomposition (MOEA/D) [17] and the Many-Objective Metaheuristic Based on the $R2$ Indicator II (MOMBI-II) [18]. Finally, our conclusions and some possible paths for future research are provided in Sect. 4.

---

[2] $POF := \{\mathbf{F}(\mathbf{x}) \in \mathbb{R}^m : \mathbf{x} \in \mathcal{S}, \nexists \mathbf{y} \in \mathcal{S}, \mathbf{y} \prec \mathbf{x}\}.$

2   Scalarizing Functions

We focus on unconstrained scalarizing functions that transform a multi-objective optimization problem (MOP) into a single-objective problem, using a predefined weight vector $\mathbf{w} := (w_1, \ldots, w_m)$ of the following form:

$$\text{minimize} \quad u(\mathbf{f}'(\mathbf{x}); \mathbf{w}) \tag{1}$$

$$\text{subject to} \quad \mathbf{x} \in \mathcal{S}, \tag{2}$$

where $\mathbf{x}$ is the decision vector, $\mathcal{S} \in \mathbb{R}^n$ is the feasible space, $\mathbf{f} \in \mathbb{R}^m$ is the vector of $m(\geq 2)$ objective functions, $\mathbf{f}'(\mathbf{x}) := \mathbf{f}(\mathbf{x}) - \mathbf{z}$ and $\mathbf{z} := (z_1, \ldots, z_m)^T$ is a reference point. Unless otherwise stated we used as reference point the ideal point, i.e., $z_i := \min\{f_i(\mathbf{x}) \mid \mathbf{x} \in \mathcal{S}\} \; \forall i \in \{1, \ldots, m\}$. Each component of $\mathbf{w}$ must satisfy $w_i \geq 0$. Although there is no particular reason beyond achieving uniformity among solutions, we also assume that $\sum_i w_i = 1$ [15].

In the remainder of this section, we present a review of fifteen aggregation functions. They differ in that they minimize some sort of a distance metric to the reference point, others combine two distance metrics, and a minority of them also consider the deviation to the weight vector. In all cases, their computational complexity is $O(m)$. Several of these scalarizing functions can generate (weakly) Pareto optimal solutions.[3] Moreover, interactive methods have coupled some of these scalarizing functions, where the reference point reflects the decision maker's preferences [13, p. 131], [14, 19–21]. Let $p \in \mathbb{N}_+$, $\alpha \in \mathbb{R}_+$ and $\theta \in \mathbb{R}$ be the model parameters. The dot product is symbolized as •, the absolute value of a real number is denoted by $|\cdot|$, and the magnitude of a vector is represented by $\|\cdot\|$.

The **Weighted Compromise Programming (WCP)** [22] is derived from the *Global Criterion Method* [23, p. 32], which includes the weight vector for modeling preferences as follows:

$$u^{wcp}(\mathbf{f}'; \mathbf{w}) := \sum_i (w_i f_i')^p. \tag{3}$$

A high value of $p \in (1, \infty)$ is preferred to obtain the complete POS [24]. In [25], the authors recommend using odd values for this parameter and coupled WCP ($p = 9$) with a metaheuristic for solving convex and concave POFs with 2 and 3 objectives.

The **Weighted Sum (WS)** [26] is one of the most commonly used scalarizing functions, which linearly combines the objectives as follows:

$$u^{ws}(\mathbf{f}'; \mathbf{w}) := \sum_i w_i f_i'. \tag{4}$$

However, WS cannot generate solutions in concave regions of the POF [25]. Some attempts have been proposed to alleviate this drawback, such as its combination with other scalarizing functions [27, 29], the use of dynamic weights coupled with

---

[3] Let be $\mathbf{x}, \mathbf{y} \in \mathcal{S}$. It is said that $\mathbf{x}$ is Pareto optimal if there is no $\mathbf{y}$ such that $\mathbf{y} \prec \mathbf{x}$. $\mathbf{x}$ is weakly Pareto optimal if there is no $\mathbf{y}$ such that $\forall i \in \{1, \ldots, m\}$, $f_i(\mathbf{y}) < f_i(\mathbf{x})$.

a secondary population [30], and the use of WS as a local search engine [31,32]. Additionally, some studies have reported that WS is an effective method for solving MaOPs [9,33]. This scalarizing function has been integrated into several evolutionary algorithms (see e.g., [17,34]).

The **Exponential Weighted Criteria (EWC)** [24] can deal with any Pareto-front shape, and is given by:

$$u^{ewc}(\mathbf{f}';\mathbf{w}) := \sum_i \left(e^{p\,w_i} - 1\right) e^{p\,f_i'}. \tag{5}$$

A large value of p is required to achieve Pareto optimality, but this can lead to numerical overflow [12]. In [35], EWC was used to solve a problem related to a voltage distribution network. To the best of our knowledge, this scalarizing function has not been integrated into any MOEA until now.

The **Weighted Power (WPO)** [36] relies on the principle that the POF can be convexified as follows:

$$u^{wpo}(\mathbf{f}';\mathbf{w}) := \sum_i w_i \left(f_i'\right)^p. \tag{6}$$

For a suitable value of $p \in [1,\infty)$, this scalarizing function can also find optimal solutions in concave POFs [12,37]. In [38], WPO was coupled with a genetic algorithm, where the weight vectors and the exponent p were updated during the evolution process according to predefined rules.

The **Weighted Product (WPR)** [39, p. 9], also called *product of powers*, is defined as follows:

$$u^{wpr}(\mathbf{f}';\mathbf{w}) := \prod_i \left(f_i'\right)^{w_i}. \tag{7}$$

This scalarizing function has been integrated with several ant colony optimization algorithms [40] and it has also been applied to solve a network design problem [41]. However, this approach has not been widely used with MOEAs.

The **Weighted Norm (WN)** [22], or *weighted L_p-metrics*, is given by:

$$u^{wn}(\mathbf{f}';\mathbf{w}) := \left(\sum_i w_i \,|f_i'|^p\right)^{\frac{1}{p}}. \tag{8}$$

WN can generate Pareto optimal solutions when $p \in [1,\infty)$ [13, p. 98]. Moreover, other scalarizing functions can be derived from it, such as WS with $p = 1$ and Least Squares [13, p. 97], [42] with $p = 2$. Moreover, concave POFs can be covered with a larger value of p. In [43], the behavior of WN was studied on MOEA/D, using continuous test problems having up to 7 objectives. The value of p was adaptively fine-tuned based on a local estimation of the Pareto front shape, taking different values from the set $\{1/2, 2/3, 1, 2, 3, \ldots, 10, 1000\}$.

The **Chebyshev or Tchebycheff function (TCH)**[4] [44], also known as the weighted min-max [12,28], is a particular case of WN with $p = \infty$, given by:

$$u^{tch}(\mathbf{f}';\mathbf{w}) := \max_i \left\{w_i \,|f_i'|\right\}. \tag{9}$$

---

[4] Although the French spelling Tchebycheff is the most preferred, the proper English transliteration is Chebyshev.

This scalarizing function can find at least weakly Pareto optimal solutions regardless of the Pareto-front shape (whenever $\mathbf{w} \in \mathbb{R}_+^m$) [13, p. 98]. By definition, f_i' is guaranteed to be always positive. Thus, the absolute value could be discarded. Even though it is possible to obtain all Pareto optimal solutions for some weight vector [13, p. 99], a recent analysis has revealed that the search ability of TCH is equivalent to Pareto-based methods [45]. Thus, in MaOPs, the probability to obtain non-dominated solutions using TCH is lower than the WN $(0 < p < \infty)$ and equivalent to Pareto-based methods. TCH has been applied to different MOEAs (see, e.g., [17,46]).

The **Augmented Chebyshev (ATCH)** [19] is a variant of TCH, where an extra term is considered in order to avoid the generation of weakly Pareto optimal solutions. It is given by:

$$u^{atch}(\mathbf{f}'; \mathbf{w}) := \max_i \left\{ w_i \, |f_i'| \right\} + \alpha \sum_i |f_i'| . \tag{10}$$

A too small value of α may result in a loss of significance of the additional term, still leading to the generation of weakly Pareto optimal solutions. However, a too large value of this parameter may cause that some non-dominated points become unreachable [47]. Although the recommendation is to use small values of α, such as $[0.001, 0.01]$ [19], some studies have shown a better performance when using large values, revealing a high sensitivity of this parameter on discrete MaOPs [27].

The **Modified Chebyshev (MTCH)** [48] is a variation of TCH, given by:

$$u^{mtch}(\mathbf{f}'; \mathbf{w}) := \max_i \left\{ w_i \left(|f_i'| + \alpha \sum_i |f_i'| \right) \right\} . \tag{11}$$

α should be a small positive value. The features of this method are discussed and illustrated in [13, p. 101]. To the best of our knowledge, this scalarizing function has not been exploited in any MOEA.

The **Achievement Scalarizing Function (ASF)** [49] can produce weakly Pareto optimal solutions, expressed as:

$$u^{asf}(\mathbf{f}; \mathbf{w}) := \max \left\{ \frac{f_i'}{w_i} \right\} . \tag{12}$$

Although this scalarizing function is similar to TCH, ASF can find an objective vector parallel to $\mathbf{w}$, improving diversity in MaOPs [18]. ASF has been employed in some MOEAs (e.g., [18,50,51]).

The **Augmented Achievement Scalarizing Function (AASF)** [13, p. 111] is a variation of ASF, where an extra term is considered for discarding weakly Pareto optimal solutions, and is defined as:

$$u^{aasf}(\mathbf{f}'; \mathbf{w}) := \max \left\{ \frac{f_i'}{w_i} \right\} + \alpha \sum_i \frac{f_i'}{w_i} . \tag{13}$$

α should take small values. In [8], it was recommended to set $\alpha \approx 10^{-4}$. There are few MOEAs that adopt this scalarizing function [8,20,52].

The **Penalty Boundary Intersection (PBI)** [17] draws ideas from the Normal-Boundary Intersection (NBI) method [53], defined as follows:

$$u^{pbi}(\mathbf{f}';\mathbf{w}) \; := \; d_1 + \theta d_2$$

$$\text{where} \quad d_1 := \left| \mathbf{f}' \cdot \frac{\mathbf{w}}{\|\mathbf{w}\|} \right| \quad \text{and} \quad d_2 := \left\| \mathbf{f}' - d_1 \frac{\mathbf{w}}{\|\mathbf{w}\|} \right\|. \tag{14}$$

θ is a penalty parameter that balances convergence (measured by d_1) and diversity (measured by d_2), both should be minimized [54]. d_2 can be seen as the distance between $\mathbf{f}'$ and its orthogonal projection on $\mathbf{w}$. When $\theta = 0$, the behavior of PBI is similar to WS. The PBI function can produce uniformly distributed solutions in objective space by setting an appropriate value for θ. Several options for setting θ have been studied considering the geometry of the POF and the number of objectives [9,54]. Recently, some attempts have been proposed for the adaptation of this parameter in MOEA/D [54].

The **Inverted Penalty Boundary Intersection (IPBI)** [55] is an extension of PBI, given by:

$$u^{ipbi}(\mathbf{f}';\mathbf{w}) \; := \; \theta d_2 - d_1$$

$$\text{where} \quad d_1 := \left| \mathbf{f}'' \cdot \frac{\mathbf{w}}{\|\mathbf{w}\|} \right| \quad \text{and} \quad d_2 := \left\| \mathbf{f}'' - d_1 \frac{\mathbf{w}}{\|\mathbf{w}\|} \right\|,$$

where $\mathbf{f}''$ is defined as:

$$\mathbf{f}''(\mathbf{x}) := \mathbf{z}^* - \mathbf{f}(\mathbf{x}), \tag{15}$$

and $\mathbf{z}^* := (z_1^*, \dots, z_m^*)^T$ is the nadir point, i.e., $z_i^* := \max\{f_i(\mathbf{x}) \mid \mathbf{x} \in POS\}$. The aim of IPBI is to enhance the spread of solutions in objective space and to improve the performance in MaOPs [55]. As in PBI, θ handles the balance between d_1 and d_2. However, a solution having a large d_1 and a small d_2 is considered as a better solution. In [11,55], a set of different values for θ were tested in MOEA/D for solving MaOPs in discrete and continuous spaces.

The **Conic Scalarization (CS)** [56] is a variant of WS, where an extra term is added for dealing with concave regions of the POF. It is defined as:

$$u^{cs}(\mathbf{f}';\mathbf{w}) := \sum_i w_i f_i' + \alpha \sum_i |f_i'|. \tag{16}$$

CS can generate weakly Pareto optimal solutions if $\alpha \in [0, w_i]$, $w_i > 0$ for all $i \in \{1, \dots, m\}$ and there exists $k \in \{1, \dots, m\}$ such that $w_k > \alpha$. There are few MOEAs that adopt CS (e.g., [57]).

The **Vector Angle Distance Scaling (VADS)** [28] can discover solutions in concavities that may appear as discontinuities in the POF, given by:

$$u^{vads}(\mathbf{f}';\mathbf{w}) := \frac{\|\mathbf{f}'\|}{\left(\frac{\mathbf{w}}{\|\mathbf{w}\|} \cdot \frac{\mathbf{f}'}{\|\mathbf{f}'\|} \right)^p}. \tag{17}$$

Here, the numerator measures convergence, whereas the denominator measures the deviation of the objective vector from the weight vector. Thus, the final solution should be lying parallel to $\mathbf{w}$. Orthogonal vectors require special care. Small values of p hinder the search of sharp concavities. Authors in [28] recommend to use $p = 100$. This scalarizing function is not compatible with any form of Pareto optimality. VADS has been implemented on a MOEA in combination with TCH [28,58].

3   Experimental Methodology

The goals of the experiments reported next are twofold: first, we want to determine which are the most suitable scalarizing functions for MOEA/D and MOMBI-II on different scenarios. These scenarios were built considering the Pareto front geometry and the number of objective functions. Second, we want to verify if this robust set of scalarizing functions can significantly improve the performance of the baseline versions of these MOEAs. To achieve the first target, we rely on a powerful tuning tool called Evolutionary Calibrator (EVOCA) [16], which can find good values for numerical and categorical parameters without requiring a strong knowledge of parameter tuning methods. The second goal is tackled by performing a comparative study using Nonparametric statistics. In the remainder of this section, we describe the test problems adopted for our experimental study, as well as the parameters settings of the MOEAs and further details of these experiments.

3.1   Test Problems and Parameters Settings

We selected the Lamé Superspheres test problems [59] since they encompass the three basic Pareto front geometries in which the scalarizing functions present challenges. Moreover, this benchmark is scalable to any number of variables and objective functions. Hence, we tested for $2, 3, 5, 7$ and 10 objectives (m). The number of decision variables was set to $n = m + 4$. We fixed the parameter $\gamma \in \{0.5, 1.0, 2.0\}$ to achieve Pareto fronts with convex, linear and concave geometries, respectively. Only unimodal problems were considered, since our aim was to determine if the scalarizing functions can handle different shapes of the Pareto front for multi- and many-objective problems. Thus, adding difficulties in the MOPs would introduce noise to the selection process of a MOEA.

The common parameters of MOEA/D and MOMBI-II adopted the same values. The population size was set to $\{100, 120, 196, 210, 276\}$ individuals for $\{2, 3, 5, 7, 10\}$ objectives, respectively. The probabilities of Polynomial-based mutation and Simulated Binary Crossover (SBX) were set to $\frac{1}{n}$ and 0.9, respectively; in both cases, the distribution index was set to 20. The weight vectors were generated using the method described in [53], looking for a cardinality analogous to the population size. The stopping criterion consisted of reaching a maximum number of 50,000 evaluations of the MOP. The neighborhood size in MOEA/D was equal to 20. The parameter values employed for MOMBI-II were: record 5, tolerance threshold 1×10^{-3} and 0.5 for the variance threshold.

Our performance indicator was the *normalized hypervolume*, which is the hypervolume indicator [1] divided by the factor $\prod_i r_i$, where $\mathbf{r} := (r_1, \dots, r_m)^T$ is a reference point, set to $(2, \dots, 2)^T$ in all our test problems.

3.2   Tuning Process

EVOCA [16] is a genetic algorithm whose aim is to find the parameter values of an optimizer that maximizes the profit for a given budget. In our case, the profit was the normalized hypervolume indicator, and the budget was limited to 10,000 executions of the optimizer.

An individual represents a calibration that involves one of the 15 scalarizing functions of Sect. 2, and the set of all the model parameters. To achieve accurate results, we consider that all the model parameters are real values. Based on our literature review, we selected the ranges: $p \in [0.1, 10.0]$ for WCP, EWC, WPO, WN and VADS; $\alpha \in [0.0001, 1.0]$ for ATCH, MTCH, AASF and CS; $\theta \in [0.1, 50.0]$ for PBI and IPBI. Our main interest was to identify the scalarizing function(s) that solve a wide variety of test instances, or at least to know in which instances these scalarizing functions perform well. For this reason, we designed the following scenarios:

- Six scenarios that considered all the combinations of the cartesian product between the Pareto-front geometries, given by $\gamma \in \{0.5, 1.0, 2.0\}$, and the {multi, many} cases. The "multi" case groups 2 and 3 objectives, whereas the "many" considers 5, 7 and 10 objectives.
- A global scenario which includes all the previous combinations.

The tuning process was independent among scenarios and between MOEAs. At the end of a tuning process, EVOCA returned 20 individuals, corresponding to the best-found calibrations. We filtered these calibrations using Wilcoxon's test (with a confidence level of 99%), executing 30 times each algorithm. The best-ranked calibrations are presented in Table 1.

Table 1. EVOCA's recommendation for each possible scenario. In every calibration, it is shown the scalarization function and its model parameter value (in parentheses).

γ	MOEA/D		MOMBI-II	
	Multi-objective	Many-objective	Multi-objective	Many-objective
0.5	AASF (0.8001)	EWC (7.6)	EWC (8.6)	TCH
			AASF (1.0)	
1.0	PBI (15.1)	AASF (0.3001)	PBI (21.4123)	AASF (0.3001)
	PBI (20.5639)			
2.0	PBI (2.6)	PBI (2.6)	PBI (2.4026)	VADS (5.6)
	PBI (7.6)			
	VADS (8.6)			VADS (2.1)
Global	ASF	AASF (0.2423)	AASF (0.0501)	ASF
	TCH		TCH	

In the multi-objective case, more than one calibration was obtained, but in the many-objective case, we found only one recommended scalarizing function.

To summarize, EVOCA found that 6 out of 15 scalarizing functions (EWC, TCH, ASF, AASF, PBI, and VADS) had an outstanding performance in the particular scenarios that we studied. In the global scenario, EVOCA determined that TCH, ASF, and AASF had the best results, remarking that ASF forms part of the original version of MOMBI-II, while TCH is used by MOEA/D for 2 objectives. Furthermore, the best options to solve problems with convex shape ($\gamma = 0.5$) were AASF, EWC, and TCH. For linear shapes ($\gamma = 1.0$) PBI and AASF. For convex shapes ($\gamma = 2.0$) the best choices were PBI and VADS. Finally, the tuning process obtained a greater accuracy for the corresponding model parameters than that provided by the values recommended in the literature.

3.3 Comparative Study

In this subsection, we examine the performance of the calibrated versions given by EVOCA using the normalized hypervolume indicator (see Sect. 3.1). We performed 30 independent runs for all scenarios. We applied the Wilcoxon rank sum test (one-tailed) to the median of this indicator, to determine whether the calibrated version performed better than the baseline algorithm at a confidence interval of 99%. The baseline version adopted in MOEA/D was TCH for 2 objectives and PBI with $\theta = 5$ for the remaining objectives [17,60]. The baseline of MOMBI-II was ASF [18]. Both MOEAs and scalarizing functions were implemented in EMO Project,[5] a framework for Evolutionary Multi-Objective Optimization. This software is implemented in C language.

For all scenarios, our experimental results are shown in Tables 2, 3 and 4. The best value between the calibrated versions and the baseline MOEA is shown in gray. A line above the median ($\bar{\cdot}$) implies that the calibrated version outperformed in a significant way the baseline algorithm. Conversely, a line under the median ($\underset{\cdot}{}$) means that the calibrated version was significantly outperformed.

Table 2. Median ($\times 10^{-1}$) and standard deviation of the normalized hypervolume indicator on the **multi-objective scenarios**.

γ	Config.	MOEA/D m 2	MOEA/D m 3	Config.	MOMBI-II m 2	MOMBI-II m 3
0.5	Baseline	9.570400 5.0e-07	9.906204 3.4e-04	Baseline	9.570404 6.0e-07	9.917509 1.2e-04
	AASF (0.8001)	9.573699 4.5e-08	9.974616 1.2e-05	EWC (8.6)	9.575178 5.4e-08	9.902034 6.8e-04
				AASF (1.0)	9.573940 9.4e-08	9.975483 1.3e-05
1.0	Baseline	8.737370 8.2e-08	9.744895 2.5e-07	Baseline	8.737369 8.0e-08	9.636011 6.3e-04
	PBI (15.1)	8.737374 4.5e-08	9.744894 7.0e-07	PBI (21.4123)	8.737374 4.5e-07	9.744881 1.8e-06
	PBI (20.5639)	8.737374 3.0e-07	9.744894 1.0e-06			
2.0	Baseline	8.025324 1.3e-07	9.277872 1.6e-06	Baseline	8.025323 9.6e-08	9.209490 1.0e-03
	PBI (2.6)	8.025325 1.7e-07	9.277875 1.1e-06	PBI (2.4026)	8.025325 1.1e-07	9.277874 1.0e-06
	PBI (7.6)	8.025326 5.1e-07	9.277867 1.4e-06			
	VADS (8.6)	8.025324 1.4e-06	9.277872 1.5e-06			

---

[5] Available at http://computacion.cs.cinvestav.mx/~rhernandez.

In the multi-objective scenarios of Table 2, we can observe a clear performance improvement over the calibrated versions for MOEA/D with 2 objectives and MOMBI-II with 2 and 3 objectives. Only the version of MOMBI-II using EWC (8.6) was outperformed by the baseline MOMBI-II. In the particular case of MOEA/D, the major gains were achieved for the convex MOPs, and the best suited scalarizing functions were AASF (0.8001) and PBI (15.1, 20.5639, 2.6, 7.6). In MOMBI-II, the major gains were in the convex MOPs and the remaining problems with 3 objectives. The best scalarizing functions for this optimizer were EWC (8.6), AASF (1.0) and PBI (21.4123, 2.4026). As it can be noticed, AASF worked very well for both MOEAs in the convex problems, while PBI performed best in the linear and concave problems. However, this scalarizing function is sensitive to its parameter value.

In the many-objective scenarios of Table 3, we can notice a clear performance improvement over the calibrated versions of both optimizers. In MOEA/D, there were only 2 ties for the concave problems, and the major gains were in the convex MOPs. The best scalarizing functions for MOEA/D were: EWC (7.6), AASF (0.3001) and PBI (2.6). In MOMBI-II, the major gains were in 5 and 7 objectives. The best scalarizing functions for this optimizer were TCH, AASF (0.3001), VADS (5.6, 2.1). In both optimizers, AASF (0.3001) worked very well on the linear problems.

Table 3. Median $(\times 10^{-1})$ and standard deviation of the normalized hypervolume indicator on the **many-objective scenarios**.

γ	Config.	m		
		5	**7**	**10**
		MOEA/D		
0.5	Baseline	9.961741 8.0e-04	9.908978 1.4e-03	9.831210 1.1e-03
	EWC (7.6)	9.999756 6.1e-08	9.999988 1.8e-08	9.999985 1.1e-16
1.0	Baseline	9.989021 2.1e-07	9.999387 1.9e-06	9.999864 1.5e-05
	AASF (0.3001)	9.989041 5.7e-07	9.999427 3.8e-07	9.999990 5.9e-07
2.0	Baseline	9.904560 2.3e-06	9.986141 4.5e-06	9.999186 4.2e-06
	PBI (2.6)	9.904578 1.8e-06	9.986145 5.4e-06	9.999160 3.8e-06
		MOMBI-II		
0.5	Baseline	9.970399 3.9e-04	9.987786 3.3e-04	9.991083 2.8e-04
	TCH	9.999357 3.2e-06	9.999964 1.5e-06	9.999942 4.1e-06
1.0	Baseline	9.937457 9.2e-04	9.977353 8.9e-04	9.993752 4.5e-04
	AASF (0.3001)	9.989052 5.6e-07	9.999427 4.8e-08	9.999991 0.0e+0
	Baseline	9.884399 3.9e-04	9.976128 2.2e-04	9.995263 1.5e-04
2.0	VADS (5.6)	9.904381 6.7e-06	9.985715 1.9e-05	9.998632 1.7e-05
	VADS (2.1)	9.904350 8.6e-06	9.985683 2.4e-05	9.998591 2.0e-05

In the global scenario of Table 4 a different pattern is observed. In the case of MOEA/D, ASF and TCH worked very well in the convex problems from 3 up to 10 objectives. However, they worsened their behavior in the linear and concave MOPs. Similarly, AASF (0.2423) performed well in the convex problems and the linear problems for 5, 7 and 10 objectives. However, its performance deteriorated

Table 4. Median $(\times 10^{-1})$ and standard deviation of the normalized hypervolume indicator on the **global scenario**.

γ	Config.	m 2	3	5	7	10
		MOEA/D				
0.5	Baseline	$9.570400_{5.0e\text{-}07}$	$9.906204_{3.4e\text{-}04}$	$9.961741_{8.0e\text{-}04}$	$9.908978_{1.4e\text{-}03}$	$9.831210_{1.1e\text{-}03}$
	ASF	$9.570399_{3.2e\text{-}07}$	$9.917019_{1.5e\text{-}04}$	$9.971177_{5.7e\text{-}04}$	$9.987843_{2.6e\text{-}04}$	$9.994325_{2.8e\text{-}04}$
	TCH	$9.570400_{5.0e\text{-}07}$	$9.953880_{2.6e\text{-}06}$	$9.999339_{1.6e\text{-}06}$	$9.999972_{5.5e\text{-}06}$	$9.999967_{6.3e\text{-}06}$
	AASF (0.2423)	$9.571983_{1.9e\text{-}07}$	$9.966615_{7.4e\text{-}05}$	$9.995188_{5.0e\text{-}05}$	$9.997490_{4.8e\text{-}05}$	$9.998646_{5.4e\text{-}05}$
1.0	Baseline	$8.737370_{8.2e\text{-}08}$	$9.744895_{2.5e\text{-}07}$	$9.989021_{2.1e\text{-}07}$	$9.999387_{1.9e\text{-}06}$	$9.999864_{1.5e\text{-}05}$
	ASF	$8.737369_{7.2e\text{-}08}$	$9.638966_{6.2e\text{-}04}$	$9.914455_{7.2e\text{-}04}$	$9.963243_{8.9e\text{-}04}$	$9.988689_{4.1e\text{-}04}$
	TCH	$8.737370_{8.2e\text{-}08}$	$9.689283_{1.8e\text{-}05}$	$9.967928_{2.4e\text{-}05}$	$9.983080_{4.6e\text{-}04}$	$9.932911_{2.2e\text{-}03}$
	AASF (0.2423)	$8.668011_{1.3e\text{-}06}$	$9.720175_{4.1e\text{-}05}$	$9.989038_{2.8e\text{-}07}$	$9.999426_{2.8e\text{-}07}$	$9.999991_{4.9e\text{-}07}$
2.0	Baseline	$8.025324_{1.3e\text{-}07}$	$9.277872_{1.6e\text{-}06}$	$9.904560_{2.3e\text{-}06}$	$9.986141_{4.5e\text{-}06}$	$9.999186_{4.2e\text{-}06}$
	ASF	$8.025323_{1.0e\text{-}07}$	$9.207519_{7.9e\text{-}04}$	$9.870073_{6.6e\text{-}04}$	$9.970228_{5.6e\text{-}04}$	$9.985247_{1.6e\text{-}03}$
	TCH	$8.025324_{1.3e\text{-}07}$	$9.228448_{5.7e\text{-}05}$	$9.854362_{1.3e\text{-}05}$	$9.807065_{2.3e\text{-}03}$	$9.561658_{6.5e\text{-}03}$
	AASF (0.2423)	$7.876793_{2.0e\text{-}06}$	$9.152424_{2.5e\text{-}03}$	$9.847468_{2.3e\text{-}03}$	$9.969067_{5.6e\text{-}04}$	$9.996733_{1.3e\text{-}04}$
		MOMBI-II				
0.5	Baseline	$9.570404_{6.0e\text{-}07}$	$9.917509_{1.2e\text{-}04}$	$9.970399_{3.9e\text{-}04}$	$9.987786_{3.3e\text{-}04}$	$9.991083_{2.8e\text{-}04}$
	TCH	$9.570404_{6.2e\text{-}07}$	$9.953905_{3.6e\text{-}06}$	$9.999357_{3.2e\text{-}06}$	$9.999964_{1.5e\text{-}06}$	$9.999942_{4.1e\text{-}06}$
	AASF (0.0501)	$9.570326_{4.4e\text{-}07}$	$9.924333_{2.7e\text{-}04}$	$9.979430_{3.4e\text{-}04}$	$9.989421_{2.1e\text{-}04}$	$9.993698_{2.2e\text{-}04}$
1.0	Baseline	$8.737369_{8.0e\text{-}08}$	$9.636011_{6.3e\text{-}04}$	$9.937457_{9.2e\text{-}04}$	$9.977353_{8.9e\text{-}04}$	$9.993752_{4.5e\text{-}04}$
	TCH	$8.737369_{7.2e\text{-}08}$	$9.689161_{1.2e\text{-}05}$	$9.968534_{1.1e\text{-}04}$	$9.984665_{6.2e\text{-}04}$	$9.931828_{2.4e\text{-}03}$
	AASF (0.0501)	$8.732288_{1.2e\text{-}05}$	$9.744887_{6.7e\text{-}07}$	$9.989033_{2.2e\text{-}07}$	$9.999427_{1.8e\text{-}08}$	$9.999991_{0.0e+0}$
2.0	Baseline	$8.025323_{9.6e\text{-}08}$	$9.209490_{1.0e\text{-}03}$	$9.884399_{3.9e\text{-}04}$	$9.976128_{2.2e\text{-}04}$	$9.995263_{1.5e\text{-}04}$
	TCH	$8.025323_{9.6e\text{-}08}$	$9.229191_{1.5e\text{-}04}$	$9.855497_{7.7e\text{-}04}$	$9.851832_{2.2e\text{-}03}$	$9.563050_{6.3e\text{-}03}$
	AASF (0.0501)	$8.003164_{2.5e\text{-}05}$	$9.269135_{1.0e\text{-}04}$	$9.898159_{1.1e\text{-}04}$	$9.985653_{1.7e\text{-}05}$	$9.999221_{5.8e\text{-}06}$

in the concave MOPs. On the other hand, for MOMBI-II, the 2 scalarizing functions were complementary to each other. For example, for 2 objectives, TCH was competitive with respect to the baseline version, while in the concave problems for 3 to 10 objectives AASF (0.0501) performed best.

These results suggest that no scalarizing function can solve effectively all the problems. Instead, there is a subset of them that can tackle in an effective manner some specific problems.

4   Conclusions and Future Work

In this work, we presented an overview and a comparative study to determine the weighted and unconstrained scalarizing functions that are the most suitable for MOEA/D and MOMBI-II. For this purpose, we designed several test scenarios considering different Pareto-front shapes and objectives. We used the tuning tool EVOCA to determine the best calibration for each of these scenarios. In almost all cases, EVOCA recommendations outperform the baseline version of these MOEAs. Our most important conclusion is that no single scalarizing function performs best in all the scenarios but a set of them, regarding the normalized hypervolume indicator. In general, we obtained good results with AASF, PBI, EWC, and VADS. These scalarizing functions deserve further study.

As part of our future work, we are interested in studying additional scenarios that consider more difficult problems including other performance indicators. We are interested in analyzing the effect of using the nadir point instead of the ideal point (as in IPBI) and using vectors formed by the inverse components of the weight vectors (as in ASF). From the obtained results, we would like to design self-adaptive models and new MOEAs that combine several scalarizing functions.

Acknowledgments. The authors gratefully acknowledge support from CONACyT project no. 221551 and PCCI140054 CONACYT/CONICYT project.

References

1. Zitzler, E.: Evolutionary algorithms for multiobjective optimization: methods and applications. Ph.D. thesis, Swiss Federal Institute of Technology (ETH), Zurich, November 1999
2. Brockhoff, D., Wagner, T., Trautmann, H.: On the properties of the $R2$ indicator. In: 2012 Genetic and Evolutionary Computation Conference (GECCO 2012), Philadelphia, pp. 465–472. ACM Press, July 2012. ISBN: 978-1-4503-1177-9
3. Li, B., Li, J., Tang, K., Yao, X.: Many-objective evolutionary algorithms: a survey. ACM Comput. Surv. **48**(1), 13:1–13:35 (2015)
4. Adra, S.F., Fleming, P.J.: Diversity management in evolutionary many-objective optimization. IEEE Trans. Evol. Comput. **15**(2), 183–195 (2011)
5. Fleischer, M.: The measure of Pareto optima applications to multi-objective meta-heuristics. In: Fonseca, C.M., Fleming, P.J., Zitzler, E., Thiele, L., Deb, K. (eds.) EMO 2003. LNCS, vol. 2632, pp. 519–533. Springer, Heidelberg (2003). doi:10.1007/3-540-36970-8_37
6. Bringmann, K., Friedrich, T.: Don't be greedy when calculating hypervolume contributions. In: Proceedings of the Tenth ACM SIGEVO Workshop on Foundations of Genetic Algorithms. FOGA 2009, Orlando, pp. 103–112. ACM, January 2009
7. Eichfelder, G.: An adaptive scalarization method in multiobjective optimization. SIAM J. Optim. **19**(4), 1694–1718 (2009)
8. Tutum, C.C., Deb, K.: A multimodal approach for evolutionary multi-objective optimization (MEMO): proof-of-principle results. In: Gaspar-Cunha, A., Henggeler Antunes, C., Coello Coello, C.A. (eds.) EMO 2015. LNCS, vol. 9018, pp. 3–18. Springer, Cham (2015). doi:10.1007/978-3-319-15934-8_1
9. Ishibuchi, H., Akedo, N., Nojima, Y.: A study on the specification of a scalarizing function in MOEA/D for many-objective knapsack problems. In: Nicosia, G., Pardalos, P. (eds.) LION 2013. LNCS, vol. 7997, pp. 231–246. Springer, Heidelberg (2013). doi:10.1007/978-3-642-44973-4_24
10. Ishibuchi, H., Akedo, N., Nojima, Y.: Behavior of multiobjective evolutionary algorithms on many-objective knapsack problems. IEEE Trans. Evol. Comput. **19**(2), 264–283 (2015)
11. Sato, H.: MOEA/D using constant-distance based neighbors designed for many-objective optimization. In: 2015 IEEE Congress on Evolutionary Computation (CEC), pp. 2867–2874, May 2015
12. Marler, R., Arora, J.: Survey of multi-objective optimization methods for engineering. Struct. Multidiscip. Optim. **26**(6), 369–395 (2004)

13. Miettinen, K.: Nonlinear Multiobjective Optimization. International Series in Operations Research & Management Science, vol. 12. Springer, New York (1998)
14. Miettinen, K., Mäkelä, M.M.: On scalarizing functions in multiobjective optimization. OR Spectr. **24**(2), 193–213 (2002)
15. Santiago, A., Huacuja, H.J.F., Dorronsoro, B., Pecero, J.E., Santillan, C.G., Barbosa, J.J.G., Monterrubio, J.C.S.: A survey of decomposition methods for multi-objective optimization. In: Castillo, O., Melin, P., Pedrycz, W., Kacprzyk, J. (eds.) Recent Advances on Hybrid Approaches for Designing Intelligent Systems, vol. 547, pp. 453–465. Springer, Cham (2014). doi:10.1007/978-3-319-05170-3_31. ISBN: 978-3-319-05170-3
16. Riff, M.C., Montero, E.: A new algorithm for reducing metaheuristic design effort. In: 2013 IEEE Congress on Evolutionary Computation, pp. 3283–3290, June 2013
17. Zhang, Q., Li, H.: MOEA/D: a multiobjective evolutionary algorithm based on decomposition. IEEE Trans. Evol. Comput. **11**(6), 712–731 (2007)
18. Hernández Gómez, R., Coello Coello, C.A.: Improved metaheuristic based on the R2 indicator for many-objective optimization. In: Proceedings of the 2015 Annual Conference on Genetic and Evolutionary Computation. GECCO 2015, pp. 679–686. ACM, New York (2015)
19. Steuer, R.E., Choo, E.U.: An interactive weighted Tchebycheff procedure for multiple objective programming. Math. Program. **26**(3), 326–344 (1983)
20. Ruiz, A.B., Saborido, R., Luque, M.: A preference-based evolutionary algorithm for multiobjective optimization: the weighting achievement scalarizing function genetic algorithm. J. Glob. Optim. **62**(1), 101–129 (2015)
21. Luque, M., Ruiz, A.B., Saborido, R., Marcenaro-Gutiérrez, O.D.: On the use of the L_p distance in reference point-based approaches for multiobjective optimization. Ann. Oper. Res. **235**(1), 559–579 (2015)
22. Zeleny, M.: Compromise Programming. Multiple Criteria Decision Making, pp. 262–301. University of South Carolina Press, Columbia (1973)
23. Coello Coello, C.A., Lamont, G.B., Van Veldhuizen, D.A.: Evolutionary Algorithms for Solving Multi-objective Problems, 2nd edn. Springer, New York (2007). ISBN: 978-0-387-33254-3
24. Athan, T.W., Papalambros, P.Y.: A note on weighted criteria methods for compromise solutions in multi-objective optimization. Eng. Optim. **27**(2), 155–176 (1996)
25. Lee, L.H., Chew, E.P., Yu, Q., Li, H., Liu, Y.: A study on multi-objective particle swarm optimization with weighted scalarizing functions. In: Proceedings of the 2014 Winter Simulation Conference. WSC 2014, Piscataway, pp. 3718–3729. IEEE Press (2014)
26. Zadeh, L.: Optimality and non-scalar-valued performance criteria. IEEE Trans. Automat. Contr. **8**(1), 59–60 (1963)
27. Ishibuchi, H., Sakane, Y., Tsukamoto, N., Nojima, Y.: Simultaneous use of different scalarizing functions in MOEA/D. In: Proceedings of the 12th Annual Conference on Genetic and Evolutionary Computation (GECCO 2010), Portland, pp. 519–526. ACM Press, 7–11 July 2010. ISBN: 978-1-4503-0072-8
28. Hughes, E.J.: Multiple single objective Pareto sampling. In: Proceedings of the 2003 Congress on Evolutionary Computation (CEC 2003), Canberra, vol. 4, pp. 2678–2684. IEEE Press, December 2003
29. Ishibuchi, H., Sakane, Y., Tsukamoto, N., Nojima, Y.: Adaptation of scalarizing functions in MOEA/D: an adaptive scalarizing function-based multiobjective evolutionary algorithm. In: Ehrgott, M., Fonseca, C.M., Gandibleux, X., Hao, J.-K., Sevaux, M. (eds.) EMO 2009. LNCS, vol. 5467, pp. 438–452. Springer, Heidelberg (2009). doi:10.1007/978-3-642-01020-0_35

30. Jin, Y., Okabe, T., Sendho, B.: Adapting weighted aggregation for multiobjective evolution strategies. In: Zitzler, E., Thiele, L., Deb, K., Coello Coello, C.A., Corne, D. (eds.) EMO 2001. LNCS, vol. 1993, pp. 96–110. Springer, Heidelberg (2001). doi:10.1007/3-540-44719-9_7

31. Paquete, L., Stützle, T.: A two-phase local search for the biobjective traveling salesman problem. In: Fonseca, C.M., Fleming, P.J., Zitzler, E., Thiele, L., Deb, K. (eds.) EMO 2003. LNCS, vol. 2632, pp. 479–493. Springer, Heidelberg (2003). doi:10.1007/3-540-36970-8_34

32. Wang, R., Zhou, Z., Ishibuchi, H., Liao, T., Zhang, T.: Localized weighted sum method for many-objective optimization. IEEE Trans. Evol. Comput. **PP**(99) (2016). (to be published)

33. Murata, T., Taki, A.: Many-objective optimization for knapsack problems using correlation-based weighted sum approach. In: Ehrgott, M., Fonseca, C.M., Gandibleux, X., Hao, J.-K., Sevaux, M. (eds.) EMO 2009. LNCS, vol. 5467, pp. 468–480. Springer, Heidelberg (2009). doi:10.1007/978-3-642-01020-0_37

34. Ishibuchi, H., Nojima, Y.: Optimization of scalarizing functions through evolutionary multiobjective optimization. In: Obayashi, S., Deb, K., Poloni, C., Hiroyasu, T., Murata, T. (eds.) EMO 2007. LNCS, vol. 4403, pp. 51–65. Springer, Heidelberg (2007). doi:10.1007/978-3-540-70928-2_8

35. Carpinelli, G., Caramia, P., Mottola, F., Proto, D.: Exponential weighted method and a compromise programming method for multi-objective operation of plug-in vehicle aggregators in microgrids. Int. J. Electr. Power Energy Syst. **56**, 374–384 (2014)

36. Li, D.: Convexification of a noninferior frontier. J. Optim. Theory Appl. **88**(1), 177–196 (1996)

37. Li, D., Yang, J.B., Biswal, M.: Quantitative parametric connections between methods for generating noninferior solutions in multiobjective optimization. Eur. J. Oper. Res. **117**(1), 84–99 (1999)

38. Dellino, G., Fedele, M., Meloni, C.: Dynamic objectives aggregation methods for evolutionary portfolio optimisation. A computational study. Int. J. BioInspir. Comput. **4**(4), 258–270 (2012)

39. Triantaphyllou, E.: Multi-criteria Decision Making Methods: A Comparative Study, vol. 44. Springer, New York (2000)

40. Angelo, J.S., Barbosa, H.J.: On ant colony optimization algorithms for multiobjective problems. In: Ostfeld, A. (ed.) Ant Colony Optimization Methods and Applications, pp. 53–74. InTech, Cham (2011)

41. Bjornson, E., Jorswieck, E.A., Debbah, M., Ottersten, B.: Multiobjective signal processing optimization: the way to balance conflicting metrics in 5G systems. IEEE Signal Process. Mag. **31**(6), 14–23 (2014)

42. Kasprzak, E., Lewis, K.: Pareto analysis in multiobjective optimization using the collinearity theorem and scaling method. Struct. Multidiscip. Optim. **22**(3), 208–218 (2001)

43. Wang, R., Zhang, Q., Zhang, T.: Pareto adaptive scalarising functions for decomposition based algorithms. In: Gaspar-Cunha, A., Henggeler Antunes, C., Coello Coello, C.A. (eds.) EMO 2015. LNCS, vol. 9018, pp. 248–262. Springer, Cham (2015). doi:10.1007/978-3-319-15934-8_17

44. Bowman Jr., V.J.: On the relationship of the Tchebyche norm and the efficient frontier of multiple-criteria objectives. In: Thiriez, H., Zionts, S. (eds.) Multiple Criteria Decision Making. LNEMS, vol. 130, pp. 76–86. Springer, Heidelberg (1973). doi:10.1007/978-3-642-87563-2_5

45. Giagkiozis, I., Fleming, P.J.: Methods for multi-objective optimization: an analysis. Inf. Sci. **293**, 338–350 (2015)
46. Hernández Gómez, R., Coello Coello, C.A.: MOMBI: a new metaheuristic for many-objective optimization based on the $R2$ indicator. In: IEEE Congress on Evolutionary Computation (CEC 2013), Cancún, pp. 2488–2495. IEEE Press, 20–23 June 2013. ISBN: 978-1-4799-0454-9
47. Ralphs, T.K., Saltzman, M.J., Wiecek, M.M.: An improved algorithm for solving biobjective integer programs. Ann. Oper. Res. **147**(1), 43–70 (2006)
48. Kaliszewski, I.: A modified weighted Tchebycheff metric for multiple objective programming. Comput. Oper. Res. **14**(4), 315–323 (1987)
49. Wierzbicki, A.P.: The use of reference objectives in multiobjective optimization. In: Fandel, G., Gal, T. (eds.) Multiple Criteria Decision Making Theory and Application. LNE, vol. 177, pp. 468–486. Springer, Heidelberg (1980)
50. Derbel, B., Brockhoff, D., Liefooghe, A.: Force-based cooperative search directions in evolutionary multi-objective optimization. In: Purshouse, R.C., Fleming, P.J., Fonseca, C.M., Greco, S., Shaw, J. (eds.) EMO 2013. LNCS, vol. 7811, pp. 383–397. Springer, Heidelberg (2013). doi:10.1007/978-3-642-37140-0_30
51. Yuan, Y., Xu, H., Wang, B., Zhang, B., Yao, X.: Balancing convergence and diversity in decomposition-based many-objective optimizers. IEEE Trans. Evol. Comput. **20**(2), 180–198 (2016)
52. Saborido, R., Ruiz, A.B., Luque, M.: Global WASF-GA: an evolutionary algorithm in multiobjective optimization to approximate the whole Pareto optimal front. Evol. Comput. 1–40 (2017, in press)
53. Das, I., Dennis, J.E.: Normal-boundary intersection: a new method for generating the Pareto surface in nonlinear multicriteria optimization problems. SIAM J. Optim. **8**(3), 631–657 (1998)
54. Yang, S., Jiang, S., Jiang, Y.: Improving the multiobjective evolutionary algorithm based on decomposition with new penalty schemes. Soft Comput. 1–15 (2017, in press)
55. Sato, H.: Inverted PBI in MOEA/D and its impact on the search performance on multi and many-objective optimization. In: Genetic and Evolutionary Computation Conference (GECCO), Vancouver, pp. 645–652. ACM Press, 12–16 July 2014. ISBN: 978-1-4503-2662-9
56. Kasimbeyli, R.: A nonlinear cone separation theorem and scalarization in nonconvex vector optimization. SIAM J. Optim. **20**(3), 1591–1619 (2010)
57. Erozan, İ., Torkul, O., Ustun, O.: Proposal of a nonlinear multi-objective genetic algorithm using conic scalarization to the design of cellular manufacturing systems. Flex. Serv. Manuf. J. **27**(1), 30–57 (2015)
58. Hughes, E.J.: MSOPS-II: a general-purpose many-objective optimiser. In: 2007 IEEE Congress on Evolutionary Computation (CEC 2007), Singapore, pp. 3944–3951. IEEE Press, September 2007
59. Emmerich, M.T.M., Deutz, A.H.: Test problems based on Lamé superspheres. In: Obayashi, S., Deb, K., Poloni, C., Hiroyasu, T., Murata, T. (eds.) EMO 2007. LNCS, vol. 4403, pp. 922–936. Springer, Heidelberg (2007). doi:10.1007/978-3-540-70928-2_68
60. Deb, K., Jain, H.: An evolutionary many-objective optimization algorithm using reference-point-based nondominated sorting approach, Part I: solving problems with box constraints. IEEE Trans. Evol. Comput. **18**(4), 577–601 (2014)

Multi-objective Representation Setups for Deformation-Based Design Optimization

Andreas Richter[1]([envelope]), Jascha Achenbach[1], Stefan Menzel[2], and Mario Botsch[1]

[1] Computer Graphics Group, Bielefeld University, Bielefeld, Germany
anrichter@techfak.uni-bielefeld.de
[2] Honda Research Institute Europe, Offenbach, Germany

Abstract. The increase of complexity in virtual product design requires high-quality optimization algorithms capable to find the global parameter solution for a given problem. The representation, which defines the encoding of the design and the mapping from parameter space to design space, is a key aspect for the performance of the optimization process. To initialize representations for a high performing optimization we utilize the concept of *evolvability*. Our interpretation of this concept consists of three performance criteria, namely *variability*, *regularity*, and *improvement potential*, where regularity and improvement potential characterize conflicting goals. In this article we address the generation of initial representation setups trading off between these two conflicting criteria for design optimization. We analyze Pareto-optimal compromises for deformation representations with radial basis functions in two test scenarios: fitting of 1D height fields and fitting of 3D face scans. We use the Pareto front as a ground-truth to show the feasibility of a single-objective optimization targeting one preference-based trade-off. Based on the results of both optimization approaches we propose two heuristic methods, Lloyd sampling and orthogonal least squares sampling, targeting representations with high regularity and high improvement potential at the two ends of the Pareto front. Thereby, we overcome the time consuming process of an evolutionary optimization to set up high-performing representations for these two cases.

1 Introduction

The increasing complexity in modern industrial design processes requires advanced optimization methods to efficiently come up with novel and high-quality solutions for successful business. In automotive product design, our target application, concurrent development processes are applied to deal with different requirements, e.g., from physical domains such as aerodynamic or structural performance criteria, from manufacturing process layout, or from design features specified by current customer demands. Moreover, since these requirements change over time, an efficient development process needs to cope with dynamic environments to allow a high degree of flexibility.

Biologically-inspired population-based evolutionary optimization algorithms are designed to handle these demands [13]. The careful construction of the representation, the encoding of a design, is one of the most important aspects for the

© Springer International Publishing AG 2017
H. Trautmann et al. (Eds.): EMO 2017, LNCS 10173, pp. 514–528, 2017.
DOI: 10.1007/978-3-319-54157-0_35

success of an optimization process because the representation defines the solution space and determines how efficient the optimizer can explore it. Although a human designer can manually set up representations for design optimization, e.g. [19], the setup might not be optimal such that even minor changes of the setup could highly improve the performance of the optimization process. Therefore, we target to develop automatic procedures to optimize the initial representation along with the designer's input or preference, which is typically given by information on the expected importance of certain design regions for the current optimization task.

Based on the concept of evolvability [17] we proposed a mathematical model for evaluating the quality of deformation representation setups in [16]. The quality of each setup is numerically quantified by three criteria, namely *variability*, *regularity*, and *improvement potential*. However, since regularity and improvement potential are conflicting targets [16] a multi-objective analysis is required to finally provide the designer the possibility to choose one trade-off setup from the set of Pareto solutions according to her/his preference.

In this article our focus is on the initial generation of optimal representation setups addressing these two conflicting targets, which we evaluate for the same test scenarios as in [16]: 1D function approximation and 3D template fitting, which both are based on RBF deformation representations. For these RBF deformations we are particularly interested in the optimal distribution of the RBF centers and their efficient computation. To this end we first perform a multi-objective optimization of center distributions to analyze the trade-off between regularity and improvement potential. Although this time-consuming optimization might be infeasible for real-word applications, it results in ground-truth solutions that we use to evaluate a weighted single-objective optimization of the center distribution. We show that such a single-objective optimization is feasible in our application and speeds up the optimization process for one particular weight, which is set according to a designer's preference. The insight gained from the multi-objective optimization furthermore allows us to derive two heuristic approaches for rapidly generating center distributions on both ends of the Pareto front, i.e., aiming solely for regularity or improvement potential, respectively. Being based on Lloyd sampling [10] or orthogonal least squares [4], both methods generate high-quality center distributions within minutes in contrast to the single-objective optimization, which runs for hours.

In Sect. 2 we discuss state-of-the-art approaches for setting up deformation representations and motivate our approach. In Sect. 3 we give the technical details for RBF deformations as the representation of our choice. In Sect. 4 we describe our model of evolvability that we use to evaluate these representations. This yields the basis to perform and analyze a multi-objective optimization in two test scenarios in Sect. 5. The Pareto front is the ground-truth for a preference-based single-objective optimization in Sect. 6. Moreover, the Pareto-optimal solutions motivate heuristics, Lloyd and OLS, which we discuss in Sect. 7.

2   Related Work

In shape optimization based on deformation methods designer-driven approaches are typically applied to set up initial deformation representations. The designer defines target regions where the design has to be varied/optimized and places control points adapted to these regions. For example, in [19] a control grid for free-form deformation (FFD) is manually constructed and handle regions for deformations with radial basis functions (RBF) are manually set up. For basic automated representation setups commercial tools provide a uniform distribution of control points, e.g., a glider optimization [5].

Originally, deformation representations are employed in scattered data approximation, e.g., for approximating a target shape. In [3] the control points of non-uniform rational B-splines are optimized by a gradient-based method to improve the approximation quality of a wing. In [22,27] a uniform setup of a control grid is refined in sensitive regions, i.e. parameters are added, resulting in an improved approximation. Amoignon [1,2] tackles the problem that uniform control grids for FFD might have empty grid cells. Instead of adjusting the grid to the design he deforms the design (e.g., wings) to completely fill out the grid. To obtain RBF setups that are adapted to a target, different basis functions are iteratively evaluated and selected at fixed locations [24] or their location is being optimized [7,14]. All these approaches are specialized to set up control points for approximating one fixed target. Thereby, they neglect numerical properties of the deformation setup which are important for, e.g., the convergence speed of an evolutionary optimization.

The representation setup of adaptive B-splines for an evolutionary design process is targeted in [15,26]. The optimization alternates between approximation of a shape and adaptation of the representation. To test whether an adjusted representation is beneficial for the optimization this process is performed for a few iterations. Thereby the performance of a representation is measured by the objective function of the actual optimization task. In [20] the representation is optimized implicitly by adding its parameters to the approximation problem. The criterion for a high-quality representation purely depends on the target of the optimization omitting further aspects of this process like convergence speed.

In contrast, we utilize quality criteria based on the concept of evolvability to evaluate representations. We include an objective-independent criterion to address the convergence speed as well as target information or human knowledge. Based on this model we set up high-quality deformation representations for evolutionary design optimization. In the next section we give the technical details for these deformations.

3   RBF Representations

In a shape optimization scenario, for instance in automotive product design, the design model to be optimized (the phenotype) is typically represented by a surface polygon mesh, where the n mesh vertices $\mathbf{x}_1, \ldots, \mathbf{x}_n \in \mathbb{R}^3$ represent points on the surface, which are connected by polygonal faces (usually trian-gles or quads). The vertex positions $\mathbf{x}_i$ could in theory be used as optimization parameters in an evolutionary optimization. However, for non-trivial models the complexity of the model easily exceeds one million vertices, thus making the direct optimization of vertex positions intractable.

Even for highly complex shapes the actual *deformations* applied during opti-mization are typically rather simple, low-frequency functions, which can there-fore be controlled by a small number of parameters. Hence we choose as rep-resentation a deformation function $\mathbf{u}(\mathbf{x})$, which maps deformation parameters (genotypes) to shape variations (phenotypes), which are then evaluated by a fitness function. Both free-form deformation (FFD) and radial basis functions (RBFs) have been successfully employed in design optimization [19]. In this paper we focus on RBF deformations, since their kernel-based setup is more flexible than lattice-based FFD representations.

The initial design $(\mathbf{x}_1, \ldots, \mathbf{x}_n)$ is deformed into a shape variant $(\mathbf{x}_1', \ldots, \mathbf{x}_n')$ by adding the displacement $\mathbf{u}(\mathbf{x}_i)$ to each vertex $\mathbf{x}_i$ (Fig. 1), which for RBF deformation has the form

$$\mathbf{u}(\mathbf{x}) \;=\; \sum_{j=1}^{m} \mathbf{w}_j \, \varphi(\|\mathbf{c}_j - \mathbf{x}\|) \;=: \sum_{j=1}^{m} \mathbf{w}_j \, \varphi_j(\mathbf{x}) \,. \tag{1}$$

Here, $\varphi_j(\mathbf{x}) = \varphi(\|\mathbf{c}_j - \mathbf{x}\|)$ denotes the j-th scalar-valued radial basis function, which is centered at $\mathbf{c}_j \in \mathbb{R}^3$ and weighted by the coefficient $\mathbf{w}_j \in \mathbb{R}^3$.

The choice of the kernel function $\varphi \colon \mathbb{R} \to \mathbb{R}$ has a significant influence on the resulting deformation and its computation complexity [18]. In this paper we employ and analyze globally-supported triharmonic thin-plate splines, φ_{tri}, as well as compactly-supported Wendland functions, φ_W, with support radii s varying from rather local to more global [25]:

$$\varphi_{tri}(r) = \begin{cases} r^2 \log(r) & \text{for 2D domains,} \\ r^3 & \text{for 3D domains.} \end{cases}$$

$$\varphi_W(r) = \begin{cases} \left(1 - \frac{r}{s}\right)^4 \left(\frac{4r}{s} + 1\right) & \text{for } r < s, \\ 0 & \text{otherwise.} \end{cases}$$

The RBF deformation (and thus the deformed shape) is linear in the RBF weights $\mathbf{w}_j$. If we write the initial and deformed shapes as $(n \times 3)$-matrices $\mathbf{X} = \left(\mathbf{x}_1^{\mathrm{T}}, \ldots, \mathbf{x}_n^{\mathrm{T}}\right)^{\mathrm{T}}$ and $\mathbf{X}' = \left(\mathbf{x}_1'^{\mathrm{T}}, \ldots, \mathbf{x}_n'^{\mathrm{T}}\right)^{\mathrm{T}}$, respectively, we can write the shape deformation in matrix notation

$$\mathbf{X}' = \mathbf{X} + \mathbf{\Phi}\mathbf{W} \tag{2}$$

using an $(n \times m)$ RBF matrix $(\mathbf{\Phi})_{i,j} = \varphi_j(\mathbf{x}_i)$ and the RBF weights $\mathbf{W} = \left(\mathbf{w}_1^{\mathrm{T}}, \ldots, \mathbf{w}_m^{\mathrm{T}}\right)^{\mathrm{T}} \in \mathbb{R}^{m \times 3}$.

In the above setting, the deformation $\mathbf{u}$ is controlled by manipulating the RBF weights $\mathbf{w}_j$, which we call *indirect manipulation*. However, it has been

Center distribution   Target

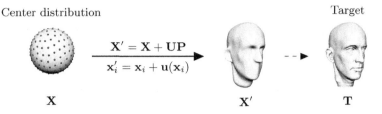

$$\mathbf{X}' = \mathbf{X} + \mathbf{UP}$$
$$\mathbf{x}'_i = \mathbf{x}_i + \mathbf{u}(\mathbf{x}_i)$$

$\mathbf{X}$ $\mathbf{X}'$ $\mathbf{T}$

Fig. 1. The RBF deformation $\mathbf{u}$ transforms the initial mesh $\mathbf{X}$ to $\mathbf{X}'$ by translating each vertex $\mathbf{x}_i$ of $\mathbf{X}$ by the displacement $\mathbf{u}(\mathbf{x}_i)$. The distribution of the RBF centers (red dots) is crucial for a high-performing fit of the mesh $\mathbf{X}$ to the target $\mathbf{T}$. (Color figure online)

shown in the context of free-form deformation that so-called *direct manipulation* is more intuitive for the human designer [8] as well as more efficient in an evolutionary optimization [12], due to the more direct and stronger causal relation between optimization parameters and the resulting shape deformation. In the RBF setting, a direct manipulation is controlled by specifying the displacement $\mathbf{d}_j$ for each center position $\mathbf{c}_j$, and then solving a linear system for the weights $\mathbf{w}_j$ that meet these interpolation constraints:

$$\mathbf{W} = \mathbf{\Psi}^{-1}\mathbf{D}\,, \qquad (3)$$

with $\mathbf{D} = (\mathbf{d}_1^{\mathrm{T}},\ldots,\mathbf{d}_m^{\mathrm{T}})^{\mathrm{T}} \in \mathbb{R}^{m\times 3}$ and $(\mathbf{\Psi})_{i,j} = \varphi_j(\mathbf{c}_i) \in \mathbb{R}^{m\times m}$. Combining Eqs. (2) and (3) leads to the matrix representation of direct RBF deformation:

$$\mathbf{X}' = \mathbf{X} + \mathbf{\Phi}\mathbf{\Psi}^{-1}\mathbf{D}. \qquad (4)$$

Note that both *indirect manipulation* (2) and *direct manipulation* (4) can be written as a linear deformation operator

$$\mathbf{X}' = \mathbf{X} + \mathbf{UP}\,, \qquad (5)$$

using a deformation matrix $\mathbf{U}$, being either $\mathbf{\Phi}$ or $\mathbf{\Phi}\mathbf{\Psi}^{-1}$, and deformation parameters $\mathbf{P}$, being either $\mathbf{W}$ or $\mathbf{D}$. The deformation matrix $\mathbf{U}$ (which is the deformation setup) depends on the employed kernel and the center distribution. These two aspects define the realizable deformations and thereby the performance of a design optimization process. Evaluating and optimizing different deformation setups, different kernels, and different center distributions, allows us to initialize a high performing design optimization. The concept of evolvability reveals quality criteria that we discuss in the next section.

4   Evolvability for Linear Deformations

The biological concept of *evolvability* is a very promising approach to measure the *expected* performance of evolutionary processes [23]. In [17] we gathered, categorized, and extensively discussed this concept not only in the biological context, but also in the context of technical engineering. In agreement with Sterelny [21], we understand evolvability as a combination of three major attributes: *variability*, *regularity*, and *improvement potential*. Based on this classification we proposed a mathematical model to quantify the quality of linear deformation representations $\mathbf{U}$ in design optimization. In this section, we give a short summary of our model of evolvability and refer to [16] for details.

Variability $V(\mathbf{U})$ measures the potential of a deformation setup to explore the design space and we define it as

$$V(\mathbf{U}) = \frac{\text{rank}\,(\mathbf{U})}{n}, \tag{6}$$

where n is the number of vertices of a design and $\text{rank}\,(\mathbf{U})$ denotes the rank of the matrix $\mathbf{U}$ [6]. We showed that variability is independent of the manipulation type, being either indirect or direct (Eqs. (2) or (4)). Furthermore, for a fixed number of RBF centers it is independent of the center distribution as long as the deformation matrix has maximal rank, which is the case if centers do not coincide. In our test scenarios we assume a fixed number of centers, which results in constant variability, which thereby does not represent a conflicting interest to regularity and improvement potential.

We define *regularity* $R(\mathbf{U})$ as

$$R(\mathbf{U}) = \kappa^{-1}(\mathbf{U}) = \frac{\sigma_{\min}}{\sigma_{\max}}, \tag{7}$$

where κ is the condition number of a matrix and $\sigma_{\min}$ and $\sigma_{\max}$ are its minimal and maximal singular value [6]. The regularity of a deformation setup characterizes the expected convergence speed of an evolutionary optimization. This criterion is an interpretation of the concept of robustness. Robust representations aim to prevent infeasible designs and thereby speed up the optimization process. Our regularity measure addresses the convergence directly. Furthermore, it corresponds to the concept of *causality*, because the condition number κ characterizes the causal relation between genotype (parameter space) and phenotype (design space). In [16] we show that RBF deformation setups with a local kernel (a kernel with a small support radius) have higher regularity than setups with a global kernel and that direct manipulation has higher regularity than indirect manipulation.

We define *improvement potential* $P(\mathbf{U})$ as the potential of a representation to improve the fitness of a design. From a local point of view the most beneficial variation of a design is according to the (estimated) fitness gradient $\mathbf{g}$. Thus, we measure the improvement potential $P(\mathbf{U})$ as the approximation error to this gradient, which leads to:

$$P(\mathbf{U}) = 1 - \left\| (\mathbf{I} - \mathbf{U}\mathbf{U}^{+})\mathbf{g} \right\|_{2}^{2}, \tag{8}$$

with $\mathbf{U}^{+}$ being the Pseudo-inverse of $\mathbf{U}$ [6]. Because for complex design optimization applications the calculation of the fitness gradient is infeasible, designer knowledge and experience offer valuable insight to derive an estimated gradient. Improvement potential is independent of the manipulation type (indirect or direct), as is variability. Furthermore, we showed that global kernels have better improvement potential than local ones thus lead to solutions with higher fitness.

Our experiments in [16] revealed that regularity and improvement potential are conflicting targets, because local kernels show high regularity but low improvement potential, whereas global kernels show low regularity but high improvement potential. These two conflicting criteria motivate the multi-objective optimization in the next section to gain insight into possible trade-off setups. Such a multi-objective optimization is very time-consuming and hence infeasible for most applications. However, it results in ground truth solutions to evaluate a more efficient weighted single-objective optimization, which we propose and discuss in Sect. 6. To further increase efficiency we analyze heuristics for generating setups in Sect. 7, which are more robust against local optima where a single-objective optimization might easily get stuck.

5   Pareto Analysis

In this section we show the results of the multi-objective optimization of deformation setups towards the two conflicting targets regularity and improvement potential. Due to the enormous computational costs of automotive design optimization we analyze two simpler test scenarios instead, namely 1D function approximation and 3D template fitting, according to our test framework in [16].

5.1   Test Scenario: Height Field Approximation

In this scenario we fit an initial plane (Fig. 2, left) to a target height field (Fig. 2, right) by minimizing the approximation error (see [16] for the details). Instead of performing the actual fit (as in [16]), our goal in this paper is to find a well-performing *deformation setup*. We employ RBF deformation and construct the matrix $\mathbf{U}$ according to either indirect or direct manipulation, Eqs. (2) or (4). To cover a variety of kernel types, from rather local to global, we employ compact Wendland kernels with support radii s of 0.25 and 0.5, and global triharmonic kernels. Given the type of manipulation, the kernel, and the support radius, the optimal center distribution with respect to the conflicting targets regularity and improvement potential, Eqs. (7) and (8), is the goal of a multi-objective optimization. Because we analyze distributions with a fixed number of centers, $m = 25$, variability is constant thus not included in the optimization. Each center has two coordinates, thus we solve a 50-dimensional optimization problem.

We realize the multi-objective optimization with the NSGA2 algorithm of the shark 2.3 library [9] with the following settings: 100 individuals, tournament selection, polynomial mutation rate with a probability of 1/50, crossover with a probability of 0.9, and 25000 iterations. We initialize the algorithm with randomized center distributions and restrict the centers to the initial plane ($[0, 1] \times [0, 2]$)

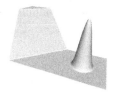

Fig. 2. Example of a (uniform) RBF center setup for function approximation. The center distribution on the initial plane is to be optimized to allow a high quality fit of the right test function (compare [16]).

during the optimization. With these settings one optimization run took approximately 2 days.

In Fig. 3 we plot the resulting Pareto front as blue dots for the three tested kernels with indirect and direct manipulation, respectively. The green dots are the values of the initial population. The tests indicate a smooth well-shaped Pareto front. For the local Wendland kernel the front almost reaches the optimal value of 1 for regularity and improvement potential, respectively. Note that the very low regularity values of the triharmonic kernel for indirect manipulation goes along with our results in [16] and theoretical results in [25].

Especially the center distributions maximizing either regularity or improvement potential, respectively, are interesting because they can be computed through a single-objective optimization. For *indirect* manipulation we obtain uniformly distributed centers (Fig. 4, top) resulting in maximal regularity, in agreement with theoretical results [25]. In contrast, the center distributions leading to optimal regularity for *direct* manipulation are unpredictable (Fig. 4, bottom), which shows the advantages of an automatic procedure for distributing centers in contrast to a purely designer-driven approach.

Center distributions with maximal improvement potential are adapted to the target height field for the compact Wendland kernels (Fig. 5). The distribution is denser in regions which have to be deformed more. In contrast, centers for the global triharmonic kernel are not placed in these regions (Fig. 5, right), which is unintuitive for a designer. This again emphasizes the demand for an automatic construction of setups instead of a purely designer driven approach.

5.2 Test Scenario: 3D Template Fitting

In the second test scenario we deform an initial sphere to closely fit the point cloud of a given face scan (see Fig. 1 and [16]). Like in the height field approximation scenario we intend to *set up* an optimal center distribution rather than performing the fitting. However, distributing centers for template fitting is more complex because the sphere and the scan are embedded in 3D such that each of the 25 centers consists of 3 coordinates, resulting in 75 parameters to be optimized. We choose the initial distributions randomly on the initial sphere, restrict the search domain to its bounding box $[-1, 1]^3$, and choose support radii of 0.5 and 1 for the Wendland kernels (since the initial domain is larger than

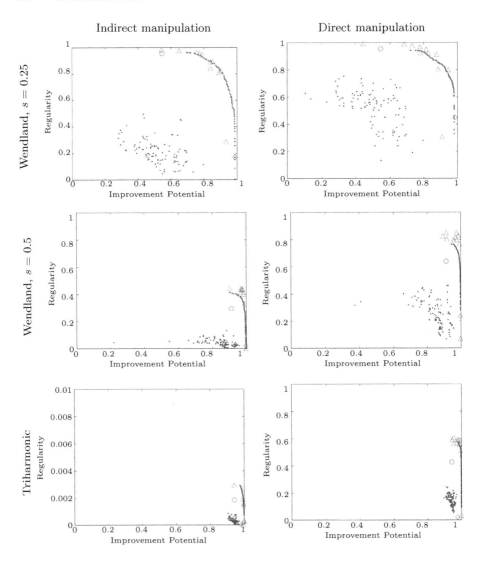

Fig. 3. The Pareto front (blue dots) and the initial random population (green dots) for the function approximation scenario. The magenta triangles are the results of the weighted single-objective optimization described in Sect. 6. Heuristic setups generated with Lloyd (orange circle) or OLS (orange diamond), discussed in Sect. 7, result in very regular setups or setups with a very good improvement potential. (Color figure online)

in the function approximation scenario). Apart from the mutation rate, which we set to 1/75 according to the 75 parameters, we perform the multi-objective optimization with identical settings as in the function approximation scenario. In Fig. 6 we plot the Pareto front for the three kernel types with direct and

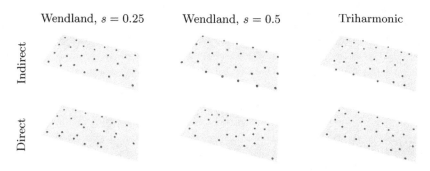

Fig. 4. Optimized center distributions towards regularity are uniform for indirect manipulation (top) but tends to be unintuitive for direct manipulation (bottom).

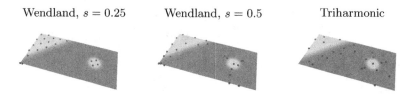

Fig. 5. Target-adapted center distributions with optimal improvement potential. The compact kernels are mainly placed in regions with locally high fitting error (yellow), whereas triharmonic kernels are placed less intuitive (blue). (Color figure online)

indirect manipulation, respectively. These plots are qualitatively equivalent to the plots of the function approximation scenario, compare to Fig. 3.

The multi-objective optimization in both test scenarios, height field approximation and template fitting, runs up to 2 days, hence it is infeasible for most real-world applications. Instead of computing the whole Pareto front we are rather interested in one particular setup trading off regularity and improvement potential according to our preference. Therefore, we employ a weighted single-objective optimization in the next section and utilize the Pareto front as a ground truth to test if this optimization is able to converge towards the front.

6   Weighted Single-Objective Optimization

The runtime of 2 days of a multi-objective optimization in our tests motivates alternative optimization approaches. Instead of computing the whole Pareto front the designer guides the construction of trade-off setups between regularity and improvement potential by setting a preference $\lambda \in [0, 1]$ based on his expertise.

By weighting Eqs. (7) and (8) we define an objective function f for a preference-based single-objective optimization:

$$f_\lambda(\mathbf{U}) = \lambda R(\mathbf{U}) + (1 - \lambda)P(\mathbf{U}). \tag{9}$$

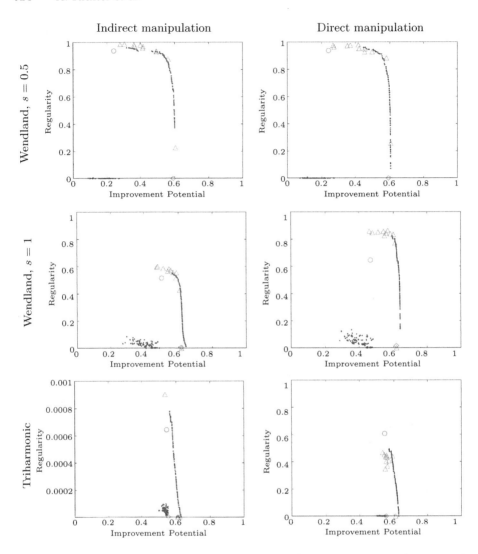

Fig. 6. The Pareto front (blue dots) and the initial random population (green dots) for the template fitting scenario. The magenta triangles are the results of the weighted single-objective optimization described in Sect. 6. Heuristic setups, described in Sect. 7, generated with Lloyd (orange circle) or OLS (orange diamond) result in very regular setups or setups with a very good improvement potential. (Color figure online)

Because such a single-objective optimization might not converge to the Pareto front, we analyze this in the following. As an optimization algorithm we choose a (25,100)-CMA-ES of the shark 2.3 library [9], we choose the preferences λ to be $0, 0.1, 0.2, \ldots, 1$ for Eq. (9) and run the optimization for 1000 generations. The optimization of a setup for one preference took approximately 2 h. The results of the single-objective optimization in Figs. 3 and 6 are depicted with the magenta triangles. The clustering of solutions, e.g., Figs. 3 and 6 middle, shows that uniformly distributed preferences λ do not result in uniformly distributed solutions along the Pareto front. Therefore, a designer has to set the preference carefully. The single-objective optimization converges towards the Pareto front and even performs slightly better because of its focus in one preferred direction, except for the triharmonic kernel in the template fitting scenario, where the optimizer gets stuck in local optima (Fig. 6, bottom right). This shows the feasibility of such an optimization for scenarios where a designer is interested in a setup for one particular preference. The runtime of 2 h and these local optima motivate efficient heuristics to distribute centers.

7 Heuristic Setup Strategies

Heuristic methods aim to generate good center distributions in a robust and efficient manner. They are analytically and geometrically motivated but lack the guarantee to be Pareto-optimal. In our test scenarios a single-objective optimization still runs for hours and might get stuck in local optima. Because we expect these drawbacks become worse for more complex scenarios, e.g., with a more complex initial design or a larger amount of parameters, we propose and analyze a geometry-motivated approach for very regular setups and an analytically motivated approach for setups with high improvement potential.

Pareto-optimal center distributions targeting regularity for indirect manipulation are uniform distributions in all our tests (Fig. 4, top). Hence, we apply Lloyd sampling which is also known as k-means clustering (see [10,11], respectively, for algorithmic details), which result in uniform center distributions similar to the Pareto-optimal solutions (compare Fig. 2, left, and Fig. 4, top). Comparing the regularity score of the resulting setup to the Pareto front (see Figs. 3 and 6, orange circles) reveals that the Lloyd sampling is close to the front for local Wendland kernels ($s = 0.25$ for the plane or $s = 0.5$ for the sphere). Even for direct manipulation uniform Lloyd sampling results in good regularity. For the triharmonic kernel in the template fitting scenario the heuristic even out-performs the multi- and single-objective optimization (Fig. 6, bottom right). According to Eq. (7) regularity is the ratio of the smallest to the largest singular value of the deformation matrix. For indirect RBF manipulation this singular value is bounded by the *separation distance*, which measures the minimal distance between any pairs of centers [25]. The uniform Lloyd sampling by construction has a good separation distance and thus results in good regularity. This sampling performs better than any tested random distributions, performs better than the evolutionary optimization in one test, is robust to local optima (Fig. 6, bottom right), and fast to set up (one minute).

In [16] we motivated improvement potential (Eq. (8)) by solving the approximation problem $\mathbf{g} = \mathbf{U}\mathbf{p}$ for an estimated fitness gradient $\mathbf{g}$, the deformation matrix $\mathbf{U}$, and the deformation parameters $\mathbf{p} = (p_1, \ldots, p_m)$. Each parameter p_j is the coefficient for $\varphi_j(\mathbf{x})$, which corresponds to a column $\mathbf{U}_j$ of the deformation matrix and to a center $\mathbf{c}_j$ for indirect manipulation. The orthogonal least

Wendland, $s = 0.25$ Wendland, $s = 0.5$ Triharmonic

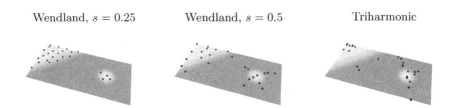

Fig. 7. Heuristic OLS setups with high improvement potential are adapted to the target. Wendland kernels are placed in regions with locally high fitting error (yellow) rather than in already optimal ones (blue). (Color figure online)

squares method (OLS, detailed description and algorithm in [4,7]) determines the influence of each parameter to minimize the approximation error to the estimated gradient in a greedy manner. According to their influence OLS ranks the parameters, thereby ranks the centers, and we select the most important ones. We initialize OLS with a large set of centers as candidates (30×30 in 2D or $30 \times 30 \times 30$ in 3D) on a uniform grid and greedily select the best 25 ones. We cannot apply this procedure for direct manipulation because the interpolation matrix $\mathbf{\Psi}^{-1}$ in Eq. (4) disbands the correspondence between parameters and centers. Since direct manipulation and indirect manipulation result in equal improvement potential for identical center distributions, we simply apply this algorithm for indirect manipulation and switch to direct manipulation afterwards. For the function approximation scenario we show that target-adapted setups for the Wendland kernels in Fig. 7 (left) are similar to the Pareto-optimal solutions in Fig. 5. But for the global triharmonic kernels OLS results in an unintuitive center placement (Fig. 7 right). Nonetheless, the OLS setups are close to the Pareto front or even hit it in both test scenarios (see Figs. 3 and 6, orange diamonds) for the compact Wendland kernels. In conjunction with the small computation time of 1 min, OLS is very efficient.

8 Summary and Future Work

The initial representation setup is crucial for the performance of an evolutionary optimization process. We analyzed the generation of RBF deformation setups for evolutionary design optimization for two test scenarios. The concept of evolvability reveals powerful criteria for setups, namely variability, regularity, and improvement potential, to measure the expected performance of a setup. Regularity and improvement potential are conflicting targets, which we therefore analyze with a multi-objective optimization. As downside this optimization process has a runtime of 2 days for our comparatively simple test scenarios.

In real-world applications we rather aim for one optimal deformation setup with respect to a user-specified preference between regularity and improvement potential. We demonstrated the feasibility of such a weighted single-objective optimization. For some tests the quality of the deformation setup is even better than the Pareto-optimal ones. This process is much faster, but it still runs for

2 h for our simple problems. Furthermore, the single-objective optimization gets stuck in local optima in some of our tests.

In order to further improve computational performance and robustness we proposed and analyzed heuristics to generate setups. The regular setups constructed by Lloyd sampling are close to the Pareto front for local Wendland kernels. Even for direct manipulation where we lack the geometrical motivation the regularity of the setup is significantly better than a random initialization. The Lloyd sampling even out-performs the evolutionary solutions in one example. Center distributions constructed with orthogonal least squares have high improvement potential and are on or very close to the Pareto front in all tests. Both methods reduce the computational effort from 2 h to 1 min.

For future work we blend between the heuristics, Lloyd and OLS, according to our preference. Moreover, we analyze the generalization of our methods to alternative deformations like free-from deformation.

Acknowledgments. Andreas Richter gratefully acknowledges the financial support from Honda Research Institute Europe (HRI-EU). Mario Botsch is supported by the Cluster of Excellence Cognitive Interaction Technology "CITEC" (EXC 277) at Bielefeld University, funded by the German Research Foundation (DFG).

References

1. Amoignon, O., Hradil, J., Navratil, J.: A numerical study of adaptive FFD in aerodynamic shape optimization. In: Proceedings of 52nd Aerospace Sciences Meeting (2014)
2. Amoignon, O., Navrátil, J., Hradil, J.: Study of parameterizations in the project CEDESA. In: Proceedings of 52nd Aerospace Sciences Meeting (2014)
3. Becker, G., Schäfer, M., Jameson, A.: An advanced NURBS fitting procedure for post-processing of grid-based shape optimizations. In: Proceedings of 49th Aerospace Sciences Meeting (2011)
4. Chen, S., Billings, S.A., Luo, W.: Orthogonal least squares methods and their application to non-linear system identification. Int. J. Control $50(5)$, 1873–1896 (1989)
5. Costa, E., Biancolini, M.E., Groth, C., Cella, U., Veble, G., Andrejasic, M.: RBF-based aerodynamic optimization of an industrial glider. In: Proceedings of International CAE Conference (2014)
6. Golub, G.H., Van Loan, C.F.: Matrix Computations. Johns Hopkins University Press, Baltimore (2012)
7. Gomm, J.B., Yu, D.L.: Selecting radial basis function network centers with recursive orthogonal least squares training. IEEE Trans. Neural Netw. $11(2)$, 306–314 (2000)
8. Hsu, W.M., Hughes, J.F., Kaufman, H.: Direct manipulation of free-form deformations. In: Proceedings of ACM SIGGRAPH, pp. 177–184 (1992)
9. Igel, C., Heidrich-Meisner, V., Glasmachers, T.: Shark. J. Mach. Learn. Res. 9, 993–996 (2008)
10. Lloyd, S.P.: Least squares quantization in PCM. IEEE Trans. Inf. Theory $28(2)$, 129–137 (1982)

11. MacQueen, J.: Some methods for classification and analysis of multivariate observations. In: Proceedings of the Fifth Berkeley Symposium on Mathematical Statistics and Probability, pp. 281–297 (1967)
12. Menzel, S., Olhofer, M., Sendhoff, B.: Direct manipulation of free form deformation in evolutionary design optimisation. In: Proceedings of the International Conference on Parallel Problem Solving From Nature, pp. 352–361 (2006)
13. Mina, A.A., Braha, D., Bar-Yam, Y.: Complex engineered systems: a new paradigm. In: Complex Engineered Systems: Science Meets Technology. Understanding Complex Systems, pp. 1–21. Springer, Heidelberg (2006)
14. Ohtake, Y., Belyaev, A., Seidel, H.P.: 3D scattered data approximation with adaptive compactly supported radial basis functions. In: Proceedings of IEEE International Conference on Shape Modeling Applications, pp. 31–39 (2004)
15. Olhofer, M., Jin, Y., Sendhoff, B.: Adaptive encoding for aerodynamic shape optimization using evolution strategies. In: Proceedings of IEEE Congress on Evolutionary Computation, pp. 576–583 (2001)
16. Richter, A., Achenbach, J., Menzel, S., Botsch, M.: Evolvability as a quality criterion for linear deformation representations in evolutionary optimization. In: Proceedings of IEEE Congress on Evolutionary Computation, pp. 901–910 (2016)
17. Richter, A., Botsch, M., Menzel, S.: Evolvability of representations in complex system engineering: a survey. In: Proceedings of IEEE Congress on Evolutionary Computation, pp. 1327–1335 (2015)
18. Sieger, D., Gaulik, S., Achenbach, J., Menzel, S., Botsch, M.: Constrained space deformation techniques for design optimization. Comput. Aided Des. **73**, 40–51 (2016)
19. Sieger, D., Menzel, S., Botsch, M.: A comprehensive comparison of shape deformation methods in evolutionary design optimization. In: Proceedings of the International Conference on Engineering Optimization (2012)
20. Simões, L.F., Izzo, D., Haasdijk, E., Eiben, A.E.: Self-adaptive genotype-phenotype maps: neural networks as a meta-representation. In: International Conference on Parallel Problem Solving from Nature, pp. 110–119 (2014)
21. Sterelny, K.: Symbiosis, evolvability and modularity. In: Modularity in Development and Evolution, pp. 490–516. University of Chicago Press (2004)
22. Vuong, A.V., Giannelli, C., Jüttler, B., Simeon, B.: A hierarchical approach to adaptive local refinement in isogeometric analysis. Comput. Methods Appl. Mech. Eng. **200**(49), 3554–3567 (2011)
23. Wagner, G.P., Altenberg, L.: Perspectives: complex adaptations and the evolution of evolvability. Evolution **50**(3), 967–976 (1996)
24. Webb, A.R., Shannon, S.: Shape-adaptive radial basis functions. IEEE Trans. Neural Netw. **9**(6), 1155–1166 (1998)
25. Wendland, H.: Scattered Data Approximation. Cambridge University Press, Cambridge (2004)
26. Yang, Z., Sendhoff, B., Tang, K., Yao, X.: Target shape design optimization by evolving B-splines with cooperative coevolution. Appl. Soft Comput. **48**, 672–682 (2016)
27. Zheng, J., Wang, Y., Seah, H.S.: Adaptive T-spline surface fitting to z-map models. In: Proceedings of the 3rd International Conference on Computer Graphics and Interactive Techniques in Australasia and South East Asia, pp. 405–411 (2005)

Design Perspectives of an Evolutionary Process for Multi-objective Molecular Optimization

Susanne Rosenthal[1]([✉]) and Markus Borschbach[1,2]([✉])

[1] Steinbeis Innovation Center 'Intelligent and Self-Optimizing Software
Assistance Systems', Paracelsusstr. 9, 51465 Bergisch Gladbach, Germany
{susanne.rosenthal,markus.borschbach}@stw.de
[2] University of Applied Sciences, FHDW, Hauptstr. 2, Bergisch Gladbach, Germany
markus.borschbach@fhdw.de

Simultaneous optimization of several physiochemical properties is an important task in the drug design process. Molecule optimization formulated as optimization problems usually provide several conflicting objectives. The number of molecular properties as well as the cost-intensive methods of molecule property prediction are stringent requirements to *in silico* aided drug design process. Numerical approximations of the physiochemical molecule properties are challenging and *in vitro* methods are still essential. The objective of an *in silico* aided drug design process is the evolution of a multi-objective evolutionary process with the potential of a selected number of improved molecule identification within a very low iteration number for a further efficient laboratory examinations. This paper presents design perspectives of a multi-objective evolutionary process that identifies a wide variety of genetic different but selected number of optimized molecules within a low number of generations. Its search behavior is compared to NSGA-II. Furthermore, limitations of the proposed algorithm are demonstrated with regard to its performance in many-objective molecular optimization and potential concepts of adaptations for this purpose are discussed.

1 Introduction

Peptides have a central role in the area of drug design for therapeutic and diagnostic interventions due to their high specificity and low toxicity profile. In the drug design process, native peptides or promising protein fragments are transformed into pharmaceutically acceptable components that dispose several physiochemical properties [1]. For this purpose a successive optimization of potential adequate molecules is required. Therefore, a computer-aided drug design process is a requested procedure due to its cost and time efficiency. In a recent work [2], a single-objective process for molecule identification has been published that revealed an exponential fitness improvement of candidate molecules within 10 generations. This advanced approach identifies optimized molecules by an evolutionary process as *in silico* methods and the fitness of the candidate solutions is accessed by an appropriate laboratory *in vitro* assay. This supports the fundamental finding of Singh [3] that improved molecules are identified in each successive generation of an evolutionary process. The potential of this early convergence behavior in single-objective evolutionary optimization is an important

© Springer International Publishing AG 2017
H. Trautmann et al. (Eds.): EMO 2017, LNCS 10173, pp. 529–544, 2017.
DOI: 10.1007/978-3-319-54157-0_36

characteristic as several physiochemical properties are not predictable by numerical approximation models and a combination of *in vitro* and *in silico* methods is required for a cost-efficient computer-aided drug design process. This raises the necessity of an evolutionary concept for simultaneous optimization of several molecular properties in an as low as possible number of iterations. More precisely, the issue is an evolutionary concept that identifies a wide variety of genetic different but selected number of molecules complying with several physiochemical properties in an optimal manner in a low number of iterations.

This paper presents design considerations of such an evolutionary process and discusses its search behavior in contrast to the basic model NSGA-II. This proposed algorithm makes use of a combination of deterministic dynamic variation operators and a selection strategy for the determination of the individuals for the succeeding generation. This selection strategy is tournament-based and a combination of fitness-proportionate and indicator-based selection. Furthermore, the limitations of the proposed algorithm are demonstrated with regard to its performance in many-objective molecular optimization.

This paper is structured as follows: in Sect. 2, related work with regard to multi-objective evolutionary molecule optimization is presented. Design perspectives for an evolutionary concept identifying a selected number of molecules complying with several physiochemical properties in an optimal manner within a low iterations number are discussed in Sect. 3. Section 4 motivates the molecular optimization problems and summarizes the simulation onsets. The simulation results are presented and discussed in Sect. 5. Section 6 gives an outlook on future work.

2   Related Work

Multi-objective Evolutionary Algorithms (MOEAs) have become established methods in the field of peptide- or protein-based drug design. Some work has been published regarding the use and the adaptations of state- of-the-art MOEAs for the purpose of multi-objective molecular optimization. This section gives an overview of this work published so far.

Hohm, Limbourg and Hoffmann published a Pareto-based MOEA for the design of effective peptide-based drugs [4]. This MOEA with problem-specific component is applied on an optimization problem for mimotope design with 3 objectives. Mutation, crossover and swapping are used as variation operators and a three-criteria based selection strategy is used to include the idea of elitism as well as genetic diversity. The mutation process makes use of a mutation pool comprising single amino acids for mutation as well as short amino acid sequences. These amino acids and amino acid sequences have different probabilities to be selected for insertion. This mutation pool undergoes a selection process: the probability of insertion depends on the times that a member of this pool is inserted and results in an improved individual. Pairwise single mutation crossover is used and the crossover points are chosen randomly. Since molecules sometimes provide good motifs in a suboptimal ordering, motives are swapped. The selection of the individuals for the succeeding generation is performed by binary tournament selection. A hierarchy of three selection criteria is used: firstly, the non-dominated solution is selected, secondly the individual in the less crowded hypercube is selected and thirdly the individual is randomly selected. Furthermore, this MOEA uses an archive pool for Pareto-optimal individuals. The test runs are performed with a population size of 30 and 2.500 generations.

NSGA-II is a commonly used and adapted evolutionary algorithm in the area of molecular optimization: Oduguwa, Tiwari, Fiorentino and Roy [5] perform a comparison of three multi-objective algorithms PAES [7], SPEA [6] and NSGA-II [8] for solving a 3-dimensional protein-ligand docking problem. The population size was set to 100, and 500 generations were performed. NSGA-II and PAES perform accurate and robust, but similar optimal solutions were identified by all three algorithms.

Lee, Shin and Zhang published the NSGA-II with constrained tournament selection for the DNA sequence optimization [9]. The DNA sequence problem is formulated as a 4-objective MOP with two constraints. A two-stage crossover process is used: the first stage is a sequence set level crossover, which is performed by an exchange of the sequences between two chromosomes. The second step is the one-point crossover with a probability of $p = 1.0$. The constrained tournament selection favors solutions, which are feasible, have less penalty or belong to a better front. Therefore, the selection process comprises three cases: firstly, the feasible solutions are selected, secondly the one with less penalty is selected and thirdly the dominating one is selected or otherwise the one with the larger crowding distance. The experiments were performed with a population size of 1000 and 200 generations.

Deb and Reddy published experiments with the traditional NSGA-II and two variants for identification of gene subsets to classify cancer samples [10]. The traditional NSGA-II as well as the two alternatives are applied on a 3-objective MOP. The NSGA-II variant termed multimodal NSGA-II identifies multimodal and non-dominated solutions in one single run by an adapted selection process. Multimodal solutions have identical objective values but different phenotypes. The alternative NSGA-II makes use of the biased dominance principle as an adaptation of the Pareto dominance principle. The biased dominance principle ensures that two solutions with identical complementary objective values are not dominating each other. Furthermore, solutions lying along the one objective axis have the potential to be non-dominated to each other. The test runs have performed with a population size of 500 with 500 generations.

Small et al. [20] apply the Indicator-Based Evolutionary Algorithm (IBEA) to optimize reagent combinations from a dynamic chemical library of 33 compounds with established or predicted targets in the regulatory network controlling IL-1β expression. IBEA identified excellent solutions within 11 generations, during which only 550 combinations out of a search space of 9 billions have been evaluated.

Sanchez-Faddev et al. [23] use a$(10 + 1)$-SMS-EMOA to identify 22-mer peptide ligand out of a search dimension of 23^{22} possible solutions. The optimization problem of ligand identification is formulated as bi-objective optimization problem. A new peptide is generated by mutation. This is performed by replacing a random amino acid of the sequence. A set of 9 Pareto optimal solutions is identified after 1189 iterations.

Sliwoski et al. [21] present a review of *in vitro*-based methods as well as computer-aided drug discovery methods used over the last three decades. Furthermore, the work discusses theory behind these methods and summarizes successful applications.

This work proposes design perspectives of a MOEA for multi-objective molecular optimization as well as for 'small' many-objective optimization problems. This algorithm is highly recommended to complement an *in vitro* drug design process as computer-assisting system since a wide variety of genetic different but selected number of highly qualified peptides are identified within a very low number of generation, more precisely within 10 generations.

3   Design Perspectives of the Evolutionary Process

Due to reported favorable convergence and diversity properties of NSGA-II at the same time in various applications [5,10,13], our proposed evolutionary process imitates the flow of NSGA-II, which is depicted in Fig. 1 and is further termed COmponent-Specific Evolutionary Algorithm for Molecule Optimization (COSEA-MO). The initial population of size N is generated by random individuals. The variation components of NSGA-II, pairwise single mutation and crossover, as well as the selection process with the critical component crowding distance operator are substituted. This section discusses the motivation of the component choice.

3.1   Individual Encoding and Search Space

The encoding scheme is aligned to the recommendations of Kershenbaum [11]. Five potentially conflicting properties are advised for an ideal encoding: the encoding schema has to present all feasible and only feasible solutions, which have an equal probability to be presented. The encoding scheme has to represent a useful scheme in a small number of genes that are close to one another in the chromosome and it has to possess locality in the way that small changes to the chromosome result in small changes in the solution.

A character string encoding is selected as encoding scheme to avoid the use of genotype-phenotype mapping. Various tools for molecular property prediction use single-letter string and consequently, a character string encoding is an established way of encoding. The individuals represent peptides and proteins consisting of commonly 20 canonical amino acids. Two alternative character string encodings are conceivable: firstly, the single-letters represent the character code of the amino acids and secondly a representation of each amino acid as nucleotide triple. The single amino acids are represented by nucleotide triples consisting of the four nucleotides adenine (A), cytosine (C), guanine (G) and uracil (U). The clear advantages of the single letter code for each amino acid

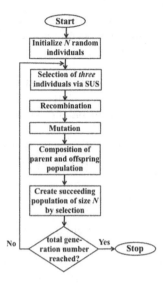

Fig. 1. Procedure of COSEA-MO for molecular optimization

in contrast to nucleotide tripe is that each molecule is represented by exactly one character string, only feasible molecules are represented, each molecule has the same probability to be represented and - more important - small changes performed by variation operators preserve similarity of the created offspring to their parents in the genetic material. This is an important property in molecular optimization as similarity in molecule structure is often related to similarity in its molecule properties [12]. The different probabilistic appearance of amino acids in a molecule which arises with nucleotide triple code is incorporable in the evolutionary process by a mutation pool containing amino acids according to their natural numerical existence. The complexity of the search space is l^{20}, where l is the molecule length and 20 is the number of the canonical amino acids.

3.2   Variation Operators

The issue of a suitable combination of mutation and recombination operator is motivated by a high explorative search behavior in the early generations and a more motifs-maintaining recombination in later generations as well as mutation method with high mutation rates in early generations to support the explorative search and therefore allow the discovery of new regions of the fitness landscape. Lower mutation rates in later generations support the exploitive search, which allows the convergence to other optima in the vicinity. Deterministic dynamic variation operators are suitable operators to achieve this purpose. The characteristic of deterministic dynamic operators is the adaptation of mutation and recombination rates by a predefined functional reduction with the iteration progress. This is implemented by deterministic functions depending on the current iteration number, the total number of iterations and the molecule length. Several deterministic dynamic operators have been investigated in their performance on a generic 3-dimensional molecular optimization problem [15,16] and the empirically optimal combination is achieved by the linear recombination operator 'LiDeRP' and an adaptation version of the deterministic dynamic mutation of Bäck and Schütz [18]. The combination of LiDeRP and the deterministic dynamic mutation provides the most successful balance of exploitation and exploration of the search process.

The recombination operator LiDeRP varies the number of recombination points over the generations via a linearly decreasing function:

$$x_R(t) = \frac{l}{2} - \frac{l/2}{T} \cdot t, \tag{1}$$

which depends on the length of the individual l, the total number of the generations T and the index of the current generation t.

The deterministic dynamic operator of Bäck and Schütz [18] determines the mutation probabilities via the following function with $a = 2$

$$p_{BS} = (a + \frac{l-2}{T-1}t)^{-1}, \tag{2}$$

The mutation rate is bounded by $(0; \frac{1}{2}]$. The mutation rate of the first generation has been adapted to a lower starting mutation rate with $a = 5$. This is due to the fact that the combination of a high mutation and recombination probability corresponds to a random creation of an individual.

3.3   Selection Strategy

There are several issues when designing an appropriate selection strategy for a MOEA in biochemical optimization. The first issue refers to the question of how to guide the search in the direction of optimal solutions. The second issue is to ensure a high spread of the non-dominated solutions. The third issue refers to the specific purpose of biochemical optimization: the selection has to ensure a high diversity of the genetic material inherited to the succeeding population. The high diversity of the genetic material supports the global search process. Ideally, the selection strategy has to comply with these three issues at the same time. Furthermore, change is another important component for the selection process especially in the field of molecular optimization. Change in the selection procedure imitates the aspect of change in a natural evolutionary process. The proposed selection strategy for COSEA-MO differs from the traditional selection process of NSGA-II in avoiding the critical component crowding distance to provide a reliably good algorithm performance independent of the problem dimension.

These considerations results in the following concept of a selection strategy: the use of tournament selection provides the subject of change in a selection process and ensures a high solution spread. Stochastic Universal Sampling (SUS) is used to ensure selection frequencies of the individuals in each front are in line with the expected frequencies. Consequently, the selection frequencies of individuals in one front are identical. Furthermore, SUS provides the opportunity for low quality solutions to find their way into the succeeding generation. Low quality solutions potentially have high quality genetic motifs, which produce high quality solutions in later generations. An indicator is used as a further selection criterium to ensure the selection of the optimal individual according to all objectives. This selection strategy and the ACV-indicator are introduced in [17]. The procedure of ACV-based Selection is depicted in Fig. 2. It starts with the tournament selection of ts individuals from the population. These individuals are ranked among themselves and the ACV value is calculated for each individual. From this ranked tournament set, the individuals with the lowest ACV-values are selected for the succeeding generation with a probability p_0, with the aim of guiding the search process in direction of high quality solutions. With a probability $1 - p_0$, the individuals are chosen from different fronts via SUS. The number of pointers in front-based SUS is equal to the number of fronts detected in the ranking process. The segments are equal in size to the number of individuals in each front. These steps repeat until the succeeding filial generation is complete. Consequently, this selection strategy has two parameters, the tournament size and the probability p_0 for choosing the individuals from the first front. Default values are $ts = 10$ and $p_0 = 50\%$.

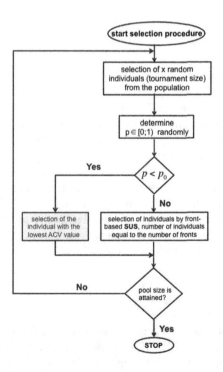

Fig. 2. ACV-based selection strategy with SUS.

4   Empirical Analysis

This section describes the simulation onsets of the experimental part. For comparison and discussion of the different algorithm search behavior, a generic 3- and 4-dimensional optimization problem predicting physiochemical molecular properties is generated. These optimization problems are generic in the sense that the physiochemical properties of molecules are calculated by descriptor values of each amino acid in the molecule sequence. The numerical approximation of physiochemical molecule properties is a commonly used method and a challenge in the area of molecule optimization. The dimension sizes of the optimization problems are selected under the aspect that evolutionary multi-objective optimization commonly concentrates on 2- and 3-dimensional problems, many-objective optimization refers to optimization problems with 4 and more objectives [14].

4.1   Four Molecular Fitness Objectives

The objective functions work on the amino acid sequence or the primary structure of molecules. Two fitness functions are provided by the BioJava library [19]: the first objective function determines the Molecular Weight (MW) of each individual, since molecules for drug design have to provide a minimized MW for a good membrane permeability. MW of a peptide sequence of the length l is determined in a most simple way, by the sum of the mass of each amino acid (a_i) plus a water molecule:

$$\sum_{i=1}^{l} mass(a_i) + 17.0073(OH) + 1.0079(H),\qquad(3)$$

where O (oxygen) and H (hydrogen) are the elements of the periodic system. The second function is the Needleman-Wunsch algorithm that is used as a method for the global sequence alignment to a pre-defined reference individual. This algorithm refers to the common hypothesis that a high similarity between molecules refers to similar molecular properties.

A common problem of drug peptides is a low solubility in aqueous solutions, especially peptides with stretches of hydrophobic amino acids. Therefore, the third fitness function calculates the average hydrophilicity via the hydrophilicity scale of Hopp and Woods with a window size equal to the peptide length l. An average hydrophilicity value is assigned to each candidate peptide using the scales of Hopp and Woods for each amino acid a_i [25]:

$$\frac{1}{l} \cdot (\sum_{i=1}^{l} hydro(a_i)).\qquad(4)$$

These three fitness functions generate the 3-dimensional problem (3D-MOP). The 4-dimensional problem (4D-MOP) is designed by the addition of the Instability Index (InstInd), also provided by the BioJava library. The use of molecules as therapeutic agents is potentially restricted by their instability. Peptides as therapeutic agents are of great interest as they are simple to synthesis, but a great problem is their potential degradation by enzymes in systemic application. The stability is addressed by the Instability Index as stability is a very important feature of drug components. The Instability Index is determined by the Dipeptide Instability Weight Values (DIWV) of each two consecutive amino acids in the peptide sequence. These values are summarized and the final sum is normalized by the peptide length l:

$$Instability\ Index = \frac{10}{l} \sum_{i=1}^{l} DIWV(x_i, x_{i+1}) \tag{5}$$

DIWV are provided by the GRP-Matrix [22].

These four objective functions act comparatively as they reflect the similarity of a particular molecule to a pre-defined reference molecule:

$$f(\text{CandidatePeptide}) := |f(\text{CandidatePeptide}) - f(\text{ReferencePeptide})| \tag{6}$$

Therefore, the four objective functions have to be minimized and the optimization problems are minimization problems.

4.2   Simulation Conditions

The molecules are short peptide sequences of the length 20 consisting of the 20 canonical amino acids. The search space has a complexity of 20^{20}. The populations are of a size of 100 and the initial population of every run is generated by random amino acid sequences. For statistical reasons, each configuration is repeated 30 times until the 10th generation. This low number of generations is generally motivated by the complement of the *in vitro* process with COSEA-MO. In this computer-assisted optimization, COSEA-MO is used to propose peptides *in silico* and the peptide fitness - constituted by the molecular objective function values - are determined in an laboratory *in vitro* assay.

These experiments are evaluated according to the three evolutionary multi-objective solution set requirements: convergence, diversity and pertinency. The Average Cuboid Volume (ACV) indicator is used as statistical convergence measure calculating the averaged covered space spanned by the solution set and an ideal reference point. The ideal reference point is chosen as the zero point. The ACV indicator has been introduced in [26]. There are two major aspects for the use of ACV as a convergence indicator: Firstly, ACV does not require the knowledge Pareto optimal solutions or a reference set of Pareto optimal solutions that are usually unknown in real-world MOPs. Secondly, ACV measures the quality of the solutions set relative to the set size. As a consequence, ACV allows the comparison of differently sized solution sets in a statistically reasonable way. The properties of the ACV indicator have been discussed in [24]: ACV is insensitive to multiple copies of equal solutions in the solution set, sensitive to refined Pareto optimal solution sets, but ACV is not usable as a diversity indicator. ACV shows a bias towards a specific part of the Pareto front with the lowest ACV-values. The ACV indicator is given by

$$ACV = \frac{1}{n} \sum_{i=1}^{n} (\prod_{j=1}^{k} (x_{ij} - r_j)), \tag{7}$$

where n is the number of individuals that are evaluated, k the number of objectives and r_j the pre-defined reference point. The lower the ACV indicator values are, the better is the global convergence behavior of the evaluated solution set.

The diversity within the population is assessed via:

$$\Delta = \sum_{i,j=1, i<j} \frac{|d_{ij} - \bar{d}|}{N} \qquad \text{with } N = \binom{n}{2} = \frac{n(n-1)}{2}, \tag{8}$$

where d_{ij} is the Euclidean distance of each possible combination of solutions. n is the number of solutions, here $n = 100$, and $\bar{d}$ is the average distance over all determined distances. The indicator values of diversity are scaled under the same criterion.

5   Simulation Results and Discussion

In a first test series, the search behavior of the traditional NSGA-II with a recombination rate of $p_r = 0.7$ and a mutation of $p_m = 0.1$ is compared against the search characteristics of COSEA-MO in the case of the 3D-MOP. The recombination and mutation rate corresponds to the findings in [10] and is motivated by a suitable balance of exploration and exploitation. Figures 3 and 4 depict the ACV results of the non-dominated solutions (NDS) in each generation identified by COSEA-MO and NSGA-II. The ACV-values are scaled for a better visualization. The stock charts show the average ACV-values as well as the standard deviations depicted by the

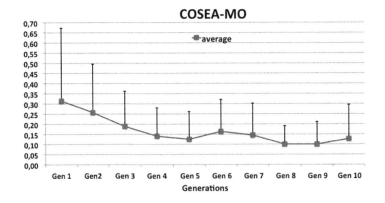

Fig. 3. ACV-values achieved by COSEA-MO within 10 generations for 3D-MOP.

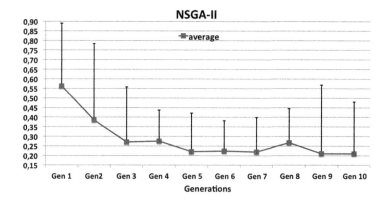

Fig. 4. ACV-values achieved by NSGA-II within 10 generations for 3D-MOP.

whiskers. Both algorithm reveal a quality improvement of the NDS over 10 generations, whereas COSEA-MO generally provides significantly lower and therefore
better convergence results compared to NSGA-II. Figure 5 provides box plots for
the average number of NDS within the 10 generations identified by COSEA-MO
and NSGA-II for the 3D-MOP. COSEA-MO detects a significantly lower number
of NDS compared to NSGA-II. Figure 7 depicts the diversity results of the NDS
identified by NSGA-II within 10 generations. The NDS diversity is slightly higher
in the case of NSGA-II compared to COSEA-MO. Therefore, it can be concluded
that the NDS identified by NSGA-II tends to extremal solutions for each objective function. Summarizing, the simulation results of COSEA-MO and NSGA-II
reveal that COSEA-MO provides a more selected number of high qualified NDS
within the first 10 generations. In the second test series, the search behavior of the
traditional NSGA-II is compared against the search characteristics of COSEA-
MO in the case of the 4D-MOP. Generally, the simulation results are comparable to those of the first test series: COSEA-MO reveals significantly better ACV-
indicator values than NSGA-II as depicted in the stock charts of Figs. 9 and 10.
The number of NDS is once again considerably higher in the case of NSGA-II
apart from some outlier (Fig. 6). Furthermore, an increase of the NDS number
is observable with the iteration number in the case of NSGA-II. The NDS diversity of COSEA-MO is generally of a broader range within the first 10 generations,
whereas the NDS diversity of NSGA-II is once again slightly higher caused by the
tendency of the NDS to extremal objective values (Fig. 8). These results allow the
conclusion that COSEA-MO provides a selected number of improved molecules in
a higher quality on average than NSGA-II. The simulation results of NSGA-II dis-
close three considerable disadvantages: firstly, a significantly lower evolutionary
progress within the first 10 generations is observable. Secondly, the manual inspection of the NDS identified a considerable lower genetic diversity of the molecule
sequences. Thirdly, the NDS reveal optimal values for only one objective function
each. The general higher NDS number makes this algorithm inefficient in a combined *in vitro* and *in silico* process. A well-know fact for the worse simulations

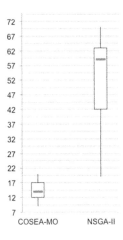

Fig. 5. Avg. no. of NDS for 3D-MOP.

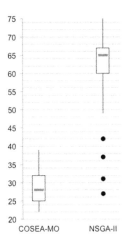

Fig. 6. Avg. no. of NDS for 4D-MOP.

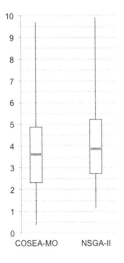

Fig. 7. Diversity of NDS for 3D-MOP.

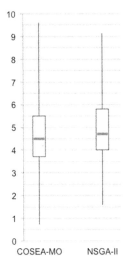

Fig. 8. Diversity of NDS for 4D-MOP.

results of NSGA-II, especially in the case of 4D-MOP, is the critical component selection of individuals for the succeeding generation by ranking and crowding distance operator, which explicitly uses the Pareto dominance principle. Pareto dominance principle usually results in increased number of NDS by a reduction of the front diversity with an increase of the problem dimension.

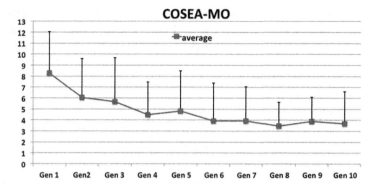

Fig. 9. ACV-values achieved by COSEA-MO within 10 generations for 4D-MOP.

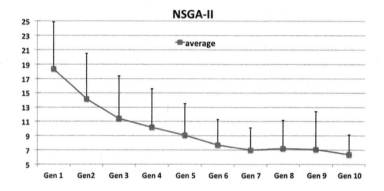

Fig. 10. ACV-values achieved by NSGA-II within 10 generations for 4D-MOP.

COSEA-MO explicitly makes use of the Pareto dominance principle by the front-based SUS in the selection strategy to determine the individuals for the succeeding generation. On the other hand, the combination of indicator-based and fitness-proportionate selection results in a selected number of improved molecules compared to NSGA-II. Due to the nearly doubled number of NDS comparing the COSEA-MO results of the 3D- and 4D-MOP allows the hypothesis that a further dimension increase results in a further increase of the NDS with the same disadvantages in the search behavior revealed by NSGA-II. Generally, two alternatives are conceivable: Firstly, an adaption of the selection parameter p_0 for the percentage selection of individuals by the ACV indicator. Secondly, another alternative is the substitution of the front-based SUS. The performance of the first alternative is limited by the property of the ACV indicator, ACV has a tendency to potentially cover one specific part of the Pareto front with the disadvantage of low diversity. Diversity is generally an important and challenging topic in many-objective optimization [27] and a method for ensuring a high spread within the solution set is necessary. This leads to the second alternative. A potential alternative for the front-based SUS is a method for ensuring biodiversity within the solution set as biodiversity is an important feature in the drug design process. A potential method to ensure biodiversity and simulation analysis of these methods are part of future research.

6   Conclusion

In this work, design perspectives of a MOEA for molecular optimization are presented with the aim of identification a selected number of improved molecules within a very low iteration number for a further efficient laboratory examination. COSEA-MO has been proposed that includes a combination of deterministic dynamic variation operators and a combined tournament-based and fitness-proportionate selection strategy for the determination of the individuals for the succeeding generation. COSEA-MO is compared to NSGA-II with regard to the search behavior investigated in the case of a generic multi-objective and a many-objective optimization problem. NSGA-II differs in its search behavior to COSEA-MO by its significant lower evolutionary progress in the first 10 generations, the general lower genetic diversity in molecule sequences and NDS which reveal optimal values for only one objective function each. The critical component of the proposed COSEA-MO related to many-objective optimization is the front-based SUS as selection criterion, which explicitly uses Pareto dominance ranking. Alternative methods for this strategy to maintain biodiversity as well as the influence of coevolution is part of future research.

References

1. Otvos, L.: Peptide-Based Drug Design: Methods and Protocols. Humana Press Inc., New York City (2000)
2. Röckendorf, N., Borschbach, M., Frey, A.: Molecular evolution of peptide ligands with custom-tailored characteristics. PLOS Comput. Biol. **8**(12), e1002800 (2012). doi:10.1371/journal.pcbi.1002800
3. Singh, J., Ator, M.A., Jaeger, E.P., et al.: Application of genetic algorithms to combinatorial synthesis: a computational approach to lead identification and lead optimization. J. Am. Chem. Soc. **118**, 1669–1676 (1996)
4. Hohm, T., Limbourg, P., Hoffmann, D.: A multiobjective evolutionary method for the design of peptidic mimotopes. J. Comput. Biol. **13**(1), 113–125 (2006)
5. Oduguwa, A., Tiwari, A., Fiorentino, S.: Multi-objective optimization of the protein-ligand docking problem in drug discovery. In: Genetic and Evolutionary Computation Conference, GECCO 2006, pp. 1793–1800 (2006)
6. Zitzler, E., Thiele, L.: Multiobjective evolutionary algorithms: a comparative case study and the strength Pareto approach. IEEE Trans. Evol. Comput. **3**(4), 257–271 (1999)
7. Knowles, J.D., Corne, D.W.: The Pareto archived evolution strategy : a new baseline algorithm for Pareto multiobjective optimisation. In: Proceedings of the Congress on Evolutionary Computation (CEC 1999), pp. 98–105 (1999)

8. Deb, K., Pratap, A., Agarwal, S.: A fast and elitist multiobjective genetic algorithm: NSGA-II. IEEE Trans. Evol. Comput. **6**(2), 182–197 (2002)
9. Lee, I.-H., Shin, S.-Y., Zhang, B.-T.: DNA sequence optimization using constrained multi-objective evolutionary algorithm. In: Evolutionary Computation, CEC 2003, pp. 2270–2276 (2003)
10. Deb, K., Reddy, A.: Reliable classification of two-class cancer data using evolutionary algorithms. BioSystems **72**, 111–129 (2003)
11. Kershenbaum, A.: When genetic algorithms work best. INFORMS J. Comput. **9**(3), 254–255 (1997)
12. Emmerich, M., Beume, N., Naujoks, B.: An EMO algorithm using the hypervolume measure as selection criterion. In: Coello Coello, C.A., Hernández Aguirre, A., Zitzler, E. (eds.) EMO 2005. LNCS, vol. 3410. Springer, Heidelberg (2005)
13. Ramesh, S., Kannan, S., Baskers, S.: Application of modified NSGA-II algorithms to multi-objective reactive power planning. Appl. Soft Comput. **12**(2), 741–753 (2012)
14. Fleming, P.J., Purshouse, R.C., Lygoe, R.J.: Many-objective optimization: an engineering design perspective. In: Coello Coello, C.A., Hernández Aguirre, A., Zitzler, E. (eds.) EMO 2005. LNCS, vol. 3410, pp. 14–32. Springer, Heidelberg (2005). doi:10.1007/978-3-540-31880-4_2
15. Rosenthal, S., El-Sourani, N., Borschbach, M.: Impact of different recombination methods in a mutation-specific MOEA for a biochemical application. In: Vanneschi, L., Bush, W.S., Giacobini, M. (eds.) EvoBIO 2013. LNCS, vol. 7833, pp. 188–199. Springer, Heidelberg (2013). doi:10.1007/978-3-642-37189-9_17
16. Rosenthal, S., Borschbach, M.: A benchmark on the interaction of basic variation operators in multi-objective peptide design evaluated by a three dimensional diversity metric and a minimized hypervolume. In: Emmerich, M., et al. (eds.) EVOLVE - A Bridge between Probability, Set Oriented Numerics and Evolutionary Computation IV, pp. 139–153. Springer, Heidelberg (2013)
17. Rosenthal, S., Borschbach, M.: Average cuboid volume as a convergence indicator and selection criterion for multi-objective biochemical optimization. In: Tantar, E., Tantar, A.-A., Bouvry, P., Del Moral, P., Legrand, P., Coello Coello, C.A., Schütze, O. (eds.) EVOLVE - A Bridge between Probability, Set Oriented Numerics and Evolutionary Computation. Springer, Heidelberg (2015)
18. Bäck, T., Schütz, M.: Intelligent mutation rate control in canonical genetic algorithm. In: Proceedings of the International Symposium on Methodology for Intelligent systems, pp. 158–167 (1996)
19. BioJava: CookBook, release 3.0. http://www.biojava.org/wiki/BioJava
20. Small, B.M., McColl, B.W., Allmendinger, R., et al.: Efficient discovery of anti-inflammatory small-molecule combinations using evolutionary computing. Nat. Chem. Biol. **7**, 902–908 (2011)
21. Sliwoski, G., Kothiwale, S., Meiler, J., Lowe, E.D.: Computational methods in drug discovery. Pharmacol Rev. **66**(1), 334–395 (2014)
22. Guruprasad, K., Reddy, B., Pandit, M.: Correlation between stability of a protein and its dipeptide composition: a novel approach for predicting in vivo stability of a protein from its primary structure. Protein Eng. **4**(2), 155–161 (1990)
23. Sanchez-Faddeev, H., Emmerich, M., Verbeek, F., et al.: Using multiobjective optimization and energy minimization to design an Isoform-selective ligand of the 14-3-3 protein. In: International Symposium on Leveraging Applications of Formal Methods, Verification and Validation, pp. 12–24 (2012)

24. Rosenthal, S., Borschbach, M.: Average cuboid volume as a convergence indicator and selection criterion for multi-objective biochemical optimization. In: Emmerich, M., Deutz, A., Schütze, O., Legrand, P., Tantar, E., Tantar, A.-A. (eds.) EVOLVE - A Bridge Between Probability. Set Oriented Numerics and Evolutionary Computation VII. Springer, Heidelberg (2017). doi:10.1007/978-3-319-49325-1_9

25. Hopp, T.P., Woods, K.R.: A computer program for predicting protein antigenic determinants. Mol. Immunol. **20**(4), 483–489 (1983)

26. Rosenthal, S., Borschbach, M.: Impact of population size and selection within a customized NSGA-II for biochemical optimization assessed on the basis of the average cuboid volume indicator. In: Proceedings of the Sixth International Conference on Bioinformatics, Biocomputational Systems and Biotechnologies, BIOTECHNO 2014, Charmonix, April 2014

27. Wang, H., Jin, Y., Yao, X.: Diversity assessment in many-objective optimization. IEEE Trans. Cybern. **PP**(99), 1–13 (2016)

Towards a Better Balance of Diversity and Convergence in NSGA-III: First Results

Haitham Seada[1(✉)], Mohamed Abouhawwash[2], and Kalyanmoy Deb[3]

[1] Department of Computer Science and Engineering, Michigan State University,
East Lansing, USA
seadahai@msu.edu
[2] Department of Mathematics, Faculty of Science, Mansoura University,
Mansoura, Egypt
saleh1284@mans.edu.eg, mhawwash@msu.edu
[3] Department of Electrical and Computer Engineering, Michigan State University,
East Lansing, USA
kdeb@egr.msu.edu
http://www.coin-laboratory.com

Abstract. Over the last few decades we have experienced a plethora of successful optimization concepts, algorithms, techniques and softwares. Each trying to excel in its own niche. Logically, combining a carefully selected subset of them may deliver a novel approach that brings together the best of some those previously independent worlds. The span of applicability of the new approach and the magnitude of improvement are completely dependent on the selected techniques and the level of perfection in weaving them together. In this study, we combine NSGA-III with local search and use the recently proposed Karush-Kuhn-Tucker Proximity Measure (KKTPM) to guide the whole process. These three carefully selected building blocks are intended to perform well on several levels. Here, we focus on Diversity and Convergence (DC-NSGA-III), hence we use Local Search and KKTPM respectively, in the course of a multi/many objective algorithm (NSGA-III). The results show how DC-NSGA-III can significantly improve performance on several standard multi- and many-objective optimization problems.

Keywords: NSGA-III, Diversity, Convergence, Local Search, KKTPM

1 Introduction

Traditionally, single objective evolutionary optimization researchers focused on the convergence abilities of their algorithms. However, since its advent, Evolutionary Mutliobjective Optimization (EMO) focused on both convergence and diversity. EMO switched the interest from only one solution to a set of competing solutions representing the trade-off among conflicting objectives, the Pareto front. Most early EMO algorithms relied on the concept of Pareto domination [1] for convergence. This practice continued until today. The major difference

© Springer International Publishing AG 2017
H. Trautmann et al. (Eds.): EMO 2017, LNCS 10173, pp. 545–559, 2017.
DOI: 10.1007/978-3-319-54157-0_37

between these algorithms was in the techniques they use to maintain diversity among solutions. Those techniques were mainly borrowed from single-objective evolutionary computation literature [2].

SPEA2 [3] and NSGA-II [4] are very good examples of successful algorithms that dominated the field for years. They were however unable to maintain diversity in more than two objectives [5]. Although SPEA2 can be considered the first considerable attempt in three dimensions, its high complexity and inability to scale further put it in the same category as NSGA-II.

Maintaining diversity in three or more objectives remained an obstacle until Zhang and Li proposed MOEA/D in 2007 [6]. Instead of the widely used crowding distance operator adopted by NSGA-II, MOEA/D used a decomposition based approach where a multiobjective optimization problem is divided into a number of different single objective optimization subproblems. Along with a population based algorithm, MOEA/D was able to solve up to four objectives. In 2014, Deb and Jain, proposed NSGA-III [7]. Their algorithm used a predefined evenly distributed set of reference directions to guide the search procedure. With a carefully designed normalization mechanism, their algorithm was shown to successfully maintain diversity up to fifteen objectives.

All these efforts remained in their majority isolated form theoretical/mathematical optimization and even from single-objective evolutionary optimization algorithms. Ishibuchi's IM-MOGLS [8] was the first study incorporating mathematical optimization as a local search in the context of EMO. His idea was simply to combine several objectives using a weighted sum approach and start a local search from each individual in his combined parents-offspring population. Subsequent researchers followed the same path [9]. Others explored achievement scalarization function (ASF) as an alternative means of combining several objectives into one in EMO, among those are Bosman [10] and Sindhya et al. [11].

Dutta et al. took another important step in bridging the gap between evolutionary and theoretical/mathematical optimization when they proposed their Karush-Kuhn-Tucker Proximity Measure (KKTPM) for single objective optimization [12]. They theoretically proved the correctness of their new metric and clearly showed how effective it can be in the context of evolutionary algorithms. The only drawback was that exactly computing KKTPM for a single solution is very expensive. The process involves solving a whole optimization sub-problem. Recently, their efforts were carried even further by Deb and Abouhawwash [13] who extended the metric to the realm of multiobjective optimization. In a later study they proposed an approximate yet computationally fast method to calculate KKTPM that avoids the internal optimization task [14].

All these advancements motivate us to explore further use of KKTPM in the design of efficient EMO methods. In this study we propose an integrated approach that combines NSGA-III, local search and KKTPM. The local search is performed using two distinct operators, one designed to enhance overall population diversity while the other puts additional emphasis on convergence. And since we are now able to quickly compute KKTPM to get an idea on the level of convergence of solutions, the second local search operator is guided by KKTPM throughout the entire optimization process.

A bipartite precursor of this study [15,16] showed that using KKTPM was very fruitful in terms of speeding up convergence, however we noticed a visible deterioration in terms of diversity. On the other hand if we completely abandoned KKTPM we will miss out on a significant convergence speed up. Hence, our proposed approach is intended to allow for improved diversity and convergence while keeping balance between them.

It is worth noting that NSGA-III follows the usual trend of continuously emphasising non-domination throughout the entire optimization process. This continuous emphasis influences the algorithm to seek convergence then diversity, in this specific order. DC-NSGA-III frequently breaks this emphasis during optimization. The reason behind this change is that the convergence-first principle that NSGA-III follows has some drawbacks which we summarize in the following points.

1. Losing diversity in early stages of the optimization process might be incurable in later generations, causing the population either to be trapped in local Pareto fronts or lose part(s) of the true Pareto front. The latter effect is usually observed with problems having non-continuous or disconnected Pareto fronts.
2. Most of the association and normalization efforts spent by NSGA-III at early generations are not very meaningful. This is true because reference directions keep changing with the least change in any of the extreme points reached so far.

In order to address all the aforementioned issues, we propose (DC-NSGA-III). The next section describes our proposed algorithm in detail. Section 3, discusses the theoretical and practical reasons behind employing each of our local search techniques in its place. Section 4, provides extensive simulation results on a set of both unconstrained and constrained, bi-objective and many-objective optimization problems. Finally, we conclude this study in Sect. 5.

2   DC-NSGA-III

In this study we keep the key characteristics of NSGA-III intact. Specifically speaking, our algorithm uses a predefined set of reference directions to keep diversity. These directions are evenly distributed over the normalized hyperplane which connects the normalized extreme point of the current population. Extreme points are extracted at each iteration from the combined parent offspring population using an ASF-based comparison along the M available extreme directions ($M =$ number of objectives). Association is performed every generation over the combined population, where each individual (solution/point) gets attached to its closest reference direction (in terms of perpendicular distance). Throughout this study we will use the notations $s.d$ and $d.surroundings$ to refer to the direction

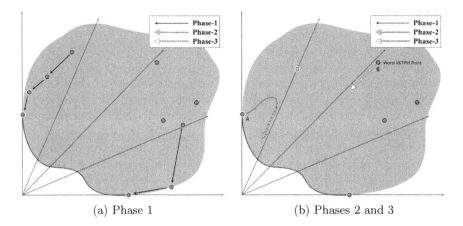

(a) Phase 1   (b) Phases 2 and 3

Fig. 1. Phases 1, 2 and 3. Part(a) shows how the algorithm continues in *phase-1* until all extreme points stagnates. Part(b) shows both *phase-2*, where a local search operation is started from the closest point to an empty reference direction, and *phase-3* where the algorithm tries to pull a poorly converged (large KKTPM) point.

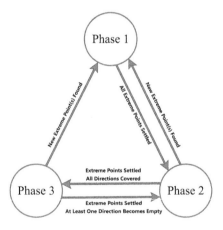

Fig. 2. Details showing how phases alternate among each other.

to which an individual s is attached and the set of individuals attached to a direction d respectively.

Having that said, our proposed approach breaks the usual NSGA-III pattern of continuously emphasising non-dominance. Every α generations our approach uses a different niching algorithm. The proposed algorithm alternates among three different phases. *Phase-1* is outlined in Algorithm 1, while *phase-2* and *phase-3* are both outlined in Algorithm 2.

2.1   Phase-1

Figure 1 shows a graphical representation of the three proposed phases. In the first phase, our goal is to find the extreme points of the Pareto-optimal front as accurately as possible. Thus, initially, the algorithm seeks better extreme points using a biased weighted-sum local search, starting from the already existing extreme points. We call this step *phase-1*. Any problem has a number of extreme points equal to the number of objectives M. If a better extreme point – than the one already existing – is found, the old one will no longer serve as an extreme point. The new point will then be used instead for later normalizations. And since The number of iterations required to find accurate extreme points of the Pareto front is unknown beforehand, *phase-1* starts in the beginning and continues for an undefined number of generations, and may even continue until the final generation if extreme points keep improving.

A stagnation triggers the transition from *phase-1* to *phase-2*. A stagnant extreme point is a one that is not improving anymore using *phase-1* local search. If the algorithm finds itself trying to repeat extreme point local search from the same starting point, it implies that the last local search was unsuccessful and consequently any further local search operations from the same starting point and using the same optimizer will be useless. Once all extreme points reach stagnation, the algorithm shifts its focus towards maintaining better diversity and moves on to *phase-2*.

2.2   Phase-2

In *phase-2*, our approach re-normalizes the current population using those extreme points attained so far. Association is then performed using the new reference directions. As opposed to the traditional front-by-front accommodation scheme adopted in NSGA-II and NSGA-III, this algorithm selects the closest individual to each reference direction. And since each reference directions represents a niche in the objective space, the algorithm tries to fill all the gaps in the current front. Consequently, regardless of an individual's rank, if it is the only representative of its niche, it will be selected at the expense of some – possibly – better ranked individuals which are already outperformed in their own niche (line 4).

The procedure explained above does not guarantee generating enough individuals to completely fill the next population. And even if it did, some of these individuals might be already dominated. Hence, the proposed algorithm makes an initial attempt to cover empty reference directions (i.e. those having no associations so far), using an ASF-based local search procedure. Since, *empty* reference directions has no associated points, local search is started from the closest point in the first front to the empty direction. Figure 1(b) shows how point A is used to start a local search along direction D.

2.3   Phase-3

After covering all reference directions, the algorithm then moves to the final phase. *Phase-3* focus completely on achieving better convergence for badly converged individuals. However, in order to detect such individuals in a population a deterministic and accurate scheme of preferability is needed. The usual domination check or a niching-based check cannot provide information about the extent of convergence of a solution in a multi-objective algorithm. However, in many cases, specially towards the end of optimization, most individuals fall in the same non-dominated front, although some of the nondominated solutions can be far away from the true Pareto-optimal front. The niching-based metrics can only provide information about their inter-spaced diversity. This is when a metric like KKTPM which has the ability to provide point-wise convergence property comes to our rescue. KKTPM can differentiate better or worse converged individuals even if they are on the same front. Such a metric was unavailable before and in this study we exploit its theoretical underpinning to isolate badly converged non-dominated solutions for a special treatment. It is worth noting that for black box optimization, using numerical gradients might significantly increase the total number of function evaluations. However this is left for a future study to investigate. *Phase-3* uses individuals having the worst KKTPM values from the entire combined feasible population as starting points for the ASF-based local search. Figure 1(b) shows how point B having the highest KKTPM is enhanced through an ASF-based local search along the direction to which it is attached.

All Function Evaluations (FEs) from the three phases contribute to the total FE count of the algorithm. In order keep this number low, the total number of local search operations performed per cycle in both phases 2 and 3 is bounded by the parameter β which is arbitrarily set to 2 in all our simulations.

Algorithm 1. Phase 1

Input: parent population (P), offspring (O) population size (N), reference directions (D), ideal point (I), intercepts (T), maximum number of function evaluations ($FeMax$), maximum number of local search operations per iteration β
Output: None
1: $All \leftarrow P \cup O$
2: $F \leftarrow getFeasible(All)$
3: $E \leftarrow getExtremePoints(F)$
4: $\hat{E}(i) \leftarrow localSearch_{BWS}(E(i), I, T, FeMax), \quad i = 1, \ldots, M$
5: $O(randomIndex) \leftarrow \hat{E}(i), \quad 1 \leq i \leq M, 1 \leq randomIndex \leq |O|$

In *phases 2 and 3*, and for the sake of saving even more function evaluations, the maximum limit of allowed function evaluation is kept proportional to the KKTPM value of the starting point. Consequently, weak points will be given extra emphasis during local search without consuming more function evaluations. Finally, the new population is completed by adding the best-ranked individuals among those overlooked during all the previous steps.

2.4   Alternating Phases

It is important to mention that although extreme points can reach a state of stagnation by the end of *phase-1*, this does not mean that these are the true extremes of the Pareto front. That is why the three phases of DC-NSGA-III are always applied in an alternating manner. During the evolutionary process, and while the algorithm is in either *phase-2* or *3*, if a better extreme point is found using the traditional genetic operators, the algorithm switches back immediately to *phase-1* and waits for another state of stagnation. Figure 2 graphically explains what triggers transition from each phase to the other.

The key idea behind this is to always find the extremes as accurately as possible. This will make the normalization of the objectives better and will help in matching the reference points with population objective vectors as close as possible so that a good distribution of points can be obtained at the end. *Phase-2* ensures finding points for empty reference lines proactively, thereby emphasizing the distribution of points. *Phase-3* focuses specifically on the convergence issue of poorly converged solutions. The reason for alternate use of these phases is the unreliability in obtaining the true desired Pareto-optimal solution in a single application of the local search. A local search using classical point-based optimization method can be useful in a problem, but it is not always sufficient with one application. It is after all a classical single-objective point by point procedure, which is prone to be trapped in a local optimum, and unable to solve the multi-objective optimization problem through decomposition specially in a single run. On the other hand, EMO algorithms are powerful, reliable but lack the ability to focus or even identify those points/sections that need more attention. Hence, we propose this approach of seamless information exchange and alternation between the three phases by the use of recent advancements in multi-objective algorithmic and theoretical fields.

3   Local Search Formulations

In this study we employ Matlab's® `fmincon()` optimization routine. First, we formulate a single-objective optimization problem out of the original multiobjective problem (scalarization), then `fmincon()` is applied to reach a solution. As mentioned before, our approach includes two types of local search, one for searching extreme points (*phase-1*) and the other for locally searching solutions along some specific reference direction (*phases-2* and *3*). Although both types of local search use the same optimizer, they differ in the way the problem is formulated.

$$
\begin{aligned}
\underset{\mathbf{x}}{\text{Minimize}} \quad & \text{BWS}_i(\mathbf{x}) = \epsilon \tilde{f}_i(x) + \sum_{j=1,\ j \neq i}^{M} w_j \tilde{f}_j(x), \\
\text{subject to} \quad & \epsilon < \min_{j=1, j \neq i}^{M} w_j
\end{aligned}
\tag{1}
$$

$$\text{Minimize} \quad \text{ASF}(\mathbf{x}, \mathbf{z}^r, \mathbf{w}) = \max_{i=1}^{M} \left(\frac{\tilde{f}_i(x) - u_i}{w_i} \right), \tag{2}$$

$$\text{subject to} \quad g_j(\mathbf{x}) \leq 0, \quad j = 1, 2, \ldots, J.$$

Extreme points local search optimizers are formulated in a Biased Weighted Sum (BWS) form. Equation 1 shows how a minimization problem aiming at finding extreme point (i) is formulated. Note that $\tilde{f}_k(x)$ is the normalized value of objective k. For all our bi-objective and many objectives simulations we use $\epsilon = 0.01$ and $\epsilon = 0.1$ respectively.

The first term in Eq. 1 is included to avoid weak extreme points. Thanks to this term, weak extreme points will have larger objective values than their true extreme counterparts.

Algorithm 2. Phases 2 and 3

Input: parent population (P), offspring (O) population size (N), reference directions (D), ideal point (I), intercepts (T), maximum number of function evaluations $(FeMax)$, maximum number of local search operations per iteration β

Output: New Population $(\hat{P})$

1: $Fronts \leftarrow nonDominatedSorting(All)$
2: $All \leftarrow P \cup O$
3: $F \leftarrow getFeasible(All)$
4: $\hat{P} \leftarrow getBestWithinNiche(d, O) \; \forall d \in D$
5: **if** $stagnant(E)$ **then**
6: $\quad D_{empty} \leftarrow \{d \in D \mid d.surroundings = \phi\},$
7: $\quad s.kktpm \leftarrow calculateKKTPM(s) \quad \forall s \in \hat{P},$
8: $\quad$ **for** i = 1 to β **do**
9: $\quad\quad$ **if** $D_{empty} = \phi$ **then**
10: $\quad\quad\quad d \leftarrow randomPick(D_{empty})$ ▶ Phase-2
11: $\quad\quad\quad s \leftarrow \{x \mid x \in F_1, \perp_d (x, d) \leq \perp_d (y, d) \; \forall y \in Front_1\}$
12: $\quad\quad$ **else**
13: $\quad\quad\quad s \leftarrow \{x \in \hat{P} \mid x.kktpm \geq y.kktpm \; \forall y \in \hat{P}\}$ ▶ Phase-3
14: $\quad\quad$ **end if**
15: $\quad\quad \hat{s} \leftarrow localSearch_{ASF}(s, s.kktpm, I, T, FeMax)$
16: $\quad\quad \hat{P} \leftarrow \hat{P} \cup x$
17: $\quad$ **end for**
18: **end if**
19: **while** $|\hat{P}| < N$ **do**
20: $\quad z \leftarrow \{x \mid x \in All, x.rank \geq y.rank \; \forall y \in All\}$
21: $\quad All \leftarrow All \setminus x$
22: $\quad$ **if** $x \notin \hat{P}$ **then**
23: $\quad\quad \hat{P} \leftarrow \hat{P} \cup x$
24: $\quad$ **end if**
25: **end while**

For directional local search, an Achievement Scalarization Function (ASF) is used, as shown in Eq. 2. Given a specific reference direction, an ASF is a means of converting a multi-objective optimization problem to a single-objective optimization problem whose optimal solution is the intersection point between the given direction and the original Pareto front.

The reason behind using two different formulations instead of one is that each of them has its merits and weaknesses. A weighted sum approach is straightforward, easy to understand and implement and – as will be shown later – easy to optimize using `fmincon()`. But, it is theoretically unable to attain any point lying on a non-convex section of the front. Consequently, it cannot be used as a general local search procedure for finding Pareto points.

However, since extreme points by their very nature can never be located on non-convex sections of the front, weighted sum can theoretically reach any extreme point. Combined with its simplicity and ease of use, BWS becomes the perfect choice for searching extreme points. On the other hand, ASF can – in general – reach any Pareto point if the appropriate reference direction is provided, which makes it the perfect choice to be used as a general local search mechanism with a reference direction based algorithm like NSGA-III. But when it comes to extreme points, ASF faces some serious problems that we summarize in the next points:

1. Unless the starting point is placed in a certain position with respect to the extreme point (inferior to the extreme point in all objectives), an ASF-based local search will never be able to reach it.
2. We observe that, the more flat/steep the provided direction is, the more difficult the problem is for `fmincon()` to solve.

For all these reasons, we use a BWS based local search in *phase-1* to seek extreme points, and an ASF-based local search in *phases-2* and *3* to cover gaps and enhance internal diversity. It is worth noting that ASF-based local searches are unable to handle even some of the easiest problems (e.g. ZDT2) due to the reasons discussed earlier. On the contrary, BWS based local search, can reach the true extreme points of easy problems in one attempt. However, for harder problems, one local search per extreme point is not sufficient. In these situations, our proposed approach with its inherent information exchange and alternation of phases shows its aptitude.

4   Results

Our simulation results involve a selected set of bi-objective and many-objective problems with and without constraints. The problems are also selected to exhibit a wide range of difficulties. The parameters used in our simulations are summarized in Table 1. Both GD and IGD are shown to compare convergence and overall performance, respectively. All the results presented in this study are for the median run out of 31 independent runs each started with a different random initial population.

Table 1. Parameters - From left to right: problem name, population size, total number of solution evaluations per run (SE), local search frequency (α), maximum local search operations per generation (β), maximum limit of solution evaluations per each local search operation

Problem	Pop. size	SE	α	β	Max. LS FE
ZDT4	48	36,288	5	2	500
ZDT6	24	4,032	5	2	500
TNK	24	2,832	5	2	500
SRN	24	6,144	5	2	500
OSY	48	7,488	5	2	500
DTLZ1(5)	128	64,128	5	2	500
DTLZ2(5)	128	31,347	5	2	500

4.1   Two Objectives

For bi-objective simulations we use TNK, SRN, OSY, ZDT4 and ZDT6 [17]. An appropriate maximum number of solution evaluations is set for each problem based on the difficulty of the problem as shown in Table 1. Note that both DC-NSGA-III and NSGA-III are allocated the same number of total SEs as mentioned in the third column of Table 1. The numbers are decided by running DC-NSGA-III for enough generations to get a reasonable performance (IGD). For easy problems (TNK and SRN) the difference in performance is minimal (figures removed due to limited space). This is because these problems are too easy to require any additional special consideration. However we emphasis that even with all these additional local search operations there is no degradation in performance even on these easy problems. The vertical lines indicate the number of solution evaluations at which each algorithm reach stagnation.

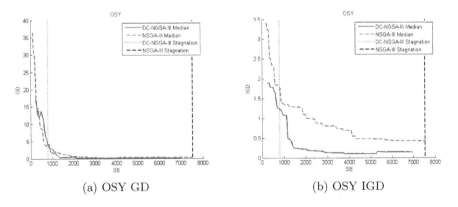

(a) OSY GD   (b) OSY IGD

Fig. 3. Variation of GD and IGD with solution evaluations for OSY.

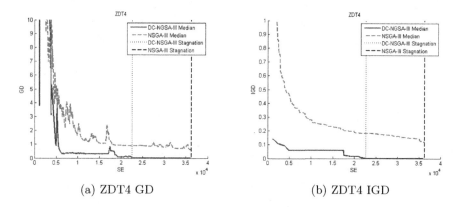

(a) ZDT4 GD   (b) ZDT4 IGD

Fig. 4. Variation of GD and IGD with solution evaluations for ZDT4.

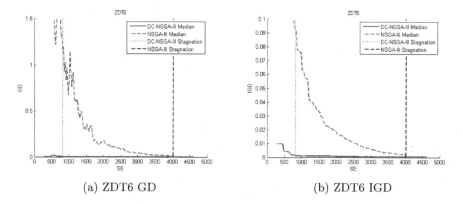

(a) ZDT6 GD   (b) ZDT6 IGD

Fig. 5. Variation of GD and IGD with solution evaluations for ZDT6.

It is clear from the figures that with more difficulties in problems the merit of our approach is more evident. Obviously, DC-NSGA-III outperforms NSGA-III using the same number of function evaluations in both convergence and diversity as shown in Figs. 3, 4 and 5 for OSY, ZDT4 and ZDT6, respectively. ZDT4 represents a real challenge to the convergence ability of any algorithm due to the presence of multiple local Pareto-optimal fronts. ZDT6 and OSY test the ability of the algorithm to maintain a diverse set of solutions due to biased density or unfavorable feasibility in the search space. Obviously, DC-NSGA-III excels in overcoming these two types of challenges compared to NSGA-III with the limited number of total solution evaluations fixed in this study (mentioned in Table 1). The final front-to-front comparison of the three problems are shown in Figs. 6 and 7. These SEs are too few for NSGA-III (without any local search or other checks and balances) to bring its population close to the true Pareto-optimal front.

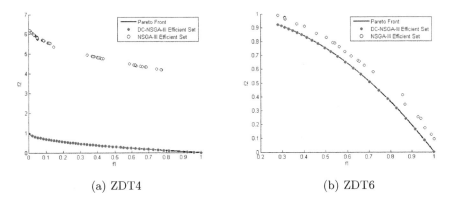

(a) ZDT4   (b) ZDT6

Fig. 6. Comparison of obtained fronts for ZDT4 and ZDT6 between DC-NSGA-III and NSGA-III median of 31 runs (final generation)

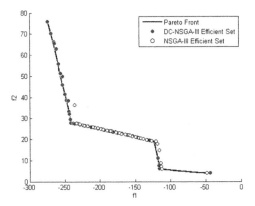

Fig. 7. Comparison of obtained fronts for OSY between DC-NSGA-III and NSGA-III for the median of 31 runs.

Figure 8 shows a typical alternation sequence recorded over a single optimization run of OSY. Although the algorithm keeps alternating between the three phases, it clearly indicates that *phase-1* is dominant in the beginning of the run yet less frequent later. That is when all extreme points tend to stagnate, giving higher priority to convergence towards the end of the run.

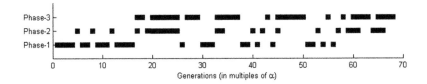

Fig. 8. Alternation among the three phases in a random OSY run

Thus, the results on bi-objective problems reveal that our proposed DC-NSGA-III approach performs equally well to a vanilla NSGA-III approach on simpler problems, but exhibits a much better performance on more difficult ones.

4.2   Many Objectives

On the many objectives side, we consider DTLZ1 and DTLZ2 [18], both using five objectives. The results show a visible improvement on DTLZ2 (5 obj) achieved by DC-NSGA-III over NSGA-III in terms of both convergence and diversity measures. However, for five-objective DTLZ1, NSGA-III performance is slightly better. This is because the BWS-based local search optimizer is faced by a difficult problem having a large number of weakly dominated false extreme points along each axis. As the number of objectives grow, the effect of the ϵ-objective gradually vanishes.

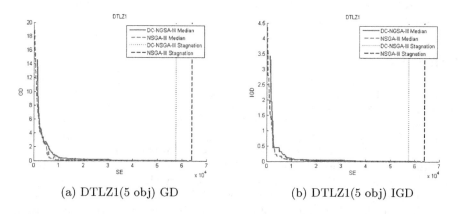

(a) DTLZ1(5 obj) GD   (b) DTLZ1(5 obj) IGD

Fig. 9. Variation of GD and IGD with solution evaluations for 5-objective DTLZ1.

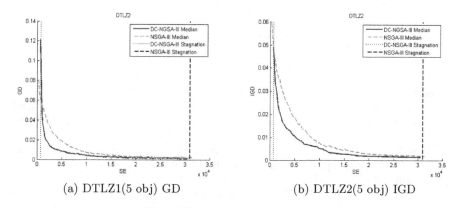

(a) DTLZ1(5 obj) GD   (b) DTLZ2(5 obj) IGD

Fig. 10. Variation of GD and IGD with solution evaluations for 5-objective DTLZ2.

5   Conclusions

In this study we have proposed a novel integrated EMO approach combining NSGA-III with a recently proposed KKTPM measure. The proposed DC-NSGA-III has used two types of local search operators. Through an alternating three-phases approach, the proposed algorithm has shown to enhance both diversity and convergence on a set of bi-objective and many-objective optimization problems. This confirms the usefulness of integrating optimization algorithms and techniques to form more powerful optimization concepts, a research direction that has started to receive attention among EMO researchers. This work can be further extended through a parametric study of the ϵ parameter used in the BWS local search operator. Also, more problems involving more objectives can be included. Another possible extension is to modify this algorithmic design to encompass single-objective optimization problems as well. Despite the need for pursuing these additional studies, these initial results have indicated a definite merit of our proposed hybrid EMO algorithm which has embraced point-based local search methods and theoretical results of optimization in a way to benefit the overall algorithm.

References

1. Deb, K.: Multi-objective Optimization Using Evolutionary Algorithms, vol. 16. John Wiley Sons, Hoboken (2001)
2. DeJong, K.A.: An analysis of the behavior of a class of genetic adaptive systems. Ph.D. thesis, Ann Arbor, MI: University of Michigan, Dissertation Abstracts International 36(10), 5140B (University Microlms No. 76–9381) (1975)
3. Zitzler, E., Laumanns, M., Thiele, L., Zitzler, E., Zitzler, E., Thiele, L., Thiele, L.: SPEA2: improving the strength pareto evolutionary algorithm (2001)
4. Deb, K., Agrawal, S., Pratap, A., Meyarivan, T.: A fast elitist non-dominated sorting genetic algorithm for multi-objective optimization: NSGA-II. In: Schoenauer, M., Deb, K., Rudolph, G., Yao, X., Lutton, E., Merelo, J.J., Schwefel, H.-P. (eds.) PPSN 2000. LNCS, vol. 1917. Springer, Heidelberg (2000)
5. Khare, V., Yao, X., Deb, K.: Performance scaling of multi-objective evolutionary algorithms. In: Fonseca, C.M., Fleming, P.J., Zitzler, E., Thiele, L., Deb, K. (eds.) EMO 2003. LNCS, vol. 2632, pp. 376–390. Springer, Heidelberg (2003). doi:10.1007/3-540-36970-8_27
6. Zhang, Q., Li, H.: MOEA/D: a multiobjective evolutionary algorithm based on decomposition. IEEE Trans. Evol. Comput. **6**, 712–731 (2007)
7. Deb, K., Jain, H.: An evolutionary many-objective optimization algorithm using reference-point-based nondominated sorting approach, part I: solving problems with box constraints. IEEE Trans. Evol. Comput. **18**(4), 577–601 (2014)
8. Ishibuchi, H., Murata, T.: A multi-objective genetic local search algorithm and its application to owshop scheduling. IEEE Trans. Syst. Man Cybern. Part C Appl. Rev. **28**(3), 392–403 (1998)
9. Knowles, J.D., Corne, D.W.: M-PAES: a memetic algorithm for multiobjective optimization. In: Proceedings of the Congress on Evolutionary Computation, vol. 1, pp. 325–332. IEEE (2000)

10. Bosman, P.A.: On gradients and hybrid evolutionary algorithms for real-valued multiobjective optimization. IEEE Trans. Evol. Comput. **1**, 51–69 (2012)

11. Sindhya, K., Deb, K., Miettinen, K.: A local search based evolutionary multi-objective optimization approach for fast and accurate convergence. In: Rudolph, G., Jansen, T., Beume, N., Lucas, S., Poloni, C. (eds.) PPSN 2008. LNCS, vol. 5199, pp. 815–824. Springer, Heidelberg (2008). doi:10.1007/978-3-540-87700-4_81

12. Dutta, J., Deb, K., Tulshyan, R., Arora, R.: Approximate KKT points and a proximity measure for termination. J. Global Optim. **4**, 1463–1499 (2013)

13. Deb, K., Abouhawwash, M.: An optimality theory based proximity measure for set based multi-objective optimization. IEEE Trans. Evol. Comput. (in press)

14. Deb, K., Abouhawwash, M.: A computationally fast and approximate method for Karush-Kuhn-Tucker proximity measure. Technical report, COIN Report No. 015, Department of Electrical and Computer Engineering, Michigan State University, East Lansing, USA (2015)

15. Abouhawwash, M., Seada, H., Deb, K.: Karush-Kuhn-Tucker optimality based local search for enhanced convergence of evolutionary multi-criterion optimization methods. Comput. Oper. Res. (in press)

16. Seada, H., Abouhawwash, M., Deb, K.: Towards a better diversity of evolutionary multi-criterion optimization algorithms using local searches. In: Proceedings of the on Genetic and Evolutionary Computation Conference Companion, pp. 77–78. ACM (2016)

17. Zitzler, E., Deb, K., Thiele, L.: Comparison of multiobjective evolutionary algorithms: empirical results. Evol. Comput. **8**(2), 173–195 (2000)

18. Deb, K., Thiele, L., Laumanns, M., Zitzler, E.: Scalable test problems for evolutionary multiobjective optimization. In: Abraham, A., Jain, L., Goldberg, R. (eds.) Evolutionary Multiobjective Optimization. Springer, Heidelberg (2005)

A Comparative Study of Fast Adaptive Preference-Guided Evolutionary Multi-objective Optimization

Florian Siegmund[1(✉)], Amos H.C. Ng[1], and Kalyanmoy Deb[2]

[1] School of Engineering Science, University of Skövde,
Högskolevägen, 54128 Skövde, Sweden
{florian.siegmund,amos.ng}@his.se
[2] Department of Electrical and Computer Engineering,
Michigan State University, 428 S. Shaw Lane, East Lansing, MI 48824, USA
kdeb@egr.msu.edu

Abstract. In Simulation-based Evolutionary Multi-objective Optimization, the number of simulation runs is very limited, since the complex simulation models require long execution times. With the help of preference information, the optimization result can be improved by guiding the optimization towards relevant areas in the objective space with, for example, the Reference Point-based NSGA-II algorithm (R-NSGA-II) [4]. Since the Pareto-relation is the primary fitness function in R-NSGA-II, the algorithm focuses on exploring the objective space with high diversity. Only after the population has converged close to the Pareto-front does the influence of the reference point distance as secondary fitness criterion increase and the algorithm converges towards the preferred area on the Pareto-front.

In this paper, we propose a set of extensions of R-NSGA-II which adaptively control the algorithm behavior, in order to converge faster towards the reference point. The adaption can be based on criteria such as elapsed optimization time or the reference point distance, or a combination thereof. In order to evaluate the performance of the adaptive extensions of R-NSGA-II, a performance metric for reference point-based EMO algorithms is used, which is based on the Hypervolume measure called the Focused Hypervolume metric [12]. It measures convergence and diversity of the population in the preferred area around the reference point. The results are evaluated on two benchmark problems of different complexity and a simplistic production line model.

Keywords: Evolutionary multi-objective optimization · Guided search · Preference-guided EMO · Reference point · Decision support · Adaptive

1 Introduction

In Simulation-based Evolutionary Multi-objective Optimization with complex simulation models, the number of solution simulations is usually limited.

© Springer International Publishing AG 2017
H. Trautmann et al. (Eds.): EMO 2017, LNCS 10173, pp. 560–574, 2017.
DOI: 10.1007/978-3-319-54157-0_38

In today's world, the preconditions for system operation and processes change frequently. For example, while an automotive production line may be optimized for the current input and customer demand, those figures may change within a few weeks, which means the car manufacturer will need to react and adapt the production promptly. Consequently, only a few days would be available for optimizing the system. Given a 15 min simulation time and half a day of available time, only 48 configuration solutions could be simulated. Running optimization in parallel with high-performance computing resources with 50 parallel evaluations will allow the execution of approximately 2,500 evaluations during those 12 h.

This number of available evaluation samples should be used in the best possible way. If a decision maker specifies an area in the objective space where he or she expects good solutions, the optimization can be guided in that direction. The exploration of the whole Pareto-front can be avoided and the preferred area can be explored better within the available optimization time. Preference articulation can be done in several ways. The Reference Point-based NSGA-II algorithm [4] asks the decision maker to specify one or several reference points in the objective space and guides the population or multiple sub-populations towards the reference points. In the Visual Steering method [14], reference points are used as attractor points. However, Pareto-dominance is not used for all operation modes. [1] use reference points (and other methods) to build a weighted hypervolume indicator that guides the optimization towards the reference point. Another way of preference articulation is to let the decision maker compare multiple pairs of solutions [5]. In their publication, the binary preferences are transformed into a value function in the objective space that increases the fitness of solutions in preferred areas. Optimization is used to build a function that matches the specified binary preferences as good as possible.

In this paper, we focus on the R-NSGA-II algorithm which features three fitness criteria: Non-domination sorting, ϵ-clustering, and the distance to the reference point. The Pareto-dominance is the primary fitness criterion with the most influence on the algorithm behavior. Only after convergence to the Pareto-front do the other fitness criteria become as important as the Pareto-dominance. This leads to R-NSGA-II spending a lot of time exploring the objective space instead of exploring the preferred area. Therefore, we investigate how R-NSGA-II can be extended to reduce the strong influence of the dominance relation. We propose several adaptive extensions of the algorithm, aiming to achieve a speedup of the convergence towards the reference point.

The paper is structured as follows. Section 2 provides an introduction to the Reference point-based NSGA-II algorithm (R-NSGA-II). Different criteria for adaptive control of the R-NSGA-II are explained in Sect. 3. Section 4 describes and proposes the new adaption strategies for R-NSGA-II to accelerate convergence towards the reference point. In Sect. 5, numerical experiments with the different adaptive R-NSGA-II variants are performed and evaluated with the Focused Hypervolume metric. Conclusions are drawn and possible future work is suggested in Sect. 6.

2   The R-NSGA-II Algorithm

In this section, the R-NSGA-II algorithm for Preference-Guided Evolutionary Multi-objective Optimization is described. The R-NSGA-II algorithm was proposed by [4] and it is based on the NSGA-II algorithm [2]. However, the crowding distance for the partially selected front is replaced by a fitness criterion that is based on the distance to the reference point. Offspring solutions are created by crossover and mutation. From the set of population and offspring solutions, the next population is selected by non-domination sorting and distance-based selection. A selection step example is shown in Fig. 2. After selection, the process continues with parental selection. This loop is called a generation and it continues until a maximum number of generations has been reached. The algorithm pseudocode is given in Algorithm 1.

In the selection step of R-NSGA-II, a clustering method is used that allows the diversity of the population to be controlled. Without this step, the population would converge towards one point that is closest to the reference point.

Two extensions of R-NSGA-II have been proposed by the authors in [10] (Constrained Pareto-fitness) and [13] (Delayed Pareto-fitness). They are described in the following section, together with two new adaptive R-NSGA-II extensions.

3   Adaptive R-NSGA-II

Our experience with the bi-objective ZDT benchmark function suite with deterministic objective functions [15] shows that R-NSGA-II in the standard configuration as in article [4], with population size $|P| = 50$, first focuses on the Ideal point [10]. After convergence, the population explores the Pareto-front, and then it focuses towards the reference point of R-NSGA-II. This leads to a delay in convergence towards the reference point. The effect is stronger if the objective functions are noisy. In this paper, we therefore introduce different R-NSGA-II modifications that can avoid delayed convergence towards the reference point, even for stochastic objective functions. These methods are adaptive and can change according to the problem settings. Therefore, in the following section, we first present criteria which can help R-NSGA-II adapt to the problem. These criteria are the elapsed optimization time in limited-time optimization, and the distance and progress towards the reference point.

3.1   Time-Based Adaptive Control

With Time-based Adaptive Control (TAC), we can define a progress value $p \in [0, 1]$ which can be used to adapt the algorithm behavior. In Eq. 1, the time criterion increases along with the elapsed optimization time B_t until the total budget B is reached. This total available time is adjusted by B_F final objective value samples which are evaluated for the last population. Each solution in the last population is sampled b_f times in order to reduce noise in the objective

---

Algorithm 1. R-NSGA-II algorithm pseudocode.

1: Generate initial population P.
2: Evaluate the solutions.
3: Perform non-domination sorting.   // Start new generation
4: Calculate solution distances to reference point r and their reference point ranks,
5: based on ASF distance $\delta_{ASF}(F(s), r) = \max_{i=1}^{H}\{F_i(s) - r\}$, where $F(s)$ is the
6: objective vector of solution s with H objectives.
7: **repeat**   // Parental tournament selection and mating
8:     **for** $i \leftarrow 1, 2$ **do**
9:         Randomly select two parent solutions out of the population.
10:         Select the solution for mating which dominates the other solution.
11:         In case of mutual non-domination, compare by reference point rank.
12:     **end for**
13:     Create offspring solution by mating of the two parents with SBX crossover.
14: **until** $|Q| = |P|$ offspring solutions have been created.
15: Apply Polynomial Mutation on all offspring solutions.
16: Evaluate each solution in Q.
17:    // Environmental selection
18: Perform non-domination sorting on $P \cup Q$.
19: Calculate reference point ranks.
20: **Variant**: Select all solutions in all fronts that fit into the next generation.
21: (non-clustered)
22: (Re-)run ϵ-CLUSTERING() on all fronts with remaining solutions (non-selected).
23: **for all** fronts **do**
24:     Determine representative solution in each cluster (randomly or closest to r).
25: **end for**
26: Select all representative solutions in all fronts that fit completely into the next
27: generation (only representatives).
28: **if** all fronts fit into the next population **then**
29:     go to line 22.
30: **else**
31:     For the front that only partially fits, step-wise select the cluster representative
32:     solutions which are closest to r, until the next population is complete.
33: **end if**
34: **if** less than B solutions have been evaluated **then**   // Termination criterion
35:     go to line 3 and start new generation.
36: **end if**
37: **return** non-dominated solutions.

38: **function** ϵ-CLUSTERING(*front*, r, ϵ)
39:     **repeat**
40:         Find the solution closest to r in *front*, based on ASF distance.
41:         Around this solution, create cluster of all solutions within ϵ-distance.
42:         (Euclidean distance)
43:         Remove all solutions within the new cluster from *front*.
44:     **until** *front* is empty.
45:     **return** set of clusters.
46: **end function**

---

values, i.e., $B_F = \sum_{i=1}^{|P|} b_f$. Parameter $a > 0$ allows to accelerate or delay the progress measure.

$$p^{\mathrm{TAC}} = \min \left\{ 1, \frac{B_t}{B - B_F} \right\}^a. \tag{1}$$

Since we have preference information available in R-NSGA-II, we can make use of it to define another adaptive control strategy. We propose Distance-based Adaptive Control based on the DDR resampling algorithm [11], which is described in the following section.

3.2 Distance-Based Adaptive Control for Unattainable Reference Points

Distance-based Adaptive Control (DAC) is based on the adaption mechanism of Distance-based Dynamic Resampling (DDR). DDR was proposed in [11] and it assigns samples to a solution based on its distance to a reference point. Since a reference point in the objective space can be unattainable by the optimization, the DDR algorithm combines the distance allocation with a progress measure. Such reference points are common, since the decision maker usually picks a reference point from an optimistic area in the objective space. Therefore, if the optimization convergence slows down, the progress measure will indicate that the best attainable solutions with the closest distance to the reference point have been found. In addition, DDR also considers the elapsed optimization time, in order to handle situations of premature convergence in a local Pareto-front [11]. This adaptive control method can be used to provide a progress measure $p \in [0, 1]$ for an adaptive R-NSGA-II algorithm. The formula for the distance-based control is given in Eq. 2. For the progress and time mechanisms please refer to [11].

$$p^{\mathrm{DAC}} = \min \left\{ 1, \left(\frac{1 - d_s}{1 - \underline{\delta}} \right)^a \right\}, \quad d_s = d_s^{1/4}, \ s \in S. \tag{2}$$

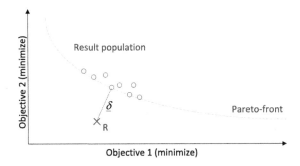

Fig. 1. Bi-objective min/min-problem scenario with infeasible reference point R. The distance of the result population to R is limited by a lower bound $\underline{\delta}$.

In Eq. 2, d_s stands for the distance of solution s to the reference point R. $\underline{\delta}$ is the minimum achievable distance to R by any solution. p is determined by calculating the DDR allocation for solution s which has the lower quartile distance $d_s^{1/4}$ among all solutions in the set of population and offspring S. The lower quartile is chosen to avoid the influence of outliers (Fig. 1).

4 Adaption Strategies for R-NSGA-II Speedup

In this section, we describe four different adaptive extensions for the R-NSGA-II algorithm which can accelerate convergence towards the reference point.

4.1 Delayed Pareto-Fitness

We proposed the Delayed Pareto-fitness (DPF) extension for R-NSGA-II in [13]. The goal was to achieve a convergence speedup as well as a better integration of Dynamic Resampling into R-NSGA-II. Here, we extend this approach by the Distance-based Adaptive Control. The DPF allows to run R-NSGA-II exclusively with the reference point distance and ϵ-diversity as fitness functions (scalar mode), until a certain progress threshold θ is reached. When this threshold is exceeded, the algorithm works as the standard R-NSGA-II. Equation 3 shows the condition for the two operation modes and the right side of Fig. 2 depicts an example of a selection step with Delayed Pareto-fitness.

$$op\_mode = \begin{cases} scalar & \text{if } p < \theta, \\ standard & \text{otherwise.} \end{cases} \tag{3}$$

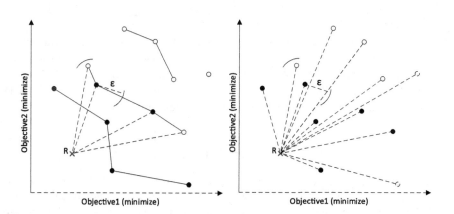

Fig. 2. R-NSGA-II algorithm – Example for selection step in objective space (left side). The population size is $|P| = 6$. Four solutions are selected from the first front, and two from the second front. For the Delayed Pareto-fitness extension (right side), the distances from the objective vectors to the reference point R are compared and the best $|P| = 6$ solutions are selected.

The standard mode is also activated as soon as a solution is found that dominates the reference point. This means that the reference point is attainable and, if the standard mode were not used in this case, then the best solutions dominating the reference point could not be found. A similar approach was used by [8,9], but it makes comparisons based on dominance and distance during the whole runtime. In [13], we designed the DPF extension to exclusively use the distance criterion, which facilitates the application of Dynamic Resampling.

A similar approach was used by [14] and is called Visual Steering. They guide the search towards an arbitrary attractor point in the objective space. However, their guided search by attractor point does not use Pareto-dominance, which leads to the described effect that for feasible reference points the best solutions cannot be found. Instead, a circle of solutions as close as possible around the reference point is found.

4.2   Constrained Pareto-Fitness

The Constrained Pareto-fitness (CPF) extension was proposed in [10]. In this paper, the selection operator of R-NSGA-II was modified by controlling the number of solutions selected from each individual front. The idea with this extension is to constrain the number of solutions from the first fronts which have the highest influence on the optimization process through mating. However, it is still important to select a large part of the next population from those fronts. Otherwise, a strong emphasis on solutions from inferior fronts would render the algorithm ineffective. Therefore, a decreasing assignment of solutions for selection is used. The number of solutions that should be selected from front F_i is calculated as in Eq. 4.

$$\#Sel(F_i) \leq \frac{|P|}{2^i}, \quad i = 1, \ldots, |P| \quad \Rightarrow \quad \#Sel \leq \sum_{i=1}^{n} \left\lfloor \frac{|P|}{2^i} \right\rfloor \leq |P|. \qquad (4)$$

Here, we propose an adaptive version which allows to control the fraction of solutions chosen from the first and following fronts, based on the progress value $p \in [0,1]$ (Eq. 5). The extent of control is determined by parameter $\Delta > 0$ which is set to 1.5 for the experiments in this paper. Pseudocode of the R-NSGA-II selection step with Adaptive CPF is listed in Algorithm 2.

$$\#Sel(p) = \sum_{i=1}^{n} \left\lfloor \frac{|P|}{(1 + \Delta p)^i} \right\rfloor. \qquad (5)$$

With $\Delta = 1.5$, the denominator can take values in $[1, 2.5]$ (the standard fixed value was 2 in the original publication [10]). With this allocation, it could happen that the assigned sum of selected solutions exceeds the population size $|P|$. We implement an upper limit which makes sure that the total number of selected solutions is exactly $|P|$. It is also possible that a front contains fewer solutions than the assigned number. In this case, the missing solutions will be transferred and added to the preferred number of solutions of the next front. A detailed description of the algorithm can be found in [10].

Algorithm 2. R-NSGA-II – Environmental selection – Adaptive Constrained Pareto-Fitness pseudocode.

16:    // Environmental selection
17: Perform non-domination sorting on $P \cup Q$.
18: Calculate reference point ranks.
19: (Re-)run ϵ-CLUSTERING() on all fronts with remaining solutions (non-selected).
20: **for** $i \leftarrow 1$, #fronts **do**
21:    Determine representative solution in each cluster (randomly or closest to r).
22:    Select max $\left\{ |P| - \#Repr(F_i), \#Repr(F_i), \left\lfloor \frac{|P|}{(1+\Delta p)^i} \right\rfloor \right\}$ representative solutions
23:    which are closest to r.
24:    **if** there are unselected representative solutions in F_i and $\exists j < i : \#Sel(F_j, p) < \left\lfloor \frac{|P|}{(1+\Delta p)^j} \right\rfloor$ **then**
25:        Select remaining solutions $\left\lceil \frac{|P|}{(1+\Delta p)^j} \right\rceil - \#Sel(F_j, p)$ from F_i.
26:    **end if**
27: **end for**
28: **if** less than $|P|$ solutions have been selected **then**
29:    go to line 19.
30: **end if**

4.3 CORE NSGA-II

We propose a Preference-based EMO algorithm, called Cone Reference point-based NSGA-II algorithm (CORE NSGA-II), which is able to focus on the reference point faster than R-NSGA-II. It is based on the cylinder filter of the Focused Hypervolume performance metric [12]. For R-NSGA-II and a bi-objective problem, solutions within the distance $r = \frac{|P|}{2}\epsilon$ from a cylinder or cone axis, with $|P|$ being the population size and ϵ the R-NSGA-II clustering parameter, are allowed to be selected into the next population. The CORE NSGA-II requires two points to be specified: A reference point R and a direction point D, defining the cylinder axis (D can be chosen, for example, as the center of the initial population). The distance d of solution s to the cylinder axis is calculated by projecting $s - R$ onto the cylinder axis (See Fig. 3). The projection vector p is obtained as $p = (s - R) \cdot (D - R) / |(D - R)|$, where the operator $\cdot$ denotes the scalar product. The distance of s towards the cylinder axis is then obtained as $d = |p - (s - R)|$ and has to be smaller than the cylinder radius, $d \leq r$, in order for solution s to be considered for selection.

The cone filter is created by making the radius dependent on the reference point distance, making CORE NSGA-II an adaptive algorithm: $r(p) = c|p| + \frac{|P|}{2}\epsilon$. The cone filter coefficient c is usually chosen in $c \in [0, 0.1]$. In order to not be too restrictive, we allow solutions outside of the cone to be selected, but they are assigned a penalty value: $Penalty(s) = \max\{0, d(s) - r(p)\}$.

The cylinder filter must be applied before the non-domination sorting is performed. Otherwise, dominated solutions are filtered out during the non-domination sorting, which would be non-dominated after the application of the cylinder filter. In our case of stochastic simulation with a limited budget, we

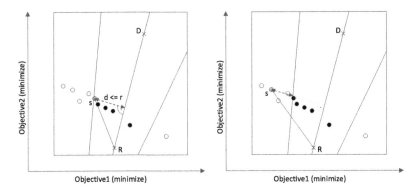

Fig. 3. CORE NSGA-II filter. On the right side, the penalty function for solutions outside of the cylinder/cone is shown.

allow the cylinder radius to be larger: $r = c\frac{|P|}{2}\epsilon$, $c \geq 1$, since it is not possible for the optimization to converge with enough solutions into the cylinder, within the available optimization time.

4.4 Adaptive Population Size

In this section, we propose an adaptive extension of R-NSGA-II which allows to change the population size dynamically during the optimization runtime. The idea behind this modification is the assumption that the R-NSGA-II algorithm needs a larger population at the beginning of the optimization runtime to explore the search space, and a smaller population when it covers the preferred area around the reference point. The changing population size is defined in Eq. 6 and can be controlled by the criteria described in Sect. 3 through the progress parameter p.

$$|P| = \lfloor (1-p)\left(|P|_{max} - |P|_{min}\right)\rceil + |P|_{min}. \tag{6}$$

5 Numerical Experiments

In this section, the described adaptive R-NSGA-II algorithm variants are evaluated on two benchmark functions of different complexity, as well as a simplistic production line model. The experiments are conducted for optimistic, unattainable reference points.

5.1 Benchmark Functions, Simulation Model, and Settings

The benchmark functions used, ZDT1 and ZDT4 [15], are deterministic in their original versions. Therefore, zero-mean normal relative noise has been added to create noisy optimization problems. The added noise 1% is $(\mathcal{N}(0, 0.01), \mathcal{N}(0, 0.1))$ for ZDT1 and $(\mathcal{N}(0, 0.01), \mathcal{N}(0, 1))$ for ZDT4 (considering the relevant objective ranges of $[0, 1] \times [0, 10]$ and $[0, 1] \times [0, 100]$ respectively). In the following, the ZDT benchmark optimization problems are called ZDT1-1% and ZDT4-1%.

The adaptive algorithm variants are also evaluated on a stochastic simulation model consisting of 6 machines (Cycle time 1 min, $\sigma = 1.5$ min, CV = 1.5, lognormal distribution) and 5 buffers with sizes $\in [1, 50]$. The source distribution is lognormal with $\mu = 1$ min and $\sigma = 1.5$ min. The basic structure is depicted in Fig. 4. The simulated warm-up time for the production line model is 3 days and the total simulated time of operation is 10 days. The conflicting objectives are to maximize the main production output, measured as Throughput (TH) (parts per hour), and to minimize the sum of the buffer sizes, TNB = Total number of buffers. This is a generalization of the lean buffering problem [6] (finding the minimal number of buffers required to obtain a certain level of system throughput). In order to consider the maintenance aspect, the machines are simulated with an Availability of 90% and a MTTR of 5 min, leading to a MTBF of 45 min. The Throughtput (TH) objective is noisy with a measured average standard deviation of approximately 0.6 which corresponds to 1% of the TH objective range. We therefore call the problem PL-1%.

Fig. 4. A simplistic production line configuration.

Since the described benchmark functions and the simulation model are low noise function scenarios, no resampling is done during the optimization runs. However, $b_f = 10$ final samples are evaluated for the final population, in order to provide accurate objective values to the decision maker.

5.2   Algorithm Parameter Settings

The limited simulation budget is chosen as $B = 2,500$ replications for both ZDT problems, which we consider to be realistic. This corresponds to a one day optimization runtime on a cluster with 100 computers/cores and a 15 min function evaluation time, which could be a realistic real-world optimization scenario. However, the benchmark functions and simplistic production line models run much faster, which allows fast experimentation. R-NSGA-II is run with a population size of $|P| = 50$, crossover rate $p_c = 0.8$, SBX crossover operator with $\eta_c = 2$, Mutation probability $p_m = 0.07$ and Polynomial Mutation operator with $\eta_m = 5$. ϵ is chosen as $\epsilon = 0.001$ as in the original R-NSGA-II publication [4]. The initial population is generated by Latin Hypercube Sampling.

The time threshold and the distance threshold of Delayed Pareto-fitness θ is set to $\theta = 0.75$. The adaptive Constrained Pareto-fitness variant uses a Δ value of 1.5, in order to realize denominator values in $[1, 2.5]$ (the standard fixed value was 2 in the original publication [10]). The cylinder filter of CORE R-NSGA-II uses an ϵ value of 0.002 for the cylinder/cone radius. This is because, in our scenario with light objective noise, we want to allow more solutions to be inside the cylinder filter. The distance-based adaption of the cone filter is controlled by the parameter $c = 0.01$. The adaptive population size R-NSGA-II extension changes the population size between $|P|_{max} = 50$ and $|P|_{min} = 20$.

5.3   Performance Measurement

In this paper, a performance metric we call Focused Hypervolume metric (F-HV) is used, which we proposed in [12]. It is based on the standard Hypervolume definition [16] and the R-HV metric [3,7]. Also, it measures convergence and diversity, but in a limited area of the objective space which is of interest to the decision maker. This limited area is defined by a cylinder (cf. CORE NSGA-II description above) which filters solutions which are outside of the cylinder. The cylinder is defined by the decision maker by setting certain points in the objective space. An example is given in Fig. 5.

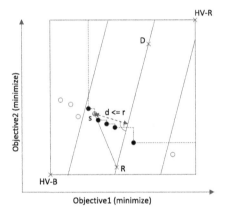

Fig. 5. The Focused Hypervolume metric.

For the Focused Hypervolume, the Hypervolume reference point $HV\text{-}R$ is chosen close to the R-NSGA-II reference point R, in order to limit the objective space that is of interest to the decision maker. Solutions that do not dominate the Hypervolume reference point are ignored by the Focused Hypervolume. In addition, a base point $HV\text{-}B$ used for normalization of the Hypervolume value is defined (Fig. 5). It also limits the objective space that is relevant to the decision maker. A direction point D is chosen in a way to define a cylinder axis possibly perpendicular to the Pareto-front. Another performance metric with the same purpose has been proposed by [1] who use a weighted hypervolume indicator which values the hypervolume contribution of those solutions higher which are close to a so-called preference point.

Table 1. F-HV configuration. Cylinder radius $r = c|P|/2\epsilon = 0.025c$.

	R	HV-R	HV-B	D	r
ZDT1-1%	0.5, 0	0.6, 1	0.4, 0	0.6, 2	0.05
ZDT4-1%	0.5, 0	0.6, 50	0.4, 0	0.5, 50	0.1
PL-1%	50, 50	60, 40	50, 50	60, 40	0.025

The F-HV metric parameter values are chosen as in Table 1. Due to the higher complexity of ZDT4, we chose a wider cylinder radius ($r = 0.1$), in order to allow more algorithm variants to find solutions in the preferred area within the available time.

5.4   Evaluation and Replication

All experiments performed in this study are replicated five times, while median performance metric values as well as standard deviation values are calculated. To follow the performance development over time, the performance metric is evaluated after every generation of the optimization algorithm (Fig. 6). The cylinder filter of the Focused Hypervolume metric is applied before the non-domination sorting is performed and the hypervolume of the filtered non-dominated solution set is calculated.

5.5   Results

The metric values for the result population on ZDT1-1%, ZDT4-1%, and PL-1% are listed in Table 2. A visual comparison during the whole optimization runtime for R-NSGA-II with the Constrained Pareto-fitness extension and with the Delayed Pareto-fitness are shown in Figs. 6, 7, and 8, respectively.

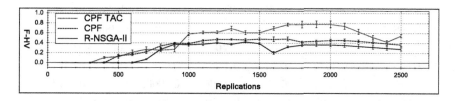

Fig. 6. Normalized F-HV results for ZDT1-1%, measured over time up to 2,500 function evaluations. The results show the R-NSGA-II algorithm with and without the Constrained Pareto-fitness extension. The values shown are median metric values. The error bars show the standard error based on 5 experiment replications. For the last population, $b_f = 10$ final samples have been executed.

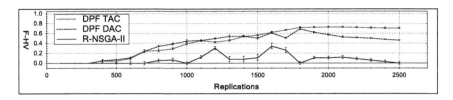

Fig. 7. Normalized F-HV results for ZDT4-1%, measured over time up to 2,500 function evaluations. The results compare R-NSGA-II with time-based and distance-based Delayed Pareto-fitness. Measurement settings as in Fig. 6.

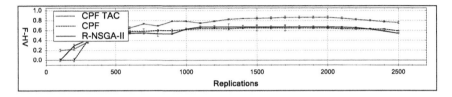

Fig. 8. Normalized F-HV results for the simplistic production line PL-1%, measured over time. The results show the R-NSGA-II algorithm with and without the Constrained Pareto-fitness extension. Measurement settings as in Fig. 6.

Many algorithm extension variants were able to achieve better values than the R-NSGA-II algorithm. However, some experiments with the Adaptive Population Size extension revealed a result that was worse than the result for the standard R-NSGA-II. At the same time, these experiments show a high standard deviation of the results. DPF and CPF showed the best results.

Table 2. F-HV performance measurement results for R-NSGA-II with different adaptive extensions. Five experiment replications have been performed. The measurement is performed on the last population where $b_f = 10$ final samples have been executed.

	ZDT1-1%	StdDev	ZDT4-1%	StdDev	PL-1%	StdDev
R-NSGA-II	0.2792	0.1682	0.0000	0.1384	0.5383	0.0437
DPF TAC	0.3883	0.0478	**0.7045**	0.0657	0.6056	0.0127
DPF DAC	0.4080	0.0366	0.4624	0.0236	0.3975	0.0541
CPF	0.3685	0.1730	0.3517	0.4000	0.5876	0.0474
CPF TAC	**0.5478**	0.1680	0.0000	0.1997	**0.7564**	0.1010
CPF DAC	0.4708	0.0980	0.0121	0.0365	0.6791	0.0652
APS TAC	0.3302	0.0620	0.0000	0.3938	0.0000	0.2640
APS DAC	0.2279	0.1404	0.0000	0.3020	0.6079	0.0407
CORE	0.3826	0.0347	0.2355	0.0991	0.4596	0.0750
CORE DAC	0.3736	0.0221	0.3552	0.2489	0.4825	0.0135

6   Conclusions

The experiments show that R-NSGA-II with the adaptive extensions can achieve better results than the standard R-NSGA-II. It is therefore recommended to use one of the proposed methods to speed up convergence towards the reference point. The Delayed and Constrained Pareto-fitness extensions showed the best results. Adapting the population size led to an unsteady performance but some algorithm variants were able to achieve a minor performance gain.

6.1   Future Work

As future work, we will run the proposed R-NSGA-II extensions on real-world industrial case studies. In addition, the algorithms will be extended for scenarios with attainable reference points, and a parameter study will be conducted.

Acknowledgments. This study was partially funded by the Knowledge Foundation, Sweden, through the IDSS project. The authors gratefully acknowledge their provision of research funding.

References

1. Brockhoff, D., Bader, J., Thiele, L., Zitzler, E.: Directed multiobjective optimization based on the weighted hypervolume indicator. J. Multi-Criteria Decis. Anal. **20**(5–6), 291–317 (2013)
2. Deb, K., Agrawal, S., Pratap, A., Meyarivan, T.: A fast and elitist multi-objective genetic algorithm: NSGA-II. IEEE Trans. Evol. Comput. **6**(2), 182–197 (2002)
3. Deb, K., Siegmund, F., Ng, A.H.C.: R-HV: a metric for computing hyper-volume for reference point based EMOs. In: Panigrahi, B.K., Suganthan, P.N., Das, S. (eds.) SEMCCO 2014. LNCS, vol. 8947, pp. 98–110. Springer, Cham (2015). doi:10.1007/978-3-319-20294-5_9. ISBN: 978-3-319-20293-8
4. Deb, K., Sundar, J., Bhaskara Rao, N., Chaudhur, S.: Reference point based multi-objective optimization using evolutionary algorithms. Int. J. Comput. Intell. Res. **2**(3), 273–286 (2006)
5. Deb, K., Sinha, A., Korhonen, P.J., Wallenius, J.: An interactive evolutionary multi-objective optimization method based on progressively approximated value functions. IEEE Trans. Evol. Comput. **14**(5), 723–739 (2010)
6. Enginarlar, E., Li, J., Meerkov, S.M.: How lean can lean buffers be? IIE Trans. **37**(4), 333–342 (2005)
7. Li, K., Deb, K.: Performance Assessment for Preference-Based Evolutionary Multi-Objective Optimization Using Reference Points (2016, submitted to journal). http://www.egr.msu.edu/kdeb/papers/c2016001.pdf
8. López-Jaimes, A., Arias Montaño, A., Coello Coello, C.A.: Preference incorporation to solve many-objective airfoil design problems. In: Proceedings of the Congress on Evolutionary Computation 2011, New Orleans, USA, pp. 1605–1612 (2011). ISBN: 978-1-4244-7834-7
9. López-Jaimes, A., Arias Montaño, A., Coello Coello, C.A.: Including preferences into a multiobjective evolutionary algorithm to deal with many-objective engineering optimization problems. Inf. Sci. **277**, 1–20 (2014). ISSN: 0020–0255

10. Siegmund, F., Ng, A.H.C., Deb, K.: Finding a preferred diverse set of pareto-optimal solutions for a limited number of function calls. In: Proceedings of the IEEE Congress on Evolutionary Computation 2012, Brisbane, Australia, pp. 2417–2424 (2012). ISBN: 978-1-4673-1508-1

11. Siegmund, F., Ng, A.H.C., Deb, K.: A comparative study of dynamic resampling strategies for guided evolutionary multi-objective optimization. In: Proceedings of the IEEE Congress on Evolutionary Computation 2013, Cancún, Mexico, pp. 1826–1835 (2013). ISBN: 978-1-4799-0454-9

12. Siegmund, F., Ng, A.H.C., Deb, K.: Hybrid dynamic resampling for guided evolutionary multi-objective optimization. In: Proceedings of the 8th International Conference on Evolutionary Multi-Criterion Optimization, Guimarães, Portugal, pp. 366–380 (2015). ISBN: 978-3-319-15934-8

13. Siegmund, F., Ng, A.H.C., Deb, K.: A ranking and selection strategy for preference-based evolutionary multi-objective optimization of variable-noise problems. In: Proceedings of the Congress on Evolutionary Computation WCCI-CEC 2016, Vancouver, Canada, pp. 3035–3044, July 2016. ISBN: 978-1-5090-0623-6

14. Stump, G., Simpson, T.W., Donndelinger, J.A., Lego, S., Yukish, M.: Visual steering commands for trade space exploration: user-guided sampling with example. J. Comput. Inf. Sci. Eng. **9**(4), 1–10 (2009). 044501, ISSN: 1530-9827

15. Zitzler, E., Deb, K., Thiele, L.: Comparison of multiobjective evolutionary algorithms: empirical results. Evol. Comput. **8**(2), 173–195 (2000)

16. Zitzler, E., Thiele, L.: Multiobjective optimization using evolutionary algorithms — a comparative case study. In: Eiben, A.E., Bäck, T., Schoenauer, M., Schwefel, H.-P. (eds.) PPSN 1998. LNCS, vol. 1498, pp. 292–301. Springer, Heidelberg (1998). doi:10.1007/BFb0056872

A Population-Based Algorithm for Learning a Majority Rule Sorting Model with Coalitional Veto

Olivier Sobrie[1,2]([✉]), Vincent Mousseau[1]([✉]), and Marc Pirlot[2]([✉])

[1] CentraleSupelec, Université Paris-Saclay, Grande Voie des Vignes,
92295 Châtenay Malabry, France
vincent.mousseau@centralesupelec.fr
[2] Université de Mons, Faculté Polytechnique, 9, Rue de Houdain,
7000 Mons, Belgium
olivier.sobrie@gmail.com, marc.pirlot@umons.ac.be

Abstract. MR-Sort (Majority Rule Sorting) is a multiple criteria sorting method which assigns an alternative a to category C^h when a is better than the lower limit of C^h on a weighted majority of criteria, and this is not true with the upper limit of C^h. We enrich the descriptive ability of MR-Sort by the addition of coalitional vetoes which operate in a symmetric way as compared to the MR-Sort rule w.r.t. to category limits, using specific veto profiles and veto weights. We describe a heuristic algorithm to learn such an MR-Sort model enriched with coalitional veto from a set of assignment examples, and show how it performs on real datasets.

1 Introduction

Multiple Criteria Sorting Problems aim at assigning alternatives to one of the predefined ordered categories $C^1, C^2, ..., C^p$, C^1 and C^p being the worst and the best category, respectively. This type of assignment method contrasts with classical supervised classification methods in that the assignments have to be monotone with respect to the scores of the alternatives. In other words, an alternative which has better scores on all criteria than another cannot be assigned to a worse category.

Many multiple criteria sorting methods have been proposed in the literature (see e.g., [1,2]). MR-Sort (Majority Rule Sorting, see [3]) is an outranking-based multiple criteria sorting method which corresponds to a simplified version of ELECTRE TRI where the discrimination and veto thresholds are omitted[1]. MR-Sort proved able to compete with state-of-the-art classification methods such as Choquistic regression [4] on real datasets.

---

[1] It is worth noting that outranking methods used for sorting are not subject to Condorcet effects (cycles in the preference relation), since alternatives are not compared in pairwise manner but only to profiles limiting the categories.

© Springer International Publishing AG 2017
H. Trautmann et al. (Eds.): EMO 2017, LNCS 10173, pp. 575–589, 2017.
DOI: 10.1007/978-3-319-54157-0_39

In the pessimistic version of ELECTRE TRI, veto effects make it possible to worsen the category to which an alternative is assigned when this alternative has very bad performances on one/several criteria. We consider a variant of MR-Sort which introduces possible veto effects. While in ELECTRE TRI, a veto involves a single criterion, we consider a more general formulation of veto (see [5]) which can involve a coalition of criteria (such a coalition can be reduced to a singleton).

The definition of such a *"coalitional veto"* exhibits a noteworthy symmetry between veto and concordance. To put it simple, in a two-category context (*Bad/Good*), an alternative is classified as *Good* when its performances are above the concordance profile on a sufficient majority of criteria, and when its performances are not below the veto profile for a sufficient majority of criteria. Hence, the veto condition can be viewed as the negation of a majority rule using a specific veto profile, and specific veto weights.

Algorithms to learn the parameters of an MR-Sort model without veto (category limits and criteria weights) have been proposed, either using linear programming involving integer variables (see [3]) or using a specific heuristic (see [6,7]). When the size of the learning set exceeds 100, only heuristic algorithms are able to provide a solution in a reasonable computing time.

Olteanu and Meyer [8] have developed a simulated annealing based algorithm to learn a MR-Sort model with classical veto (not coalitional ones).

In this paper, we propose a new heuristic algorithm to learn the parameters of a MR-Sort model with coalitional veto (called MR-Sort-CV) which makes use of the symmetry between the concordance and the coalitional veto conditions. Preliminary results obtained using an initial version of the algorithm can be found in [9]. The present work describes an improved version of the algorithm and the results of tests on real datasets involving two or more categories (while the preliminary version was only tested for classifying in two categories).

The paper is organized as follows. In Sect. 2, we recall MR-Sort the present work describes and define its extension when considering monocriterion veto and coalitional veto. After a brief reminder of the heuristic algorithm to learn an MR-Sort model, Sect. 3 is devoted to the presentation of the algorithm to learn an MR-Sort model with coalitional veto. Section 4 presents experimentations of this algorithm and Sect. 5 groups conclusions and directions for further research.

2   Considering Vetoes in MR-Sort

2.1   MR-Sort Model

MR-Sort is a method for assigning objects to ordered categories. It is a simplified version of ELECTRE TRI, another MCDA method [10,11].

The MR-Sort rule works as follows. Formally, let X be a set of objects evaluated on n ordered attributes (or criteria), $F = \{1, ..., n\}$. We assume that X is the Cartesian product of the criteria scales, $X = \prod_{j=1}^{n} X_j$, each scale X_j being completely ordered by the relation $\geq_j$. An object $a \in X$ is a vector $(a_1, \ldots, a_j, \ldots, a_n)$, where $a_j \in X_j$ for all j. The ordered categories which the objects are assigned to by the MR-Sort model are denoted by C^h, with $h = 1, \ldots, p$. Category C^h is delimited by its lower limit profile b^{h-1} and its upper limit profile b^h, which is also the lower limit profile of category C^{h+1} (provided $0 < h < p$). The profile b^h is the vector of criterion values $(b_1^h, \ldots, b_j^h, \ldots, b_n^h)$, with $b_j^h \in X_j$ for all j. We denote by $P = \{1, \ldots, p\}$ the list of category indices. By convention, the best category, C^p, is delimited by a fictive upper profile, b^p, and the worst one, C^1, by a fictive lower profile, b^0. It is assumed that the profiles dominate one another, i.e.: $b_j^h \geq_j b_j^{h-1}$, for $h = \{1, \ldots, p\}$ and $j = \{1, \ldots, n\}$.

Using the MR-Sort procedure, an object is assigned to a category if its criterion values are at least as good as the category lower profile values on a weighted majority of criteria while this condition is not fulfilled when the object's criterion values are compared to the category upper profile values. In the former case, we say that the object is *weakly preferred* to the profile, while, in the latter, it is not[2]. Formally, if an object $a \in X$ is *weakly preferred* to a profile b^h, we denote this by $a \succcurlyeq b^h$. Object a is preferred to profile b^h whenever the following condition is met:

$$a \succcurlyeq b^h \Leftrightarrow \sum_{j:a_j \geq_j b_j^h} w_j \geq \lambda, \tag{1}$$

where w_j is the nonnegative weight associated with criterion j, for all j and λ sets a majority level. The weights satisfy the normalization condition $\sum_{j \in F} w_j = 1$; λ is called the *majority threshold*.

The preference relation $\succcurlyeq$ defined by (1) is called an *outranking* relation without veto or a *concordance* relation ([11]; see also [12,13] for an axiomatic description of such relations). Consequently, the condition for an object $a \in X$ to be assigned to category C^h reads:

$$\sum_{j:a_j \geq_j b_j^{h-1}} w_j \geq \lambda \quad \text{and} \quad \sum_{j:a_j \geq_j b_j^h} w_j < \lambda. \tag{2}$$

The MR-Sort assignment rule described above involves $pn + 1$ parameters, i.e. n weights, $(p-1)n$ profiles evaluations and one majority threshold.

A *learning set* A is a subset of objects $A \subseteq X$ for which an assignment is known. For $h \in P$, A_h denotes the subset of objects $a \in A$ which are assigned to category C^h. The subsets A_h are disjoint; some of them may be empty.

2.2 MR-Sort-MV

In this section, we recall the traditional monocriterion veto rule as defined by [14,15]. In a MR-Sort model with *monocriterion veto*, an alternative a is "at least as good as" a profile b^h if it has at least equal to or better performances than b^h on a weighted majority of criteria and if it is not strongly worse than the profile on any criterion. In the sequel, we call b^h a *concordance profile* and we define "strongly worse than the profile" b^h by means of a veto profile

---

[2] "weak preference" means being "at least as good as".

$v^h = (v_1^h, v_2^h, ..., v_n^h)$, with $v_j^h \leq_j b_j^h$. It represents a vector of performances such that any alternative having a performance worse than or equal to this profile on any criterion would be excluded from category C^{h+1}. Formally, the assignment rule is described by the following condition:

$$a \succcurlyeq b^h \iff \sum_{j:a_j \geq_j b_j^h} w_j \geq \lambda \text{ and not } a V b^h,$$

with $a V b^h \iff \exists j \in F : a_j \leq_j v_j^h$. Note that non-veto condition is frequently presented in the literature using a veto threshold (see e.g. [16]), i.e. a maximal difference w.r.t. the concordance profile in order to be assigned to the category above the profile. Using veto profiles instead of veto thresholds better suits the context of multicriteria sorting. We recall that a profile b^h delimits the category C^h from C^{h+1}, with $C^{h+1} \succ C^h$; with monocriterion veto, the MR-Sort assignment rule reads as follows:

$$a \in C^h \iff \left[\sum_{j:a_j \geq_j b_j^{h-1}} w_j \geq \lambda \text{ and } \nexists j \in F : a_j \leq v_j^{h-1} \right]$$

$$\text{and } \left[\sum_{j:a_j \geq_j b_j^h} w_j < \lambda \text{ or } \exists j \in F : a_j \leq v_j^h \right]. \tag{3}$$

We remark that a MR-Sort model with more than 2 categories remains consistent only if veto profiles v_j^h do not overlap, i.e., are chosen such that $v_j^h \geq v_j^{h'}$ for all $\{h, h'\}$ s.t. $h > h'$. Otherwise, an alternative might be on the one hand in veto against a profile b^h, which prevents it to be assigned to C^{h+1} and, on the other hand, not in veto against b^{h+1}, which does not prevent it to be assigned to C^{h+2}.

2.3   MR-Sort-CV

We introduce here a new veto rule considering vetoes w.r.t. coalitions of criteria, which we call "*coalitional veto*". With this rule, the veto applies and forbids an alternative a to be assigned to category C^{h+1} when the performance of an alternative a is not better than v_j^h on a weighted majority of criteria.

As for the monocriterion veto, the veto profiles are vectors of performances $v^h = (v_1^h, v_2^h, ..., v_n^h)$, for all $h = \{1, .., p\}$. Coalitional veto also involves a set of nonnegative *veto weights* denoted z_j, for all $j \in F$. Without loss of generality, the sum of z_j is set to 1. Furthermore, a veto cutting threshold Λ is also involved and determines whether a coalition of criteria is sufficient to impose a veto. Formally, we express the coalitional veto rule $a V b^h$, as follows:

$$a V b^h \iff \sum_{j:a_j \leq_j v_j^h} z_j \geq \Lambda. \tag{4}$$

Using coalitional veto, the outranking relation of MR-Sort (2.2) is modified as follows:

$$a \succcurlyeq b^h \iff \sum_{j:a_j \geq_j b_j^h} w_j \geq \lambda \text{ and } \sum_{j:a_j \leq_j v_j^h} z_j < \Lambda. \tag{5}$$

Using coalitional veto with MR-Sort modifies the assignment rule as follows:

$$a \in C^h \iff \left[\sum_{j:a_j \geq_j b_j^{h-1}} w_j \geq \lambda \text{ and } \sum_{j:a_j \leq_j v_j^{h-1}} z_j < \Lambda \right]$$

$$\text{and } \left[\sum_{j:a_j \geq_j b_j^h} w_j < \lambda \text{ or } \sum_{j:a_j \leq_j v_j^h} z_j \geq \Lambda \right] \tag{6}$$

In MR-Sort, the coalitional veto can be interpreted as a combination of performances preventing the assignment of an alternative to a category. We call this new model, MR-Sort-CV.

The coalitional veto rule given in Eq. (5) is a generalization of the monocriterion rule. Indeed, if the veto cut threshold Λ is equal to $\frac{1}{n}$ (n being the number of criteria), and each veto weight z_j is set to $\frac{1}{n}$, then the veto rule defined in Eq. (4) corresponds to a monocriterion veto for each criterion.

2.4 The Non Compensatory Sorting (NCS) Model

In this subsection, we recall the non compensatory sorting (NCS) rule as defined by [14,15], which will be used in the experimental part (Sect. 4) for comparison purposes. These rules allow to model criteria interactions. MR-Sort is a particular case of these, in which criteria do not interact.

In order to take criteria interactions into account, it has been proposed to modify the definition of the global outranking relation, $a \succcurlyeq b^h$, given in (1). We introduce the notion of capacity. A *capacity* is a function $\mu : 2^F \to [0, 1]$ such that:

- $\mu(B) \geq \mu(A)$, for all $A \subseteq B \subseteq F$ (monotonicity);
- $\mu(\emptyset) = 0$ and $\mu(F) = 1$ (normalization).

The Möbius transform allows to express the capacity in another form:

$$\mu(A) = \sum_{B \subseteq A} m(B), \tag{7}$$

for all $A \subseteq F$, with $m(B)$ defined as:

$$m(B) = \sum_{C \subseteq B} (-1)^{|B|-|C|} \mu(C) \tag{8}$$

The value $m(B)$ can be interpreted as the weight that is exclusively allocated to B as a whole. A capacity can be defined directly by its Möbius transform also called "interaction". An interaction m is a set function $m : 2^F \to [-1, 1]$ satisfying the following conditions:

$$\sum_{j \in K \subseteq J \cup \{j\}} m(K) \geq 0, \quad \forall j \in F, J \subseteq F \setminus \{i\} \tag{9}$$

and

$$\sum_{K \subseteq F} m(K) = 1.$$

If m is an interaction, the set function defined by $\mu(A) = \sum_{B \subseteq A} m(B)$ is a capacity. Conditions (9) guarantee that μ is monotone [17].

Using a capacity to express the weight of the coalition in favor of an object, we transform the outranking rule as follows:

$$a \succcurlyeq b^h \Leftrightarrow \mu(A) \geq \lambda \text{ with } A = \{j : a_j \geq_j b_j^h\}$$
$$\text{and } \mu(A) = \sum_{B \subseteq A} m(B) \tag{10}$$

Computing the value of $\mu(A)$ with the Möbius transform induces the evaluation of $2^{|A|}$ parameters. In a model composed of n criteria, it implies the elicitation of 2^n parameters, with $\mu(\emptyset) = 0$ and $\mu(F) = 1$. To reduce the number of parameters to elicit, we use a 2-additive capacity in which all the interactions involving more than 2 criteria are equal to zero. Inferring a 2-additive capacity for a model having n criteria requires the determination of $\frac{n(n+1)}{2} - 1$ parameters.

Finally, the condition for an object $a \in X$ to be assigned to category C^h can be expressed as follows:

$$\mu(F_{a,h-1}) \geq \lambda \quad \text{and} \quad \mu(F_{a,h}) < \lambda \tag{11}$$

with $F_{a,h-1} = \{j : a_j \geq_j b_j^{h-1}\}$ and $F_{a,h} = \{j : a_j \geq_j b_j^h\}$.

3   Learning MR-Sort

Learning the parameters of MR-Sort and ELECTRE TRI models has been studied in several articles [3, 18–25]. In this section, we recall how to learn the parameters of a MR-Sort model using respectively an exact method [3] and a heuristic algorithm [18]. We then extend the heuristic algorithm to MR-Sort-CV.

3.1   Learning a Simple MR-Sort Model

It is possible to learn a MR-Sort model from a learning set using Mixed Integer Programming (MIP), see [3]. Such a MIP formulation is not suitable for large data sets because of the high computing time required to infer the MR-Sort parameters. In order to learn MR-Sort models in the context of large data sets, a heuristic algorithm has been proposed in [18]. As for the MIP, the heuristic algorithm takes as input a set of assignment examples and their vectors of performances. The algorithm returns the parameters of a MR-Sort model.

The heuristic algorithm proposed in [18] works as follows. First a population of N_{mod} MR-Sort models is initialized. Thereafter, the following two steps are repeated iteratively on each model in the population:

$$\min \sum_{a \in A} (x'_a + y'_a)$$
$$\text{s.t.}$$
$$\sum_{j:a_j \geq_j b_j^{h-1}} w_j - x_a + x'_a = \lambda \qquad \forall a \in A_h, h = \{2, ..., p\}$$
$$\sum_{j:a_j \geq_j b_j^{h}} w_j + y_a - y'_a = \lambda - \epsilon \quad \forall a \in A_h, h = \{1, ..., p-1\}$$
$$\sum_{j=1}^{n} w_j = 1 \qquad\qquad (12)$$
$$w_j \in [0; 1] \quad \forall j \in F$$
$$\lambda \in [0; 1]$$
$$x_a, y_a, x'_a, y'_a \in \mathbb{R}_0^+$$
ε a small positive number.

1. A linear program optimizes the weights and the majority threshold on the basis of assignment examples while keeping the profiles fixed.
2. Given the inferred weights and the majority threshold, a heuristic adjusts the profiles of the model on the basis of the assignment examples.

After applying these two steps to all the models in the population, the $\lfloor \frac{N_{\text{mod}}}{2} \rfloor$ models restoring the least numbers of examples are reinitialized. These steps are repeated until the heuristic finds a model that fully restores all the examples or after a number of iterations specified a priori. This approach can be viewed as a sort of evolutionary metaheuristic (without crossover) in which a population of models is evolved.

The linear program designed to learn the weights and the majority threshold is given by (12). It minimizes a sum of slack variables, x'_a and y'_a, that is equal to 0 when all the objects are correctly assigned, i.e. assigned to the category defined in the input data set. We remark that the objective function of the linear program does not explicitly minimize the 0/1 loss but a sum of slacks. This implies that compensatory effects might appear, with undesirable consequences on the 0/1 loss. However in this heuristic, we consider that these effects are acceptable. The linear program doesn't involve binary variables. Therefore, the computing time remains reasonable when the size of the problem increases.

The objective function of the heuristic varying the profiles maximizes the number of examples compatible with the model. To do so, it iterates over each profile b^h and each criterion j and identifies a set of candidate moves for the profile, which correspond to the performances of the examples on criterion j located between profiles b^{h-1} and b^{h+1}. Each candidate move is evaluated as a function of the probability to improve the classification accuracy of the model. To assess whether a candidate move is likely to improve the classification of one or several objects, the examples which have an evaluation on criterion j located between the current value of the profile, b_j^h, and the candidate move, $b_j^h + \delta$ (resp. $b_j^h - \delta$), are grouped in different subsets:

$V_{h,j}^{+\delta}$ (**resp.** $V_{h,j}^{-\delta}$): the sets of objects misclassified in C^{h+1} instead of C^h (resp. C^h instead of C^{h+1}), for which moving the profile b^h by $+\delta$ (resp. $-\delta$) on j results in a correct assignment.

$W_{h,j}^{+\delta}$ (**resp.** $W_{h,j}^{-\delta}$): the sets of objects misclassified in C^{h+1} instead of C^h (resp. C^h instead of C^{h+1}), for which moving the profile b^h by $+\delta$ (resp. $-\delta$) on j strengthens the criteria coalition in favor of the correct classification but will not by itself result in a correct assignment.

$Q_{h,j}^{+\delta}$ (**resp.** $Q_{h,j}^{-\delta}$): the sets of objects correctly classified in C^{h+1} (resp. C^{h+1}) for which moving the profile b^h by $+\delta$ (resp. $-\delta$) on j results in a misclassification.

$R_{h,j}^{+\delta}$ (**resp.** $R_{h,j}^{-\delta}$): the sets of objects misclassified in C^{h+1} instead of C^h (resp. C^h instead of C^{h+1}), for which moving the profile b^h by $+\delta$ (resp. $-\delta$) on j weakens the criteria coalition in favor of the correct classification but does not induce misclassification by itself.

$T_{h,j}^{+\delta}$ (**resp.** $T_{h,j}^{-\delta}$): the sets of objects misclassified in a category higher than C^h (resp. in a category lower than C^{h+1}) for which the current profile evaluation weakens the criteria coalition in favor of the correct classification.

A formal definition of these sets can be found in [18]. The evaluation of the candidate moves is done by aggregating the number of elements in each subset. Finally, the choice to move or not the profile on the criterion is determined by comparing the candidate move evaluation to a random number drawn uniformly. These operations are repeated multiple times on each profile and each criterion.

3.2   Learning a MR-Sort-CV Model

As compared with MR-Sort, a MR-Sort-CV model requires the elicitation of additional parameters: a veto profile, veto weights and a veto threshold. In (2), the MR-Sort condition $\sum_{j:a_j \geq_j b_j^{h-1}} w_j \geq \lambda$ is a necessary condition for an alternative to be assigned to a category at least as good as C^h. Basically, the coalitional veto rule can be viewed as a dual version of the majority rule. It provides a sufficient condition for being assigned to a category worse than C^h. An alternative a is assigned to such a category as soon as $\sum_{j:a_j \leq_j v_j^{h-1}} z_j \geq \lambda$. This condition has essentially the same form as the MR-Sort rule except that the sum is over the criteria on which the alternative's performance is at *most* as good as the profile (instead of "at least as good", for the MR-Sort rule).

In order to learn a MR-Sort-CV model, we modify the heuristic presented in the previous subsection as shown in Algorithm 1. The main differences with the MR-Sort heuristic are highlighted in grey.

In the rest of this section, we give some detail about the changes that have been brought to the heuristic.

Concordance weights optimization. The concordance profiles being given, the weights are optimized using the linear program (12). The sum of the error variables $x'_a + y'_a$ was the objective to be minimized. In the linear program,

Algorithm 1. Metaheuristic to learn the parameters of an MR-Sort-CV model.

Generate a population of N_{model} models with concordance profiles initialized with a heuristic

repeat

    for all model M of the set **do**

        Learn the concordance weights and majority threshold with a linear program, using the current concordance profiles

        Apply the heuristic N_{it} times to adjust the concordance profiles, using the current concordance weights and threshold.

        Initialize a set of veto profiles, taking into account the concordance profiles (in the first step and in case the coalitional veto rule was discarded in the previous step)

        Learn the veto weights and majority threshold with a linear program, using the current profiles

        Adjust the veto profiles by applying the heuristic N_{it} times, using the current veto weights and threshold

        Discard the coalitional veto if it does not improve classification accuracy

    end for

    Reinitialize the $\left\lfloor \frac{N_{model}}{2} \right\rfloor$ models giving the worst CA

until Stopping criterion is met

---

x'_a is set to a positive value whenever it is not possible to satisfy the condition which assigns a to a category at least as good as C^h, while a actually belongs to C^h. Impeding the assignment of positive values to x'_a amounts to favor false positive assignments. Hence, positive values of x'_a should be heavily penalized. In contrast, positive values of y'_a correspond to the case in which the conditions for assigning a to the categories above the profile are met while a belongs to the category below the profile. Positive values of y'_a need not be discouraged as much as those of x'_a and therefore we changed the objective function of the linear program into $\min \sum_{a \in A} 10x'_a + y'_a$.

Adjustment of the concordance profiles. In order to select moves by a quantity $\pm\delta$ applied to the profile level on a criterion, we compute a probability which takes into account the sizes of the sets listed at the end of Section 3.1. To ensure the consistency of the model, the candidate moves are located in the interval $[\max(b_j^{h-1}, v_j^h), b_j^{h+1}]$. In all cases, the movements which lower the profile $(-\delta)$ are more favorable to false positive than the opposite movements. Therefore, all other things being equal (i.e. the sizes of the sets), the probability of choosing a downward move $-\delta$ should be larger than that of an upward move $+\delta$. The probability of an upward move is thus computed by the following formula

$$P(b_j^h + \delta) = \frac{2|V_{h,j}^{-\delta}| + 1|W_{h,j}^{-\delta}| + 0.1|T_{h,j}^{-\delta}|}{|V_{h,j}^{-\delta}| + |W_{h,j}^{-\delta}| + |T_{h,j}^{-\delta}| + 5|Q_{h,j}^{-\delta}| + |R_{h,j}^{-\delta}|}, \tag{13}$$

while that of a downward move is

$$P(b_j^h - \delta) = \frac{4|V_{h,j}^{+\delta}| + 2|W_{h,j}^{+\delta}| + 0.1|T_{h,j}^{+\delta}|}{|V_{h,j}^{+\delta}| + |W_{h,j}^{+\delta}| + |T_{h,j}^{+\delta}| + 5|Q_{h,j}^{+\delta}| + |R_{h,j}^{+\delta}|}. \tag{14}$$

The values appearing in (13) and (14) were determined empirically.

Initialization of veto profiles. A randomly selected veto profile is associated with each concordance profile. The veto profiles are initialized in ascending order, i.e. from the profile delimiting the worst category to the one delimiting the best category. The generation of a veto profile is done by drawing a random number in the interval $[b_j^h, v_j^{h-1}]$ on each criterion j. For the profile delimiting the worst categories, v_j^{h-1} corresponds to the worst possible performance on criterion j. Veto rules that do not prove useful at the end of the improvement phase (i.e., that do not contribute to reduce misclassification) are discarded and new veto profiles are generated during the next loop.

Veto weights optimization. The same type of linear program is used as for the concordance weights. One difference lies in the objective function in which we give the same importance to the positive and negative slack variables. Another difference lies in the constraints ensuring that an alternative $a \in C^h$ should not outrank the profile b^{h+1}. In order not to outrank the profile b^{h+1}, an alternative should either fulfill the condition $\sum_{j=1}^n w_j < \lambda$ or the condition $\sum_{j=1}^n z_j \geq \Lambda$ Since this is a disjunction of conditions, it is not necessary to ensure both. Therefore we do not consider the alternatives $a \in C^h$ that satisfy the first condition ($\sum_{j=1}^n w_j < \lambda$) in the linear program.

Adjustment of the veto profiles. The same heuristic as for the concordance profiles is used. The coefficients in (13) and (14) are modified in order to treat equally upward and downward moves.

4 Experiments

4.1 Datasets

In view of assessing the performance of the heuristic algorithm designed for learning the parameters of a MR-Sort-CV model, we use the algorithm to learn MR-Sort-CV models from several real data sets available at http://www.uni-marburg.de/fb12/kebi/research/repository/monodata, which serve as benchmark to assess monotone classification algorithms [4]. They involve from 120 to 1728 instances, from 4 to 8 monotone attributes and from 2 to 36 categories (see Table 1).

In our experiments, categories were initially binarized by thresholding at the median. We split the datasets in a twofold 50/50 partition: a learning set and a test set. Models are learned on the first set and evaluated on the test set; this is done 100 times on learning sets drawn at random.

4.2 Results Obtained with the Binarized Datasets

In a previous experimental study [6] performed on the same datasets, we compared the classification accuracy on the test sets obtained with MR-Sort and NCS. These results are reproduced in columns 2 and 3 of Table 2. No significant improvement in classification accuracy was observed when comparing NCS to MR-Sort. We have added the results for the new MR-Sort-CV heuristic in the fourth column of this table. In four cases, no improvement is observed as compared with MR-Sort. A slight improvement (of the order of 1%) is obtained in three cases (CPU, ESL, LEV). In one case (BCC), the results are slightly worse (2%).

Table 1. Data sets characteristics

Data set	#instances	#attributes	#categories
DBS	120	8	2
CPU	209	6	4
BCC	286	7	2
MPG	392	7	36
ESL	488	4	9
MMG	961	5	2
ERA	1000	4	4
LEV	1000	4	5
CEV	1728	6	4

4.3 Results Obtained with the Original Datasets

We also checked the algorithm for assigning alternatives to the original classes, in case there are more than two classes. Datasets with more than two classes

Table 2. Average and standard deviation of the classification accuracy on the test sets obtained with three different heuristics

Data set	MR-Sort	NCS	MR-Sort-CV
DBS	0.8377 ± 0.0469	0.8312 ± 0.0502	0.8390 ± 0.0476
CPU	0.9325 ± 0.0237	0.9313 ± 0.0272	0.9429 ± 0.0244
BCC	0.7250 ± 0.0379	0.7328 ± 0.0345	0.7044 ± 0.0299
MPG	0.8219 ± 0.0237	0.8180 ± 0.0247	0.8240 ± 0.0391
ESL	0.8996 ± 0.0185	0.8970 ± 0.0173	0.9024 ± 0.0179
MMG	0.8268 ± 0.0151	0.8335 ± 0.0138	0.8267 ± 0.0119
ERA	0.7944 ± 0.0173	0.7944 ± 0.0156	0.7959 ± 0.0270
LEV	0.8408 ± 0.0122	0.8508 ± 0.0188	0.8551 ± 0.0171
CEV	0.8516 ± 0.0091	0.8662 ± 0.0095	0.8516 ± 0.0665

are CPU, MPG, ESL, ERA, LEV and CEV. We did not consider MPG and ESL in which alternatives are respectively assigned to 36 and 9 categories. It is not reasonable indeed to aim at learning 35 (resp. 9) concordance and veto profiles on the basis of 392 (resp. 488) assignment examples in the dataset, half of them being reserved for testing purposes. The results obtained with the four remaining datasets are reported in Table 3.

Table 3. Average and standard deviation of the accuracy of classification in more than two categories obtained for the test sets

Data set	MR-Sort	MR-SortCV
CPU	0.8039 ± 0.0354	0.8469 ± 0.0426
ERA	0.5123 ± 0.0233	0.5230 ± 0.0198
LEV	0.5662 ± 0.0258	0.5734 ± 0.0213
CEV	0.7664 ± 0.0193	0.7832 ± 0.0130

We observe improvements w.r.t. MR-Sort on all four datasets. The gain ranges between a little less than 1% for ERA till 4% for CPU.

4.4   Results Obtained Using Randomly Generated MR-Sort-CV models

In view of the slight improvements w.r.t. MR-Sort obtained on the benchmark datasets, one may wonder whether this result should not be ascribed to the design of our heuristic algorithm. In order to check whether our algorithm is able to learn a MR-Sort-CV model in case the objects have been assigned to categories according to a hidden MR-Sort-CV rule, we performed the following experiments.

For a number of criteria ranging from 4 to 7, we generate at random a MR-Sort-CV model with two categories. We generate a learning set composed of 1000 random vectors of alternatives. Then we assign the alternatives to the two categories using the MR-Sort-CV model. We use the algorithm to learn a MR-Sort-CV model that reproduces as accurately as possible the assignments of the alternatives in the learning set. Having generated 10000 additional alternatives at random and having assigned them to a category using the generated MR-Sort-CV model, we compare these assignments with those produced by the learned model. We repeat this 100 times for each number of criteria. The average classification accuracy for the learning and the test sets is displayed in Table 4.

We observe that, on average, the learned model correctly restores more than 98.5% of the assignments in the learning set and more than 97.5% of the assignments in the test set. This means that our heuristic algorithm is effective in learning a MR-Sort-CV model when the assignments are actually made according to such a model.

Table 4. Average and standard deviation of the classification accuracy of MR-Sort-CV models learned on data generated by random MR-Sort-CV models with 2 categories and 4 to 7 criteria. The learning set is composed of 1000 alternatives and the test set is composed of 10000 alternatives.

#criteria	Learning set	Test set
4	0.9908 ± 0.01562	0.98517 ± 0.01869
5	0.9904 ± 0.01447	0.98328 ± 0.01677
6	0.9860 ± 0.01560	0.97547 ± 0.02001
7	0.9827 ± 0.01766	0.96958 ± 0.02116

An alternative explanation of the modest improvements yielded by using the MR-Sort-CV model is that the latter has limited additional descriptive power as compared to MR-Sort. We check this hypothesis by running the MR-Sort learning algorithm on the same artificial datasets generated by random MR-Sort-CV rules as in Table 4. The experimental results are reported in Table 5. By comparing Tables 4 and 5, we see that MR-Sort models are able to approximate very well MR-Sort-CV models. Indeed, the classification accuracy obtained with MR-Sort models is quite high on the test sets and only about 2% below that obtained with a learned MR-Sort-CV model.

Table 5. Average and standard deviation of the classification accuracy of MR-Sort models learned on data generated by random MR-Sort-CV models with 2 categories and 4 to 7 criteria. The learning set is composed of 1000 alternatives and the test set is composed of 10000 alternatives.

#criteria	Learning set	Test set
4	0.9760 ± 0.0270	0.9700 ± 0.0309
5	0.9713 ± 0.0275	0.9627 ± 0.0318
6	0.9645 ± 0.0248	0.9525 ± 0.0307
7	0.9639 ± 0.0264	0.9518 ± 0.0301

5   Conclusion

We have presented MR-Sort-CV, a new original extension of the MR-Sort ordered classification model. This model introduces a new and more general form of veto condition which applies on coalitions of criteria rather than a single criterion. This coalitional veto condition can be expressed as a "reversed" MR-Sort rule. Such a symmetry enables us to design a heuristic algorithm to learn an MR-Sort-CV model, derived from an algorithm used to learn MR-Sort.

The experimental results obtained on benchmark datasets show that there is no significant improvement in classification accuracy as compared with MR-Sort in the case of two categories. The new model and the proposed learning algorithm lead to modest but definite gains in classification accuracy in the case of several categories. We checked that the learning algorithm was not responsible for the weak improvement of the assignment accuracy on the test sets. Therefore, the conclusion should be that the introduction of coalitional veto only adds limited descriptive ability to the MR-Sort model. This was also checked empirically.

The fact that coalitional veto (and this holds *a fortiori* for ordinary, single-criterion, veto) adds little descriptive power to the MR-Sort model is an important information by itself. In a learning context, the present study indicates that there is little hope to substantially improve classification accuracy by moving from a MR-Sort to a MR-Sort-CV model (as was also the case with the NCS model [6]). Note that small improvements in classification accuracy may be valuable, for instance, in medical applications (see e.g. [26]). Therefore it may be justified to consider MR-Sort-CV in spite of the increased complexity of the algorithm and of the model interpretation as compared to MR-Sort. It should be emphasized that the MR-Sort models lean themselves to easy interpretation in terms of rules [26]. The MR-Sort-CV model, although more complex, inherits this property since the coalitional veto condition is a reversed MR-Sort rule. Therefore, MR-Sort-CV models may be useful in specific applications.

References

1. Doumpos, M., Zopounidis, C.: Multicriteria Decision Aid Classification Methods. Kluwer Academic Publishers, Dordrecht (2002)
2. Zopounidis, C., Doumpos, M.: Multicriteria classification and sorting methods: a literature review. Eur. J. Oper. Res. **138**(2), 229–246 (2002)
3. Leroy, A., Mousseau, V., Pirlot, M.: Learning the parameters of a multiple criteria sorting method. In: Brafman, R.I., Roberts, F.S., Tsoukiàs, A. (eds.) ADT 2011. LNCS (LNAI), vol. 6992, pp. 219–233. Springer, Heidelberg (2011). doi:10.1007/978-3-642-24873-3_17
4. Tehrani, A.F., Cheng, W., Dembczyński, K., Hüllermeier, E.: Learning monotone nonlinear models using the Choquet integral. Mach. Learn. **89**(1–2), 183–211 (2012)
5. Sobrie, O., Pirlot, M., Mousseau, V.: New veto relations for sorting models. Technical report, Laboratoire Génie Industriel, Ecole Centrale Paris Cahiers de recherche 2014–04, October 2014
6. Sobrie, O., Mousseau, V., Pirlot, M.: Learning the parameters of a non compensatory sorting model. In: Walsh, T. (ed.) ADT 2015. LNCS (LNAI), vol. 9346, pp. 153–170. Springer, Cham (2015). doi:10.1007/978-3-319-23114-3_10
7. Sobrie, O., Mousseau, V., Pirlot, M.: Learning the parameters of a majority rule sorting model taking attribute interactions into account. In: DA2PL 2014 Workshop From Multiple Criteria Decision Aid to Preference Learning, pp. 22–30, Paris, France (2014)
8. Olteanu, A.L., Meyer, P.: Inferring the parameters of a majority rule sorting model with vetoes on large datasets. In: DA2PL 2014 : From Multicriteria Decision Aid to Preference Learning, pp. 87–94 (2014)

9. Sobrie, O., Mousseau, V., Pirlot, M.: Learning MR-sort rules with coalitional veto. In: DA2PL 2016: From Multicriteria Decision Aid to Preference Learning, p. 7 (2016)

10. Yu, W.: Aide multicritère à la décision dans le cadre de la problématique du tri: méthodes et applications. Ph.D. thesis, LAMSADE, Université Paris Dauphine, Paris (1992)

11. Roy, B., Bouyssou, D.: Aide multicritère à la décision: méthodes et cas. Economica Paris (1993)

12. Bouyssou, D., Pirlot, M.: A characterization of concordance relations. Eur. J. Oper. Res. **167**(2), 427–443 (2005)

13. Bouyssou, D., Pirlot, M.: Further results on concordance relations. Eur. J. Oper. Res. **181**, 505–514 (2007)

14. Bouyssou, D., Marchant, T.: An axiomatic approach to noncompensatory sorting methods in MCDM, I: the case of two categories. Eur. J. Oper. Res. **178**(1), 217–245 (2007)

15. Bouyssou, D., Marchant, T.: An axiomatic approach to noncompensatory sorting methods in MCDM, II: more than two categories. Eur. J. Oper. Res. **178**(1), 246–276 (2007)

16. Roy, B.: The outranking approach and the foundations of ELECTRE methods. Theor. Decis. **31**, 49–73 (1991)

17. Chateauneuf, A., Jaffray, J.-Y.: Derivation of some results on monotone capacities by Mobius inversion. In: Bouchon, B., Yager, R.R. (eds.) IPMU 1986. LNCS, vol. 286, pp. 95–102. Springer, Heidelberg (1987). doi:10.1007/3-540-18579-8_8

18. Sobrie, O., Mousseau, V., Pirlot, M.: Learning a majority rule model from large sets of assignment examples. In: Perny, P., Pirlot, M., Tsoukiàs, A. (eds.) ADT 2013. LNCS (LNAI), vol. 8176, pp. 336–350. Springer, Heidelberg (2013). doi:10.1007/978-3-642-41575-3_26

19. Mousseau, V., Słowiński, R.: Inferring an ELECTRE TRI model from assignment examples. J. Global Optim. **12**(1), 157–174 (1998)

20. Mousseau, V., Figueira, J.R., Naux, J.P.: Using assignment examples to infer weights for ELECTRE TRI method: some experimental results. Eur. J. Oper. Res. **130**(1), 263–275 (2001)

21. The, A.N., Mousseau, V.: Using assignment examples to infer category limits for the ELECTRE TRI method. J. Multi-criteria Decis. Anal. **11**(1), 29–43 (2002)

22. Dias, L., Mousseau, V., Figueira, J.R., Clímaco, J.: An aggregation/disaggregation approach to obtain robust conclusions with ELECTRE TRI. Eur. J. Oper. Res. **138**(1), 332–348 (2002)

23. Doumpos, M., Marinakis, Y., Marinaki, M., Zopounidis, C.: An evolutionary approach to construction of outranking models for multicriteria classification: the case of the ELECTRE TRI method. Eur. J. Oper. Res. **199**(2), 496–505 (2009)

24. Cailloux, O., Meyer, P., Mousseau, V.: Eliciting ELECTRE TRI category limits for a group of decision makers. Eur. J. Oper. Res. **223**(1), 133–140 (2012)

25. Zheng, J., Metchebon, S.A., Mousseau, V., Pirlot, M.: Learning criteria weights of an optimistic ELECTRE TRI sorting rule. Comput. OR **49**, 28–40 (2014)

26. Sobrie, O., Lazouni, M.E.A., Mahmoudi, S., Mousseau, V., Pirlot, M.: A new decision support model for preanesthetic evaluation. Comput. Methods Programs Biomed. **133**, 183–193 (2016)

Injection of Extreme Points in Evolutionary Multiobjective Optimization Algorithms

A.K.M. Khaled Ahsan Talukder[1]([✉]), Kalyanmoy Deb[1],
and Shahryar Rahnamayan[2]

[1] Computational Optimization and Innovation Laboratory,
Michigan State University, East Lansing, MI 48824, USA
talukde1@msu.edu, kdeb@egr.msu.edu
[2] Department of Electrical, Computer, and Software Engineering,
University of Ontario Institute of Technology, Oshawa, Canada
shahryar.rahnamayan@uoit.ca
https://www.coin-laboratory.com

Abstract. This paper investigates a curious case of *informed initialization* technique to solve difficult multi-objective optimization (MOP) problems. The initial population was injected with non-exact (i.e. approximated) nadir objective vectors, which are the boundary solutions of a Pareto optimal front (PF). The algorithm then successively improves those boundary solutions and utilizes them to generate non-dominated solutions targeted to the vicinity of the PF along the way. The proposed technique was ported to a standard Evolutionary Multi-objective Optimization (EMO) algorithm and tested on a wide variety of benchmark MOP problems. The experimental results suggest that the proposed approach is very helpful in achieving extremely fast convergence, especially if an experimenter's goal is to find a set of well distributed trade-off solutions within a fix-budgeted solution evaluations (SEs). The proposed approach also ensures a more focused exploration of the underlying search space.

1 Introduction

Since the past two decades, the algorithmic techniques to solve multi-objective optimization problems (MOPs) have been developed mainly by the two communities independently. Centrally, by the classical numerical optimization practitioners and also by the Evolutionary Multi-objective Optimization (EMO) community. However, in many occasions, they both address the same problem in two completely different perspectives. For example, a canonical algorithm like Normal Boundary Intersection (NBI) [1] assumes that the bound (i.e. affine subspace of the lowest dimension that contains *convex hull of individual minima* or $CHIM_+$) of the Pareto-optimal front (PF) is already known, on the other hand, such assumption is not necessary in many standard EMO algorithms [2].

However, bounding/bracketing of the search space is the first step in many classical numerical optimization algorithms [3]. As EMO algorithms are stochastic global search algorithms, they do not require such a measure. Moreover, the

© Springer International Publishing AG 2017
H. Trautmann et al. (Eds.): EMO 2017, LNCS 10173, pp. 590–605, 2017.
DOI: 10.1007/978-3-319-54157-0_40

concept of *bound* might not be understood as it has been done in the single function optimization problems. In MOPs, bounds can be considered as the set of *boundary* solutions beyond which the objective function is undefined (or in constrained problems, infeasible). However, if those bounding solutions are available before any EMO run, they will be very useful in finding the complete PF in a more efficient way. This paper will try to explore this possibility.

Interestingly, we can find some examples in the recent EMO literature where such *bracketing* techniques have been used. For example, in [4] the algorithm simultaneously searches for the boundary solutions and inserts them into the population on different stages. In another study [5], a high performance single objective search algorithm is used to seed the initial population with solutions to a scalarized version of the problem, and reported to be useful with standard EMO algorithms. From all these examples, there are some interesting questions that we think are not answered yet:

- Is it computationally feasible[1] to find the boundary solutions first and later use them in an EMO algorithm?
- Instead of only depending on random local search operations [4,5,7], can these boundary solutions be *explicitly* utilized in a more intuitive and deterministic way?
- How many bounding (or a minimum number of bounding) solutions are *sufficient enough* to be utilized efficiently?

In this paper, we would like to address these questions. The paper is organized as follows: in Sect. 2 we will review some basic definitions related to MOPs, then in Sect. 3 we will discuss two specific motivations for this study. After that we will describe our approach in Sect. 4. The proposed algorithm can explicitly construct (i.e. explores) the rest of the Pareto optimal solutions within (or, the vicinity of) the approximate bounding solutions through an intuitive heuristic. Moreover, our approach also improves the approximations using a careful *update mechanism* to ensure that they can reach to the true PF bounding points. By this way, we can facilitate a more *informed* exploration and save a huge number of solution evaluations. Our scheme is then applied to a standard EMO algorithms (i.e. NSGA-II [2]). Then we will discuss our experiment results on different benchmark MOP problem sets in Sect. 5 and then we conclude the paper in Sect. 6.

2   Preliminaries

We consider multi-objective optimization problems (MOPs) involving m conflicting objectives ($f_i : \mathcal{S} \subset \mathbb{R}^n \to \mathcal{F} \subset \mathbb{R}^m$) as functions of decision vector

---

[1] PF bounds are no easier to find than any other solution in the PF, this has been proved numerous times [6].

$\mathbf{x} \in \mathcal{S}$. We also assume that the readers are already familiar with the basic terminologies[2] concerning an MOP. Specific to this paper, we are more interested in the boundary of the set $\mathcal{F}$, the boundary can also be defined in terms of convex hull of finite point set [1]:

Definition 1 (*Convex Hull of Inidividual Minima (CHIM)*). *Given an ideal objective vector* $\mathbf{z}^* = [f_1^*, f_2^*, \ldots, f_m^*]^T$, *let us assume* Φ *be an* $m \times m$ *matrix[3] whose* i-th *column is* f_i^*. *Then the set of points in* $\mathbb{R}^m$ *which are convex combinations of* f_i^*, *i.e.* $\Phi\beta : \beta \in \mathbb{R}^m$, $\sum_{i=1}^n \beta_i = 1$, $\beta_i \geq 0$, *are referred as the CHIM.*

From the above definition, a PF can be understood in terms of so-called $CHIM_+$:

Definition 2 (CHIM$_+$). *Let* $CHIM_\infty$ *be the affine subspace of the lowest dimension that contains the CHIM, i.e. the set* $\{\Phi\beta : \beta \in \mathbb{R}^m, \sum_{i=1}^m \beta_i = 1\}$. *Moreover, denote* $\delta\mathcal{F}$ *as the boundary of the set* $\mathcal{F}$. *Then* $CHIM_+$ *is defined as the convex hull of the points in the set* $\mathcal{F} \cap CHIM_\infty$. *More informally, if we extend (or withdraw) the boundary of the CHIM simplex to touch* $\delta\mathcal{F}$; *the "extension" of CHIM thus obtained is defined as* $CHIM_+$.

Therefore, the Pareto-optimal front $\mathcal{PF}$ can also be defined as a set of intersection points found from the normals emanating from the $CHIM_+$ (towards the *ideal objective vector* $\mathbf{z}^*$) on to $\delta\mathcal{F}$. The bounding points of the PF is the end point solutions of $CHIM_+$, they can be approximated using either $\mathbf{z}^*$ or the *nadir objective vector* $\mathbf{z}^{\text{nad}}$.

As we have discussed before, the bounding points to the PF is a topic of special interest in the classical numerical optimization community since, in many cases, these points are the first step to model the PF. And therefore, it is the starting point of many non-stochastic MOP solvers [1,10,11]. Whereas in the EMO community, the bounding points are generally overlooked; just because: (i) bounding points do not represent the trade-off – being minimum at all objectives except one, (ii) they are not of much interest in Decision Maker's (DM's) perspective, (iii) since all EMO algorithms employ a population based parallel search, the bounds can be found out along with an EA run. As a result, finding boundary solutions[4] is taken as granted. Or such points can be found with a careful change in the original algorithm [6].

In this paper, we will see how the performance of an EMO can be drastically improved if we first approximate such bounding solutions, inject them into the initial population and thereafter take necessary measures to improve the approximations to the true PF bounds. To approximate them, we will use

---

[2] Weak/Strong Pareto dominance, non-dominated (Pareto) set ($\mathcal{PS}$), local/global Pareto-optimal front ($\mathcal{PF}$), critical point and ideal/Nadir objective vector etc. They are also discussed in [8].

[3] Also, known as the *pay-off* matrix [9].

[4] The terms "bounding solutions" and "boundary solutions" will be used interchangeably throughout the paper.

classical optimization methods like *Interior point method* (IPM) [12] and *generalized pattern-search* (GPS) [13] approach by optimizing m single objective achievement-scalarizing functions [14]. Given the function evaluations spent to find those bounding solutions, our approach demonstrates that it can still save a lot of computational effort compared to that of a standard EMO algorithms. Due to the space constraints, other relevant discussions and the results are included in the supplementary document [8] accompanied with this paper.

3   Motivations

The main motivating guide of this study stems from two aspects: (i) we wanted to take the advantages of the determinism in the classical numerical optimization methods and, (ii) we wanted to guide the search trajectory in a more non-skewed way so that all the trade-off regions will be explored with a similar pace.

3.1   Inspirations from the Classical Methods

In the classical numerical optimization discipline, the first thing we learn is the concept of *bounding (or bracketing)* of the minima (or maxima) [3]. Otherwise, the optimization algorithms will spend a lot of time figuring out a suitable place to start the actual search process.

The first example is the famous Normal Boundary Intersection (NBI) algorithm [1], where the primary theory is developed based on the assumption that we already have the end-point solutions of the $CHIM_+$. Another example is presented in [10], where the algorithm first finds the boundary solutions, then discretizes the PF using triangulation algorithm and then enumerates the entire PF. In [11], the idea is to find the z^{nad} (although the points are not referred as "nadir" objective vector in the paper), then construct the $CHIM_+$ from it. After that, the algorithms divides the space into smaller parts to find the rest of the solutions on the true PF. The authors mentioned that the approach is not suitable for more than 2 objectives. A more detailed survey on such non-population/stochastic MOP approaches to "bound first, then optimize" can be found in [15].

Unfortunately, in most cases the algorithms presented are not simple and easy to implement. Moreover, many examples of such strategies are specialized to a particular problem domain. According to [15], such approaches are termed as the *first order approximation algorithms*. In that sense, our approach can be considered as an example of such category, however simpler and more intuitive than the existing approaches. Moreover, our method is applicable to wide range of MOPs.

3.2   Search Trajectory Bias

Most of the standard EMO algorithms are elitist by design. They are also "opportunistic" in a sense that the population always try to converge to a particular portion of the PF which seems to be easier to solve at a particular moment. Therefore, they tend to generate more solutions on a certain portion of the objective space which is easier to explore. For example, we can see such bias when we try to solve the ZDT4 problem using NSGA-II. In this case, the first objective is easier to minimize than the second one. In ZDT4, reaching the actual PF is difficult because of the local fronts, however finding the weakly dominated solutions along the f_2 axis is comparatively easy. This kind of non-symmetric search behaviour, is what we think, impedes the optimization procedure – especially, if the goal is to find *well-distributed* solutions within a budgeted function evaluation. In addition, such bias might form in multiple spaces if the number of objective increases. Moreover, this can also lead to a stagnation on a locally optimal front.

4   The Algorithm Description

The algorithm works in two phases. The first phase approximates the boundary of the $CHIM_+$ using a fixed budget. In the next phase, an EMO algorithm will explore the entire PF. And most importantly, during the same time, the algorithm needs to successfully update (or improve) the bounds, given the fact that the bounds found in the first phase might not be exact. Our approach nicely fits into a standard elitist EMO algorithm like NSGA-II [2].

4.1   Approximating the $CHIM_+$ Bounds

In m-dimensional objective space if the true PF is a surface, then it can have more than m number of $CHIM_+$ end points. For example, in the three-objective DTLZ7 problem [16], $CHIM_+$ has eight end points. However, we do not want to trace all the solutions that encompasses the entire PF boundary (as it has been done in [10]), we are only interested in the $CHIM_+$ end-points. To find the bounds, we keep these criteria in mind:

1. Finding the exact PF bound is non-trivial, so the goal is to get an approximation that is "good enough". Which is also difficult because, to measure the closeness, we need to know the true PF.
2. Use the "best possible" effort to approximate the PF bounds, by spending a fixed number (as small as possible) of solution evaluations[5] (SEs).
3. Use a deterministic classical numerical single-objective optimization algorithm.
4. Try to approximate at most m bounding solutions, but not more than that: running the single objective optimizer for at most m times.

For a given EA run, let's assume we start with N_p individuals for N_{gen} generations. Therefore, the total number of function evaluations will be $T_e = N_p N_{\text{gen}}$.

---

[5] Instead of *Function Evaluation* we refer it as *Solution Evaluation*, because a fitness function might be evaluated from a series of multiple mathematical function evaluations.

Algorithm 1. Approximate All $CHIM_+$ Bounds

1: $m \leftarrow$ no. of objectives
2: $N_p \leftarrow$ population size
3: $N_{gen} \leftarrow$ maximum generation
4: $T_b \leftarrow \frac{1}{4} N_p N_{gen}$
5: $\mathcal{Z}_b^* \leftarrow \emptyset$, an empty solution set
6: **for** i from 1 **to** m **do**
7:      $f_i \leftarrow i$-th objective function
8:      $\mathbf{x}_i^* \leftarrow$ random initial vector
9:      **repeat**
10:          $\mathbf{x}_i^* \leftarrow$ minimize f_i
11:      **until** $\frac{T_b}{2m}$ solution evaluation reached
12:      $\mathbf{z}^{ref} \leftarrow [f_1(\mathbf{x}_i^*), f_2(\mathbf{x}_i^*), \ldots, f_m(\mathbf{x}_i^*)]^T$
13:      $f_{aasf} \leftarrow$ construct AASF according to Eq. 1
14:      **repeat**
15:          $\mathbf{x}_i^* \leftarrow$ minimize f_{aasf} w.r.t. $\mathbf{z}^{ref}$
16:      **until** $\frac{T_b}{2m}$ solution evaluation reached
17:      $\mathcal{Z}_b^* \leftarrow \{\mathcal{Z}_b^* \cup \mathbf{x}_i^*\}$
18: **end for**
19: **return** $\mathcal{Z}_b^*$

For the fixed budget, we will allocate $T_b = T_e/4$ function evaluations for the bound search and $T_{ea} = 3T_e/4$ for the actual EA run. To approximate the bound, we will separately optimize m objective functions, we will spend T_b/m function evaluation for each bounding points.

The actual $CHIM_+$ bound computation algorithm was conducted in two steps – (i) given a particular objective function f_i, first we minimize it by spending $T_b/2m$ number of function evaluations and assume a solution $\mathbf{x}_i^*$ is found. Moreover, assume $\mathbf{z}^{ref} = \mathbf{f}(\mathbf{x}_i^*)$. This step is done as a "bootstrap" that will help us to reach close to the f_i^*. (ii) then we construct an Augmented Achievement Scalarizing Function (AASF) [14] by taking $\mathbf{z}^{ref}$ as a reference point:

$$f_{aasf} = \rho \sum_{j=1}^{m} w_j(f_j(\mathbf{x}) - z_j^{ref}) + \max_{j \in \{1,2,\ldots,m\}} w_j(f_j(\mathbf{x}) - z_j^{ref}) \tag{1}$$

and solve it again for $\frac{T_b}{2m}$ iterations. Here, we set $w_i = 0.9$, $w_{j \neq i} = \frac{1}{10(m-1)}$. We set $\rho = 0.0001$, the value of ρ should be kept small enough so that we can avoid the weakly dominated solutions. For the single objective solver, we use classical methods like Interior Point Method (IPM) [12] or Generalized Pattern Search (GPS) [13] etc.

A basic listing for this routine is presented in Algorithm 1 and the concept is illustrated in the Fig. 1. It should be noted that, this algorithm does not always find the all the unique bounds correctly, especially when the problem is hard, or when the PF is composed of multiple disconnected fronts.

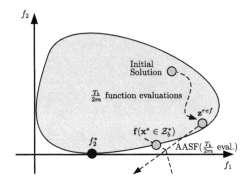

Fig. 1. The illustration of Algorithm 1 while finding a bounding solution that minimizes f_2. The algorithm spends $T_b/2m$ function evaluations to find $\mathbf{z}^{\text{ref}}$ and another $T_b/2m$ function evaluations to approximate one of the $CHIM_+$ bounds. The figure also shows that the discovered solution $\mathbf{x}^*$ might not be the exact bounding point.

4.2   The Exploration Step

The basic exploration is conducted as follows: after estimating the $CHIM_+$ bounds, we will include the set $\mathcal{Z}_b^*$ into the initial population and then the EA execution will start. On each generation, we will select 25% of the best (i.e. the first front) individuals from the current population, then we pick one arbitrary decision vector $\mathbf{x}_s$ (we call it *source vector*) from the selected 25% solutions. Next we pick one random solution from $\mathcal{Z}_b^*$ as a *target vector* $\mathbf{x}_t$. Here, the decision vector $\mathbf{x}_t$ is the *promising solution* that we want to get close to. Given $\mathbf{x}_s$ and $\mathbf{x}_t \in \mathcal{Z}_b^*$, the variation operation is achieved using a simple linear translation –

$$\mathbf{x}_c = \mathbf{x}_s + \mathbf{U}\left[\left(\frac{3d}{4}, \frac{5d}{4}\right)\right] \circ \left[\frac{\mathbf{x}_s - \mathbf{x}_t}{||\mathbf{x}_s - \mathbf{x}_t||}\right] \qquad (2)$$

Here, $\mathbf{U}[(l, u)]$ is a uniform random vector where each element is within the range $[l, u]$, $d = ||\mathbf{x}_s - \mathbf{x}_t||$ and $\circ$ is the *Hadamard product*. $\mathbf{x}_c$ is the generated solution. Basically, we are trying to generate a solution that is within an n-dimensional sphere with radius $d/4$ centered at $\mathbf{x}_t$.

In an ideal case, if the set $\mathcal{Z}_b^*$ contains the true $CHIM_+$ bounds, then the later operation will ensure a very fast (almost immediate) convergence. However this does not address the issue of so-called *Search Trajectory Bias*, to alleviate such a degeneracy, we will include more decision vectors in the set $\mathcal{Z}_b^*$. For this treatment, we will also include solutions with the *maximal crowding distances* from the current best front. If we compute the crowding distances as have been done in the NSGA-II algorithm, the extreme solutions will have the crowding distance of ∞ and the solutions that are rendered isolated *within* the current best front will have the maximal values. The *gaps* in the PF are enclosed by such isolated solutions.

Let us assume the solutions that reside on the edge of the *gaps* are denoted as $\mathcal{E}_g$ and the solutions with ∞ crowding distances are denoted as $\mathcal{E}_\infty$. Obviously, if we could generate more points to the vicinity of $\mathcal{E}_g$, the search trajectory bias could be repaired long before they become more severe. Therefore, we also include $\mathcal{E}_g$ in $\mathcal{Z}_b^*$ and we denote this set as $\mathcal{V}$, i.e. $\mathcal{V} = \{\mathcal{Z}_b^* \cup \mathcal{E}_g\}$. To do this, we will pick m solutions with the maximal crowding distances (excluding the solutions in $\mathcal{E}_\infty$) from the current population, as $\mathcal{E}_g$. Moreover, this inclusion needs to be done in every generation, because we do not know if the current front is the PF or not. Furthermore, $\mathcal{E}_g$ is never identical across the generations.

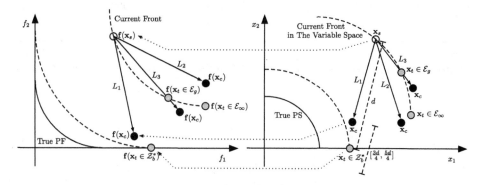

Fig. 2. The exploration step, i.e. the illustration of lines 9–12 in Algorithm 3. The right axes are the variable space and the left axes are the corresponding objective space. The point $\mathbf{x}_c$ is the child (black circle) and $\mathbf{x}_s$ is the *source* (white circle) decision vectors. The vectors $\mathbf{x}_t$ are the *target* points (grey circles). The operation will arbitrarily choose one of the directions denoted by L_1, L_2 or L_3. If $\mathbf{x}_c$ violates the variable bound then it is reverted back to the corresponding *target* point $\mathbf{x}_t$.

Now, instead of taking the target decision vectors $\mathbf{x}_t$ from $\mathcal{Z}_b^*$, we will consider them from $\mathcal{V}$. We call all the solutions in $\mathcal{V}$ as "pivot points". Along with the search trajectory bias corrections, we also consider the case when $\mathcal{Z}_b^*$ is not the

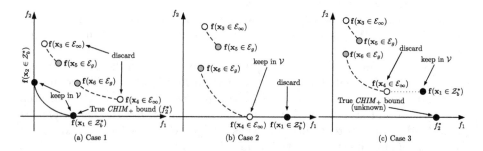

Fig. 3. Different configurations and the relative positioning of the solutions in $\mathcal{V}$. The first two cases ((a), (b)) are ideal, since the target solutions can be retained in the set $\mathcal{V}$ by a simple non-dominated sorting. For the third case (c), a special swapping step (line 7–9, Algorithm 2) is required to maintain a solution set to encourage diversity.

true $CHIM_+$ bounds (in fact, this is the most likely case). Which follows that we need to improve the bounds in $\mathcal{Z}_b^*$ during the EA generations. This procedure is explained in the next section.

4.3   Successive Bound Correction

The procedure described in the previous section only considers an ideal case, where the positions of $\mathcal{Z}_b^*$, $\mathcal{E}_g$ and $\mathcal{E}_\infty$ never coincide with any other. However during the evolutionary runs, different configurations may arise –

Algorithm 2. Update Pivot Points

Require: approx. $CHIM_+$ bounds $\mathcal{Z}_b^*$ from Algorithm 1 or the set $\mathcal{Z}_b^*$ found from the
    previous update.
1: $\mathcal{E}_\infty \leftarrow$ boundary solutions from the current front
2: $\mathcal{V} \leftarrow \{\mathcal{Z}_b^* \cup \mathcal{E}_\infty\}$
3: apply non-dominated sorting on $\mathcal{V}$
4: rank $\mathcal{V}$: $\mathcal{V} \rightarrow \{\mathcal{F}_1, \mathcal{F}_2, \ldots, \mathcal{F}_n\}$
5: take the best front in $\mathcal{V}'$: $\mathcal{V}' \leftarrow \mathcal{F}_1$
6: **for** all points $\mathbf{x}_i$ in $(\mathcal{V} - \mathcal{V}')$ **do**
7:     **if** $\mathbf{x}_i$ weakly dominates any $\mathbf{x}_j \in \mathcal{V}'$ **then**
8:         replace $\mathbf{x}_j$ by $\mathbf{x}_i$
9:     **end if**
10: **end for**
11: update $\mathcal{Z}_b^*$: $\mathcal{Z}_b^* \leftarrow \mathcal{V}'$
12: $\mathcal{E}_g \leftarrow$ find m solutions with maximal crowding distances.
13: $\mathcal{V} \leftarrow \{\mathcal{V}' \cup \mathcal{E}_g\}$
14: **return** $\mathcal{V}$

- Case 1 *(Ideal Case)*: where $\mathcal{Z}_b^*$ is close to (or coincident with) the true $CHIM_+$ bounds. The current front is far from the true PF, therefore $\mathcal{E}_g$ and $\mathcal{E}_\infty$ do not coincide with $\mathcal{Z}_b^*$. Refer to the Fig. 3a.
- Case 2 *(Ideal Case)*: where $\mathcal{Z}_b^*$ is weakly dominated by the corresponding solutions in $\mathcal{E}_\infty$ (Fig. 3b). In this case, we need to discard the solution $\mathbf{x}_1$ and keep $\mathbf{x}_4$ in $\mathcal{V}$.
- Case 3 *(Degenerate Case)*: where $\mathcal{Z}_b^*$ does not coincide with the true PF bound (or both non-dominated), and one of the solutions in $\mathcal{E}_\infty$ weakly dominates the corresponding solution in $\mathcal{Z}_b^*$ (Fig. 3c). In this case, we need to discard the solution $\mathbf{x}_4$ and keep $\mathbf{x}_1$ in $\mathcal{V}$. Because the solution $\mathbf{x}_1$ has the higher chance of extending the current/future front, so that it does not loose the diversity in terms of the objective function trade-off.

    The above three configurations indicates that we also need to include $\mathcal{E}_\infty$ into the pivot set $\mathcal{V}$. In order to address the case 1 and 2, we first merge the sets $\mathcal{Z}_b^* \cup \mathcal{E}_\infty$ into $\mathcal{V}$ and apply non-dominated sorting. Then we will take out the non-dominated solutions $\mathcal{V}'$ from $\mathcal{V}$, i.e. $(\mathcal{V} - \mathcal{V}')$. After that, we will check if any point $\mathbf{x}_i \in (\mathcal{V} - \mathcal{V}')$ weakly dominates a point $\mathbf{x}_j \in \mathcal{V}'$. If yes, we will replace $\mathbf{x}_j$ by $\mathbf{x}_i$. This step will handle the case 3 as presented in the Fig. 3c. The procedural description of this step is presented in the Algorithm 2.

If we run Algorithm 2 in the case 1 (Fig. 3a), the resultant pivot set will be $\mathcal{V} = \{\mathbf{x}_1, \mathbf{x}_2, \mathbf{x}_5, \mathbf{x}_6\}$. In case 2 (Fig. 3b), it will be $\mathcal{V} = \{\mathbf{x}_2, \mathbf{x}_4, \mathbf{x}_5, \mathbf{x}_6\}$ and in the case 3 (Fig. 3c), the result will be $\mathcal{V} = \{\mathbf{x}_1, \mathbf{x}_2, \mathbf{x}_5, \mathbf{x}_6\}$, etc. Also, $|\mathcal{V}'| \leq |\mathcal{V}|$ is trivially true. On the line 14 of Algorithm 2, it is always the case that $|\mathcal{V}| > m$. The integration into NSGA-II is straight-forward, which is discussed in the next section.

4.4 Integrating into an Elitist EMO Algorithm: NSGA-II

The integration listing is presented in the Algorithm 3. The only difference with NSGA-II is that we approximate the $CHIM_+$ bounds and inject them into the initial individual. Then we take 25% of arbitrary individuals and project them on to the pivot solutions, and all these projected solutions will be merged into the child population on every generation. On the same iteration, the successive improvement on the bounds is done on line 7.

Moreover, upon generating the vector $\mathbf{x}_c$, if one of the variable values go beyond the variable bounds (i.e. $x_j > x_{j_H}$ or $x_j < x_{j_L}$), then we replace the overshot with the corresponding *target* vector value y_j, i.e. $y_j \in \mathbf{x}_t$. Therefore, if a certain vector $\mathbf{x}_c$ can't make a successful translation, then $\mathbf{x}_c$ is reverted back to the vicinity of $\mathbf{x}_t$. Thus, we assure a local best estimated translation of the *source* vector $\mathbf{x}_s$. This process is done on the line 13 of the Algorithm 3, also explained in the Fig. 2. Again, if we look at the line 18, it may seem that the pivot points $\mathcal{V}$ are inserted into the current parent population on every generation. If the $\mathcal{Z}_b^*$ are true PF bounds, the line 18 will inject the same copies of solutions for multiple times. This can easily be avoided by keeping a global pointers to the solutions in $\mathcal{V}$. Another interesting aspect of this approach: it is "pluggable" – in a sense that this can be ported to any other elitist EMO algorithm.

5 Experiments and Comparisons

Before describing the main experimental results, in the following subsection we are going to review a special comparison technique called *Relative Speed-up Ratio (RSR)* that has been utilized in this study. We also see how the benchmark problems' specifications are also modified in order to ensure a fair comparative analysis. The benchmark problems are taken from the existing EMO literature [16–18].

5.1 Performance Measure: The Relative Speed-Up Ratio (RSR)

In this experiment, we are interested to see if our approach can offer a better convergence speed while maintaining a good diversity in terms of objective trade-offs. For the spread and convergence measure, we have used the Hypervolume (HV) metric using the procedure discussed in [19]. For the speed, we have formulated a metric called *Relative Speed-up Ratio (RSR)*, which also depends on the HV metric of the current front in each generation.

We ran the Algorithm 3 to solve a particular problem P and then the same problem was solved using the original NSGA-II. At each generation, the HV measure was recorded from both algorithms and compared side by side. To understand the statistics, all the experiment data were collated from 31 independent runs, therefore the HV metric is calculated as the mean of those 31 runs at each generation.

Algorithm 3. NSGA-II with Bracketing

Require: construct $\mathcal{Z}_b^*$ using Algorithm 1
1: $N_p \leftarrow$ population size $|P_t|$
2: $N_{\text{gen}} \leftarrow$ maximum generation
3: initial child population: $Q_1 \leftarrow \emptyset$
4: $t \leftarrow 1$
5: **while** $t \leq N_{\text{gen}}$ **do**
6:     $P_t' \leftarrow$ select uniquely random $\frac{N_p}{4}$ solutions from P_t
7:     $\mathcal{V}_t \leftarrow$ construct pivot set $\mathcal{V}$ using Algorithm 2
8:     $S_t \leftarrow \emptyset$
9:     **for** each solution $\mathbf{x}_s \in P_t'$ **do**
10:         $\mathbf{x}_\tau \leftarrow$ pick an arbitrary point from $\mathcal{V}_t$
11:         $d \leftarrow ||\mathbf{x}_\tau - \mathbf{x}_s||$
12:         $\mathbf{x}_c \leftarrow \mathbf{x}_s + \mathbf{U}\left[\left(\frac{3d}{4}, \frac{5d}{4}\right)\right] \circ \left(\frac{1}{d}(\mathbf{x}_s - \mathbf{x}_\tau)\right)$
13:         **if** $((x_j > x_{j_H}) \vee (x_j < x_{j_L}) \,|\, \exists x_j \in \mathbf{x}_c)$ **then**
14:             correct overshoot: $x_j \leftarrow y_j \,|\, y_j \in \mathbf{x}_\tau$
15:         **end if**
16:         $S_t \leftarrow \{S_t \cup \mathbf{x}_c\}$
17:     **end for**
18:     include the bounds: $P_t \leftarrow \{P_t \cup \mathcal{Z}_b^*\}$
19:     $R_t \leftarrow \{P_t \cup Q_t\}$
20:     rank R_t into fronts: $R_t \rightarrow \{\mathcal{F}_1, \mathcal{F}_2, \ldots, \mathcal{F}_n\}$
21:     $P_{t+1} \leftarrow \emptyset$
22:     $i \leftarrow 1$
23:     **while** $|P_{t+1}| + |\mathcal{F}_i| \leq N_p$ **do**
24:         assign crowding distances on the front $\mathcal{F}_i$
25:         $P_{t+1} \leftarrow \{P_t \cup \mathcal{F}_i\}$
26:         $i \leftarrow i + 1$
27:     **end while**
28:     sort $\mathcal{F}_i$ in descending order using $\preceq_n$
29:     $P_{t+1} \leftarrow$ the first $N_p - |P_{t+1}|$ solutions from $\mathcal{F}_i$
30:     $Q_{t+1} \leftarrow$ select, crossover and mutate P_{t+1}
31:     $Q_{t+1} \leftarrow Q_{t+1} \cup S_t$ and randomly shuffle Q_{t+1}
32:     $t \leftarrow t + 1$
33: **end while**

Given a pair of EMO algorithms A and B, that solve a particular problem P, the RSR measure is computed as follows:

$$RSR(A, B) = \frac{SE_A|_{HV_A \geq r\overline{HV}_{B_{\max}}}}{SE_B|_{\overline{HV}_{B_{\max}}}} \qquad (3)$$

In the above equation, SE_A and SE_B denotes the total number of solution evaluations (SE) for algorithms A and B, respectively. And the subscripted expression after $|$ denotes the limit when the SE values need to be recorded. $\overline{HV}_{B_{\max}}$ denotes the maxmium mean-HV measure for the algorithm B (in the convergence plot); and r is a value $0.0 < r \leq 1.0$. Hence, the above expression computes the ratio of two quantities –

- The total number of SE of algorithm A to reach *at-least* a certain r-portion of the maximum of the mean-hypervolumes of B and
- The total number of SE of algorithm B to reach its maximum of the mean-hypervolumes.

Therefore, if the algorithm A has a slower convergence rate than that of B, then $RSR(A, B) > 0.0$, otherwise it will be equal to 0.0. For all the experiments, we have set the value of r within $0.8 \leq r \leq 0.9$. All the benchmark test problems used in this paper are well defined and their shape of the PF is also known. Therefore, for a given reference objective vector $\mathbf{z}$, an *ideal HV (IHV)* is also computable analytically. For the RSR results (presented in Table 1), we have used the IHV values instead of $\overline{HV}_{B_{\max}}$. So, in all our experiments, the RSR values are computed as:

$$RSR_{IHV}(A, B) = \frac{SE_A|_{HV_A \geq rIHV}}{SE_B|_{IHV}} \qquad (4)$$

5.2   Experiments with the Benchmark Problem Set

First we have tested the performance of Algorithm 3 on five 2-objective problems [17], namely ZDT1, ZDT2, ZDT3, ZDT4 and ZDT6, and we have set NSGA-II results as the *control groups*. To maintain a fair comparison, we have compensated the extra solution evaluations by the Algorithm 1 for the NSGA-II runs, and compared NSGA-II and Algorithm 3 side by side. We are interested to see which algorithm can reach to a desired IHV within less solution evaluations (SE). For all the problems, we have seen our algorithm can demonstrate a very steep convergence to the true PF, given that the extra SE from Algorithm 1 are compensated for NSGA-II. The experiment with ZDT1 is illustrated in Fig. 4, here we can see that the Algorithm 1 takes up to around 2K of solution evaluations. Given that, NSGA-II still lags behind with a multiple factors to reach the desired PF, we have seen a similar effect on the problems ZDT2, ZDT4 and ZDT6; except for ZDT3, for which we saw some fluctuations due to the disconnected nature of the true PF. We only presented two of our results in the Fig. 5,

Table 1. The ideal HV (IHV), their corresponding reference points and RSR_{IHV}(NSGA-II, Algorithm 3) values.

Problem	Ideal HV (IHV)	Reference Points ($\mathbf{z}$)	SE at 90%-IHV (NSGA-II)	SE at 90%-IHV (Algorithm 3)	RSR_{IHV} (NSGA-II, Algorithm 3)
ZDT1	3.67	$(2.0, 2.0)$	3600	**1689**	**2.13**
ZDT2	3.34	$(2.0, 2.0)$	6000	**1877**	**3.20**
ZDT3	4.82	$(2.0, 2.0)$	3400	**2340**	**1.45**
ZDT4	3.67	$(2.0, 2.0)$	14500	**5442**	**2.66**
ZDT6	15.35	$(4.0, 4.0)$	20000	**3036**	**6.59**
DTLZ1	999.98	$(10.0, 10.0, 10.0)$	24800	**11402**	**2.18**
DTLZ2	7.48	$(2.0, 2.0, 2.0)$	4000	5998	0.00
DTLZ3	3374.98	$(15.0, 15.0, 15.0)$	30800	**14732**	**2.09**
DTLZ4	7.48	$(2.0, 2.0, 2.0)$	4400	4422	0.00
DTLZ5	6.1	$(2.0, 2.0, 2.0)$	4400	**3282**	**1.34**
DTLZ6	55.6	$(4.0, 4.0, 4.0)$	40000	**16802**	**2.38**
DTLZ7	134.20	$(10.0, 10.0, 10.0)$	6400	**3891**	**1.64**

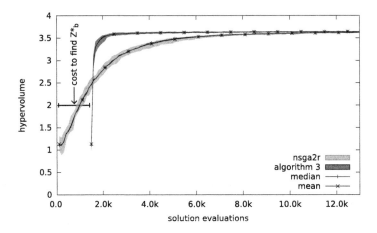

Fig. 4. The convergence test of Algorithm 3 vs. NSGA-II on problem ZDT1

due to the space constraint. To see the figures corresponding to the results of the rest of the problems, readers are referred to the accompanying supplementary material [8] with this paper.

In the next experiment, we have carried out similar tests with the scalable problem sets – DTLZ1, DTLZ2, DTLZ3, DTLZ4, DTLZ5, DTLZ6 and DTLZ7 [16]. For all cases, we have considered 3-objectives. Similarly, we compensate the measure by the extra SE spent to find bounds. All the results are collated from 31 independent runs. In DTLZ6, our approach shows a noticeable improvement, and also for DTLZ1 (Fig. 6). Figures for the results on other problems can be

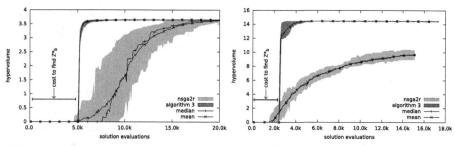

(a) Convergence test for ZDT4 problem   (b) Convergence test for ZDT6 problem

Fig. 5. These plots illustrates the comparative analysis of the convergence rates for different 2-objective problems, the curves are actually consisted of box-plots. Here Algorithm 3 denotes our algorithm and nsga2r is NSGA-II.

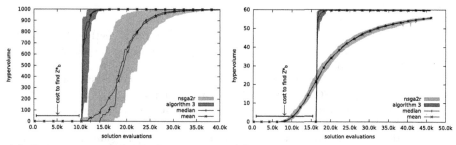

(a) Convergence test for DTLZ1 problem   (b) Convergence test for DTLZ6 problem

Fig. 6. These plots illustrates the comparative analysis of the convergence rates for different 3-objective problems, the curves are actually consisted of box-plots. Here Algorithm 3 denotes our algorithm and nsga2r is NSGA-II.

seen in [8]. There are some problems (i.e. DTLZ2, DTLZ4 and DTLZ5) for which the Algorithm 3 does not demonstrate any improvement. The reason behind this is that NSGA-II does not face much difficulty in reaching the true PF, as a result the outcome stays the same even if we introduce extreme points to guide the search. For example, DTLZ6 is harder than DTLZ5[6], as a result, our approach shows even better speed-up in solving harder problems.

6   Conclusions and Future Works

The main contribution of this paper is the utilization of the concept of *bracketing* in the MOP. Although in many classical optimization algorithms, they are assumed to be known. In EMO algorithms, the bounding solutions can play a very important step in the actual optimization run. Moreover, our approach does

---

[6] DTLZ5 and DTLZ6 are basically the same problem except an exponential growth added to the g function, and DTLZ7 has a disconnected PF.

not assume that the initial approximations are true $CHIM_+$ bounds. To overcome this limitation, it uses an intuitive and a simple way to improve the bounds during the evolutionary run – which is more intuitive than many approaches [4, 7] that use an archive or a population to improve/utilize the bounds.

Most importantly, our study has demonstrated that, even after spending a portion of the total allocated solution evaluations to find the $CHIM_+$ bounds, they are so useful that the original EMO can achieve an extremely fast convergence. Our technique is also easy to implement and understand. We have also seen that the efficacy of our algorithm becomes more salient with the increasing level of problem difficulty. Even though we have carried out an extensive study of our model on a variety of benchmark problems, we did not conduct our study on *many-objective* cases yet. We hope to investigate along this line in the future.

References

1. Das, I., Dennis, J.E.: Normal-boundary intersection: a new method for generating the pareto surface in nonlinear multicriteria optimization problems. SIAM J. Optim. **8**(3), 631–657 (1998)
2. Deb, K., Pratap, A., Agarwal, S., Meyarivan, T.: A fast and elitist multiobjective genetic algorithm: NSGA-II. IEEE Trans. Evol. Comput. **6**(2), 182–197 (2002)
3. Kiefer, J.: Sequential minimax search for a maximum. Proc. Am. Math. Soc. **4**(3), 502–502 (1953)
4. Freire, H., de Moura Oliveira, P.B., Pires, E.J.S., Bessa, M.: Many-objective optimization with corner-based search. Memetic Comput. **7**(2), 105–118 (2015)
5. Dasgupta, D., Hernandez, G., Romero, A., Garrett, D., Kaushal, A., Simien, J.: On the use of informed initialization and extreme solutions sub-population in multi-objective evolutionary algorithms. In: 2009 IEEE Symposium on Computational Intelligence in Multi-Criteria Decision-Making, pp. 58–65 (2009)
6. Ehrgott, M., Tenfelde-Podehl, D.: Computation of ideal and nadir values and implications for their use in MCDM methods. Eur. J. Oper. Res. **151**(1), 119–139 (2003)
7. Singh, H.K., Isaacs, A., Ray, T.: A pareto corner search evolutionary algorithm and dimensionality reduction in many-objective optimization problems. IEEE Trans. Evol. Comput. **15**(4), 539–556 (2011)
8. Talukder, A.K.A., Deb, K., Rahnamayan, S.: Injection of extreme points in evolutionary multiobjective optimization algorithms (supplimentary discussions and results). http://www.cse.msu.edu/talukde1/papers/chimps-emo17.pdf
9. Benayoun, R., de Montgolfier, J., Tergny, J., Laritchev, O.: Linear programming with multiple objective functions: step method (stem). Math. Program. **1**(1), 366–375 (1971)
10. Mueller-Gritschneder, D., Graeb, H., Schlichtmann, U.: A successive approach to compute the bounded pareto front of practical multiobjective optimization problems. SIAM J. Optim. **20**(2), 915–934 (2009)
11. Schandl, B., Klamroth, K., Wiecek, M.: Norm-based approximation in multicriteria programming. Comput. Math. Appl. **44**(7), 925–942 (2002)
12. Powell, M.J.D.: A fast algorithm for nonlinearly constrained optimization calculations. In: Numerical Analysis, pp. 144–157. Springer, Heidelberg (1978)
13. Audet, C., Dennis, J.E.: Analysis of generalized pattern searches. SIAM J. Optim. **13**(3), 889–903 (2002)

14. Wierzbicki, A.P.: The use of reference objectives in multiobjective optimization. In: Fandel, G., Gal, T. (eds.) Multiple Criteria Decision Making Theory and Application, Lecture Notes in Economics and Mathematical Systems, vol. 177, pp. 468–486. Springer, Heidelberg (1980)
15. Ruzika, S., Wiecek, M.M.: Approximation methods in multiobjective programming. J. Optim. Theor. Appl. **126**(3), 473–501 (2005)
16. Deb, K., Thiele, L., Laumanns, M., Zitzler, E.: Scalable multi-objective optimization test problems. In: Proceedings of the 2002 Congress on Evolutionary Computation, CEC 2002, vol. 1, pp. 825–830 (2002)
17. Zitzler, E., Deb, K., Thiele, L.: Comparison of multiobjective evolutionary algorithms: empirical results. Evol. Comput. **8**(2), 173–195 (2000)
18. Osyczka, A., Kundu, S.: A new method to solve generalized multicriteria optimization problems using the simple genetic algorithm. Struct. Optim. **10**(2), 94–99 (1995)
19. While, L., Bradstreet, L., Barone, L.: A fast way of calculating exact hypervolumes. IEEE Trans. Evol. Comput. **16**(1), 86–95 (2012)

The Impact of Population Size, Number of Children, and Number of Reference Points on the Performance of NSGA-III

Ryoji Tanabe[✉] and Akira Oyama

Institute of Space and Astronautical Science,
Japan Aerospace Exploration Agency, Sagamihara, Kanagawa, Japan
{tanabe,oyama}@flab.isas.jaxa.jp

Abstract. We investigate the impact of three control parameters (the population size μ, the number of children λ, and the number of reference points H) on the performance of Nondominated Sorting Genetic Algorithm III (NSGA-III). In the past few years, many efficient Multi-Objective Evolutionary Algorithms (MOEAs) for Many-Objective Optimization Problems (MaOPs) have been proposed, but their control parameters have been poorly analyzed. The recently proposed NSGA-III is one of most promising MOEAs for MaOPs. It is widely believed that NSGA-III is almost parameter-less and requires setting only one control parameter (H), and the value of μ and λ can be set to $\mu = \lambda \approx H$ as described in the original NSGA-III paper. However, the experimental results in this paper show that suitable parameter settings of μ, λ, and H values differ from each other as well as their widely used parameter settings. Also, the performance of NSGA-III significantly depends on them. Thus, the usually used parameter settings of NSGA-III (i.e., $\mu = \lambda \approx H$) might be unsuitable in many cases, and μ, λ, and H require a particular parameter tuning to realize the best performance of NSGA-III.

1 Introduction

A Multi-Objective continuous Optimization Problem (MOP) which frequently appears in engineering problems can be formulated as follows:

$$\text{minimize} \quad \boldsymbol{f}(\boldsymbol{x}) = (f_1(\boldsymbol{x}), ..., f_M(\boldsymbol{x}))^{\mathrm{T}}$$
$$\text{subject to} \quad \boldsymbol{x} \in \mathbb{S} \subseteq \mathbb{R}^D \tag{1}$$

where, $\boldsymbol{f} : \mathbb{S} \to \mathbb{R}^M$ is an objective function vector which consists of M conflicting objective functions, and $\mathbb{R}^M$ is the objective function space. $\boldsymbol{x} = (x_1, ..., x_D)^{\mathrm{T}}$ is a D-dimensional solution vector, and $\mathbb{S} = \Pi_{j=1}^D [a_j, b_j]$ is the bound-constrained search space, where $a_j \le x_j \le b_j$ for each variable index $j \in \{1, ..., D\}$.

We say that $\boldsymbol{x}^1$ dominates $\boldsymbol{x}^2$ and denote $\boldsymbol{x}^1 \prec \boldsymbol{x}^2$ if and only if $f_i(\boldsymbol{x}^1) \le f_i(\boldsymbol{x}^2)$ for all $i \in \{1, ..., M\}$ and $f_i(\boldsymbol{x}^1) < f_i(\boldsymbol{x}^2)$ for at least one index i. $\boldsymbol{x}^*$ is a Pareto optimal solution if there exists no $\boldsymbol{x} \in \mathbb{S}$ such that $\boldsymbol{x} \prec \boldsymbol{x}^*$. $\boldsymbol{f}(\boldsymbol{x}^*)$ is also called a Pareto optimal objective function vector. The set of all $\boldsymbol{x}^*$ in $\mathbb{S}$

© Springer International Publishing AG 2017
H. Trautmann et al. (Eds.): EMO 2017, LNCS 10173, pp. 606–621, 2017.
DOI: 10.1007/978-3-319-54157-0_41

is the Pareto optimal solution Set (PS), and the set of all $f(x^*)$ is the Pareto Frontier (PF). The goal of MOPs is finding a set of well-distributed nondominated solutions which close to the PF in the objective function space.

A Multi-Objective Evolutionary Algorithm (MOEA) is one of most promising approaches for solving MOPs [14]. Since MOEAs use a set of individuals (solutions of a given problem) for the search, they can find good nondominated solutions in one single run. NSGA-II [5] proposed in the early 2000's is a representative MOEA. However, NSGA-II performs relatively well on MOPs with $M \leq 3$, but its performance significantly degrades on MOPs with $M \geq 4$ [13,22], where an MOP with $M \geq 4$ is called a Many-Objective continuous Optimization Problem (MaOP). Since MaOPs frequently appear in real-world applications [14], the poor performance of MOEAs is a critical problem. Thus, in the past few years, researchers in the Evolutionary Computation community have studied on the development of MOEAs which can handle a large number of objectives [14].

To the best of our knowledge, while many efficient MOEAs for MaOPs have been proposed in previous work, their control parameters have been poorly analyzed. Almost all previous work only proposed a novel MOEA and evaluated its performance on some benchmark problems, but analysis of their control parameters have been minimal. In particular, effects of the population size μ and the number of children λ on the performance of MOEAs are poorly understood. It is well-known that the performance of EAs and MOEAs significantly depends on control parameter settings [8]. Therefore, users of MOEAs need to appropriately set their parameter values for a target application. General rules of thumb for parameter settings (e.g., the population size should be set to about 20 on six-objective multimodal MOPs) help users with such tedious tasks. In addition, since appropriate parameter settings might differ from each MOEA, the performance comparison between MOEAs with inappropriate ones is unfair [10,12]. An analysis of the impact of control parameter settings is also useful for understanding the performance behavior of MOEAs and designing a novel, efficient MOEA. For example, a small population size leads NSGA-II to the fast convergence [3] and is effective for expensive optimization problems.

In this paper, we analyze the impact of three control parameters (the population size μ, the number of children λ, and the number of reference points H) on the performance of Nondominated Sorting Genetic Algorithm III (NSGA-III) [6]. We exhaustively investigate the performance of NSGA-III with various parameter settings of μ, λ, and H on the WFG benchmarks [9] with $M \in \{2, 4, 6, 8\}$. NSGA-III, one of most promising MOEAs for MaOPs, is an improved version of NSGA-II [5] by replacing the crowding distance-based selection with the reference vectors-based niching selection. NSGA-III can be categorized into the $(\mu + \lambda)$-MOEA framework. Therefore, NSGA-III essentially has the three control parameters (μ, λ, and H), except for settings of variation operators (e.g., the crossover probability of SBX), and μ, λ, and H can be set independently of each other. However, it is widely believed that NSGA-III is almost parameter-less and requires setting only one control parameter (H), and the value of μ and λ can be set to $\mu = \lambda \approx H$ as described in [6]. In fact, μ, λ, and H of NSGA-III were set to same values as $\mu = \lambda \approx H$ in almost all previous work (exceptions will be discussed in Sect. 3). Thus, the performance of NSGA-III with $\mu \neq \lambda \neq H$ has never been examined while parameter studies of MOEAs are important for practical uses as well as algorithm designs as discussed above. In this paper, we show that suitable parameter settings of μ, λ, and H values differ from each other as well as widely used parameter settings, and the performance of NSGA-III significantly depends on them.

This paper is organized as follows: Sect. 2 introduces the NSGA-III algorithm. We describe related work in Sect. 3. Section 4 describes experimental settings, and Sect. 5 presents the experimental analysis of μ, λ, and H of NSGA-III. Finally, Sect. 6 concludes this paper and discusses our future work.

2   NSGA-III

This section briefly introduces NSGA-III [6]. A NSGA-III population $\boldsymbol{P}^t = \{\boldsymbol{x}^{1,t}, ..., \boldsymbol{x}^{\mu,t}\}$ is a set of individuals $\boldsymbol{x}^{i,t} = (x_1^{i,t}, ..., x_D^{i,t})^{\mathrm{T}}$, $i \in \{1, ..., \mu\}$. After the initialization of the population $\boldsymbol{P}^t$, the following steps are repeated until the termination criteria are met.

At the beginning of each iteration t, children $\boldsymbol{Q}^t = \{\boldsymbol{u}^{1,t}, ..., \boldsymbol{u}^{\lambda,t}\}$ are generated by applying variation operators (e.g., crossover and mutation operators) to randomly selected individuals from $\boldsymbol{P}^t$, where $\boldsymbol{u}^{i,t} = (u_1^{i,t}, ..., u_D^{i,t})^{\mathrm{T}}$, $i \in \{1, ..., \lambda\}$. After the generation of $\boldsymbol{Q}^t$, μ individuals composing the population in the next iteration $\boldsymbol{P}^{t+1}$ are selected from a union $\boldsymbol{R}^t = \boldsymbol{P}^t \cup \boldsymbol{Q}^t$, where $|\boldsymbol{R}^t| = \mu + \lambda$. The nondominated sorting [5] is applied to $\boldsymbol{R}^t$, and the individuals in $\boldsymbol{R}^t$ are grouped as $\{\boldsymbol{F}^{1,t}, \boldsymbol{F}^{2,t}, ...\}$ according to their nondomination levels. Starting from $i = 1$, the individuals in $\boldsymbol{F}^{i,t}$ is added to a new temporal population $\boldsymbol{S}^t$, and the index counter i is incremented as $i = i + 1$. This procedure is repeated until the size of $\boldsymbol{S}^t$ equals or exceeds μ. The index counter i at the end of the nondomination levels-based selection is recorded as l. When $|\boldsymbol{S}^t| = \mu$, $\boldsymbol{P}^{t+1} = \boldsymbol{S}^t$, and this environmental selection procedure is finished. Otherwise (i.e., $|\boldsymbol{S}^t| > \mu$), all individuals in $\boldsymbol{S}^t \backslash \boldsymbol{F}^{l,t}$ are added to $\boldsymbol{P}^{t+1}$, and remaining K individuals ($K = \mu - |\boldsymbol{P}^{t+1}|$) are selected from $\boldsymbol{F}^{l,t}$ so that $|\boldsymbol{P}^{t+1}| = \mu$ according to the following reference vectors-based niching selection.

First, for each f_i ($i \in \{1, ..., M\}$), for each $\boldsymbol{x} \in \boldsymbol{S}^t$, the objective function value $f_i(\boldsymbol{x})$ is normalized into $[0, 1]$ by an originally defined normalization rule in [6]. After the normalization, the niching procedure with the reference vectors is applied to each $\boldsymbol{x} \in \boldsymbol{S}^t$. The reference vector consists of the reference point $\boldsymbol{w} = (w_1, ..., w_M)^{\mathrm{T}}$, $\sum_{j=1}^{M} w_j = 1$ and the ideal point which is identical to the origin $(0, ..., 0)^{\mathrm{T}}$. NSGA-III requires a set of the reference points $\boldsymbol{W} = \{\boldsymbol{w}^1, ..., \boldsymbol{w}^H\}$. $\boldsymbol{W}$ can be generated by using any method, but it is desirable that $\boldsymbol{W}$ consist of uniformly distributed reference points when a user's preference information cannot be utilized for the search.

For each $\boldsymbol{x} \in \boldsymbol{S}^t \backslash \boldsymbol{F}^{l,t}$ and each reference vector, the perpendicular distance between the normalized objective function vector $\boldsymbol{f}'(\boldsymbol{x})$ and the reference vector is calculated. Then, $\boldsymbol{x}$ is associated with a reference vector which has the minimum distance. For each $\boldsymbol{w}^i$ ($i \in \{1, ..., H\}$), the number of associated individuals is counted as $\rho_{i,t}$. After the association procedure of all members in $\boldsymbol{S}^t \backslash \boldsymbol{F}^{l,t}$, the other members in $\boldsymbol{F}^{l,t} \in \boldsymbol{S}^t$ are also associated in the same manner, but the niche count ρ is not counted at this time.

Here, K individuals in $\boldsymbol{F}^{l,t}$ ($K = \mu - |\boldsymbol{P}^{t+1}|$) is selected for the population in the next iteration $\boldsymbol{P}^{t+1}$. First, $\boldsymbol{w}^{\min,t}$ having the minimum niche count $\rho_{\min,t} = \min_{i=1}^{H}\{\rho_{i,t}\}$ is selected. When there are more than two such reference points, one of them is randomly selected as $\boldsymbol{w}^{\min,t}$. Let $\boldsymbol{I}^t$ be a set of the individuals in $\boldsymbol{F}^{l,t}$ associated with $\boldsymbol{w}^{\min,t}$. In this case, the following three situations (i)–(iii) can be considered: (i) $|\boldsymbol{I}^t| = 0$, (ii) $|\boldsymbol{I}^t| \geq 1$ and $\rho_{\min,t} = 0$, and (iii) $|\boldsymbol{I}^t| \geq 1$ and $\rho_{\min,t} \geq 1$. In the case of (i), $\boldsymbol{w}^{\min,t}$ is temporarily removed from W at the iteration t. In the case of (ii), an individual having the minimum perpendicular distance between the normalized objective function vector $\boldsymbol{f}'(\boldsymbol{x})$ and the reference vector is selected from $\boldsymbol{I}^t$ and added to $\boldsymbol{P}^{t+1}$. In the case of (iii), a randomly selected individual in $\boldsymbol{I}^t$ is added to $\boldsymbol{P}^{t+1}$. In any of the situations (i)–(iii), the selected individual is removed from $\boldsymbol{I}^t$, and $\rho_{\min,t} = \rho_{\min,t} + 1$. The niching procedure described above is repeatedly applied until $|\boldsymbol{P}^{t+1}| = \mu$.

3   Previous Parameter Studies of NSGA-III

Here, we review previous parameter studies of three control parameters (the population size μ, the number of children λ, and the number of reference points H) of NSGA-III. In [6], Deb and Jain visually verified that NSGA-III with small μ and H values can find good nondominated solutions on three-objective functions. Seada and Deb evaluated the convergence performance of U-NSGA-III with $\mu > H$ on two-objective MOPs [18], where U-NSGA-III is a unified version of NSGA-III for single, multi, and many-objective optimization. The experimental results in [18] show that the larger μ value than H is suitable for U-NSGA-III when solving a multi-modal MOP. Andersson et. al. presented a parameter tuning study of NSGA-II and NSGA-III [1]. μ ($\approx H$) and four control parameters of the SBX crossover and polynomial mutation (i.e., the crossover/mutation probability p_c and p_m, the distribution index η_c and η_m) of NSGA-II and NSGA-III were tuned using a meta-GA. However, the work [1] mainly focused on the latter four control parameters (p_c, p_m, η_c, and η_m) of the variation operators, and the discussion on the tuned μ ($\approx H$) value was minimal. The experiments in the two previous work [1,18] were performed only on two-objective MOPs, and the impact of μ, λ, and H was not investigated on MaOPs with $M \geq 4$. Also, λ was set as $\lambda = \mu$ in [6,18], and three parameters were set as $\mu = \lambda \approx H$ in [1].

Very recently, Ishibuchi et al. [12] examined the performance of various MOEAs, including NSGA-III, with various μ values on the three- and five-objective functions. Their experimental results show that the performance of the MOEAs significantly depends on the settings of μ, and a suitable μ value differs from each MOEA. However, they set the three parameter values of NSGA-III as $\mu = \lambda \approx H$ and did not examine the performance of NSGA-III (and the other MOEAs) with $\mu \neq \lambda \neq H$ in the same way as [1].

In summary, the performance of NSGA-III with different μ, λ, and H values has been poorly understood. Note that the situation is not unique to NSGA-III. Whereas there is much previous work on parameter studies of variation operators [21,25], the impact of μ and λ on the performance of MOEAs has been hardly investigated. While there are several parameter studies on μ such as [3,11,17,23], they set μ and λ to the same value, and the performance of MOEAs with $\mu \neq \lambda$ has been poorly discussed (the only one exception [7] will be discussed in Sect. 5.2).

4   Experimental Settings

We used the nine WFG functions [9] with $M \in \{2, 4, 6, 8\}$. The shapes of the PFs of the WFG1, WFG2, and WFG3 functions are complicated, discontinuous, and degenerate respectively, and those of the others WFG functions are nonconvex. The WFG4 and WFG9 functions are multimodal, and the WFG2, WFG6, WFG8, and WFG9 functions are also nonseparable. As suggested in [9], the position parameter k was set to $k = 2(M - 1)$, and the distance parameter l was set to $l = 20$, where the number of variables D is $D = k + l$.

We used the source code of NSGA-III implemented by Yuan[1]. We used the SBX crossover and polynomial mutation as in the original NSGA-III [6]. As suggested in [6], we set the control parameters of the variation operators as follows: $p_c = 1.0$, $\eta_c = 30$, $p_m = 1/D$, $\eta_m = 20$. The number of independent runs was set to 31 for $M \in \{2, 4, 6\}$ and only 11 for $M = 8$ due to expensive computational cost of the hypervolume calculation. The number of maximum function evaluations was set to 1×10^4 as [2]. We set the population size μ, the number of children λ, and the number of reference points H as follows: $\mu \in \{10, 20, 40, 80, 160, 320\}$, $\lambda \in \{1, 10, 20, 40, 80, 160, 320, 640\}$, and $H \in \{10, 20, 40, 80, 160\}$ (i.e., $6 \times 8 \times 5 = 240$ configurations were verified).

We used the hypervolume (HV) indicator [28] for evaluating the quality of a set of obtained nondominated solutions $\boldsymbol{A}$[2]. Before calculating the HV value (as well as the MinSum and SumMin values [13] used in Sect. 5.3), the objective function vector $\boldsymbol{f}(\boldsymbol{x})$ of each $\boldsymbol{x} \in \boldsymbol{A}$ was normalized using the ideal point $(0, ..., 0)^{\mathrm{T}}$ and the nadir point $(2, ..., 2M)^{\mathrm{T}}$ as in [26]. The reference point for calculating HV was set to $(1.1, ..., 1.1)^{\mathrm{T}}$. Almost all previous work (e.g., [2,18,25–27]) used nondominated solutions in the bounded archive (= the population $\boldsymbol{P}$) for calculating the HV value. The bounded archive maintains (nondominated) solutions obtained during the search process, but its size is limited. As pointed out in [10], when using the bounded archive and the HV as the performance indicator,

---

[1] The code was downloaded from http://www.cs.bham.ac.uk/~xin/journal_papers.html.

[2] Since the IGD indicator [28] used in [6] is unsuitable for comparing nondominated solution sets of different size as pointed out in [10], we did not use it.

Algorithm 1. Generation method of a set of reference points $\boldsymbol{W}$

---

1   Generate $\boldsymbol{W}^{\mathrm{tmp}} = \{\boldsymbol{w}^{\mathrm{tmp},1}, ..., \boldsymbol{w}^{\mathrm{tmp},H^{\mathrm{tmp}}}\}$ using SLD so that $H^{\mathrm{tmp}} \gg H$;

2   $\boldsymbol{W} \leftarrow \emptyset$, $\boldsymbol{w}^c = (1/M, ..., 1/M)^{\mathrm{T}}$, $\boldsymbol{W} \leftarrow \boldsymbol{W} \cup \{\boldsymbol{w}^c\}$;

3   **for** $i = 1$ **to** H^{tmp} **do**

4     $d_i \leftarrow$ EuclideanDistance$(\boldsymbol{w}^{\mathrm{tmp},i}, \boldsymbol{w}^c)$, $V_i \leftarrow TRUE$;

5   **while** $|\boldsymbol{W}| < H$ **do**

6     $k \leftarrow \underset{j | V_j = TRUE}{\arg \max} \; d_j$, $V_k \leftarrow FALSE$, $\boldsymbol{W} \leftarrow \boldsymbol{W} \cup \{\boldsymbol{w}^{\mathrm{tmp},k}\}$;

7   **for** $l = 1$ **to** H^{tmp} **do**

8   If $V_l = TRUE$, then $d_l \leftarrow \min($EuclideanDistance$(\boldsymbol{w}^{\mathrm{tmp},l}, \boldsymbol{w}^{\mathrm{tmp},k}), d_l)$;

---

the large μ values are more beneficial than the small μ values. Furthermore, a monotonic increase of the hypervolume over time (= the number of function evaluations) cannot be ensured when using the bounded archive [15,17]. Thus, we cannot perform a fair comparison of MOEAs with various μ values using the bounded archive. As suggested in [1,3,12,15,17], for the fair comparison of NSGA-III with various μ values, we used the *unbounded* archive that stores *all* nondominated solutions found during the search process.

As described in Sect. 2, NSGA-III requires a set of reference points $\boldsymbol{W} = \{\boldsymbol{w}^1, ..., \boldsymbol{w}^H\}$. In the original NSGA-III paper [6], Simplex-Lattice Design (SLD) [4] and Two-layered SLD (TSLD) [6] were used for generating $\boldsymbol{W}$. SLD and TSLD are the widely used weight/reference points generation methods, but they cannot generate $\boldsymbol{W}$ with arbitrary size [20,24]. That is, they are not suitable for our scale-up studies of H and M. In addition to SLD and TSLD, generation methods of $\boldsymbol{W}$ with arbitrary size were proposed in [16,20,27]. In this paper, we used a generation method of $\boldsymbol{W}$ inspired by the proposed methods in [16, 27]. Algorithm 1 shows the procedure. First, a set of candidate reference points $\boldsymbol{W}^{\mathrm{tmp}} = \{\boldsymbol{w}^{\mathrm{tmp},1}, ..., \boldsymbol{w}^{\mathrm{tmp},H^{\mathrm{tmp}}}\}$ are generated using SLD so that $H^{\mathrm{tmp}} \gg H$. For each $M \in \{2, 4, 6, 8\}$, we set the division number of SLD to $10^5, 83, 24$, and 14 respectively (i.e., $H^{\mathrm{tmp}} = 100\,001, 102\,340, 118\,755$, and $116\,280$). Then, after adding a central point $\boldsymbol{w}^c = (1/M, ..., 1/M)^{\mathrm{T}}$ to $\boldsymbol{W}$, a reference point in $\boldsymbol{W}^{\mathrm{tmp}}$ having the largest Euclidean distance value from ones already belonging to $\boldsymbol{W}$ is selected and added to $\boldsymbol{W}$ (lines 2–4 and 5–8 in Algorithm 1 respectively). This trial is repeated until the size of $\boldsymbol{W}$ is equal to H.

5   Experimental Results and Discussion

We analyze the impact of the parameter settings of H, λ, and μ on the performance of NSGA-III. The results are shown in Sects. 5.1, 5.2, and 5.3 respectively. We also discuss how much the performance of NSGA-III can be improved by tuning the μ, λ, and H values in Sect. 5.4.

5.1   Influence of H and μ Values on the Performance of NSGA-III

Figure 1 shows the influence of the H values on the performance of NSGA-III
with the various μ values on the WFG1, WFG2, and WFG4 functions with
$M \in \{2, 4, 6, 8\}$ at the number of function evaluations $= 1 \times 10^4$. We can see
that the choice of μ and H values significantly affects the performance of NSGA-
III, and their appropriate values also depend on the function, M, and the other
parameter values. Due to space constraints, we show only the results on the
selected functions, but the results on the WFG3, 7, 9 functions and the WFG5,
6, 8 functions are similar to the WFG1 and WFG4 functions respectively. Note
that some HV values in Fig. 1 exceed 1 since the reference point for the HV
calculation was set to $(1.1, ..., 1.1)^{\mathrm{T}}$.

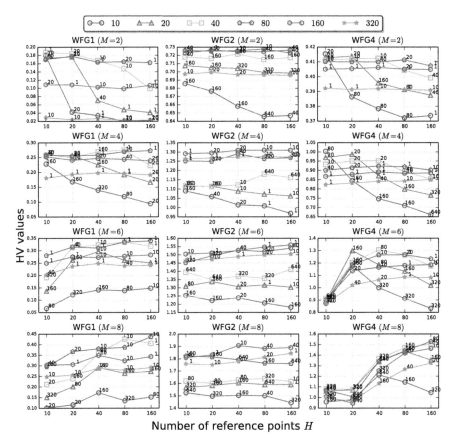

Fig. 1. Influence of the H values on the performance of NSGA-III with the various μ
values on the WFG1, WFG2, and WFG4 functions with $M = \{2, 4, 6, 8\}$. We show the
results of NSGA-III with the best configuration of μ, H, and λ values which achieved
the best median HV value at the number of function evaluations $= 1 \times 10^4$. The numbers
in the figure represent the corresponding λ values.

First, we discuss the impact of the μ values on the performance of NSGA-III. NSGA-III with $\mu \in \{40, 80\}$ and $\{80, 160\}$ achieves the relatively higher HV values for the WFG1 and WFG2 functions respectively. Large μ values might be suitable for NSGA-III when solving a MOP with the disconnected PF (e.g., WFG2 function). The results on the multimodal WFG4 function with $M \in \{2, 4\}$ is counterintuitive – NSGA-III with $\mu = 10$ performs best. With an increasing M, an appropriate μ value is also increased as $\mu \in \{20, 40\}$ for $M = 6$ and $\mu \in \{40, 80\}$ for $M = 8$. The reason for this is simply that the objective function space is increasing according to M, and large μ values become suitable for NSGA-III. The reason why NSGA-III with the smallest population size $\mu = 10$ performs well on the WFG4 function with $M \in \{2, 4\}$ can be considered as follows: The WFG4 function is multimodal, but its shape of the PF is not so complex such as that of the WFG1 and WFG2 functions. Therefore, NSGA-III with the smallest μ value can find good nondominated solutions. NSGA-III with $\mu = 320$ performs poorly in many cases but outperforms NSGA-III using the small μ values when M increases. Therefore, large μ values might be suitable for solving MaOPs with a very large M (i.e., $M \geq 10$).

Then, we analyze the impact of the H values on the performance of NSGA-III. As shown in Fig. 1, for the WFG functions with $M \in \{2, 4\}$, the increase of the H values deteriorates the performance of NSGA-III with the small population size $\mu \in \{10, 20\}$ while it does not significantly influence the performance of NSGA-III with the large μ values. On the other hand, for the WFG functions with $M \in \{6, 8\}$, there exists a particular H value which is suitable for NSGA-III with each μ value. For example, NSGA-III with $\mu = 80$ achieves the highest HV values on the eight-objective WFG1, WFG2 and WFG4 functions when using $H = 160$. Since large H values increase the number of niches in the environmental selection and promote diversity in the population (see Sect. 2), they might be suitable for NSGA-III when solving MaOPs as shown Fig. 1. Seada and Deb [18] reports that the larger μ value than H is suitable for U-NSGA-III when solving a two-objective multimodal MOP (see Sect. 3). Although the examined version of NSGA-III is slightly different from [18], the experimental results of this study show that the use of the smaller μ value than H can improve the performance of NSGA-III on MaOPs with $M \geq 6$.

5.2   Influence of λ and μ Values on the Performance of NSGA-III

Figure 2 shows the influence of the λ values on the performance of NSGA-III with the various μ values. For $M = 2$, the large λ values deteriorate the performance of NSGA-III. This is consistent with the results reported in the previous work [7]. In [7], Durillo et al. investigated the performance of the steady state version of NSGA-II and SPEA2 (i.e., $\lambda = 1$) on two-objective MOPs, and the results show their promising performance. This trend can also be seen in the results of NSGA-III with the large μ values ($\mu \in \{160, 320\}$) on the WFG functions with $M \in \{4, 6, 8\}$. On the other hand, for the WFG functions with $M \in \{6, 8\}$, there is a particular λ value for each μ value which makes NSGA-III with the small μ

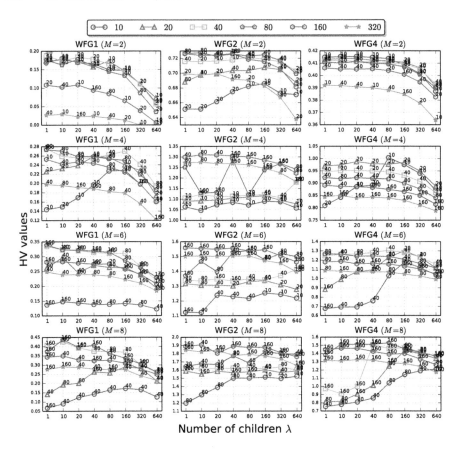

Fig. 2. Influence of the λ values on the performance of NSGA-III with the various μ values on the WFG1, WFG2, and WFG4 functions with $M = \{2, 4, 6, 8\}$. We show the results of NSGA-III with the best configuration of μ, H, and λ values which achieved the best median HV value at the number of function evaluations $= 1 \times 10^4$. The numbers in the figure represent the corresponding H values.

values ($\mu \in \{10, 20, 40, 80\}$) efficient. For example, NSGA-III with $\mu = 10$ and $\lambda = 80$ performs best on the WFG4 function with $M = 4$.

In [7], the performance comparison of $(\mu + \mu)$-MOEAs and $(\mu + 1)$-MOEAs was performed only on MOPs with $M = 2$. Our results show that the larger λ value than μ might be suitable for NSGA-III with small μ values on MaOPs with $M \geq 4$. Unfortunately, we cannot explain the reason why NSGA-III with $\lambda > \mu$ performs well on MaOPs. A further investigation of the impact of λ on the performance of NSGA-III is our future work.

5.3   Convergence Behavior of NSGA-III with Various μ Values

We here analyze the convergence behavior of NSGA-III with the various μ values. For the analysis, in addition to the HV indicator, we also measured two convergence metrics (MinSum and SumMin [13]) and the number of nondominated solutions stored in the unbounded archive $\boldsymbol{A}^3$. MinSum indicates only the convergence quality of $\boldsymbol{A}$ to the center of the PF. The MinSum value of $\boldsymbol{A}$ is calculated as follows: $\text{MinSum}(\boldsymbol{A}) = \min_{\boldsymbol{x} \in \boldsymbol{A}}\{\sum_{j=1}^{M} f_j(\boldsymbol{x})\}$. Also, SumMin evaluates only the convergence performance of $\boldsymbol{A}$ toward the PF around its M edges. The SumMin value of $\boldsymbol{A}$ is obtained as follows: $\text{SumMin}(\boldsymbol{A}) = \sum_{j=1}^{M} \min_{\boldsymbol{x} \in \boldsymbol{A}}\{f_j(\boldsymbol{x})\}$. We also report the number of nondominated solutions stored in the unbounded archive $\boldsymbol{A}$.

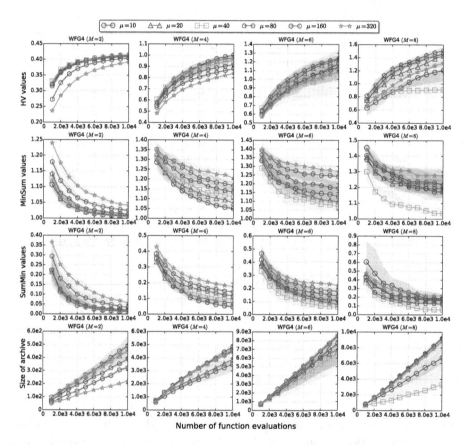

Fig. 3. Convergence behavior of NSGA-III with the various μ values. The figures show the HV, MinSum, SumMin, and $|\boldsymbol{A}|$ over the number of function evaluations on the WFG4 function with $M \in \{2, 4, 6, 8\}$. We plot the median and quartile values of each indicator across all runs. We show the data of NSGA-III with the best configuration of μ, H, and λ values which achieved the best median HV value at the number of function evaluations $= 1 \times 10^4$.

---

[3] Since, as far as we know, there is no good diversity indicator for the *unbounded* archive, we could not measure the diversity of the obtained nondominated solutions.

Figure 3 shows the HV, MinSum, SumMin, and $|\boldsymbol{A}|$ over the number of function evaluations on the WFG4 function with $M \in \{2, 4, 6, 8\}$. We here only discuss the results on the WFG4 function, but note that the obtained results highly depend on the function instance. The analysis of the convergence behavior of NSGA-III on the other WFG functions is our future work.

Figure 3 shows that the large μ values make the convergence speed of NSGA-III slow for $M \in \{2, 4, 6\}$. For example, NSGA-III with $\mu = 320$ achieves the worst HV value at any time, and its convergence speed towards the PF is slow as shown in the MinSum and SumMin values. The results are consistent with the previous work by Brockhoff et. al. [3] which investigated the anytime behavior of NSGA-II with various μ values on two-objective MOPs. NSGA-III with the small μ values ($\mu \in \{10, 20\}$) show the good anytime behavior on the WFG4 function with $M \in \{2, 4\}$ and the strong convergence to the PF (see the MinSum and SumMin values). On the other hand, the performance of NSGA-III with $\mu = 10$ is poor on the WFG4 function with $M \in \{6, 8\}$, and $\mu = 80$ seems to be suitable. For $M = 8$, NSGA-III with $\mu = 10$ performs worse than NSGA-III with $\mu = 320$. Although NSGA-III with $\mu = 40$ achieves the best MinSum and SumMin value, its HV value is worst. This might be due to its low number of obtained nondominated solutions $|\boldsymbol{A}|$. In summary, the results show that large μ values make the convergence speed of NSGA-III slow but might be suitable for MaOPs with $M \geq 6$.

5.4   Generalist vs. Specialist: How Much Can the Performance of NSGA-III be Improved by Tuning μ, λ, and H Values?

So far in this paper, we have analyzed the impact of parameter settings on the performance of NSGA-III for each WFG function instance. We here investigate how much the performance of NSGA-III can be improved by tuning the μ, λ, and H values. We consider the following two parameter tuning scenarios: (1) finding the parameter settings that robustly perform well on a given set of functions, and (2) finding ones that perform well only on a specific function. We say that the parameter settings obtained by the parameter tuning on the scenarios (1) and (2) are *generalist* and *specialist* respectively, where they were named by Smit and Eiben in [19].

First, we deal with the parameter tuning scenario (1) for finding the generalist. We used the Average Performance Score (APS) [2]. Recall that we verified the 240 configurations of μ, λ, and H in this paper (see Sect. 4). Suppose that we compare the performance of n algorithms $\{A_1, ..., A_n\}$ on a given problem instance using the HV values obtained in multiple runs ($n = 240$ in this study). For each $i \in \{1, ..., n\}$ and $j \in \{1, ..., n\}\backslash\{i\}$, let $\delta_{i,j} = 1$, if A_j outperforms A_i based on the Wilcoxon rank-sum test with $p < 0.05$, otherwise $\delta_{i,j} = 0$. Then, the performance score $P(A_i)$ is defined as follows: $P(A_i) = \sum_{j\in\{1,...,n\}\backslash\{i\}}^{n} \delta_{i,j}$. $P(A_i)$ represents the number of compared algorithms that outperforms A_i. The APS is an average of the $P(A_i)$ values on all problem instances.

Table 1 provides the combination of μ, λ, and H values which achieved the smallest APS value for each M on all WFG functions, i.e., the generalist for

Table 1. Generalist for each M: Combination of (μ, λ, H) values that achieved the smallest APS value for each M at the number of function evaluations $= 1 \times 10^4$.

$M = 2$	$M = 4$	$M = 6$	$M = 8$
$(40, 10, 20)$	$(40, 10, 40)$	$(40, 80, 40)$	$(80, 40, 80)$

Table 2. Specialist for each WFG function with $M = 8$: Combination of (μ, λ, H) values that achieved the highest median HV value for each eight-objective WFG function at the number of function evaluations $= 1 \times 10^4$.

WFG1	WFG2	WFG3	WFG4	WFG5	WFG6	WFG7	WFG8	WFG9
$(80, 10, 160)$	$(160, 10, 40)$	$(160, 80, 20)$	$(80, 80, 160)$	$(40, 80, 40)$	$(40, 10, 40)$	$(40, 80, 40)$	$(40, 80, 40)$	$(80, 20, 40)$

each M. We have no intention of claiming that NSGA-III with the generalist shown in Table 1 performs best on all possible problems (it performs well only on the WFG benchmarks). As discussed in [19], obtaining such "best" parameter settings is impossible in practice. We can see that $\mu = 40$ is relatively suitable for NSGA-III for $M \in \{2, 4, 6\}$. On the other hand, NSGA-III with $\mu = 80$ performs well on the WFG functions with $M = 8$. The results also show that the appropriate H values depend on M as discussed in Sect. 5.1. Unfortunately, general rules of thumb for the appropriate parameter settings of λ cannot be found in Table 1.

Then, we deal with the parameter tuning scenario (2) for finding the specialist. Table 2 shows the combination of (μ, λ, H) values which achieved the best median HV value for each eight-objective WFG function, i.e., the specialist for each WFG function with $M = 8$. The results show that the appropriate parameter settings of μ, λ, and H significantly depend on a given function. When solving the WFG1 and WFG4 functions, the large H value is suitable, but NSGA-III with the small H value ($H = 40$) performs best on the other WFG functions, except for the WFG3 function. The large μ value is also suitable for the WFG3 and WFG4 functions.

Although the parameter tuning is not the main concern of this paper, it is interesting to compare the performance of NSGA-III with different parameter configurations. Figure 4 shows the convergence behavior of NSGA-III with the generalist, specialist, and default parameter settings of (μ, λ, H) on the WFG functions with $M = 8$. For the default parameter settings, we set $\mu = \lambda = H = 156$ as suggested in the original NSGA-III paper [6]. We also show the results of NSGA-III with default parameter settings using a set of reference points generated by TSLD (see Sect. 4), i.e., the original NSGA-III. For the generalist and specialist, we used the parameter settings shown in Tables 1 and 2 respectively.

Figure 4 shows that NSGA-III with the specialist clearly outperforms NSGA-III with the three other parameter settings on all WFG functions with $M = 8$. NSGA-III with the generalist performs better than NSGA-III with the default settings, except for the WFG2 and WFG3 functions. Since the shapes of the PFs

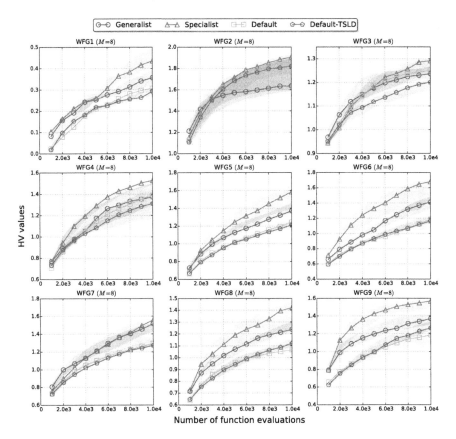

Fig. 4. Convergence behavior of NSGA-III with the generalist, specialist, and default settings of (μ, λ, H) on the WFG functions with $M = 8$. We plot the median and quartile HV values across all 11 runs. For the default settings, we set $\mu = \lambda = H = 156$ as suggested in the original NSGA-III paper [6]. We also show the results of NSGA-III with default settings using a set of reference points generated by TSLD, i.e., the original NSGA-III. For the generalist and specialist, see Tables 1 and 2 respectively.

of the WFG2 and WFG3 functions are complex, NSGA-III requires particular parameter settings (see Table 2). Due to this reason, we consider that NSGA-III with the generalist did not work well on these WFG functions. We cannot find a clear winner on the comparison between NSGA-III with the default parameter settings using a set of reference points $\boldsymbol{W}$ generated by Algorithm 1 and TSLD (labeled as "Default" and "Default-TSLD" in Fig. 4 respectively). A further performance evaluation of reference points generation methods is an avenue for future work. However, Algorithm 1 has an advantage compared to TSLD – it can generate a set of reference points $\boldsymbol{W}$ with arbitrary size.

In summary, the results show that (i) the performance of NSGA-III can be greatly improved by tuning the parameter settings of μ, λ, and H individually, and (ii) their appropriate settings depend on the function instance.

6   Conclusion

We analyzed the impact of the population size μ, the number of children λ, and the number of reference points H on the performance of NSGA-III [6]. It was widely believed that NSGA-III was almost parameter-less and requires setting only one control parameter (H), and the value of μ and λ could be set to $\mu = \lambda \approx H$ [6]. In contrast, our experimental results showed that appropriate parameter settings of μ, λ, and H differ from each other as well as widely used parameter settings. Also, the performance of NSGA-III significantly depends on the settings of μ, λ, and H. In fact, NSGA-III with a carefully tuned parameter setting (i.e., the specialist) performs significantly better than NSGA-III with the widely used settings. Thus, the usually used parameter settings of NSGA-III (i.e., $\mu = \lambda \approx H$) might be unsuitable in many cases, and NSGA-III requires a particular parameter tuning to realize its best performance.

Future work will analyze the behavior of NSGA-III with different values of μ, λ, and H. We will also investigate the relationship between good parameter combinations of NSGA-III and the shape of the PF by using some statistical approaches (e.g., main effect plots). While the focus of this study was limited to the bound-constrained WFG functions, an analysis on constrained MOPs is an avenue for future work. We believe that the effect of the parameter settings of μ, λ, and H is applicable to other reference vectors-based MOEAs such as U-NSGA-III [18], θ-DEA [26], and VaEA [24], and its investigation is also our future direction.

Acknowledgments. This research is supported by the HPCI System Research Project "Research and development of multiobjective design exploration and high-performance computing technologies for design innovation" (Project ID:hp160203).

References

1. Andersson, M., Bandaru, S., Ng, A., Syberfeldt, A.: Parameter tuning of MOEAs using a bilevel optimization approach. In: Gaspar-Cunha, A., Henggeler Antunes, C., Coello, C.C. (eds.) EMO 2015. LNCS, vol. 9018, pp. 233–247. Springer, Cham (2015). doi:10.1007/978-3-319-15934-8_16
2. Bader, J., Zitzler, E.: HypE: an algorithm for fast hypervolume-based many-objective optimization. Evol. Comput. **19**(1), 45–76 (2011)
3. Brockhoff, D., Tran, T., Hansen, N.: Benchmarking numerical multiobjective optimizers revisited. In: GECCO, pp. 639–646 (2015)
4. Das, I., Dennis, J.E.: Normal-boundary intersection: a new method for generating the pareto surface in nonlinear multicriteria optimization problems. SIAM J. Optim **8**(3), 631–657 (1998)
5. Deb, K., Agrawal, S., Pratap, A., Meyarivan, T.: A fast and elitist multiobjective genetic algorithm: NSGA-II. IEEE TEVC **6**(2), 182–197 (2002)
6. Deb, K., Jain, H.: An evolutionary many-objective optimization algorithm using reference-point-based nondominated sorting approach, part I: solving problems with box constraints. IEEE TEVC **18**(4), 577–601 (2014)

7. Durillo, J.J., Nebro, A.J., Luna, F., Alba, E.: On the effect of the steady-state selection scheme in multi-objective genetic algorithms. In: Ehrgott, M., Fonseca, C.M., Gandibleux, X., Hao, J.-K., Sevaux, M. (eds.) EMO 2009. LNCS, vol. 5467, pp. 183–197. Springer, Heidelberg (2009). doi:10.1007/978-3-642-01020-0_18
8. Eiben, A.E., Hinterding, R., Michalewicz, Z.: Parameter control in evolutionary algorithms. IEEE TEVC **3**(2), 124–141 (1999)
9. Huband, S., Hingston, P., Barone, L., While, R.L.: A review of multiobjective test problems and a scalable test problem toolkit. IEEE TEVC **10**(5), 477–506 (2006)
10. Ishibuchi, H., Masuda, H., Nojima, Y.: Comparing solution sets of different size in evolutionary many-objective optimization. In: IEEE CEC, pp. 2859–2866 (2015)
11. Ishibuchi, H., Sakane, Y., Tsukamoto, N., Nojima, Y.: Evolutionary many-objective optimization by NSGA-II and MOEA/D with large populations. In: IEEE SMC, pp. 1758–1763 (2009)
12. Ishibuchi, H., Setoguchi, Y., Masuda, H., Nojima, Y.: How to compare many-objective algorithms under different settings of population and archive sizes. In: IEEE CEC, pp. 1149–1156 (2016)
13. Ishibuchi, H., Tsukamoto, N., Nojima, Y.: Evolutionary many-objective optimization: a short review. In: IEEE CEC, pp. 2419–2426 (2008)
14. Li, B., Li, J., Tang, K., Yao, X.: Many-objective evolutionary algorithms: a survey. ACM Comput. Surv. **48**(1), 13 (2015)
15. López-Ibáñez, M., Knowles, J., Laumanns, M.: On sequential online archiving of objective vectors. In: Takahashi, R.H.C., Deb, K., Wanner, E.F., Greco, S. (eds.) EMO 2011. LNCS, vol. 6576, pp. 46–60. Springer, Heidelberg (2011). doi:10.1007/978-3-642-19893-9_4
16. Martínez, S., Aguirre, H.E., Tanaka, K., Coello, C.A.C.: On the low-discrepancy sequences and their use in MOEA/D for high-dimensional objective spaces. In: IEEE CEC, pp. 2835–2842 (2015)
17. Radulescu, A., López-Ibáñez, M., Stützle, T.: Automatically improving the anytime behaviour of multiobjective evolutionary algorithms. In: Purshouse, R.C., Fleming, P.J., Fonseca, C.M., Greco, S., Shaw, J. (eds.) EMO 2013. LNCS, vol. 7811, pp. 825–840. Springer, Heidelberg (2013). doi:10.1007/978-3-642-37140-0_61
18. Seada, H., Deb, K.: A unified evolutionary optimization procedure for single, multiple, and many objectives. IEEE TEVC **20**(3), 358–369 (2016)
19. Smit, S.K., Eiben, A.E.: Parameter tuning of evolutionary algorithms: generalist vs. specialist. In: Chio, C., Cagnoni, S., Cotta, C., Ebner, M., Ekárt, A., Esparcia-Alcazar, A.I., Goh, C.-K., Merelo, J.J., Neri, F., Preuß, M., Togelius, J., Yannakakis, G.N. (eds.) EvoApplications 2010. LNCS, vol. 6024, pp. 542–551. Springer, Heidelberg (2010). doi:10.1007/978-3-642-12239-2_56
20. Tan, Y., Jiao, Y., Li, H., Wang, X.: MOEA/D + uniform design: a new version of MOEA/D for optimization problems with many objectives. Comput. OR **40**(6), 1648–1660 (2013)
21. Tušar, T., Filipič, B.: Differential evolution versus genetic algorithms in multiobjective optimization. In: Obayashi, S., Deb, K., Poloni, C., Hiroyasu, T., Murata, T. (eds.) EMO 2007. LNCS, vol. 4403, pp. 257–271. Springer, Heidelberg (2007). doi:10.1007/978-3-540-70928-2_22
22. Wagner, T., Beume, N., Naujoks, B.: Pareto-, aggregation-, and indicator-based methods in many-objective optimization. In: Obayashi, S., Deb, K., Poloni, C., Hiroyasu, T., Murata, T. (eds.) EMO 2007. LNCS, vol. 4403, pp. 742–756. Springer, Heidelberg (2007). doi:10.1007/978-3-540-70928-2_56

23. Wessing, S., Beume, N., Rudolph, G., Naujoks, B.: Parameter tuning boosts performance of variation operators in multiobjective optimization. In: Schaefer, R., Cotta, C., Kołodziej, J., Rudolph, G. (eds.) PPSN 2010. LNCS, vol. 6238, pp. 728–737. Springer, Heidelberg (2010). doi:10.1007/978-3-642-15844-5_73
24. Xiang, Y., Zhou, Y., Li, M., Chen, Z.: A vector angle based evolutionary algorithm for unconstrained many-objective optimization. IEEE TEVC (2016, in press)
25. Yuan, Y., Xu, H., Wang, B.: An experimental investigation of variation operators in reference-point based many-objective optimization. In: GECCO, pp. 775–782 (2015)
26. Yuan, Y., Xu, H., Wang, B., Yao, X.: A new dominance relation-based evolutionary algorithm for many-objective optimization. IEEE TEVC **20**(1), 16–37 (2016)
27. Zhang, Q., Liu, W., Li, H.: The performance of a new version of MOEA/D on CEC09 unconstrained MOP test instances. In: IEEE CEC, pp. 203–208 (2009)
28. Zitzler, E., Thiele, L., Laumanns, M., Fonseca, C.M., da Fonseca, V.G.: Performance assessment of multiobjective optimizers: an analysis and review. IEEE TEVC **7**(2), 117–132 (2003)

Multi-objective Optimization for Liner Shipping Fleet Repositioning

Kevin Tierney[1(✉)], Joshua Handali[2], Christian Grimme[2],
and Heike Trautmann[2]

[1] Decision Support and Operations Research Lab,
University of Paderborn, Paderborn, Germany
tierney@dsor.de
[2] Information Systems and Statistics Group,
University of Münster, Münster, Germany
j_hand02@uni-muenster.de,
{christian.grimme,heike.trautmann}@wi.uni-muenster.de

Abstract. The liner shipping fleet repositioning problem (LSFRP) is a central optimization problem within the container shipping industry. Several approaches exist for solving this problem using exact and heuristic techniques, however all of them use a single objective function for determining an optimal solution. We propose a multi-objective approach based on a simulated annealing heuristic so that repositioning coordinators can better balance profit making with cost-savings and environmental sustainability. As the first multi-objective approach in the area of liner shipping routing, we show that giving more options to decision makers need not be costly. Indeed, our approach requires no extra runtime than a weighted objective heuristic and provides a rich set of solutions along the Pareto front.

1 Introduction

Liner shipping networks form the backbone of world trade, providing efficient, cheap and secure transportation of containerized goods. The container shipping industry carried over 171 million twenty foot equivalent units (TEU)[1] in 2015 on over 5,000 ships [18]. As container volumes steadily increase each year, the challenge of maintaining liner shipping networks becomes ever more difficult for liner carriers to handle.

Liner shipping networks consist of regularly scheduled, cyclical routes between ports, called *services*. Services are regularly added, removed and modified in a network to adapt to seasonal cargo demands and macro economic changes. To carry out such changes, ships must be moved, or *repositioned*, between services. Liner carriers must balance the conflicting objectives of minimizing their sailing costs and maximizing their cargo intake (profit). Carriers

---

[1] A single TEU represents one twenty foot container, with two TEU representing the commonly found forty foot container.

H. Trautmann et al. (Eds.): EMO 2017, LNCS 10173, pp. 622–638, 2017.
DOI: 10.1007/978-3-319-54157-0_42

are particularly interested in reducing their carbon footprint through improved routing and slow-steaming (see, e.g., [9]), which also results in reduced costs due to lower fuel consumption.

Experienced planners create repositioning plans often with little to no decision support [17]. Recently, however, several algorithms have been developed for solving the liner shipping fleet repositioning problem (LSFRP), both to optimality [12,13,15] and heuristically [12,14]. In addition, a decision support system (DSS) with an integrated repositioning visualization has been developed [10] to assist planners making repositioning decisions. However, all of the research in this area is based on a single objective function consisting of weighted costs and revenues. The solutions found with this function may not be desirable for liner carriers. Moreover, trade-offs between objectives are ignored and cannot be adequately assessed by the decision maker.

We propose a multi-objective simulated annealing (MOSA) algorithm for the LSFRP that allows repositioning coordinators to find LSFRP solutions under varying sailing costs, cargo profit and empty container transports. Our algorithm provides solutions across a wide expanse of the Pareto front, ensuring that coordinators can easily choose the solution that best fits the current business environment. We provide experimental results and a brief case study on LSFRP instances based on data from Maersk Line, the world's largest liner carrier.

This paper is organized as follows. We first briefly describe fleet repositioning in Sect. 2, followed by an overview of related work in Sect. 3. Next, we describe MOSA for the LSFRP in Sect. 4 along with detailed experimental results in Sect. 5. Finally, we conclude and discuss future work in Sect. 6.

2   Liner Shipping Fleet Repositioning

We briefly describe the LSFRP and provide a mathematical model, referring readers to [17] for a more detailed description of the problem and the liner shipping domain.

The process of repositioning a set of container ships has several phases. Before a repositioning begins, ships sail along their regularly scheduled route(s). In contrast to other forms of shipping, such as tramp or industrial, liner shipping is completely schedule based, like a bus or train, with ships arriving at ports on a periodic (usually weekly or bi-weekly) basis. At a time point determined by the repositioning coordinator the vessel *phases out* from its service and begins repositioning towards its new service, called the *goal service*. During this time, the vessel may undertake any number of activities to complete its repositioning as described below. Once the vessel reaches its goal service, it *phases in*, which indicates the start of regular service.

Vessel repositioning is expensive due to the cost of fuel (on the order of hundreds of thousands of dollars) and the revenue lost due to cargo flow disruptions. Liner carriers reposition hundreds of vessels each year, meaning that optimizing vessel movements can significantly reduce the economic and environmental burdens of containerized shipping, and furthermore allow shippers to better utilize repositioning vessels to transport cargo. The aim of the LSFRP is to maximize the profit earned when repositioning a number of vessels from their initial services to a goal service being added or expanded.

Liner Shipping Services. Services are composed of multiple *slots*, each of which represents a cycle assigned to a particular vessel. A slot consists of a number of *visits*, or port calls, i.e., a specific time when a vessel is scheduled to arrive at a port. A vessel that is assigned to a particular slot sequentially sails to each visit in the slot.

Vessel sailing speeds can be adjusted throughout repositioning to balance cost savings with punctuality, as vessel fuel consumption increases approximately cubically with the speed of the vessel. *Slow steaming*, in which vessels sail near or at their minimum speed, allows vessels to sail more cheaply between two ports than at higher speeds, albeit with a longer duration. We linearize the bunker consumption of each repositioning vessel in order to more easily model the LSFRP.

Phase-Out. The repositioning period for each vessel starts at a specific time when the vessel may cease normal operations. Each vessel is assigned a different starting time, or phase-out time, for its repositioning. The repositioning coordinator specifies these times based on customer requirements, leasing agreements, etc. However, the phase out time does not require a ship to leave its service immediately. Ships may continue sailing on their initial service if it is profitable to do so in the context of the required repositioning. Note that the ships involved in a repositioning often do not all share the same initial service. After ships leave their initial service, they may undertake a number of repositioning activities.

Phase-In. Each vessel must phase in to a slot on the goal service before a time set by the repositioning coordinator. After this time, normal operations on the goal service must begin. Some services start with all ships arriving one week after the other at the same port, while large services may begin simultaneously in multiple ports. The repositioning of each vessel and optimization of its activities thus takes place in the period between two fixed times, the vessel's earliest phase-out time and the latest phase-in time of all vessels.

Cargo and Equipment. Revenue is earned through delivering *cargo* and *equipment* (typically empty containers). We take a detailed view of cargo flows in which a demand is represented with a source port, destination port, container type, latest delivery time, amount of TEU available and revenue per TEU delivered. There are typically multiple customers shipping containers between a given pair of two ports, all at different prices due to differing agreements between the carrier and the shipper. We aggregate all of these demands into a single demand and compute an average revenue per container for the port to port pair. Furthermore, we do not require that all demands are delivered. Demands can be ignored or partially delivered with no penalty.

For each TEU of cargo transported, the liner carrier must pay for the cost of loading and unloading the containers from the ship from the revenue to determine the profit per TEU of each demand. Vessels have a maximum capacity in terms of both *dry* and *reefer* (refrigerated) cargo. Reefer containers must be placed in positions with plugs or refrigeration hookups, which are limited on each vessel. Dry containers can be placed anywhere, however if one is placed in a reefer position it prevents a reefer container from being loaded there. We note that container stowage is in its own right a difficult optimization problem [16], which is why we only constrain the overall capacity of the vessel. The cargo flows can be seen as a multi-commodity flow where each demand is a commodity with a start and end port. Equipment, in contrast to cargo, can be sent from any port where it is in surplus to any port where it is in demand and the revenue represents the cost saved by the carrier by moving the equipment through a repositioning rather than a more expensive method.

Some ports have equipment, but are not on any service visited by repositioning vessels. These ports are called *flexible* ports, and are associated with flexible visits. The repositioning coordinator may choose the time a vessel arrives at such visits, if at all. All other visits are called *inflexible*, because the time a vessel arrives is fixed.

Sail-on-Service (SoS) Opportunities. While repositioning, vessels may use certain services to cheaply sail between different parts of the network called *SoS opportunities*. There are two vessels involved in SoS opportunities, referred to as the *repositioning vessel*, which is the vessel under the control of a repositioning coordinator, and the *on-service vessel*, which is the vessel assigned to a slot on the service being offered as an SoS opportunity. Repositioning vessels use SoS opportunities by replacing the on-service vessel and sailing in its place for a portion of the service. SoS opportunities save significant amounts of money on fuel, since the on-service vessel no longer has to sail anywhere and can be laid up or leased out. Utilizing an SoS is subject to a number of constraints, which are described in [17].

Asia-CA3 Case Study. Figure 1 shows a subset of a real repositioning scenario in which a vessel must be repositioned from its initial service (the phase-out service), the Chennai-Express, to the goal service (the phase-in service), the Intra-WCSA. The Asia-CA3 service is offered as a SoS opportunity to the vessel repositioning from Chennai Express to Intra-WCSA. One possible repositioning plan (marked in red) could involve a vessel leaving the Chennai Express at TPP, and sailing to HKG, where it can pick up the Asia-CA3, thus replacing the on-service vessel. The repositioning vessel would then sail along the Asia-CA3 until it gets to BLB, where it can join the Intra-WCSA. Note that no vessel sails on the backhaul of the Asia-CA3, and this is allowed because very little cargo travels on the Asia-CA3 towards Asia.

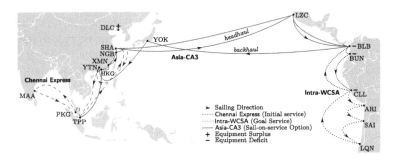

Fig. 1. A subset of a case study of the LSFRP, from [13].

2.1 Mathematical Model

We present the mixed-integer linear arc-flow formulation of the LSFRP from [12]. Several other models exist for special cases of the LSFRP that are detailed in [17].

Graph. The LSFRP can be modeled using a graph in which each visit is represented by a node, and valid sailings between visits are arcs between nodes in the graph. This means that graph nodes represent both a place and a time, except in the case of flexible nodes, where the time a ship arrives must be determined. We note that the LSFRP graph is rather detailed and embeds several LSFRP constraints (see [17]).

Parameters. The parameters of the model are as follows.

S	Set of ships.
V'	Set of visits minus the graph sink.
V^i, V^f	Set of inflexible and flexible visits, respectively.
A^i, A^f	Set of inflexible and flexible arcs, respectively.
A'	Set of arcs minus those arcs connecting to the graph sink, i.e., $(i,j) \in A$, $i,j \in V'$.
Q	Set of equipment types. $Q = \{dc, rf\}$.
M	Set of demand triplets of the form (o,d,q), where $o \in V', d \subseteq V'$ and $q \in Q$ are the origin visit, destination visits and the cargo type, respectively.
$V^{q+} \subseteq V'$	Set of visits with an equipment surplus of type q.
$V^{q-} \subseteq V'$	Set of visits with an equipment deficit of type q.
$V^{q^*} \subseteq V'$	Set of visits with an equipment surplus or deficit of type q ($V^{q^*} = V^{q+} \cup V^{q-}$).
$u_s^q \in \mathbb{R}^+$	Capacity of vessel s for cargo type $q \in Q$.
$M_i^{Orig}, (M_i^{Dest}) \subseteq M$	Set of demands with an origin (destination) visit $i \in V$.
$v_s \in V'$	Starting visit of ship $s \in S$.
$t_{si}^{Mv} \in \mathbb{R}$	Move time per TEU for vessel s at visit $i \in V'$.

$t_i^E \in \mathbb{R}$	Enter time at inflexible visit $i \in V'$.
$t_i^X \in \mathbb{R}$	Exit time at inflexible visit $i \in V'$.
$t_i^P \in \mathbb{R}$	Pilot time at visit $i \in V'$.
$r^{(o,d,q)} \in \mathbb{R}^+$	Amount of revenue gained per TEU for the demand triplet.
$c_{sij}^{Sail} \in \mathbb{R}^+$	Fixed cost of vessel s utilizing arc $(i,j) \in A'$.
$c_{sij}^{VarSail} \in \mathbb{R}^+$	Variable hourly cost of vessel $s \in S$ utilizing arc $(i,j) \in A'$.
$c_i^{Mv} \in \mathbb{R}^+$	Cost of a TEU move at visit $i \in V'$.
$c_{si}^{Port} \in \mathbb{R}$	Port fee associated with vessel s at visit $i \in V'$.
$d_{ijs}^{Min} \in \mathbb{R}^+$	Minimum duration for vessel s to sail on flexible arc (i,j).
$d_{ijs}^{Max} \in \mathbb{R}^+$	Maximum duration for vessel s to sail on flexible arc (i,j).
$a^{(o,d,q)} \in \mathbb{R}^+$	Amount of demand available for the demand triplet.
$In(i) \subseteq V'$	Set of visits with an arc connecting to visit $i \in V$.
$Out(i) \subseteq V'$	Set of visits receiving an arc from $i \in V$.
$\tau \in V$	Graph sink, which is not an actual visit.
$\phi^{Sail}, \phi^{Cargo}, \phi^{Eqp}$	Weights for the objective functions.

Variables. We now list the decision variables of the model.

$w_{ij}^s \in \mathbb{R}_0^+$	The duration that vessel $s \in S$ sails on flexible arc $(i,j) \in A^f$
$x_{ij}^{(o,d,q)} \in \mathbb{R}_0^+$	Amount of flow of demand triplet $(o,d,q) \in M$ on $(i,j) \in A'$
$x_{ij}^q \in \mathbb{R}_0^+$	Amount of equipment of type $q \in Q$ flowing on $(i,j) \in A'$
$y_{ij}^s \in \{0,1\}$	Indicates whether vessel s is sailing on arc $(i,j) \in A$
$z_i^E \in \mathbb{R}_0^+$	Defines the enter time of a vessel at visit i
$z_i^X \in \mathbb{R}_0^+$	Defines the exit time of a vessel at visit i.

Objective and Constraints

$$\max \; - \phi^{Sail} \left[\sum_{s \in S} \left(\sum_{(i,j) \in A'} c_{sij}^{Sail} y_{ij}^s + \sum_{(i,j) \in A^f} c_{sij}^{VarSail} w_{ij}^s \right) \right.$$

$$\left. + \sum_{j \in V'} \sum_{i \in In(j)} \sum_{s \in S} c_{sj}^{Port} y_{ij}^s \right] \tag{1}$$

$$+ \phi^{Cargo} \sum_{(o,d,q) \in M} \left(\sum_{j \in d} \sum_{i \in In(j)} \left(r^{(o,d,q)} - c_o^{Mv} - c_j^{Mv} \right) x_{ij}^{(o,d,q)} \right) \tag{2}$$

$$+ \phi^{Eqp} \sum_{q \in Q} \sum_{i \in V^{q+}} \sum_{j \in Out(i)} x_{ij}^q \tag{3}$$

subject to $\displaystyle\sum_{s\in S}\sum_{i\in In(j)} y_{ij}^s \leq 1$ $\hspace{3em} \forall j \in V'$ $\hspace{3em}$ (4)

$$\sum_{j\in Out(i)} y_{ij}^s = 1 \hspace{2em} \forall s \in S, i = v_s \hspace{2em} (5)$$

$$\sum_{i\in In(\tau)}\sum_{s\in S} y_{i\tau}^s = |S| \hspace{2em} (6)$$

$$\sum_{i\in In(j)} y_{ij}^s - \sum_{i\in Out(j)} y_{ji}^s = 0 \hspace{1em} \forall j \in \{V'\setminus\bigcup_{s\in S} v_s\}, s \in S \hspace{1em} (7)$$

$$\sum_{(o,d,rf)\in M} x_{ij}^{(o,d,rf)} \leq \sum_{s\in S} u_s^{rf} y_{ij}^s \hspace{2em} \forall (i,j) \in A' \hspace{2em} (8)$$

$$\sum_{(o,d,q)\in M} x_{ij}^{(o,d,q)} + \sum_{q'\in Q} x_{ij}^{q'} \leq \sum_{s\in S} u_s^{dc} y_{ij}^s \hspace{1em} \forall (i,j) \in A' \hspace{1em} (9)$$

$$\sum_{i\in Out(o)} x_{oi}^{(o,d,q)} \leq a^{(o,d,q)} \sum_{i\in Out(o)}\sum_{s\in S} y_{oi}^s \hspace{1em} \forall (o,d,q) \in M \hspace{1em} (10)$$

$$\sum_{i\in In(j)} x_{ij}^{(o,d,q)} - \sum_{k\in Out(j)} x_{jk}^{(o,d,q)} = 0 \; \forall (o,d,q) \in M, j \in V'\setminus(o\cup d) \hspace{0.5em} (11)$$

$$\sum_{i\in In(j)} x_{ij}^q - \sum_{k\in Out(j)} x_{jk}^q = 0 \hspace{2em} \forall q \in Q, j \in V'\setminus V^{q^*} \hspace{1em} (12)$$

$$d_{ijs}^{Min} y_{ij}^s \leq w_{ij}^s \leq d_{ijs}^{Max} y_{ij}^s \hspace{2em} \forall (i,j) \in A^f, s \in S \hspace{1em} (13)$$

$$z_i^E = t_i^E \sum_{s\in S}\sum_{j\in In(i)} y_{ij}^s \hspace{2em} \forall i \in V^i \hspace{2em} (14)$$

$$z_i^X = t_i^X \sum_{s\in S}\sum_{j\in Out(i)} y_{ij}^s \hspace{2em} \forall i \in V^i \hspace{2em} (15)$$

$$z_i^X + \sum_{s\in S} w_{ij}^s \leq z_j^E \hspace{2em} \forall (i,j) \in A^f \hspace{2em} (16)$$

$$\sum_{(o,d,q)\in M_i^{Orig}}\sum_{j\in Out(o)} t_{so}^{Mv} x_{oj}^{(o,d,q)} + \sum_{(o,d,q)\in M_i^{Dest}}\sum_{d'\in d}\sum_{j\in In(d')} t_{sd}^{Mv} x_{jd'}^{(o,d,q)}$$

$$+ \sum_{q\in Q}\left(\sum_{i'\in\{V^{q^+}\cap\{i\}\}}\sum_{j\in Out(i')} t_{si'}^{Mv} x_{i'j}^q + \sum_{i'\in\{V^{q^-}\cap\{i\}\}}\sum_{j\in In(i')} t_{sj}^{Mv} x_{ji'}^q \right)$$

$$- z_i^X + z_i^E + t_i^P \sum_{j\in In(i)} y_{ij}^s \leq 0 \hspace{2em} \forall i \in V^f, s \in S \hspace{1em} (17)$$

The domains of the variables are described above. The objective function consists of three components. The sailing cost objective (1) takes into account the precomputed sailing costs on arcs between inflexible visits, the variable cost for sailings to and from flexible visits, and port fees. Note that the fixed sailing cost on an arc may also include canal fees or other revenue related to SoS opportunities. The profit from delivering cargo (2) earns revenue from delivering cargo subtracted by the cost to load and unload the cargo from the vessel. The equipment profit is taken into account in (3). Equipment is handled similar to cargo, except that equipment can flow from any port where it is in supply to any port where it is in demand. When $\phi^{Cargo} = \phi^{Sail} = 1$ and $\phi^{Equip} = 150$ the objective function is nearly equivalent to the one used in previous work on the LSFRP.

The node-disjointness of the paths is enforced by Constraints (4), and the flow of each vessel from its source node to the graph sink is handled by (5), (6) and (7), with (6) ensuring that all vessels arrive at the sink. The vessel capacity is modeled in Constraints (8) (for reefer containers) and (9) (for all containers), and these are only enforced if a vessel uses an arc. Note that Constraints (8) do not take into account empty reefer equipment, since empty containers do not need power, and can therefore be placed anywhere on the vessel. Cargo is only allowed to flow on arcs where a vessel is sailing in Constraints (10). The flow of cargo from its source to its destination, through intermediate nodes, is handled by Constraints (11). Constraints (12) balance the flow of equipment in to and out of nodes.

Flexible arcs have a duration that is constrained by the minimum and maximum sailing time of the vessel on the arc in Constraints (13). The enter and exit time of a vessel at inflexible ports is handled by Constraints (14) and (15). In practice, these constraints are only necessary if one of the outgoing arcs from an inflexible visit ends at a flexible visit. Constraints (16) set the enter time of a visit to be the duration of a vessel on a flexible arc plus the exit time of the vessel at the start of the arc. Constraints (17) set the amount of time a vessel spends at a flexible visit. The first part of the constraint computes the time required to load and unload cargo and equipment, with the final term of the constraint adding the piloting time to the duration only if one of the incoming arcs is enabled (i.e., the flexible visit is being used).

This mathematical model has been shown in the literature to be effective for solving small LSFRP instances (less than 4 or 5 ships), but runs out of time or memory on larger instances [17]. To support decision makers in industry, heuristics are needed that can solve even large LSFRP instances in little time.

3   Related Work

Despite the relevance of the LSFRP to daily liner carrier operations, it was not mentioned in the most recent survey of work in the liner shipping domain [4]. Work in liner shipping optimization has mainly focused on strategic problems such as network design (e.g., [3]) and fleet deployment (e.g., [11]), as well as tactical problems such as cargo allocation/routing [6]. The LSFRP is different due to its relatively short operational timespan and detailed view of shipping operations. However, a similarity to the literature on the LSFRP and these other liner shipping optimization problems is that they have all been solved using only a single objective function.

As identified in a review of multi-objective DSSs in maritime transportation by Mansouri, Lee and Aluko [8], very few papers in maritime transportation consider multiple objectives, other than as components of a single, monetary-based objective function. In fact, the only articles we found solving an operational liner shipping problem with multi-objective techniques consider empty container repositioning with two objectives in [19,20]. The total cost of repositioning the containers within a network is balanced with the amount of unsatisfied demand.

Several DSSs exist within maritime transportation for assisting companies. The TurboRouter system [5] supports fleet scheduling planners, but does not explicitly provide a way of viewing a tradeoff between objectives. A DSS designed specifically for the LSFRP [10] also does not explicitly consider multi-objective optimization, but could be easily augmented with our proposed approach in this work. The StowMan[s] system [7] assists in the stowage of a container vessel. The system does not explicitly generate a Pareto front, but it does allow users to interact with objective weightings and hard constraints, as well as compare the solutions found with various performance indicators.

Models in the liner shipping literature often assign a dollar value to all portions of the objective function. In many cases this makes sense. In the worst case, however, the dollar values given are difficult to compute or even arbitrary, for example, the value of repositioning an empty container in an operational liner shipping network. We suspect the dearth of literature in this area is due to the relative youth of the area of maritime optimization in comparison to other problem domains and large scale of the optimization problems found. In this work, we report good news, as exploring the Pareto front of the LSFRP turns out to not be significantly more difficult than optimizing a weighted objective function.

4   Multi-objective Simulated Annealing Approach

We propose a multi-objective simulated annealing algorithm (MOSA) for solving the LSFRP, based heavily on the existing simulated annealing algorithm in [12]. The intuition behind the algorithm is that we simply conduct a standard SA with reheating using a weighted objective function, saving solutions along the way using an archiving strategy. This works surprisingly well at finding a variety of solutions to the LSFRP. We first describe the solution representation used for the MOSA, the local search neighborhoods used, then the objective functions we consider, and finally we formalize and discuss the MOSA method for the LSFRP.

4.1   Solution Representation

The solution to the LSFRP can be represented as a sequence of visits for each vessel starting from the vessel's initial visit and ending at the graph sink, which is connected to the final phase-in visit of each slot of the goal service. A key requirement of the LSFRP is that the vessel paths are *node disjoint*, as in most ports there is neither the space at the quay nor the necessity of having multiple repositioning vessels at the same place at the same time. However, we relax this constraint in the solution representation and allow vessels to have the same node on their paths and introduce a penalty function to discourage such solutions. Additionally, we allow and penalize infeasible sailing times to flexible visits.

4.2 Neighborhoods

We use three "basic" neighborhoods and two compound neighborhoods within the MOSA. Our basic neighborhoods add, remove or swap visits between the paths of the vessels, while the compound neighborhoods reconstruct large portions of the paths. These neighborhoods were computationally investigated in [12].

Visit Addition. A ship is selected uniformly at random along with an arc (u, v) on its path. A new visit is chosen such that the sailing from u to the new visit and back to v is feasible (in the sense of the penalized objective function). If no such visit exists, the neighborhood simply makes no changes to the solution.

Visit Removal. A ship is selected uniformly at random as is a visit on its path. This visit may not be the first or last visit of the path. The visit is removed from the path if there is an arc that can be added from the previous node to the successor node of the removed node. Otherwise, nothing is changed.

Visit Swap. Two ships are chosen uniformly at random and a visit is chosen from the path of one of the ships. If there is a visit on the path of the other ship, such that swapping the visits between the paths is possible, then these are swapped. Otherwise the solution is not changed.

Random Path Completion (RPC). A ship is selected uniformly at random along with a visit on its path. All visits subsequent to this visit are removed from its path, and a random path to the graph sink is generated. We avoid loops by requiring that no visit appears in the vessel's path more than once. If it is impossible to complete a random path without a loop, the neighborhood is abandoned and the solution is not changed, however we never encountered this situation on real data.

Demand Destination Completion (DDC). This neighborhood attempts to make a small change to a vessel path to increase the amount of containers carried. A vessel is selected at random along with a visit on the vessel's path. A demand is chosen that could be loaded at the visit, but is not carried because the destination of the demand is not on the vessel's path. Using a breadth first search, we attempt to find a path from any visit on the path before the selected visit to the destination of the demand. If one is found, we then search for a path back to a successive visit on the path in the same way. All nodes between the chosen visit and where the path is rejoined are dropped from the path.

Additional Neighborhoods. We note that in [1] a reactive tabu search for the LSFRP is proposed that contains two additional neighborhoods to those that we propose here. These neighborhoods can be seamlessly integrated into our MO-LSFRP approach, and could potentially find even better solutions.

4.3   Objective Functions

We consider three objective functions in this work that in previous work were converted into dollar amounts and summed together (see (1), (2) and (3) in Sect. 2.1.).

Sailing Cost (Minimize). This objective function sums the cost of sailing for a vessel on all of its arcs, both those with fixed sailing times and those where the sailing duration must be determined during the optimization. Only fuel costs and port fees are considered here; ship leasing and crew/hotel costs are ignored, as they are fixed over the timespan of the repositioning, regardless of the route taken. We note that low values of this objective function correspond to less CO_2 and SO_x emissions, and larger values may be associated with higher eco-efficiency depending on the amount of containers transported on the ships.

Cargo Profit (Maximize). The first objective is the cargo profit from transporting customers' containers. This function is made up of the revenue earned per TEU transferred from one port to another minus the cost of loading and unloading the containers from the vessel. Note that transshipment is not necessary in this model.

Equipment Profit (Maximize). This objective counts the number of empty containers transported between ports. This objective function is particularly hard to convert into a dollar amount, as the value of having an empty container is not easily quantified. In cases where no empty container is available in a place where one is needed, carriers might be able to lease a container at market rates. However, these rates vary significantly. Furthermore, carrying empty containers on a repositioning saves money a carrier may have otherwise spent moving the empty containers either on other ships in the network or on extra ships that are chartered just to carry empty containers.

4.4   MOSA

Algorithm 1 extends the SA algorithm in [12] with our MOSA approach. The algorithm accepts the following parameters. The set of functions f represents the three objective functions, while f' is the weighted function with arbitrary weights defined by the algorithm user. The initial temperature of the SA is given by t^{Init}, the cooling factor by α and the reheating factor with β. A single iteration of the SA is assumed to have converged if the temperature falls below t^{Min}, and r^{Itrs} gives the maximum number of iterations without an improvement before convergence is assumed. After $r^{Restart}$ reheats, the search is restarted, but converges after a maximum of r^{Reheat} SA reheats without any improvement.

The algorithm starts by creating an initial solution through one of three construction heuristics proposed in [12]. We then begin the main reheat loop, followed by the SA loop. On line 11, one of the neighborhoods described above is

Algorithm 1. The MOSA algorithm with reheating and restarts.

```
 1: function SA(f, f', t^{Init}, α, β, t^{Min}, r^{ltrs}, r^{Restart}, r^{Reheat})
 2:     s^{Init} ← CREATESOLUTION();  s* ← s^{Init};  s*_{prev} ← s^{Init}
 3:     t ← t^{Init};  reheats ← 0;  nonImprovingReheats ← 0
 4:     A ← ∅                                                              ▷ Solution archive
 5:     repeat                                                             ▷ Reheat/restart loop.
 6:         nonImprovingItr ← 0;
 7:         if reheats ≥ r^{Restart} then
 8:             s ← s^{Init}
 9:             reheats ← 0
10:         repeat                                                         ▷ SA loop.
11:             s' ← SELECTNEIGHBOR(s)
12:             if f(s') ≻ f(s) then
13:                 UPDATEARCHIVE(A, s')
14:                 s ← s'
15:             else
16:                 if exp(\frac{f'(s')-f'(s)}{t}) > RANDOM(0, 1) then s ← s'   ▷ Metropolis criterion.
17:                 nonImprovingItr ← nonImprovingItr + 1
18:             if f'(s') > f'(s*) then s* ← s'
19:             t ← tα
20:         until t < t^{Min} or nonImprovingItr ≥ r^{ltrs}
21:         t ← t^{Init}β                              ▷ Reheat to a factor β of the initial temperature.
22:         reheats ← reheats + 1
23:         if s* ≤ s*_{prev} then nonImprovingReheats ← nonImprovingReheats + 1
24:         s*_{prev} ← max{s*, s*_{prev}}
25:     until Time limit reached or nonImprovingReheats ≥ r^{Reheat}
26: return (A, s*)
```

chosen uniformly at random and applied to the current candidate solution s. If the neighboring solution s' Pareto dominates the current solution s, it is added to the solution archive (UPDATEARCHIVE) and the candidate is updated. We define the domination operator $\succ$ as follows. We save solutions that are better than the solutions in the archive in at least one objective function value. Note that f includes the penalty function as an objective, however any solution with no penalty violations dominates all solutions with penalty violations, meaning that as soon as a solution is found that is feasible, the archive is purged of infeasible solutions. This ensures that MOSA remains an anytime algorithm if a planner has to stop it before it has found feasible solutions, but in the case that feasible solutions are available (as is the case for every instance we have encountered), these are automatically preferred to any infeasible ones.

When no dominating solution is observed, the metropolis acceptance criterion is consulted to determine whether to accept the neighboring solution. Afterwards, the current best solution according to the weighted objective, s^*, is updated. Although not strictly necessary, we keep track of s^* because this solution may be of particular interest to the repositioning coordinator.

If a reheat is necessary, the temperature is reset on line 21, and the number of non improving reheats is updated. The algorithm continues until a maximum time is reached, or the number of non-improving reheats is exceeded. The solution archive containing the Pareto front, along with s^*, is returned.

5   Experimental Evaluation

We test our multi-objective approach experimentally on the dataset of public LSFRP instances from [12]. Several of these instances are anonymized versions of real repositioning scenarios from a large liner carrier, with the remaining instances being crafted scenarios using real (anonymized) data. We conduct our experiments on a PC with an Intel i7-2600K CPU at 3.4 GHz running Ubuntu 14.04. We allow up to 8 GB of RAM per execution of our approach, but note that our heuristic is not memory bound. For each of the 44 instances in the dataset, we run MOSA 25 times for 60 s each and record the solutions in the Pareto front each time. In this section, we show representations of the Pareto front for two-dimensional cases. For more objectives we apply so-called level diagrams, proposed by [2]. Level diagrams allow for analysis and decision making considering n-dimensional Pareto fronts. The infinity norm over all normalized objective values for a given solution is plotted against the original objective value under consideration. This implicitly leads to a v-shaped plot where points are placed according to their proximity to an ideal point. A decision maker can then determine a level of tolerated loss according to the normalized objectives (up from the peak of the v-shape), and see the possible range of values for a given objective.

Case Study 1: Intra-WCSA. We first investigate the optimization of a small instance, repos2p, in which three ships must be sent to the Intra-WCSA service shown previously in Fig. 1. Figure 2a shows the Pareto front across all runs of the instance. Although this is a small instance with only 36 graph nodes, 150 arcs, and 28 demands, the planner has a number of options he or she can carry out. The solution with the lowest sailing costs (i.e., the lowest fuel consumption) also has the lowest amount of cargo profit. This is not so surprising, as fuel savings come through visiting fewer ports. Indeed, two out of three vessels sail directly from their phase-out port to the phase-in, and the third vessel utilizes the Suez canal (followed by the panama canal) to reach the goal service.

At the opposite end of the solution spectrum, the highest profit generating solution (on the Pareto front) is also associated with the highest sailing costs. In this solution, one vessel sails directly to its phase-in, while the other two vessels go out of their way to deliver cargo in both the Middle East and the east coast of the United States. The two solutions on the Pareto front with roughly 5e6 in sailing costs show how taking a slightly different route can vastly change the cargo profit. These are visualized in Fig. 2b, with dashed lines showing the changes necessary to move from the low profit solution to the high profit solution. The boxes provide the "UN Locode" of each visited port, consisting of a two letter country code and a 3 character port identifier. Adding a visit to JOAQB to the path of the green vessel (with a slightly different goal service start for one vessel) results in the solution with profit 8e5. If for some reason the planner cannot send a ship to JOAQB (no space at the quay, etc.), an alternative is available by extending the phase-out on CHX to INMAA before leaving for PABLB.

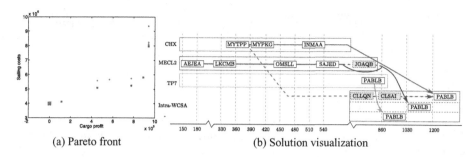

(a) Pareto front   (b) Solution visualization

Fig. 2. Pareto front computed over all 25 runs for repos2p, with the meta-Pareto front shown as blue squares, the best weighted solution as a large, red square (weights set as described in Sect. 2.1), and (meta)-dominated solutions as green dots. No empty equipment is present in this instance, and the x-axis is measured in USD. The figure on the right shows the repositioning at point (4e5, 5e6) with the x-axis representing the time in hours. (Color figure online)

Case Study 2: SAECS. Our second case study involves instance repos29p, which repositions seven Panamax sized ships, with three starting on an Asia-Europe service (AE), three starting on a Middle East to Europe service (ME), and one starting on a transatlantic route (TA). The goal service is a Europe-Africa line called SAECS that connects northern Europe, Spain and South Africa. The repositioning coordinator is presented with a rich variety of options for solving this instance. Even on only just one execution of MOSA, we see several potential paths for every vessel involved in the repositioning, including at least one "interesting" path that does more than just sail from the phase out to the phase in. The level diagrams in Fig. 3 show how the solutions found compare for the three objectives. Figure 3(c) shows that there are only three different options for equipment: either none is taken or equipment is taken at the levels of 5e3 and under 6e3.

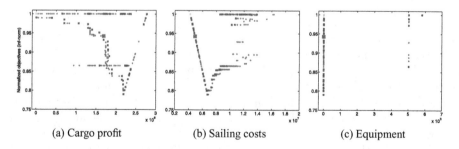

(a) Cargo profit   (b) Sailing costs   (c) Equipment

Fig. 3. Level diagrams computed over all 25 runs for repos29p, with the meta-Pareto front shown as blue squares, the best weighted solution as a large, red square (weights set as described in Sect. 2.1), and the ten solutions with the lowest sailing cost as light blue squares. (Color figure online)

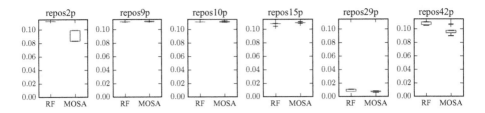

Fig. 4. Comparison of the (normalized) hypervolume of the solutions found by MOSA to a reference front on selected instances.

Comparison of MOSA to a Reference Front. An alternative to generating the Pareto front with MOSA in a single run is to run MOSA several times with different objective weights and store the best solution found in each run. A reference front can then be assembled out of the solutions found. We perform this experiment to see how well MOSA performs against this reference front. The weights we choose for MOSA are 1, 150 and 1 for the cargo, equipment and sailing objectives, respectively, resulting in 3^3 weight configurations. These weights correspond to the values used in the literature for the objective function and were suggested by an industrial partner [17]. We repeat this analysis 10 times, creating 10 reference fronts and compare this against 10 executions of MOSA. We note that computing a reference front is 27 times as expensive as a single execution of MOSA, nonetheless the resulting MOSA fronts are comparable in terms of the amount of hypervolume given the extensive saving in computation time. Figure 4 shows box plots of the normalized hypervolume on several instances for the reference front (RF) and MOSA. We further note that the size of the front is roughly 3.9 times larger using MOSA than the reference front, meaning decision makers are provided with a wider range of non-dominated options.

6   Conclusion and Future Work

This work presents a multiobjective extension of the liner shipping fleet repositioning problem. The problem involves a number of constraints as well as three objectives. We show clear benefits from presenting multiple solutions that balance the fulfillment of the (single-objective) optimization. Furthermore, we avoid requiring decision makers to specify arbitrary weight values for the objectives, while nonetheless ensuring that comparisons between alternative solutions can be made. We also show that level diagrams greatly facilitate the analysis of the generated solutions.

In future work we will extend the embedding of the multiobjective approach within a single-objective optimizer by designing a specialized multi-objective strategy and possibly include further objective functions.

Acknowledgements. Christian Grimme and Heike Trautmann acknowledge support from the European Center for Information Systems (ERCIS).

References

1. Becker, M., Tierney, K.: A hybrid reactive tabu search for liner shipping fleet repositioning. In: Corman, F., Voß, S., Negenborn, R.R. (eds.) ICCL 2015. LNCS, vol. 9335, pp. 123–138. Springer, Cham (2015). doi:10.1007/978-3-319-24264-4_9

2. Blasco, X., Herrero, J., Sanchis, J., Martínez, M.: A new graphical visualization of n-dimensional pareto front for decision-making in multiobjective optimization. Inf. Sci. **178**(20), 3908–3924 (2008)

3. Brouer, B., Alvarez, J., Plum, C., Pisinger, D., Sigurd, M.: A base integer programming model and benchmark suite for liner-shipping network design. Transp. Sci. **48**(2), 281–312 (2013)

4. Christiansen, M., Fagerholt, K., Nygreen, B., Ronen, D.: Ship routing and scheduling in the new millennium. Eur. J. Oper. Res. **228**(3), 467–483 (2013)

5. Fagerholt, K.: A computer-based decision support system for vessel fleet schedulingexperience and future research. Decis. Support Syst. **37**(1), 35–47 (2004)

6. Guericke, S., Tierney, K.: Liner shipping cargo allocation with service levels and speed optimization. Transp. Res. Part E Logistics Transp. Rev. **84**, 40–60 (2015)

7. INTERSCHALT Maritime Systems AG: StowMan[s]: Efficient stowage planning for higher cargo intake. A case study, November 2015. http://www.interschalt. com/fileadmin/user_upload/StowManS_Case_Study.pdf. Accessed 26 Sep 2016

8. Mansouri, S.A., Lee, H., Aluko, O.: Multi-objective decision support to enhance environmental sustainability in maritime shipping: a review and future directions. Transp. Res. Part E Logistics Transp. Rev. **78**, 3–18 (2015)

9. Meyer, J., Stahlbock, R., Voß, S.: Slow steaming in container shipping. In: 45th Hawaii International Conference on System Science (HICSS 2012), pp. 1306–1314. IEEE (2012)

10. Müller, D., Tierney, K.: Decision support and data visualization for liner shipping fleet repositioning. Inf. Technol. Manage. 1–19 (2016)

11. Perakis, A., Jaramillo, D.: Fleet deployment optimization for liner shipping Part 1. Background, problem formulation and solution approaches. Marit. Policy Manage. **18**(3), 183–200 (1991)

12. Tierney, K., Askelsdóttir, B., Jensen, R., Pisinger, D.: Solving the liner shipping fleet repositioning problem with cargo flows. Transp. Sci. **49**(3), 652–674 (2014)

13. Tierney, K., Coles, A., Coles, A., Kroer, C., Britt, A., Jensen, R.: Automated planning for liner shipping fleet repositioning. In: McCluskey, L., Williams, B., Silva, J., Bonet, B. (eds.) Proceedings of the 22nd International Conference on Automated Planning and Scheduling, pp. 279–287 (2012)

14. Tierney, K., Jensen, R.M.: The liner shipping fleet repositioning problem with cargo flows. In: Hu, H., Shi, X., Stahlbock, R., Voß, S. (eds.) ICCL 2012. LNCS, vol. 7555, pp. 1–16. Springer, Heidelberg (2012). doi:10.1007/978-3-642-33587-7_1

15. Tierney, K., Jensen, R.M.: A node flow model for the inflexible visitation liner shipping fleet repositioning problem with cargo flows. In: Pacino, D., Voß, S., Jensen, R.M. (eds.) ICCL 2013. LNCS, vol. 8197, pp. 18–34. Springer, Heidelberg (2013). doi:10.1007/978-3-642-41019-2_2

16. Tierney, K., Pacino, D., Jensen, R.: On the complexity of container stowage planning problems. Discrete Appl. Math. **169**, 225–230 (2014)

17. Tierney, K.: Optimizing Liner Shipping Fleet Repositioning Plans. Springer, Cham (2015)

18. United Nations Conference on Trade and Development (UNCTAD): Review of maritime transport (2015)

19. Wong, E., Lau, H., Mak, K.: Immunity-based evolutionary algorithm for optimal global container repositioning in liner shipping. OR Spectr. **32**(3), 739–763 (2010)
20. Wong, E., Yeung, H., Lau, H.: Immunity-based hybrid evolutionary algorithm for multi-objective optimization in global container repositioning. Eng. Appl. Artif. Intell. **22**(6), 842–854 (2009)

Surrogate-Assisted Partial Order-Based Evolutionary Optimisation

Vanessa Volz[1][(✉)], Günter Rudolph[1], and Boris Naujoks[2]

[1] TU Dortmund University, Dortmund, Germany
{vanessa.volz,guenter.rudolph}@tu-dortmund.de
[2] TH Köln - University of Applied Sciences, Cologne, Germany
boris.naujoks@th-koeln.de

Abstract. In this paper, we propose a novel approach (SAPEO) to support the survival selection process in evolutionary multi-objective algorithms with surrogate models. The approach dynamically chooses individuals to evaluate exactly based on the model uncertainty and the distinctness of the population. We introduce multiple SAPEO variants that differ in terms of the uncertainty they allow for survival selection and evaluate their anytime performance on the BBOB bi-objective benchmark. In this paper, we use a Kriging model in conjunction with an SMS-EMOA for SAPEO. We compare the obtained results with the performance of the regular SMS-EMOA, as well as another surrogate-assisted approach. The results open up general questions about the applicability and required conditions for surrogate-assisted evolutionary multi-objective algorithms to be tackled in the future.

Keywords: Partial order · Multi-objective · Surrogates · Evolutionary algorithms · BBOB

1 Introduction

Surrogate model-assisted evolutionary multi-objective algorithms (SA-EMOAs) are a group of fairly recent but popular approaches[1] to solve multi-objective problems with expensive fitness functions. Using surrogate model predictions of the function values instead of/to complement exact evaluations within an evolutionary algorithm (EA) can save computational time and in some cases make the problem tractable at all.

Many EMOAs only consider objective values for the purpose of sorting and then selecting the best individuals in a population. Assuming that individuals can confidently be distinguished based on surrogate model predictions, knowing the individuals' exact objective values is not necessary. Under this assumption, the algorithm and its evolutionary path would not be affected at all by trusting the predicted sorting, and the computational budget could be reduced.

---

[1] Workshop on the topic in 2016: http://samco.gforge.inria.fr/doku.php.

© Springer International Publishing AG 2017
H. Trautmann et al. (Eds.): EMO 2017, LNCS 10173, pp. 639–653, 2017.
DOI: 10.1007/978-3-319-54157-0_43

In this paper, we present a novel approach to integrate surrogate models and evolutionary (multi-objective) algorithms (dubbed SAPEO for **S**urrogate-**A**ssisted **P**artial Order-Based **E**volutionary **O**ptimisation[2]) that seeks to reduce the number of function evaluations while simultaneously controlling the probability of detrimental effects on the solution quality. The idea is to choose the individuals for exact evaluation dynamically based on the model uncertainty and the distinctness of the population. Preliminary experiments on single-objective problems showed promising results, so in this paper, we investigate the approach in the currently sought-after multi-objective context. We also present different versions that allow differing levels and types of uncertainties for the survival selection process, which, in turn, can potentially effect the solution quality.

In the following, we describe our extensive analysis of the anytime performance of SAPEO using the BBOB-BIOBJ benchmark (refer to Sect. 2.2), focusing on use cases with low budgets to simulate applications with expensive functions. Our SAPEO implementation[3] uses a Kriging surrogate model [14] in conjunction with the SMS-EMOA [2]. We further compare the algorithm and its variants to the underlying SMS-EMOA and an SA-EMOA approach called *pre-selection* [4] (SA-SMS in the following).

We specifically investigate if and under which conditions SAPEO outperforms the SMS-EMOA and SA-SMS in terms of the hypervolume indicator that all algorithms use to evaluate populations. Surprisingly, none of the surrogate-assisted algorithms can convincingly beat out the baseline SMS-EMOA, even for small function budgets. This result opens up questions about SA-EMOAs in general and about the necessary quality of the integrated surrogate models.

A potential explanation for this performance is the increased uncertainty of the surrogate model predictions when compared to the single-objective experiments. Thus, we analyse the effects of prediction uncertainty on the overall performance of the SA-EMOAs. In the future, the resulting insights could become important when (1) deciding whether using a surrogate model is beneficial on a given problem at all and (2) when choosing the sample size and further parameters for the model. This is especially crucial for multi- and many-objective problems, where learning an accurate surrogate model becomes increasingly expensive and thus renders analysing the trade-off between surrogate model computation and function evaluations critical. Additionally, the stated questions and insights are also relevant for noisy optimisation problems where uncertainty can be reduced by repeated evaluations (although not eliminated).

In the following, we present related work in Sect. 2 and introduce the proposed SAPEO algorithm in Sect. 3. The description and visualisation of the benchmarking results are found in Sect. 4. Section 5 concludes the paper with an analysis of the results and lists open research problems.

---

[2] Acknowledgement: The SAPEO concept was developed during the SAMCO Workshop in March 2016 at the Lorentz Center (Leiden, NL). https://www.lorentzcenter.nl This work is part of a project that has received funding from the European Unions Horizon 2020 research and innovation program under grant agreement No 692286.
[3] Code and visualisations available at: http://url.tu-dortmund.de/volz.

2   Background and Related Work

2.1   Surrogate-Assisted Evolutionary Multi-objective Optimisation

Let $X_1, \ldots, X_\lambda \in \mathbb{R}^n$ be a population and the corresponding fitness function $f : \mathbb{R}^n \rightarrow \mathbb{R}^d$. General concepts of multi-objective optimisation will not be discussed here (refer to e.g. [16]). We will be referring to Pareto-dominance as $\preceq$, and to its weak and strong versions as $\precsim$ and $\prec$, respectively. We use the same notation to compare vectors in objective space $\mathbb{R}^d$, i.e. let $a, b \in \mathbb{R}^d$, then $a \preceq b \iff \forall k \in \{1 \ldots d\} : a_k \leq b_k \wedge \exists k \in \{1 \ldots d\} : a_k < b_k$.

Surrogate-assisted evolutionary multi-objective optimisation is surveyed in [7,8], where several approaches for the integration of surrogates and EMOAs are described. According to the surveys, the selection approaches can generally be divided into individual-based (e.g. SAPEO), generation-based, and population-based strategies. Additionally, there is pre-selection (e.g. SA-SMS [4]), which is similar to individual-based strategies but does not retain any individuals with uncertain fitness values.

However, neither of the cited surveys, nor [9], features any algorithm that chooses to propagate uncertainty instead of assuming a distribution and using aggregated metrics such as the expected value. In [10], uncertainty propagation is implemented for noisy optimisation problems using a partial order based on confidence intervals (or hypercubes in higher dimensions) as examined in [13]. The only work transferring this approach to SA-EMOAs we are aware of is GP-DEMO [11] and the authors' previous publications who use a differential evolution algorithm.

Apart from the differences owed to the underlying optimisation algorithms (e.g. crowding distance vs. hypervolume), GP-DEMO is very similar to the SAPEO variant that allows the least uncertainty (SAPEO-uf-ho, cf. Sect. 4.1). An important difference, however, is SAPEOs dynamic adaptation of allowed uncertainty throughout the runtime of the algorithm and executing different partial orders in sequence. In addition to the partial orders inspired by [13] used in both [11] and SAPEO, we propose another order that interprets the confidence interval bounds as objectives and thus deals differently with overlapping intervals. A further difference is the choice of samples for the surrogate models: GP-DEMO uses the Pareto front, whereas SAPEO always uses a local model relative to the solution in question.

2.2   Benchmarking with BBOB

BBOB-BIOBJ is a bi-objective Black-Box Optimisation Benchmarking test suite [15]. It consists of 55 bi-objective functions that are a combination of 10 of the 24 single-objective functions in the BBOB test suite established in 2009 [6]. In order to measure general algorithm performance across function types, single-objective functions were chosen such that the resulting benchmark would be diverse in terms of separability, conditioning, modality and global structure [6]. Based on these properties, the single-objective functions are divided into 5

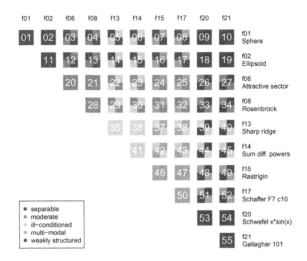

Fig. 1. The 55 BBOB-BIOBJ functions are combinations of 10 single-objective functions (on the top and right). The groups the single-objective and the resulting bi-objective functions belong to are colour-coded according to the legend. (Colour figure online)

function groups, from which 2 functions are chosen each. The resulting problems and corresponding properties are visualised in Fig. 1.

In an effort to measure performance on the different function types more accurately, each of the functions in the bi-objective test suite has 10 instances that differ in terms of some properties, e.g. the location of optima. As an effect, the scale of the optima and achievable absolute improvement of the objective values also vary significantly across instances (thus also between objectives). The robustness of an algorithm's performance on a function group can be evaluated with a higher confidence by testing on multiple members of that group.

All of the functions in the test suites are defined for search spaces of multiple dimensions, of which we will be considering dimensions $2, 3, 5, 10$ and 20 in order to be able to evaluate a wide range of problem sizes. The search space of each function is limited to $[-100, 100] \subset \mathbb{R}$ per dimension in BBOB-BIOBJ.

The performance of an algorithm on the benchmarking suite is measured using a quality indicator expressing both the size of the obtained Pareto set and the proximity to a reference front. Since the true Pareto front is not known for the functions in the test suite, an approximation is obtained by combining all known solutions from popular algorithms. The ideal and nadir points are known, however, and used to normalise the quality indicator to enable comparisons across functions [3]. The metric reported as a performance measure for the algorithm is called precision. It is the difference of the quality indicator of the reference set I_{ref} and the indicator value of the obtained set. 58 target precisions are fixed and the number of function evaluations needed to achieve them is reported during a benchmark run. This way, the COCO platform enables an anytime comparison of algorithms, i.e. an evaluation of algorithm performance for each target precision and number of function evaluations [3].

3   Surrogate-Assisted Partial Order-Based Evolutionary Optimisation

3.1   Formal Description

Let $\tilde{f}(X_i) \in \mathbb{R}^d$ be the predicted fitness for individual X_i as computed by a local surrogate model with uncertainty $\tilde{\sigma}_i$ modelled by

$$\tilde{f}_k(X_i) = f_k(X_i) + e_i, \quad e_i \sim \mathcal{N}(0, \tilde{\sigma}_i), \; k \in \{1 \ldots d\}.$$

Assuming (Assumption **A1**) that the assumptions made by Kriging models [14] hold and $\tilde{\sigma}_i$ was estimated correctly, it follows that

$$\mathbb{P}\Big(f_k(x_i) \in [\tilde{f}_k(x_i) - u_i, \tilde{f}_k(x_i) + u_i]\Big) = 1 - \alpha \quad \text{with}$$

$$u_i = \tilde{\sigma}_i z \left(1 - \frac{\alpha}{2}\right) \tag{1}$$

since $P(|e_i| \leq u_i) = 1 - \alpha$. Here, z denotes the quantile function of the standard normal distribution. In case the objective values are stochastically independent, which is true for the BBOB-BIOBJ benchmark, we can therefore conclude that $f(X_i)$ lies within the hypercube (or bounding box) bounded by the described confidence interval in each dimension with probability $(1 - \alpha)^d$.

Assuming the function values lie within the defined hypercubes (Assumption **A2**), we can distinguish individuals confidently just based on the predicted hypercubes. Of course, in case of large uncertainties $\tilde{\sigma}_i$, a distinction can rarely be meaningful. To combat this problem, SAPEO introduces a threshold ε_g for the uncertainty that is adapted in each iteration g of the EMOA and decreased over the runtime of the algorithm depending on the distinctness of the population (more details in Sect. 3.2). Individuals X_i with an uncertainty higher than the threshold ($u_i > \varepsilon_g$) are evaluated exactly in generation g.

To distinguish between individuals, we propose binary relations incorporating the information on confidence bounds to varying degrees. All of these relations induce a strict partial order (irreflexivity, transitivity) on a population, including and akin to the Pareto dominance commonly used in EMOAs for the first part of the selection process. We analyse the proposed relations in terms of the probability and magnitude of a **sorting error** e_o that per our definition are the pairwise differences between the different orders induced by Pareto dominance and the proposed relations, respectively. We define the probability $\mathbb{P}(e_{o,r}^{i,j})$ of a sorting error made by relation $\preceq_r$ on individuals X_i, X_j and the magnitude of the error $e_{o,r}^{i,j}$:

$$\mathbb{P}(e_{o,r}^{i,j}) = \mathbb{P}\Big(X_i \npreceq X_j \mid X_i \preceq_r X_j\Big)$$

$$e_{o,r}^{i,j} = \begin{cases} |f(X_i) - f(X_j)| & \text{if } (X_i \preceq_r X_j) \wedge (X_i \npreceq X_j) \\ 0 & \text{else} \end{cases}$$

A single or more sorting errors can but do not have to lead to **selection errors** e_s, where the individuals selected differ from the baseline comparison. This type of error will not be analysed in this paper, but it is bounded by e_o.

We define:

$\preceq_f$: *Pareto dominance on function values* This relation is the standard in EMOAs and, since only f is considered, it is obvious that $\mathbb{P}(e_{o,f}^{i,j}) = 0$.

$$X_i \preceq_f X_j := f(X_i) \preceq f(X_j)$$

$\preceq_u$: *Confidence interval dominance* (cf. [11,13]) Assuming **A2**, if

$$X_i \preceq_u X_j := \bigwedge_{k \in [1\ldots d]} \tilde{f}_k(X_i) + u_i < \tilde{f}_k(X_j) - u_j$$

holds, it is guaranteed that $X_i \prec X_j$. Assuming stochastic independence of the errors on predicted uncertainty, we can compute an upper bound for the probability of sorting errors per dimension:

$$\mathbb{P}(e_{o,u}^{i,j}(k)) = \mathbb{P}\left(f_k(X_i) \geq f_k(X_j) \mid X_i \preceq_u X_j \right)$$

$$\leq \mathbb{P}\left(f_k(X_i) \geq \tilde{f}_k(X_i) + u_i \ \vee \ f_k(X_j) \leq \tilde{f}_k(X_j) - u_j \right)$$

$$\leq \frac{\alpha}{2} + \frac{\alpha}{2} = \alpha$$

Only if the confidence hypercubes of two individuals intersect, the probability of them being incomparable is greater than 0. Because of the way $X_i \preceq_u X_j$ is defined, this is only possible if a sorting error is made in every dimension. It follows that $\mathbb{P}(e_{o,u}^{i,j}) \leq \alpha^d$ assuming **A1**, making sorting errors controllable.

$\preceq_c$: *Confidence interval bounds as objectives* Another way of limiting the prediction errors potentially perpetuated through the algorithm is to limit the magnitude of the sorting error. For this reason we define:

$$X_i \preceq_c X_j := \bigwedge_{k \in [1\ldots d]} \left(\begin{smallmatrix} \tilde{f}_k(X_i)-u_i \\ \tilde{f}_k(X_i)+u_i \end{smallmatrix} \right) \precsim \left(\begin{smallmatrix} \tilde{f}_k(X_j)-u_j \\ \tilde{f}_k(X_j)+u_j \end{smallmatrix} \right)$$

$$\wedge \exists k \in [1\ldots d] : \left(\begin{smallmatrix} \tilde{f}_k(X_i)-u_i \\ \tilde{f}_k(X_i)+u_i \end{smallmatrix} \right) \prec \left(\begin{smallmatrix} \tilde{f}_k(X_j)-u_j \\ \tilde{f}_k(X_j)+u_j \end{smallmatrix} \right)$$

Under assumption **A2**, the error per dimension is bounded by the length of intersection of the confidence intervals, which is in turn bounded by the width of the smaller interval. Therefore, it holds that $e_{o,c}^{i,j} \leq 2^d \min(u_i, u_j)^n$.

$\preceq_p$: *Pareto dominance on predicted values* This relation is the most straightforward, but it does not take the uncertainties of the predictions into account.

$$X_i \preceq_p X_j := \tilde{f}(X_i) \preceq \tilde{f}(X_j)$$

Assuming **A2** again, a sorting error can only be committed if the confidence intervals intersect. Because of the symmetric nature of the interval, it holds that

$e_{o,p}^{i,j} \leq (u_i + u_j)^d$, as the magnitude of the sorting error is again bounded by the confidence interval widths.

$\preceq_o$: *Pareto dominance on lower bounds* This optimistic relation was motivated by [4] where SA-EMOAs performed better using $\preceq_o$ instead of $\preceq_p$.

$$X_i \preceq_o X_j := \bigwedge_{k \in [1\ldots d]} \tilde{f}_k(X_i) - u_i \leq \tilde{f}_k(X_j) - u_j$$

$$\wedge \exists k \in [1\ldots d] : \tilde{f}_k(X_i) - u_i < \tilde{f}_k(X_j) - u_j$$

The maximum error occurs when the lower confidence interval bounds are close together, but in the wrong order, making $e_{o,o}^{i,j} \leq 2^n \max(u_i, u_j)^d$.

Now assume we have obtained a strict partial order based on any of the given binary relations. Let r_c be the rank of the μ-th individual. Then, all individuals with rank less than r_c can confidently (with maximum errors as described above) be selected. In case a selection has to be made from the individuals with the critical rank r_c, one option is to apply another dominance relation to the individuals in question in hopes that the required distinction can be made. In case of $\prec_f$, this always means evaluating uncertain individuals until a confident distinction can be made according to the previous relation.

Another option is to use a relation inducing a total preorder (transitivity, totality) as a secondary selection criterion (and random choice in case of further ties) as most EMOAs do. Again incorporating different information, we have tested the following hypervolume-based (hv) relations for this purpose:

- Hypervolume contribution of objective values:

$$X_i \leq_{ho} X_j := hv(f_o(X_i)) \geq hv(f_o(X_j)), \text{ where } f_o(X_i) = \begin{cases} f(X_i) & u_i = 0 \\ \tilde{f}(X_i) & \text{else} \end{cases}$$

- Hypervolume contribution of confidence interval bounds:

$$X_i \leq_{hc} X_j := \prod_{k \in [1\ldots d]} hv\left(\frac{\tilde{f}_k(X_i) - u_i}{\tilde{f}_k(X_i) + u_i}\right) \geq \prod_{k \in [1\ldots d]} hv\left(\frac{\tilde{f}_k(X_j) - u_j}{\tilde{f}_k(X_j) + u_j}\right)$$

3.2 SAPEO Algorithm

Algorithm 1 describes the basic SAPEO algorithm, which any EMOA and surrogate model with uncertainty estimates can be plugged into. As inputs, the algorithm receives the fitness function **fun**, the number of points considered for the surrogate model **local_size** and the **budget** of function evaluations. The order of dominance relations (**strategies**) and the secondary criterion (**scnd_crit**) are used for selection (cf. Sect. 3.1). The output is the final population.

The algorithm starts with mandatory data structures; the population is initialised randomly (line 1) and evaluated using the considered fitness function **fun** (line 2), the EMOA is set up (line 3) and the error tolerance ε as well as the generation counter are set to their initial values (line 4).

The core optimisation loop starts in line 5 and stops if either the considered optimiser terminates (due to the allocated budget or convergence), but not while both the error tolerance ε is larger than 0 and there are function evaluations left

Algorithm 1. SAPEO

Input: fun, local_size, budget, strategies, scnd_crit
Output: X_{final} ▷ final population
1: $X_0.genome \Leftarrow [random(n) : i \in 1\ldots pop\_size]$ ▷ random initialisation
2: $X_0.f \Leftarrow [fun(x) : x \in X_0]$; $X_0.e = 0$ ▷ evaluate sampled individuals
3: $O \Leftarrow init(X_0, budget)$ ▷ init optimiser with initial population
4: $\varepsilon_0 \Leftarrow \infty$; $g \Leftarrow 1$ ▷ init error tolerance, generation counter
5: **while** $(\neg O.stop()) \vee (\varepsilon > 0 \wedge budget > 0)$ **do**
6:     $X_g.phenome \Leftarrow O.evolve(X_{g-1})$ ▷ get new population
7:     $X_g.f, X_g.e \Leftarrow [model(x, knearest(x, X[X.e == 0], local\_size)) : x \in X_g]$
8:         ▷ predict value, error with surrogate from evaluated neighbours
9:     $\varepsilon_g \Leftarrow \min(e_{g-1}, \alpha\text{-percentiles}(\text{diff}(X_g.f)))$ ▷ update error tolerance
10:    **for** $x \in X_g$ **do**
11:        **if** $x.e > \varepsilon_g \vee O.select(X_g, strategy, scnd\_crit) ==$ NULL **then**
12:            $x.f = fun(x)$ ▷ evaluate individual
13:            $bbob.recommend(X[X.e > 0, \text{last}])$ ▷ recommend solution
14:        **end if**
15:    **end for**
16:    $X_g = O.select(X_g, strategy, scnd\_crit)$ ▷ SAPEO survival selection
17:    $g = g + 1$ ▷ increase generation counter
18: **end while**
19: $X_{\text{final}} = X_{g-1}$; $X_{\text{final}}.f = [fun(x) : x \in X_{\text{final}}]$ ▷ evaluate final population

---

to avoid convergence on imprecise values. Within the loop, new candidate solutions X are first generated by the optimisation algorithm (line 6) and evaluated based on a local surrogate model trained from the `local_size` evaluated individuals closest in design space (line 8). The predicted function and the expected model errors (cf. Eq. 1) are stored. The error tolerance threshold is then adapted (line 9). We reduce the threshold during the course of the algorithm in order to limit the probability of sorting errors with large effects on the final population. Therefore, ε_g is the minimum of the previous threshold ε_{g-1} and the α-percentiles of the euclidian distances in objective space per dimension. The distances are a way to measure the distinctness of a population and thus the potential of overlapping confidence intervals. By adapting ϵ_g accordingly, we reduce the number and magnitude of potential sorting errors.

If any of the individuals in the population need to be evaluated - either because the predicted uncertainty is above the threshold or because the individuals cannot be distinguished (see line 11) - they are evaluated in line 12 and updated accordingly. In order to simulate anytime behaviour of the algorithm, each time a solution is evaluated, an individual is recommended to the BBOB framework in line 13. This serves the purpose of measuring the solution quality of the algorithm had it been stopped at the time more accurately.

The set of candidate solutions, along with the (predicted but reasonably certain) function values and the expected prediction errors are then passed to the optimiser in line 16. Depending on the selected strategy, the optimiser then selects the succeeding population as described above with regard to the predicted function values and uncertainties and resumes its regular process.

Finally, after the optimisation loop terminates, the function values of the individuals in the final population are computed using the real fitness function in line 19, in case there are any individuals left that have not been evaluated.

4 Evaluation

4.1 Experimental Setup

Each experiment was run with 550 parallel jobs that took less than 3 hours each with specifications according to Table 1. Since the performance is strictly measured in terms of function evaluations (target precision reached per function evaluation, cf. Sect. 2.2), the runtime does not influence it.

Table 1. Experiment specifications and parameters

Budget	1000 per dimension (usecase: expensive function)
Variation operators	Standard for all algorithms (cf. [2])
Populations size	100 (as suggested in [4])
Sample size for surrogate	15 (due to computational concerns)[a]
Number of candidate offspring	15 (for SA-SMS, same as sample size)
Correlation assumption	Squared exponential
Trend assumption	Constant
Regression weights	Maximum likelihood using COBYLA
	Start: 10^{-2}, bounds: $[10^{-4}, 10^{1}]$

[a]In a real-world application, the sample size should be chosen considering the tradeoff between computation times for the model and the fitness function

We compare the performances with a standard [2] and surrogate-assisted SMS-EMOA with pre-selection as proposed by [4], since we are not aware of any other SA-EMOAs using the SMS-EMOA with individual-based surrogate management strategies. Specifically, we look at the following algorithms:

SMS-EMOA. Standard SMS-EMOA as baseline comparison.

SA-SMS-p. Surrogate assisted SMS-EMOA using $\preceq_p$ for pre-selection.

SA-SMS-o. Surrogate assisted SMS-EMOA using $\preceq_o$ instead (experimentally shown to improve the performance of pre-selection for the NSGA-II [4]).

SAPEO-uf-ho. SAPEO using $\preceq_u$ to rank the offspring, thus accepting a risk of sorting errors of only α^2 (cf. Sect. 3.1). For as long as the population cannot be distinguished by $\preceq_u$, the invididuals are evaluated according to $\preceq_f$, thus avoiding making any further sorting errors. The hypervolume relation $\leq_{ho}$ is used as secondary criterion. This algorithm should therefore only take small risks and behave like the SMS-EMOA while saving function evaluations.

SAPEO-ucp-ho. SAPEO using increasingly risky relations $\preceq_u, \preceq_c, \preceq_p$ to avoid evaluations completely if not forced by the uncertainty threshold ε, taking the opposite approach as SAPEO-uf-ho. $\leq_{ho}$ is used as secondary criterion.

SAPEO-uc-hc. SAPEO using multi-objectification of the confidence interval boundaries fully. It uses $\preceq_u$ as a first safer way of ranking, followed by $\preceq_c$ on critical individuals. Secondary criterion is $\leq_{hc}$.

4.2   Visualisation and Interpretation of Results

There are two main angles to evaluating the anytime performance of algorithms: (1) measuring the performance indicator after a predefined number of function evaluations (*fixed budget*) and (2) recording the function evaluation when target performances are reached (*fixed target*) [6]. In the following, we use the latter.

For a detailed depiction of an algorithm's performance for a fixed target, we use heatmaps (cf. Fig. 2) that show the percentage of budget used per dimension until a target was reached according to the colour scale on the right. If the target is not reached within the allocated budget, the corresponding square is white. The dimensions and instances of each function are shown separately to enable analysis of the generalisation of algorithm performance across function instances and dimensions. This is very important to justify the aggregation of performance measures across instances. The functions have colour codes according to the legend in Fig. 1 that specify their function groups.

From the plot, it is apparent that for the selected target 10^0, the algorithm SAPEO-uf-ho has trouble with a number of functions even in small dimensions.

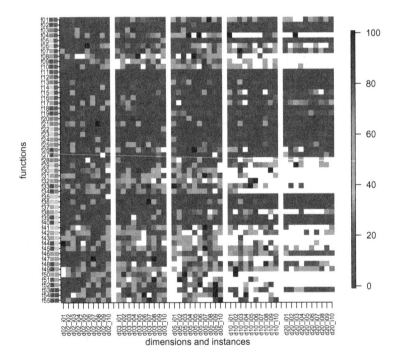

Fig. 2. SAPEO-uf-ho performance in terms of the percentage of the budget per dimension used to reach target 10^0 for all function instances and dimensions. (Colour figure online)

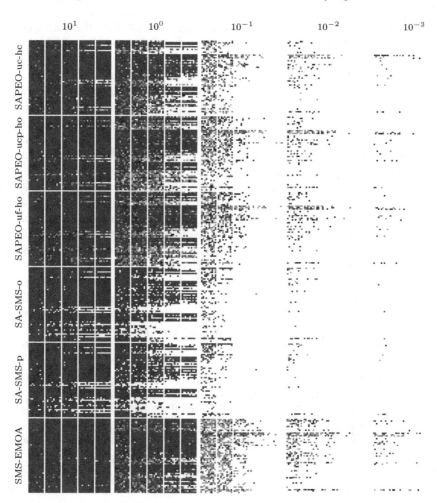

Fig. 3. Heatmaps visualising target performances for all algorithms (rows) across multiple targets (columns). Refer to Fig. 2 for a detailed explanation. (Colour figure online)

Additionally, for those functions, the algorithm's performance seems to drop with increasing dimension of the search space. Especially the Rosenbrock function seems to be problematic for the algorithm: SAPEO-uf-ho rarely reaches the target for dimensions 10 or 20 when the Rosenbrock function is part of the problem (f 04, 13, 21, 28-34). A potential cause is an inaccurate representation of its narrow valley containing the optimum with the surrogate model. Another explanation could be a mismatch of the Rosenbrock function and the variation operators, causing difficulty for the underlying SMS-EMOA. As expected, some of the weakly structured problems were difficult for SAPEO-uf-ho as well.

A discussion of the potential causes of the performances is only possible with reference to other algorithms. In order to get a better overview of the performances of all algorithms and to detect patterns, we have compiled Fig. 3, which is an assembly of 30 heatmaps like the one in Fig. 2 for all algorithms and different targets. The same colour scale as in Fig. 2 is used. Recall that white spaces signify targets that were not reached within the allocated budget.

In Fig. 3, the most obvious trend is the declining performance for each target, which was of course expected. It is also apparent that the SMS-EMOA performs better in general than all other algorithms for each target precision. We can also see that all SAPEO versions are an improvement when compared to the SA-SMS algorithms. Interestingly, we also see similar patterns in terms of which functions are more difficult for all algorithms, indicating that the added surrogate models do not influence the underlying optimisation behaviour significantly.

Unfortunately, while providing a good overview, Fig. 3 is not well suited to interpret the performance of each algorithm per function. While very detailed, the plots are not easy to interpret due to the abundance of information displayed at once. In order to analyse the circumstances of different performance patterns, we compile a plot that aggregates the different instances of a function. This way, the general performance of an algorithm per function can be expressed without risk of overfitting, as intended by the COCO framework. To do that, we use the *expected runtime* (expected number of function evaluations) to reach a target [5] as a performance measure. The measure is estimated for a restart algorithm with 1000 samples. The results are again displayed in a heatmap (Fig. 4). The colour visualises the estimated expected runtime per dimension according to the scale on the right in log10-scale. Higher values than the maximum budget (>3) occur if a target is not reached in all instances. White spaces occur if the target was never reached by the algorithm in all instances.

The plot displays expected runtime for all dimensions in different columns according to the labels above. Each of these columns is again divided into 3, displaying the results for different algorithms according to the labels on the bottom. There are two algorithms per column, whose results are displayed on top of each other for each row corresponding to a function. For each algorithm, the expected runtimes for targets $10^1, 10^0, 10^{-1}, 10^{-2}, 10^{-3}$ are depicted in that order. In case a target was never reached for all algorithms in a column, it is omitted. For example, the expected runtime for SAPEO-uf-ho on function 01, target 10^1 and dimension 2 is on the top left corner and encoded in a light blue. Therefore, the expected runtime to reach target 10^1 is around $10^0 * 2 = 2$. The SMS-EMOA is directly below that and a shade lighter, so has a slightly higher expected runtime. Like in Fig. 2, the groups each function belongs to are encoded according to the colour scheme in the legend of Fig. 1.

The general trends as seen in Fig. 3 can be observed here as well. However, we can also see that SAPEO-uf-ho beats the SMS-EMOA in terms of precision reached on very rare occasions, for example on functions f03 and f41 in dimension 2 and function f20 in dimension 10. Still, the SMS-EMOA generally reaches the same or more precision targets than the other algorithms. However, in most cases where a higher precision target is reached by only a single algorithm, the corresponding colour indicates a very high expected runtime. This means that the algorithm did not reach the higher target for most instances, which speaks

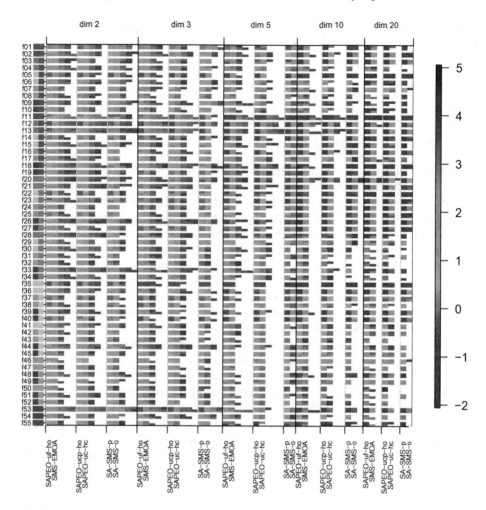

Fig. 4. BBOB-BIOBJ performance results for all algorithms regarding expected runtime (colour coded in log-scale) for targets $10^1, 10^0, 10^{-1}, 10^{-2}, 10^{-3}$ (Colour figure online)

against a robust performance of that algorithm. More importantly, the SA-SMS variants often reach less targets than the other algorithms, especially in higher dimensional problems, meaning they are clearly outperformed.

The colour gradients in most functions are remarkably alike, indicating similar behaviour and difficulties experienced with each problem. This is expected, as the intention of SA-EMOAs is to avoid function evaluations with only controlled effects on the evolutionary path. Possibly due to the aggregating nature of the expected runtime measure, the performance contrast does not appear to be as stark as in Fig. 3. The gradient and number of performance targets reached per function is in fact relatively similar for all algorithms. In most cases, differences occur towards the end of the gradient, indicating that the precision improvement of the surrogate-assisted algorithms is less steep than for the SMS-EMOA.

However, in order to analyse the algorithms' behaviour appropriately in that regard, a more thorough analysis of the separate selection steps is required.

Regarding the different functions, there seems to be no clear performance pattern. The different SAPEO versions vary rarely. The performance of all algorithms seems to be more closely tied to the single-objective functions, e.g., Rosenbrock seems to pose problems whereas Schwefel seems more manageable.

5   Conclusions and Future Work

In this paper, we have proposed a novel approach to surrogate-assisted multi-objective evolutionary algorithms called SAPEO. An extensive analysis of its anytime performance using the BBOB-BIOBJ benchmark showed that it was outperformed by its underlying algorithm in this study, the SMS-EMOA. This fact is quite surprising, since the SAPEO-uf-ho variant allows minimal uncertainties and should rarely make different decisions. However, SAPEO still beats another SA-EMOA based on the SMS-EMOA [4] on the benchmark.

One potential source of error is the surrogate model, e.g. assumptions **A1, A2** (Sect. 3.1) could be wrong. A large error in the predicted uncertainty could have a tremendous influence on the algorithm. However, this is controllable through the adaptation of the uncertainty threshold ε. Additionally, the uncertainties during the start of the SAPEOs were relatively large, which could also send the algorithm into a wrong direction. The uncertainties could be mitigated by using surrogate ensembles instead, distributing the samples better, increasing the sample size or selecting a fitting kernel. Additionally, the performances of local vs. global surrogates should be analysed more thoroughly. It is apparent that the quality of the surrogate model is a major concern for SA-EMOAs, which could be problematic for black-box optimisation in general.

Apart from the model, there are possible improvements regarding the binary relations used. For one, $\preceq_u$ could be defined without forcing strict Pareto dominance of the hypercubes. Furthermore, using hypercubes for the potential location of the fitness values is a simplification. Perhaps a binary relation on hyperellipsoids could provide better results.

Furthermore, while the SAPEO approach worked well for single-objective problems, the corresponding multi-objective problems pose an incomparably larger difficulty for a surrogate model. Additionally, even slightly overestimated function values could lead to an incorrect identification of dominated individuals. This is because a large number of critical values would need to be distinguished with less certain relations or expensive function evaluations.

Notice that previous studies [1,12] have shown that EAs with a larger number of offspring are less vulnerable to noisy fitness functions. Therefore, it may be conjectured that the $\mu + 1$ selection scheme of the SMS-EMOA causes the poor performance under noise induced by the surrogate. This hypothesis, however, remains a question for future research. In general, the influence of the quality of surrogate models on SA-EMOAs should be analysed more carefully. With the proper noise-robust optimisation algorithm and parametrisation, SAPEO should be able to beat its underlying algorithm as it does on single-objective problems.

References

1. Arnold, D., Beyer, H.-G.: On the benefits of populations for noisy optimization. Evol. Comput. **11**(2), 111–127 (2003)
2. Beume, N., Naujoks, B., Emmerich, M.: SMS-EMOA: multiobjective selection based on dominated hypervolume. Eur. J. Oper. Res. **181**(3), 1653–1669 (2007)
3. Brockhoff, D., Tušar, T., Tušar, D., Wagner, T., Hansen, N., Auger, A.: Biobjective performance assessment with the COCO platform. CoRR abs/1605.01746 (2016). Accessed 22 Dec 2016
4. Emmerich, M., Giannakoglou, K., Naujoks, B.: Single- and multi-objective evolutionary optimization assisted by Gaussian random field metamodels. IEEE Trans. Evol. Comput. **10**(4), 421–439 (2006)
5. Hansen, N., Auger, A., Ros, R., Finck, S., Pošík, P.: Comparing results of 31 algorithms from the black-box optimization benchmarking BBOB-2009. In: Companion of Genetic and Evolutionary Computation Conference (GECCO 2010), pp. 1689–1696. ACM Press, New York (2010)
6. Hansen, N., Finck, S., Ros, R., Auger, A.: Real-parameter black-box optimization benchmarking 2009: noiseless functions definitions. Research report RR-6829, INRIA (2009). Accessed 22 Dec 2016
7. Jin, Y.: A comprehensive survey of fitness approximation in evolutionary computation. Softw. Comput. **9**(1), 3–12 (2005)
8. Jin, Y.: Surrogate-assisted evolutionary computation: recent advances and future challenges. Swarm Evol. Comput. **1**(2), 61–70 (2011)
9. Knowles, J., Nakayama, H.: Meta-modeling in multiobjective optimization. In: Branke, J., Deb, K., Miettinen, K., Słowiński, R. (eds.) Multiobjective Optimization. LNCS, vol. 5252, pp. 245–284. Springer, Heidelberg (2008). doi:10.1007/978-3-540-88908-3_10
10. Limbourg, P., Aponte, D.E.S.: An optimization algorithm for imprecise multiobjective problem functions. In: IEEE Congress on Evolutionary Computation (CEC 2005). IEEE Press, Piscataway (2005)
11. Mlakar, M., Petelin, D., Tušar, T., Filipič, B.: GP-DEMO: differential evolution for multiobjective optimization based on Gaussian process models. Eur. J. Oper. Res. **243**(2), 347–361 (2015)
12. Nissen, V., Propach, J.: Optimization with noisy function evaluations. In: Eiben, A.E., Bäck, T., Schoenauer, M., Schwefel, H.-P. (eds.) PPSN 1998. LNCS, vol. 1498, pp. 159–168. Springer, Heidelberg (1998). doi:10.1007/BFb0056859
13. Rudolph, G.: A partial order approach to noisy fitness functions. In: IEEE Congress on Evolutionary Computation (CEC 2001), pp. 318–325. IEEE Press, Piscataway (2001)
14. Sacks, J., Welch, W.J., Mitchell, T.J., Wynn, H.P.: Design and analysis of computer experiments. Stat. Sci. **4**(4), 409–423 (1989)
15. Tušar, T., Brockhoff, D., Hansen, N., Auger, A.: COCO: the bi-objective black box optimization benchmarking (BBOB-BIOBJ) test suite. CoRR abs/1604.00359 (2016). Accessed 22 Dec 2016
16. Zitzler, E., Knowles, J., Thiele, L.: Quality assessment of pareto set approximations. In: Branke, J., Deb, K., Miettinen, K., Słowiński, R. (eds.) Multiobjective Optimization. LNCS, vol. 5252, pp. 373–404. Springer, Heidelberg (2008). doi:10.1007/978-3-540-88908-3_14

Hypervolume Indicator Gradient Ascent Multi-objective Optimization

Hao Wang[(⊠)], André Deutz, Thomas Bäck, and Michael Emmerich

Leiden Institute of Advanced Computer Science, Leiden University,
Niels Bohrweg 1, 2333 CA Leiden, The Netherlands
{h.wang,a.h.deutz,t.h.w.baeck,m.t.m.emmerich}@liacs.leidenuniv.nl

Abstract. Many evolutionary algorithms are designed to solve black-box multi-objective optimization problems (MOPs) using stochastic operators, where neither the form nor the gradient information of the problem is accessible. In some real-world applications, e.g. surrogate-based global optimization, the gradient of the objective function is accessible. In this case, it is straightforward to use a gradient-based multi-objective optimization algorithm to achieve fast convergence speed and the stability of the solution. In a relatively recent approach, the hypervolume indicator gradient in the decision space is derived, which paves the way for the method for maximizing the hypervolume indicator of a fixed size population. In this paper, several mechanisms which originated in the field of evolutionary computation are proposed to make this gradient ascent method applicable. Specifically, the well-known non-dominated sorting is used to help steering the dominated points. The principle of the so-called cumulative step-size control that is originally proposed for evolution strategies is adapted to control the step-size dynamically. The resulting algorithm is called Hypervolume Indicator Gradient Ascent Multi-objective Optimization (HIGA-MO). The proposed algorithm is tested on ZDT problems and its performance is compared to other methods of moving the dominated points as well as to some evolutionary multi-objective optimization algorithms that are commonly used.

Keywords: Set-based scalarization · Hypervolume indicator · Gradient ascent · Non-dominated sorting · Cumulative step-size control

1 Introduction

Different multi-objective optimization algorithms have been proposed and exploited in real-world problem over the years, e.g. NSGA-II [2], SPEA2 [20] and SMS-EMOA [1]. These evolutionary multi-criteria optimization (EMO) algorithms employ heuristic operators (e.g. random variation and selection operators), instead of using the gradient information of the objective functions. For a large subclass of such problems, that is the **continuous** multi-objective optimization problem, gradient-based algorithms are of interest due to the fact that they are generally fast, precise and stable with respect to local convergence.

© Springer International Publishing AG 2017
H. Trautmann et al. (Eds.): EMO 2017, LNCS 10173, pp. 654–669, 2017.
DOI: 10.1007/978-3-319-54157-0_44

Various gradient-based approaches have been proposed for the multi-objective optimization task [5,7,9,13]. A relatively new idea is proposed by [3,4], in which the gradient of the hypervolume indicator with respect to a set of decision vectors is computed. In this paper, we adopt the definition and the computation of the *hypervolume indicator gradient* to steer the search points within the decision space. By using the hypervolume indicator gradient [19], the search points are moved in the direction of steepest ascent in the hypervolume indicator. Therefore, the proposed algorithm is termed *hypervolume indicator gradient ascent multi-objective optimization* (HIGA-MO). The major benefit of exploiting hypervolume gradients are (1) the points in the objective space will be well distributed on the Pareto front, (2) it is almost free of control parameters, and (3) the algorithm has a high precision of convergence to the Pareto front.

However, the first implementation of this idea showed numerical problems. As a remedy, ideas that were developed in the field of evolutionary multi-criterion optimization are adopted in this paper. Firstly, the hypervolume indicator may have zero gradient components at some decision vectors, e.g., the dominated points. The well-known non-dominated sorting technique is adopted and combined with the hypervolume indicator gradient computation, in order to equip each decision vector with a multi-layered gradient. Secondly, the normalization of the hypervolume indicator *sub*-gradient is used to overcome the "creepiness" phenomenon observed in earlier versions of hypervolume gradient ascent, and caused by an imbalance in the length of sub-gradients which leads to a slow convergence speed [14]. Thirdly, the usage of constant step-sizes is no longer appropriate if precise convergence to the Pareto front is aimed for. Instead, a cumulative step-size control inspired by the optimal gradient ascent is proposed to dynamically adapt the step-size. Such a cumulative step-size control resembles the step-size adaptation mechanism in the well-known CMA-ES [6], an evolutionary algorithm for single objective continuous optimization. The resulting algorithm is tested on problems named ZDT1-4 and ZDT6 from [18]. Its performance is compared to three evolutionary algorithms: NSGA-II [2], SPEA2 [20] and SMS-EMOA [1], as well as the other methods for steering the dominated points.

This paper is organized as follows. In Sect. 2, the multi-objective optimization problem and some notations used in this paper are introduced. In Sect. 3, derivations of the hypervolume indicator gradient are revisited with simplification of notations. The method for steering the dominated points is discussed in Sect. 4. The cumulative step-size control method is illustrated in Sect. 5. In Sect. 7, an experimental study of the resulting algorithm is conducted on the ZDT problems. Finally, we conclude the paper and suggest potential future improvements on HIGA-MO.

2   Background and Notations

In this section, a brief introduction of the problem and terminology that will be used later is given. In multi-objective optimization problems (MOPs), a collection of functions, represented as the m-tuple below, are optimized simultaneously:

$$(f_1 : S_1 \rightarrow \mathbb{R}, f_2 : S_2 \rightarrow \mathbb{R}, \ldots, f_m : S_m \rightarrow \mathbb{R}), \quad S_1, S_2, \ldots, S_m \subseteq \mathbb{R}^d$$

where d denotes the dimension of the domain of each function and m denotes the number of objective functions. Without loss of generality, we assume all the functions above are to be maximized (minimization problem can be transformed from maximization). In this work, it is assumed that each objective function f_i is continuous differentiable *almost everywhere* in S_i. Thus, the MOP can be formulated as follows:

$$\max_{\mathbf{x} \in S} \mathbf{f}(\mathbf{x}), \quad S = \bigcap_{i=1}^{m} S_i \subseteq \mathbb{R}^d,$$

where $\mathbf{f}(\mathbf{x}) = [f_1(\mathbf{x}), f_2(\mathbf{x}), \ldots, f_m(\mathbf{x})]^\top$ is a vector-valued function composed of m objective functions:

$$\mathbf{f} : S \rightarrow \mathbb{R}^m$$

Due to the continuous differentiability assumption on each objective function, $\mathbf{f}$ is again continuous differentiable almost everywhere in S. The gradient information is expressed as transpose of the Jacobian matrix as follows:

$$\frac{\partial \mathbf{f}(\mathbf{x})}{\partial \mathbf{x}} = [\nabla f_1(\mathbf{x}), \nabla f_2(\mathbf{x}), \ldots, \nabla f_m(\mathbf{x})], \quad \nabla f_i(\mathbf{x}) : S \rightarrow \mathbb{R}^d, \quad i = 1, 2, \ldots m$$

In addition, it is assumed that each gradient vector above (column vector) can be computed either analytically or numerically. In MOPs, a set of decision vectors are moved in *decision space* S to approximate the Pareto efficient set, which is the so-called Pareto efficient set approximation:

$$X = \left\{ \mathbf{x}^{(1)}, \mathbf{x}^{(2)}, \ldots, \mathbf{x}^{(\mu)} \right\}, \quad \mathbf{x}^{(i)} \in S, \ i = 1, 2, \ldots, \mu$$

with corresponding Pareto front approximation set (objective vectors) in the *objective space*:

$$Y = \left\{ \mathbf{y}^{(1)}, \mathbf{y}^{(2)}, \ldots, \mathbf{y}^{(\mu)} \right\}, \quad \mathbf{y}^{(i)} = \mathbf{f}(\mathbf{x}^{(i)}) \in \mathbb{R}^m, \ i = 1, 2, \ldots, \mu$$

In order to measure and compare the quality among Pareto front approximation sets Y, one approach is to quantify the quality by constructing a proper indicator. The most common one is the hypervolume indicator H [21,22]. Given a reference point $\mathbf{r} \in \mathbb{R}^m$, the hypervolume indicator of the Pareto front approximation set Y can be expressed as:

$$H(Y; \mathbf{r}) = \lambda \left(\bigcup_{\mathbf{y} \in Y} [\mathbf{r}, \mathbf{y}] \right),$$

where λ denotes the Lebesgue measure on $\mathbb{R}^m$, which is the size of the hypervolume dominated by the approximation set Y with respect to the reference space. Note that the reference point $\mathbf{r}$ will be assumed to be a given constant and thus omitted in the following notations for brevity.

3   Hypervolume Indicator Gradient

The hypervolume indicator gradient is defined as the gradient of the hypervolume indicator with respect to the approximation of the Pareto efficient set, which is proposed in [3,4]. In this work, the derivation of the hypervolume indicator gradient is reformulated and the notation is simplified. In the following, we shall use matrix calculus notations with denominator layout, meaning that the derivative of a vector/matrix is laid out according to the denominator.

Intuitively, the hypervolume indicator can be expressed as a function of the Pareto efficient set approximation X, which allows for the differentiation of hypervolume indicator with respect to decision vectors. More specifically, by concatenation of all the vectors in this set, we obtain a so-called $\mu \cdot d$-vector:

$$\mathbf{X} = \left[\mathbf{x}^{(1)^\top}, \mathbf{x}^{(2)^\top}, \ldots, \mathbf{x}^{(\mu)^\top}\right]^\top \in \mathbb{S}^\mu \subseteq \mathbb{R}^{\mu \cdot d}$$

and its corresponding Pareto front approximation vector can be written as a $\mu \cdot m$-vector:

$$\mathbf{Y} = \left[\mathbf{y}^{(1)^\top}, \mathbf{y}^{(2)^\top}, \ldots, \mathbf{y}^{(\mu)^\top}\right]^\top \in \mathbb{R}^{\mu \cdot m}$$

In order to establish a connection between $\mu \cdot d$-vectors and $\mu \cdot m$-vectors, we define a mapping $\mathbf{F} : \mathbb{S}^\mu \to \mathbb{R}^{\mu \cdot m}$:

$$\mathbf{F}(\mathbf{X}) := \left[\mathbf{f}(\mathbf{x}^{(1)})^\top, \mathbf{f}(\mathbf{x}^{(1)})^\top, \ldots, \mathbf{f}(\mathbf{x}^{(\mu)})^\top\right]^\top$$

Now consider that the hypervolume indicator, that is normally defined in the objective space, can be re-written as a function of $\mu \cdot d$-vectors by composition:

$$\mathcal{H}_\mathbf{F}(\mathbf{X}) := H(\mathbf{F}(\mathbf{X})),$$

which is a continuous mapping from $\mathbb{S}^\mu$ to $\mathbb{R}$, for which under certain regularity conditions the gradient is defined (in case of differentiable objective functions only for a zero measure subset of $\mathbb{R}^{\mu \cdot d}$ the gradient is undefined, in which case one-sided derivatives still exist). Given $\mathcal{H}_\mathbf{F}$, its derivatives are defined (given they exist) by:

$$\frac{\partial \mathcal{H}_\mathbf{F}(\mathbf{X})}{\partial \mathbf{X}} = \left[\frac{\partial \mathcal{H}_\mathbf{F}(\mathbf{X})}{\partial \mathbf{x}^{(1)}}^\top, \ldots, \frac{\partial \mathcal{H}_\mathbf{F}(\mathbf{X})}{\partial \mathbf{x}^{(\mu)}}^\top\right]^\top, \tag{1}$$

where each of the term in the RHS of the equation above is called *sub-gradient*, which is the local hypervolume change rate by moving each decision vector infinitesimally. It has been shown in [3] that the hypervolume indicator gradient is the concatenation of the hypervolume contribution gradients. Moreover, the sub-gradients can be calculated by applying the chain rule:

$$\frac{\partial \mathcal{H}_\mathbf{F}(\mathbf{X})}{\partial \mathbf{x}^{(i)}} = \frac{\partial \mathbf{y}^{(i)}}{\partial \mathbf{x}^{(i)}} \frac{\partial \mathcal{H}_\mathbf{F}(\mathbf{X})}{\partial \mathbf{y}^{(i)}} \tag{2}$$

$$= \sum_{k=1}^m \frac{\partial \mathcal{H}_\mathbf{F}(\mathbf{X})}{\partial f_k(\mathbf{x}^{(i)})} \nabla f_k(\mathbf{x}^{(i)}) \tag{3}$$

The first partial derivative in Eq. 2 is the gradient of $\mathcal{H}_{\mathbf{F}}$ in the objective space while the second one is the transpose of the Jacobian matrix of the mapping $\mathbf{F}$. Eq. 3 is the detailed form. From it, it is clear that the hypervolume indicator gradient is a *linear combination of gradient vectors of objective functions*, where the weight for an objective function is the partial derivative of the hypervolume indicator at this objective value. We omit the calculation for gradients of $\mathcal{H}_{\mathbf{F}}$ in the objective space for simplicity, noting that in the bi-objective case they correspond to the length of the steps of the attainment curve. For the high dimensional case and efficient computation, see [3].

Note that in practice the length of the sub-gradients usually differ by orders of magnitude, leading to the "creepiness" behavior [14] that some decision vectors move much faster than the rest, Such a behavior results in a very slow convergence speed and points might get dominated by others. As a remedy, it is suggested to normalize all the sub-gradients.

4   Steering Dominated Points

The difficulty increases when applying the hypervolume indicator gradient direction for steering the decision vectors: the hypervolume indicator can either be zero or only one-sided at decision vectors. For example, at every strictly dominated search point, the hypervolume indicator sub-gradient is zero, because the Pareto front and thus the hypervolume indicator remain unchanged if it is moved locally in an infinitesimally small neighborhood. For every weakly dominated point, the hypervolume indicator sub-gradient at this point, even does not exist due to the fact that only one-sided partial derivatives exist. Consequently, such decision vectors will become stationary in gradient ascent method. One obvious solution to such a problem is to apply evolutionary operators (mutation and crossover) on those search points (decision vectors) until they become non-dominated. However, as we are aiming for a fully deterministic multi-objective optimization algorithm, randomized operators are not adopted in this work.

Some methods have been proposed to steer dominated points [9,11,17]. The most prominent one, proposed by Lara [9], computes the gradient at dominated points as follows (for bi-objective problems):

$$\frac{\nabla f_1(\mathbf{x}^{(i)})}{\|\nabla f_1(\mathbf{x}^{(i)})\|} + \frac{\nabla f_2(\mathbf{x}^{(i)})}{\|\nabla f_2(\mathbf{x}^{(i)})\|}, \quad \mathbf{x}^{(i)} \text{ is dominated}$$

which is a sum of normalized gradients of each objective function. It guarantees that dominated decision vectors move into the *dominance cone* [17]. However, such a method only considers the movement of single points, instead of a set of search points and it does not generalize to more than two dimensions. We shall call this method **Lara's direction** in the following experiments, where it is compared with the method proposed in this work. Another method for steering the dominated points is proposed by the authors in [17]. It steers dominated points towards the nearest gap on the non-dominated set. The search direction is determined as the gradient of the distance of the dominated objective vector to the center of its nearest gap. Again, this method steers dominated points independently and is termed as **Gap-filling** in this paper. In the above methods, dominated points are steered widely independent of each other, which might result in diversity loss.

In this work, we propose to use the *non-dominated sorting* technique that is developed in the NSGA-II algorithm [15], in order to compute the hypervolume indicator gradients of multiple layers of non-dominated sets. In detail, the decision and objective vectors are partitioned into q subsets, or *layers* according to their dominance rank in the objective space:

$$\mathbf{X} \rightarrow \left\{\mathbf{X}^1, \mathbf{X}^2, \ldots, \mathbf{X}^q\right\},$$

$$\mathbf{X}^i = \left[\mathbf{x}^{(i_1)^\top}, \mathbf{x}^{(i_2)^\top}, \ldots, \mathbf{x}^{(i_\mu)^\top}\right]^\top,$$

where X^i indicates a layer of order i and i_μ denotes the number of decision vectors in the ith rank layer. The layers can be recursively defined as (given ND as the operator that selects the non-dominated subset from approximation set):

$$X^1 = ND(X), \quad X^{i+1} = ND(X - \bigcup_{j=1}^{i} X^j),$$

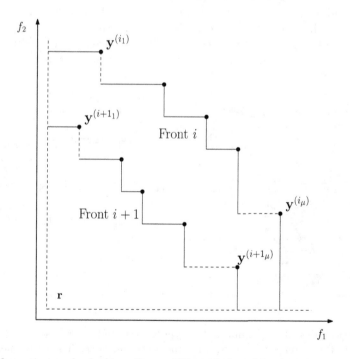

Fig. 1. Schematic graph showing the partition of the objective vectors using non-dominated sorting. For each partition (layer), a hypervolume indicator is defined and thus its gradient can be computed.

where q is the highest index i such that $X^i \neq \emptyset$. Note that the $\mu \cdot m$-vector is also partitioned as above. In principle, it is possible to compute the hypervolume indicator gradient for any layer by ignoring all the layers that dominate it (have a lower rank) temporarily. This partition is illustrated in Fig. 1. In this manner, the hypervolume volume indicator gradient on the whole approximation set $\mathbf{X}$ can be (re-)defined as the concatenation of the hypervolume indicator gradient on each layer:

$$\frac{\partial \mathcal{H}_\mathbf{F}(\mathbf{X})}{\partial \mathbf{X}} := \left[\frac{\partial \mathcal{H}_\mathbf{F}(\mathbf{X}^1)}{\partial \mathbf{X}^1}^\top, \frac{\partial \mathcal{H}_\mathbf{F}(\mathbf{X}^2)}{\partial \mathbf{X}^2}^\top, \ldots, \frac{\partial \mathcal{H}_\mathbf{F}(\mathbf{X}^q)}{\partial \mathbf{X}^q}^\top \right]^\top \qquad (4)$$

Note that again q is the number of layers obtained from non-dominated sorting techniques. The gradient computation given in Eq. 3 can be used to compute each gradient term above. Thus, each decision vector is associated with a steepest ascent direction that maximizes its hypervolume contribution on each layer.

There are two advantages of using the non-dominated sorting procedure. Firstly, maximizing the hypervolume will not only steer the points towards the Pareto front, but also spread out the points across the intermediate Pareto front approximation. By applying the hypervolume indicator gradient direction on each layer, the decision vectors on each layer will be well distributed before a dominated layer merges into the global Pareto front and thus the additional cost to spread out points after the merging is small. Moreover, when the Pareto efficient set is disconnected in the decision space, the proposed approach will increase the convergence speed due to the fact that each connected efficient set is treated as one layer and the decision vectors on it are spread quickly over

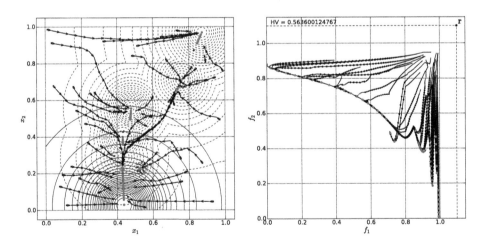

Fig. 2. Trajectories of 50 points under hypervolume indicator gradient direction to approximate the Pareto front using 10^3 function evaluations. The experiment is conducted on a bi-objective problem MPM2 (from R *smoof* package) in the 2-D decision space. All five disconnected components of the Pareto front are obtained with well distributed points. **Left:** the decision space. **Right:** the objective space.

the efficient sets. This effect can be shown by visualizing the trajectories of the approximation set on a simple objective landscape. In Fig. 2, trajectories of the approximation set are illustrated in both decision and objective space, on MPM2 functions (from the R *smoof* package[1]). In the decision space, it is clear that our layering approach (Fig. 2) manages to approximate five disconnected efficient sets with a good distribution of points.

Secondly, on the real landscape, it is possible that local Pareto fronts exist (e.g. consider the well-known ZDT4 problem [18]). Using the non-dominated sorting, it is more likely to identify those local Pareto fronts, which could be helpful to balance global and local search. This advantage of the proposed approach is exploited by the authors in multi-objective multi-modal landscape analysis [8].

5 Step-Size Adaptation

The constant step-size setting that is common in gradient descent (ascent) for the single objective optimization task, is no longer appropriate. Usually, the length of the gradient vector (in the gradient field) gradually goes to zero when approaching the local optimum. In this case, a properly set constant step-size will lead to the local optimum in a stable manner. However, in our case, due to the normalization, the length of the search steps is always 1 when decision vectors are approaching the Pareto efficient set. If a constant step-size is applied here, the decision vector will *overshoot* its optimal position and begin to oscillate (even diverge). In order to tackle this issue, the step-size of the decision vectors needs to (1) gradually decrease when approaching the Pareto efficient set and (2) increase quickly when the decision vectors are far away from the efficient set. In addition, it is reasonable to use individual step-sizes that are controlled independently for each decision vector because their optimal step-size differs largely.

A cumulative step-size adaptation mechanism is proposed to approximate the optimal step-size in the optimization process. It is inspired by the following observation: in single objective gradient optimization, if the step-size is set optimally, then consecutive search direction are perpendicular to each other. In order to approximate the optimal step-size setting, the inner product of consecutive normalized hypervolume indicator gradients is calculated. If such an inner product is positive, it indicates the current step-size is smaller than the optimal one and vice versa:

$$I_t^{(i)} = \left\langle \frac{\partial \mathcal{H}_{\mathbf{F}}(\mathbf{X})}{\partial \mathbf{x}^{(i)}}^{(t-1)}, \frac{\partial \mathcal{H}_{\mathbf{F}}(\mathbf{X})}{\partial \mathbf{x}^{(i)}}^{(t)} \right\rangle, \quad i = 1, \ldots, \mu, \quad t = 1, 2, \ldots.$$

Note that superscripts $(t), (t-1)$ are iteration indices. In addition, such an inner product computed in each iteration fluctuates hugely and direct use of it leads to unstable adaptation behavior. Therefore, the inner product is cumulated using exponentially decreasing weights through the iterations to get a more stable

---

[1] https://github.com/jakobbossek/smoof.

indicator for the step-size adaptation. The cumulative rule for the inner product is written as follows:

$$p_t^{(i)} \leftarrow (1 - c) \cdot p_{t-1}^{(i)} + c \cdot I_t^{(i)}, \quad i = 1, \ldots, \mu, \quad t = 1, 2, \ldots. \tag{5}$$

Note that $p_t^{(i)}$ denotes the cumulated inner product for search point i at iteration t and c $(0 < c < 1)$ is the accumulation coefficient. Such an inner product accumulation rule is similar to the cumulative step-size adaptation mechanism in the Covariance Matrix Adaptation Evolution Strategy (CMA-ES) [6], where consecutive mutation steps are accumulated for step-size adaptation. Based on the cumulated inner product, a simple control rule is designed to adapt the step-size online:

$$\sigma_{t+1}^{(i)} = \begin{cases} \sigma_t^{(i)} \cdot \alpha & \text{if } p_t^{(i)} < 0, \\ \sigma_t^{(i)} & \text{if } p_t^{(i)} = 0, \quad 0 < \alpha < 1 \\ \sigma_t^{(i)}/\alpha & \text{if } p_t^{(i)} > 0. \end{cases} \tag{6}$$

where $\sigma_t^{(i)}$ is the individual step-size for search point i at iteration t. This rule dictates that (1) if the inner product accumulation is positive, then the step-size is increased by a factor of α, (2) if the inner product accumulation is negative, the the step-size is decrease by a factor of α, and (3) otherwise, the step-size remains unchanged. In this work, the settings of $c = 0.7, \alpha = 0.8$ are suggested by tuning the algorithmic performance on MPM2 functions.

The backtracking line search [10], which is a common technique to approximate the optimal step-size in single objective gradient ascent, is not suitable for the proposed algorithm. It requires additional function evaluations for each search point to estimate the optimal step-size setting. Such additional costs are no longer acceptable for the set-based algorithm. In contrast, the proposed cumulative step-size adaptation mechanism does not bring any additional overheads.

6   Hypervolume Indicator Gradient Ascent Algorithm

In this section, the algorithmic components developed in the previous sections are combined into the hypervolume indicator gradient ascent algorithm.

6.1   Handling Non-differentiable Points

In practice, the continuous objective function can be non-differentiable at some points, even the function is almost everywhere differentiable (e.g. on the constraint boundary of the ZDT1 problem). To overcome this issue, it is suggested to *mutate* those points in the decision space. Given a point $\mathbf{x} \in \mathbb{R}^d$, it is mutated in the decision space S when the gradient of objective functions at $\mathbf{x}$ contains invalid values (e.g. infinite). The mutation of $\mathbf{x}$ should be local but large enough to escape from the non-differentiable regions. For this purpose, then mutation operator in Differential Evolution [16] is adopted here because it is adaptive and only contains a single parameter. Suppose $\mathbf{x}$ is in the ith ranked layer ($\mathbf{x} \in \mathbf{X}^i$), then it is mutated as follows:

$$\mathbf{x} \leftarrow \mathbf{x} + F \cdot (\mathbf{x}^{(a)} - \mathbf{x}^{(b)}), \tag{7}$$

where $\mathbf{x}^{(a)}, \mathbf{x}^{(b)}$ are randomly picked from $\mathbf{X}^i$. $F \in [0, 2]$ is the differential weight that is set according to the literature. It is necessary to compute the differential vector within the same layer of $\mathbf{x}$ because the Pareto efficient set is possibly disconnected in the decision space and differential vectors computed across layers possibly create non-local mutations.

6.2 Pseudo-code

The resulting algorithm is presented in Algorithm 1. In line 4, the non-dominated sorting procedure is called to partition the approximation set. In line 7 the hypervolume indicator gradient is computed for every decision vector on each layer. If a decision vector has either zero gradient or is not differentiable, it is mutated in line 9 according to Eq. 7. In line 11, the hypervolume indicator sub-gradient is normalized before decision vectors are moved in the steepest ascent manner (line 12). The cumulative step-size adaptation (Eqs. 5 and 6) is then applied in line 13. In addition to the common usage of the function evaluation budget for the termination criterion, it is suggested here to check stationarity of search points: a decision vector is considered stationary if the norm of its sub-gradient multiplied by the step-size is close to zero.

Algorithm 1. Hypervolume Indicator Gradient Ascent Multi-Objective Optimization

---

1 Initialize the search points $\mathbf{X}$ uniformly in the search space
2 **while** the termination criteria are not satisfied **do**
3   Evaluation: $\mathbf{Y} \leftarrow \mathbf{X}$.
4      $\{\mathbf{X}^1, \mathbf{X}^2, \ldots, \mathbf{X}^q\} \leftarrow$ non-dominated-sorting($\mathbf{X}$)
5   **for** $k = 1$ to q **do**
6   **for** every $\mathbf{x}^{(i)}$ in $\mathbf{X}^k$ **do**
7   Compute the sub-gradient $\frac{\partial \mathcal{H}_{\mathbf{F}}(\mathbf{X})}{\partial \mathbf{x}^{(i)}}$ according to Eq. 2
8   **if** $\frac{\partial \mathcal{H}_{\mathbf{F}}(\mathbf{X})}{\partial \mathbf{x}_j}$ is undefined **then**
9                  $\mathbf{x}^{(i)} \leftarrow \mathbf{x}^{(i)} + F \cdot (\mathbf{x}^{(a)} - \mathbf{x}^{(b)})$, where $\mathbf{x}^{(a)}, \mathbf{x}^{(b)}$ are randomly picked from $\mathbf{X}^k$
10   **else**
11   Sub-gradient normalization: $\mathbf{g}^{(i)} \leftarrow \frac{\partial \mathcal{H}_{\mathbf{F}}(\mathbf{X})}{\partial \mathbf{x}^{(i)}} / \| \frac{\partial \mathcal{H}_{\mathbf{F}}(\mathbf{X})}{\partial \mathbf{x}^{(i)}} \|$
12   Gradient ascent: $\mathbf{x}^{(i)} \leftarrow \mathbf{x}^{(i)} + \sigma^{(i)} \mathbf{g}^{(i)}$
13   Apply the step-size control according to Eq. 5 and 6.
14   **end if**
15   **end for**
16   **end for**
17 **end while**
18 **return** $\mathbf{X}, \mathbf{Y}$

---

7   Experiments

Experiment settings. To test the performance of HIGA-MO, the well-known ZDT problems [2] are selected as benchmark problem set. The proposed algorithm is compared to three well-established evolutionary multi-objective optimization algorithm: NSGA-II, SPEA2 and SMS-EMOA. The parameters in those two algorithms are set according to the literature [1, 2, 20]. In addition, other methods for steering the dominated point (Sect. 4), Lara's direction and Gap-filling, are tested against HIGA-MO. For these two methods, the non-dominated points are moved using the hypervolume indicator gradient.

The hypervolume indicator and convergence measure used in [1], are adopted here as the performance metrics. The convergence measure is calculated numerically by discretizing the Pareto front into 1000 points. For the hypervolume indicator computation, the reference point $[11, 11]^\top$ is used for the test problems ZDT1-4 and ZDT6. Two experiments are conducted: one with a relatively small population setting $\mu = 40$ while the other uses a large population, $\mu = 100$. A relatively small function evaluation budget, 100μ is chosen here due the reason that in long runs, all deterministic methods stagnate to local optima. All the algorithms terminate if the maximal function evaluation budget is reached. For each algorithm, 15 independent runs are conducted to obtain average performance measures. The initial step-size of the proposed HIGA-MO algorithm is set to 0.05 multiplied by the maximum range of the decision space. The internal reference point to compute the hypervolume indicator gradient is set to $[11, 11]^\top$ to ensure every objective vector is within the reference space.

Results. The test results are shown in Table 1 for $\mu = 40$ and Table 2 for $\mu = 100$. The hypervolume of the non-dominated set after termination is used to compute the performance measures. For the small population setting, HIGA-MO outperforms the evolutionary algorithms (NSGA-II, SPEA2 and SMS-EMOA) on ZDT1-3 and ZDT6 problems, both in terms of hypervolume indicator and convergence measure. By checking the standard deviation, it is obvious that HIGA-MO generates more stable results compared to evolutionary algorithms and such deviations are only affected by the initialization of the approximation set and the technique to handle the non-differentiable points (Eq. 7). Comparing it to the other two methods, namely, Lara's direction and Gap-filling, that steer the dominated points independently, HIGA-MO gives a higher hypervolume indicator value on ZDT1-3 while Lara's method performs better on ZDT6. In terms of the convergence measure, Lara's direction always outperforms HIGA-MO on ZDT1-3 and 6. Lara's direction moves the dominated points toward the Pareto front without considering the distribution of them while HIGA-MO is designed to achieve both. Thus, HIGA-MO requires more efforts to approach the Pareto front than Lara's direction, in terms of the convergence measure. On ZDT4, which has a highly multi-modal landscape, none of the gradient-based methods (HIGA-MO, Lara's direction and Gap-filling) achieves comparable results to evolutionary algorithms. The gradient-based methods easily stagnate in the local Pareto-front and fail to move towards the global one. For such a highly multi-modal optimization problem, a restart heuristic could improve the performance of gradient-based algorithms. For the large population setting, Table 2 shows roughly the same results for algorithm comparisons as for the small population setting.

Table 1. $\mu = 40$: performance measures on ZDT1-4 and ZDT6 problems.

Test-function	Algorithm	Convergence measure			Hypervolume indicator		
		Average	Std. dev.	Rank	Average	Std. dev.	Rank
ZDT1	HIGA-MO	**0.00500490**	1.3075e−02	1	**120.62948062**	4.0750e−03	1
	Lara's direction	0.07747718	6.4031e−02	3	120.33761711	1.2309e−01	2
	Gap-filling	0.06061863	1.2352e−01	2	120.22307239	4.6840e−01	3
	NSGA-II	0.10960371	3.2542e−02	5	119.33541376	3.7345e−01	4
	SMS-EMOA	0.09376444	3.5934e−02	4	119.20965862	4.8101e−01	5
	SPEA2	0.32006024	5.9788e−02	6	116.27370195	1.6826e+00	6
ZDT2	HIGA-MO	0.00036082	3.6233e−05	3	**120.31634691**	9.8307e−04	1
	Lara's direction	**0.00011253**	5.0289e−05	1	118.92812930	3.5019e+00	3
	Gap-filling	0.00015973	2.0645e−04	2	119.45871166	2.5324e+00	2
	NSGA-II	0.16511979	7.7092e−02	4	114.03423180	3.7806e+00	4
	SMS-EMOA	0.24929199	8.4178e−02	5	109.17629732	3.2584e+00	5
	SPEA2	0.67688451	1.5708e−01	6	104.54506810	3.3537e+00	6
ZDT3	HIGA-MO	0.00031903	5.0492e−05	2	128.55259300	7.9970e−01	2
	Lara's direction	**0.00028076**	5.0842e−05	1	125.78304061	3.5114e+00	6
	Gap-filling	0.00034568	5.4557e−05	3	**128.75911576**	9.2658e−03	1
	NSGA-II	0.00228282	5.9689e−03	4	126.56081625	2.8857e+00	3
	SMS-EMOA	0.00405046	5.7238e−03	5	125.88966563	2.9289e+00	5
	SPEA2	0.00635668	1.0852e−02	6	126.55026001	2.5895e+00	4
ZDT4	HIGA-MO	38.13060527	7.6780e+00	4	0.00000000	0.0000e+00	6
	Lara's direction	43.19742796	1.1544e+01	5	0.00000000	0.0000e+00	5
	Gap-filling	52.35972878	1.2465e+01	6	1.16325406	4.3525e+00	4
	NSGA-II	4.07411956	1.6869e+00	2	75.28344930	1.8038e+01	2
	SMS-EMOA	**3.52099683**	1.7386e+00	1	**78.04608227**	1.8555e+01	1
	SPEA2	11.17677922	4.9514e+00	3	19.34577362	2.2000e+01	3
ZDT6	HIGA-MO	3.83694298	1.3668e+00	6	113.28359226	1.3577e+00	2
	Lara's direction	**0.00010409**	4.3909e−05	1	**116.86127498**	1.6820e+00	1
	Gap-filling	3.02249489	2.7090e+00	5	106.81768735	2.0573e+01	3
	NSGA-II	1.28139859	3.0071e−01	2	97.53535725	3.8143e+00	4
	SMS-EMOA	1.36426329	3.1163e−01	3	96.84386232	4.2309e+00	5
	SPEA2	2.22799304	7.2398e−01	4	86.25780584	7.9570e+00	6

Table 2. $\mu = 100$: performance measures on ZDT1-4 and ZDT6 problems.

Test-function	Algorithm	Convergence measure			Hypervolume indicator		
		Average	Std. dev.	Rank	Average	Std. dev.	Rank
ZDT1	HIGA-MO	**0.00031201**	4.1269e−05	1	**120.64580412**	1.7718e−03	1
	Lara's direction	0.02103585	4.7314e−02	5	120.48926778	5.2474e−02	2
	Gap-filling	0.02091304	6.1387e−02	4	120.42616648	2.7937e−01	5
	NSGA-II	0.01769266	4.6048e−03	3	120.45030137	4.5135e−02	4
	SMS-EMOA	0.01234011	2.6377e−03	2	120.48071780	3.6130e−02	3
	SPEA2	0.06017346	1.7966e−02	6	119.86686583	2.1615e−01	6
ZDT2	HIGA-MO	0.00028335	3.3303e−05	3	**120.31710222**	2.3560e−03	1
	Lara's direction	**0.00005498**	1.2085e−05	1	120.30338190	2.9998e−03	2
	Gap-filling	0.00007857	8.7094e−05	2	120.14758158	1.5778e−01	3
	NSGA-II	0.02834448	4.4153e−03	5	119.16220851	1.0985e+00	4
	SMS-EMOA	0.02338094	7.0938e−03	4	118.40070248	2.7352e+00	5
	SPEA2	0.08566545	4.8472e−02	6	114.48551919	4.4285e+00	6
ZDT3	HIGA-MO	0.00047505	7.5997e−05	3	128.77154126	8.5828e−03	3
	Lara' direction	0.00046485	5.9553e−05	2	128.77257561	5.2596e−03	2
	Gap-filling	**0.00039660**	4.9392e−05	1	128.77099724	3.3611e−03	4
	NSGA-II	0.00063823	5.1880e−05	5	**128.77436195**	1.1318e−03	1
	SMS-EMOA	0.00055256	3.5594e−05	4	128.34841609	1.0889e+00	6
	SPEA2	0.00243258	6.6391e−03	6	128.55447469	7.9741e−01	5
ZDT4	HIGA-MO	31.34155544	3.9090e+00	4	0.00000000	0.0000e+00	6
	Lara's direction	40.35930710	1.1041e+01	5	0.00000000	0.0000e+00	5
	Gap-filling	43.47103886	1.5933e+01	6	5.23444012	1.5425e+01	4
	NSGA-II	**0.80498648**	5.0038e−01	1	**109.60569075**	5.4368e+00	1
	SMS-EMOA	1.01209147	6.3095e−01	2	107.14186469	7.1460e+00	2
	SPEA2	2.80155378	1.3959e+00	3	83.82023960	1.5461e+01	3
ZDT6	HIGA-MO	3.54689504	1.2985e+00	5	113.79978098	8.8488e−01	2
	Lara's direction	**0.00004369**	1.2553e−05	1	**116.49314419**	1.4990e+00	1
	Gap-filling	4.12388484	2.9230e+00	6	86.58598768	3.4123e+01	6
	NSGA-II	0.43202530	7.1773e−02	3	109.28079070	1.2513e+00	4
	SMS-EMOA	0.40028650	1.1394e−01	2	109.87049482	1.8951e+00	3
	SPEA2	0.49692387	1.2882e−01	4	108.17997611	1.9177e+00	5

8   Conclusions

In this paper, a full gradient-based multi-objective optimization algorithm is proposed. The gradient direction is derived by differentiating the hypervolume indicator with respect to the concatenation of decision vectors. Moreover, several techniques are devised to solve difficulties in applying the hypervolume indicator gradient to the approximation set: (1) the non-dominated sorting procedure is used to steer the dominated points using the hypervolume indicator gradient. (2) a cumulative step-size adaptation mechanism is developed to approximate the optimal step-size in gradient ascent search. The algorithm is tested on 5 ZDT problems, and its performance is compared to evolutionary algorithms and some other gradient-based approaches. The proposed algorithm shows a fast convergence speed in terms of the hypervolume indicator.

As shown in the experimental results on ZDT4, the proposed algorithm fails to approach the global Pareto front and gets stuck in local ones instead. In practice, such an issue can be tackled by using restart heuristics to re-sample the stagnated points. In addition, it is possible to hybridize HIGA-MO with an evolutionary multi-objective (EMO) algorithm, where the global search ability of an EMO helps the algorithm to escape from a deceptive, local Pareto front and HIGA-MO could achieve fast convergence speed when approaching the global Pareto front. Such an approach has been proposed in [9] and the optimal way to combine HIGA-MO with EMOs should be investigated.

The experiments conducted in this paper are on a small number of problems. In future research, the proposed algorithm should be investigated on more multi-objective problems. When a using large number of search points, the objective vectors on the Pareto front are close to each other, which might result in relatively slow movement. In this case, its performance needs to be further tested. In addition, it is of interest to compare HIGA-MO empirically to other set-based scalarization method [12].

For the proposed method for steering the dominated points, it should be also be empirically compared to alternative methods that are proposed in [17]. Those methods should be thoroughly compared to characterize their performance in terms of convergence measure and the hypervolume indicator value. In addition, as described in Sect. 5, the parameter tuning for the step-size adaptation is merely tested on a simple test problem (MPM2 function). A most rigorous parameter tuning procedure should be performed to get a reliable and robust parameter setting.

Acknowledgments. This work presented in this paper is financially supported by the Dutch Research Project (NWO) PROMIMOOC (project number: 650.002.001).

References

1. Beume, N., Naujoks, B., Emmerich, M.: SMS-EMOA: multiobjective selection based on dominated hypervolume. Eur. J. Oper. Res. **181**(3), 1653–1669 (2007)
2. Deb, K., Agrawal, S., Pratap, A., Meyarivan, T.: A fast elitist non-dominated sorting genetic algorithm for multi-objective optimization: NSGA-II. In: Schoenauer, M., Deb, K., Rudolph, G., Yao, X., Lutton, E., Merelo, J.J., Schwefel, H.-P. (eds.) PPSN 2000. LNCS, vol. 1917, pp. 849–858. Springer, Heidelberg (2000). doi:10. 1007/3-540-45356-3_83
3. Emmerich, M., Deutz, A.: Time complexity and zeros of the hypervolume indicator gradient field. In: Schütze, O., Coello, C.A.C., Tantar, A.-A., Tantar, E., Bouvry, P., Moral, P.D., Legrand, P. (eds.) EVOLVE - A Bridge between Probability, Set Oriented Numerics, and Evolutionary Computation III. SCI, vol. 500, pp. 169–193. Springer (2014)
4. Emmerich, M., Deutz, A., Beume, N.: Gradient-based/evolutionary relay hybrid for computing pareto front approximations maximizing the S-metric. In: Bartz-Beielstein, T., Blesa Aguilera, M.J., Blum, C., Naujoks, B., Roli, A., Rudolph, G., Sampels, M. (eds.) HM 2007. LNCS, vol. 4771, pp. 140–156. Springer, Heidelberg (2007). doi:10.1007/978-3-540-75514-2_11

5. Fliege, J., Svaiter, B.F.: Steepest descent methods for multicriteria optimization. Math. Meth. Oper. Res. **51**(3), 479–494 (2000)
6. Hansen, N., Ostermeier, A.: Completely derandomized self-adaptation in evolution strategies. Evol. Comput. **9**(2), 159–195 (2001)
7. Hillermeier, C.: Generalized homotopy approach to multiobjective optimization. J. Optim. Theor. Appl. **110**(3), 557–583 (2001)
8. Kerschke, P., Wang, H., Preuss, M., Grimme, C., Deutz, A., Trautmann, H., Emmerich, M.: Towards analyzing multimodality of continuous multiobjective landscapes. In: Handl, J., Hart, E., Lewis, P.R., López-Ibáñez, M., Ochoa, G., Paechter, B. (eds.) PPSN 2016. LNCS, vol. 9921, pp. 962–972. Springer, Cham (2016). doi:10.1007/978-3-319-45823-6_90
9. López, A.L., Coello, C.A.C., Schütze, O.: Using gradient based information to build hybrid multi-objective evolutionary algorithms. Ph.D. thesis, CINVESTAV-IPN, Mexico city, May 2012
10. Nocedal, J., Wright, S.: Numerical Optimization. Operations Research and Financial Engineering. Springer, New York (2000)
11. Ren, Y., Deutz, A., Emmerich, M.: On steering dominated points in hypervolume gradient ascent for bicriteria continuous optimization (extended abstract). In: Numerical and Evolutionary Optimization, NEO (2015), Tijuana, Mexico (Book of abstracts) (2015)
12. Schütze, O., Domínguez-Medina, C., Cruz-Cortés, N., Gerardo de la Fraga, L., Sun, J.-Q., Toscano, G., Landa, R.: A scalar optimization approach for averaged hausdorff approximations of the pareto front. Eng. Optim. **48**(9), 1593–1617 (2016)
13. Schütze, O., Lara, A., Coello, C.A.C.: The directed search method for unconstrained multi-objective optimization problems. In: Proceedings of the EVOLVE-A Bridge Between Probability, Set Oriented Numerics, and Evolutionary Computation, pp. 1–4 (2011)
14. Hernández, V.A.S., Schütze, O., Emmerich, M.: Hypervolume maximization via set based Newton's method. In: Tantar, A.-A., et al. (eds.) EVOLVE - A Bridge between Probability, Set Oriented Numerics, and Evolutionary Computation V, pp. 15–28. Springer, Cham (2014)
15. Srinivas, N., Deb, K.: Muiltiobjective optimization using nondominated sorting in genetic algorithms. Evol. Comput. **2**(3), 221–248 (1994)
16. Storn, R., Price, K.: Differential evolution-a simple and efficient heuristic for global optimization over continuous spaces. J. Glob. Optim. **11**(4), 341–359 (1997)
17. Wang, H., Ren, Y., Deutz, A., Emmerich, M.: On steering dominated points in hypervolume indicator gradient ascent for Bi-objective optimization. In: Schütze, O., Trujillo, L., Legrand, P., Maldonado, Y. (eds.) NEO 2015: Results of the Numerical and Evolutionary Optimization Workshop NEO 2015, 23–25 September 2015, Tijuana, Mexico, pp. 175–203. Springer, Cham (2017)
18. Zitzler, E., Deb, K., Thiele, L.: Comparison of multiobjective evolutionary algorithms: empirical results. Evol. Comput. **8**(2), 173–195 (2000)
19. Zitzler, E., Künzli, S.: Indicator-based selection in multiobjective search. In: Yao, X., et al. (eds.) PPSN 2004. LNCS, vol. 3242, pp. 832–842. Springer, Heidelberg (2004). doi:10.1007/978-3-540-30217-9_84
20. Zitzler, E., Laumanns, M., Thiele, L., et al.: SPEA2: improving the strength pareto evolutionary algorithm. Eurogen **3242**, 95–100 (2001)

21. Zitzler, E., Thiele, L.: Multiobjective optimization using evolutionary algorithms — a comparative case study. In: Eiben, A.E., Bäck, T., Schoenauer, M., Schwefel, H.-P. (eds.) PPSN 1998. LNCS, vol. 1498, pp. 292–301. Springer, Heidelberg (1998). doi:10.1007/BFb0056872

22. Zitzler, E., Thiele, L., Laumanns, M., Fonseca, C.M., Da Fonseca, V.G.: Performance assessment of multiobjective optimizers: an analysis and review. IEEE Trans. Evol. Comput. **7**(2), 117–132 (2003)

Toward Step-Size Adaptation in Evolutionary Multiobjective Optimization

Simon Wessing[(✉)], Rosa Pink, Kai Brandenbusch, and Günter Rudolph

Computer Science Department, Technische Universität Dortmund,
Otto-Hahn-Str. 14, 44221 Dortmund, Germany
{simon.wessing,rosa.pink,kai.brandenbusch,guenter.rudolph}@tu-dortmund.de

Abstract. We give a definition for step size optimality in multiobjective optimization and visualize the optimal step sizes for a few two-dimensional example constellations. After that, we try to engineer a step size adaptation mechanism that also works in the real world. For this mechanism, we employ the self-adaptation of mutation strength, which is simple and well-known from single-objective optimization. The resulting approach obtains better results than simulated binary crossover and polynomial mutation on the bi-objective BBOB testbed.

1 Introduction

In contrast to selection, variation is a comparatively little explored topic in evolutionary multiobjective optimization (EMO). Especially the adaptation of the mutation strength is an open question even for simple problems, since the situation is quite different from the well understood situation in the single-objective case. For example, the optimal step size[1] s^* of the $(1 + 1)$ evolutionary algorithm (EA) for minimizing $f(x) = x^\top x$ is known if the mutation vector Z is uniformly distributed on a hypersphere surface [25] or multinormally distributed [23], yielding a mutation $X + s \cdot Z$ of individual X. The step size is optimal, if the expected improvement $\mathsf{E}[\max\{0, f(X) - f(X + s \cdot Z)\}]$ gets maximal, or equivalently, if the expected objective function value $\mathsf{E}[\min\{f(X), f(X + s \cdot Z)\}]$ gets minimal.

In multiobjective optimization the optimal solution is not a single point with minimal objective function value, but a set of points (the Pareto set) with mutually incomparable objective function vectors. A broadly accepted scalar measure to assess the quality of an approximation of the Pareto set is the dominated hypervolume: the larger is the dominated hypervolume the better is the approximation. It is also used in the $(\mu + 1)$ S-metric selection EMO algorithm (SMS-EMOA) as decision criterion whether a new point is accepted and which old point from the population is removed [4]. Therefore, an apparently natural choice for an optimal step size might be that step size for which the improvement of the expected hypervolume becomes maximal. In the bicriteria case

---

[1] We are using *step size* and *mutation strength* synonymously in this work.

© Springer International Publishing AG 2017
H. Trautmann et al. (Eds.): EMO 2017, LNCS 10173, pp. 670–684, 2017.
DOI: 10.1007/978-3-319-54157-0_45

$f(x) = (f_1(x), f_2(x))^\top$ it has been proven [2] that the $(1+1)$-SMS-EMOA is algorithmically equivalent to a single-objective $(1+1)$-EA minimizing the surrogate function $f^S(x) = \frac{1}{2}(f_1(x) + f_2(x))$. If $f_1(x) = x^\top x$ and $f_2(x) = (x-a)^\top(x-a)$ with $a \neq 0$ the resulting surrogate problem $f^S(x) = (x-a/2)^\top(x-a/2) + a^\top a/4$ is a displaced version of the problem $f(x) = x^\top x$ with constant offset. As a consequence, we know the optimal step size from the theory of the single-objective case: it is the step size for which the expected value of the surrogate function gets minimal, or equivalently, for which the expected hypervolume gets maximal. Unfortunately, the algorithmic equivalence between single-objective and multi-objective EAs breaks if $\mu > 1$, even in the bicriteria case (cf. Theorem 11 in [3]). Therefore, optimal step sizes for the $(\mu+1)$-SMS-EMOA and related algorithms with $\mu > 1$ are unknown. But without knowledge about optimal step sizes for simple multiobjective problems as that above, it is difficult to devise adaptive mechanisms for a step size control in the general case.

The mutative self-adaptation of step sizes is a simple approach for such a task in single-objective optimization. According to Eiben et al. [14], the term *self-adaptive* describes a parameter control mechanism in EAs that treats parameters as part of an individual. This of course only makes sense for individual-related parameters such as the mutation strength. Together with the individual, the parameters are subject to recombination and mutation. The survivor selection eliminating unfit individuals indirectly also eliminates the corresponding mutation strengths, which created these unfit individuals. Thus, this system typically provides enough feedback to adapt a single mutation strength parameter for a single-objective local search. Due to the emergent behavior, it is one of the simplest mechanisms for controlling the step size of the search [6]. It is also possible to have a separate step size for every decision variable, which is helpful on separable ill-conditioned problems [6]. However, the approach has never been widely adopted in EMO, because the $(\mu + 1)$-selection frequently used here is quite different from the (μ, λ)-selection known to be necessary for self-adaptive step size control to work well in single-objective optimization. In EMO, instead simulated binary crossover (SBX) and polynomial mutation (PM) of Deb and Agrawal [12] have become a popular choice for variation operators in continuous search spaces, since they became the standard variation operator of NSGA2 [13]. Beyer and Deb [5] argued that SBX has quasi-self-adaptive properties, because the offspring's distance to the parents is proportional to the distance between the parents. However, in practice quite severe performance differences may appear between SBX/PM and other variation operators.

The first appearance of Gaussian distributions in multiobjective EAs was probably in 1991 in an approach by Kursawe [20]. However, the mutation strength was not the focus of investigation. A theoretical analysis of Gaussian and hypersphere distributions in the multiobjective setting was due to Rudolph in 1998 [24]. Then, a first hint towards the virtues of self-adaptive mutation for the SMS-EMOA was found in 2006 by Naujoks et al. [21]. They coupled Gaussian mutation with discrete recombination and applied it to an airfoil design problem. The approach yielded better results than SBX/PM in terms of dominated hypervolume. Also the more sophisticated approach of covariance matrix adaptation has been transferred to multiobjective optimization [17]. However, the topic of self-adaptation apparently was only revisited once, by Klinkenberg et al. [19], who generated λ offspring from the same parent and used a metamodel to preselect one of them for exact evaluation. The low interest in Gaussian mutations may have to do with the kind of test problems commonly used at the time, which were largely separable. For such problems, SBX/PM has proven to be quite high-performing. However, [28] showed that for aerodynamic optimization problems, also the variation of differential evolution is mostly preferable to SBX/PM. This was confirmed by [1] for a modern set of benchmark problems.

In this work, we will focus on the SMS-EMOA, which is a simple algorithm for the maximization of dominated hypervolume [4]. At first, we explore the characteristics of supposedly optimal step sizes for a simple bi-objective problem yielding some surprising insights, before we apply a standard version of self-adaptation to a $(\mu + \lambda)$-SMS-EMOA with $\mu > 1$ and $\lambda > 1$.

2   Optimal Step Sizes for a Simple Bi-objective Problem

Let $V = \{v^{(1)}, v^{(2)}, \ldots, v^{(\mu)}\}$ be μ lexicographically ordered, mutually incomparable objective vectors in $\mathbb{R}^2$ and $r \in \mathbb{R}^2$ a reference point that is dominated by each of the objective vectors. The *dominated hypervolume* of the objective vectors in V relative to r is given by

$$H(v^{(1)}, v^{(2)}, \ldots, v^{(\mu)}; r) = (r_1 - v_1^{(1)}) \cdot ((r_2 - v_2^{(1)}) + \sum_{i=2}^{\mu} (r_1 - v_1^{(i)}) \cdot (v_2^{(i-1)} - v_2^{(i)})$$

whereas the *dominated hypervolume contribution* of objective vector $v^{(k)} \in V$ is

$$H_\Delta(v^{(k)}; V, r) = H(V; r) - H(V \backslash \{v^{(k)}\}; r)$$

for $k = 1, \ldots, \mu$. Suppose that some individual x has been chosen as parent from the current population, so that mutation leads to the new individual $x + s\,Z$ with objective vector $f(x + s\,Z)$, which is added to the current set $V^{(t)}$ of objective vectors. Before selection, the enlarged population is $V^+ = V^{(t)} \cup \{v(s)\}$ where $v(s) = f(x + s\,Z)$. The SMS-EMOA removes that individual with least hypervolume contribution, i. e., $w = \operatorname{argmin}\{H_\Delta(v; V^+, r) : v \in V^+\}$. Thus, after selection the new set of objective vectors is $V^{(t+1)} = V^+ \backslash \{w\}$. As a consequence, the *expected hypervolume improvement (EHVI) after survivor selection* is

$$\mathsf{E}[\,H(V^{(t+1)}) - H(V^{(t)})\,] = \mathsf{E}[\,H(V^{(t+1)})\,] - H(V^{(t)})\,. \tag{1}$$

This formulation hides the fact that the EHVI is actually a function of the step size $s > 0$. In detailed notation, the left part of the EHVI in the right hand side of (1) reads

$$\mathsf{E}\left[H(V^{(t)} \cup \{v(s)\} \backslash \operatorname*{argmin}_{v \in V^{(t)} \cup \{v(s)\}} \{H_\Delta(v; V^{(t)} \cup \{v(s)\}, r)\}; r) \right], \tag{2}$$

revealing that the choice of s determines which individual is added and which individual is removed. Note that this is a crucial difference from the common EHVI definition in model-based multiobjective optimization, where no solutions are removed at all [16]. The kind of interrelationship encountered here requires a case-by-case analysis that hampers an analytical treatment with bearable effort. Due to this fact, we resort to a numerical analysis. The objective function is $f(x) = (x^\top x, (x - a)^\top (x - a))^\top$ whose Pareto set is the line segment between 0 and a, in our case $a = (10, 10)^\top$. We restrict the study to the two-dimensional decision space, enabling a graphical illustration and comparison of the mutations for each individual of the population.

In the numerical analysis we assume that a reference point and some population $x^{(1)}, \ldots, x^{(\mu)} \in \mathbb{R}^2$ with objective vectors $v^{(k)} = f(x^{(k)})$ are given $(k = 1, \ldots, \mu)$. Three examples P_1, P_2, P_3 with two different reference points each are depicted in Fig. 1. To determine the optimal step size for each individual, let angle $\omega_i = i \cdot 2\pi/360$ and direction $d_i = (\cos \omega_i, \sin \omega_i)^\top$ with length $\|d_i\| = 1$ for $i = 0, 1, \ldots, 359$. Similarly, let step size $s_j = j \cdot \Delta s$ for $j = 1, \ldots, s_{\max}$ for some appropriate $\Delta s > 0$. Thus, we are going to approximate the expected hypervolume improvement in case of step size s_j by simulating a uniform distribution on the unit circle with radius s_j. We average the hypervolume improvement in direction d_i by advancing the angle of the directions in steps of 1 degree. In detail, the hypervolume improvement h_{ij} of some individual x with step size s_j in direction d_i relative to a population V with reference point r is

$$h_{ij} = H(V \cup \{v_{ij}\} \setminus \operatorname*{argmin}_{v \in V \cup \{v_{ij}\}} \{H_\Delta(v; V \cup \{v_{ij}\}, r)\}; r)$$

where $v_{ij} = v(s_j, d_i) = f(x + s_j \cdot d_i)$ for $i = 0, 1, \ldots, 359$ and $j = 1, \ldots, s_{\max}$. These values are illustrated in the second row of Figs. 2 and 3 as a gray-scale heatmap. The estimator for the expected hypervolume improvement of individual x with step size s_j is

$$\hat{h}_j = \frac{1}{360} \sum_{i=0}^{359} h_{ij}.$$

These values are depicted in the last row of Figs. 2 and 3. In the top row we depict the probability of a hypervolume improvement for given step size; the value is estimated by the relative frequency of a success. Unfortunately, the maximal EHVI is not always in good agreement with the probability of improvement, which indicates that $s^* = \operatorname{argmax}_{s_j} \{\hat{h}_j\}$ is sometimes not very robust (see Fig. 2, solution 4). In these cases, the bulk of the contribution to $\hat{h}_j$ comes from a small area with very large improvement. In Fig. 2, there are also small areas where the EHVI gets negative. This is possible because the selection of the SMS-EMOA with its dynamically constructed reference point is not completely in accordance with the evaluation using a fixed reference point r, as discovered by [18].

In Fig. 1, one can see from the areas with many circle crossings that with a far away reference point of $r = (900, 900)^\top$, the boundary points of the Pareto

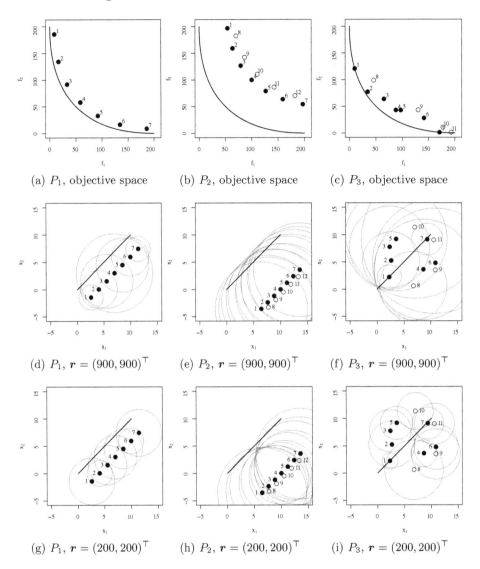

(a) P_1, objective space

(b) P_2, objective space

(c) P_3, objective space

(d) P_1, $\boldsymbol{r} = (900, 900)^\top$

(e) P_2, $\boldsymbol{r} = (900, 900)^\top$

(f) P_3, $\boldsymbol{r} = (900, 900)^\top$

(g) P_1, $\boldsymbol{r} = (200, 200)^\top$

(h) P_2, $\boldsymbol{r} = (200, 200)^\top$

(i) P_3, $\boldsymbol{r} = (200, 200)^\top$

Fig. 1. Example populations in different constellations. The two lower rows depict the decision space with optimal step sizes s^* as red circles. White markers indicate dominated solutions. (Color figure online)

set are much preferred. For $\boldsymbol{r} = (200, 200)^\top$, which is actually the nadir point, solutions near the interior of the line segment are favoured. Similar observations can be made in Figs. 2 and 3, where two pronounced peaks exist in the case of $\boldsymbol{r} = (900, 900)^\top$, corresponding to the two boundary points of the Pareto set. Another important observation is that for dominated individuals, improvement is only possible with quite large step sizes (see Fig. 3).

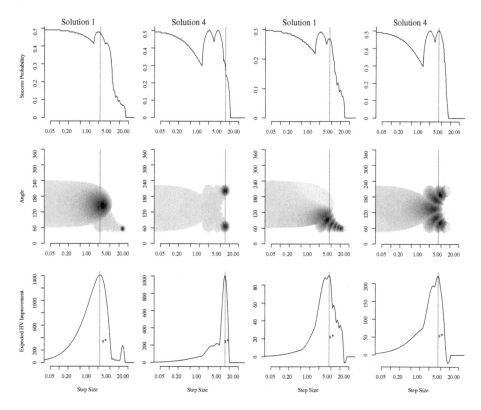

Fig. 2. Success probability and EHVI of the first and fourth individual of P_1 vs. step size. The left two columns show $r = (900, 900)^\top$, the right two columns $r = (200, 200)^\top$. The optimal step size s^* is marked with a vertical line. Cf. [22].

3   Experiments with Self-adaptation

In lack of an immediate, theoretically supported alternative, we are now focusing on the basic mutation operator of evolution strategies using Gaussian random numbers, to obtain step size control for the real world. As we already mentioned, the applicability of the operator to the SMS-EMOA was already shown in [21]. It creates an offspring y from an individual x by calculating

$$y = x + \sigma \mathcal{N}(0, I) \,, \tag{3}$$

where I is the identity matrix. In (3), the simplest variant with one step size parameter σ is presented. This variant requires one exogenous learning parameter τ in its adaptation rule, which is typically recommended to be chosen as $\tau = 1/\sqrt{n}$ for unimodal single-objective problems [6]. In this case, the step size is varied by a multiplicative mutation

$$\sigma = \tilde{\sigma} \cdot \exp(\tau \mathcal{N}(0, 1)) \,, \tag{4}$$

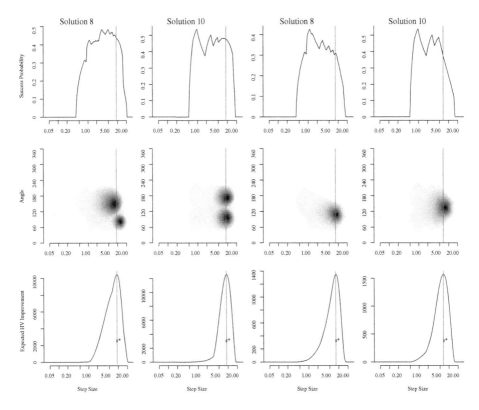

Fig. 3. Success probability and EHVI of some dominated individuals of P_2 against step size. The left two columns show $r = (900, 900)^\top$, the right two columns $r = (200, 200)^\top$. The optimal step size s^* is marked with a vertical line.

where $\tilde{\sigma}$ is the step size of the parent individual. It is important to note that for the adaptation to work, (4) has to be carried out before the actual mutation of the decision variable in (3).

3.1 Analysis on Unimodal Problems

Research question. What is the best configuration for an SMS-EMOA with self-adaptive mutation on unimodal bi-objective problems?

Pre-experimental planning. Aiming for a basic setup, we do not incorporate restart mechanisms in our algorithm. Correspondingly, we restrict the set of test problems to unimodal problems from the bi-objective black-box optmization benchmarking (BBOB) suite [26], to avoid interference from the additional difficulty of multimodal problems.

As the recommended choice of $1/\sqrt{n}$ for τ only pertains to single-objective optimization, we introduce a factor c to vary the learning rate $\tau = c/\sqrt{n}$.

Task. Following the performance measurement guidelines of the BBOB workshop [9,10], we record all non-dominated points produced during an algorithm run to use them for the assessment. This approach is gaining more interest recently, as it presumably reduces the dependence on μ, in comparison to only evaluating the final population. Furthermore, powerful post-processing algorithms have now become available, at least for bi-objective problems [8].

The calculation of dominated hypervolume slightly deviates from the BBOB setting. We do not normalize the hypervolume between ideal and nadir point, but simply calculate a reference point for each problem instance by combining the non-dominating fronts of all runs and adding an offset of one to the worst objective value in each dimension. To sensibly aggregate results over several problem instances and dimensions, we carry out a rank transformation (RT-2 in [11]) of the hypervolume values first.

Table 1. Experimental factors

Factor	Type	Symbol	Levels
Number variables	Observable	n	$\{2, 3, 5, 10, 20\}$
Learning param. constant	Control	c	$\{2^{-2}, 2^{-1}, 2^0, 2^1, 2^2, 2^3\}$
Population size	Control	μ	$\{10, 50\}$
Number offspring	Control	λ	$\{1, \mu, 5\mu\}$
Recombination	Control		$\{$discrete, intermediate, arithmetic, none$\}$

Setup. Table 1 summarizes the investigated factors in this experiment. Configurations with $\lambda > 1$ are included in the setup to better approximate promising configurations from the single-objective scenario. To enable them, we use a greedy selection that iteratively removes the worst individual, until the desired population size is restored. The number of parents is fixed at $\rho = 2$ and the initial mutation strength is $\sigma_{\text{init}} = 0.025$. Recombination variants are chosen and named with an orientation towards [15, p. 50]:

- Arithmetic recombination: ρ weights $w_i \in [0,1]$ are chosen randomly, so that $\sum_{i=1}^{\rho} w_i = 1$. The offspring $\boldsymbol{y}$ is chosen as convex combination of the parents, i.e., $\boldsymbol{y} = \sum_{i=1}^{\rho} w_i \boldsymbol{x}_i$. The offspring therefore is in the convex hull of the parents.
- Intermediate recombination: A special case of arithmetic recombination with $w_i = 1/\rho$.
- Discrete recombination: Each decision variable of the offspring is chosen random uniformly from the parents' values at the same position.
- No recombination: The offspring is just a clone of one individual.

The budget is set to $10^4 n$ function evaluations per SMS-EMOA run. We carry out a full factorial experiment and thus investigate $6 \cdot 2 \cdot 3 \cdot 4 = 144$ different algorithm configurations. Furthermore, there are 15 unimodal problems available in the bi-objective BBOB testbed (all combinations of sphere, separable

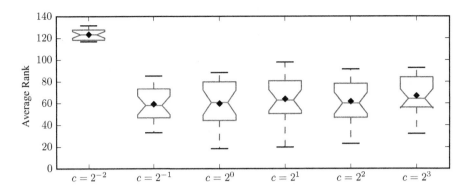

Fig. 4. Performance comparison of learning parameters $\tau = c/\sqrt{n}$ with different numerators c. Black diamonds indicate mean values.

ellipsoid, attractive sector, sharp ridge, and sum of different powers functions). As a concession to this costly setup, we only use one instance/replication per problem. The search space $[-100, 100]^n$ of each problem [26] is normalized to the unit hypercube. To always stay inside the bound constraints, we use Lamarckian reflection as the repair method [27].

Results and observations. Figure 4 shows that a setting of $c = 2^{-2}$ in the learning parameter produces outliers with clearly worse performance than all the other c values. Thus, we exclude this setting from the remaining analysis. Figure 5 shows performance data for the remaining factors. It suggests that $\mu = 50$, $\lambda = 5\mu$, and not using recombination are the best settings regarding mean and median performance. Actually, these factor levels in combination together with $c = 2^0$ are also the best performing configuration. However, we also want to investigate if there are interactions between the factors. This is done in Tables 2, 3, and 4, where mean ranks are depicted as in the figures. In Table 2, we see that in contrast to the other recombination variants, arithmetic recombination works better with a small μ and the effect of λ on performance is quite weak. No recombination is also very dependent on μ, but with a clear advantage for the larger value. In Table 3 we see that the influence of the learning parameter is the largest for discrete recombination.

Figure 6 gives us some insight into why $c = 2^{-2}$ performs so bad: of the tested settings, $c = 2^{-2}$ is the only one where the step size is hardly adapted at all. In all the other cases, the step size decreases by about two orders of magnitude after a brief phase of increase. This is shown here by averaging the step size parameters σ of all individuals in the population at each time step, and then averaging the resulting $\bar{\sigma}$ values over all 15 test problems.

Discussion. Figure 6 indicates that self-adaptation of step size on unimodal bi-objective problems works surprisingly similar as in single-objective optimization and has a positive influence on performance. (Note that even the original learning parameter obtained the best performance.) However, as the step size is

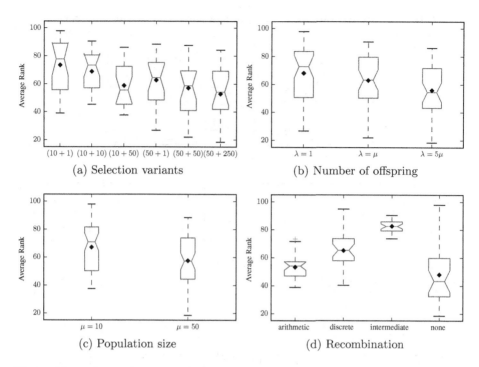

Fig. 5. Main effects of controllable factors except learning parameter. In these figures, the level $c = 2^{-2}$ for the learning parameter has been excluded from the data. Black diamonds indicate mean values.

Table 2. Selection variant against recombination. Lower is better.

	arithmetic	discrete	intermediate	none
$(10+1)$	46.97	85.43	82.53	78.95
$(10+10)$	51.29	72.55	83.48	68.34
$(10+50)$	47.69	62.90	82.25	42.50
$(50+1)$	61.93	63.21	84.93	40.95
$(50+50)$	58.23	55.88	84.06	30.43
$(50+250)$	53.77	51.34	78.82	27.14

Table 3. Learning parameter against recombination. Lower is better.

	arithmetic	discrete	intermediate	none
$2^{-1}/\sqrt{n}$	49.96	66.60	79.90	40.82
$2^{0}/\sqrt{n}$	57.01	53.97	83.87	44.49
$2^{1}/\sqrt{n}$	55.65	65.43	82.33	52.42
$2^{2}/\sqrt{n}$	48.70	66.57	80.38	50.98
$2^{3}/\sqrt{n}$	55.25	73.53	86.90	51.54

Table 4. Selection variant against learning parameter. Lower is better..

	$2^{-1}/\sqrt{n}$	$2^0/\sqrt{n}$	$2^1/\sqrt{n}$	$2^2/\sqrt{n}$	$2^3/\sqrt{n}$
$(10+1)$	62.93	74.85	80.58	74.45	74.55
$(10+10)$	61.57	68.47	73.43	69.07	72.04
$(10+50)$	57.47	54.80	55.29	60.12	66.49
$(50+1)$	57.52	58.83	67.85	63.23	66.34
$(50+50)$	55.84	53.60	57.41	56.51	62.40
$(50+250)$	60.61	48.46	49.18	46.57	59.01

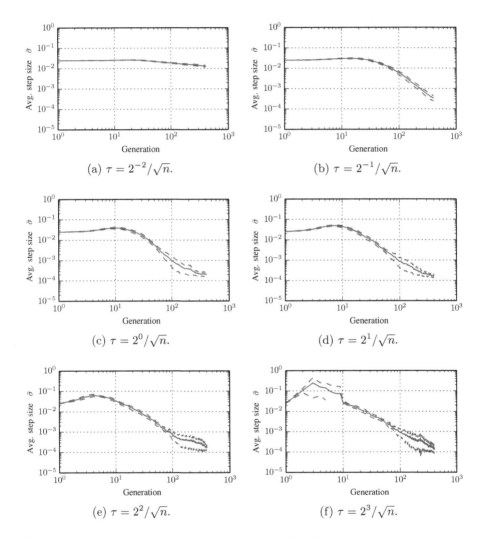

(a) $\tau = 2^{-2}/\sqrt{n}$.

(b) $\tau = 2^{-1}/\sqrt{n}$.

(c) $\tau = 2^0/\sqrt{n}$.

(d) $\tau = 2^1/\sqrt{n}$.

(e) $\tau = 2^2/\sqrt{n}$.

(f) $\tau = 2^3/\sqrt{n}$.

Fig. 6. Development of the average step size $\bar{\sigma}$ with different learning parameters τ. The shown configuration is $(50+250)$ selection, no recombination, and $n = 20$. Dashed lines indicate point-wise 95 % confidence intervals for the mean.

increasing at first, a larger initial step size σ_{init} might have been more suitable. Unfortunately, σ_{init} and the repair method for constraint violations could not be varied due to restrictions on the computational resources. These two factors are known to interact strongly in single-objective optimization [27], so they also should be investigated in the future.

Regarding the recombination variants, it seems advisable to concentrate on arithmetic and no recombination in the future, as they yield the best performance. Further improvements might be possible by implementing some special partner selection, e.g. by mating individuals with their nearest neighbors.

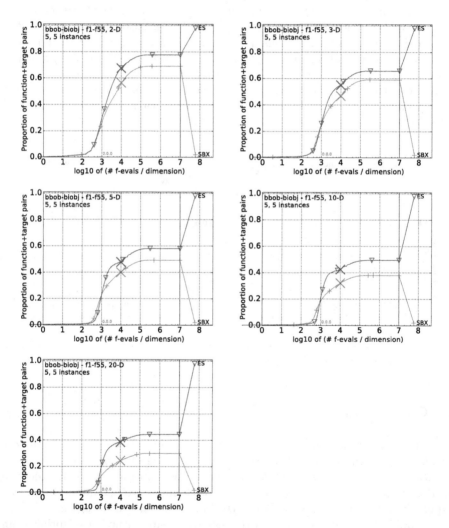

Fig. 7. Comparison of the SMS-EMOA with ES and SBX-variation on the whole BBOB testbed. Larger is better. Cf. [7].

3.2   Comparison of Self-adaptive Variation with SBX/PM

After having found a high-quality SMS-EMOA configuration with ES-variation, we compare this configuration with the standard SBX/PM variation. For this comparison, we are using all 55 available BBOB problems with 10 instances per problem, so the self-adaptive ES-variation has to prove its competitiveness also on multimodal problems. For the budget and n, the same values as in Sect. 3.1 are chosen. For the ES-variation, the best variant on unimodal problems is chosen (no recombination with $\tau = 2^0/\sqrt{n}$). The selection is $(50 + 250)$ for both algorithms. The used SBX and PM parameters are $\eta_c = 10$, $p_c = 0.7$, $\eta_m = 10$, and $p_m = 0.1$. Figure 7 summarizes the results by plotting the algorithms' empirical cumulative distribution functions. It shows that ES-variation has an advantage in all tested search space dimensions. Although the ES configuration has not been adapted to them, it also seems to work relatively well on multimodal problems. However, as is well known [28], SBX/PM has its strengths on separable problems. Correspondingly, in this test suite it achieves a better performance than ES especially on those problems which contain the separable ellipsoid function (see Fig. 8).

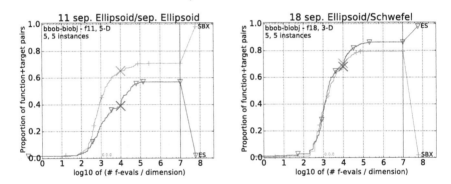

Fig. 8. Examples of separable problems where SBX/PM obtains better or comparable performance. Larger is better.

4   Conclusion

From our preliminary investigations, we can conclude that even on unimodal problems, the situation is much more complicated than in single-objective optimization. Nonetheless, a success-dependent step size control as in single-objective optimization seems to be possible, even with mutative self-adaptation. The experimental results on the BBOB testbed are generally encouraging.

A sufficiently large step size is required to create surviving offspring from dominated solutions, indicating that a bias towards larger step sizes is advisable in this case. Thus, a more sophisticated mechanism may be custom-tailored to the multiobjective case by exploiting this fact.

Furthermore, the experiments on unimodal multiobjective problems reveal that the recombination operators commonly used in single-objective EAs should be questioned, as best results have been achieved without recombination. The interaction between recombination and the geometry of a multiobjective landscape deserve careful scrutiny in future.

References

1. Auger, A., Brockhoff, D., Hansen, N., Tušar, D., Tušar, T., Wagner, T.: The impact of variation operators on the performance of SMS-EMOA on the bi-objective BBOB-2016 test suite. In: Companion Publication of the 2016 Conference on Genetic and Evolutionary Computation, pp. 1225–1232. ACM (2016)
2. Beume, N., Laumanns, M., Rudolph, G.: Convergence rates of $(1 + 1)$ evolutionary multiobjective optimization algorithms. In: Schaefer, R., Cotta, C., Kołodziej, J., Rudolph, G. (eds.) PPSN 2010. LNCS, vol. 6238, pp. 597–606. Springer, Heidelberg (2010). doi:10.1007/978-3-642-15844-5_60
3. Beume, N., Laumanns, M., Rudolph, G.: Convergence rates of SMS-EMOA on continuous bi-objective problem classes. In: Proceedings of the 11th Workshop on Foundations of Genetic Algorithms, FOGA 2011, pp. 243–252. ACM (2011)
4. Beume, N., Naujoks, B., Emmerich, M.: SMS-EMOA: multiobjective selection based on dominated hypervolume. Eur. J. Oper. Res. **181**(3), 1653–1669 (2007)
5. Beyer, H.G., Deb, K.: On self-adaptive features in real-parameter evolutionary algorithms. IEEE Trans. Evol. Comput. **5**(3), 250–270 (2001)
6. Beyer, H.G., Schwefel, H.P.: Evolution strategies - a comprehensive introduction. Nat. Comput. **1**(1), 3–52 (2002)
7. Brandenbusch, K.: Experimentelle Analyse selbst-adaptiver Schrittweitenanpassung bei mehrkriteriellen Evolutionären Algorithmen. Bachelor's thesis. TU Dortmund University, Department of Computer Science, Dortmund, September 2016. (in German)
8. Bringmann, K., Friedrich, T., Klitzke, P.: Generic postprocessing via subset selection for hypervolume and epsilon-indicator. In: Bartz-Beielstein, T., Branke, J., Filipič, B., Smith, J. (eds.) PPSN 2014. LNCS, vol. 8672, pp. 518–527. Springer, Heidelberg (2014). doi:10.1007/978-3-319-10762-2_51
9. Brockhoff, D., Tran, T.D., Hansen, N.: Benchmarking numerical multiobjective optimizers revisited. In: Proceedings of the 2015 Annual Conference on Genetic and Evolutionary Computation, GECCO 2015, pp. 639–646. ACM (2015)
10. Brockhoff, D., Tušar, T., Tušar, D., Wagner, T., Hansen, N., Auger, A.: Biobjective performance assessment with the COCO platform. e-print 1605.01746, arXiv (2016). https://arxiv.org/abs/1605.01746
11. Conover, W.J., Iman, R.L.: Rank transformations as a bridge between parametric and nonparametric statistics. Am. Stat. **35**(3), 124–129 (1981)
12. Deb, K., Agrawal, R.B.: Simulated binary crossover for continuous search space. Complex Syst. **9**, 115–148 (1995)
13. Deb, K., Pratap, A., Agarwal, S., Meyarivan, T.: A fast and elitist multiobjective genetic algorithm: NSGA-II. IEEE Trans. Evol. Comput. **6**(2), 182–197 (2002)
14. Eiben, A.E., Hinterding, R., Michalewicz, Z.: Parameter control in evolutionary algorithms. IEEE Trans. Evol. Comput. **3**(2), 124–141 (1999)
15. Eiben, A.E., Smith, J.E.: Introduction to Evolutionary Computing. Springer, Heidelberg (2003)

16. Hupkens, I., Deutz, A., Yang, K., Emmerich, M.: Faster exact algorithms for computing expected hypervolume improvement. In: Gaspar-Cunha, A., Henggeler Antunes, C., Coello, C.C. (eds.) EMO 2015. LNCS, vol. 9019, pp. 65–79. Springer, Cham (2015). doi:10.1007/978-3-319-15892-1_5

17. Igel, C., Hansen, N., Roth, S.: Covariance matrix adaptation for multi-objective optimization. Evol. Comput. **15**(1), 1–28 (2007)

18. Judt, L., Mersmann, O., Naujoks, B.: Non-monotonicity of observed hypervolume in 1-greedy S-metric selection. J. Multi-Criteria Decis. Anal. **20**(5–6), 277–290 (2013)

19. Klinkenberg, J.W., Emmerich, M.T.M., Deutz, A.H., Shir, O.M., Bäck, T.: A reduced-cost SMS-EMOA using kriging, self-adaptation, and parallelization. In: Ehrgott, M., Naujoks, B., Stewart, J.T., Wallenius, J. (eds.) Multiple Criteria Decision Making for Sustainable Energy and Transportation Systems. LNE, vol. 634, pp. 301–311. Springer, Heidelberg (2010). doi:10.1007/978-3-642-04045-0_26

20. Kursawe, F.: A variant of evolution strategies for vector optimization. In: Schwefel, H.-P., Männer, R. (eds.) PPSN 1990. LNCS, vol. 496, pp. 193–197. Springer, Heidelberg (1991). doi:10.1007/BFb0029752

21. Naujoks, B., Quagliarella, D., Bartz-Beielstein, T.: Sequential parameter optimisation of evolutionary algorithms for airfoil design. In: Winter, G. (ed.) Proceedings of Design and Optimization: Methods and Application (ERCOFTAC 2006), pp. 231–235. University of Las Palmas de Gran Canaria (2006)

22. Pink, R.: Optimale Schrittweiten für einen bikriteriellen Evolutionären Algorithmus mit S-Metrik Selektion. Bachelor's thesis. TU Dortmund University, Department of Computer Science, Dortmund, September 2016. (in German)

23. Rechenberg, I.: Evolutionsstrategie - Optimierung technischer Systeme nach Prinzipien der biologischen Evolution. Frommann Holzboog, Stuttgart (1973)

24. Rudolph, G.: On a multi-objective evolutionary algorithm and its convergence to the Pareto set. In: IEEE International Conference on Evolutionary Computation Proceedings, pp. 511–516 (1998)

25. Schumer, M., Steiglitz, K.: Adaptive step size random search. IEEE Trans. Autom. Control **13**(3), 270–276 (1968)

26. Tušar, T., Brockhoff, D., Hansen, N., Auger, A.: COCO: the bi-objective black box optimization benchmarking (bbob-biobj) test suite. e-print 1604.00359, arXiv (2016). https://www.arxiv.org/abs/1604.00359v2

27. Wessing, S.: Repair methods for box constraints revisited. In: Esparcia-Alcázar, A.I. (ed.) EvoApplications 2013. LNCS, vol. 7835, pp. 469–478. Springer, Heidelberg (2013). doi:10.1007/978-3-642-37192-9_47

28. Wessing, S., Beume, N., Rudolph, G., Naujoks, B.: Parameter tuning boosts performance of variation operators in multiobjective optimization. In: Schaefer, R., Cotta, C., Kołodziej, J., Rudolph, G. (eds.) PPSN 2010. LNCS, vol. 6238, pp. 728–737. Springer, Heidelberg (2010). doi:10.1007/978-3-642-15844-5_73

Computing 3-D Expected Hypervolume Improvement and Related Integrals in Asymptotically Optimal Time

Kaifeng Yang[1]([✉]), Michael Emmerich[1], André Deutz[1], and Carlos M. Fonseca[2]

[1] LIACS, Leiden University, Niels Bohrweg 1, 2333 CA Leiden, The Netherlands
{k.yang,m.t.m.emmerich,a.h.deutz}@liacs.leidenuniv.nl
[2] CISUC, Department of Informatics Engineering, University of Coimbra, Polo II, Pinhal de Marrocos, 3030-290 Coimbra, Portugal
cmfonsec@dei.uc.pt
http://moda.liacs.nl

Abstract. The Expected Hypervolume Improvement (EHVI) is a frequently used infill criterion in surrogate-assisted multi-criterion optimization. It needs to be frequently called during the execution of such algorithms. Despite recent advances in improving computational efficiency, its running time for three or more objectives has remained in $O(n^d)$ for $d \geq 3$, where d is the number of objective functions and n is the size of the incumbent Pareto-front approximation. This paper proposes a new integration scheme, which makes it possible to compute the EHVI in $\Theta(n \log n)$ optimal time for the important three-objective case ($d = 3$). The new scheme allows for a generalization to higher dimensions and for computing the Probability of Improvement (PoI) integral efficiently. It is shown, both theoretically and empirically, that the hidden constant in the asymptotic notation is small. Empirical speed comparisons were designed between the C++ implementations of the new algorithm (which will be in the public domain) and those recently published by competitors, on randomly-generated non-dominated fronts of size 10, 100, and 1000. The experiments include the analysis of batch computations, in which only the parameters of the probability distribution change but the incumbent Pareto-front approximation stays the same. Experimental results show that the new algorithm is always faster than the other algorithms, sometimes over 10^4 times faster.

Keywords: Expected hypervolume improvement · Time complexity · Surrogate-assisted multi-criterion optimization · Efficient global optimization · Probability of improvement

1 Introduction

Surrogate-assisted multi-criterion optimization (SAMCO) uses approximations to the objective functions in order to quickly assess the potential quality of a candidate solution. In the context of SAMCO, the *Expected Hypervolume Improvement* (EHVI) is frequently used as an infill or pre-selection criterion [1–8],

© Springer International Publishing AG 2017
H. Trautmann et al. (Eds.): EMO 2017, LNCS 10173, pp. 685–700, 2017.
DOI: 10.1007/978-3-319-54157-0_46

and is a straightforward generalization of the single-objective *expected improvement* (EI). It is called multiple times because, in each iteration of SAMCO, either an optimum of the EI needs to be found, as in the case of *Bayesian global optimization*[1] [10], or, in *surrogate-assisted evolutionary algorithms*, it is used to pre-assess the quality of the individuals of a larger population.

Informally, the EHVI for a problem with d objectives is defined as follows [4]: Given an incumbent approximation to a Pareto front of size n, the EHVI is an integral that measures the expected increase of the *Hypervolume Indicator*, given a predictive multivariate Gaussian distribution on the d-dimensional objective space stemming, for instance, from a Gaussian process or Kriging Approximation, the outcome at a yet unevaluated design point. It, therefore, measures the expected quantity by which the Pareto-front approximation would improve, if that design point were evaluated. Both the mean values and variances of the objective function value predictions are considered.

Although the EHVI is interesting due to its theoretical properties, as shown by Wagner et al. [11], it has been critizised in the same paper for the computational effort required for its exact computation. This is a problem, as many SAMCO algorithms require a large number of EHVI computations to be performed in every iteration. Since the announcement of exact computation schemes [12], significant progress has been made, resulting in faster computation schemes [8,13,14]. Notably, in 2-D, the computation time could be improved from $O(n^3 \log n)$–using a straightforward integration scheme, with a grid partitioning and repeated hypervolume computations for each grid cell – to asymptotically optimal $\Theta(n \log n)$ in [13] using a stripe partitioning and a multi-layered integration scheme. However, despite significant efficiency improvements [8,14], the running time of the best-known algorithms for $d \geq 3$ has remained in $O(n^d)$.

This paper shows that the new integration technique proposed in [13] can be generalized to $d \geq 3$, yielding a new algorithm with asymptotically optimal time complexity in $\Theta(n \log n)$. This improves existing upper bounds by a factor of $O(n^2/\log n)$. Moreover, it is shown that similar approaches provide asymptotically optimal time algorithms for the *Probability of Improvement*, i.e. the probability that a point will be non-dominated, and the *Truncated Hypervolume Improvement* (TEHVI) [15].

The paper is structured as follows: Sect. 2 discusses the problem in the wider context of surrogate-assisted multi-criterion optimization, and Sect. 3 provides the necessary definitions. In Sect. 4 the new algorithm for 3-D EHVI calculation is discussed, including the analysis of its complexity. Experimental results on test data are reported in Sect. 5, including a speed comparison to other algorithms proposed in the literature [8,14]. Section 6 discusses how the new algorithm can be modified for solving the related problems of PoI and TEHVI. Section 7 draws the main conclusions and discusses future work.

---

[1] Also called Efficient Global Optimization [9].

2   Relevance and Related Work

In the context of single-objective optimization, the expected improvement was firstly proposed by Mockus et al. [10]. They assumed that the objective function, say $f : \mathbb{R}^q \to \mathbb{R}$, $q \geq 1$, is a realization of a Gaussian process with known (or estimated *a priori*) mean and covariance structure. Given that the values $f(x^{(i)})$ at sites $x^{(1)}, \ldots, x^{(t)}$ are known, e.g., from previous exact evaluations, the conditional mean $\mu(x)$ and standard deviations $\sigma(x)$ at a yet unevaluated point x can be computed. The 1-D Gaussian distribution with mean $\mu(x)$ and standard deviation $\sigma(x)$ assigns to $y \in \mathbb{R}$ the likelihood of the event $f(x) = y$. Consider the problem of maximizing $f(x)$, and a corresponding sequence of points $(x_i)_{i \in \mathbb{N}}$. Then, for a given point in time, t, we can express the improvement of a given $y \in \mathbb{R}$ as $I(y, f_{\max,t}) = \max\{0, y - f_{\max,t}\}$, where $f_{\max,t} = \max\{f(x_i) | i \in \{1, \ldots, t\}\}$. The expected improvement of x is then defined as $\mathrm{EI}(x) = \int_{y \in \mathbb{R}} I(y, f_{\max,t}) \, \xi_{\mu(x),\sigma(x)}(y) \, \mathrm{d}y$, where $\xi_{\mu(x),\sigma(x)}$ is the probability density function of the 1-D Gaussian distribution conditioned on $(x_i, f(x_i))_{i \in \{1, \ldots, t\}}$.

The EI is widely used to pre-assess the quality of solutions in evolutionary and deterministic optimization with expensive black box evaluations. It was popularized by Jones et al. [9] as an infill criterion in the so-called Efficient Global Optimization (EGO) algorithm. In each iteration EGO evaluates the design point with maximal EI. Its convergence properties are discussed in [16], where a proof of global convergence under mild assumptions on the global covariance and the smoothness of the function is given. Roughly speaking, global convergence is due to the fact that EI rewards high variance and also high mean values.

Various generalizations of EI to multi-criterion optimization have been discussed in the literature, e.g., [7,17–20]. See also [11] for an overview. In the case of multiple objectives, it is possible to consider a Gaussian process model for each objective function separately and independently, resulting in a multivariate distribution with d mean values $\mu_i(x)$ and standard deviation $\sigma_i(x)$. A key question when generalizing the expected improvement is how to define improvement of a given Pareto-front approximation. In indicator-based multi-criterion optimization, the performance of a Pareto-front approximation is assessed by a unary indicator, typically the *Hypervolume Indicator*, which allows for a simple generalization of the *Expected Improvement*–the EHVI. Given $(x_i, \mathbf{f}(x_i))$, $i = 1, \cdots, t$, the incumbent Pareto-front approximation becomes $\mathbf{P}_{\max,t} = $ Non-Dominated subset of $\{\mathbf{f}(x_i) | i \in \{1, \ldots, t\}\}$ and its hypervolume replaces $f_{\max,t}$ in the definition of the EI. The EHVI was first proposed in [21], and since then it has been used in Evolutionary Algorithms for airfoil optimization [4] and quantum control [22], as well as in multi-criterion generalizations of Efficient Global Optimization for applications in fluid dynamics [23], event controllers in wastewater treatment [1], efficient algorithm tuning [5], electrical component design [8], and bio-fuel power-generation [3]. In all of these applications, the bi-objective EHVI was used. Due to its high computation time for problems with three and more objectives, using it as an infill criterion was not advised in such settings, and fast, but imprecise, alternatives were sought [24].

So far it has remained unknown whether the integration algorithms used in the literature achieved optimal performance. Hence, it is an important question to study whether, and to what extent, the computational efficiency of the exact computation of the EHVI can be further improved. In early work, the EHVI integral was computed by Monte Carlo simulation [21]. The first algorithm for the exact calculation of the 2-D EHVI was introduced by Emmerich et al. [12], with a computational complexity of $O(n^3 \log n)$, where n is the number of non-dominated points in the front. Couckuyt et al. [8] provided an exact EHVI calculation algorithm for $d > 2$, and, according to experimental data, this algorithm was typically much faster than the one introduced in [12], but it still had a high worst-case time complexity. Hupkens et al. [14] found algorithms for computing EHVI with the then lowest worst-case time complexity of $O(n^2)$ and $O(n^3)$, for two and three objectives, respectively. Recently, Emmerich et al. [13] proposed an asymptotically optimal algorithm for the bi-objective case, with a computational time complexity of $\Theta(n \log n)$. In 3-D, the time complexity of this problem and those of the related problems of computing the *Probability of Improvement* [8] and *Truncated Expected Hypervolume Improvement* [2] have so far remained cubic. The following sections outline algorithms that can solve these problems in optimal time $\Theta(n \log n)$.

3   Expected Hypervolume Improvement

This section formally introduces the *Expected Hypervolume Improvement* and related definitions. In the following, we consider problems with three objectives. Most of the discussion will be on mathematical objects defined in the objective space. We will denote the axes of the objective space spanned by $f_1, f_2, ..., f_d$, where $f_i : \mathbb{R}^q \to \mathbb{R}$ are the objective functions to be maximized. Points in the objective space will be denoted by $\mathbf{y}$, in case we are not interested in their preimage. We define the usual Pareto dominance order on the objective space, i.e. $\mathbf{y}^{(1)} \succ \mathbf{y}^{(2)}$, iff $\mathbf{y}^{(1)} \geq \mathbf{y}^{(2)}$ (componentwise) and $\mathbf{y}^{(1)} \neq \mathbf{y}^{(2)}$.

The *Hypervolume Indicator*, introduced in [25], is one of the most important unary indicators for evaluating the quality of a Pareto-front approximation. Its theoretical properties are discussed in [26,27]. Notably, it does not require the Pareto front to be known in advance, and its maximization leads to high quality and diverse Pareto-front approximation sets. The *Hypervolume Indicator* is defined as follows:

Definition 1 (Hypervolume Indicator). *Given a finite approximation to a Pareto front, say* $\mathbf{P} = \{\mathbf{y}^{(1)}, \dots, \mathbf{y}^{(n)}\} \subset \mathbb{R}^d$, *the Hypervolume Indicator (HV) of* $\mathbf{P}$ *is defined as the d-dimensional Lebesgue measure of the subspace dominated by* $\mathbf{P}$ *and bounded below by a reference point* $\mathbf{r}$:

$$HV(\mathbf{P}) = \lambda_d(\cup_{\mathbf{y} \in \mathbf{P}}[\mathbf{r}, \mathbf{y}]) \tag{1}$$

with λ_d *being the Lebesgue measure on* $\mathbb{R}^d$.

The hypervolume measures the size of the dominated subspace bounded below by a reference point $\mathbf{r}$ – we consider maximization. This reference point needs to be provided by the user, and it should, if possible, be chosen such that it is dominated by all elements of the Pareto-front approximation sets $\mathbf{P}$ that might occur during the optimization process.

In order to generalize the *Expected Improvement* (EI) criterion, we will first generalize the concept of improvement to the multiobjective case:

Definition 2 (Hypervolume Improvement). *Given a finite collection of vectors $\mathbf{P} \subset \mathbb{R}^d$, the Hypervolume Improvement of a vector $\mathbf{y} \in \mathbb{R}^d$ is defined as:*

$$HVI(\mathbf{y}, \mathbf{P}) = HV(\mathbf{P} \cup \{\mathbf{y}\}) - HV(\mathbf{P}) \tag{2}$$

In case we want to emphasize the reference point $\mathbf{r}$, the notation $HVI(\mathbf{y}, \mathbf{P}, \mathbf{r})$ will be used to denote the Hypervolume Improvement.

The *Expected Hypervolume Improvement* is based on the theory of the *Hypervolume Indicator*, and is a measure of how much hypervolume may improve, considering the uncertainty of the prediction. The prediction is a probability distribution that measures, for a given input vector in the decision space, the likelihood of outcomes in the objective space. Typically, Gaussian process models or Kriging models are used for prediction, delivering for each input vector an independent, multivariate probability distribution with a mean vector $\boldsymbol{\mu} \in \mathbb{R}^d$ and a standard deviation vector $\boldsymbol{\sigma} \in \mathbb{R}^d$, each component of which corresponding to a particular objective function prediction.

Definition 3 (Expected Hypervolume Improvement). *Given a mean vector $\boldsymbol{\mu} \in \mathbb{R}^d$ and a standard deviation vector $\boldsymbol{\sigma} \in \mathbb{R}^d$, an incumbent Pareto-front approximation $\mathbf{P} \subset \mathbb{R}^d$, and a reference point $\mathbf{r} \in \mathbb{R}^d$ the expected improvement is given by:*

$$EHVI(\boldsymbol{\mu}, \boldsymbol{\sigma}, \mathbf{P}, \mathbf{r}) = \int_{\mathbf{y} \in \mathbb{R}^d} HVI(\mathbf{y}, \mathbf{P}, \mathbf{r}) \xi_{\boldsymbol{\mu},\boldsymbol{\sigma}}(\mathbf{y}) d\mathbf{y} \tag{3}$$

Example 1. Figure 1(left) depicts the dominated hypervolume for a small approximation set $\mathbf{P} = (\mathbf{y}^{(1)} = (4, 4, 1), \mathbf{y}^{(2)} = (1, 2, 4), \mathbf{y}^{(3)} = (2, 1, 3))$. The volume of all slices is the 3-D *Hypervolume Indicator* of $\mathbf{P}$, with $\mathbf{r}$ being the origin of the coordinate system. The *Hypervolume Improvement* of $\mathbf{y}^{(+)} = (3, 3, 2)$ relative to $\mathbf{P}$ is given by the joint volume covered by the red slices. The *Expected Hypervolume Improvement* would average over different realizations of $\mathbf{y}^{(+)}$ following a 3-D normal distribution with mean vector (μ_1, μ_2, μ_3) and standard deviation σ_1, σ_2, and σ_3.

4   Asymptotically Optimal Algorithm for 3-D EHVI Calculation

For the convenience of illustration, the EHVI calculation in this section only involves the maximization case. The main idea of the proposed algorithm is

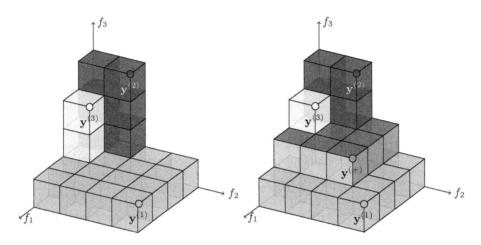

Fig. 1. The left figure shows a given 3-D Pareto-front approximation consisting of the points $\mathbf{y}^{(1)} = (4,4,1)$, $\mathbf{y}^{(2)} = (1,2,4)$, and $\mathbf{y}^{(3)} = (2,1,3)$. The right figure shows how the *Hypervolume Indicator* increases when the red point $\mathbf{y}^{(+)} = (3,3,2)$ is added. The volume of the increment (red blocks) is the *Hypervolume Improvement*. The *Expected Hypervolume Improvement* is the mean value of the *Hypervolume Improvement*, if $\mathbf{y}^{+}$ would be sampled from a 3-D normal distribution. (Color figure online)

separating the integration volume into integration slices, as few as possible, and then per integration slice compute all contributions to the EHVI integral that are related to the area of the bases of the integration slices.

4.1 Partitioning into $O(n)$ Integration Slices

Example 2. An illustration of integration slices is shown in Fig. 2. A Pareto front set is composed by 4 points ($\mathbf{y}^{(1)} = (1,3,4), \mathbf{y}^{(2)} = (4,2,3), \mathbf{y}^{(3)} = (2,4,2)$ and $\mathbf{y}^{(4)} = (3,5,1)$), and this Pareto front is shown in the left figure. The right figure illustrates the projection onto $y_1 y_2$-plane with rectangle slices and $\mathbf{l}, \mathbf{u}$. The rectangular slices, which share the similar color but different opacity, represent integration slices with the same value of y_3 in their lower bound. The lower bound of the 3-D integration slice B_4 is $\mathbf{l}^{(4)} = (1,2,2)$, and the upper bound of the slice is $\mathbf{u}^{(4)} = (2,4,\infty)$.

For maximization problems, the upper bound of each integration slice is always ∞ in the y_3 axis, therefore we can describe each integration slice by its lower bound ($\mathbf{l}$) and upper bound ($\mathbf{u}$) as follows.

$$B_i = (\mathbf{l}, \mathbf{u}) = \left(\begin{pmatrix} l_1^{(i)} \\ l_2^{(i)} \\ l_3^{(i)} \end{pmatrix}, \begin{pmatrix} u_1^{(i)} \\ u_2^{(i)} \\ \infty \end{pmatrix} \right), \qquad i = 1, \ldots, 2n+1 \qquad (4)$$

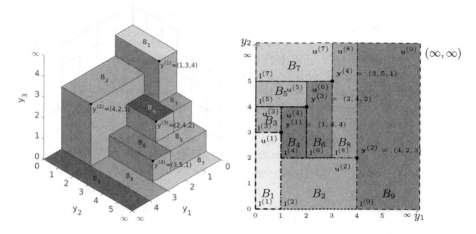

Fig. 2. Left: 3-D Pareto front. Right: 3-D Pareto front in 2-D, each slice can be described by lower bound and upper bound

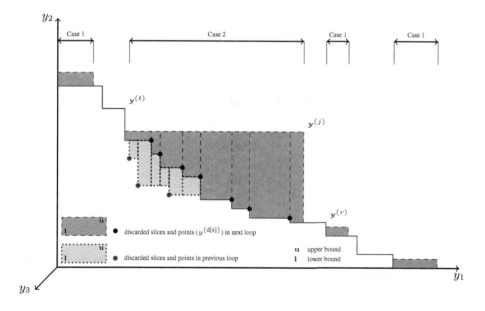

Fig. 3. Boundary search for slices in 3-D case

Algorithm 1 describes how to obtain the slices $B_1, \ldots, B_i, \ldots, B_{2n+1}$ with the corresponding lower and upper bounds ($\mathbf{l}^{(i)}$ and $\mathbf{u}^{(i)}$) and compute the integrals for them. The partitioning algorithm is similar to the sweep line algorithm described in [28]. The basic idea of this algorithm is using an AVL tree, structured by y_1 and y_2 coordinates, to process the points in a descending order of the y_3 coordinate. For each such point, say $\mathbf{y}^{(i)}$, add this point to the AVL tree and find all the points $(\mathbf{y}^{(d[1])}, \ldots, \mathbf{y}^{(d[s])})$ which are dominated by it in the y_1y_2-plane and discard them from the AVL tree. See Fig. 3 for describing one such iteration. In each iteration, $s + 1$ slices are created by coordinates of the points $\mathbf{y}^{(t)}, \mathbf{y}^{(d[1])}, \ldots, \mathbf{y}^{(d[s])}, \mathbf{y}^{(r)}$, and $\mathbf{y}^{(i)}$ as illustrated in Fig. 3. The integration of the slices (calculation_3D) will be described in Subsect. 4.2.

Here, the number of the integration slices is $2n + 1$ when all points are in general position. (Otherwise $2n + 1$ provides an upper bound for the obtained number of slices.) The reason is as follows: In the algorithm each point $\mathbf{y}^{(i)}, i = 1, \ldots, n$ creates a slice, say slice $A^{(i)}$, when it is created and a slice, say slice $B^{(i)}$, when it is discarded from the AVL tree due to domination by another point, say $\mathbf{y}^{(s)}$, in the y_1y_2-plane. The two slices are defined as follows $A^{(i)} = ((y_1^{(t)}, y_2^{(l2)}, y_3^{(i)}), (y_1^{(u1)}, y_2^{(i)}, \infty))$ whereas $y_2^{(l2)}$ is either $y_2^{(r)}$ if no points are dominated by $\mathbf{y}^{(i)}$ in the y_1y_2-plane or $y_2^{(d[1])}$, otherwise. Moreover, $B^{(i)} = ((y_1^{(i)}, y_2^{(r)}, y_3^{(s)}), (y_1^{(u)}, y_2^{(s)}, \infty))$, and $\mathbf{y}^{(u)}$ denotes either the right neighbour among the newly dominated points in the y_1y_2-plane, or $\mathbf{y}^{(s)}$ if $\mathbf{y}^{(i)}$ is the rightmost point among all newly dominated points. This way each slice can be attributed to exactly one point in $\mathbf{P}$, except for one slice that is created in the final iteration. In the final iteration one additional point $\mathbf{y}^{(n+1)} = (\infty, \infty, \infty)$ is added in the y_1y_2-plane. This point leads to the creation of a slice when it is added, but because it is never discarded it adds only a single slice. In total, therefore $2n + 1$ slices are created.

4.2 Computing Slice-Based Parts of the Integral

Given a partitioning of the non-dominated space into integration slices $B_1, \ldots, B_i, \ldots, B_{2n+1}$ the part of the integral related to each of the integration slices can be computed separately. To see how this can be done, the *Hypervolume Improvement* of a point $\mathbf{y} \in \mathbb{R}^3$ is rewritten as:

$$\mathrm{HVI}_3(\mathbf{y}, \mathbf{P}, \mathbf{r}) = \sum_{i=1}^{2n+1} \lambda_3[B_i \cap \Delta(\mathbf{y})] \tag{5}$$

where $\Delta \mathbf{y}$ is the part of the objective space that is dominated by $\mathbf{y}$. The HVI expression in the definition of EHVI in Eq. (3) can be replaced by HVI_3 in Eq. (5):

$$\mathrm{EHVI}(\boldsymbol{\mu}, \boldsymbol{\sigma}, \mathbf{P}, \mathbf{r}) = \sum_{i=1}^{2n+1} \int_{y_1=l_1^{(i)}}^{\infty} \int_{y_2=l_2^{(i)}}^{\infty} \int_{y_3=l_3^{(i)}}^{\infty} \lambda_3[B_i \cap \Delta(\mathbf{y})] \cdot \xi_{\boldsymbol{\mu},\boldsymbol{\sigma}}(\mathbf{y})d\mathbf{y} \tag{6}$$

In Eq. (6), the summation can be done after the integration, because integration is a linear mapping. Then we can divide the integration interval $\int_{y_1=l_1^{(i)}}^{\infty}$ and $\int_{y_2=l_2^{(i)}}^{\infty}$ into $(\int_{y_1=l_1^{(i)}}^{u_1^{(i)}} + \int_{y_1=u_1^{(i)}}^{\infty})$ and $(\int_{y_2=l_2^{(i)}}^{u_2^{(i)}} + \int_{y_2=u_2^{(i)}}^{\infty})$, respectively. Based on this division, Eq. (6) can be expressed by:

$$\text{Eq.6} = \sum_{i=1}^{2n+1} \int_{y_1=l_1^{(i)}}^{u_1^{(i)}} \int_{y_2=l_2^{(i)}}^{u_2^{(i)}} \int_{y_3=l_3^{(i)}}^{\infty} \lambda_3[B_i \cap \Delta(\mathbf{y})] \cdot \xi_{\boldsymbol{\mu},\boldsymbol{\sigma}}(\mathbf{y})d\mathbf{y} \tag{7}$$

$$+ \sum_{i=1}^{2n+1} \int_{y_1=l_1^{(i)}}^{u_1^{(i)}} \int_{y_2=u_2^{(i)}}^{\infty} \int_{y_3=l_3^{(i)}}^{\infty} \lambda_3[B_i \cap \Delta(\mathbf{y})] \cdot \xi_{\boldsymbol{\mu},\boldsymbol{\sigma}}(\mathbf{y})d\mathbf{y} \tag{8}$$

$$+ \sum_{i=1}^{2n+1} \int_{y_1=u_1^{(i)}}^{\infty} \int_{y_2=l_2^{(i)}}^{u_2^{(i)}} \int_{y_3=l_3^{(i)}}^{\infty} \lambda_3[B_i \cap \Delta(\mathbf{y})] \cdot \xi_{\boldsymbol{\mu},\boldsymbol{\sigma}}(\mathbf{y})d\mathbf{y} \tag{9}$$

$$+ \sum_{i=1}^{2n+1} \int_{y_1=u_1^{(i)}}^{\infty} \int_{y_2=u_2^{(i)}}^{\infty} \int_{y_3=l_3^{(i)}}^{\infty} \lambda_3[B_i \cap \Delta(\mathbf{y})] \cdot \xi_{\boldsymbol{\mu},\boldsymbol{\sigma}}(\mathbf{y})d\mathbf{y} \tag{10}$$

The idea of this decomposition is that the improvement for $\int_{y_1=u_1^{(i)}}^{\infty}(\cdots)$ and $\int_{y_2=u_2^{(i)}}^{\infty}(\cdots)$ are constant, that is: $\int_{y_k=u_k^{(i)}}^{\infty} \lambda_1[B_i \cap \Delta(y_k)] \cdot \xi_{\mu_k,\sigma_k}(y_k)dy_k = \int_{y_k=u_k^{(i)}}^{\infty}(u_k^{(i)} - l_k^{(i)}) \cdot \xi_{\mu_k,\sigma_k}(y_k)dy_k = (u_k^{(i)} - l_k^{(i)}) \cdot (1 - \Phi(\frac{u_k^{(i)}-\mu_k}{\sigma_k})) =: \vartheta(l_k^{(i)}, u_k^{(i)}, \sigma_k, \mu_k)$, where $k = 1, 2$ and $\lambda_1[B_i \cap \Delta(y_k)]$ is the *Hypervolume Improvement* in dimension k, i.e., a 1-D *Hypervolume Improvement*. We introduce the following abbreviations: $\lambda_1[B_i \cap \Delta(y_k)] = ||[l_k^{(i)}, u_k^{(i)}] \cap [l_k^{(i)}, y_k]|| = \min\{u_k^{(i)}, y_k\} - l_k^{(i)} =: \ell_{l_k^{(i)}}^{(u_k^{(i)},y_k)}, \xi_i = \xi_{\mu_i,\sigma_i}(y_i)$, where $i = 1, 2, 3$. Based on this abbreviation, Eq. (7) can be written as

$$\text{Eq.7} = \sum_{i=1}^{2n+1} \int_{l_1^{(i)}}^{u_1^{(i)}} \ell_{l_1^{(i)}}^{(u_1^{(i)},y_1)} \xi_1 dy_1 \int_{l_2^{(i)}}^{u_2^{(i)}} \ell_{l_2^{(i)}}^{(u_2^{(i)},y_2)} \xi_2 dy_2 \int_{l_3^{(i)}}^{\infty} \ell_{l_3^{(i)}}^{(u_3^{(i)},y_3)} \xi_3 dy_3$$

$$= \sum_{i=1}^{2n+1} \left(\int_{l_1^{(i)}}^{\infty} \ell_{l_1^{(i)}}^{(u_1^{(i)},y_1)} \xi_1 dy_1 - \int_{u_1^{(i)}}^{\infty} \ell_{l_1^{(i)}}^{(u_1^{(i)},y_1)} \xi_1 dy_1 \right) \cdot$$

$$\left(\int_{l_2^{(i)}}^{\infty} \ell_{l_2^{(i)}}^{(u_2^{(i)},y_2)} \xi_2 dy_2 - \int_{u_2^{(i)}}^{\infty} \ell_{l_2^{(i)}}^{(u_2^{(i)},y_2)} \xi_2 y_2 \right) \cdot \int_{l_3^{(i)}}^{\infty} \ell_{l_3^{(i)}}^{(u_3^{(i)},y_3)} \xi_3 dy_3$$

$$= \sum_{i=1}^{2n+1} \left(\int_{l_1^{(i)}}^{\infty} (y_1 - l_1^{(i)})\xi_1 dy_1 - \int_{u_1^{(i)}}^{\infty} (u_1^{(i)} - l_1^{(i)})\xi_1 dy_1 \right) \cdot$$

$$\left(\int_{l_2^{(i)}}^{\infty} (y_2 - l_2^{(i)})\xi_2 dy_2 - \int_{u_2^{(i)}}^{\infty} (u_1^{(i)} - l_2^{(i)})\xi_2 dy_2 \right) \cdot \int_{l_3^{(i)}}^{\infty} (y_3 - l_3^{(i)})\xi_3 dy_3$$

$$= \sum_{i=1}^{2n+1} \left(\Psi_{\infty}(l_1^{(i)}, l_1^{(i)}, \mu_1, \sigma_1) - \vartheta(l_1^{(i)}, u_1^{(i)}, \sigma_1, \mu_1) \right) \cdot$$

$$\left(\Psi_{\infty}(l_2^{(i)}, l_2^{(i)}, \mu_2, \sigma_2) - \vartheta(l_2^{(i)}, u_2^{(i)}, \sigma_2, \mu_2) \right) \cdot \Psi_{\infty}(l_3^{(i)}, l_3^{(i)}, \mu_3, \sigma_3) \tag{11}$$

The Ψ_∞ functions that are used here, are integrals of the 1-D cumulative Gaussian distribution function (Φ) and the 1-D Gaussian probability density function ξ, these two functions are discussed in the appendix. Similar to the derivation of Eqs. (7), (8), (9) and (10) can be written as follows:

$$
\text{Eq.8} = \sum_{i=1}^{2n+1} \left(\Psi_\infty(l_1^{(i)}, l_1^{(i)}, \mu_1, \sigma_1) - \vartheta(l_1^{(i)}, u_1^{(i)}, \sigma_1, \mu_1) \right) \cdot \vartheta(l_2^{(i)}, u_2^{(i)}, \sigma_2, \mu_2) \cdot
$$
$$
\Psi_\infty(l_3^{(i)}, l_3^{(i)}, \mu_3, \sigma_3) \tag{12}
$$

$$
\text{Eq.9} = \sum_{i=1}^{2n+1} \vartheta(l_1^{(i)}, u_1^{(i)}, \sigma_1, \mu_1) \cdot \left(\Psi_\infty(l_2^{(i)}, l_2^{(i)}, \mu_2, \sigma_2) - \vartheta(l_2^{(i)}, u_2^{(i)}, \sigma_2, \mu_2) \right) \cdot
$$
$$
\Psi_\infty(l_3^{(i)}, l_3^{(i)}, \mu_3, \sigma_3) \tag{13}
$$

$$
\text{Eq.10} = \sum_{i=1}^{2n+1} \vartheta(l_1^{(i)}, u_1^{(i)}, \sigma_1, \mu_1) \cdot \vartheta(l_2^{(i)}, u_2^{(i)}, \sigma_2, \mu_2) \cdot \Psi_\infty(l_3^{(i)}, l_3^{(i)}, \mu_3, \sigma_3) \tag{14}
$$

The final EHVI formula is the sum of Eqs. (11), (12), (13) and (14). The Pseudo code of the proposed algorithm is shown in Algorithm 1.

During the EHVI calculation, as the $y_1 y_2$-projections are mutually non-dominated, the points are also sorted by the y_2 coordinate in the AVL tree, identifying a neighbouring point or a discard point takes time $O(\log n)$. Then the EHVI for these integration slices will be calculated by the *calculation_3d* function in Algorithm 1 (line 13 and 23), which is the summation of Eqs. (11), (12), (13) and (14) with the parameters of μ, σ and B_{2n+1}. The EHVI computational complexity for each slice is $O(1)$. Moreover, the dominated points ($\mathbf{y}^{(d[s])}$) will be removed from the AVL tree, and the new points ($\mathbf{y}^{(j)}$) will be inserted in the AVL tree. Since the points that are dominated by the new point $\mathbf{y}^{(j)}$ will be deleted at the end of the current loop, they will not occur again in the later computations. Hence the total number of open slices does not exceed $2n + 1$, as mentioned before, and the total computation costs $O(n \log n)$.

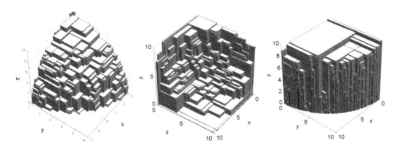

Fig. 4. Randomly generated fronts of type CONVEXSPHERICAL, CONCAVESPHERICAL, and CLIFF3D from [28] with $|P| = 100$ (left, middle and right).

Algorithm 1. EHVI calculation algorithm in 3-D

Input: $(\mathbf{y}^{(1)}, \cdots, \mathbf{y}^{(n)})$: mutually non-dominated $\mathbb{R}^3$-points sorted by third coordinate (y_3) in descending order

Output: EHVI value

1 $\mathbf{y}^{(n+1)} = (\infty, \infty, r_3)$;

2 Initialize AVL tree T for 3-D points

   Insert $\mathbf{y}^{(1)}$, $(\infty, r_2, \infty)^T$ and $(r_1, \infty, \infty)^T$ into T;

3 Initialize the number of integration slices $n_b = 1$;

4 Initialize $EHVI = 0$;

5 **for** $i = 2$ **to** $n + 1$ **do**   /* Main loop */

6   Retrieve the following information from tree T:

7   r: index of the successor of $\mathbf{y}^{(i)}$ in x-coordinate (right neighbour);

8   t: index of the successor of $\mathbf{y}^{(i)}$ in y-coordinate (left neighbour);

9   d[1], $\cdots$, d[s]: indices of points dominated by $\mathbf{y}^{(i)}$ in $y_1 y_2$-plane, sorted ascendingly in the first coordinate(y_1);

10     $B_{n_b}.l_3 = y_3^{(i)}$, $B_{n_b}.u_2 = y_2^{(i)}$, $B_{n_b}.u_3 = \infty$;

11   **if** $s == 0$ **then**   /* Case 1 */

12        $B_{n_b}.l_1 = y_1^{(t)}$, $B_{n_b}.l_2 = y_2^{(r)}$, $B_{n_b}.u_1 = y_1^{(i)}$;

13        $EHVI = EHVI+ $ calculation_3d$(\boldsymbol{\mu}, \boldsymbol{\sigma}, B_{n_b})$;

14        $n_b = n_b + 1$;

15   **else**   /* Case 2 */

16   **for** $j = 1$ **to** $s + 1$ **do**

17   **if** $j == 1$ **then**

18              $B_{n_b}.l_1 = y_1^{(t)}$, $B_{n_b}.l_2 = y_2^{(d[1])}$, $B_{n_b}.u_1 = y_1^{(d[1])}$;

19   **else if** $j == s + 1$ **then**

20              $B_{n_b}.l_1 = y_1^{(d[s])}$, $B_{n_b}.l_2 = y_2^{(r)}$, $B_{n_b}.u_1 = y_1^{(i)}$;

21   **else**

22              $B_{n_b}.l_1 = y_1^{(d[j-1])}$, $B_{n_b}.l_2 = y_2^{(d[j])}$, $B_{n_b}.u_1 = y_1^{(d[j])}$;

23           $EHVI = EHVI+ $ calculation_3d$(\boldsymbol{\mu}, \boldsymbol{\sigma}, B_{n_b})$;

24           $n_b = n_b + 1$;

25   Discard $\mathbf{y}^{(d[1])}, \cdots, \mathbf{y}^{(d[s])}$ from tree T;

26   Insert $\mathbf{y}^{(i)}$ in tree T.

The C++ source-code and MATLAB .mex file for computing the EHVI is available on http://moda.liacs.nl or on request from the authors.

5 Empirical Comparison

Three algorithms, IRS_fast [14], CDD13 [8] and KMAC, which is short for the authors given name, in this paper are compared via the same benchmarks. The test benchmarks from Emmerich and Fonseca [28] are used to generated the Pareto fronts. The Pareto fronts and evaluated points are randomly generated based on CONVEXSPHERICAL, CONCAVESPHERICAL, and CLIFF3D functions.

Table 1. Empirical comparisons of strategies for 3-D EHVI calculation

Type	$|P|$	Batch Size	Time Average (s)		
			CDD13 [8]	IRS_fast [14]	KMAC
CONVEX	10	1	0.13785	0.00037	**0.00005**
CONVEX	10	10	0.14090	0.00056	**0.00021**
CONVEX	10	100	0.16500	0.00304	**0.00095**
CONVEX	10	1000	0.69104	0.02778	**0.00754**
CONVEX	100	1	13.97556	0.05337	**0.00038**
CONVEX	100	10	17.05551	0.13730	**0.00099**
CONVEX	100	100	45.90095	0.93196	**0.00831**
CONVEX	100	1000	422.31263	8.38585	**0.06462**
CONVEX	1000	1	>3 h	94.72402	**0.00390**
CONVEX	1000	10	>3 h	155.77306	**0.01067**
CONVEX	1000	100	>3 h	795.11319	**0.06517**
CONVEX	1000	1000	>3 h	2838.31854	**0.53801**
CONCAVE	10	1	0.11209	0.00026	**0.00007**
CONCAVE	10	10	0.12790	0.00054	**0.00014**
CONCAVE	10	100	0.14002	0.00294	**0.00077**
CONCAVE	10	1000	0.36697	0.02597	**0.00840**
CONCAVE	100	1	10.62329	0.04895	**0.00031**
CONCAVE	100	10	12.63582	0.12927	**0.00146**
CONCAVE	100	100	27.51827	0.85124	**0.00768**
CONCAVE	100	1000	314.32314	7.67280	**0.06285**
CONCAVE	1000	1	>3 h	91.51055	**0.00332**
CONCAVE	1000	10	>3 h	149.58491	**0.01079**
CONCAVE	1000	100	>3 h	744.46691	**0.06696**
CONCAVE	1000	1000	>3 h	2499.29737	**0.50981**
CLIFF3D	10	1	0.12514	0.00026	**0.00007**
CLIFF3D	10	10	0.13222	0.00055	**0.00013**
CLIFF3D	10	100	0.14432	0.00278	**0.00075**
CLIFF3D	10	1000	0.44964	0.02725	**0.00761**
CLIFF3D	100	1	10.90605	0.04730	**0.00029**
CLIFF3D	100	10	12.85031	0.12709	**0.00112**
CLIFF3D	100	100	44.79395	0.80735	**0.00689**
CLIFF3D	100	1000	679.51368	7.46205	**0.06099**
CLIFF3D	1000	1	>3 h	136.37944	**0.00344**
CLIFF3D	1000	10	>3 h	165.34537	**0.01007**
CLIFF3D	1000	100	>3 h	731.03794	**0.06480**
CLIFF3D	1000	1000	>3 h	2543.16864	**0.51032**

The parameters: $\boldsymbol{\sigma} = (2.5, 2.5, 2.5)$, $\boldsymbol{\mu} = (10, 10, 10)$ were used in the experiments. Pareto front sizes $|P| \in \{10, 100, 1000\}$ and the number of predictions (candidate points) or Batch Size $\in \{1, 10, 100, 1000\}$ are used together with $\boldsymbol{\sigma}$ and $\boldsymbol{\mu}$. Ten trials were randomly generated by the same parameters, and average runtimes (10 runs) for the whole 10 trails with the same parameters were computed. The data for 3-D case with $|P| = 100$ are visualized in Fig. 4, and these figures are originally from [28]. All the experiments were run on the same hardware: Intel(R) Xeon(R) CPU E5-2667 v2 3.30 GHz, RAM 48 GB. The operating system was Ubuntu 12.04 LTS (64 bit), and the compiler was gcc 4.9.2 with compiler flag -Ofast, except for SUMO code, MATLAB 8.4.0.150421 (R2014b), 64 bit. The experiments were set to halt if the algorithms would not finish the EHVI computation within 3 hours. The results are shown in Table 1. While the speed-up gained by batch processing in IRS_fast is considerable, in KMAC the idea is not the case and we just use repeated computation. The results show that the proposed algorithm, KMAC, is the fastest one for all the test problems. Empirical comparisons on randomly generated Pareto fronts of different shape show that the new algorithm is by a factor of 7 to 3.9×10^4 faster than previously published implementations.

6 Related Problems

It is now straightforward to compute other integrals in a similar manner. For instance the *Probability of Improvement* [8,21], that is the probability that a point is non-dominated w.r.t. $\mathbf{P}$, can be computed simply by integration in parts over the $2n + 1$ integration slices, which yields an algorithm with running time in $O(n \log n)$ algorithm for $d = 3$:

$$\mathrm{PoI}(\boldsymbol{\mu}, \boldsymbol{\sigma}, \mathbf{P}) = \sum_{i=1}^{2n+1} \prod_{j=1}^{d} \Phi(\frac{u_j^{(i)} - \mu_j}{\sigma_j}) - \Phi(\frac{l_j^{(i)} - \mu_j}{\sigma_j}) \tag{15}$$

Moreover, we can compute the *Truncated Expected Hypervolume Improvement* (TEHVI) [2], by replacing the formulas of Ψ_∞, Φ, and ξ by corresponding formulas of the truncated Gaussian distribution (see [2] for details) and discard irrelevant integration slices that do not intersect with the reference slice [2].

7 Conclusions and Future Research

In this paper, an asymptotically optimal algorithm with a computational complexity of $O(n \log n)$, for the 3-D EHVI exact calculation, was proposed. Compared to [14], the computational complexity is improved by the factor $n^2 / \log n$. This meets the lower bound for the time complexity of the 3-D EHVI computation for $d = 3$, shown by reduction to the *Hypervolume Indicator* problem, see [14]. Thus the algorithm is asymptotically optimal and the time complexity of 3-D EHVI computation is in $\Theta(n \log n)$. As opposed to previous techniques, which required grid decomposition of the non-dominated subspace into $O(n^3)$ integration slices, the new integration technique can make use of efficient partitioning of the dominated space into only $2n+1$ axis-aligned integration slices. In practice, the new computation scheme will be of great advantage for making the EHVI and related integrals applicable in multiobjective optimization with three objectives, especially in Bayesian Optimization and surrogate-assisted multi-criterion evolutionary algorithms. Empirical comparisons are in line with this theoretical improvement and show the proposed algorithm is by a factor of 7 to 3.9×10^4 faster than previous existing implementations.

It is expected that this technique can be generalized to higher dimensional objective spaces, using for instance the partitioning technique for the non-dominated space discussed in [29]. The time complexity per slice is expected to be $O(2^{d-1})$.

Acknowledgements. Kaifeng Yang acknowledges financial support from the China Scholarship Council (CSC), CSC No. 201306370037. Carlos M. Fonseca was supported by national funds through the Portuguese Foundation for Science and Technology (FCT), and by the European Regional Development Fund (FEDER) through COMPETE 2020 – Operational Program for Competitiveness and Internationalization (POCI).

Appendix

Definition 4 (Ψ_∞ function (see also [14])). *Let $\phi(s) = 1/\sqrt{2\pi}e^{-\frac{1}{2}s^2}(s \in \mathbb{R})$ denote the probability density function (PDF) of the standard normal distribution. Moreover, let $\Phi(s) = \frac{1}{2}\left(1 + erf\left(\frac{s}{\sqrt{2}}\right)\right)$ denote its cumulative probability distribution function (CDF), and erf is Gaussian error function. The general normal distribution with mean μ and standard deviation σ has as PDF, $\xi_{\mu,\sigma}(s) = \phi_{\mu,\sigma}(s) = \frac{1}{\sigma}\phi(\frac{s-\mu}{\sigma})$ and its CDF is $\Phi_{\mu,\sigma}(s) = \Phi(\frac{s-\mu}{\sigma})$. Then the function $\Psi_\infty(a,b,\mu,\sigma)$ is defined as:*

$$\Psi_\infty(a,b,\mu,\sigma) = \int_b^\infty (z-a)\frac{1}{\sigma}\phi\left(\frac{z-\mu}{\sigma}\right)dz$$

$$= \sigma\phi\left(\frac{b-\mu}{\sigma}\right) + (\mu-a)\left[1 - \Phi\left(\frac{b-\mu}{\sigma}\right)\right]$$

References

1. Zaefferer, M., Bartz-Beielstein, T., Naujoks, B., Wagner, T., Emmerich, M.: A case study on multi-criteria optimization of an event detection software under limited budgets. In: Purshouse, R.C., Fleming, P.J., Fonseca, C.M., Greco, S., Shaw, J. (eds.) EMO 2013. LNCS, vol. 7811, pp. 756–770. Springer, Heidelberg (2013). doi:10.1007/978-3-642-37140-0_56

2. Yang, K., Deutz, A., Yang, Z., Bäck, T., Emmerich, M.: Truncated expected hypervolume improvement: exact computation and application. In: IEEE Congress on Evolutionary Computation (CEC). IEEE (2016)

3. Yang K, Gaida D, Bäck T, Emmerich M.: Expected hypervolume improvement algorithm for PID controller tuning and the multiobjective dynamical control of a biogas plant. In: 2015 IEEE Congress on Evolutionary Computation (CEC), pp. 1934–1942, May 2015

4. Michael, T.M., Giannakoglou, K.C., Naujoks, B.: Single-and multiobjective evolutionary optimization assisted by Gaussian random field metamodels. IEEE Trans. Evol. Comput. **10**(4), 421–439 (2006)

5. Koch, P., Wagner, T., Emmerich, M.T., Bäck, T., Konen, W.: Efficient multicriteria optimization on noisy machine learning problems. Appl. Soft Comput. **29**, 357–370 (2015)

6. Shimoyama, K., Jeong, S., Obayashi, S.: Kriging-surrogate-based optimization considering expected hypervolume improvement in non-constrained many-objective test problems. In: IEEE Congress on Evolutionary Computation (CEC), pp. 658–665. IEEE (2013)

7. Shimoyama, K., Sato, K., Jeong, S., Obayashi, S.: Comparison of the criteria for updating kriging response surface models in multi-objective optimization. In: IEEE Congress on Evolutionary Computation, pp. 1–8. IEEE (2012)

8. Couckuyt, I., Deschrijver, D., Dhaene, T.: Fast calculation of multiobjective probability of improvement and expected improvement criteria for pareto optimization. J. Global Optim. **60**(3), 575–594 (2014)

9. Jones, D.R., Schonlau, M., Welch, W.J.: Efficient global optimization of expensive black-box functions. J. Global Optim. **13**(4), 455–492 (1998)

10. Mockus, J., Tiešis, V., Žilinskas, A.: The application of Bayesian methods for seeking the extremum. In: Towards Global Optimization, vol. 2, pp. 117–131. North-Holland, Amsterdam (1978)

11. Wagner, T., Emmerich, M., Deutz, A., Ponweiser, W.: On expected-improvement criteria for model-based multi-objective optimization. In: Schaefer, R., Cotta, C., Kołodziej, J., Rudolph, G. (eds.) PPSN 2010. LNCS, vol. 6238, pp. 718–727. Springer, Heidelberg (2010). doi:10.1007/978-3-642-15844-5_72

12. Emmerich, M.T., Deutz, A.H., Klinkenberg, J.W.: Hypervolume-based expected improvement: monotonicity properties and exact computation. In: IEEE Congress on Evolutionary Computation (CEC), pp. 2147–2154. IEEE (2011)

13. Emmerich, M., Yang, K., Deutz, A., Wang, H., Fonseca, C.M.: A multicriteria generalization of bayesian global optimization. In: Pardalos, P.M., Zhigljavsky, A., Žilinskas, J. (eds.) Advances in Stochastic and Deterministic Global Optimization. SOIA, vol. 107, pp. 229–242. Springer, Cham (2016). doi:10.1007/978-3-319-29975-4_12

14. Hupkens, I., Deutz, A., Yang, K., Emmerich, M.: Faster exact algorithms for computing expected hypervolume improvement. In: Gaspar-Cunha, A., Henggeler Antunes, C., Coello, C.C. (eds.) EMO 2015. LNCS, vol. 9019, pp. 65–79. Springer, Cham (2015). doi:10.1007/978-3-319-15892-1_5

15. Yang, K., Li, L., Deutz, A., Bäck, T., Emmerich, M.: Preference-based multiobjective optimization using truncated expected hypervolume improvement. In: 12th International Conference on Natural Computation, Fuzzy Systems and Knowledge Discovery. IEEE (2016)

16. Vazquez, E., Bect, J.: Convergence properties of the expected improvement algorithm with fixed mean and covariance functions. J. Stat. Plan. Infer. **140**(11), 3088–3095 (2010)

17. Knowles, J., Hughes, E.J.: Multiobjective optimization on a budget of 250 evaluations. In: Coello Coello, C.A., Hernández Aguirre, A., Zitzler, E. (eds.) EMO 2005. LNCS, vol. 3410, pp. 176–190. Springer, Heidelberg (2005). doi:10.1007/978-3-540-31880-4_13

18. Keane, A.J.: Statistical improvement criteria for use in multiobjective design optimization. AIAA J. **44**(4), 879–891 (2006)

19. Shimoyama, K., Sato, K., Jeong, S., Obayashi, S.: Updating kriging surrogate models based on the hypervolume indicator in multi-objective optimization. J. Mech. Des. **135**(9), 094503–094503-7 (2013)

20. Svenson, J., Santner, T.: Multiobjective optimization of expensive-to-evaluate deterministic computer simulator models. Comput. Stat. Data Anal. **94**, 250–264 (2016)

21. Emmerich, M.T.M.: Single-and multi-objective evolutionary design optimization assisted by Gaussian random field metamodels. Ph.D. thesis, FB Informatik, University of Dortmund, ELDORADO, Dortmund, 10 (2005)

22. Shir, O.M., Emmerich, M., Bäckck, T., Vrakking, M.J.: The application of evolutionary multi-criteria optimization to dynamic molecular alignment. In: IEEE Congress on Evolutionary Computation, pp. 4108–4115. IEEE (2007)

23. Laniewski-Wołłk, P, Obayashi S, Jeong S.: Development of expected improvement for multi-objective problems. In: Proceedings of 42nd Fluid Dynamics Conference/Aerospace Numerical, Simulation Symposium (CD ROM), Varna, Bulgaria (2010)

24. Luo, C., Shimoyama, K., Obayashi, S.: Kriging model based many-objective optimization with efficient calculation of expected hypervolume improvement. In: IEEE Congress on Evolutionary Computation (CEC), pp. 1187–1194. IEEE (2014)

25. Zitzler, E., Thiele, L.: Multiobjective evolutionary algorithms: a comparative case study and the strength pareto approach. IEEE Trans. Evol. Comput. **3**(4), 257–271 (1999)

26. Zitzler, E., Thiele, L., Laumanns, M., Fonseca, C.M., Fonseca, V.G.D.: Performance assessment of multiobjective optimizers: an analysis and review. IEEE Trans. Evol. Comput. **7**(2), 117–132 (2003)

27. Auger, A., Bader, J., Brockhoff, D., Zitzler, E.: Hypervolume-based multiobjective optimization: theoretical foundations and practical implications. Theor. Comput. Sci. **425**, 75–103 (2012)

28. Emmerich, M.T.M., Fonseca, C.M.: Computing hypervolume contributions in low dimensions: asymptotically optimal algorithm and complexity results. In: Takahashi, R.H.C., Deb, K., Wanner, E.F., Greco, S. (eds.) EMO 2011. LNCS, vol. 6576, pp. 121–135. Springer, Heidelberg (2011). doi:10.1007/978-3-642-19893-9_9

29. Lacour, R., Klamroth, K., Fonseca, C.M.: A box decomposition algorithm to compute the hypervolume indicator. Comput. Oper. Res. **79**, 347–360 (2016)

Author Index

Printed in the United States
By Bookmasters